Available in MyStatLab™ for Your Statistics Courses

MyStatLab is the market-leading online resource for learning and teaching statistics.

Leverage the Power of StatCrunch

MyStatLab leverages the power of StatCrunch—powerful, web-based statistics software. Integrated into MyStatLab, students can easily analyze data from their exercises and etext. In addition, access to the full online community allows users to take advantage of a wide variety of resources and applications at www.statcrunch.com.

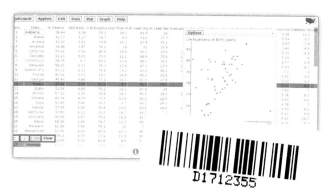

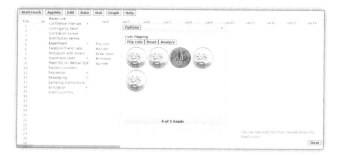

Bring Statistics to Life

Virtually flip coins, roll dice, draw cards, and interact with animations on your mobile device with the extensive menu of experiments and applets in StatCrunch. Offering a number of ways to practice resampling procedures, such as permutation tests and bootstrap confidence intervals, StatCrunch is a complete and modern solution.

Real-World Statistics

MyStatLab video resources help foster conceptual understanding. StatTalk Videos, hosted by fun-loving statistician Andrew Vickers, demonstrate important statistical concepts through interesting stories and real-life events. This series of 24 videos includes assignable questions built in MyStatLab and an instructor's guide.

www.mystatlab.com

Probability & Statistics for Engineers & Scientists

NINTH EDITION
MyStatLab™ Update

Ronald E. Walpole
Roanoke College

Raymond H. Myers
Virginia Tech

Sharon L. Myers
Radford University

Keying Ye
University of Texas at San Antonio

PEARSON

Boston   Columbus   Indianapolis   New York   San Francisco
Amsterdam   Cape Town   Dubai   London   Madrid   Milan   Munich   Paris   Montréal   Toronto
Delhi   Mexico City   São Paulo   Sydney   Hong Kong   Seoul   Singapore   Taipei   Tokyo

Editorial Director: *Chris Hoag*
Editor in Chief: *Deirdre Lynch*
Acquisitions Editor: *Patrick Barbera*
Editorial Assistant: *Justin Billing*
Program Manager: *Chere Bemelmans*
Project Manager: *Christine Whitlock*
Program Management Team Lead: *Karen Wernholm*
Project Management Team Lead: *Peter Silvia*
Media Producer: *Aimee Thorne*
MathXL Content Manager: *Bob Carroll*
Product Marketing Manager: *Tiffany Bitzel*
Field Marketing Manager: *Evan St. Cyr*
Marketing Coordinator: *Brooke Smith*
Senior Author Support/Technology Specialist: *Joe Vetere*
Rights and Permissions Project Manager: *Gina Cheselka*
Procurement Specialist: *Carol Melville*
Associate Director of Design USHE EMSS/HSC/EDU: *Andrea Nix*
Program Design Lead: *Heather Scott*
Composition: *Keying Ye*
Cover Image: *Marjory Dressler/Dressler Photo-Graphics*

Copyright © 2017, 2012, 2007 by Pearson Education, Inc. All Rights Reserved. Printed in the United States of America. This publication is protected by copyright, and permission should be obtained from the publisher prior to any prohibited reproduction, storage in a retrieval system, or transmission in any form or by any means, electronic, mechanical, photocopying, recording, or otherwise. For information regarding permissions, request forms and the appropriate contacts within the Pearson Education Global Rights & Permissions department, please visit www.pearsoned.com/permissions/.

Acknowledgements of third party content appear on page xvii, which constitutes an extension of this copyright page.

PEARSON, ALWAYS LEARNING and MYSTATLAB are exclusive trademarks owned by Pearson Education, Inc. or its affiliates in the U.S. and/or other countries.

Unless otherwise indicated herein, any third-party trademarks that may appear in this work are the property of their respective owners and any references to third-party trademarks, logos or other trade dress are for demonstrative or descriptive purposes only. Such references are not intended to imply any sponsorship, endorsement, authorization, or promotion of Pearson's products by the owners of such marks, or any relationship between the owner and Pearson Education, Inc. or its affiliates, authors, licensees or distributors.

Library of Congress Cataloging-in-Publication Data
Names: Walpole, Ronald E. | Myers, Raymond H. | Myers, Sharon L., 1944- | Ye, Keying.
Title: Probability & statistics for engineers & scientists.
Other titles: Probability and statistics for engineers and scientists
Description: 9th edition, MyStatLab update / Ronald E. Walpole, Roanoke College, Raymond H. Myers, Virginia Tech, Sharon L. Myers, Radford University, Keying Ye, University of Texas at San Antonio. | Boston : Pearson, [2017] | Includes bibliographical references and index.
Identifiers: LCCN 2015016078 | ISBN 0134115856
Subjects: LCSH: Engineering–Statistical methods. | Probabilities. Classification: LCC TA340 .P738 2017 | DDC 519.5–dc23 LC record available at http://lccn.loc.gov/2015016078

2 2019

ISBN 10: 0-13-411585-6
ISBN 13: 978-0-13-411585-6

This book is dedicated to

Billy and Julie
R.H.M. and S.L.M.
Limin, Carolyn and Emily
K.Y.

Contents

Preface xiii

1 Introduction to Statistics and Data Analysis 1
- 1.1 Overview: Statistical Inference, Samples, Populations, and the Role of Probability 1
- 1.2 Sampling Procedures; Collection of Data 7
- 1.3 Measures of Location: The Sample Mean and Median 11
  - Exercises 13
- 1.4 Measures of Variability 14
  - Exercises 17
- 1.5 Discrete and Continuous Data 17
- 1.6 Statistical Modeling, Scientific Inspection, and Graphical Diagnostics 18
- 1.7 General Types of Statistical Studies: Designed Experiment, Observational Study, and Retrospective Study 27
  - Exercises 30

2 Probability 35
- 2.1 Sample Space 35
- 2.2 Events 38
  - Exercises 42
- 2.3 Counting Sample Points 44
  - Exercises 51
- 2.4 Probability of an Event 52
- 2.5 Additive Rules 56
  - Exercises 59
- 2.6 Conditional Probability, Independence, and the Product Rule ... 62
  - Exercises 69
- 2.7 Bayes' Rule 72
  - Exercises 76
  - Review Exercises 77

	2.8	Potential Misconceptions and Hazards; Relationship to Material in Other Chapters..	79

3 Random Variables and Probability Distributions — 81

	3.1	Concept of a Random Variable....................................	81
	3.2	Discrete Probability Distributions................................	84
	3.3	Continuous Probability Distributions............................	87
		Exercises	91
	3.4	Joint Probability Distributions...................................	94
		Exercises	104
		Review Exercises...	107
	3.5	Potential Misconceptions and Hazards; Relationship to Material in Other Chapters..	109

4 Mathematical Expectation — 111

	4.1	Mean of a Random Variable.....................................	111
		Exercises	117
	4.2	Variance and Covariance of Random Variables...................	119
		Exercises	127
	4.3	Means and Variances of Linear Combinations of Random Variables	128
	4.4	Chebyshev's Theorem...	135
		Exercises	137
		Review Exercises...	139
	4.5	Potential Misconceptions and Hazards; Relationship to Material in Other Chapters..	142

5 Some Discrete Probability Distributions — 143

	5.1	Introduction and Motivation	143
	5.2	Binomial and Multinomial Distributions.........................	143
		Exercises	150
	5.3	Hypergeometric Distribution	152
		Exercises	157
	5.4	Negative Binomial and Geometric Distributions	158
	5.5	Poisson Distribution and the Poisson Process....................	161
		Exercises	164
		Review Exercises...	166
	5.6	Potential Misconceptions and Hazards; Relationship to Material in Other Chapters..	169

6 Some Continuous Probability Distributions ... **171**

- 6.1 Continuous Uniform Distribution ... 171
- 6.2 Normal Distribution ... 172
- 6.3 Areas under the Normal Curve ... 176
- 6.4 Applications of the Normal Distribution ... 182
  - Exercises ... 185
- 6.5 Normal Approximation to the Binomial ... 187
  - Exercises ... 193
- 6.6 Gamma and Exponential Distributions ... 194
- 6.7 Chi-Squared Distribution ... 200
- 6.8 Beta Distribution ... 201
- 6.9 Lognormal Distribution ... 201
- 6.10 Weibull Distribution (Optional) ... 203
  - Exercises ... 206
  - Review Exercises ... 207
- 6.11 Potential Misconceptions and Hazards; Relationship to Material in Other Chapters ... 209

7 Functions of Random Variables (Optional) ... **211**

- 7.1 Introduction ... 211
- 7.2 Transformations of Variables ... 211
- 7.3 Moments and Moment-Generating Functions ... 218
  - Exercises ... 222

8 Fundamental Sampling Distributions and Data Descriptions ... **225**

- 8.1 Random Sampling ... 225
- 8.2 Some Important Statistics ... 227
  - Exercises ... 230
- 8.3 Sampling Distributions ... 232
- 8.4 Sampling Distribution of Means and the Central Limit Theorem ... 233
  - Exercises ... 241
- 8.5 Sampling Distribution of S^2 ... 243
- 8.6 t-Distribution ... 246
- 8.7 F-Distribution ... 251
- 8.8 Quantile and Probability Plots ... 254
  - Exercises ... 259
  - Review Exercises ... 260
- 8.9 Potential Misconceptions and Hazards; Relationship to Material in Other Chapters ... 262

9 One- and Two-Sample Estimation Problems ... **265**

- 9.1 Introduction ... 265
- 9.2 Statistical Inference ... 265
- 9.3 Classical Methods of Estimation ... 266
- 9.4 Single Sample: Estimating the Mean ... 269
- 9.5 Standard Error of a Point Estimate ... 276
- 9.6 Prediction Intervals ... 277
- 9.7 Tolerance Limits ... 280
  - Exercises ... 282
- 9.8 Two Samples: Estimating the Difference between Two Means ... 285
- 9.9 Paired Observations ... 291
  - Exercises ... 294
- 9.10 Single Sample: Estimating a Proportion ... 296
- 9.11 Two Samples: Estimating the Difference between Two Proportions ... 300
  - Exercises ... 302
- 9.12 Single Sample: Estimating the Variance ... 303
- 9.13 Two Samples: Estimating the Ratio of Two Variances ... 305
  - Exercises ... 307
- 9.14 Maximum Likelihood Estimation (Optional) ... 307
  - Exercises ... 312
  - Review Exercises ... 313
- 9.15 Potential Misconceptions and Hazards; Relationship to Material in Other Chapters ... 316

10 One- and Two-Sample Tests of Hypotheses ... **319**

- 10.1 Statistical Hypotheses: General Concepts ... 319
- 10.2 Testing a Statistical Hypothesis ... 321
- 10.3 The Use of P-Values for Decision Making in Testing Hypotheses ... 331
  - Exercises ... 334
- 10.4 Single Sample: Tests Concerning a Single Mean ... 336
- 10.5 Two Samples: Tests on Two Means ... 342
- 10.6 Choice of Sample Size for Testing Means ... 349
- 10.7 Graphical Methods for Comparing Means ... 354
  - Exercises ... 356
- 10.8 One Sample: Test on a Single Proportion ... 360
- 10.9 Two Samples: Tests on Two Proportions ... 363
  - Exercises ... 365
- 10.10 One- and Two-Sample Tests Concerning Variances ... 366
  - Exercises ... 369
- 10.11 Goodness-of-Fit Test ... 370
- 10.12 Test for Independence (Categorical Data) ... 373

	10.13	Test for Homogeneity	376
	10.14	Two-Sample Case Study	379
		Exercises	382
		Review Exercises	384
	10.15	Potential Misconceptions and Hazards; Relationship to Material in Other Chapters	386

11 Simple Linear Regression and Correlation **389**

11.1	Introduction to Linear Regression	389
11.2	The Simple Linear Regression Model	390
11.3	Least Squares and the Fitted Model	394
	Exercises	398
11.4	Properties of the Least Squares Estimators	400
11.5	Inferences Concerning the Regression Coefficients	403
11.6	Prediction	408
	Exercises	411
11.7	Choice of a Regression Model	414
11.8	Analysis-of-Variance Approach	414
11.9	Test for Linearity of Regression: Data with Repeated Observations	416
	Exercises	421
11.10	Data Plots and Transformations	424
11.11	Simple Linear Regression Case Study	428
11.12	Correlation	430
	Exercises	435
	Review Exercises	436
11.13	Potential Misconceptions and Hazards; Relationship to Material in Other Chapters	442

12 Multiple Linear Regression and Certain Nonlinear Regression Models **443**

12.1	Introduction	443
12.2	Estimating the Coefficients	444
12.3	Linear Regression Model Using Matrices	447
	Exercises	450
12.4	Properties of the Least Squares Estimators	453
12.5	Inferences in Multiple Linear Regression	455
	Exercises	461
12.6	Choice of a Fitted Model through Hypothesis Testing	462
12.7	Special Case of Orthogonality (Optional)	467
	Exercises	471
12.8	Categorical or Indicator Variables	472

	Exercises	476
12.9	Sequential Methods for Model Selection	476
12.10	Study of Residuals and Violation of Assumptions (Model Checking)	482
12.11	Cross Validation, C_p, and Other Criteria for Model Selection	487
	Exercises	494
12.12	Special Nonlinear Models for Nonideal Conditions	496
	Exercises	500
	Review Exercises	501
12.13	Potential Misconceptions and Hazards; Relationship to Material in Other Chapters	506

13 One-Factor Experiments: General 507

13.1	Analysis-of-Variance Technique	507
13.2	The Strategy of Experimental Design	508
13.3	One-Way Analysis of Variance: Completely Randomized Design (One-Way ANOVA)	509
13.4	Tests for the Equality of Several Variances	516
	Exercises	518
13.5	Single-Degree-of-Freedom Comparisons	520
13.6	Multiple Comparisons	523
	Exercises	529
13.7	Comparing a Set of Treatments in Blocks	532
13.8	Randomized Complete Block Designs	533
13.9	Graphical Methods and Model Checking	540
13.10	Data Transformations in Analysis of Variance	543
	Exercises	545
13.11	Random Effects Models	547
13.12	Case Study	551
	Exercises	553
	Review Exercises	555
13.13	Potential Misconceptions and Hazards; Relationship to Material in Other Chapters	559

14 Factorial Experiments (Two or More Factors) 561

14.1	Introduction	561
14.2	Interaction in the Two-Factor Experiment	562
14.3	Two-Factor Analysis of Variance	565
	Exercises	575
14.4	Three-Factor Experiments	579
	Exercises	586

	14.5	Factorial Experiments for Random Effects and Mixed Models....	588
		Exercises...	592
		Review Exercises...	594
	14.6	Potential Misconceptions and Hazards; Relationship to Material in Other Chapters..	596

15 2^k Factorial Experiments and Fractions **597**

	15.1	Introduction	597
	15.2	The 2^k Factorial: Calculation of Effects and Analysis of Variance	598
	15.3	Nonreplicated 2^k Factorial Experiment..........................	604
		Exercises...	609
	15.4	Factorial Experiments in a Regression Setting..................	612
	15.5	The Orthogonal Design	617
		Exercises...	625
	15.6	Fractional Factorial Experiments...................................	626
	15.7	Analysis of Fractional Factorial Experiments	632
		Exercises...	634
	15.8	Higher Fractions and Screening Designs	636
	15.9	Construction of Resolution III and IV Designs with 8, 16, and 32 Design Points...	637
	15.10	Other Two-Level Resolution III Designs; The Plackett-Burman Designs...	638
	15.11	Introduction to Response Surface Methodology.................	639
	15.12	Robust Parameter Design	643
		Exercises...	652
		Review Exercises...	653
	15.13	Potential Misconceptions and Hazards; Relationship to Material in Other Chapters..	654

16 Nonparametric Statistics **655**

	16.1	Nonparametric Tests...	655
	16.2	Signed-Rank Test	660
		Exercises...	663
	16.3	Wilcoxon Rank-Sum Test	665
	16.4	Kruskal-Wallis Test	668
		Exercises...	670
	16.5	Runs Test..	671
	16.6	Tolerance Limits	674
	16.7	Rank Correlation Coefficient	674
		Exercises...	677
		Review Exercises...	679

17 Statistical Quality Control ... **681**

- 17.1 Introduction ... 681
- 17.2 Nature of the Control Limits ... 683
- 17.3 Purposes of the Control Chart ... 683
- 17.4 Control Charts for Variables ... 684
- 17.5 Control Charts for Attributes ... 697
- 17.6 Cusum Control Charts ... 705
- Review Exercises ... 706

18 Bayesian Statistics ... **709**

- 18.1 Bayesian Concepts ... 709
- 18.2 Bayesian Inferences ... 710
- 18.3 Bayes Estimates Using Decision Theory Framework ... 717
- Exercises ... 718

Bibliography ... **721**

Appendix A: Statistical Tables and Proofs ... **725**

Appendix B: Answers to Odd-Numbered Non-Review Exercises ... **769**

Index ... **785**

Preface

General Approach and Mathematical Level

Our emphasis in creating the ninth edition MyStatLab Update is less on adding new material and more on providing clarity and deeper understanding. This objective was accomplished in part by including new end-of-chapter material that adds connective tissue between chapters. We affectionately call these comments at the end of the chapter "Pot Holes." They are very useful to remind students of the big picture and how each chapter fits into that picture, and they aid the student in learning about limitations and pitfalls that may result if procedures are misused. A deeper understanding of real-world use of statistics is made available through class projects, which were added in several chapters. These projects provide the opportunity for students alone, or in groups, to gather their own experimental data and draw inferences. In some cases, the work involves a problem whose solution will illustrate the meaning of a concept or provide an empirical understanding of an important statistical result. Some existing examples were expanded and new ones were introduced to create "case studies," in which commentary is provided to give the student a clear understanding of a statistical concept in the context of a practical situation.

In this edition, we continue to emphasize a balance between theory and applications. Calculus and other types of mathematical support (e.g., linear algebra) are used at about the same level as in previous editions. The coverage of analytical tools in statistics is enhanced with the use of calculus when discussion centers on rules and concepts in probability. Probability distributions and statistical inference are highlighted in Chapters 2 through 10. Linear algebra and matrices are very lightly applied in Chapters 11 through 15, where linear regression and analysis of variance are covered. Students using this text should have had the equivalent of one semester of differential and integral calculus. Linear algebra is helpful but not necessary so long as the section in Chapter 12 on multiple linear regression using matrix algebra is not covered by the instructor. As in previous editions, a large number of exercises that deal with real-life scientific and engineering applications are available to challenge the student. The many data sets associated with the exercises are available for download from the website http://www.pearsonhighered.com/datasets or in MyStatLab.

Summary of the Changes in the Ninth Edition MyStatLab Update

- We've added MyStatLab, a course management system that delivers proven results in helping individual students succeed. MyStatLab provides engaging experiences that personalize, stimulate, and measure learning for each student.

To learn more about how MyStatLab combines proven learning applications with powerful assessment, visit www.mystatlab.com or contact your Pearson representative.

- Class projects were added in several chapters to provide a deeper understanding of the real-world use of statistics. Students are asked to produce or gather their own experimental data and draw inferences from these data.
- More case studies were added and others expanded to help students understand the statistical methods being presented in the context of a real-life situation.
- "Pot Holes" were added at the end of some chapters and expanded in others. These comments are intended to present each chapter in the context of the big picture and discuss how the chapters relate to one another. They also provide cautions about the possible misuse of statistical techniques.
- Chapter 1 has been enhanced to include more on single-number statistics as well as graphical techniques. New fundamental material on sampling and experimental design is presented.
- Examples added to Chapter 8 on sampling distributions are intended to motivate P-values and hypothesis testing. This prepares the student for the more challenging material on these topics that will be presented in Chapter 10.
- Chapter 12 contains additional development regarding the effect of a single regression variable in a model in which collinearity with other variables is severe.
- Chapter 15 now introduces material on the important topic of response surface methodology (RSM). The use of noise variables in RSM allows the illustration of mean and variance (dual response surface) modeling.
- The central composite design (CCD) is introduced in Chapter 15.
- More examples are given in Chapter 18, and the discussion of using Bayesian methods for statistical decision making has been enhanced.

Content and Course Planning

This text is designed for either a one- or a two-semester course. A reasonable plan for a one-semester course might include Chapters 1 through 10. This would result in a curriculum that concluded with the fundamentals of both estimation and hypothesis testing. Instructors who desire that students be exposed to simple linear regression may wish to include a portion of Chapter 11. For instructors who desire to have analysis of variance included rather than regression, the one-semester course may include Chapter 13 rather than Chapters 11 and 12. Chapter 13 features one-factor analysis of variance. Another option is to eliminate portions of Chapters 5 and/or 6 as well as Chapter 7. With this option, one or more of the discrete or continuous distributions in Chapters 5 and 6 may be eliminated. These distributions include the negative binomial, geometric, gamma, Weibull, beta, and log normal distributions. Other features that one might consider removing from a one-semester curriculum include maximum likelihood estimation, prediction, and/or tolerance limits in Chapter 9. A one-semester curriculum has built-in flexibility, depending on the relative interest of the instructor in regression, analysis of variance, experimental design, and response surface methods (Chapter 15). There are several discrete and continuous distributions (Chapters 5 and 6) that have applications in a variety of engineering and scientific areas.

Chapters 11 through 18 contain substantial material that can be added for the second semester of a two-semester course. The material on simple and multiple linear regression is in Chapters 11 and 12, respectively. Chapter 12 alone offers a substantial amount of flexibility. Multiple linear regression includes such "special topics" as categorical or indicator variables, sequential methods of model selection such as stepwise regression, the study of residuals for the detection of violations of assumptions, cross validation and the use of the PRESS statistic as well as C_p, and logistic regression. The use of orthogonal regressors, a precursor to the experimental design in Chapter 15, is highlighted. Chapters 13 and 14 offer a relatively large amount of material on analysis of variance (ANOVA) with fixed, random, and mixed models. Chapter 15 highlights the application of two-level designs in the context of full and fractional factorial experiments (2^k). Special screening designs are illustrated. Chapter 15 also features a new section on response surface methodology (RSM) to illustrate the use of experimental design for finding optimal process conditions. The fitting of a second order model through the use of a central composite design is discussed. RSM is expanded to cover the analysis of robust parameter design type problems. Noise variables are used to accommodate dual response surface models. Chapters 16, 17, and 18 contain a moderate amount of material on nonparametric statistics, quality control, and Bayesian inference.

Chapter 1 is an overview of statistical inference presented on a mathematically simple level. It has been expanded from the eighth edition to more thoroughly cover single-number statistics and graphical techniques. It is designed to give students a preliminary presentation of elementary concepts that will allow them to understand more involved details that follow. Elementary concepts in sampling, data collection, and experimental design are presented, and rudimentary aspects of graphical tools are introduced, as well as a sense of what is garnered from a data set. Stem-and-leaf plots and box-and-whisker plots have been added. Graphs are better organized and labeled. The discussion of uncertainty and variation in a system is thorough and well illustrated. There are examples of how to sort out the important characteristics of a scientific process or system, and these ideas are illustrated in practical settings such as manufacturing processes, biomedical studies, and studies of biological and other scientific systems. A contrast is made between the use of discrete and continuous data. Emphasis is placed on the use of models and the information concerning statistical models that can be obtained from graphical tools.

Chapters 2, 3, and 4 deal with basic probability as well as discrete and continuous random variables. Chapters 5 and 6 focus on specific discrete and continuous distributions as well as relationships among them. These chapters also highlight examples of applications of the distributions in real-life scientific and engineering studies. Examples, case studies, and a large number of exercises edify the student concerning the use of these distributions. Projects bring the practical use of these distributions to life through group work. Chapter 7 is the most theoretical chapter in the text. It deals with transformation of random variables and will likely not be used unless the instructor wishes to teach a relatively theoretical course. Chapter 8 contains graphical material, expanding on the more elementary set of graphical tools presented and illustrated in Chapter 1. Probability plotting is discussed and illustrated with examples. The very important concept of sampling distributions is presented thoroughly, and illustrations are given that involve the central limit theorem and the distribution of a sample variance under normal, independent (i.i.d.) sampling. The t and F distributions are introduced to motivate their use in chapters to follow. New material in Chapter 8 helps the student to visualize the importance of hypothesis testing, motivating the concept of a P-value.

Chapter 9 contains material on one- and two-sample point and interval estimation. A thorough discussion with examples points out the contrast between the different types of intervals—confidence intervals, prediction intervals, and tolerance intervals. A case study illustrates the three types of statistical intervals in the context of a manufacturing situation. This case study highlights the differences among the intervals, their sources, and the assumptions made in their development, as well as what type of scientific study or question requires the use of each one. A new approximation method has been added for the inference concerning a proportion. Chapter 10 begins with a basic presentation on the pragmatic meaning of hypothesis testing, with emphasis on such fundamental concepts as null and alternative hypotheses, the role of probability and the P-value, and the power of a test. Following this, illustrations are given of tests concerning one and two samples under standard conditions. The two-sample t-test with paired observations is also described. A case study helps the student to develop a clear picture of what interaction among factors really means as well as the dangers that can arise when interaction between treatments and experimental units exists. At the end of Chapter 10 is a very important section that relates Chapters 9 and 10 (estimation and hypothesis testing) to Chapters 11 through 16, where statistical modeling is prominent. It is important that the student be aware of the strong connection.

Chapters 11 and 12 contain material on simple and multiple linear regression, respectively. Considerably more attention is given in this edition to the effect that collinearity among the regression variables plays. A situation is presented that shows how the role of a single regression variable can depend in large part on what regressors are in the model with it. The sequential model selection procedures (forward, backward, stepwise, etc.) are then revisited in regard to this concept, and the rationale for using certain P-values with these procedures is provided. Chapter 12 offers material on nonlinear modeling with a special presentation of logistic regression, which has applications in engineering and the biological sciences. The material on multiple regression is quite extensive and thus provides considerable flexibility for the instructor, as indicated earlier. At the end of Chapter 12 is commentary relating that chapter to Chapters 14 and 15. Several features were added that provide a better understanding of the material in general. For example, the end-of-chapter material deals with cautions and difficulties one might encounter. It is pointed out that there are types of responses that occur naturally in practice (e.g. proportion responses, count responses, and several others) with which standard least squares regression should not be used because standard assumptions do not hold and violation of assumptions may induce serious errors. The suggestion is made that data transformation on the response may alleviate the problem in some cases. Flexibility is again available in Chapters 13 and 14, on the topic of analysis of variance. Chapter 13 covers one-factor ANOVA in the context of a completely randomized design. Complementary topics include tests on variances and multiple comparisons. Comparisons of treatments in blocks are highlighted, along with the topic of randomized complete blocks. Graphical methods are extended to ANOVA to aid the student in supplementing the formal inference with a pictorial type of inference that can aid scientists and engineers in presenting material. A new project is given in which students incorporate the appropriate randomization into each plan and use graphical techniques and P-values in reporting the results. Chapter 14 extends the material in Chapter 13 to accommodate two or more factors that are in a factorial structure. The ANOVA presentation in Chapter 14 includes work in both random and fixed effects models. Chapter 15 offers material associated with 2^k factorial designs; examples and case studies present the use of screening designs and special higher fractions of the 2^k. Two new and special features are the presentations of response surface methodology (RSM) and robust parameter design. These topics are linked in a case study that describes and illustrates a dual response surface design and analysis featuring the use of process mean and variance response surfaces.

Computer Software

Case studies, beginning in Chapter 8, feature computer printout and graphical material generated using both SAS and MINITAB. The inclusion of the computer reflects our belief that students should have the experience of reading and interpreting computer printout and graphics, even if the software in the text is not that which is used by the instructor. Exposure to more than one type of software can broaden the experience base for the student. There is no reason to believe that the software used in the course will be that which the student will be called upon to use in practice following graduation. Examples and case studies in the text are supplemented, where appropriate, by various types of residual plots, quantile plots, normal probability plots, and other plots. Such plots are particularly prevalent in Chapters 11 through 15.

Acknowledgments

We are indebted to those colleagues who reviewed the previous editions of this book and provided many helpful suggestions for this edition. They are Julie Couton, *University of Nebraska-Lincoln*; David Groggel, *Miami University*; Lance Hemlow, *Raritan Valley Community College*; Ying Ji, *University of Texas at San Antonio*; Thomas Kline, *University of Northern Iowa*; Sheila Lawrence, *Rutgers University*; Luis Moreno, *Broome County Community College*; David Nembhard, *Pennsylvania State University*; Joon Jin Song, *Baylor University*; Donald Waldman, *University of Colorado—Boulder*; Marlene Will, *Spalding University*; and Connie Wilson, *Old Dominion University*. We would also like to thank Delray Schulz, *Millersville University*; Roxane Burrows, *Hocking College*; and Frank Chmely for ensuring the accuracy of this text.

We would like to thank the editorial and production services provided by numerous people from Pearson, especially editor in chief Deirdre Lynch, acquisitions editor Patrick Barbera, project manager Christine Whitlock, editorial assistant Justin Billing, and copyeditor Sally Lifland. Many useful comments and suggestions by proofreader Gail Magin are greatly appreciated. We thank the Virginia Tech Statistical Consulting Center, which was the source of many real-life data sets.

R.H.M.
S.L.M.
K.Y.

Get the Most Out of MyStatLab™

BREAKTHROUGH
To improving results

MyStatLab is the world's leading online resource for teaching and learning statistics. MyStatLab helps students and instructors improve results and provides engaging experiences and personalized learning for each student so learning can happen in any environment. Plus, it offers flexible and time-saving course management features to allow instructors to easily manage their classes while remaining in complete control, regardless of course format.

Personalized Support for Students

- MyStatLab comes with many learning resources—eText, applets, videos, and more—all designed to support your students as they progress through their course.

- The Adaptive Study Plan acts as a personal tutor, updating in real time based on student performance to provide personalized recommendations on what to work on next. With the new Companion Study Plan assignments, instructors can now assign the Study Plan as a prerequisite to a test or quiz, helping to guide students through concepts they need to master.

- Personalized Homework allows instructors to create homework assignments tailored to each student's specific needs, focused on just the topics they have not yet mastered.

Used by nearly 4 million students each year, the MyStatLab and MyMathLab family of products delivers consistent, measurable gains in student learning outcomes, retention, and subsequent course success.

www.mystatlab.com

BREAKTHROUGH
To improving results

Resources for Success

Instructor's Solutions Manual

The Instructor's Solutions Manual contains worked-out solutions to all text exercises and is available for download from Pearson Education's Instructor's Resource Center **(www.pearsonhighered.com/irc)** and in MyStatLab.

PowerPoint Slides

The PowerPoint slides include most of the figures and tables from the text. Slides are available to download from Pearson Education's Instructor Resource Center **(www.pearsonhighered.com/irc)** and in MyStatLab.

Student Solutions Manual

The Student Solutions Manual features complete solutions to selected exercises. This is a great tool for students as they study and work through the problems.

ISBN-10: 0-13-411623-2
ISBN-13: 978-0-13-411623-5

MyStatLab™ Online Course (access code required)

MyStatLab from Pearson is the world's leading online resource for teaching and learning statistics; it integrates interactive homework, assessment, and media in a flexible, easy to use format. MyStatLab is a course management system that helps individual students succeed. It provides engaging experiences that personalize, stimulate, and measure learning for each student. Tools are embedded to make it easy to integrate statistical software into the course. And, it comes from an experienced partner with educational expertise and an eye on the future. MyStatLab leverages the power of the web-based statistical software, StatCrunch™, and includes access to **www.StatCrunch.com.** To learn more about how MyStatLab combines proven learning applications with powerful assessment, visit **www.mystatlab.com** or contact your Pearson representative.

Chapter 1

Introduction to Statistics and Data Analysis

1.1 Overview: Statistical Inference, Samples, Populations, and the Role of Probability

Beginning in the 1980s and continuing into the 21st century, an inordinate amount of attention has been focused on *improvement of quality* in American industry. Much has been said and written about the Japanese "industrial miracle," which began in the middle of the 20th century. The Japanese were able to succeed where we and other countries had failed–namely, to create an atmosphere that allows the production of high-quality products. Much of the success of the Japanese has been attributed to the use of *statistical methods* and statistical thinking among management personnel.

Use of Scientific Data

The use of statistical methods in manufacturing, development of food products, computer software, energy sources, pharmaceuticals, and many other areas involves the gathering of information or **scientific data**. Of course, the gathering of data is nothing new. It has been done for well over a thousand years. Data have been collected, summarized, reported, and stored for perusal. However, there is a profound distinction between collection of scientific information and **inferential statistics**. It is the latter that has received rightful attention in recent decades.

The offspring of inferential statistics has been a large "toolbox" of statistical methods employed by statistical practitioners. These statistical methods are designed to contribute to the process of making scientific judgments in the face of **uncertainty** and **variation**. The product density of a particular material from a manufacturing process will not always be the same. Indeed, if the process involved is a batch process rather than continuous, there will be not only variation in material density among the batches that come off the line (batch-to-batch variation), but also within-batch variation. Statistical methods are used to analyze data from a process such as this one in order to gain more sense of where in the process changes may be made to improve the **quality** of the process. In this process, quality may well be defined in relation to closeness to a target density value in harmony with *what portion of the time* this closeness criterion is met. An engineer may be concerned with a specific instrument that is used to measure sulfur monoxide in the air during pollution studies. If the engineer has doubts about the effectiveness of the instrument, there are two **sources of variation** that must be dealt with. The first is the variation in sulfur monoxide values that are found at the same locale on the same day. The second is the variation between values observed and the **true** amount of sulfur monoxide that is in the air at the time. If either of these two sources of variation is exceedingly large (according to some standard set by the engineer), the instrument may need to be replaced. In a biomedical study of a new drug that reduces hypertension, 85% of patients experienced relief, while it is generally recognized that the current drug, or "old" drug, brings relief to 80% of patients that have chronic hypertension. However, the new drug is more expensive to make and may result in certain side effects. Should the new drug be adopted? This is a problem that is encountered (often with much more complexity) frequently by pharmaceutical firms in conjunction with the FDA (Federal Drug Administration). Again, the consideration of variation needs to be taken into account. The "85%" value is based on a certain number of patients chosen for the study. Perhaps if the study were repeated with new patients the observed number of "successes" would be 75%! It is the natural variation from study to study that must be taken into account in the decision process. Clearly this variation is important, since variation from patient to patient is endemic to the problem.

Variability in Scientific Data

In the problems discussed above the statistical methods used involve dealing with variability, and in each case the variability to be studied is that encountered in scientific data. If the observed product density in the process were always the same and were always on target, there would be no need for statistical methods. If the device for measuring sulfur monoxide always gives the same value and the value is accurate (i.e., it is correct), no statistical analysis is needed. If there were no patient-to-patient variability inherent in the response to the drug (i.e., it either always brings relief or not), life would be simple for scientists in the pharmaceutical firms and FDA and no statistician would be needed in the decision process. Statistics researchers have produced an enormous number of analytical methods that allow for analysis of data from systems like those described above. This reflects the true nature of the science that we call inferential statistics, namely, using techniques that allow us to go beyond merely reporting data to drawing conclusions (or inferences) about the scientific system. Statisticians make use of fundamental laws of probability and statistical inference to draw conclusions about scientific systems. Information is gathered in the form of **samples**, or collections of **observations**. The process of sampling is introduced in Chapter 2, and the discussion continues throughout the entire book.

Samples are collected from **populations**, which are collections of all individuals or individual items of a particular type. At times a population signifies a scientific system. For example, a manufacturer of computer boards may wish to eliminate defects. A sampling process may involve collecting information on 50 computer boards sampled randomly from the process. Here, the population is all computer boards manufactured by the firm over a specific period of time. If an improvement is made in the computer board process and a second sample of boards is collected, any conclusions drawn regarding the effectiveness of the change in process should extend to the entire population of computer boards produced under the "improved process." In a drug experiment, a sample of patients is taken and each is given a specific drug to reduce blood pressure. The interest is focused on drawing conclusions about the population of those who suffer from hypertension.

Often, it is very important to collect scientific data in a systematic way, with planning being high on the agenda. At times the planning is, by necessity, quite limited. We often focus only on certain properties or characteristics of the items or objects in the population. Each characteristic has particular engineering or, say, biological importance to the "customer," the scientist or engineer who seeks to learn about the population. For example, in one of the illustrations above the quality of the process had to do with the product density of the output of a process. An engineer may need to study the effect of process conditions, temperature, humidity, amount of a particular ingredient, and so on. He or she can systematically move these **factors** to whatever levels are suggested according to whatever prescription or **experimental design** is desired. However, a forest scientist who is interested in a study of factors that influence wood density in a certain kind of tree cannot necessarily design an experiment. This case may require an **observational study** in which data are collected in the field but **factor levels** can not be preselected. Both of these types of studies lend themselves to methods of statistical inference. In the former, the quality of the inferences will depend on proper planning of the experiment. In the latter, the scientist is at the mercy of what can be gathered. For example, it is sad if an agronomist is interested in studying the effect of rainfall on plant yield and the data are gathered during a drought.

The importance of statistical thinking by managers and the use of statistical inference by scientific personnel is widely acknowledged. Research scientists gain much from scientific data. Data provide understanding of scientific phenomena. Product and process engineers learn a great deal in their off-line efforts to improve the process. They also gain valuable insight by gathering production data (on-line monitoring) on a regular basis. This allows them to determine necessary modifications in order to keep the process at a desired level of quality.

There are times when a scientific practitioner wishes only to gain some sort of summary of a set of data represented in the sample. In other words, inferential statistics is not required. Rather, a set of single-number statistics or **descriptive statistics** is helpful. These numbers give a sense of center of the location of the data, variability in the data, and the general nature of the distribution of observations in the sample. Though no specific statistical methods leading to **statistical inference** are incorporated, much can be learned. At times, descriptive statistics are accompanied by graphics. Modern statistical software packages allow for computation of **means**, **medians**, **standard deviations**, and other single-number statistics as well as production of graphs that show a "footprint" of the nature of the sample. Definitions and illustrations of the single-number statistics and graphs, including histograms, stem-and-leaf plots, scatter plots, dot plots, and box plots, will be given in sections that follow.

The Role of Probability

In this book, Chapters 2 to 6 deal with fundamental notions of probability. A thorough grounding in these concepts allows the reader to have a better understanding of statistical inference. Without some formalism of probability theory, the student cannot appreciate the true interpretation from data analysis through modern statistical methods. It is quite natural to study probability prior to studying statistical inference. Elements of probability allow us to quantify the strength or "confidence" in our conclusions. In this sense, concepts in probability form a major component that supplements statistical methods and helps us gauge the strength of the statistical inference. The discipline of probability, then, provides the transition between descriptive statistics and inferential methods. Elements of probability allow the conclusion to be put into the language that the science or engineering practitioners require. An example follows that will enable the reader to understand the notion of a P-value, which often provides the "bottom line" in the interpretation of results from the use of statistical methods.

Example 1.1: Suppose that an engineer encounters data from a manufacturing process in which 100 items are sampled and 10 are found to be defective. It is expected and anticipated that occasionally there will be defective items. Obviously these 100 items represent the sample. However, it has been determined that in the long run, the company can only tolerate 5% defective in the process. Now, the elements of probability allow the engineer to determine how conclusive the sample information is regarding the nature of the process. In this case, the **population** conceptually represents all possible items from the process. Suppose we learn that *if the process is acceptable*, that is, if it does produce items no more than 5% of which are defective, there is a probability of 0.0282 of obtaining 10 or more defective items in a random sample of 100 items from the process. This small probability suggests that the process does, indeed, have a long-run rate of defective items that exceeds 5%. In other words, under the condition of an acceptable process, the sample information obtained would rarely occur. However, it did occur! Clearly, though, it would occur with a much higher probability if the process defective rate exceeded 5% by a significant amount.

From this example it becomes clear that the elements of probability aid in the translation of sample information into something conclusive or inconclusive about the scientific system. In fact, what was learned likely is alarming information to the engineer or manager. Statistical methods, which we will actually detail in Chapter 10, produced a P-value of 0.0282. The result suggests that the process **very likely is not acceptable**. The concept of a **P-value** is dealt with at length in succeeding chapters. The example that follows provides a second illustration.

Example 1.2: Often the nature of the scientific study will dictate the role that probability and deductive reasoning play in statistical inference. Exercise 9.40 on page 294 provides data associated with a study conducted at the Virginia Polytechnic Institute and State University on the development of a relationship between the roots of trees and the action of a fungus. Minerals are transferred from the fungus to the trees and sugars from the trees to the fungus. Two samples of 10 northern red oak seedlings were planted in a greenhouse, one containing seedlings treated with nitrogen and the other containing seedlings with no nitrogen. All other environmental conditions were held constant. All seedlings contained the fungus *Pisolithus tinctorus*. More details are supplied in Chapter 9. The stem weights in grams were recorded after the end of 140 days. The data are given in Table 1.1.

Table 1.1: Data Set for Example 1.2

No Nitrogen	Nitrogen
0.32	0.26
0.53	0.43
0.28	0.47
0.37	0.49
0.47	0.52
0.43	0.75
0.36	0.79
0.42	0.86
0.38	0.62
0.43	0.46

Figure 1.1: A dot plot of stem weight data.

In this example there are two samples from two **separate populations**. The purpose of the experiment is to determine if the use of nitrogen has an influence on the growth of the roots. The study is a comparative study (i.e., we seek to compare the two populations with regard to a certain important characteristic). It is instructive to plot the data as shown in the dot plot of Figure 1.1. The ○ values represent the "nitrogen" data and the × values represent the "no-nitrogen" data.

Notice that the general appearance of the data might suggest to the reader that, on average, the use of nitrogen increases the stem weight. Four nitrogen observations are considerably larger than any of the no-nitrogen observations. Most of the no-nitrogen observations appear to be below the center of the data. The appearance of the data set would seem to indicate that nitrogen is effective. But how can this be quantified? How can all of the apparent visual evidence be summarized in some sense? As in the preceding example, the fundamentals of probability can be used. The conclusions may be summarized in a probability statement or *P*-value. We will not show here the statistical inference that produces the summary probability. As in Example 1.1, these methods will be discussed in Chapter 10. The issue revolves around the "probability that data like these could be observed" *given that nitrogen has no effect*, in other words, given that both samples were generated from the same population. Suppose that this probability is small, say 0.03. That would certainly be strong evidence that the use of nitrogen does indeed influence (apparently increases) average stem weight of the red oak seedlings.

How Do Probability and Statistical Inference Work Together?

It is important for the reader to understand the clear distinction between the discipline of probability, a science in its own right, and the discipline of inferential statistics. As we have already indicated, the use or application of concepts in probability allows real-life interpretation of the results of statistical inference. As a result, it can be said that statistical inference makes use of concepts in probability. One can glean from the two examples above that the sample information is made available to the analyst and, with the aid of statistical methods and elements of probability, conclusions are drawn about some feature of the population (the process does not appear to be acceptable in Example 1.1, and nitrogen does appear to influence average stem weights in Example 1.2). Thus for a statistical problem, **the sample along with inferential statistics allows us to draw conclusions about the population, with inferential statistics making clear use of elements of probability**. This reasoning is *inductive* in nature. Now as we move into Chapter 2 and beyond, the reader will note that, unlike what we do in our two examples here, we will not focus on solving statistical problems. Many examples will be given in which no sample is involved. There will be a population clearly described with all features of the population known. Then questions of importance will focus on the nature of data that might hypothetically be drawn from the population. Thus, one can say that **elements in probability allow us to draw conclusions about characteristics of hypothetical data taken from the population, based on known features of the population**. This type of reasoning is *deductive* in nature. Figure 1.2 shows the fundamental relationship between probability and inferential statistics.

Figure 1.2: Fundamental relationship between probability and inferential statistics.

Now, in the grand scheme of things, which is more important, the field of probability or the field of statistics? They are both very important and clearly are complementary. The only certainty concerning the pedagogy of the two disciplines lies in the fact that if statistics is to be taught at more than merely a "cookbook" level, then the discipline of probability must be taught first. This rule stems from the fact that nothing can be learned about a population from a sample until the analyst learns the rudiments of uncertainty in that sample. For example, consider Example 1.1. The question centers around whether or not the population, defined by the process, is no more than 5% defective. In other words, the conjecture is that **on the average** 5 out of 100 items are defective. Now, the sample contains 100 items and 10 are defective. Does this support the conjecture or refute it? On the surface it would appear to be a refutation of the conjecture because 10 out of 100 seem to be "a bit much." But without elements of probability, how do we know? Only through the study of material in future chapters will we learn the conditions under which the process is acceptable (5% defective). The probability of obtaining 10 or more defective items in a sample of 100 is 0.0282.

We have given two examples where the elements of probability provide a summary that the scientist or engineer can use as evidence on which to build a decision. The bridge between the data and the conclusion is, of course, based on foundations of statistical inference, distribution theory, and sampling distributions discussed in future chapters.

1.2 Sampling Procedures; Collection of Data

In Section 1.1 we discussed very briefly the notion of sampling and the sampling process. While sampling appears to be a simple concept, the complexity of the questions that must be answered about the population or populations necessitates that the sampling process be very complex at times. While the notion of sampling is discussed in a technical way in Chapter 8, we shall endeavor here to give some common-sense notions of sampling. This is a natural transition to a discussion of the concept of variability.

Simple Random Sampling

The importance of proper sampling revolves around the degree of confidence with which the analyst is able to answer the questions being asked. Let us assume that only a single population exists in the problem. Recall that in Example 1.2 two populations were involved. **Simple random sampling** implies that any particular sample of a specified *sample size* has the same chance of being selected as any other sample of the same size. The term **sample size** simply means the number of elements in the sample. Obviously, a table of random numbers can be utilized in sample selection in many instances. The virtue of simple random sampling is that it aids in the elimination of the problem of having the sample reflect a different (possibly more confined) population than the one about which inferences need to be made. For example, a sample is to be chosen to answer certain questions regarding political preferences in a certain state in the United States. The sample involves the choice of, say, 1000 families, and a survey is to be conducted. Now, suppose it turns out that random sampling is not used. Rather, all or nearly all of the 1000 families chosen live in an urban setting. It is believed that political preferences in rural areas differ from those in urban areas. In other words, the sample drawn actually confined the population and thus the inferences need to be confined to the "limited population," and in this case confining may be undesirable. If, indeed, the inferences need to be made about the state as a whole, the sample of size 1000 described here is often referred to as a **biased sample**.

As we hinted earlier, simple random sampling is not always appropriate. Which alternative approach is used depends on the complexity of the problem. Often, for example, the sampling units are not homogeneous and naturally divide themselves into nonoverlapping groups that are homogeneous. These groups are called *strata*, and a procedure called *stratified random sampling* involves random selection of a sample *within* each stratum. The purpose is to be sure that each of the strata is neither over- nor underrepresented. For example, suppose a sample survey is conducted in order to gather preliminary opinions regarding a bond referendum that is being considered in a certain city. The city is subdivided into several ethnic groups which represent natural strata. In order not to disregard or overrepresent any group, separate random samples of families could be chosen from each group.

Experimental Design

The concept of randomness or random assignment plays a huge role in the area of **experimental design**, which was introduced very briefly in Section 1.1 and is an important staple in almost any area of engineering or experimental science. This will be discussed at length in Chapters 13 through 15. However, it is instructive to give a brief presentation here in the context of random sampling. A set of so-called **treatments** or **treatment combinations** becomes the populations to be studied or compared in some sense. An example is the nitrogen versus no-nitrogen treatments in Example 1.2. Another simple example would be "placebo" versus "active drug," or in a corrosion fatigue study we might have treatment combinations that involve specimens that are coated or uncoated as well as conditions of low or high humidity to which the specimens are exposed. In fact, there are four treatment or factor combinations (i.e., 4 populations), and many scientific questions may be asked and answered through statistical and inferential methods. Consider first the situation in Example 1.2. There are 20 diseased seedlings involved in the experiment. It is easy to see from the data themselves that the seedlings are different from each other. Within the nitrogen group (or the no-nitrogen group) there is considerable **variability** in the stem weights. This variability is due to what is generally called the **experimental unit**. This is a very important concept in inferential statistics, in fact one whose description will not end in this chapter. The nature of the variability is very important. If it is too large, stemming from a condition of excessive nonhomogeneity in experimental units, the variability will "wash out" any detectable difference between the two populations. Recall that in this case that did not occur.

The dot plot in Figure 1.1 and *P*-value indicated a clear distinction between these two conditions. What role do those experimental units play in the data-taking process itself? The common-sense and, indeed, quite standard approach is to assign the 20 seedlings or experimental units **randomly to the two treatments or conditions**. In the drug study, we may decide to use a total of 200 available patients, patients that clearly will be different in some sense. They are the experimental units. However, they all may have the same chronic condition for which the drug is a potential treatment. Then in a so-called **completely randomized design**, 100 patients are assigned randomly to the placebo and 100 to the active drug. Again, it is these experimental units within a group or treatment that produce the variability in data results (i.e., variability in the measured result), say blood pressure, or whatever drug efficacy value is important. In the corrosion fatigue study, the experimental units are the specimens that are the subjects of the corrosion.

Why Assign Experimental Units Randomly?

What is the possible negative impact of not randomly assigning experimental units to the treatments or treatment combinations? This is seen most clearly in the case of the drug study. Among the characteristics of the patients that produce variability in the results are age, gender, and weight. Suppose merely by chance the placebo group contains a sample of people that are predominately heavier than those in the treatment group. Perhaps heavier individuals have a tendency to have a higher blood pressure. This clearly biases the result, and indeed, any result obtained through the application of statistical inference may have little to do with the drug and more to do with differences in weights among the two samples of patients.

We should emphasize the attachment of importance to the term **variability**. Excessive variability among experimental units "camouflages" scientific findings. In future sections, we attempt to characterize and quantify measures of variability. In sections that follow, we introduce and discuss specific quantities that can be computed in samples; the quantities give a sense of the nature of the sample with respect to center of location of the data and variability in the data. A discussion of several of these single-number measures serves to provide a preview of what statistical information will be important components of the statistical methods that are used in future chapters. These measures that help characterize the nature of the data set fall into the category of **descriptive statistics**. This material is a prelude to a brief presentation of pictorial and graphical methods that go even further in characterization of the data set. The reader should understand that the statistical methods illustrated here will be used throughout the text. In order to offer the reader a clearer picture of what is involved in experimental design studies, we offer Example 1.3.

Example 1.3: A corrosion study was made in order to determine whether coating an aluminum metal with a corrosion retardation substance reduced the amount of corrosion. The coating is a protectant that is advertised to minimize fatigue damage in this type of material. Also of interest is the influence of humidity on the amount of corrosion. A corrosion measurement can be expressed in thousands of cycles to failure. Two levels of coating, no coating and chemical corrosion coating, were used. In addition, the two relative humidity levels are 20% relative humidity and 80% relative humidity.

The experiment involves four treatment combinations that are listed in the table that follows. There are eight experimental units used, and they are aluminum specimens prepared; two are assigned randomly to each of the four treatment combinations. The data are presented in Table 1.2.

The corrosion data are averages of two specimens. A plot of the averages is pictured in Figure 1.3. A relatively large value of cycles to failure represents a small amount of corrosion. As one might expect, an increase in humidity appears to make the corrosion worse. The use of the chemical corrosion coating procedure appears to reduce corrosion.

In this experimental design illustration, the engineer has systematically selected the four treatment combinations. In order to connect this situation to concepts with which the reader has been exposed to this point, it should be assumed that the

Table 1.2: Data for Example 1.3

Coating	Humidity	Average Corrosion in Thousands of Cycles to Failure
Uncoated	20%	975
	80%	350
Chemical Corrosion	20%	1750
	80%	1550

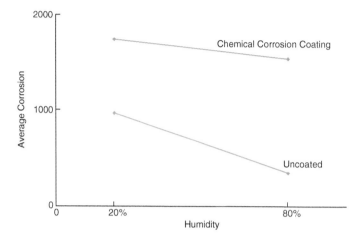

Figure 1.3: Corrosion results for Example 1.3.

conditions representing the four treatment combinations are four separate populations and that the two corrosion values observed for each population are important pieces of information. The importance of the average in capturing and summarizing certain features in the population will be highlighted in Section 1.3. While we might draw conclusions about the role of humidity and the impact of coating the specimens from the figure, we cannot truly evaluate the results from an analytical point of view without taking into account the *variability around* the average. Again, as we indicated earlier, if the two corrosion values for each treatment combination are close together, the picture in Figure 1.3 may be an accurate depiction. But if each corrosion value in the figure is an average of two values that are widely dispersed, then this variability may, indeed, truly "wash away" any information that appears to come through when one observes averages only. The foregoing example illustrates these concepts:

(1) random assignment of treatment combinations (coating, humidity) to experimental units (specimens)

(2) the use of sample averages (average corrosion values) in summarizing sample information

(3) the need for consideration of measures of variability in the analysis of any sample or sets of samples

1.3 Measures of Location: The Sample Mean and Median

This example suggests the need for what follows in Sections 1.3 and 1.4, namely, descriptive statistics that indicate measures of center of location in a set of data, and those that measure variability.

Measures of location are designed to provide the analyst with some quantitative values of where the center, or some other location, of data is located. In Example 1.2, it appears as if the center of the nitrogen sample clearly exceeds that of the no-nitrogen sample. One obvious and very useful measure is the **sample mean**. The mean is simply a numerical average.

Definition 1.1: Suppose that the observations in a sample are $x_1, x_2, \ldots, x_n$. The **sample mean**, denoted by $\bar{x}$, is

$$\bar{x} = \sum_{i=1}^{n} \frac{x_i}{n} = \frac{x_1 + x_2 + \cdots + x_n}{n}.$$

There are other measures of central tendency that are discussed in detail in future chapters. One important measure is the **sample median**. The purpose of the sample median is to reflect the central tendency of the sample in such a way that it is uninfluenced by extreme values or outliers.

Definition 1.2: Given that the observations in a sample are $x_1, x_2, \ldots, x_n$, arranged in **increasing order** of magnitude, the sample median is

$$\tilde{x} = \begin{cases} x_{(n+1)/2}, & \text{if } n \text{ is odd}, \\ \frac{1}{2}(x_{n/2} + x_{n/2+1}), & \text{if } n \text{ is even}. \end{cases}$$

As an example, suppose the data set is the following: 1.7, 2.2, 3.9, 3.11, and 14.7. The sample mean and median are, respectively,

$$\bar{x} = 5.12, \quad \tilde{x} = 3.11.$$

Clearly, the mean is influenced considerably by the presence of the extreme observation, 14.7, whereas the median places emphasis on the true "center" of the data set. In the case of the two-sample data set of Example 1.2, the two measures of central tendency for the individual samples are

$$\begin{aligned}
\bar{x} \text{ (no nitrogen)} &= 0.399 \text{ gram}, \\
\tilde{x} \text{ (no nitrogen)} &= \frac{0.38 + 0.42}{2} = 0.400 \text{ gram}, \\
\bar{x} \text{ (nitrogen)} &= 0.565 \text{ gram}, \\
\tilde{x} \text{ (nitrogen)} &= \frac{0.49 + 0.52}{2} = 0.505 \text{ gram}.
\end{aligned}$$

Clearly there is a difference in concept between the mean and median. It may be of interest to the reader with an engineering background that the sample mean is the **centroid of the data** in a sample. In a sense, it is the point at which a fulcrum can be placed to balance a system of "weights" which are the locations of the individual data. This is shown in Figure 1.4 with regard to the with-nitrogen sample.

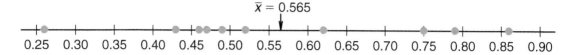

Figure 1.4: Sample mean as a centroid of the with-nitrogen stem weight.

In future chapters, the basis for the computation of $\bar{x}$ is that of an **estimate** of the **population mean**. As we indicated earlier, the purpose of statistical inference is to draw conclusions about population characteristics or **parameters** and **estimation** is a very important feature of statistical inference.

The median and mean can be quite different from each other. Note, however, that in the case of the stem weight data the sample mean value for no-nitrogen is quite similar to the median value.

Other Measures of Locations

There are several other methods of quantifying the center of location of the data in the sample. We will not deal with them at this point. For the most part, alternatives to the sample mean are designed to produce values that represent compromises between the mean and the median. Rarely do we make use of these other measures. However, it is instructive to discuss one class of estimators, namely the class of **trimmed means**. A trimmed mean is computed by "trimming away" a certain percent of both the largest and the smallest set of values. For example, the 10% trimmed mean is found by eliminating the largest 10% and smallest 10% and computing the average of the remaining values. For example, in the case of the stem weight data, we would eliminate the largest and smallest since the sample size is 10 for each sample. So for the without-nitrogen group the 10% trimmed mean is given by

$$\bar{x}_{\text{tr}(10)} = \frac{0.32 + 0.37 + 0.47 + 0.43 + 0.36 + 0.42 + 0.38 + 0.43}{8} = 0.39750,$$

and for the 10% trimmed mean for the with-nitrogen group we have

$$\bar{x}_{\text{tr}(10)} = \frac{0.43 + 0.47 + 0.49 + 0.52 + 0.75 + 0.79 + 0.62 + 0.46}{8} = 0.56625.$$

Note that in this case, as expected, the trimmed means are close to both the mean and the median for the individual samples. The trimmed mean is, of course, more insensitive to outliers than the sample mean but not as insensitive as the median. On the other hand, the trimmed mean approach makes use of more information than the sample median. Note that the sample median is, indeed, a special case of the trimmed mean in which all of the sample data are eliminated apart from the middle one or two observations.

Exercises

1.1 The following measurements were recorded for the drying time, in hours, of a certain brand of latex paint.

3.4	2.5	4.8	2.9	3.6
2.8	3.3	5.6	3.7	2.8
4.4	4.0	5.2	3.0	4.8

Assume that the measurements are a simple random sample.

(a) What is the sample size for the above sample?
(b) Calculate the sample mean for these data.
(c) Calculate the sample median.
(d) Plot the data by way of a dot plot.
(e) Compute the 20% trimmed mean for the above data set.
(f) Is the sample mean for these data more or less descriptive as a center of location than the trimmed mean?

1.2 According to the journal *Chemical Engineering*, an important property of a fiber is its water absorbency. A random sample of 20 pieces of cotton fiber was taken and the absorbency on each piece was measured. The following are the absorbency values:

18.71 21.41 20.72 21.81 19.29 22.43 20.17
23.71 19.44 20.50 18.92 20.33 23.00 22.85
19.25 21.77 22.11 19.77 18.04 21.12

(a) Calculate the sample mean and median for the above sample values.
(b) Compute the 10% trimmed mean.
(c) Do a dot plot of the absorbency data.
(d) Using only the values of the mean, median, and trimmed mean, do you have evidence of outliers in the data?

1.3 A certain polymer is used for evacuation systems for aircraft. It is important that the polymer be resistant to the aging process. Twenty specimens of the polymer were used in an experiment. Ten were assigned randomly to be exposed to an accelerated batch aging process that involved exposure to high temperatures for 10 days. Measurements of tensile strength of the specimens were made, and the following data were recorded on tensile strength in psi:

No aging: 227 222 218 217 225
          218 216 229 228 221
Aging: 219 214 215 211 209
          218 203 204 201 205

(a) Do a dot plot of the data.
(b) From your plot, does it appear as if the aging process has had an effect on the tensile strength of this polymer? Explain.
(c) Calculate the sample mean tensile strength of the two samples.
(d) Calculate the median for both. Discuss the similarity or lack of similarity between the mean and median of each group.

1.4 In a study conducted by the Department of Mechanical Engineering at Virginia Tech, the steel rods supplied by two different companies were compared. Ten sample springs were made out of the steel rods supplied by each company, and a measure of flexibility was recorded for each. The data are as follows:

Company A: 9.3 8.8 6.8 8.7 8.5
            6.7 8.0 6.5 9.2 7.0
Company B: 11.0 9.8 9.9 10.2 10.1
            9.7 11.0 11.1 10.2 9.6

(a) Calculate the sample mean and median for the data for the two companies.
(b) Plot the data for the two companies on the same line and give your impression regarding any apparent differences between the two companies.

1.5 Twenty adult males between the ages of 30 and 40 participated in a study to evaluate the effect of a specific health regimen involving diet and exercise on the blood cholesterol. Ten were randomly selected to be a control group, and ten others were assigned to take part in the regimen as the treatment group for a period of 6 months. The following data show the reduction in cholesterol experienced for the time period for the 20 subjects:

Control group: 7   3 −4 14   2
                  5 22 −7   9   5
Treatment group: −6   5   9   4   4
                 12 37   5   3   3

(a) Do a dot plot of the data for both groups on the same graph.
(b) Compute the mean, median, and 10% trimmed mean for both groups.
(c) Explain why the difference in means suggests one conclusion about the effect of the regimen, while the difference in medians or trimmed means suggests a different conclusion.

1.6 The tensile strength of silicone rubber is thought to be a function of curing temperature. A study was carried out in which samples of 12 specimens of the rubber were prepared using curing temperatures of $20°C$ and $45°C$. The data below show the tensile strength values in megapascals.

20°C: 2.07 2.14 2.22 2.03 2.21 2.03
       2.05 2.18 2.09 2.14 2.11 2.02
45°C: 2.52 2.15 2.49 2.03 2.37 2.05
       1.99 2.42 2.08 2.42 2.29 2.01

(a) Show a dot plot of the data with both low and high temperature tensile strength values.

(b) Compute sample mean tensile strength for both samples.

(c) Does it appear as if curing temperature has an influence on tensile strength, based on the plot? Comment further.

(d) Does anything else appear to be influenced by an increase in curing temperature? Explain.

1.4 Measures of Variability

Sample variability plays an important role in data analysis. Process and product variability is a fact of life in engineering and scientific systems: The control or reduction of process variability is often a source of major difficulty. More and more process engineers and managers are learning that product quality and, as a result, profits derived from manufactured products are very much a function of **process variability**. As a result, much of Chapters 9 through 15 deals with data analysis and modeling procedures in which sample variability plays a major role. Even in small data analysis problems, the success of a particular statistical method may depend on the magnitude of the variability among the observations in the sample. Measures of location in a sample do not provide a proper summary of the nature of a data set. For instance, in Example 1.2 we cannot conclude that the use of nitrogen enhances growth without taking sample variability into account.

While the details of the analysis of this type of data set are deferred to Chapter 9, it should be clear from Figure 1.1 that variability among the no-nitrogen observations and variability among the nitrogen observations are certainly of some consequence. In fact, it appears that the variability within the nitrogen sample is larger than that of the no-nitrogen sample. Perhaps there is something about the inclusion of nitrogen that not only increases the stem height ($\bar{x}$ of 0.565 gram compared to an $\bar{x}$ of 0.399 gram for the no-nitrogen sample) but also increases the variability in stem height (i.e., renders the stem height more inconsistent).

As another example, contrast the two data sets below. Each contains two samples and the difference in the means is roughly the same for the two samples, but data set B seems to provide a much sharper contrast between the two populations from which the samples were taken. If the purpose of such an experiment is to detect differences between the two populations, the task is accomplished in the case of data set B. However, in data set A the large variability *within* the two samples creates difficulty. In fact, it is not clear that there is a distinction *between* the two populations.

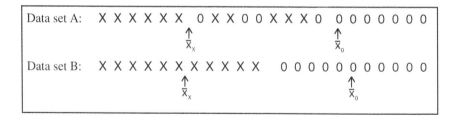

1.4 Measures of Variability

Sample Range and Sample Standard Deviation

Just as there are many measures of central tendency or location, there are many measures of spread or variability. Perhaps the simplest one is the **sample range** $X_{max} - X_{min}$. The range can be very useful and is discussed at length in Chapter 17 on *statistical quality control*. The sample measure of spread that is used most often is the **sample standard deviation**. We again let $x_1, x_2, \ldots, x_n$ denote sample values.

Definition 1.3: The **sample variance**, denoted by s^2, is given by

$$s^2 = \sum_{i=1}^{n} \frac{(x_i - \bar{x})^2}{n-1}.$$

The **sample standard deviation**, denoted by s, is the positive square root of s^2, that is,

$$s = \sqrt{s^2}.$$

It should be clear to the reader that the sample standard deviation is, in fact, a measure of variability. Large variability in a data set produces relatively large values of $(x - \bar{x})^2$ and thus a large sample variance. The quantity $n - 1$ is often called the **degrees of freedom associated with the variance** estimate. In this simple example, the degrees of freedom depict the number of independent pieces of information available for computing variability. For example, suppose that we wish to compute the sample variance and standard deviation of the data set (5, 17, 6, 4). The sample average is $\bar{x} = 8$. The computation of the variance involves

$$(5-8)^2 + (17-8)^2 + (6-8)^2 + (4-8)^2 = (-3)^2 + 9^2 + (-2)^2 + (-4)^2.$$

The quantities inside parentheses sum to zero. In general, $\sum_{i=1}^{n}(x_i - \bar{x}) = 0$ (see Exercise 1.16 on page 31). Then the computation of a sample variance does not involve n **independent squared deviations** from the mean $\bar{x}$. In fact, since the last value of $x - \bar{x}$ is determined by the initial $n - 1$ of them, we say that these are $n - 1$ "pieces of information" that produce s^2. Thus, there are $n - 1$ degrees of freedom rather than n degrees of freedom for computing a sample variance.

Example 1.4: In an example discussed extensively in Chapter 10, an engineer is interested in testing the "bias" in a pH meter. Data are collected on the meter by measuring the pH of a neutral substance (pH = 7.0). A sample of size 10 is taken, with results given by

7.07 7.00 7.10 6.97 7.00 7.03 7.01 7.01 6.98 7.08.

The sample mean $\bar{x}$ is given by

$$\bar{x} = \frac{7.07 + 7.00 + 7.10 + \cdots + 7.08}{10} = 7.0250.$$

The sample variance s^2 is given by

$$s^2 = \frac{1}{9}[(7.07 - 7.025)^2 + (7.00 - 7.025)^2 + (7.10 - 7.025)^2$$
$$+ \cdots + (7.08 - 7.025)^2] = 0.001939.$$

As a result, the sample standard deviation is given by

$$s = \sqrt{0.001939} = 0.044.$$

So the sample standard deviation is 0.0440 with $n - 1 = 9$ degrees of freedom.

Units for Standard Deviation and Variance

It should be apparent from Definition 1.3 that the variance is a measure of the average squared deviation from the mean $\bar{x}$. We use the term *average squared deviation* even though the definition makes use of a division by degrees of freedom $n - 1$ rather than n. Of course, if n is large, the difference in the denominator is inconsequential. As a result, the sample variance possesses units that are the square of the units in the observed data whereas the sample standard deviation is found in linear units. As an example, consider the data of Example 1.2. The stem weights are measured in grams. As a result, the sample standard deviations are in grams and the variances are measured in grams^2. In fact, the individual standard deviations are 0.0728 gram for the no-nitrogen case and 0.1867 gram for the nitrogen group. Note that the standard deviation does indicate considerably larger variability in the nitrogen sample. This condition was displayed in Figure 1.1.

Which Variability Measure Is More Important?

As we indicated earlier, the sample range has applications in the area of statistical quality control. It may appear to the reader that the use of both the sample variance and the sample standard deviation is redundant. Both measures reflect the same concept in measuring variability, but the sample standard deviation measures variability in linear units whereas the sample variance is measured in squared units. Both play huge roles in the use of statistical methods. Much of what is accomplished in the context of statistical inference involves drawing conclusions about characteristics of populations. Among these characteristics are constants which are called **population parameters**. Two important parameters are the **population mean** and the **population variance**. The sample variance plays an explicit role in the statistical methods used to draw inferences about the population variance. The sample standard deviation has an important role along with the sample mean in inferences that are made about the population mean. In general, the variance is considered more in inferential theory, while the standard deviation is used more in applications.

Exercises

1.7 Consider the drying time data for Exercise 1.1 on page 13. Compute the sample variance and sample standard deviation.

1.8 Compute the sample variance and standard deviation for the water absorbency data of Exercise 1.2 on page 13.

1.9 Exercise 1.3 on page 13 showed tensile strength data for two samples, one in which specimens were exposed to an aging process and one in which there was no aging of the specimens.
(a) Calculate the sample variance as well as standard deviation in tensile strength for both samples.
(b) Does there appear to be any evidence that aging affects the variability in tensile strength? (See also the plot for Exercise 1.3 on page 13.)

1.10 For the data of Exercise 1.4 on page 13, compute both the mean and the variance in "flexibility" for both company A and company B. Does there appear to be a difference in flexibility between company A and company B?

1.11 Consider the data in Exercise 1.5 on page 13. Compute the sample variance and the sample standard deviation for both control and treatment groups.

1.12 For Exercise 1.6 on page 13, compute the sample standard deviation in tensile strength for the samples separately for the two temperatures. Does it appear as if an increase in temperature influences the variability in tensile strength? Explain.

1.5 Discrete and Continuous Data

Statistical inference through the analysis of observational studies or designed experiments is used in many scientific areas. The data gathered may be **discrete** or **continuous**, depending on the area of application. For example, a chemical engineer may be interested in conducting an experiment that will lead to conditions where yield is maximized. Here, of course, the yield may be in percent or grams/pound, measured on a continuum. On the other hand, a toxicologist conducting a combination drug experiment may encounter data that are binary in nature (i.e., the patient either responds or does not).

Great distinctions are made between discrete and continuous data in the probability theory that allow us to draw statistical inferences. Often applications of statistical inference are found when the data are *count data*. For example, an engineer may be interested in studying the number of radioactive particles passing through a counter in, say, 1 millisecond. Personnel responsible for the efficiency of a port facility may be interested in the properties of the number of oil tankers arriving each day at a certain port city. In Chapter 5, several distinct scenarios, leading to different ways of handling data, are discussed for situations with count data.

Special attention even at this early stage of the textbook should be paid to some details associated with binary data. Applications requiring statistical analysis of binary data are voluminous. Often the measure that is used in the analysis is the *sample proportion*. Obviously the binary situation involves two categories. If there are n units involved in the data and x is defined as the number that fall into category 1, then $n - x$ fall into category 2. Thus, x/n is the sample proportion in category 1, and $1 - x/n$ is the sample proportion in category 2. In the biomedical application, 50 patients may represent the sample units, and if 20 out of 50 experienced an improvement in a stomach ailment (common to all 50) after all were given the drug, then $\frac{20}{50} = 0.4$ is the sample proportion for which the drug was a success and $1 - 0.4 = 0.6$ is the sample proportion for which the drug was not successful. Actually the basic numerical measurement for binary data is generally denoted by either 0 or 1. For example, in our medical example, a successful result is denoted by a 1 and a nonsuccess a 0. As a result, the sample proportion is actually a sample mean of the ones and zeros. For the successful category,

$$\frac{x_1 + x_2 + \cdots + x_{50}}{50} = \frac{1 + 1 + 0 + \cdots + 0 + 1}{50} = \frac{20}{50} = 0.4.$$

What Kinds of Problems Are Solved in Binary Data Situations?

The kinds of problems facing scientists and engineers dealing in binary data are not a great deal unlike those seen where continuous measurements are of interest. However, different techniques are used since the statistical properties of sample proportions are quite different from those of the sample means that result from averages taken from continuous populations. Consider the example data in Exercise 1.6 on page 13. The statistical problem underlying this illustration focuses on whether an intervention, say, an increase in curing temperature, will alter the population mean tensile strength associated with the silicone rubber process. On the other hand, in a quality control area, suppose an automobile tire manufacturer reports that a shipment of 5000 tires selected randomly from the process results in 100 of them showing blemishes. Here the sample proportion is $\frac{100}{5000} = 0.02$. Following a change in the process designed to reduce blemishes, a second sample of 5000 is taken and 90 tires are blemished. The sample proportion has been reduced to $\frac{90}{5000} = 0.018$. The question arises, "Is the decrease in the sample proportion from 0.02 to 0.018 substantial enough to suggest a real improvement in the population proportion?" Both of these illustrations require the use of the statistical properties of sample averages—one from samples from a continuous population, and the other from samples from a discrete (binary) population. In both cases, the sample mean is an **estimate** of a population parameter, a population mean in the first illustration (i.e., mean tensile strength), and a population proportion in the second case (i.e., proportion of blemished tires in the population). So here we have sample estimates used to draw scientific conclusions regarding population parameters. As we indicated in Section 1.3, this is the general theme in many practical problems using statistical inference.

1.6 Statistical Modeling, Scientific Inspection, and Graphical Diagnostics

Often the end result of a statistical analysis is the estimation of parameters of a **postulated model**. This is natural for scientists and engineers since they often deal in modeling. A statistical model is not deterministic but, rather, must entail some probabilistic aspects. A model form is often the foundation of **assumptions** that are made by the analyst. For example, in Example 1.2 the scientist may wish to draw some level of distinction between the nitrogen and no-nitrogen populations through the sample information. The analysis may require a certain model for

1.6 Statistical Modeling, Scientific Inspection, and Graphical Diagnostics

the data, for example, that the two samples come from **normal** or **Gaussian distributions**. See Chapter 6 for a discussion of the normal distribution.

Obviously, the user of statistical methods cannot generate sufficient information or experimental data to characterize the population totally. But sets of data are often used to learn about certain properties of the population. Scientists and engineers are accustomed to dealing with data sets. The importance of characterizing or *summarizing* the nature of collections of data should be obvious. Often a summary of a collection of data via a graphical display can provide insight regarding the system from which the data were taken. For instance, in Sections 1.1 and 1.3, we have shown dot plots.

In this section, the role of sampling and the display of data for enhancement of **statistical inference** is explored in detail. We merely introduce some simple but often effective displays that complement the study of statistical populations.

Scatter Plot

At times the model postulated may take on a somewhat complicated form. Consider, for example, a textile manufacturer who designs an experiment where cloth specimen that contain various percentages of cotton are produced. Consider the data in Table 1.3.

Table 1.3: Tensile Strength

Cotton Percentage	Tensile Strength
15	7, 7, 9, 8, 10
20	19, 20, 21, 20, 22
25	21, 21, 17, 19, 20
30	8, 7, 8, 9, 10

Five cloth specimens are manufactured for each of the four cotton percentages. In this case, both the model for the experiment and the type of analysis used should take into account the goal of the experiment and important input from the textile scientist. Some simple graphics can shed important light on the clear distinction between the samples. See Figure 1.5; the sample means and variability are depicted nicely in the scatter plot. One possible goal of this experiment is simply to determine which cotton percentages are truly distinct from the others. In other words, as in the case of the nitrogen/no-nitrogen data, for which cotton percentages are there clear distinctions between the populations or, more specifically, between the population means? In this case, perhaps a reasonable model is that each sample comes from a normal distribution. Here the goal is very much like that of the nitrogen/no-nitrogen data except that more samples are involved. The formalism of the analysis involves notions of hypothesis testing discussed in Chapter 10. Incidentally, this formality is perhaps not necessary in light of the diagnostic plot. But does this describe the real goal of the experiment and hence the proper approach to data analysis? It is likely that the scientist anticipates the existence of a *maximum population mean tensile strength* in the range of cotton concentration in the experiment. Here the analysis of the data should revolve around a different type of model, one that postulates a type of structure relating the population mean tensile strength to the cotton concentration. In other words, a model may be written

$$\mu_{t,c} = \beta_0 + \beta_1 C + \beta_2 C^2,$$

where $\mu_{t,c}$ is the population mean tensile strength, which varies with the amount of cotton in the product C. The implication of this model is that for a fixed cotton level, there is a population of tensile strength measurements and the population mean is $\mu_{t,c}$. This type of model, called a **regression model**, is discussed in Chapters 11 and 12. The functional form is chosen by the scientist. At times the data analysis may suggest that the model be changed. Then the data analyst "entertains" a model that may be altered after some analysis is done. The use of an empirical model is accompanied by **estimation theory**, where β_0, β_1, and β_2 are estimated by the data. Further, statistical inference can then be used to determine model adequacy.

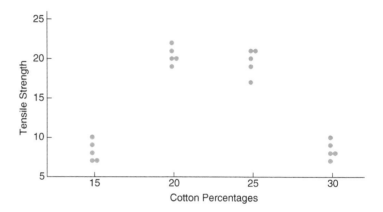

Figure 1.5: Scatter plot of tensile strength and cotton percentages.

Two points become evident from the two data illustrations here: (1) The type of model used to describe the data often depends on the goal of the experiment; and (2) the structure of the model should take advantage of nonstatistical scientific input. A selection of a model represents a **fundamental assumption** upon which the resulting statistical inference is based. It will become apparent throughout the book how important graphics can be. Often, plots can illustrate information that allows the results of the formal statistical inference to be better communicated to the scientist or engineer. At times, plots or **exploratory data analysis** can teach the analyst something not retrieved from the formal analysis. Almost any formal analysis requires assumptions that evolve from the model of the data. Graphics can nicely highlight **violation of assumptions** that would otherwise go unnoticed. Throughout the book, graphics are used extensively to supplement formal data analysis. The following sections reveal some graphical tools that are useful in exploratory or descriptive data analysis.

Stem-and-Leaf Plot

Statistical data, generated in large masses, can be very useful for studying the behavior of the distribution if presented in a combined tabular and graphic display called a **stem-and-leaf plot**.

To illustrate the construction of a stem-and-leaf plot, consider the data of Table 1.4, which specifies the "life" of 40 similar car batteries recorded to the nearest tenth of a year. The batteries are guaranteed to last 3 years. First, split each observation into two parts consisting of a stem and a leaf such that the stem represents the digit preceding the decimal and the leaf corresponds to the decimal part of the number. In other words, for the number 3.7, the digit 3 is designated the stem and the digit 7 is the leaf. The four stems 1, 2, 3, and 4 for our data are listed vertically on the left side in Table 1.5; the leaves are recorded on the right side opposite the appropriate stem value. Thus, the leaf 6 of the number 1.6 is recorded opposite the stem 1; the leaf 5 of the number 2.5 is recorded opposite the stem 2; and so forth. The number of leaves recorded opposite each stem is summarized under the frequency column.

Table 1.4: Car Battery Life

2.2	4.1	3.5	4.5	3.2	3.7	3.0	2.6
3.4	1.6	3.1	3.3	3.8	3.1	4.7	3.7
2.5	4.3	3.4	3.6	2.9	3.3	3.9	3.1
3.3	3.1	3.7	4.4	3.2	4.1	1.9	3.4
4.7	3.8	3.2	2.6	3.9	3.0	4.2	3.5

Table 1.5: Stem-and-Leaf Plot of Battery Life

Stem	Leaf	Frequency
1	69	2
2	25669	5
3	0011112223334445567778899	25
4	11234577	8

The stem-and-leaf plot of Table 1.5 contains only four stems and consequently does not provide an adequate picture of the distribution. To remedy this problem, we need to increase the number of stems in our plot. One simple way to accomplish this is to write each stem value twice and then record the leaves 0, 1, 2, 3, and 4 opposite the appropriate stem value where it appears for the first time, and the leaves 5, 6, 7, 8, and 9 opposite this same stem value where it appears for the second time. This modified double-stem-and-leaf plot is illustrated in Table 1.6, where the stems corresponding to leaves 0 through 4 have been coded by the symbol ⋆ and the stems corresponding to leaves 5 through 9 by the symbol ·.

In any given problem, we must decide on the appropriate stem values. This decision is made somewhat arbitrarily, although we are guided by the size of our sample. Usually, we choose between 5 and 20 stems. The smaller the number of data available, the smaller is our choice for the number of stems. For example, if the data consist of numbers from 1 to 21 representing the number of people in a cafeteria line on 40 randomly selected workdays and we choose a double-stem-and-leaf plot, the stems will be 0⋆, 0·, 1⋆, 1·, and 2⋆ so that the smallest observation 1 has stem 0⋆ and leaf 1, the number 18 has stem 1· and leaf 8, and the largest observation 21 has stem 2⋆ and leaf 1. On the other hand, if the data consist of numbers from $18,800 to $19,600 representing the best possible deals on 100 new automobiles from a certain dealership and we choose a single-stem-and-leaf plot, the stems will be 188, 189, 190, ..., 196 and the leaves will now each contain two digits. A car that sold for $19,385 would have a stem value of 193 and the two-digit leaf 85. Multiple-digit leaves belonging to the same stem are usually separated by commas in the stem-and-leaf plot. Decimal points in the data are generally ignored when all the digits to the right of the decimal represent the leaf. Such was the case in Tables 1.5 and 1.6. However, if the data consist of numbers ranging from 21.8 to 74.9, we might choose the digits 2, 3, 4, 5, 6, and 7 as our stems so that a number such as 48.3 would have a stem value of 4 and a leaf of 8.3.

Table 1.6: Double-Stem-and-Leaf Plot of Battery Life

Stem	Leaf	Frequency
1·	69	2
2⋆	2	1
2·	5669	4
3⋆	001111222333444	15
3·	5567778899	10
4⋆	11234	5
4·	577	3

The stem-and-leaf plot represents an effective way to summarize data. Another way is through the use of the **frequency distribution**, where the data, grouped into different classes or intervals, can be constructed by counting the leaves belonging to each stem and noting that each stem defines a class interval. In Table 1.5, the stem 1 with 2 leaves defines the interval 1.0–1.9 containing 2 observations; the stem 2 with 5 leaves defines the interval 2.0–2.9 containing 5 observations; the stem 3 with 25 leaves defines the interval 3.0–3.9 with 25 observations; and the stem 4 with 8 leaves defines the interval 4.0–4.9 containing 8 observations. For the double-stem-and-leaf plot of Table 1.6, the stems define the seven class intervals 1.5–1.9, 2.0–2.4, 2.5–2.9, 3.0–3.4, 3.5–3.9, 4.0–4.4, and 4.5–4.9 with frequencies 2, 1, 4, 15, 10, 5, and 3, respectively.

Histogram

Dividing each class frequency by the total number of observations, we obtain the proportion of the set of observations in each of the classes. A table listing relative frequencies is called a **relative frequency distribution**. The relative frequency distribution for the data of Table 1.4, showing the midpoint of each class interval, is given in Table 1.7.

The information provided by a relative frequency distribution in tabular form is easier to grasp if presented graphically. Using the midpoint of each interval and the

Table 1.7: Relative Frequency Distribution of Battery Life

Class Interval	Class Midpoint	Frequency, f	Relative Frequency
1.5–1.9	1.7	2	0.050
2.0–2.4	2.2	1	0.025
2.5–2.9	2.7	4	0.100
3.0–3.4	3.2	15	0.375
3.5–3.9	3.7	10	0.250
4.0–4.4	4.2	5	0.125
4.5–4.9	4.7	3	0.075

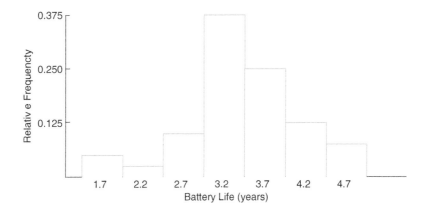

Figure 1.6: Relative frequency histogram.

corresponding relative frequency, we construct a **relative frequency histogram** (Figure 1.6).

Many continuous frequency distributions can be represented graphically by the characteristic bell-shaped curve of Figure 1.7. Graphical tools such as what we see in Figures 1.6 and 1.7 aid in the characterization of the nature of the population. In Chapters 5 and 6 we discuss a property of the population called its **distribution**. While a more rigorous definition of a distribution or **probability distribution** will be given later in the text, at this point one can view it as what would be seen in Figure 1.7 in the limit as the size of the sample becomes larger.

A distribution is said to be **symmetric** if it can be folded along a vertical axis so that the two sides coincide. A distribution that lacks symmetry with respect to a vertical axis is said to be **skewed**. The distribution illustrated in Figure 1.8(a) is said to be skewed to the right since it has a long right tail and a much shorter left tail. In Figure 1.8(b) we see that the distribution is symmetric, while in Figure 1.8(c) it is skewed to the left.

If we rotate a stem-and-leaf plot counterclockwise through an angle of 90°, we observe that the resulting columns of leaves form a picture that is similar to a histogram. Consequently, if our primary purpose in looking at the data is to determine the general shape or form of the distribution, it will seldom be necessary

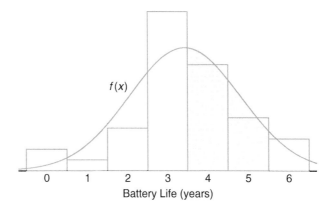

Figure 1.7: Estimating frequency distribution.

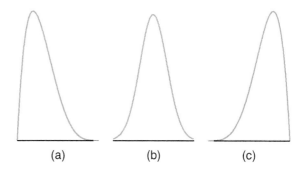

Figure 1.8: Skewness of data.

to construct a relative frequency histogram.

Box-and-Whisker Plot or Box Plot

Another display that is helpful for reflecting properties of a sample is the **box-and-whisker plot**. This plot encloses the *interquartile range* of the data in a box that has the median displayed within. The interquartile range has as its extremes the 75th percentile (upper quartile) and the 25th percentile (lower quartile). In addition to the box, "whiskers" extend, showing extreme observations in the sample. For reasonably large samples, the display shows center of location, variability, and the degree of asymmetry.

In addition, a variation called a **box plot** can provide the viewer with information regarding which observations may be **outliers**. Outliers are observations that are considered to be unusually far from the bulk of the data. There are many statistical tests that are designed to detect outliers. Technically, one may view an outlier as being an observation that represents a "rare event" (there is a small probability of obtaining a value that far from the bulk of the data). The concept of outliers resurfaces in Chapter 12 in the context of regression analysis.

1.6 Statistical Modeling, Scientific Inspection, and Graphical Diagnostics

The visual information in the box-and-whisker plot or box plot is not intended to be a formal test for outliers. Rather, it is viewed as a diagnostic tool. While the determination of which observations are outliers varies with the type of software that is used, one common procedure is to use a **multiple of the interquartile range**. For example, if the distance from the box exceeds 1.5 times the interquartile range (in either direction), the observation may be labeled an outlier.

Example 1.5: Nicotine content was measured in a random sample of 40 cigarettes. The data are displayed in Table 1.8.

Table 1.8: Nicotine Data for Example 1.5

1.09	1.92	2.31	1.79	2.28	1.74	1.47	1.97
0.85	1.24	1.58	2.03	1.70	2.17	2.55	2.11
1.86	1.90	1.68	1.51	1.64	0.72	1.69	1.85
1.82	1.79	2.46	1.88	2.08	1.67	1.37	1.93
1.40	1.64	2.09	1.75	1.63	2.37	1.75	1.69

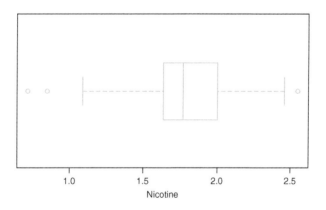

Figure 1.9: Box-and-whisker plot for Example 1.5.

Figure 1.9 shows the box-and-whisker plot of the data, depicting the observations 0.72 and 0.85 as mild outliers in the lower tail, whereas the observation 2.55 is a mild outlier in the upper tail. In this example, the interquartile range is 0.365, and 1.5 times the interquartile range is 0.5475. Figure 1.10, on the other hand, provides a stem-and-leaf plot.

Example 1.6: Consider the data in Table 1.9, consisting of 30 samples measuring the thickness of paint can "ears" (see the work by Hogg and Ledolter, 1992, in the Bibliography). Figure 1.11 depicts a box-and-whisker plot for this asymmetric set of data. Notice that the left block is considerably larger than the block on the right. The median is 35. The lower quartile is 31, while the upper quartile is 36. Notice also that the extreme observation on the right is farther away from the box than the extreme observation on the left. There are no outliers in this data set.

```
The decimal point is 1 digit(s) to the left of the |
  7 | 2
  8 | 5
  9 |
 10 | 9
 11 |
 12 | 4
 13 | 7
 14 | 07
 15 | 18
 16 | 3447899
 17 | 045599
 18 | 2568
 19 | 0237
 20 | 389
 21 | 17
 22 | 8
 23 | 17
 24 | 6
 25 | 5
```

Figure 1.10: Stem-and-leaf plot for the nicotine data.

Table 1.9: Data for Example 1.6

Sample	Measurements	Sample	Measurements
1	29 36 39 34 34	16	35 30 35 29 37
2	29 29 28 32 31	17	40 31 38 35 31
3	34 34 39 38 37	18	35 36 30 33 32
4	35 37 33 38 41	19	35 34 35 30 36
5	30 29 31 38 29	20	35 35 31 38 36
6	34 31 37 39 36	21	32 36 36 32 36
7	30 35 33 40 36	22	36 37 32 34 34
8	28 28 31 34 30	23	29 34 33 37 35
9	32 36 38 38 35	24	36 36 35 37 37
10	35 30 37 35 31	25	36 30 35 33 31
11	35 30 35 38 35	26	35 30 29 38 35
12	38 34 35 35 31	27	35 36 30 34 36
13	34 35 33 30 34	28	35 30 36 29 35
14	40 35 34 33 35	29	38 36 35 31 31
15	34 35 38 35 30	30	30 34 40 28 30

There are additional ways that box-and-whisker plots and other graphical displays can aid the analyst. Multiple samples can be compared graphically. Plots of data can suggest relationships between variables. Graphs can aid in the detection of anomalies or outlying observations in samples.

There are other types of graphical tools and plots that are used. These are discussed in Chapter 8 after we introduce additional theoretical details.

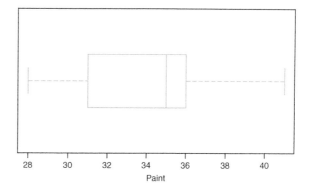

Figure 1.11: Box-and-whisker plot for thickness of paint can "ears."

Other Distinguishing Features of a Sample

There are features of the distribution or sample other than measures of center of location and variability that further define its nature. For example, while the median divides the data (or distribution) into two parts, there are other measures that divide parts or pieces of the distribution that can be very useful. Separation is made into four parts by *quartiles*, with the third quartile separating the upper quarter of the data from the rest, the second quartile being the median, and the first quartile separating the lower quarter of the data from the rest. The distribution can be even more finely divided by computing percentiles of the distribution. These quantities give the analyst a sense of the so-called *tails* of the distribution (i.e., values that are relatively extreme, either small or large). For example, the 95th percentile separates the highest 5% from the bottom 95%. Similar definitions prevail for extremes on the lower side or *lower tail* of the distribution. The 1st percentile separates the bottom 1% from the rest of the distribution. The concept of percentiles will play a major role in much that will be covered in future chapters.

1.7 General Types of Statistical Studies: Designed Experiment, Observational Study, and Retrospective Study

In the foregoing sections we have emphasized the notion of sampling from a population and the use of statistical methods to learn or perhaps affirm important information about the population. The information sought and learned through the use of these statistical methods can often be influential in decision making and problem solving in many important scientific and engineering areas. As an illustration, Example 1.3 describes a simple experiment in which the results may provide an aid in determining the kinds of conditions under which it is not advisable to use a particular aluminum alloy that may have a dangerous vulnerability to corrosion. The results may be of use not only to those who produce the alloy, but also to the customer who may consider using it. This illustration, as well as many more that appear in Chapters 13 through 15, highlights the concept of designing or controlling experimental conditions (combinations of coating conditions and humidity) of interest to learn about some characteristic or measurement (level of corrosion) that results from these conditions. Statistical methods that make use of measures of central tendency in the corrosion measure, as well as measures of variability, are employed. As the reader will observe later in the text, these methods often lead to a statistical model like that discussed in Section 1.6. In this case, the model may be used to estimate (or predict) the corrosion measure as a function of humidity and the type of coating employed. Again, in developing this kind of model, descriptive statistics that highlight central tendency and variability become very useful.

The information supplied in Example 1.3 illustrates nicely the types of engineering questions asked and answered by the use of statistical methods that are employed through a designed experiment and presented in this text. They are

(i) What is the nature of the impact of relative humidity on the corrosion of the aluminum alloy within the range of relative humidity in this experiment?

(ii) Does the chemical corrosion coating reduce corrosion levels and can the effect be quantified in some fashion?

(iii) Is there **interaction** between coating type and relative humidity that impacts their influence on corrosion of the alloy? If so, what is its interpretation?

What Is Interaction?

The importance of questions (i) and (ii) should be clear to the reader, as they deal with issues important to both producers and users of the alloy. But what about question (iii)? The concept of *interaction* will be discussed at length in Chapters 14 and 15. Consider the plot in Figure 1.3. This is an illustration of the detection of interaction between two **factors** in a simple designed experiment. Note that the lines connecting the sample means are not parallel. **Parallelism** would have indicated that the effect (seen as a result of the slope of the lines) of relative humidity is the same, namely a negative effect, for both an uncoated condition and the chemical corrosion coating. Recall that the negative slope implies that corrosion becomes more pronounced as humidity rises. Lack of parallelism implies an interaction between coating type and relative humidity. The nearly "flat" line for the corrosion coating as opposed to a steeper slope for the uncoated condition suggests that *not only is the chemical corrosion coating beneficial (note the displacement between the lines), but the presence of the coating renders the effect of humidity negligible*. Clearly all these questions are very important to the effect of the two individual factors and to the interpretation of the interaction, if it is present.

Statistical models are extremely useful in answering questions such as those listed in (i), (ii), and (iii), where the data come from a designed experiment. But one does not always have the luxury or resources to employ a designed experiment. For example, there are many instances in which the conditions of interest to the scientist or engineer cannot be implemented simply because the *important factors cannot be controlled*. In Example 1.3, the relative humidity and coating type (or lack of coating) are quite easy to control. This of course is the defining feature of a designed experiment. In many fields, factors that need to be studied cannot be controlled for any one of various reasons. Tight control as in Example 1.3 allows the analyst to be confident that any differences found (for example, in corrosion levels) are due to the factors under control. As a second illustration, consider Exercise 1.6 on page 13. Suppose in this case 24 specimens of silicone rubber are selected and 12 assigned to each of the curing temperature levels. The temperatures are controlled carefully, and thus this is an example of a designed experiment with a **single factor** being curing temperature. Differences found in the mean tensile strength would be assumed to be attributed to the different curing temperatures.

What If Factors Are Not Controlled?

Suppose there are no factors controlled and *no random assignment* of fixed treatments to experimental units and yet there is a need to glean information from a data set. As an illustration, consider a study in which interest centers around the relationship between blood cholesterol levels and the amount of sodium measured in the blood. A group of individuals were monitored over time for both blood cholesterol and sodium. Certainly some useful information can be gathered from such a data set. However, it should be clear that there certainly can be no strict control of blood sodium levels. Ideally, the subjects should be divided randomly into two groups, with one group assigned a specific high level of blood sodium and the other a specific low level of blood sodium. Obviously this cannot be done. Clearly changes in cholesterol can be experienced because of changes in one of a number of other factors that were not controlled. This kind of study, without factor control, is called an **observational study**. Much of the time it involves a situation in which subjects are observed across time.

Biological and biomedical studies are often by necessity observational studies. However, observational studies are not confined to those areas. For example, consider a study that is designed to determine the influence of ambient temperature on the electric power consumed by a chemical plant. Clearly, levels of ambient temperature cannot be controlled, and thus the data structure can only be a monitoring of the data from the plant over time.

It should be apparent that the striking difference between a well-designed experiment and observational studies is the difficulty in determination of true cause and effect with the latter. Also, differences found in the fundamental response (e.g., corrosion levels, blood cholesterol, plant electric power consumption) may be due to other underlying factors that were not controlled. Ideally, in a designed experiment the *nuisance factors* would be equalized via the randomization process. Certainly changes in blood cholesterol could be due to fat intake, exercise activity, and so on. Electric power consumption could be affected by the amount of product produced or even the purity of the product produced.

Another often ignored disadvantage of an observational study when compared to carefully designed experiments is that, unlike the latter, observational studies are at the mercy of nature, environmental or other uncontrolled circumstances that impact the ranges of factors of interest. For example, in the biomedical study regarding the influence of blood sodium levels on blood cholesterol, it is possible that there is indeed a strong influence but the particular data set used did not involve enough observed variation in sodium levels because of the nature of the subjects chosen. Of course, in a designed experiment, the analyst chooses and controls ranges of factors.

A third type of statistical study which can be very useful but has clear disadvantages when compared to a designed experiment is a **retrospective study**. This type of study uses strictly **historical data**, data taken over a specific period of time. One obvious advantage of retrospective data is that there is reduced cost in collecting the data. However, as one might expect, there are clear disadvantages.

(i) Validity and reliability of historical data are often in doubt.

(ii) If time is an important aspect of the structure of the data, there may be data missing.

(iii) There may be errors in collection of the data that are not known.

(iv) Again, as in the case of observational data, there is no control on the ranges of the measured variables (the factors in a study). Indeed, the ranges found in historical data may not be relevant for current studies.

In Section 1.6, some attention was given to modeling of relationships among variables. We introduced the notion of regression analysis, which is covered in Chapters 11 and 12 and is illustrated as a form of data analysis for designed experiments discussed in Chapters 14 and 15. In Section 1.6, a model relating population mean tensile strength of cloth to percentages of cotton was used for illustration, where 20 specimens of cloth represented the experimental units. In that case, the data came from a simple designed experiment where the individual cotton percentages were selected by the scientist.

Often both observational data and retrospective data are used for the purpose of observing relationships among variables through model-building procedures discussed in Chapters 11 and 12. While the advantages of designed experiments certainly apply when the goal is statistical model building, there are many areas in which designing of experiments is not possible. Thus, *observational or historical data must be used*. We refer here to a historical data set that is found in Exercise 12.5 on page 450. The goal is to build a model that will result in an equation or relationship that relates monthly electric power consumed to average ambient temperature x_1, the number of days in the month x_2, the average product purity x_3, and the tons of product produced x_4. The data are the past year's historical data.

Exercises

1.13 A manufacturer of electronic components is interested in determining the lifetime of a certain type of battery. A sample, in hours of life, is as follows:

123, 116, 122, 110, 175, 126, 125, 111, 118, 117.

(a) Find the sample mean and median.
(b) What feature in this data set is responsible for the substantial difference between the two?

1.14 A tire manufacturer wants to determine the inner diameter of a certain grade of tire. Ideally, the diameter would be 570 mm. The data are as follows:

572, 572, 573, 568, 569, 575, 565, 570.

(a) Find the sample mean and median.
(b) Find the sample variance, standard deviation, and range.
(c) Using the calculated statistics in parts (a) and (b), can you comment on the quality of the tires?

1.15 Five independent coin tosses result in $HHHHH$. It turns out that if the coin is fair the probability of this outcome is $(1/2)^5 = 0.03125$. Does this produce strong evidence that the coin is not fair? Comment and use the concept of *P*-value discussed in Section 1.1.

Exercises

1.16 Show that the n pieces of information in $\sum_{i=1}^{n}(x_i - \bar{x})^2$ are not independent; that is, show that

$$\sum_{i=1}^{n}(x_i - \bar{x}) = 0.$$

1.17 A study of the effects of smoking on sleep patterns is conducted. The measure observed is the time, in minutes, that it takes to fall asleep. These data are obtained:

Smokers:	69.3	56.0	22.1	47.6
	53.2	48.1	52.7	34.4
	60.2	43.8	23.2	13.8
Nonsmokers:	28.6	25.1	26.4	34.9
	29.8	28.4	38.5	30.2
	30.6	31.8	41.6	21.1
	36.0	37.9	13.9	

(a) Find the sample mean for each group.
(b) Find the sample standard deviation for each group.
(c) Make a dot plot of the data sets A and B on the same line.
(d) Comment on what kind of impact smoking appears to have on the time required to fall asleep.

1.18 The following scores represent the final examination grades for an elementary statistics course:

```
23  60  79  32  57  74  52  70  82
36  80  77  81  95  41  65  92  85
55  76  52  10  64  75  78  25  80
98  81  67  41  71  83  54  64  72
88  62  74  43  60  78  89  76  84
48  84  90  15  79  34  67  17  82
69  74  63  80  85  61
```

(a) Construct a stem-and-leaf plot for the examination grades in which the stems are $1, 2, 3, \ldots, 9$.
(b) Construct a relative frequency histogram, draw an estimate of the graph of the distribution, and discuss the skewness of the distribution.
(c) Compute the sample mean, sample median, and sample standard deviation.

1.19 The following data represent the length of life in years, measured to the nearest tenth, of 30 similar fuel pumps:

```
2.0  3.0  0.3  3.3  1.3  0.4
0.2  6.0  5.5  6.5  0.2  2.3
1.5  4.0  5.9  1.8  4.7  0.7
4.5  0.3  1.5  0.5  2.5  5.0
1.0  6.0  5.6  6.0  1.2  0.2
```

(a) Construct a stem-and-leaf plot for the life in years of the fuel pumps, using the digit to the left of the decimal point as the stem for each observation.
(b) Set up a relative frequency distribution.
(c) Compute the sample mean, sample range, and sample standard deviation.

1.20 The following data represent the length of life, in seconds, of 50 fruit flies subject to a new spray in a controlled laboratory experiment:

```
17  20  10   9  23  13  12  19  18  24
12  14   6   9  13   6   7  10  13   7
16  18   8  13   3  32   9   7  10  11
13   7  18   7  10   4  27  19  16   8
 7  10   5  14  15  10   9   6   7  15
```

(a) Construct a double-stem-and-leaf plot for the life span of the fruit flies using the stems 0⋆, 0·, 1⋆, 1·, 2⋆, 2·, and 3⋆ such that stems coded by the symbols ⋆ and · are associated, respectively, with leaves 0 through 4 and 5 through 9.
(b) Set up a relative frequency distribution.
(c) Construct a relative frequency histogram.
(d) Find the median.

1.21 The lengths of power failures, in minutes, are recorded in the following table.

```
22   18  135   15   90   78   69   98  102
83   55   28  121  120   13   22  124  112
70   66   74   89  103   24   21  112   21
40   98   87  132  115   21   28   43   37
50   96  118  158   74   78   83   93   95
```

(a) Find the sample mean and sample median of the power-failure times.
(b) Find the sample standard deviation of the power-failure times.

1.22 The following data are the measures of the diameters of 36 rivet heads in 1/100 of an inch.

```
6.72  6.77  6.82  6.70  6.78  6.70  6.62  6.75
6.66  6.66  6.64  6.76  6.73  6.80  6.72  6.76
6.76  6.68  6.66  6.62  6.72  6.76  6.70  6.78
6.76  6.67  6.70  6.72  6.74  6.81  6.79  6.78
6.66  6.76  6.76  6.72
```

(a) Compute the sample mean and sample standard deviation.
(b) Construct a relative frequency histogram of the data.
(c) Comment on whether or not there is any clear indication that the sample came from a population that has a bell-shaped distribution.

1.23 The hydrocarbon emissions at idling speed in parts per million (ppm) for automobiles of 1980 and 1990 model years are given for 20 randomly selected cars.

1980 models:
```
141  359  247  940  882  494  306  210  105  880
200  223  188  940  241  190  300  435  241  380
```
1990 models:
```
140  160   20   20  223   60   20   95  360   70
220  400  217   58  235  380  200  175   85   65
```

(a) Construct a dot plot as in Figure 1.1.

(b) Compute the sample means for the two years and superimpose the two means on the plots.

(c) Comment on what the dot plot indicates regarding whether or not the population emissions changed from 1980 to 1990. Use the concept of variability in your comments.

1.24 The following are historical data on staff salaries (dollars per pupil) for 30 schools sampled in the eastern part of the United States in the early 1970s.

```
3.79  2.99  2.77  2.91  3.10  1.84  2.52  3.22
2.45  2.14  2.67  2.52  2.71  2.75  3.57  3.85
3.36  2.05  2.89  2.83  3.13  2.44  2.10  3.71
3.14  3.54  2.37  2.68  3.51  3.37
```

(a) Compute the sample mean and sample standard deviation.

(b) Construct a relative frequency histogram of the data.

(c) Construct a stem-and-leaf display of the data.

1.25 The following data set is related to that in Exercise 1.24. It gives the percentages of the families that are in the upper income level, for the same individual schools in the same order as in Exercise 1.24.

```
72.2  31.9  26.5  29.1  27.3   8.6  22.3  26.5
20.4  12.8  25.1  19.2  24.1  58.2  68.1  89.2
55.1   9.4  14.5  13.9  20.7  17.9   8.5  55.4
38.1  54.2  21.5  26.2  59.1  43.3
```

(a) Calculate the sample mean.

(b) Calculate the sample median.

(c) Construct a relative frequency histogram of the data.

(d) Compute the 10% trimmed mean. Compare with the results in (a) and (b) and comment.

1.26 Suppose it is of interest to use the data sets in Exercises 1.24 and 1.25 to derive a model that would predict staff salaries as a function of percentage of families in a high income level for current school systems. Comment on any disadvantage in carrying out this type of analysis.

1.27 A study is done to determine the influence of the wear, y, of a bearing as a function of the load, x, on the bearing. A designed experiment is used for this study. Three levels of load were used, 700 lb, 1000 lb, and 1300 lb. Four specimens were used at each level, and the sample means were, respectively, 210, 325, and 375.

(a) Plot average wear against load.

(b) From the plot in (a), does it appear as if a relationship exists between wear and load?

(c) Suppose we look at the individual wear values for each of the four specimens at each load level (see the data that follow). Plot the wear results for all specimens against the three load values.

(d) From your plot in (c), does it appear as if a clear relationship exists? If your answer is different from that in (b), explain why.

	x		
	700	1000	1300
y_1	145	250	150
y_2	105	195	180
y_3	260	375	420
y_4	330	480	750
	$\bar{y}_1 = 210$	$\bar{y}_2 = 325$	$\bar{y}_3 = 375$

1.28 Many manufacturing companies in the United States and abroad use molded parts as components of a process. Shrinkage is often a major problem. Thus, a molded die for a part is built larger than nominal size to allow for part shrinkage. In an injection molding study it is known that the shrinkage is influenced by many factors, among which are the injection velocity in ft/sec and mold temperature in °C. The following two data sets show the results of a designed experiment in which injection velocity was held at two levels (low and high) and mold temperature was held constant at a low level. The shrinkage is measured in cm × 10^4.

Shrinkage values at low injection velocity:
```
72.68  72.62  72.58  72.48  73.07
72.55  72.42  72.84  72.58  72.92
```
Shrinkage values at high injection velocity:
```
71.62  71.68  71.74  71.48  71.55
71.52  71.71  71.56  71.70  71.50
```

(a) Construct a dot plot of both data sets on the same graph. Indicate on the plot both shrinkage means, that for low injection velocity and high injection velocity.

(b) Based on the graphical results in (a), using the location of the two means and your sense of variability, what do you conclude regarding the effect of injection velocity on shrinkage at low mold temperature?

1.29 Use the data in Exercise 1.24 to construct a box plot.

1.30 Below are the lifetimes, in hours, of fifty 40-watt, 110-volt internally frosted incandescent lamps, taken from forced life tests:

919	1196	785	1126	936	918
1156	920	948	1067	1092	1162
1170	929	950	905	972	1035
1045	855	1195	1195	1340	1122
938	970	1237	956	1102	1157
978	832	1009	1157	1151	1009
765	958	902	1022	1333	811
1217	1085	896	958	1311	1037
702	923				

Construct a box plot for these data.

1.31 Consider the situation of Exercise 1.28. But now use the following data set, in which shrinkage is measured once again at low injection velocity and high injection velocity. However, this time the mold temperature is raised to a high level and held constant.

Shrinkage values at low injection velocity:

76.20 76.09 75.98 76.15 76.17
75.94 76.12 76.18 76.25 75.82

Shrinkage values at high injection velocity:

93.25 93.19 92.87 93.29 93.37
92.98 93.47 93.75 93.89 91.62

(a) As in Exercise 1.28, construct a dot plot with both data sets on the same graph and identify both means (i.e., mean shrinkage for low injection velocity and for high injection velocity).

(b) As in Exercise 1.28, comment on the influence of injection velocity on shrinkage for high mold temperature. Take into account the position of the two means and the variability around each mean.

(c) Compare your conclusion in (b) with that in (b) of Exercise 1.28 in which mold temperature was held at a low level. Would you say that there is an interaction between injection velocity and mold temperature? Explain.

1.32 Use the results of Exercises 1.28 and 1.31 to create a plot that illustrates the interaction evident from the data. Use the plot in Figure 1.3 in Example 1.3 as a guide. Could the type of information found in Exercises 1.28 and 1.31 have been found in an observational study in which there was no control on injection velocity and mold temperature by the analyst? Explain why or why not.

1.33 Group Project: Collect the shoe size of everyone in the class. Use the sample means and variances and the types of plots presented in this chapter to summarize any features that draw a distinction between the distributions of shoe sizes for males and females. Do the same for the height of everyone in the class.

Chapter 2

Probability

2.1 Sample Space

In the study of statistics, we are concerned basically with the presentation and interpretation of **chance outcomes** that occur in a planned study or scientific investigation. For example, we may record the number of accidents that occur monthly at the intersection of Driftwood Lane and Royal Oak Drive, hoping to justify the installation of a traffic light; we might classify items coming off an assembly line as "defective" or "nondefective"; or we may be interested in the volume of gas released in a chemical reaction when the concentration of an acid is varied. Hence, the statistician is often dealing with either numerical data, representing counts or measurements, or **categorical data**, which can be classified according to some criterion.

We shall refer to any recording of information, whether it be numerical or categorical, as an **observation**. Thus, the numbers 2, 0, 1, and 2, representing the number of accidents that occurred for each month from January through April during the past year at the intersection of Driftwood Lane and Royal Oak Drive, constitute a set of observations. Similarly, the categorical data N, D, N, N, and D, representing the items found to be defective or nondefective when five items are inspected, are recorded as observations.

Statisticians use the word **experiment** to describe any process that generates a set of data. A simple example of a statistical experiment is the tossing of a coin. In this experiment, there are only two possible outcomes, heads or tails. Another experiment might be the launching of a missile and observing of its velocity at specified times. The opinions of voters concerning a new sales tax can also be considered as observations of an experiment. We are particularly interested in the observations obtained by repeating the experiment several times. In most cases, the outcomes will depend on chance and, therefore, cannot be predicted with certainty. If a chemist runs an analysis several times under the same conditions, he or she will obtain different measurements, indicating an element of chance in the experimental procedure. Even when a coin is tossed repeatedly, we cannot be certain that a given toss will result in a head. However, we know the entire set of possibilities for each toss.

Given the discussion in Section 1.7, we should deal with the breadth of the term *experiment*. Three types of statistical studies were reviewed, and several examples were given of each. In each of the three cases, *designed experiments*, *observational studies*, and *retrospective studies*, the end result was a set of *data* that of course is subject to **uncertainty**. Though only one of these has the word *experiment* in its description, the process of generating the data or the process of observing the data is part of an experiment. The corrosion study discussed in Section 1.2 certainly involves an experiment, with measures of corrosion representing the data. The example given in Section 1.7 in which blood cholesterol and sodium were observed on a group of individuals represented an observational study (as opposed to a *designed experiment*), and yet the process generated data and the outcome is subject to uncertainty. Thus, it is an experiment. A third example in Section 1.7 represented a retrospective study in which historical data on monthly electric power consumption and average monthly ambient temperature were observed. Even though the data may have been in the files for decades, the process is still referred to as an experiment.

Definition 2.1: The set of all possible outcomes of a statistical experiment is called the **sample space** and is represented by the symbol S.

Each outcome in a sample space is called an **element** or a **member** of the sample space, or simply a **sample point**. If the sample space has a finite number of elements, we may *list* the members separated by commas and enclosed in braces. Thus, the sample space S, of possible outcomes when a coin is flipped, may be written

$$S = \{H, T\},$$

where H and T correspond to heads and tails, respectively.

Example 2.1: Consider the experiment of tossing a die. If we are interested in the number that shows on the top face, the sample space is

$$S_1 = \{1, 2, 3, 4, 5, 6\}.$$

If we are interested only in whether the number is even or odd, the sample space is simply

$$S_2 = \{\text{even, odd}\}.$$

Example 2.1 illustrates the fact that more than one sample space can be used to describe the outcomes of an experiment. In this case, S_1 provides more information than S_2. If we know which element in S_1 occurs, we can tell which outcome in S_2 occurs; however, a knowledge of what happens in S_2 is of little help in determining which element in S_1 occurs. In general, it is desirable to use the sample space that gives the most information concerning the outcomes of the experiment. In some experiments, it is helpful to list the elements of the sample space systematically by means of a **tree diagram**.

Example 2.2: An experiment consists of flipping a coin and then flipping it a second time if a head occurs. If a tail occurs on the first flip, then a die is tossed once. To list the elements of the sample space providing the most information, we construct the tree diagram of Figure 2.1. The various paths along the branches of the tree give the distinct sample points. Starting with the top left branch and moving to the right along the first path, we get the sample point *HH*, indicating the possibility that heads occurs on two successive flips of the coin. Likewise, the sample point *T3* indicates the possibility that the coin will show a tail followed by a 3 on the toss of the die. By proceeding along all paths, we see that the sample space is

$$S = \{HH,\ HT,\ T1,\ T2,\ T3,\ T4,\ T5,\ T6\}.$$

2.1 Sample Space

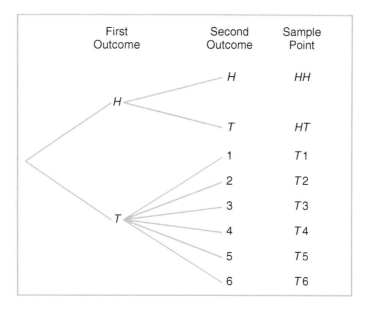

Figure 2.1: Tree diagram for Example 2.2.

Many of the concepts in this chapter are best illustrated with examples involving the use of dice and cards. These are particularly important applications to use early in the learning process, to facilitate the flow of these new concepts into scientific and engineering examples such as the following.

Example 2.3: Suppose that three items are selected at random from a manufacturing process. Each item is inspected and classified defective, D, or nondefective, N. To list the elements of the sample space providing the most information, we construct the tree diagram of Figure 2.2. Now, the various paths along the branches of the tree give the distinct sample points. Starting with the first path, we get the sample point DDD, indicating the possibility that all three items inspected are defective. As we proceed along the other paths, we see that the sample space is

$$S = \{DDD, DDN, DND, DNN, NDD, NDN, NND, NNN\}.$$

Sample spaces with a large or infinite number of sample points are best described by a **statement** or **rule method**. For example, if the possible outcomes of an experiment are the set of cities in the world with a population over 1 million, our sample space is written

$$S = \{x \mid x \text{ is a city with a population over 1 million}\},$$

which reads "S is the set of all x such that x is a city with a population over 1 million." The vertical bar is read "such that." Similarly, if S is the set of all points (x, y) on the boundary or the interior of a circle of radius 2 with center at the origin, we write the **rule**

$$S = \{(x, y) \mid x^2 + y^2 \leq 4\}.$$

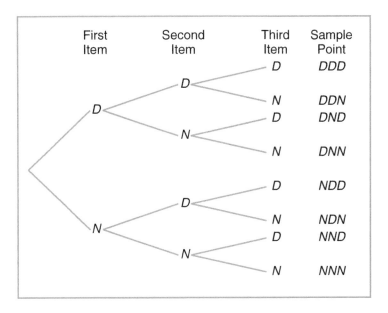

Figure 2.2: Tree diagram for Example 2.3.

Whether we describe the sample space by the rule method or by listing the elements will depend on the specific problem at hand. The rule method has practical advantages, particularly for many experiments where listing becomes a tedious chore.

Consider the situation of Example 2.3 in which items from a manufacturing process are either D, defective, or N, nondefective. There are many important statistical procedures called sampling plans that determine whether or not a "lot" of items is considered satisfactory. One such plan involves sampling until k defectives are observed. Suppose the experiment is to sample items randomly until one defective item is observed. The sample space for this case is

$$S = \{D, ND, NND, NNND, \dots\}.$$

2.2 Events

For any given experiment, we may be interested in the occurrence of certain **events** rather than in the occurrence of a specific element in the sample space. For instance, we may be interested in the event A that the outcome when a die is tossed is divisible by 3. This will occur if the outcome is an element of the subset $A = \{3, 6\}$ of the sample space S_1 in Example 2.1. As a further illustration, we may be interested in the event B that the number of defectives is greater than 1 in Example 2.3. This will occur if the outcome is an element of the subset

$$B = \{DDN, DND, NDD, DDD\}$$

of the sample space S.

To each event we assign a collection of sample points, which constitute a subset of the sample space. That subset represents all of the elements for which the event is true.

2.2 Events

Definition 2.2: An **event** is a subset of a sample space.

Example 2.4: Given the sample space $S = \{t \mid t \geq 0\}$, where t is the life in years of a certain electronic component, then the event A that the component fails before the end of the fifth year is the subset $A = \{t \mid 0 \leq t < 5\}$.

It is conceivable that an event may be a subset that includes the entire sample space S or a subset of S called the **null set** and denoted by the symbol ϕ, which contains no elements at all. For instance, if we let A be the event of detecting a microscopic organism by the naked eye in a biological experiment, then $A = \phi$. Also, if

$$B = \{x \mid x \text{ is an even factor of } 7\},$$

then B must be the null set, since the only possible factors of 7 are the odd numbers 1 and 7.

Consider an experiment where the smoking habits of the employees of a manufacturing firm are recorded. A possible sample space might classify an individual as a nonsmoker, a light smoker, a moderate smoker, or a heavy smoker. Let the subset of smokers be some event. Then all the nonsmokers correspond to a different event, also a subset of S, which is called the **complement** of the set of smokers.

Definition 2.3: The **complement** of an event A with respect to S is the subset of all elements of S that are not in A. We denote the complement of A by the symbol A'.

Example 2.5: Let R be the event that a red card is selected from an ordinary deck of 52 playing cards, and let S be the entire deck. Then R' is the event that the card selected from the deck is not a red card but a black card.

Example 2.6: Consider the sample space

$$S = \{\text{book, cell phone, mp3, paper, stationery, laptop}\}.$$

Let $A = \{\text{book, stationery, laptop, paper}\}$. Then the complement of A is $A' = \{\text{cell phone, mp3}\}$.

We now consider certain operations with events that will result in the formation of new events. These new events will be subsets of the same sample space as the given events. Suppose that A and B are two events associated with an experiment. In other words, A and B are subsets of the same sample space S. For example, in the tossing of a die we might let A be the event that an even number occurs and B the event that a number greater than 3 shows. Then the subsets $A = \{2, 4, 6\}$ and $B = \{4, 5, 6\}$ are subsets of the same sample space

$$S = \{1, 2, 3, 4, 5, 6\}.$$

Note that *both* A and B will occur on a given toss if the outcome is an element of the subset $\{4, 6\}$, which is just the **intersection** of A and B.

Definition 2.4: The **intersection** of two events A and B, denoted by the symbol $A \cap B$, is the event containing all elements that are common to A and B.

Example 2.7: Let E be the event that a person selected at random in a classroom is majoring in engineering, and let F be the event that the person is female. Then $E \cap F$ is the event of all female engineering students in the classroom.

Example 2.8: Let $V = \{a, e, i, o, u\}$ and $C = \{l, r, s, t\}$; then it follows that $V \cap C = \phi$. That is, V and C have no elements in common and, therefore, cannot both simultaneously occur.

For certain statistical experiments it is by no means unusual to define two events, A and B, that cannot both occur simultaneously. The events A and B are then said to be **mutually exclusive**. Stated more formally, we have the following definition:

Definition 2.5: Two events A and B are **mutually exclusive**, or **disjoint**, if $A \cap B = \phi$, that is, if A and B have no elements in common.

Example 2.9: A cable television company offers programs on eight different channels, three of which are affiliated with ABC, two with NBC, and one with CBS. The other two are an educational channel and the ESPN sports channel. Suppose that a person subscribing to this service turns on a television without first selecting the channel. Let A be the event that the program belongs to the NBC network and B the event that it belongs to the CBS network. Since a television program cannot belong to more than one network, the events A and B have no programs in common. Therefore, the intersection $A \cap B$ contains no programs, and consequently the events A and B are mutually exclusive.

Often one is interested in the occurrence of at least one of two events associated with an experiment. Thus, in the die-tossing experiment, if

$$A = \{2, 4, 6\} \text{ and } B = \{4, 5, 6\},$$

we might be interested in either A or B occurring or both A and B occurring. Such an event, called the **union** of A and B, will occur if the outcome is an element of the subset $\{2, 4, 5, 6\}$.

Definition 2.6: The **union** of the two events A and B, denoted by the symbol $A \cup B$, is the event containing all the elements that belong to A or B or both.

Example 2.10: Let $A = \{a, b, c\}$ and $B = \{b, c, d, e\}$; then $A \cup B = \{a, b, c, d, e\}$.

Example 2.11: Let P be the event that an employee selected at random from an oil drilling company smokes cigarettes. Let Q be the event that the employee selected drinks alcoholic beverages. Then the event $P \cup Q$ is the set of all employees who either drink or smoke or do both.

Example 2.12: If $M = \{x \mid 3 < x < 9\}$ and $N = \{y \mid 5 < y < 12\}$, then

$$M \cup N = \{z \mid 3 < z < 12\}.$$

The relationship between events and the corresponding sample space can be illustrated graphically by means of **Venn diagrams**. In a Venn diagram we let the sample space be a rectangle and represent events by circles drawn inside the rectangle. Thus, in Figure 2.3, we see that

$$A \cap B = \text{regions 1 and 2},$$
$$B \cap C = \text{regions 1 and 3},$$

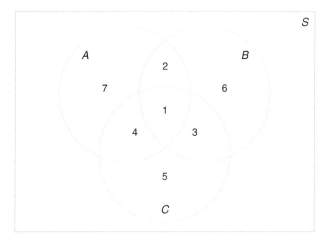

Figure 2.3: Events represented by various regions.

$$
\begin{aligned}
A \cup C &= \text{regions 1, 2, 3, 4, 5, and 7,} \\
B' \cap A &= \text{regions 4 and 7,} \\
A \cap B \cap C &= \text{region 1,} \\
(A \cup B) \cap C' &= \text{regions 2, 6, and 7,}
\end{aligned}
$$

and so forth.

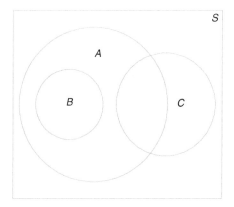

Figure 2.4: Events of the sample space S.

In Figure 2.4, we see that events A, B, and C are all subsets of the sample space S. It is also clear that event B is a subset of event A; event $B \cap C$ has no elements and hence B and C are mutually exclusive; event $A \cap C$ has at least one element; and event $A \cup B = A$. Figure 2.4 might, therefore, depict a situation where we select a card at random from an ordinary deck of 52 playing cards and observe whether the following events occur:

A: the card is red,

B: the card is the jack, queen, or king of diamonds,

C: the card is an ace.

Clearly, the event $A \cap C$ consists of only the two red aces.

Several results that follow from the foregoing definitions, which may easily be verified by means of Venn diagrams, are as follows:

1. $A \cap \phi = \phi$.
2. $A \cup \phi = A$.
3. $A \cap A' = \phi$.
4. $A \cup A' = S$.
5. $S' = \phi$.
6. $\phi' = S$.
7. $(A')' = A$.
8. $(A \cap B)' = A' \cup B'$.
9. $(A \cup B)' = A' \cap B'$.

Exercises

2.1 List the elements of each of the following sample spaces:
(a) the set of integers between 1 and 50 divisible by 8;
(b) the set $S = \{x \mid x^2 + 4x - 5 = 0\}$;
(c) the set of outcomes when a coin is tossed until a tail or three heads appear;
(d) the set $S = \{x \mid x \text{ is a continent}\}$;
(e) the set $S = \{x \mid 2x - 4 \geq 0 \text{ and } x < 1\}$.

2.2 Use the rule method to describe the sample space S consisting of all points in the first quadrant inside a circle of radius 3 with center at the origin.

2.3 Which of the following events are equal?
(a) $A = \{1, 3\}$;
(b) $B = \{x \mid x \text{ is a number on a die}\}$;
(c) $C = \{x \mid x^2 - 4x + 3 = 0\}$;
(d) $D = \{x \mid x \text{ is the number of heads when six coins are tossed}\}$.

2.4 An experiment involves tossing a pair of dice, one green and one red, and recording the numbers that come up. If x equals the outcome on the green die and y the outcome on the red die, describe the sample space S
(a) by listing the elements (x, y);
(b) by using the rule method.

2.5 An experiment consists of tossing a die and then flipping a coin once if the number on the die is even. If the number on the die is odd, the coin is flipped twice. Using the notation $4H$, for example, to denote the outcome that the die comes up 4 and then the coin comes up heads, and $3HT$ to denote the outcome that the die comes up 3 followed by a head and then a tail on the coin, construct a tree diagram to show the 18 elements of the sample space S.

2.6 Two jurors are selected from 4 alternates to serve at a murder trial. Using the notation $A_1 A_3$, for example, to denote the simple event that alternates 1 and 3 are selected, list the 6 elements of the sample space S.

2.7 Four students are selected at random from a chemistry class and classified as male or female. List the elements of the sample space S_1, using the letter M for male and F for female. Define a second sample space S_2 where the elements represent the number of females selected.

2.8 For the sample space of Exercise 2.4,
(a) list the elements corresponding to the event A that the sum is greater than 8;
(b) list the elements corresponding to the event B that a 2 occurs on either die;
(c) list the elements corresponding to the event C that a number greater than 4 comes up on the green die;
(d) list the elements corresponding to the event $A \cap C$;
(e) list the elements corresponding to the event $A \cap B$;
(f) list the elements corresponding to the event $B \cap C$;
(g) construct a Venn diagram to illustrate the intersections and unions of the events A, B, and C.

2.9 For the sample space of Exercise 2.5,
(a) list the elements corresponding to the event A that a number less than 3 occurs on the die;
(b) list the elements corresponding to the event B that two tails occur;
(c) list the elements corresponding to the event A';

(d) list the elements corresponding to the event $A' \cap B$;
(e) list the elements corresponding to the event $A \cup B$.

2.10 An engineering firm is hired to determine if certain waterways in Virginia are safe for fishing. Samples are taken from three rivers.
(a) List the elements of a sample space S, using the letters F for safe to fish and N for not safe to fish.
(b) List the elements of S corresponding to event E that at least two of the rivers are safe for fishing.
(c) Define an event that has as its elements the points
$$\{FFF, NFF, FFN, NFN\}.$$

2.11 The resumés of two male applicants for a college teaching position in chemistry are placed in the same file as the resumés of two female applicants. Two positions become available, and the first, at the rank of assistant professor, is filled by selecting one of the four applicants at random. The second position, at the rank of instructor, is then filled by selecting at random one of the remaining three applicants. Using the notation M_2F_1, for example, to denote the simple event that the first position is filled by the second male applicant and the second position is then filled by the first female applicant,
(a) list the elements of a sample space S;
(b) list the elements of S corresponding to event A that the position of assistant professor is filled by a male applicant;
(c) list the elements of S corresponding to event B that exactly one of the two positions is filled by a male applicant;
(d) list the elements of S corresponding to event C that neither position is filled by a male applicant;
(e) list the elements of S corresponding to the event $A \cap B$;
(f) list the elements of S corresponding to the event $A \cup C$;
(g) construct a Venn diagram to illustrate the intersections and unions of the events A, B, and C.

2.12 Exercise and diet are being studied as possible substitutes for medication to lower blood pressure. Three groups of subjects will be used to study the effect of exercise. Group 1 is sedentary, while group 2 walks and group 3 swims for 1 hour a day. Half of each of the three exercise groups will be on a salt-free diet. An additional group of subjects will not exercise or restrict their salt, but will take the standard medication. Use Z for sedentary, W for walker, S for swimmer, Y for salt, N for no salt, M for medication, and F for medication free.
(a) Show all of the elements of the sample space S.
(b) Given that A is the set of nonmedicated subjects and B is the set of walkers, list the elements of $A \cup B$.
(c) List the elements of $A \cap B$.

2.13 Construct a Venn diagram to illustrate the possible intersections and unions for the following events relative to the sample space consisting of all automobiles made in the United States.
F : Four door, S : Sun roof, P : Power steering.

2.14 If $S = \{0, 1, 2, 3, 4, 5, 6, 7, 8, 9\}$ and $A = \{0, 2, 4, 6, 8\}$, $B = \{1, 3, 5, 7, 9\}$, $C = \{2, 3, 4, 5\}$, and $D = \{1, 6, 7\}$, list the elements of the sets corresponding to the following events:
(a) $A \cup C$;
(b) $A \cap B$;
(c) C';
(d) $(C' \cap D) \cup B$;
(e) $(S \cap C)'$;
(f) $A \cap C \cap D'$.

2.15 Consider the sample space $S = \{$copper, sodium, nitrogen, potassium, uranium, oxygen, zinc$\}$ and the events
$$A = \{\text{copper, sodium, zinc}\},$$
$$B = \{\text{sodium, nitrogen, potassium}\},$$
$$C = \{\text{oxygen}\}.$$
List the elements of the sets corresponding to the following events:
(a) A';
(b) $A \cup C$;
(c) $(A \cap B') \cup C'$;
(d) $B' \cap C'$;
(e) $A \cap B \cap C$;
(f) $(A' \cup B') \cap (A' \cap C)$.

2.16 If $S = \{x \mid 0 < x < 12\}$, $M = \{x \mid 1 < x < 9\}$, and $N = \{x \mid 0 < x < 5\}$, find
(a) $M \cup N$;
(b) $M \cap N$;
(c) $M' \cap N'$.

2.17 Let A, B, and C be events relative to the sample space S. Using Venn diagrams, shade the areas representing the following events:
(a) $(A \cap B)'$;
(b) $(A \cup B)'$;
(c) $(A \cap C) \cup B$.

2.18 Which of the following pairs of events are mutually exclusive?
(a) A golfer scoring the lowest 18-hole round in a 72-hole tournament and losing the tournament.
(b) A poker player getting a flush (all cards in the same suit) and 3 of a kind on the same 5-card hand.
(c) A mother giving birth to a baby girl and a set of twin daughters on the same day.
(d) A chess player losing the last game and winning the match.

2.19 Suppose that a family is leaving on a summer vacation in their camper and that M is the event that they will experience mechanical problems, T is the event that they will receive a ticket for committing a traffic violation, and V is the event that they will arrive at a campsite with no vacancies. Referring to the Venn diagram of Figure 2.5, state in words the events represented by the following regions:
(a) region 5;
(b) region 3;
(c) regions 1 and 2 together;
(d) regions 4 and 7 together;
(e) regions 3, 6, 7, and 8 together.

2.20 Referring to Exercise 2.19 and the Venn diagram of Figure 2.5, list the numbers of the regions that represent the following events:
(a) The family will experience no mechanical problems and will not receive a ticket for a traffic violation but will arrive at a campsite with no vacancies.
(b) The family will experience both mechanical problems and trouble in locating a campsite with a vacancy but will not receive a ticket for a traffic violation.
(c) The family will either have mechanical trouble or arrive at a campsite with no vacancies but will not receive a ticket for a traffic violation.
(d) The family will not arrive at a campsite with no vacancies.

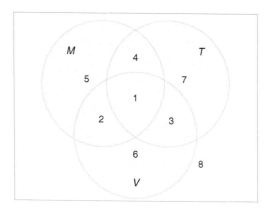

Figure 2.5: Venn diagram for Exercises 2.19 and 2.20.

2.3 Counting Sample Points

One of the problems that the statistician must consider and attempt to evaluate is the element of chance associated with the occurrence of certain events when an experiment is performed. These problems belong in the field of probability, a subject to be introduced in Section 2.4. In many cases, we shall be able to solve a probability problem by counting the number of points in the sample space without actually listing each element. The fundamental principle of counting, often referred to as the **multiplication rule**, is stated in Rule 2.1.

2.3 Counting Sample Points

Rule 2.1: If an operation can be performed in n_1 ways, and if for each of these ways a second operation can be performed in n_2 ways, then the two operations can be performed together in $n_1 n_2$ ways.

Example 2.13: How many sample points are there in the sample space when a pair of dice is thrown once?

Solution: The first die can land face-up in any one of $n_1 = 6$ ways. For each of these 6 ways, the second die can also land face-up in $n_2 = 6$ ways. Therefore, the pair of dice can land in $n_1 n_2 = (6)(6) = 36$ possible ways.

Example 2.14: A developer of a new subdivision offers prospective home buyers a choice of Tudor, rustic, colonial, and traditional exterior styling in ranch, two-story, and split-level floor plans. In how many different ways can a buyer order one of these homes?

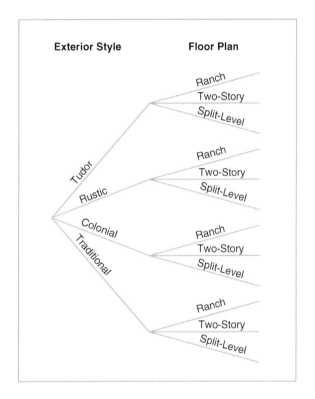

Figure 2.6: Tree diagram for Example 2.14.

Solution: Since $n_1 = 4$ and $n_2 = 3$, a buyer must choose from

$$n_1 n_2 = (4)(3) = 12 \text{ possible homes.}$$

The answers to the two preceding examples can be verified by constructing tree diagrams and counting the various paths along the branches. For instance, in Example 2.14 there will be $n_1 = 4$ branches corresponding to the different exterior styles, and then there will be $n_2 = 3$ branches extending from each of these 4 branches to represent the different floor plans. This tree diagram yields the $n_1 n_2 = 12$ choices of homes given by the paths along the branches, as illustrated in Figure 2.6.

Example 2.15: If a 22-member club needs to elect a chair and a treasurer, how many different ways can these two to be elected?

Solution: For the chair position, there are 22 total possibilities. For each of those 22 possibilities, there are 21 possibilities to elect the treasurer. Using the multiplication rule, we obtain $n_1 \times n_2 = 22 \times 21 = 462$ different ways.

The multiplication rule, Rule 2.1 may be extended to cover any number of operations. Suppose, for instance, that a customer wishes to buy a new cell phone and can choose from $n_1 = 5$ brands, $n_2 = 5$ sets of capability, and $n_3 = 4$ colors. These three classifications result in $n_1 n_2 n_3 = (5)(5)(4) = 100$ different ways for a customer to order one of these phones. The **generalized multiplication rule** covering k operations is stated in the following.

Rule 2.2: If an operation can be performed in n_1 ways, and if for each of these a second operation can be performed in n_2 ways, and for each of the first two a third operation can be performed in n_3 ways, and so forth, then the sequence of k operations can be performed in $n_1 n_2 \cdots n_k$ ways.

Example 2.16: Sam is going to assemble a computer by himself. He has the choice of chips from two brands, a hard drive from four, memory from three, and an accessory bundle from five local stores. How many different ways can Sam order the parts?

Solution: Since $n_1 = 2$, $n_2 = 4$, $n_3 = 3$, and $n_4 = 5$, there are

$$n_l \times n_2 \times n_3 \times n_4 = 2 \times 4 \times 3 \times 5 = 120$$

different ways to order the parts.

Example 2.17: How many even four-digit numbers can be formed from the digits 0, 1, 2, 5, 6, and 9 if each digit can be used only once?

Solution: Since the number must be even, we have only $n_1 = 3$ choices for the units position. However, for a four-digit number the thousands position cannot be 0. Hence, we consider the units position in two parts, 0 or not 0. If the units position is 0 (i.e., $n_1 = 1$), we have $n_2 = 5$ choices for the thousands position, $n_3 = 4$ for the hundreds position, and $n_4 = 3$ for the tens position. Therefore, in this case we have a total of

$$n_1 n_2 n_3 n_4 = (1)(5)(4)(3) = 60$$

even four-digit numbers. On the other hand, if the units position is not 0 (i.e., $n_1 = 2$), we have $n_2 = 4$ choices for the thousands position, $n_3 = 4$ for the hundreds position, and $n_4 = 3$ for the tens position. In this situation, there are a total of

$$n_1 n_2 n_3 n_4 = (2)(4)(4)(3) = 96$$

even four-digit numbers.

Since the above two cases are mutually exclusive, the total number of even four-digit numbers can be calculated as $60 + 96 = 156$.

Frequently, we are interested in a sample space that contains as elements all possible orders or arrangements of a group of objects. For example, we may want to know how many different arrangements are possible for sitting 6 people around a table, or we may ask how many different orders are possible for drawing 2 lottery tickets from a total of 20. The different arrangements are called **permutations**.

Definition 2.7: A **permutation** is an arrangement of all or part of a set of objects.

Consider the three letters a, b, and c. The possible permutations are abc, acb, bac, bca, cab, and cba. Thus, we see that there are 6 distinct arrangements. Using Rule 2.2, we could arrive at the answer 6 without actually listing the different orders by the following arguments: There are $n_1 = 3$ choices for the first position. No matter which letter is chosen, there are always $n_2 = 2$ choices for the second position. No matter which two letters are chosen for the first two positions, there is only $n_3 = 1$ choice for the last position, giving a total of

$$n_1 n_2 n_3 = (3)(2)(1) = 6 \text{ permutations}$$

by Rule 2.2. In general, n distinct objects can be arranged in

$$n(n-1)(n-2)\cdots(3)(2)(1) \text{ ways.}$$

There is a notation for such a number.

Definition 2.8: For any non-negative integer n, $n!$, called "n factorial," is defined as

$$n! = n(n-1)\cdots(2)(1),$$

with special case $0! = 1$.

Using the argument above, we arrive at the following theorem.

Theorem 2.1: The number of permutations of n objects is $n!$.

The number of permutations of the four letters a, b, c, and d will be $4! = 24$. Now consider the number of permutations that are possible by taking two letters at a time from four. These would be ab, ac, ad, ba, bc, bd, ca, cb, cd, da, db, and dc. Using Rule 2.1 again, we have two positions to fill, with $n_1 = 4$ choices for the first and then $n_2 = 3$ choices for the second, for a total of

$$n_1 n_2 = (4)(3) = 12$$

permutations. In general, n distinct objects taken r at a time can be arranged in

$$n(n-1)(n-2)\cdots(n-r+1)$$

ways. We represent this product by the symbol

$$_nP_r = \frac{n!}{(n-r)!}.$$

As a result, we have the theorem that follows.

Theorem 2.2: The number of permutations of n distinct objects taken r at a time is

$$_nP_r = \frac{n!}{(n-r)!}.$$

Example 2.18: In one year, three awards (research, teaching, and service) will be given to a class of 25 graduate students in a statistics department. If each student can receive at most one award, how many possible selections are there?

Solution: Since the awards are distinguishable, it is a permutation problem. The total number of sample points is

$$_{25}P_3 = \frac{25!}{(25-3)!} = \frac{25!}{22!} = (25)(24)(23) = 13,800.$$

Example 2.19: A president and a treasurer are to be chosen from a student club consisting of 50 people. How many different choices of officers are possible if

(a) there are no restrictions;

(b) A will serve only if he is president;

(c) B and C will serve together or not at all;

(d) D and E will not serve together?

Solution: (a) The total number of choices of officers, without any restrictions, is

$$_{50}P_2 = \frac{50!}{48!} = (50)(49) = 2450.$$

(b) Since A will serve only if he is president, we have two situations here: (i) A is selected as the president, which yields 49 possible outcomes for the treasurer's position, or (ii) officers are selected from the remaining 49 people without A, which has the number of choices $_{49}P_2 = (49)(48) = 2352$. Therefore, the total number of choices is $49 + 2352 = 2401$.

(c) The number of selections when B and C serve together is 2. The number of selections when both B and C are not chosen is $_{48}P_2 = 2256$. Therefore, the total number of choices in this situation is $2 + 2256 = 2258$.

(d) The number of selections when D serves as an officer but not E is $(2)(48) = 96$, where 2 is the number of positions D can take and 48 is the number of selections of the other officer from the remaining people in the club except E. The number of selections when E serves as an officer but not D is also $(2)(48) = 96$. The number of selections when both D and E are not chosen is $_{48}P_2 = 2256$. Therefore, the total number of choices is $(2)(96) + 2256 = 2448$. This problem also has another short solution: Since D and E can only serve together in 2 ways, the answer is $2450 - 2 = 2448$.

2.3 Counting Sample Points

Permutations that occur by arranging objects in a circle are called **circular permutations**. Two circular permutations are not considered different unless corresponding objects in the two arrangements are preceded or followed by a different object as we proceed in a clockwise direction. For example, if 4 people are playing bridge, we do not have a new permutation if they all move one position in a clockwise direction. By considering one person in a fixed position and arranging the other three in 3! ways, we find that there are 6 distinct arrangements for the bridge game.

Theorem 2.3: The number of permutations of n objects arranged in a circle is $(n-1)!$.

So far we have considered permutations of distinct objects. That is, all the objects were completely different or distinguishable. Obviously, if the letters b and c are both equal to x, then the 6 permutations of the letters a, b, and c become axx, axx, xax, xax, xxa, and xxa, of which only 3 are distinct. Therefore, with 3 letters, 2 being the same, we have $3!/2! = 3$ distinct permutations. With 4 different letters a, b, c, and d, we have 24 distinct permutations. If we let $a = b = x$ and $c = d = y$, we can list only the following distinct permutations: $xxyy$, $xyxy$, $yxxy$, $yyxx$, $xyyx$, and $yxyx$. Thus, we have $4!/(2!\,2!) = 6$ distinct permutations.

Theorem 2.4: The number of distinct permutations of n things of which n_1 are of one kind, n_2 of a second kind, ..., n_k of a kth kind is
$$\frac{n!}{n_1!n_2!\cdots n_k!}.$$

Example 2.20: In a college football training session, the defensive coordinator needs to have 10 players standing in a row. Among these 10 players, there are 1 freshman, 2 sophomores, 4 juniors, and 3 seniors. How many different ways can they be arranged in a row if only their class level will be distinguished?

Solution: Directly using Theorem 2.4, we find that the total number of arrangements is
$$\frac{10!}{1!\,2!\,4!\,3!} = 12,600.$$

Often we are concerned with the number of ways of partitioning a set of n objects into r subsets called **cells**. A partition has been achieved if the intersection of every possible pair of the r subsets is the empty set ϕ and if the union of all subsets gives the original set. The order of the elements within a cell is of no importance. Consider the set $\{a, e, i, o, u\}$. The possible partitions into two cells in which the first cell contains 4 elements and the second cell 1 element are

$$\{(a,e,i,o),(u)\}, \{(a,i,o,u),(e)\}, \{(e,i,o,u),(a)\}, \{(a,e,o,u),(i)\}, \{(a,e,i,u),(o)\}.$$

We see that there are 5 ways to partition a set of 4 elements into two subsets, or cells, containing 4 elements in the first cell and 1 element in the second.

The number of partitions for this illustration is denoted by the symbol

$$\binom{5}{4,1} = \frac{5!}{4!\,1!} = 5,$$

where the top number represents the total number of elements and the bottom numbers represent the number of elements going into each cell. We state this more generally in Theorem 2.5.

Theorem 2.5: The number of ways of partitioning a set of n objects into r cells with n_1 elements in the first cell, n_2 elements in the second, and so forth, is

$$\binom{n}{n_1, n_2, \ldots, n_r} = \frac{n!}{n_1!n_2!\cdots n_r!},$$

where $n_1 + n_2 + \cdots + n_r = n$.

Example 2.21: In how many ways can 7 graduate students be assigned to 1 triple and 2 double hotel rooms during a conference?

Solution: The total number of possible partitions would be

$$\binom{7}{3,2,2} = \frac{7!}{3!\,2!\,2!} = 210.$$

In many problems, we are interested in the number of ways of selecting r objects from n without regard to order. These selections are called **combinations**. A combination is actually a partition with two cells, the one cell containing the r objects selected and the other cell containing the $(n-r)$ objects that are left. The number of such combinations, denoted by

$$\binom{n}{r, n-r}, \text{ is usually shortened to } \binom{n}{r},$$

since the number of elements in the second cell must be $n - r$.

Theorem 2.6: The number of combinations of n distinct objects taken r at a time is

$$\binom{n}{r} = \frac{n!}{r!(n-r)!}.$$

Example 2.22: A young boy asks his mother to get 5 Game-Boy™ cartridges from his collection of 10 arcade and 5 sports games. How many ways are there that his mother can get 3 arcade and 2 sports games?

Solution: The number of ways of selecting 3 cartridges from 10 is

$$\binom{10}{3} = \frac{10!}{3!\,(10-3)!} = 120.$$

The number of ways of selecting 2 cartridges from 5 is

$$\binom{5}{2} = \frac{5!}{2!\,3!} = 10.$$

Using the multiplication rule (Rule 2.1) with $n_1 = 120$ and $n_2 = 10$, we have $(120)(10) = 1200$ ways.

Example 2.23: How many different letter arrangements can be made from the letters in the word *STATISTICS*?

Solution: Using the same argument as in the discussion for Theorem 2.6, in this example we can actually apply Theorem 2.5 to obtain

$$\binom{10}{3,3,2,1,1} = \frac{10!}{3!\ 3!\ 2!\ 1!\ 1!} = 50,400.$$

Here we have 10 total letters, with 2 letters (S, T) appearing 3 times each, letter I appearing twice, and letters A and C appearing once each. On the other hand, this result can be directly obtained by using Theorem 2.4.

Exercises

2.21 Registrants at a large convention are offered 6 sightseeing tours on each of 3 days. In how many ways can a person arrange to go on a sightseeing tour planned by this convention?

2.22 In a medical study, patients are classified in 8 ways according to whether they have blood type AB^+, AB^-, A^+, A^-, B^+, B^-, O^+, or O^-, and also according to whether their blood pressure is low, normal, or high. Find the number of ways in which a patient can be classified.

2.23 If an experiment consists of throwing a die and then drawing a letter at random from the English alphabet, how many points are there in the sample space?

2.24 Students at a private liberal arts college are classified as being freshmen, sophomores, juniors, or seniors, and also according to whether they are male or female. Find the total number of possible classifications for the students of that college.

2.25 A certain brand of shoes comes in 5 different styles, with each style available in 4 distinct colors. If the store wishes to display pairs of these shoes showing all of its various styles and colors, how many different pairs will the store have on display?

2.26 A California study concluded that following 7 simple health rules can extend a man's life by 11 years on the average and a woman's life by 7 years. These 7 rules are as follows: no smoking, get regular exercise, use alcohol only in moderation, get 7 to 8 hours of sleep, maintain proper weight, eat breakfast, and do not eat between meals. In how many ways can a person adopt 5 of these rules to follow

(a) if the person presently violates all 7 rules?

(b) if the person never drinks and always eats breakfast?

2.27 A developer of a new subdivision offers a prospective home buyer a choice of 4 designs, 3 different heating systems, a garage or carport, and a patio or screened porch. How many different plans are available to this buyer?

2.28 A drug for the relief of asthma can be purchased from 5 different manufacturers in liquid, tablet, or capsule form, all of which come in regular and extra strength. How many different ways can a doctor prescribe the drug for a patient suffering from asthma?

2.29 In a fuel economy study, each of 3 race cars is tested using 5 different brands of gasoline at 7 test sites located in different regions of the country. If 2 drivers are used in the study, and test runs are made once under each distinct set of conditions, how many test runs are needed?

2.30 In how many different ways can a true-false test consisting of 9 questions be answered?

2.31 A witness to a hit-and-run accident told the police that the license number contained the letters RLH followed by 3 digits, the first of which was a 5. If the witness cannot recall the last 2 digits, but is certain that all 3 digits are different, find the maximum number of automobile registrations that the police may have to check.

2.32 (a) In how many ways can 6 people be lined up to get on a bus?

(b) If 3 specific persons, among 6, insist on following each other, how many ways are possible?

(c) If 2 specific persons, among 6, refuse to follow each other, how many ways are possible?

2.33 If a multiple-choice test consists of 5 questions, each with 4 possible answers of which only 1 is correct,

(a) in how many different ways can a student check off one answer to each question?

(b) in how many ways can a student check off one answer to each question and get all the answers wrong?

2.34 (a) How many distinct permutations can be made from the letters of the word $COLUMNS$?

(b) How many of these permutations start with the letter M?

2.35 A contractor wishes to build 9 houses, each different in design. In how many ways can he place these houses on a street if 6 lots are on one side of the street and 3 lots are on the opposite side?

2.36 (a) How many three-digit numbers can be formed from the digits 0, 1, 2, 3, 4, 5, and 6 if each digit can be used only once?

(b) How many of these are odd numbers?

(c) How many are greater than 330?

2.37 In how many ways can 4 boys and 5 girls sit in a row if the boys and girls must alternate?

2.38 Four married couples have bought 8 seats in the same row for a concert. In how many different ways can they be seated

(a) with no restrictions?

(b) if each couple is to sit together?

(c) if all the men sit together to the right of all the women?

2.39 In a regional spelling bee, the 8 finalists consist of 3 boys and 5 girls. Find the number of sample points in the sample space S for the number of possible orders at the conclusion of the contest for

(a) all 8 finalists;

(b) the first 3 positions.

2.40 In how many ways can 5 starting positions on a basketball team be filled with 8 men who can play any of the positions?

2.41 Find the number of ways that 6 teachers can be assigned to 4 sections of an introductory psychology course if no teacher is assigned to more than one section.

2.42 Three lottery tickets for first, second, and third prizes are drawn from a group of 40 tickets. Find the number of sample points in S for awarding the 3 prizes if each contestant holds only 1 ticket.

2.43 In how many ways can 5 different trees be planted in a circle?

2.44 In how many ways can a caravan of 8 covered wagons from Arizona be arranged in a circle?

2.45 How many distinct permutations can be made from the letters of the word $INFINITY$?

2.46 In how many ways can 3 oaks, 4 pines, and 2 maples be arranged along a property line if one does not distinguish among trees of the same kind?

2.47 How many ways are there to select 3 candidates from 8 equally qualified recent graduates for openings in an accounting firm?

2.48 How many ways are there that no two students will have the same birth date in a class of size 60?

2.4 Probability of an Event

Perhaps it was humankind's unquenchable thirst for gambling that led to the early development of probability theory. In an effort to increase their winnings, gamblers called upon mathematicians to provide optimum strategies for various games of chance. Some of the mathematicians providing these strategies were Pascal, Leibniz, Fermat, and James Bernoulli. As a result of this development of probability theory, statistical inference, with all its predictions and generalizations, has branched out far beyond games of chance to encompass many other fields associated with chance occurrences, such as politics, business, weather forecasting,

2.4 Probability of an Event

and scientific research. For these predictions and generalizations to be reasonably accurate, an understanding of basic probability theory is essential.

What do we mean when we make the statement "John will probably win the tennis match," or "I have a fifty-fifty chance of getting an even number when a die is tossed," or "The university is not likely to win the football game tonight," or "Most of our graduating class will likely be married within 3 years"? In each case, we are expressing an outcome of which we are not certain, but owing to past information or from an understanding of the structure of the experiment, we have some degree of confidence in the validity of the statement.

Throughout the remainder of this chapter, we consider only those experiments for which the sample space contains a finite number of elements. The likelihood of the occurrence of an event resulting from such a statistical experiment is evaluated by means of a set of real numbers, called **weights** or **probabilities**, ranging from 0 to 1. To every point in the sample space we assign a probability such that the sum of all probabilities is 1. If we have reason to believe that a certain sample point is quite likely to occur when the experiment is conducted, the probability assigned should be close to 1. On the other hand, a probability closer to 0 is assigned to a sample point that is not likely to occur. In many experiments, such as tossing a coin or a die, all the sample points have the same chance of occurring and are assigned equal probabilities. For points outside the sample space, that is, for simple events that cannot possibly occur, we assign a probability of 0.

To find the probability of an event A, we sum all the probabilities assigned to the sample points in A. This sum is called the **probability** of A and is denoted by $P(A)$.

Definition 2.9: The **probability** of an event A is the sum of the weights of all sample points in A. Therefore,

$$0 \leq P(A) \leq 1, \quad P(\phi) = 0, \quad \text{and} \quad P(S) = 1.$$

Furthermore, if $A_1, A_2, A_3, \ldots$ is a sequence of mutually exclusive events, then

$$P(A_1 \cup A_2 \cup A_3 \cup \cdots) = P(A_1) + P(A_2) + P(A_3) + \cdots.$$

Example 2.24: A coin is tossed twice. What is the probability that at least 1 head occurs?
Solution: The sample space for this experiment is

$$S = \{HH, HT, TH, TT\}.$$

If the coin is balanced, each of these outcomes is equally likely to occur. Therefore, we assign a probability of ω to each sample point. Then $4\omega = 1$, or $\omega = 1/4$. If A represents the event of at least 1 head occurring, then

$$A = \{HH, HT, TH\} \text{ and } P(A) = \frac{1}{4} + \frac{1}{4} + \frac{1}{4} = \frac{3}{4}.$$

Example 2.25: A die is loaded in such a way that an even number is twice as likely to occur as an odd number. If E is the event that a number less than 4 occurs on a single toss of the die, find $P(E)$.

Solution: The sample space is $S = \{1, 2, 3, 4, 5, 6\}$. We assign a probability of w to each odd number and a probability of $2w$ to each even number. Since the sum of the probabilities must be 1, we have $9w = 1$ or $w = 1/9$. Hence, probabilities of $1/9$ and $2/9$ are assigned to each odd and even number, respectively. Therefore,

$$E = \{1, 2, 3\} \text{ and } P(E) = \frac{1}{9} + \frac{2}{9} + \frac{1}{9} = \frac{4}{9}.$$

Example 2.26: In Example 2.25, let A be the event that an even number turns up and let B be the event that a number divisible by 3 occurs. Find $P(A \cup B)$ and $P(A \cap B)$.

Solution: For the events $A = \{2, 4, 6\}$ and $B = \{3, 6\}$, we have

$$A \cup B = \{2, 3, 4, 6\} \text{ and } A \cap B = \{6\}.$$

By assigning a probability of $1/9$ to each odd number and $2/9$ to each even number, we have

$$P(A \cup B) = \frac{2}{9} + \frac{1}{9} + \frac{2}{9} + \frac{2}{9} = \frac{7}{9} \quad \text{and} \quad P(A \cap B) = \frac{2}{9}.$$

If the sample space for an experiment contains N elements, all of which are equally likely to occur, we assign a probability equal to $1/N$ to each of the N points. The probability of any event A containing n of these N sample points is then the ratio of the number of elements in A to the number of elements in S.

Rule 2.3: If an experiment can result in any one of N different equally likely outcomes, and if exactly n of these outcomes correspond to event A, then the probability of event A is

$$P(A) = \frac{n}{N}.$$

Example 2.27: A statistics class for engineers consists of 25 industrial, 10 mechanical, 10 electrical, and 8 civil engineering students. If a person is randomly selected by the instructor to answer a question, find the probability that the student chosen is (a) an industrial engineering major and (b) a civil engineering or an electrical engineering major.

Solution: Denote by I, M, E, and C the students majoring in industrial, mechanical, electrical, and civil engineering, respectively. The total number of students in the class is 53, all of whom are equally likely to be selected.

(a) Since 25 of the 53 students are majoring in industrial engineering, the probability of event I, selecting an industrial engineering major at random, is

$$P(I) = \frac{25}{53}.$$

(b) Since 18 of the 53 students are civil or electrical engineering majors, it follows that

$$P(C \cup E) = \frac{18}{53}.$$

Example 2.28: In a poker hand consisting of 5 cards, find the probability of holding 2 aces and 3 jacks.

Solution: The number of ways of being dealt 2 aces from 4 cards is

$$\binom{4}{2} = \frac{4!}{2!\,2!} = 6,$$

and the number of ways of being dealt 3 jacks from 4 cards is

$$\binom{4}{3} = \frac{4!}{3!\,1!} = 4.$$

By the multiplication rule (Rule 2.1), there are $n = (6)(4) = 24$ hands with 2 aces and 3 jacks. The total number of 5-card poker hands, all of which are equally likely, is

$$N = \binom{52}{5} = \frac{52!}{5!\,47!} = 2{,}598{,}960.$$

Therefore, the probability of getting 2 aces and 3 jacks in a 5-card poker hand is

$$P(C) = \frac{24}{2{,}598{,}960} = 0.9 \times 10^{-5}.$$

If the outcomes of an experiment are not equally likely to occur, the probabilities must be assigned on the basis of prior knowledge or experimental evidence. For example, if a coin is not balanced, we could estimate the probabilities of heads and tails by tossing the coin a large number of times and recording the outcomes. According to the **relative frequency** definition of probability, the true probabilities would be the fractions of heads and tails that occur in the long run. Another intuitive way of understanding probability is the **indifference** approach. For instance, if you have a die that you believe is balanced, then using this indifference approach, you determine that the probability that each of the six sides will show up after a throw is 1/6.

To find a numerical value that represents adequately the probability of winning at tennis, we must depend on our past performance at the game as well as that of the opponent and, to some extent, our belief in our ability to win. Similarly, to find the probability that a horse will win a race, we must arrive at a probability based on the previous records of all the horses entered in the race as well as the records of the jockeys riding the horses. Intuition would undoubtedly also play a part in determining the size of the bet that we might be willing to wager. The use of intuition, personal beliefs, and other indirect information in arriving at probabilities is referred to as the **subjective** definition of probability.

In most of the applications of probability in this book, the relative frequency interpretation of probability is the operative one. Its foundation is the statistical experiment rather than subjectivity, and it is best viewed as the **limiting relative frequency**. As a result, many applications of probability in science and engineering must be based on experiments that can be repeated. Less objective notions of probability are encountered when we assign probabilities based on prior information and opinions, as in "There is a good chance that the Giants will lose the Super

Bowl." When opinions and prior information differ from individual to individual, subjective probability becomes the relevant resource. In Bayesian statistics (see Chapter 18), a more subjective interpretation of probability will be used, based on an elicitation of prior probability information.

2.5 Additive Rules

Often it is easiest to calculate the probability of some event from known probabilities of other events. This may well be true if the event in question can be represented as the union of two other events or as the complement of some event. Several important laws that frequently simplify the computation of probabilities follow. The first, called the **additive rule**, applies to unions of events.

Theorem 2.7: If A and B are two events, then

$$P(A \cup B) = P(A) + P(B) - P(A \cap B).$$

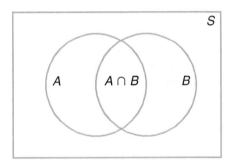

Figure 2.7: Additive rule of probability.

Proof: Consider the Venn diagram in Figure 2.7. The $P(A \cup B)$ is the sum of the probabilities of the sample points in $A \cup B$. Now $P(A) + P(B)$ is the sum of all the probabilities in A plus the sum of all the probabilities in B. Therefore, we have added the probabilities in $(A \cap B)$ twice. Since these probabilities add up to $P(A \cap B)$, we must subtract this probability once to obtain the sum of the probabilities in $A \cup B$.

Corollary 2.1: If A and B are mutually exclusive, then

$$P(A \cup B) = P(A) + P(B).$$

Corollary 2.1 is an immediate result of Theorem 2.7, since if A and B are mutually exclusive, $A \cap B = 0$ and then $P(A \cap B) = P(\phi) = 0$. In general, we can write Corollary 2.2.

2.5 Additive Rules

Corollary 2.2: If $A_1, A_2, \ldots, A_n$ are mutually exclusive, then
$$P(A_1 \cup A_2 \cup \cdots \cup A_n) = P(A_1) + P(A_2) + \cdots + P(A_n).$$

A collection of events $\{A_1, A_2, \ldots, A_n\}$ of a sample space S is called a **partition** of S if $A_1, A_2, \ldots, A_n$ are mutually exclusive and $A_1 \cup A_2 \cup \cdots \cup A_n = S$. Thus, we have

Corollary 2.3: If $A_1, A_2, \ldots, A_n$ is a partition of sample space S, then
$$P(A_1 \cup A_2 \cup \cdots \cup A_n) = P(A_1) + P(A_2) + \cdots + P(A_n) = P(S) = 1.$$

As one might expect, Theorem 2.7 extends in an analogous fashion.

Theorem 2.8: For three events A, B, and C,
$$P(A \cup B \cup C) = P(A) + P(B) + P(C) \\ - P(A \cap B) - P(A \cap C) - P(B \cap C) + P(A \cap B \cap C).$$

Example 2.29: John is going to graduate from an industrial engineering department in a university by the end of the semester. After being interviewed at two companies he likes, he assesses that his probability of getting an offer from company A is 0.8, and his probability of getting an offer from company B is 0.6. If he believes that the probability that he will get offers from both companies is 0.5, what is the probability that he will get at least one offer from these two companies?

Solution: Using the additive rule, we have
$$P(A \cup B) = P(A) + P(B) - P(A \cap B) = 0.8 + 0.6 - 0.5 = 0.9.$$

Example 2.30: What is the probability of getting a total of 7 or 11 when a pair of fair dice is tossed?

Solution: Let A be the event that 7 occurs and B the event that 11 comes up. Now, a total of 7 occurs for 6 of the 36 sample points, and a total of 11 occurs for only 2 of the sample points. Since all sample points are equally likely, we have $P(A) = 1/6$ and $P(B) = 1/18$. The events A and B are mutually exclusive, since a total of 7 and 11 cannot both occur on the same toss. Therefore,
$$P(A \cup B) = P(A) + P(B) = \frac{1}{6} + \frac{1}{18} = \frac{2}{9}.$$

This result could also have been obtained by counting the total number of points for the event $A \cup B$, namely 8, and writing
$$P(A \cup B) = \frac{n}{N} = \frac{8}{36} = \frac{2}{9}.$$

Theorem 2.7 and its three corollaries should help the reader gain more insight into probability and its interpretation. Corollaries 2.1 and 2.2 suggest the very intuitive result dealing with the probability of occurrence of at least one of a number of events, no two of which can occur simultaneously. The probability that at least one occurs is the sum of the probabilities of occurrence of the individual events. The third corollary simply states that the highest value of a probability (unity) is assigned to the entire sample space S.

Example 2.31: If the probabilities are, respectively, 0.09, 0.15, 0.21, and 0.23 that a person purchasing a new automobile will choose the color green, white, red, or blue, what is the probability that a given buyer will purchase a new automobile that comes in one of those colors?

Solution: Let G, W, R, and B be the events that a buyer selects, respectively, a green, white, red, or blue automobile. Since these four events are mutually exclusive, the probability is

$$P(G \cup W \cup R \cup B) = P(G) + P(W) + P(R) + P(B)$$
$$= 0.09 + 0.15 + 0.21 + 0.23 = 0.68.$$

Often it is more difficult to calculate the probability that an event occurs than it is to calculate the probability that the event does not occur. Should this be the case for some event A, we simply find $P(A')$ first and then, using Theorem 2.7, find $P(A)$ by subtraction.

Theorem 2.9: If A and A' are complementary events, then

$$P(A) + P(A') = 1.$$

Proof: Since $A \cup A' = S$ and the sets A and A' are disjoint,

$$1 = P(S) = P(A \cup A') = P(A) + P(A').$$

Example 2.32: If the probabilities that an automobile mechanic will service 3, 4, 5, 6, 7, or 8 or more cars on any given workday are, respectively, 0.12, 0.19, 0.28, 0.24, 0.10, and 0.07, what is the probability that he will service at least 5 cars on his next day at work?

Solution: Let E be the event that at least 5 cars are serviced. Now, $P(E) = 1 - P(E')$, where E' is the event that fewer than 5 cars are serviced. Since

$$P(E') = 0.12 + 0.19 = 0.31,$$

it follows from Theorem 2.9 that

$$P(E) = 1 - 0.31 = 0.69.$$

Example 2.33: Suppose the manufacturer's specifications for the length of a certain type of computer cable are 2000 ± 10 millimeters. In this industry, it is known that small cable is just as likely to be defective (not meeting specifications) as large cable. That is, the probability of randomly producing a cable with length exceeding 2010 millimeters is equal to the probability of producing a cable with length smaller than 1990 millimeters. The probability that the production procedure meets specifications is known to be 0.99.

(a) What is the probability that a cable selected randomly is too large?

(b) What is the probability that a randomly selected cable is larger than 1990 millimeters?

Solution: Let M be the event that a cable meets specifications. Let S and L be the events that the cable is too small and too large, respectively. Then

(a) $P(M) = 0.99$ and $P(S) = P(L) = (1 - 0.99)/2 = 0.005$.

(b) Denoting by X the length of a randomly selected cable, we have

$$P(1990 \leq X \leq 2010) = P(M) = 0.99.$$

Since $P(X \geq 2010) = P(L) = 0.005$,

$$P(X \geq 1990) = P(M) + P(L) = 0.995.$$

This also can be solved by using Theorem 2.9:

$$P(X \geq 1990) + P(X < 1990) = 1.$$

Thus, $P(X \geq 1990) = 1 - P(S) = 1 - 0.005 = 0.995$.

Exercises

2.49 Find the errors in each of the following statements:

(a) The probabilities that an automobile salesperson will sell 0, 1, 2, or 3 cars on any given day in February are, respectively, 0.19, 0.38, 0.29, and 0.15.

(b) The probability that it will rain tomorrow is 0.40, and the probability that it will not rain tomorrow is 0.52.

(c) The probabilities that a printer will make 0, 1, 2, 3, or 4 or more mistakes in setting a document are, respectively, 0.19, 0.34, −0.25, 0.43, and 0.29.

(d) On a single draw from a deck of playing cards, the probability of selecting a heart is 1/4, the probability of selecting a black card is 1/2, and the probability of selecting both a heart and a black card is 1/8.

2.50 Assuming that all elements of S in Exercise 2.8 on page 42 are equally likely to occur, find

(a) the probability of event A;

(b) the probability of event C;

(c) the probability of event $A \cap C$.

2.51 A box contains 500 envelopes, of which 75 contain \$100 in cash, 150 contain \$25, and 275 contain \$10. An envelope may be purchased for \$25. What is the sample space for the different amounts of money? Assign probabilities to the sample points and then find the probability that the first envelope purchased contains less than \$100.

2.52 Suppose that in a senior college class of 500 students it is found that 210 smoke, 258 drink alcoholic beverages, 216 eat between meals, 122 smoke and drink alcoholic beverages, 83 eat between meals and drink alcoholic beverages, 97 smoke and eat between meals, and 52 engage in all three of these bad health practices. If a member of this senior class is selected at random, find the probability that the student

(a) smokes but does not drink alcoholic beverages;

(b) eats between meals and drinks alcoholic beverages but does not smoke;

(c) neither smokes nor eats between meals.

2.53 The probability that an American industry will locate in Shanghai, China, is 0.7, the probability that it will locate in Beijing, China, is 0.4, and the probability that it will locate in either Shanghai or Beijing or both is 0.8. What is the probability that the industry will locate

(a) in both cities?

(b) in neither city?

2.54 From past experience, a stockbroker believes that under present economic conditions a customer will invest in tax-free bonds with a probability of 0.6, will invest in mutual funds with a probability of 0.3, and will invest in both tax-free bonds and mutual funds with a probability of 0.15. At this time, find the probability that a customer will invest

(a) in either tax-free bonds or mutual funds;

(b) in neither tax-free bonds nor mutual funds.

2.55 If each coded item in a catalog begins with 3 distinct letters followed by 4 distinct nonzero digits, find the probability of randomly selecting one of these coded items with the first letter a vowel and the last digit even.

2.56 An automobile manufacturer is concerned about a possible recall of its best-selling four-door sedan. If there were a recall, there is a probability of 0.25 of a defect in the brake system, 0.18 of a defect in the transmission, 0.17 of a defect in the fuel system, and 0.40 of a defect in some other area.

(a) What is the probability that the defect is the brakes or the fueling system if the probability of defects in both systems simultaneously is 0.15?

(b) What is the probability that there are no defects in either the brakes or the fueling system?

2.57 If a letter is chosen at random from the English alphabet, find the probability that the letter

(a) is a vowel exclusive of y;

(b) is listed somewhere ahead of the letter j;

(c) is listed somewhere after the letter g.

2.58 A pair of fair dice is tossed. Find the probability of getting

(a) a total of 8;

(b) at most a total of 5.

2.59 In a poker hand consisting of 5 cards, find the probability of holding

(a) 3 aces;

(b) 4 hearts and 1 club.

2.60 If 3 books are picked at random from a shelf containing 5 novels, 3 books of poems, and a dictionary, what is the probability that

(a) the dictionary is selected?

(b) 2 novels and 1 book of poems are selected?

2.61 In a high school graduating class of 100 students, 54 studied mathematics, 69 studied history, and 35 studied both mathematics and history. If one of these students is selected at random, find the probability that

(a) the student took mathematics or history;

(b) the student did not take either of these subjects;

(c) the student took history but not mathematics.

2.62 Dom's Pizza Company uses taste testing and statistical analysis of the data prior to marketing any new product. Consider a study involving three types of crusts (thin, thin with garlic and oregano, and thin with bits of cheese). Dom's is also studying three sauces (standard, a new sauce with more garlic, and a new sauce with fresh basil).

(a) How many combinations of crust and sauce are involved?

(b) What is the probability that a judge will get a plain thin crust with a standard sauce for his first taste test?

2.63 The likely location of a mobile device in the home is as follows:

Adult bedroom:	0.03
Child bedroom:	0.15
Other bedroom:	0.14
Office or den:	0.40
Other rooms:	0.28

(a) What is the probability that a mobile device is in a bedroom?

(b) What is the probability that it is not in a bedroom?

(c) Suppose a household is selected at random from households with a mobile device, in what room would you expect to find a mobile device?

2.64 Interest centers around the life of an electronic component. Suppose it is known that the probability that the component survives for more than 6000 hours is 0.42. Suppose also that the probability that the component survives *no longer than* 4000 hours is 0.04.

(a) What is the probability that the life of the component is less than or equal to 6000 hours?

(b) What is the probability that the life is greater than 4000 hours?

2.65 Consider the situation of Exercise 2.64. Let A be the event that the component fails a particular test and B be the event that the component displays strain but does not actually fail. Event A occurs with probability 0.20, and event B occurs with probability 0.35.

(a) What is the probability that the component does not fail the test?

(b) What is the probability that the component works perfectly well (i.e., neither displays strain nor fails the test)?

(c) What is the probability that the component either fails or shows strain in the test?

2.66 Factory workers are constantly encouraged to practice zero tolerance when it comes to accidents in factories. Accidents can occur because the working environment or conditions themselves are unsafe. On the other hand, accidents can occur due to carelessness or so-called human error. In addition, the worker's shift, 7:00 A.M.–3:00 P.M. (day shift), 3:00 P.M.–11:00 P.M. (evening shift), or 11:00 P.M.–7:00 A.M. (graveyard shift), may be a factor. During the last year, 300 accidents have occurred. The percentages of the accidents for the condition combinations are as follows:

Shift	Unsafe Conditions	Human Error
Day	5%	32%
Evening	6%	25%
Graveyard	2%	30%

If an accident report is selected randomly from the 300 reports,

(a) what is the probability that the accident occurred on the graveyard shift?

(b) what is the probability that the accident occurred due to human error?

(c) what is the probability that the accident occurred due to unsafe conditions?

(d) what is the probability that the accident occurred on either the evening or the graveyard shift?

2.67 Consider the situation of Example 2.32 on page 58.

(a) What is the probability that no more than 4 cars will be serviced by the mechanic?

(b) What is the probability that he will service fewer than 8 cars?

(c) What is the probability that he will service either 3 or 4 cars?

2.68 Interest centers around the nature of an oven purchased at a particular department store. It can be either a gas or an electric oven. Consider the decisions made by six distinct customers.

(a) Suppose that the probability is 0.40 that at most two of these individuals purchase an electric oven. What is the probability that at least three purchase the electric oven?

(b) Suppose it is known that the probability that all six purchase the electric oven is 0.007 while 0.104 is the probability that all six purchase the gas oven. What is the probability that at least one of each type is purchased?

2.69 It is common in many industrial areas to use a filling machine to fill boxes full of product. This occurs in the food industry as well as other areas in which the product is used in the home, for example, detergent. These machines are not perfect, and indeed they may A, fill to specification, B, underfill, and C, overfill. Generally, the practice of underfilling is that which one hopes to avoid. Let $P(B) = 0.001$ while $P(A) = 0.990$.

(a) Give $P(C)$.

(b) What is the probability that the machine does not underfill?

(c) What is the probability that the machine either overfills or underfills?

2.70 Consider the situation of Exercise 2.69. Suppose 50,000 boxes of detergent are produced per week and suppose also that those underfilled are "sent back," with customers requesting reimbursement of the purchase price. Suppose also that the cost of production is known to be $4.00 per box while the purchase price is $4.50 per box.

(a) What is the weekly profit under the condition of no defective boxes?

(b) What is the loss in profit expected due to underfilling?

2.71 As the situation of Exercise 2.69 might suggest, statistical procedures are often used for control of quality (i.e., industrial quality control). At times, the *weight* of a product is an important variable to control. Specifications are given for the weight of a certain packaged product, and a package is rejected if it is either too light or too heavy. Historical data suggest that 0.95 is the probability that the product meets weight specifications whereas 0.002 is the probability that the product is too light. For each single packaged product, the manufacturer invests $20.00 in production and the purchase price for the consumer is $25.00.

(a) What is the probability that a package chosen randomly from the production line is too heavy?

(b) For each 10,000 packages sold, what profit is received by the manufacturer if all packages meet weight specification?

(c) Assuming that all defective packages are rejected

and rendered worthless, how much is the profit reduced on 10,000 packages due to failure to meet weight specification?

2.72 Prove that

$$P(A' \cap B') = 1 + P(A \cap B) - P(A) - P(B).$$

2.6 Conditional Probability, Independence, and the Product Rule

One very important concept in probability theory is conditional probability. In some applications, the practitioner is interested in the probability structure under certain restrictions. For instance, in epidemiology, rather than studying the chance that a person from the general population has diabetes, it might be of more interest to know this probability for a distinct group such as Asian women in the age range of 35 to 50 or Hispanic men in the age range of 40 to 60. This type of probability is called a conditional probability.

Conditional Probability

The probability of an event B occurring when it is known that some event A has occurred is called a **conditional probability** and is denoted by $P(B|A)$. The symbol $P(B|A)$ is usually read "the probability that B occurs given that A occurs" or simply "the probability of B, given A."

Consider the event B of getting a perfect square when a die is tossed. The die is constructed so that the even numbers are twice as likely to occur as the odd numbers. Based on the sample space $S = \{1, 2, 3, 4, 5, 6\}$, with probabilities of $1/9$ and $2/9$ assigned, respectively, to the odd and even numbers, the probability of B occurring is $1/3$. Now suppose that it is known that the toss of the die resulted in a number greater than 3. We are now dealing with a reduced sample space $A = \{4, 5, 6\}$, which is a subset of S. To find the probability that B occurs, relative to the space A, we must first assign new probabilities to the elements of A proportional to their original probabilities such that their sum is 1. Assigning a probability of w to the odd number in A and a probability of $2w$ to the two even numbers, we have $5w = 1$, or $w = 1/5$. Relative to the space A, we find that B contains the single element 4. Denoting this event by the symbol $B|A$, we write $B|A = \{4\}$, and hence

$$P(B|A) = \frac{2}{5}.$$

This example illustrates that events may have different probabilities when considered relative to different sample spaces.

We can also write

$$P(B|A) = \frac{2}{5} = \frac{2/9}{5/9} = \frac{P(A \cap B)}{P(A)},$$

where $P(A \cap B)$ and $P(A)$ are found from the original sample space S. In other words, a conditional probability relative to a subspace A of S may be calculated directly from the probabilities assigned to the elements of the original sample space S.

2.6 Conditional Probability, Independence, and the Product Rule

Definition 2.10: The conditional probability of B, given A, denoted by $P(B|A)$, is defined by

$$P(B|A) = \frac{P(A \cap B)}{P(A)}, \quad \text{provided} \quad P(A) > 0.$$

As an additional illustration, suppose that our sample space S is the population of adults in a small town who have completed the requirements for a college degree. We shall categorize them according to gender and employment status. The data are given in Table 2.1.

Table 2.1: Categorization of the Adults in a Small Town

	Employed	Unemployed	Total
Male	460	40	500
Female	140	260	400
Total	600	300	900

One of these individuals is to be selected at random for a tour throughout the country to publicize the advantages of establishing new industries in the town. We shall be concerned with the following events:

M: a man is chosen,

E: the one chosen is employed.

Using the reduced sample space E, we find that

$$P(M|E) = \frac{460}{600} = \frac{23}{30}.$$

Let $n(A)$ denote the number of elements in any set A. Using this notation, since each adult has an equal chance of being selected, we can write

$$P(M|E) = \frac{n(E \cap M)}{n(E)} = \frac{n(E \cap M)/n(S)}{n(E)/n(S)} = \frac{P(E \cap M)}{P(E)},$$

where $P(E \cap M)$ and $P(E)$ are found from the original sample space S. To verify this result, note that

$$P(E) = \frac{600}{900} = \frac{2}{3} \quad \text{and} \quad P(E \cap M) = \frac{460}{900} = \frac{23}{45}.$$

Hence,

$$P(M|E) = \frac{23/45}{2/3} = \frac{23}{30},$$

as before.

Example 2.34: The probability that a regularly scheduled flight departs on time is $P(D) = 0.83$; the probability that it arrives on time is $P(A) = 0.82$; and the probability that it departs and arrives on time is $P(D \cap A) = 0.78$. Find the probability that a plane

(a) arrives on time, given that it departed on time, and (b) departed on time, given that it has arrived on time.

Solution: Using Definition 2.10, we have the following.

(a) The probability that a plane arrives on time, given that it departed on time, is

$$P(A|D) = \frac{P(D \cap A)}{P(D)} = \frac{0.78}{0.83} = 0.94.$$

(b) The probability that a plane departed on time, given that it has arrived on time, is

$$P(D|A) = \frac{P(D \cap A)}{P(A)} = \frac{0.78}{0.82} = 0.95.$$

The notion of conditional probability provides the capability of reevaluating the idea of probability of an event in light of additional information, that is, when it is known that another event has occurred. The probability $P(A|B)$ is an updating of $P(A)$ based on the knowledge that event B has occurred. In Example 2.34, it is important to know the probability that the flight arrives on time. One is given the information that the flight did not depart on time. Armed with this additional information, one can calculate the more pertinent probability $P(A|D')$, that is, the probability that it arrives on time, given that it did not depart on time. In many situations, the conclusions drawn from observing the more important conditional probability change the picture entirely. In this example, the computation of $P(A|D')$ is

$$P(A|D') = \frac{P(A \cap D')}{P(D')} = \frac{0.82 - 0.78}{0.17} = 0.24.$$

As a result, the probability of an on-time arrival is diminished severely in the presence of the additional information.

Example 2.35: The concept of conditional probability has countless uses in both industrial and biomedical applications. Consider an industrial process in the textile industry in which strips of a particular type of cloth are being produced. These strips can be defective in two ways, length and nature of texture. For the case of the latter, the process of identification is very complicated. It is known from historical information on the process that 10% of strips fail the length test, 5% fail the texture test, and only 0.8% fail both tests. If a strip is selected randomly from the process and a quick measurement identifies it as failing the length test, what is the probability that it is texture defective?

Solution: Consider the events

L: length defective, T: texture defective.

Given that the strip is length defective, the probability that this strip is texture defective is given by

$$P(T|L) = \frac{P(T \cap L)}{P(L)} = \frac{0.008}{0.1} = 0.08.$$

Thus, knowing the conditional probability provides considerably more information than merely knowing $P(T)$.

Independent Events

In the die-tossing experiment discussed on page 62, we note that $P(B|A) = 2/5$ whereas $P(B) = 1/3$. That is, $P(B|A) \neq P(B)$, indicating that B depends on A. Now consider an experiment in which 2 cards are drawn in succession from an ordinary deck, with replacement. The events are defined as

A: the first card is an ace,

B: the second card is a spade.

Since the first card is replaced, our sample space for both the first and the second draw consists of 52 cards, containing 4 aces and 13 spades. Hence,

$$P(B|A) = \frac{13}{52} = \frac{1}{4} \quad \text{and} \quad P(B) = \frac{13}{52} = \frac{1}{4}.$$

That is, $P(B|A) = P(B)$. When this is true, the events A and B are said to be **independent**.

Although conditional probability allows for an alteration of the probability of an event in the light of additional material, it also enables us to understand better the very important concept of **independence** or, in the present context, independent events. In the airport illustration in Example 2.34, $P(A|D)$ differs from $P(A)$. This suggests that the occurrence of D influenced A, and this is certainly expected in this illustration. However, consider the situation where we have events A and B and

$$P(A|B) = P(A).$$

In other words, the occurrence of B had no impact on the odds of occurrence of A. Here the occurrence of A is independent of the occurrence of B. The importance of the concept of independence cannot be overemphasized. It plays a vital role in material in virtually all chapters in this book and in all areas of applied statistics.

Definition 2.11: Two events A and B are **independent** if and only if

$$P(B|A) = P(B) \quad \text{or} \quad P(A|B) = P(A),$$

assuming the existences of the conditional probabilities. Otherwise, A and B are **dependent**.

The condition $P(B|A) = P(B)$ implies that $P(A|B) = P(A)$, and conversely. For the card-drawing experiments, where we showed that $P(B|A) = P(B) = 1/4$, we also can see that $P(A|B) = P(A) = 1/13$.

The Product Rule, or the Multiplicative Rule

Multiplying the formula in Definition 2.10 by $P(A)$, we obtain the following important **multiplicative rule** (or **product rule**), which enables us to calculate the probability that two events will both occur.

Theorem 2.10: If in an experiment the events A and B can both occur, then
$$P(A \cap B) = P(A)P(B|A), \text{ provided } P(A) > 0.$$

Thus, the probability that both A and B occur is equal to the probability that A occurs multiplied by the conditional probability that B occurs, given that A occurs. Since the events $A \cap B$ and $B \cap A$ are equivalent, it follows from Theorem 2.10 that we can also write

$$P(A \cap B) = P(B \cap A) = P(B)P(A|B).$$

In other words, it does not matter which event is referred to as A and which event is referred to as B.

Example 2.36: Suppose that we have a fuse box containing 20 fuses, of which 5 are defective. If 2 fuses are selected at random and removed from the box in succession without replacing the first, what is the probability that both fuses are defective?

Solution: We shall let A be the event that the first fuse is defective and B the event that the second fuse is defective; then we interpret $A \cap B$ as the event that A occurs and then B occurs after A has occurred. The probability of first removing a defective fuse is 1/4; then the probability of removing a second defective fuse from the remaining 4 is 4/19. Hence,

$$P(A \cap B) = \left(\frac{1}{4}\right)\left(\frac{4}{19}\right) = \frac{1}{19}.$$

Example 2.37: One bag contains 4 white balls and 3 black balls, and a second bag contains 3 white balls and 5 black balls. One ball is drawn from the first bag and placed unseen in the second bag. What is the probability that a ball now drawn from the second bag is black?

Solution: Let B_1, B_2, and W_1 represent, respectively, the drawing of a black ball from bag 1, a black ball from bag 2, and a white ball from bag 1. We are interested in the union of the mutually exclusive events $B_1 \cap B_2$ and $W_1 \cap B_2$. The various possibilities and their probabilities are illustrated in Figure 2.8. Now

$$P[(B_1 \cap B_2) \text{ or } (W_1 \cap B_2)] = P(B_1 \cap B_2) + P(W_1 \cap B_2)$$
$$= P(B_1)P(B_2|B_1) + P(W_1)P(B_2|W_1)$$
$$= \left(\frac{3}{7}\right)\left(\frac{6}{9}\right) + \left(\frac{4}{7}\right)\left(\frac{5}{9}\right) = \frac{38}{63}.$$

If, in Example 2.36, the first fuse is replaced and the fuses thoroughly rearranged before the second is removed, then the probability of a defective fuse on the second selection is still 1/4; that is, $P(B|A) = P(B)$ and the events A and B are independent. When this is true, we can substitute $P(B)$ for $P(B|A)$ in Theorem 2.10 to obtain the following special multiplicative rule.

2.6 Conditional Probability, Independence, and the Product Rule

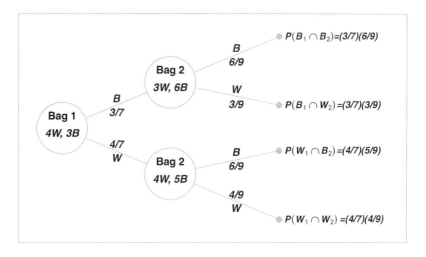

Figure 2.8: Tree diagram for Example 2.37.

Theorem 2.11: Two events A and B are independent if and only if

$$P(A \cap B) = P(A)P(B).$$

Therefore, to obtain the probability that two independent events will both occur, we simply find the product of their individual probabilities.

Example 2.38: A small town has one fire engine and one ambulance available for emergencies. The probability that the fire engine is available when needed is 0.98, and the probability that the ambulance is available when called is 0.92. In the event of an injury resulting from a burning building, find the probability that both the ambulance and the fire engine will be available, assuming they operate independently.

Solution: Let A and B represent the respective events that the fire engine and the ambulance are available. Then

$$P(A \cap B) = P(A)P(B) = (0.98)(0.92) = 0.9016.$$

Example 2.39: An electrical system consists of four components as illustrated in Figure 2.9. The system works if components A and B work and either of the components C or D works. The reliability (probability of working) of each component is also shown in Figure 2.9. Find the probability that (a) the entire system works and (b) the component C does not work, given that the entire system works. Assume that the four components work independently.

Solution: In this configuration of the system, A, B, and the subsystem C and D constitute a serial circuit system, whereas the subsystem C and D itself is a parallel circuit system.

(a) Clearly the probability that the entire system works can be calculated as follows:

$$P[A \cap B \cap (C \cup D)] = P(A)P(B)P(C \cup D) = P(A)P(B)[1 - P(C' \cap D')]$$
$$= P(A)P(B)[1 - P(C')P(D')]$$
$$= (0.9)(0.9)[1 - (1 - 0.8)(1 - 0.8)] = 0.7776.$$

The equalities above hold because of the independence among the four components.

(b) To calculate the conditional probability in this case, notice that

$$P = \frac{P(\text{the system works but } C \text{ does not work})}{P(\text{the system works})}$$
$$= \frac{P(A \cap B \cap C' \cap D)}{P(\text{the system works})} = \frac{(0.9)(0.9)(1 - 0.8)(0.8)}{0.7776} = 0.1667.$$

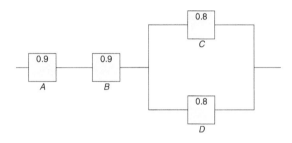

Figure 2.9: An electrical system for Example 2.39.

The multiplicative rule can be extended to more than two-event situations.

Theorem 2.12: If, in an experiment, the events $A_1, A_2, \ldots, A_k$ can occur, then

$$P(A_1 \cap A_2 \cap \cdots \cap A_k)$$
$$= P(A_1)P(A_2|A_1)P(A_3|A_1 \cap A_2) \cdots P(A_k|A_1 \cap A_2 \cap \cdots \cap A_{k-1}).$$

If the events $A_1, A_2, \ldots, A_k$ are independent, then

$$P(A_1 \cap A_2 \cap \cdots \cap A_k) = P(A_1)P(A_2) \cdots P(A_k).$$

Example 2.40: Three cards are drawn in succession, without replacement, from an ordinary deck of playing cards. Find the probability that the event $A_1 \cap A_2 \cap A_3$ occurs, where A_1 is the event that the first card is a red ace, A_2 is the event that the second card is a 10 or a jack, and A_3 is the event that the third card is greater than 3 but less than 7.

Solution: First we define the events

A_1: the first card is a red ace,

A_2: the second card is a 10 or a jack,

A_3: the third card is greater than 3 but less than 7.

Now

$$P(A_1) = \frac{2}{52}, \quad P(A_2|A_1) = \frac{8}{51}, \quad P(A_3|A_1 \cap A_2) = \frac{12}{50},$$

and hence, by Theorem 2.12,

$$P(A_1 \cap A_2 \cap A_3) = P(A_1)P(A_2|A_1)P(A_3|A_1 \cap A_2)$$
$$= \left(\frac{2}{52}\right)\left(\frac{8}{51}\right)\left(\frac{12}{50}\right) = \frac{8}{5525}.$$

The property of independence stated in Theorem 2.11 can be extended to deal with more than two events. Consider, for example, the case of three events A, B, and C. It is not sufficient to only have that $P(A \cap B \cap C) = P(A)P(B)P(C)$ as a definition of independence among the three. Suppose $A = B$ and $C = \phi$, the null set. Although $A \cap B \cap C = \phi$, which results in $P(A \cap B \cap C) = 0 = P(A)P(B)P(C)$, events A and B are not independent. Hence, we have the following definition.

Definition 2.12: A collection of events $\mathcal{A} = \{A_1, \ldots, A_n\}$ are mutually independent if for any subset of $\mathcal{A}$, $A_{i_1}, \ldots, A_{i_k}$, for $k \leq n$, we have

$$P(A_{i_1} \cap \cdots \cap A_{i_k}) = P(A_{i_1}) \cdots P(A_{i_k}).$$

Exercises

2.73 If R is the event that a convict committed armed robbery and D is the event that the convict pushed dope, state in words what probabilities are expressed by
(a) $P(R|D)$;
(b) $P(D'|R)$;
(c) $P(R'|D')$.

2.74 A class in advanced physics is composed of 10 juniors, 30 seniors, and 10 graduate students. The final grades show that 3 of the juniors, 10 of the seniors, and 5 of the graduate students received an A for the course. If a student is chosen at random from this class and is found to have earned an A, what is the probability that he or she is a senior?

2.75 A random sample of 200 adults are classified below by sex and their level of education attained.

Education	Male	Female
Elementary	38	45
Secondary	28	50
College	22	17

If a person is picked at random from this group, find the probability that
(a) the person is a male, given that the person has a secondary education;

(b) the person does not have a college degree, given that the person is a female.

2.76 In an experiment to study the relationship of hypertension and smoking habits, the following data are collected for 180 individuals:

	Nonsmokers	Moderate Smokers	Heavy Smokers
H	21	36	30
NH	48	26	19

where H and NH in the table stand for *Hypertension* and *Nonhypertension*, respectively. If one of these individuals is selected at random, find the probability that the person is
(a) experiencing hypertension, given that the person is a heavy smoker;
(b) a nonsmoker, given that the person is experiencing no hypertension.

2.77 In the senior year of a high school graduating class of 100 students, 42 studied mathematics, 68 studied psychology, 54 studied history, 22 studied both mathematics and history, 25 studied both mathematics and psychology, 7 studied history but neither mathematics nor psychology, 10 studied all three subjects, and 8 did not take any of the three. Randomly select a student from the class and find the probabilities of the following events.

(a) A person enrolled in psychology takes all three subjects.
(b) A person not taking psychology is taking both history and mathematics.

2.78 A manufacturer of a flu vaccine is concerned about the quality of its flu serum. Batches of serum are processed by three different departments having rejection rates of 0.10, 0.08, and 0.12, respectively. The inspections by the three departments are sequential and independent.

(a) What is the probability that a batch of serum survives the first departmental inspection but is rejected by the second department?
(b) What is the probability that a batch of serum is rejected by the third department?

2.79 In *USA Today* (Sept. 5, 1996), the results of a survey involving the use of sleepwear while traveling were listed as follows:

	Male	Female	Total
Underwear	0.220	0.024	0.244
Nightgown	0.002	0.180	0.182
Nothing	0.160	0.018	0.178
Pajamas	0.102	0.073	0.175
T-shirt	0.046	0.088	0.134
Other	0.084	0.003	0.087

(a) What is the probability that a traveler is a female who sleeps in the nude?
(b) What is the probability that a traveler is male?
(c) Assuming the traveler is male, what is the probability that he sleeps in pajamas?
(d) What is the probability that a traveler is male if the traveler sleeps in pajamas or a T-shirt?

2.80 The probability that an automobile being filled with gasoline also needs an oil change is 0.25; the probability that it needs a new oil filter is 0.40; and the probability that both the oil and the filter need changing is 0.14.

(a) If the oil has to be changed, what is the probability that a new oil filter is needed?
(b) If a new oil filter is needed, what is the probability that the oil has to be changed?

2.81 The probability that a married man watches a certain television show is 0.4, and the probability that a married woman watches the show is 0.5. The probability that a man watches the show, given that his wife does, is 0.7. Find the probability that

(a) a married couple watches the show;
(b) a wife watches the show, given that her husband does;
(c) at least one member of a married couple will watch the show.

2.82 For married couples living in a certain suburb, the probability that the husband will vote on a bond referendum is 0.21, the probability that the wife will vote on the referendum is 0.28, and the probability that both the husband and the wife will vote is 0.15. What is the probability that

(a) at least one member of a married couple will vote?
(b) a wife will vote, given that her husband will vote?
(c) a husband will vote, given that his wife will not vote?

2.83 The probability that a vehicle entering the Luray Caverns has Canadian license plates is 0.12; the probability that it is a camper is 0.28; and the probability that it is a camper with Canadian license plates is 0.09. What is the probability that

(a) a camper entering the Luray Caverns has Canadian license plates?
(b) a vehicle with Canadian license plates entering the Luray Caverns is a camper?
(c) a vehicle entering the Luray Caverns does not have Canadian plates or is not a camper?

2.84 The probability that you are home when a telemarketing representative calls is 0.4. Given that you are home, the probability that goods will be bought from the company is 0.3. Find the probability that you are home and goods are bought from the company.

2.85 The probability that a doctor correctly diagnoses a particular illness is 0.7. Given that the doctor makes an incorrect diagnosis, the probability that the patient files a lawsuit is 0.9. What is the probability that the doctor makes an incorrect diagnosis and the patient sues?

2.86 In 1970, 11% of Americans completed four years of college; 43% of them were women. In 1990, 22% of Americans completed four years of college; 53% of them were women (*Time*, Jan. 19, 1996).

(a) Given that a person completed four years of college in 1970, what is the probability that the person was a woman?
(b) What is the probability that a woman finished four years of college in 1990?
(c) What is the probability that a man had not finished college in 1990?

Exercises

2.87 A real estate agent has 8 master keys to open several new homes. Only 1 master key will open any given house. If 40% of these homes are usually left unlocked, what is the probability that the real estate agent can get into a specific home if the agent selects 3 master keys at random before leaving the office?

2.88 Before the distribution of certain statistical software, every fourth compact disk (CD) is tested for accuracy. The testing process consists of running four independent programs and checking the results. The failure rates for the four testing programs are, respectively, 0.01, 0.03, 0.02, and 0.01.

(a) What is the probability that a CD was tested and failed any test?

(b) Given that a CD was tested, what is the probability that it failed program 2 or 3?

(c) In a sample of 100, how many CDs would you expect to be rejected?

(d) Given that a CD was defective, what is the probability that it was tested?

2.89 A town has two fire engines operating independently. The probability that a specific engine is available when needed is 0.96.

(a) What is the probability that neither is available when needed?

(b) What is the probability that a fire engine is available when needed?

2.90 Pollution of the rivers in the United States has been a problem for many years. Consider the following events:

A: the river is polluted,
B: a sample of water tested detects pollution,
C: fishing is permitted.

Assume $P(A) = 0.3$, $P(B|A) = 0.75$, $P(B|A') = 0.20$, $P(C|A \cap B) = 0.20$, $P(C|A' \cap B) = 0.15$, $P(C|A \cap B') = 0.80$, and $P(C|A' \cap B') = 0.90$.

(a) Find $P(A \cap B \cap C)$.
(b) Find $P(B' \cap C)$.
(c) Find $P(C)$.
(d) Find the probability that the river is polluted, given that fishing is permitted and the sample tested did not detect pollution.

2.91 Find the probability of randomly selecting 4 good quarts of milk in succession from a cooler containing 20 quarts of which 5 have spoiled, by using

(a) the first formula of Theorem 2.12 on page 68;

(b) the formulas of Theorem 2.6 and Rule 2.3 on pages 50 and 54, respectively.

2.92 Suppose the diagram of an electrical system is as given in Figure 2.10. What is the probability that the system works? Assume the components fail independently.

2.93 A circuit system is given in Figure 2.11. Assume the components fail independently.

(a) What is the probability that the entire system works?

(b) Given that the system works, what is the probability that the component A is not working?

2.94 In the situation of Exercise 2.93, it is known that the system does not work. What is the probability that the component A also does not work?

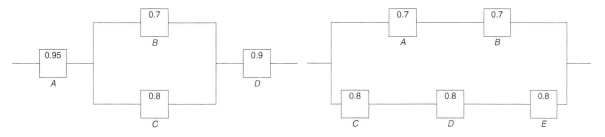

Figure 2.10: Diagram for Exercise 2.92.   Figure 2.11: Diagram for Exercise 2.93.

2.7 Bayes' Rule

Bayesian statistics is a collection of tools that is used in a special form of statistical inference which applies in the analysis of experimental data in many practical situations in science and engineering. Bayes' rule is one of the most important rules in probability theory. It is the foundation of Bayesian inference, which will be discussed in Chapter 18.

Total Probability

Let us now return to the illustration of Section 2.6, where an individual is being selected at random from the adults of a small town to tour the country and publicize the advantages of establishing new industries in the town. Suppose that we are now given the additional information that 36 of those employed and 12 of those unemployed are members of the Rotary Club. We wish to find the probability of the event A that the individual selected is a member of the Rotary Club. Referring to Figure 2.12, we can write A as the union of the two mutually exclusive events $E \cap A$ and $E' \cap A$. Hence, $A = (E \cap A) \cup (E' \cap A)$, and by Corollary 2.1 of Theorem 2.7, and then Theorem 2.10, we can write

$$P(A) = P[(E \cap A) \cup (E' \cap A)] = P(E \cap A) + P(E' \cap A)$$
$$= P(E)P(A|E) + P(E')P(A|E').$$

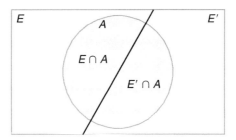

Figure 2.12: Venn diagram for the events A, E, and E'.

The data of Section 2.6, together with the additional data given above for the set A, enable us to compute

$$P(E) = \frac{600}{900} = \frac{2}{3}, \quad P(A|E) = \frac{36}{600} = \frac{3}{50},$$

and

$$P(E') = \frac{1}{3}, \quad P(A|E') = \frac{12}{300} = \frac{1}{25}.$$

If we display these probabilities by means of the tree diagram of Figure 2.13, where the first branch yields the probability $P(E)P(A|E)$ and the second branch yields

2.7 Bayes' Rule

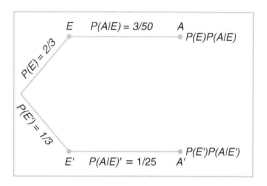

Figure 2.13: Tree diagram for the data on page 63, using additional information on page 72.

the probability $P(E')P(A|E')$, it follows that

$$P(A) = \left(\frac{2}{3}\right)\left(\frac{3}{50}\right) + \left(\frac{1}{3}\right)\left(\frac{1}{25}\right) = \frac{4}{75}.$$

A generalization of the foregoing illustration to the case where the sample space is partitioned into k subsets is covered by the following theorem, sometimes called the **theorem of total probability** or the **rule of elimination**.

Theorem 2.13: If the events $B_1, B_2, \ldots, B_k$ constitute a partition of the sample space S such that $P(B_i) \neq 0$ for $i = 1, 2, \ldots, k$, then for any event A of S,

$$P(A) = \sum_{i=1}^{k} P(B_i \cap A) = \sum_{i=1}^{k} P(B_i) P(A|B_i).$$

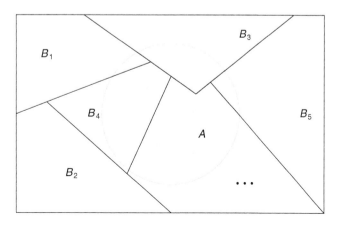

Figure 2.14: Partitioning the sample space S.

Proof: Consider the Venn diagram of Figure 2.14. The event A is seen to be the union of the mutually exclusive events

$$B_1 \cap A, \ B_2 \cap A, \ \ldots, \ B_k \cap A;$$

that is,

$$A = (B_1 \cap A) \cup (B_2 \cap A) \cup \cdots \cup (B_k \cap A).$$

Using Corollary 2.2 of Theorem 2.7 and Theorem 2.10, we have

$$\begin{aligned} P(A) &= P[(B_1 \cap A) \cup (B_2 \cap A) \cup \cdots \cup (B_k \cap A)] \\ &= P(B_1 \cap A) + P(B_2 \cap A) + \cdots + P(B_k \cap A) \\ &= \sum_{i=1}^{k} P(B_i \cap A) \\ &= \sum_{i=1}^{k} P(B_i) P(A|B_i). \end{aligned}$$

Example 2.41: In a certain assembly plant, three machines, B_1, B_2, and B_3, make 30%, 45%, and 25%, respectively, of the products. It is known from past experience that 2%, 3%, and 2% of the products made by each machine, respectively, are defective. Now, suppose that a finished product is randomly selected. What is the probability that it is defective?

Solution: Consider the following events:

A: the product is defective,

B_1: the product is made by machine B_1,

B_2: the product is made by machine B_2,

B_3: the product is made by machine B_3.

Applying the rule of elimination, we can write

$$P(A) = P(B_1)P(A|B_1) + P(B_2)P(A|B_2) + P(B_3)P(A|B_3).$$

Referring to the tree diagram of Figure 2.15, we find that the three branches give the probabilities

$$\begin{aligned} P(B_1)P(A|B_1) &= (0.3)(0.02) = 0.006, \\ P(B_2)P(A|B_2) &= (0.45)(0.03) = 0.0135, \\ P(B_3)P(A|B_3) &= (0.25)(0.02) = 0.005, \end{aligned}$$

and hence

$$P(A) = 0.006 + 0.0135 + 0.005 = 0.0245.$$

2.7 Bayes' Rule

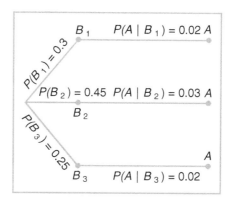

Figure 2.15: Tree diagram for Example 2.41.

Bayes' Rule

Instead of asking for $P(A)$ in Example 2.41, by the rule of elimination, suppose that we now consider the problem of finding the conditional probability $P(B_i|A)$. In other words, suppose that a product was randomly selected and it is defective. What is the probability that this product was made by machine B_i? Questions of this type can be answered by using the following theorem, called **Bayes' rule**:

Theorem 2.14: (**Bayes' Rule**) If the events $B_1, B_2, \ldots, B_k$ constitute a partition of the sample space S such that $P(B_i) \neq 0$ for $i = 1, 2, \ldots, k$, then for any event A in S such that $P(A) \neq 0$,

$$P(B_r|A) = \frac{P(B_r \cap A)}{\sum_{i=1}^{k} P(B_i \cap A)} = \frac{P(B_r)P(A|B_r)}{\sum_{i=1}^{k} P(B_i)P(A|B_i)} \quad \text{for } r = 1, 2, \ldots, k.$$

Proof: By the definition of conditional probability,

$$P(B_r|A) = \frac{P(B_r \cap A)}{P(A)},$$

and then using Theorem 2.13 in the denominator, we have

$$P(B_r|A) = \frac{P(B_r \cap A)}{\sum_{i=1}^{k} P(B_i \cap A)} = \frac{P(B_r)P(A|B_r)}{\sum_{i=1}^{k} P(B_i)P(A|B_i)},$$

which completes the proof.

Example 2.42: With reference to Example 2.41, if a product was chosen randomly and found to be defective, what is the probability that it was made by machine B_3?

Solution: Using Bayes' rule to write

$$P(B_3|A) = \frac{P(B_3)P(A|B_3)}{P(B_1)P(A|B_1) + P(B_2)P(A|B_2) + P(B_3)P(A|B_3)},$$

and then substituting the probabilities calculated in Example 2.41, we have

$$P(B_3|A) = \frac{0.005}{0.006 + 0.0135 + 0.005} = \frac{0.005}{0.0245} = \frac{10}{49}.$$

In view of the fact that a defective product was selected, this result suggests that it probably was not made by machine B_3.

Example 2.43: A manufacturing firm employs three analytical plans for the design and development of a particular product. For cost reasons, all three are used at varying times. In fact, plans 1, 2, and 3 are used for 30%, 20%, and 50% of the products, respectively. The defect rate is different for the three procedures as follows:

$$P(D|P_1) = 0.01, \qquad P(D|P_2) = 0.03, \qquad P(D|P_3) = 0.02,$$

where $P(D|P_j)$ is the probability of a defective product, given plan j. If a random product was observed and found to be defective, which plan was most likely used and thus responsible?

Solution: From the statement of the problem

$$P(P_1) = 0.30, \quad P(P_2) = 0.20, \quad \text{and} \quad P(P_3) = 0.50,$$

we must find $P(P_j|D)$ for $j = 1, 2, 3$. Bayes' rule (Theorem 2.14) shows

$$P(P_1|D) = \frac{P(P_1)P(D|P_1)}{P(P_1)P(D|P_1) + P(P_2)P(D|P_2) + P(P_3)P(D|P_3)}$$
$$= \frac{(0.30)(0.01)}{(0.3)(0.01) + (0.20)(0.03) + (0.50)(0.02)} = \frac{0.003}{0.019} = 0.158.$$

Similarly,

$$P(P_2|D) = \frac{(0.03)(0.20)}{0.019} = 0.316 \text{ and } P(P_3|D) = \frac{(0.02)(0.50)}{0.019} = 0.526.$$

The conditional probability of a defect given plan 3 is the largest of the three; thus a defective for a random product is most likely the result of the use of plan 3.

Using Bayes' rule, a statistical methodology called the Bayesian approach has attracted a lot of attention in applications. An introduction to the Bayesian method will be discussed in Chapter 18.

Exercises

2.95 In a certain region of the country it is known from past experience that the probability of selecting an adult over 40 years of age with cancer is 0.05. If the probability of a doctor correctly diagnosing a person with cancer as having the disease is 0.78 and the probability of incorrectly diagnosing a person without cancer as having the disease is 0.06, what is the probability that an adult over 40 years of age is diagnosed as having cancer?

2.96 Police plan to enforce speed limits by using radar traps at four different locations within the city limits. The radar traps at each of the locations L_1, L_2, L_3, and L_4 will be operated 40%, 30%, 20%, and 30% of the time. If a person who is speeding on her way to work has probabilities of 0.2, 0.1, 0.5, and 0.2, respectively, of passing through these locations, what is the probability that she will receive a speeding ticket?

2.97 Referring to Exercise 2.95, what is the probability that a person diagnosed as having cancer actually has the disease?

2.98 If the person in Exercise 2.96 received a speeding ticket on her way to work, what is the probability that she passed through the radar trap located at L_2?

2.99 Suppose that the four inspectors at a chocolate factory are supposed to stamp the expiration date on each package of chocolate at the end of the assembly line. John, who stamps 20% of the packages, fails to stamp the expiration date once in every 200 packages; Tom, who stamps 60% of the packages, fails to stamp the expiration date once in every 100 packages; Jeff, who stamps 15% of the packages, fails to stamp the expiration date once in every 90 packages; and Pat, who stamps 5% of the packages, fails to stamp the expiration date once in every 200 packages. If a customer complains that her package of chocolate does not show the expiration date, what is the probability that it was inspected by John?

2.100 A regional telephone company operates three identical relay stations at different locations. During a one-year period, the number of malfunctions reported by each station and the causes are shown below.

Station	A	B	C
Problems with electricity supplied	2	1	1
Computer malfunction	4	3	2
Malfunctioning electrical equipment	5	4	2
Caused by other human errors	7	7	5

Suppose that a malfunction was reported and it was found to be caused by other human errors. What is the probability that it came from station C?

2.101 A paint-store chain produces and sells latex and semigloss paint. Based on long-range sales, the probability that a customer will purchase latex paint is 0.75. Of those that purchase latex paint, 60% also purchase rollers. But only 30% of semigloss paint buyers purchase rollers. A randomly selected buyer purchases a roller and a can of paint. What is the probability that the paint is latex?

2.102 Denote by A, B, and C the events that a grand prize is behind doors A, B, and C, respectively. Suppose you randomly picked a door, say A. The game host opened a door, say B, and showed there was no prize behind it. Now the host offers you the option of either staying at the door that you picked (A) or switching to the remaining unopened door (C). Use probability to explain whether you should switch or not.

Review Exercises

2.103 A truth serum has the property that 90% of the guilty suspects are properly judged while, of course, 10% of the guilty suspects are improperly found innocent. On the other hand, innocent suspects are misjudged 1% of the time. If the suspect was selected from a group of suspects of which only 5% have ever committed a crime, and the serum indicates that he is guilty, what is the probability that he is innocent?

2.104 An allergist claims that 50% of the patients she tests are allergic to some type of weed. What is the probability that

(a) exactly 3 of her next 4 patients are allergic to weeds?

(b) none of her next 4 patients is allergic to weeds?

2.105 By comparing appropriate regions of Venn diagrams, verify that

(a) $(A \cap B) \cup (A \cap B') = A$;

(b) $A' \cap (B' \cup C) = (A' \cap B') \cup (A' \cap C)$.

2.106 The probabilities that a service station will pump gas into 0, 1, 2, 3, 4, or 5 or more cars during a certain 30-minute period are 0.03, 0.18, 0.24, 0.28, 0.10, and 0.17, respectively. Find the probability that in this 30-minute period

(a) more than 2 cars receive gas;

(b) at most 4 cars receive gas;

(c) 4 or more cars receive gas.

2.107 How many bridge hands are possible containing 4 spades, 6 diamonds, 1 club, and 2 hearts?

2.108 If the probability is 0.1 that a person will make a mistake on his or her state income tax return, find the probability that

(a) four totally unrelated persons each make a mistake;

(b) Mr. Jones and Ms. Clark both make mistakes, and Mr. Roberts and Ms. Williams do not make a mistake.

2.109 A large industrial firm uses three local motels to provide overnight accommodations for its clients. From past experience it is known that 20% of the clients are assigned rooms at the Ramada Inn, 50% at the Sheraton, and 30% at the Lakeview Motor Lodge. If the plumbing is faulty in 5% of the rooms at the Ramada Inn, in 4% of the rooms at the Sheraton, and in 8% of the rooms at the Lakeview Motor Lodge, what is the probability that

(a) a client will be assigned a room with faulty plumbing?

(b) a person with a room having faulty plumbing was assigned accommodations at the Lakeview Motor Lodge?

2.110 The probability that a patient recovers from a delicate heart operation is 0.8. What is the probability that

(a) exactly 2 of the next 3 patients who have this operation survive?

(b) all of the next 3 patients who have this operation survive?

2.111 In a certain federal prison, it is known that 2/3 of the inmates are under 25 years of age. It is also known that 3/5 of the inmates are male and that 5/8 of the inmates are female or 25 years of age or older. What is the probability that a prisoner selected at random from this prison is female and at least 25 years old?

2.112 From 4 red, 5 green, and 6 yellow apples, how many selections of 9 apples are possible if 3 of each color are to be selected?

2.113 From a box containing 6 black balls and 4 green balls, 3 balls are drawn in succession, each ball being replaced in the box before the next draw is made. What is the probability that

(a) all 3 are the same color?

(b) each color is represented?

2.114 A shipment of 12 televisions contains 3 defective ones. In how many ways can a hotel purchase 5 of these ones and receive at least 2 of the defective ones?

2.115 A certain federal agency employs three consulting firms (A, B, and C) with probabilities 0.40, 0.35, and 0.25, respectively. From past experience it is known that the probability of cost overruns for the firms are 0.05, 0.03, and 0.15, respectively. Suppose a cost overrun is experienced by the agency.

(a) What is the probability that the consulting firm involved is company C?

(b) What is the probability that it is company A?

2.116 A manufacturer is studying the effects of cooking temperature, cooking time, and type of cooking oil for making potato chips. Three different temperatures, 4 different cooking times, and 3 different oils are to be used.

(a) What is the total number of combinations to be studied?

(b) How many combinations will be used for each type of oil?

(c) Discuss why permutations are not an issue in this exercise.

2.117 Consider the situation in Exercise 2.116, and suppose that the manufacturer can try only two combinations in a day.

(a) What is the probability that any given set of two runs is chosen?

(b) What is the probability that the highest temperature is used in either of these two combinations?

2.118 A certain form of cancer is known to be found in women over 60 with probability 0.07. A blood test exists for the detection of the disease, but the test is not infallible. In fact, it is known that 10% of the time the test gives a false negative (i.e., the test incorrectly gives a negative result) and 5% of the time the test gives a false positive (i.e., incorrectly gives a positive result). If a woman over 60 is known to have taken the test and received a favorable (i.e., negative) result, what is the probability that she has the disease?

2.119 A producer of a certain type of electronic component ships to suppliers in lots of twenty. Suppose that 60% of all such lots contain no defective components, 30% contain one defective component, and 10% contain two defective components. A lot is picked, two components from the lot are randomly selected and tested, and neither is defective.

(a) What is the probability that zero defective components exist in the lot?

(b) What is the probability that one defective exists in the lot?

(c) What is the probability that two defectives exist in the lot?

2.120 A rare disease exists with which only 1 in 500 is affected. A test for the disease exists, but of course it is not infallible. A correct positive result (patient actually has the disease) occurs 95% of the time, while a false positive result (patient does not have the disease) occurs 1% of the time. If a randomly selected individual is tested and the result is positive, what is the probability that the individual has the disease?

2.121 A construction company employs two sales engineers. Engineer 1 does the work of estimating cost for 70% of jobs bid by the company. Engineer 2 does the work for 30% of jobs bid by the company. It is known that the error rate for engineer 1 is such that 0.02 is the probability of an error when he does the work, whereas the probability of an error in the work of engineer 2 is 0.04. Suppose a bid arrives and a serious error occurs in estimating cost. Which engineer would you guess did the work? Explain and show all work.

2.122 In the field of quality control, the science of statistics is often used to determine if a process is "out of control." Suppose the process is, indeed, out of control and 20% of items produced are defective.
(a) If three items arrive off the process line in succession, what is the probability that all three are defective?
(b) If four items arrive in succession, what is the probability that three are defective?

2.123 An industrial plant is conducting a study to determine how quickly injured workers are back on the job following injury. Records show that 10% of all injured workers are admitted to the hospital for treatment and 15% are back on the job the next day. In addition, studies show that 2% are both admitted for hospital treatment and back on the job the next day. If a worker is injured, what is the probability that the worker will either be admitted to a hospital or be back on the job the next day or both?

2.124 A firm is accustomed to training operators who do certain tasks on a production line. Those operators who attend the training course are known to be able to meet their production quotas 90% of the time. New operators who do not take the training course only meet their quotas 65% of the time. Fifty percent of new operators attend the course. Given that a new operator meets her production quota, what is the probability that she attended the program?

2.125 A survey of those using a particular statistical software system indicated that 10% were dissatisfied. Half of those dissatisfied purchased the system from vendor A. It is also known that 20% of those surveyed purchased from vendor A. Given that the software was purchased from vendor A, what is the probability that that particular user is dissatisfied?

2.126 During bad economic times, industrial workers are dismissed and are often replaced by machines. The history of 100 workers whose loss of employment is attributable to technological advances is reviewed. For each of these individuals, it is determined if he or she was given an alternative job within the same company, found a job with another company in the same field, found a job in a new field, or has been unemployed for 1 year. In addition, the union status of each worker is recorded. The following table summarizes the results.

	Union	Nonunion
Same Company	40	15
New Company (same field)	13	10
New Field	4	11
Unemployed	2	5

(a) If the selected worker found a job with a new company in the same field, what is the probability that the worker is a union member?
(b) If the worker is a union member, what is the probability that the worker has been unemployed for a year?

2.127 There is a 50-50 chance that the queen carries the gene of hemophilia. If she is a carrier, then each prince has a 50-50 chance of having hemophilia independently. If the queen is not a carrier, the prince will not have the disease. Suppose the queen has had three princes without the disease. What is the probability the queen is a carrier?

2.128 Group Project: Give each student a bag of chocolate M&Ms. Divide the students into groups of 5 or 6. Calculate the relative frequency distribution for color of M&Ms for each group.
(a) What is your estimated probability of randomly picking a yellow? a red?
(b) Redo the calculations for the whole classroom. Did the estimates change?
(c) Do you believe there is an equal number of each color in a process batch? Discuss.

2.8 Potential Misconceptions and Hazards; Relationship to Material in Other Chapters

This chapter contains the fundamental definitions, rules, and theorems that provide a foundation that renders probability an important tool for evaluating scientific and engineering systems. The evaluations are often in the form of probability computations, as is illustrated in examples and exercises. Concepts such as independence, conditional probability, Bayes' rule, and others tend to mesh nicely to solve practical problems in which the bottom line is to produce a probability value. Illustrations in exercises are abundant. See, for example, Exercises 2.100 and 2.101. In these and many other exercises, an evaluation of a scientific system is being made judiciously from a probability calculation, using rules and definitions discussed in the chapter.

Now, how does the material in this chapter relate to that in other chapters? It is best to answer this question by looking ahead to Chapter 3. Chapter 3 also deals with the type of problems in which it is important to calculate probabilities. We illustrate how system performance depends on the value of one or more probabilities. Once again, conditional probability and independence play a role. However, new concepts arise which allow more structure based on the notion of a random variable and its probability distribution. Recall that the idea of frequency distributions was discussed briefly in Chapter 1. The probability distribution displays, in equation form or graphically, the total information necessary to describe a probability structure. For example, in Review Exercise 2.122 the random variable of interest is the number of defective items, a discrete measurement. Thus, the probability distribution would reveal the probability structure for the number of defective items out of the number selected from the process. As the reader moves into Chapter 3 and beyond, it will become apparent that assumptions will be required in order to determine and thus make use of probability distributions for solving scientific problems.

Chapter 3

Random Variables and Probability Distributions

3.1 Concept of a Random Variable

Statistics is concerned with making inferences about populations and population characteristics. Experiments are conducted with results that are subject to chance. The testing of a number of electronic components is an example of a **statistical experiment**, a term that is used to describe any process by which several chance observations are generated. It is often important to allocate a numerical description to the outcome. For example, the sample space giving a detailed description of each possible outcome when three electronic components are tested may be written

$$S = \{NNN, NND, NDN, DNN, NDD, DND, DDN, DDD\},$$

where N denotes nondefective and D denotes defective. One is naturally concerned with the number of defectives that occur. Thus, each point in the sample space will be *assigned a numerical value* of 0, 1, 2, or 3. These values are, of course, random quantities *determined by the outcome of the experiment*. They may be viewed as values assumed by the *random variable* X, the number of defective items when three electronic components are tested.

Definition 3.1: A **random variable** is a function that associates a real number with each element in the sample space.

We shall use a capital letter, say X, to denote a random variable and its corresponding small letter, x in this case, for one of its values. In the electronic component testing illustration above, we notice that the random variable X assumes the value 2 for all elements in the subset

$$E = \{DDN, DND, NDD\}$$

of the sample space S. That is, each possible value of X represents an event that is a subset of the sample space for the given experiment.

Example 3.1: Two balls are drawn in succession without replacement from an urn containing 4 red balls and 3 black balls. The possible outcomes and the values y of the random variable Y, where Y is the number of red balls, are

Sample Space	y
RR	2
RB	1
BR	1
BB	0

Example 3.2: A stockroom clerk returns three safety helmets at random to three steel mill employees who had previously checked them. If Smith, Jones, and Brown, in that order, receive one of the three hats, list the sample points for the possible orders of returning the helmets, and find the value m of the random variable M that represents the number of correct matches.

Solution: If S, J, and B stand for Smith's, Jones's, and Brown's helmets, respectively, then the possible arrangements in which the helmets may be returned and the number of correct matches are

Sample Space	m
SJB	3
SBJ	1
BJS	1
JSB	1
JBS	0
BSJ	0

In each of the two preceding examples, the sample space contains a finite number of elements. On the other hand, when a die is thrown until a 5 occurs, we obtain a sample space with an unending sequence of elements,

$$S = \{F, NF, NNF, NNNF, \ldots\},$$

where F and N represent, respectively, the occurrence and nonoccurrence of a 5. But even in this experiment, the number of elements can be equated to the number of whole numbers so that there is a first element, a second element, a third element, and so on, and in this sense can be counted.

There are cases where the random variable is categorical in nature. Variables, often called *dummy* variables, are used. A good illustration is the case in which the random variable is binary in nature, as shown in the following example.

Example 3.3: Consider the simple condition in which components are arriving from the production line and they are stipulated to be defective or not defective. Define the random variable X by

$$X = \begin{cases} 1, & \text{if the component is defective,} \\ 0, & \text{if the component is not defective.} \end{cases}$$

Clearly the assignment of 1 or 0 is arbitrary though quite convenient. This will become clear in later chapters. The random variable for which 0 and 1 are chosen to describe the two possible values is called a **Bernoulli random variable**.

Further illustrations of random variables are revealed in the following examples.

Example 3.4: Statisticians use **sampling plans** to either accept or reject batches or lots of material. Suppose one of these sampling plans involves sampling independently 10 items from a lot of 100 items in which 12 are defective.

Let X be the random variable defined as the number of items found defective in the sample of 10. In this case, the random variable takes on the values $0, 1, 2, \ldots, 9, 10$.

Example 3.5: Suppose a sampling plan involves sampling items from a process until a defective is observed. The evaluation of the process will depend on how many consecutive items are observed. In that regard, let X be a random variable defined by the number of items observed before a defective is found. With N a nondefective and D a defective, sample spaces are $S = \{D\}$ given $X = 1$, $S = \{ND\}$ given $X = 2$, $S = \{NND\}$ given $X = 3$, and so on.

Example 3.6: Interest centers around the proportion of people who respond to a certain mail order solicitation. Let X be that proportion. X is a random variable that takes on all values x for which $0 \leq x \leq 1$.

Example 3.7: Let X be the random variable defined by the waiting time, in hours, between successive speeders spotted by a radar unit. The random variable X takes on all values x for which $x \geq 0$.

Definition 3.2: If a sample space contains a finite number of possibilities or an unending sequence with as many elements as there are whole numbers, it is called a **discrete sample space**.

The outcomes of some statistical experiments may be neither finite nor countable. Such is the case, for example, when one conducts an investigation measuring the distances that a certain make of automobile will travel over a prescribed test course on 5 liters of gasoline. Assuming distance to be a variable measured to any degree of accuracy, then clearly we have an infinite number of possible distances in the sample space that cannot be equated to the number of whole numbers. Or, if one were to record the length of time for a chemical reaction to take place, once again the possible time intervals making up our sample space would be infinite in number and uncountable. We see now that all sample spaces need not be discrete.

Definition 3.3: If a sample space contains an infinite number of possibilities equal to the number of points on a line segment, it is called a **continuous sample space**.

A random variable is called a **discrete random variable** if its set of possible outcomes is countable. The random variables in Examples 3.1 to 3.5 are discrete random variables. But a random variable whose set of possible values is an entire interval of numbers is not discrete. When a random variable can take on values on a continuous scale, it is called a **continuous random variable**. Often the possible values of a continuous random variable are precisely the same values that are contained in the continuous sample space. Obviously, the random variables described in Examples 3.6 and 3.7 are continuous random variables.

In most practical problems, continuous random variables represent *measured* data, such as all possible heights, weights, temperatures, distance, or life periods, whereas discrete random variables represent *count* data, such as the number of defectives in a sample of k items or the number of highway fatalities per year in a given state. Note that the random variables Y and M of Examples 3.1 and 3.2 both represent count data, Y the number of red balls and M the number of correct hat matches.

3.2 Discrete Probability Distributions

A discrete random variable assumes each of its values with a certain probability. In the case of tossing a coin three times, the variable X, representing the number of heads, assumes the value 2 with probability 3/8, since 3 of the 8 equally likely sample points result in two heads and one tail. If one assumes equal weights for the simple events in Example 3.2, the probability that no employee gets back the right helmet, that is, the probability that M assumes the value 0, is 1/3. The possible values m of M and their probabilities are

m	0	1	3
$P(M=m)$	$\frac{1}{3}$	$\frac{1}{2}$	$\frac{1}{6}$

Note that the values of m exhaust all possible cases and hence the probabilities add to 1.

Frequently, it is convenient to represent all the probabilities of a random variable X by a formula. Such a formula would necessarily be a function of the numerical values x that we shall denote by $f(x)$, $g(x)$, $r(x)$, and so forth. Therefore, we write $f(x) = P(X = x)$; that is, $f(3) = P(X = 3)$. The set of ordered pairs $(x, f(x))$ is called the **probability function, probability mass function**, or **probability distribution** of the discrete random variable X.

Definition 3.4: The set of ordered pairs $(x, f(x))$ is a **probability function, probability mass function**, or **probability distribution** of the discrete random variable X if, for each possible outcome x,

1. $f(x) \geq 0$,
2. $\sum_{x} f(x) = 1$,
3. $P(X = x) = f(x)$.

Example 3.8: A shipment of 20 similar laptop computers to a retail outlet contains 3 that are defective. If a school makes a random purchase of 2 of these computers, find the probability distribution for the number of defectives.

Solution: Let X be a random variable whose values x are the possible numbers of defective computers purchased by the school. Then x can only take the numbers 0, 1, and

2. Now

$$f(0) = P(X=0) = \frac{\binom{3}{0}\binom{17}{2}}{\binom{20}{2}} = \frac{68}{95}, \quad f(1) = P(X=1) = \frac{\binom{3}{1}\binom{17}{1}}{\binom{20}{2}} = \frac{51}{190},$$

$$f(2) = P(X=2) = \frac{\binom{3}{2}\binom{17}{0}}{\binom{20}{2}} = \frac{3}{190}.$$

Thus, the probability distribution of X is

x	0	1	2
$f(x)$	$\frac{68}{95}$	$\frac{51}{190}$	$\frac{3}{190}$

Example 3.9: If a car agency sells 50% of its inventory of a certain foreign car equipped with side airbags, find a formula for the probability distribution of the number of cars with side airbags among the next 4 cars sold by the agency.

Solution: Since the probability of selling an automobile with side airbags is 0.5, the $2^4 = 16$ points in the sample space are equally likely to occur. Therefore, the denominator for all probabilities, and also for our function, is 16. To obtain the number of ways of selling 3 cars with side airbags, we need to consider the number of ways of partitioning 4 outcomes into two cells, with 3 cars with side airbags assigned to one cell and the model without side airbags assigned to the other. This can be done in $\binom{4}{3} = 4$ ways. In general, the event of selling x models with side airbags and $4-x$ models without side airbags can occur in $\binom{4}{x}$ ways, where x can be 0, 1, 2, 3, or 4. Thus, the probability distribution $f(x) = P(X = x)$ is

$$f(x) = \frac{1}{16}\binom{4}{x}, \quad \text{for } x = 0, 1, 2, 3, 4.$$

There are many problems where we may wish to compute the probability that the observed value of a random variable X will be less than or equal to some real number x. Writing $F(x) = P(X \leq x)$ for every real number x, we define $F(x)$ to be the **cumulative distribution function** of the random variable X.

Definition 3.5: The **cumulative distribution function** $F(x)$ of a discrete random variable X with probability distribution $f(x)$ is

$$F(x) = P(X \leq x) = \sum_{t \leq x} f(t), \quad \text{for } -\infty < x < \infty.$$

For the random variable M, the number of correct matches in Example 3.2, we have

$$F(2) = P(M \leq 2) = f(0) + f(1) = \frac{1}{3} + \frac{1}{2} = \frac{5}{6}.$$

The cumulative distribution function of M is

$$F(m) = \begin{cases} 0, & \text{for } m < 0, \\ \frac{1}{3}, & \text{for } 0 \leq m < 1, \\ \frac{5}{6}, & \text{for } 1 \leq m < 3, \\ 1, & \text{for } m \geq 3. \end{cases}$$

Example 3.10: Find the cumulative distribution function of the random variable X in Example 3.9. Using $F(x)$, verify that $f(2) = 3/8$.

Solution: Direct calculations of the probability distribution of Example 3.9 give $f(0) = 1/16$, $f(1) = 1/4$, $f(2) = 3/8$, $f(3) = 1/4$, and $f(4) = 1/16$. Therefore,

$$F(0) = f(0) = \frac{1}{16},$$
$$F(1) = f(0) + f(1) = \frac{5}{16},$$
$$F(2) = f(0) + f(1) + f(2) = \frac{11}{16},$$
$$F(3) = f(0) + f(1) + f(2) + f(3) = \frac{15}{16},$$
$$F(4) = f(0) + f(1) + f(2) + f(3) + f(4) = 1.$$

Hence,

$$F(x) = \begin{cases} 0, & \text{for } x < 0, \\ \frac{1}{16}, & \text{for } 0 \leq x < 1, \\ \frac{5}{16}, & \text{for } 1 \leq x < 2, \\ \frac{11}{16}, & \text{for } 2 \leq x < 3, \\ \frac{15}{16}, & \text{for } 3 \leq x < 4, \\ 1 & \text{for } x \geq 4. \end{cases}$$

Now

$$f(2) = F(2) - F(1) = \frac{11}{16} - \frac{5}{16} = \frac{3}{8}.$$

It is often helpful to look at a probability distribution in graphic form. One might plot the points $(x, f(x))$ of Example 3.9 to obtain Figure 3.1. By joining the points to the x axis either with a dashed or with a solid line, we obtain a probability mass function plot. Figure 3.1 makes it easy to see what values of X are most likely to occur, and it also indicates a perfectly symmetric situation in this case.

Instead of plotting the points $(x, f(x))$, we more frequently construct rectangles, as in Figure 3.2. Here the rectangles are constructed so that their bases of equal width are centered at each value x and their heights are equal to the corresponding probabilities given by $f(x)$. The bases are constructed so as to leave no space between the rectangles. Figure 3.2 is called a **probability histogram**.

Since each base in Figure 3.2 has unit width, $P(X = x)$ is equal to the area of the rectangle centered at x. Even if the bases were not of unit width, we could adjust the heights of the rectangles to give areas that would still equal the probabilities of X assuming any of its values x. This concept of using areas to represent

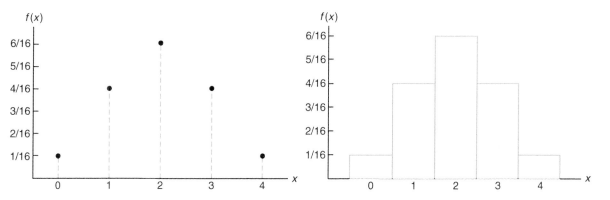

Figure 3.1: Probability mass function plot.   Figure 3.2: Probability histogram.

probabilities is necessary for our consideration of the probability distribution of a continuous random variable.

The graph of the cumulative distribution function of Example 3.9, which appears as a step function in Figure 3.3, is obtained by plotting the points $(x, F(x))$.

Certain probability distributions are applicable to more than one physical situation. The probability distribution of Example 3.9, for example, also applies to the random variable Y, where Y is the number of heads when a coin is tossed 4 times, or to the random variable W, where W is the number of red cards that occur when 4 cards are drawn at random from a deck in succession with each card replaced and the deck shuffled before the next drawing. Special discrete distributions that can be applied to many different experimental situations will be considered in Chapter 5.

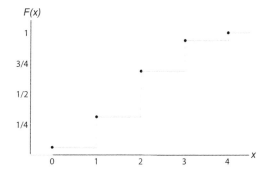

Figure 3.3: Discrete cumulative distribution function.

3.3 Continuous Probability Distributions

A continuous random variable has a probability of 0 of assuming *exactly* any of its values. Consequently, its probability distribution cannot be given in tabular form.

At first this may seem startling, but it becomes more plausible when we consider a particular example. Let us discuss a random variable whose values are the heights of all people over 21 years of age. Between any two values, say 163.5 and 164.5 centimeters, or even 163.99 and 164.01 centimeters, there are an infinite number of heights, one of which is 164 centimeters. The probability of selecting a person at random who is exactly 164 centimeters tall and not one of the infinitely large set of heights so close to 164 centimeters that you cannot humanly measure the difference is remote, and thus we assign a probability of 0 to the event. This is not the case, however, if we talk about the probability of selecting a person who is at least 163 centimeters but not more than 165 centimeters tall. Now we are dealing with an interval rather than a point value of our random variable.

We shall concern ourselves with computing probabilities for various intervals of continuous random variables such as $P(a < X < b)$, $P(W \geq c)$, and so forth. Note that when X is continuous,

$$P(a < X \leq b) = P(a < X < b) + P(X = b) = P(a < X < b).$$

That is, it does not matter whether we include an endpoint of the interval or not. This is not true, though, when X is discrete.

Although the probability distribution of a continuous random variable cannot be presented in tabular form, it can be stated as a formula. Such a formula would necessarily be a function of the numerical values of the continuous random variable X and as such will be represented by the functional notation $f(x)$. In dealing with continuous variables, $f(x)$ is usually called the **probability density function**, or simply the **density function**, of X. Since X is defined over a continuous sample space, it is possible for $f(x)$ to have a finite number of discontinuities. However, most density functions that have practical applications in the analysis of statistical data are continuous and their graphs may take any of several forms, some of which are shown in Figure 3.4. Because areas will be used to represent probabilities and probabilities are positive numerical values, the density function must lie entirely above the x axis.

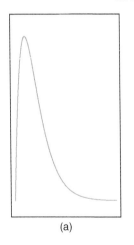
(a)

(b)
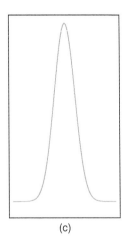
(c)
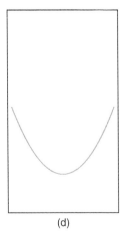
(d)

Figure 3.4: Typical density functions.

A probability density function is constructed so that the area under its curve

3.3 Continuous Probability Distributions

bounded by the x axis is equal to 1 when computed over the range of X for which $f(x)$ is defined. Should this range of X be a finite interval, it is always possible to extend the interval to include the entire set of real numbers by defining $f(x)$ to be zero at all points in the extended portions of the interval. In Figure 3.5, the probability that X assumes a value between a and b is equal to the shaded area under the density function between the ordinates at $x = a$ and $x = b$, and from integral calculus is given by

$$P(a < X < b) = \int_a^b f(x)\, dx.$$

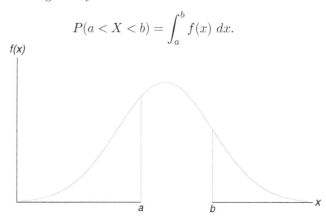

Figure 3.5: $P(a < X < b)$.

Definition 3.6: The function $f(x)$ is a **probability density function** (pdf) for the continuous random variable X, defined over the set of real numbers, if

1. $f(x) \geq 0$, for all $x \in R$.
2. $\int_{-\infty}^{\infty} f(x)\, dx = 1$.
3. $P(a < X < b) = \int_a^b f(x)\, dx$.

Example 3.11: Suppose that the error in the reaction temperature, in °C, for a controlled laboratory experiment is a continuous random variable X having the probability density function

$$f(x) = \begin{cases} \frac{x^2}{3}, & -1 < x < 2, \\ 0, & \text{elsewhere.} \end{cases}$$

(a) Verify that $f(x)$ is a density function.

(b) Find $P(0 < X \leq 1)$.

Solution: We use Definition 3.6.

(a) Obviously, $f(x) \geq 0$. To verify condition 2 in Definition 3.6, we have

$$\int_{-\infty}^{\infty} f(x)\, dx = \int_{-1}^{2} \frac{x^2}{3}\, dx = \frac{x^3}{9}\Big|_{-1}^{2} = \frac{8}{9} + \frac{1}{9} = 1.$$

(b) Using formula 3 in Definition 3.6, we obtain

$$P(0 < X \leq 1) = \int_0^1 \frac{x^2}{3} dx = \left.\frac{x^3}{9}\right|_0^1 = \frac{1}{9}.$$

Definition 3.7: The **cumulative distribution function** $F(x)$ of a continuous random variable X with density function $f(x)$ is

$$F(x) = P(X \leq x) = \int_{-\infty}^{x} f(t)\, dt, \quad \text{for } -\infty < x < \infty.$$

As an immediate consequence of Definition 3.7, one can write the two results

$$P(a < X < b) = F(b) - F(a) \text{ and } f(x) = \frac{dF(x)}{dx},$$

if the derivative exists.

Example 3.12: For the density function of Example 3.11, find $F(x)$, and use it to evaluate $P(0 < X \leq 1)$.

Solution: For $-1 < x < 2$,

$$F(x) = \int_{-\infty}^{x} f(t)\, dt = \int_{-1}^{x} \frac{t^2}{3} dt = \left.\frac{t^3}{9}\right|_{-1}^{x} = \frac{x^3 + 1}{9}.$$

Therefore,

$$F(x) = \begin{cases} 0, & x < -1, \\ \frac{x^3+1}{9}, & -1 \leq x < 2, \\ 1, & x \geq 2. \end{cases}$$

The cumulative distribution function $F(x)$ is expressed in Figure 3.6. Now

$$P(0 < X \leq 1) = F(1) - F(0) = \frac{2}{9} - \frac{1}{9} = \frac{1}{9},$$

which agrees with the result obtained by using the density function in Example 3.11.

Example 3.13: The Department of Energy (DOE) puts projects out on bid and generally estimates what a reasonable bid should be. Call the estimate b. The DOE has determined that the density function of the winning (low) bid is

$$f(y) = \begin{cases} \frac{5}{8b}, & \frac{2}{5}b \leq y \leq 2b, \\ 0, & \text{elsewhere.} \end{cases}$$

Find $F(y)$ and use it to determine the probability that the winning bid is less than the DOE's preliminary estimate b.

Solution: For $2b/5 \leq y \leq 2b$,

$$F(y) = \int_{2b/5}^{y} \frac{5}{8b} dy = \left.\frac{5t}{8b}\right|_{2b/5}^{y} = \frac{5y}{8b} - \frac{1}{4}.$$

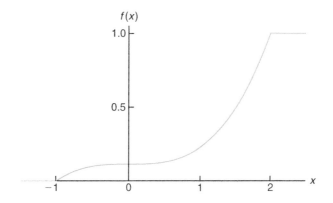

Figure 3.6: Continuous cumulative distribution function.

Thus,

$$F(y) = \begin{cases} 0, & y < \frac{2}{5}b, \\ \frac{5y}{8b} - \frac{1}{4}, & \frac{2}{5}b \leq y < 2b, \\ 1, & y \geq 2b. \end{cases}$$

To determine the probability that the winning bid is less than the preliminary bid estimate b, we have

$$P(Y \leq b) = F(b) = \frac{5}{8} - \frac{1}{4} = \frac{3}{8}.$$

Exercises

3.1 Classify the following random variables as discrete or continuous:

X: the number of automobile accidents per year in Virginia.

Y: the length of time to play 18 holes of golf.

M: the amount of milk produced yearly by a particular cow.

N: the number of eggs laid each month by a hen.

P: the number of building permits issued each month in a certain city.

Q: the weight of grain produced per acre.

3.2 An overseas shipment of 5 foreign automobiles contains 2 that have slight paint blemishes. If an agency receives 3 of these automobiles at random, list the elements of the sample space S, using the letters B and N for blemished and nonblemished, respectively; then to each sample point assign a value x of the random variable X representing the number of automobiles with paint blemishes purchased by the agency.

3.3 Let W be a random variable giving the number of heads minus the number of tails in three tosses of a coin. List the elements of the sample space S for the three tosses of the coin and to each sample point assign a value w of W.

3.4 A coin is flipped until 3 heads in succession occur. List only those elements of the sample space that require 6 or less tosses. Is this a discrete sample space? Explain.

3.5 Determine the value c so that each of the following functions can serve as a probability distribution of the discrete random variable X:

(a) $f(x) = c(x^2 + 4)$, for $x = 0, 1, 2, 3$;

(b) $f(x) = c\binom{2}{x}\binom{3}{3-x}$, for $x = 0, 1, 2$.

3.6 The shelf life, in days, for bottles of a certain prescribed medicine is a random variable having the density function

$$f(x) = \begin{cases} \frac{20{,}000}{(x+100)^3}, & x > 0, \\ 0, & \text{elsewhere.} \end{cases}$$

Find the probability that a bottle of this medicine will have a shell life of

(a) at least 200 days;

(b) anywhere from 80 to 120 days.

3.7 The total number of hours, measured in units of 100 hours, that a family runs a vacuum cleaner over a period of one year is a continuous random variable X that has the density function

$$f(x) = \begin{cases} x, & 0 < x < 1, \\ 2 - x, & 1 \le x < 2, \\ 0, & \text{elsewhere.} \end{cases}$$

Find the probability that over a period of one year, a family runs their vacuum cleaner

(a) less than 120 hours;

(b) between 50 and 100 hours.

3.8 Find the probability distribution of the random variable W in Exercise 3.3, assuming that the coin is biased so that a head is twice as likely to occur as a tail.

3.9 The proportion of people who respond to a certain mail-order solicitation is a continuous random variable X that has the density function

$$f(x) = \begin{cases} \frac{2(x+2)}{5}, & 0 < x < 1, \\ 0, & \text{elsewhere.} \end{cases}$$

(a) Show that $P(0 < X < 1) = 1$.

(b) Find the probability that more than 1/4 but fewer than 1/2 of the people contacted will respond to this type of solicitation.

3.10 Find a formula for the probability distribution of the random variable X representing the outcome when a single die is rolled once.

3.11 A shipment of 7 televisions contains 2 that are defective. A hotel makes a random purchase of 3 televisions. If x is the number of defective televisions purchased by the hotel, find the probability distribution of X. Express the results graphically as a probability histogram.

3.12 An investment firm offers its customers municipal bonds that mature after varying numbers of years. Given that the cumulative distribution function of T, the number of years to maturity for a randomly selected bond, is

$$F(t) = \begin{cases} 0, & t < 1, \\ \frac{1}{4}, & 1 \le t < 3, \\ \frac{1}{2}, & 3 \le t < 5, \\ \frac{3}{4}, & 5 \le t < 7, \\ 1, & t \ge 7, \end{cases}$$

find

(a) $P(T = 5)$;

(b) $P(T > 3)$;

(c) $P(1.4 < T < 6)$;

(d) $P(T \le 5 \mid T \ge 2)$.

3.13 The probability distribution of X, the number of imperfections per 10 meters of a synthetic fabric in continuous rolls of uniform width, is given by

x	0	1	2	3	4
$f(x)$	0.41	0.37	0.16	0.05	0.01

Construct the cumulative distribution function of X.

3.14 The waiting time, in hours, between successive speeders spotted by a radar unit is a continuous random variable with cumulative distribution function

$$F(x) = \begin{cases} 0, & x < 0, \\ 1 - e^{-8x}, & x \ge 0. \end{cases}$$

Find the probability of waiting less than 12 minutes between successive speeders

(a) using the cumulative distribution function of X;

(b) using the probability density function of X.

3.15 Find the cumulative distribution function of the random variable X representing the number of defectives in Exercise 3.11. Then using $F(x)$, find

(a) $P(X = 1)$;

(b) $P(0 < X \le 2)$.

3.16 Construct a graph of the cumulative distribution function of Exercise 3.15.

3.17 A continuous random variable X that can assume values between $x = 1$ and $x = 3$ has a density function given by $f(x) = 1/2$.

(a) Show that the area under the curve is equal to 1.

(b) Find $P(2 < X < 2.5)$.

(c) Find $P(X \le 1.6)$.

Exercises

3.18 A continuous random variable X that can assume values between $x = 2$ and $x = 5$ has a density function given by $f(x) = 2(1+x)/27$. Find
(a) $P(X < 4)$;
(b) $P(3 \leq X < 4)$.

3.19 For the density function of Exercise 3.17, find $F(x)$. Use it to evaluate $P(2 < X < 2.5)$.

3.20 For the density function of Exercise 3.18, find $F(x)$, and use it to evaluate $P(3 \leq X < 4)$.

3.21 Consider the density function
$$f(x) = \begin{cases} k\sqrt{x}, & 0 < x < 1, \\ 0, & \text{elsewhere.} \end{cases}$$

(a) Evaluate k.
(b) Find $F(x)$ and use it to evaluate
$$P(0.3 < X < 0.6).$$

3.22 Three cards are drawn in succession from a deck without replacement. Find the probability distribution for the number of spades.

3.23 Find the cumulative distribution function of the random variable W in Exercise 3.8. Using $F(w)$, find
(a) $P(W > 0)$;
(b) $P(-1 \leq W < 3)$.

3.24 Find the probability distribution for the number of comic books when 4 books are selected at random from a collection consisting of 5 comic books, 2 art books, and 3 math books. Express your results by means of a formula.

3.25 From a box containing 4 dimes and 2 nickels, 3 coins are selected at random without replacement. Find the probability distribution for the total T of the 3 coins. Express the probability distribution graphically as a probability histogram.

3.26 From a box containing 4 black balls and 2 green balls, 3 balls are drawn in succession, each ball being replaced in the box before the next draw is made. Find the probability distribution for the number of green balls.

3.27 The time to failure in hours of an important piece of electronic equipment used in a manufactured DVD player has the density function
$$f(x) = \begin{cases} \frac{1}{2000} \exp(-x/2000), & x \geq 0, \\ 0, & x < 0. \end{cases}$$

(a) Find $F(x)$.
(b) Determine the probability that the component (and thus the DVD player) lasts more than 1000 hours before the component needs to be replaced.
(c) Determine the probability that the component fails before 2000 hours.

3.28 A cereal manufacturer is aware that the weight of the product in the box varies slightly from box to box. In fact, considerable historical data have allowed the determination of the density function that describes the probability structure for the weight (in ounces). Letting X be the random variable weight, in ounces, the density function can be described as
$$f(x) = \begin{cases} \frac{2}{5}, & 23.75 \leq x \leq 26.25, \\ 0, & \text{elsewhere.} \end{cases}$$

(a) Verify that this is a valid density function.
(b) Determine the probability that the weight is smaller than 24 ounces.
(c) The company desires that the weight exceeding 26 ounces be an extremely rare occurrence. What is the probability that this rare occurrence does actually occur?

3.29 An important factor in solid missile fuel is the particle size distribution. Significant problems occur if the particle sizes are too large. From production data in the past, it has been determined that the particle size (in micrometers) distribution is characterized by
$$f(x) = \begin{cases} 3x^{-4}, & x > 1, \\ 0, & \text{elsewhere.} \end{cases}$$

(a) Verify that this is a valid density function.
(b) Evaluate $F(x)$.
(c) What is the probability that a random particle from the manufactured fuel exceeds 4 micrometers?

3.30 Measurements of scientific systems are always subject to variation, some more than others. There are many structures for measurement error, and statisticians spend a great deal of time modeling these errors. Suppose the measurement error X of a certain physical quantity is decided by the density function
$$f(x) = \begin{cases} k(3-x^2), & -1 \leq x \leq 1, \\ 0, & \text{elsewhere.} \end{cases}$$

(a) Determine k that renders $f(x)$ a valid density function.
(b) Find the probability that a random error in measurement is less than $1/2$.
(c) For this particular measurement, it is undesirable if the *magnitude* of the error (i.e., $|x|$) exceeds 0.8. What is the probability that this occurs?

3.31 Based on extensive testing, it is determined by the manufacturer of a washing machine that the time Y (in years) before a major repair is required is characterized by the probability density function

$$f(y) = \begin{cases} \frac{1}{4}e^{-y/4}, & y \geq 0, \\ 0, & \text{elsewhere.} \end{cases}$$

(a) Critics would certainly consider the product a bargain if it is unlikely to require a major repair before the sixth year. Comment on this by determining $P(Y > 6)$.

(b) What is the probability that a major repair occurs in the first year?

3.32 The proportion of the budget for a certain type of industrial company that is allotted to environmental and pollution control is coming under scrutiny. A data collection project determines that the distribution of these proportions is given by

$$f(y) = \begin{cases} 5(1-y)^4, & 0 \leq y \leq 1, \\ 0, & \text{elsewhere.} \end{cases}$$

(a) Verify that the above is a valid density function.

(b) What is the probability that a company chosen at random expends less than 10% of its budget on environmental and pollution controls?

(c) What is the probability that a company selected at random spends more than 50% of its budget on environmental and pollution controls?

3.33 Suppose a certain type of small data processing firm is so specialized that some have difficulty making a profit in their first year of operation. The probability density function that characterizes the proportion Y that make a profit is given by

$$f(y) = \begin{cases} ky^4(1-y)^3, & 0 \leq y \leq 1, \\ 0, & \text{elsewhere.} \end{cases}$$

(a) What is the value of k that renders the above a valid density function?

(b) Find the probability that at most 50% of the firms make a profit in the first year.

(c) Find the probability that at least 80% of the firms make a profit in the first year.

3.34 Magnetron tubes are produced on an automated assembly line. A sampling plan is used periodically to assess quality of the lengths of the tubes. This measurement is subject to uncertainty. It is thought that the probability that a random tube meets length specification is 0.99. A sampling plan is used in which the lengths of 5 random tubes are measured.

(a) Show that the probability function of Y, the number out of 5 that meet length specification, is given by the following discrete probability function:

$$f(y) = \frac{5!}{y!(5-y)!}(0.99)^y(0.01)^{5-y},$$

for $y = 0, 1, 2, 3, 4, 5$.

(b) Suppose random selections are made off the line and 3 are outside specifications. Use $f(y)$ above either to support or to refute the conjecture that the probability is 0.99 that a single tube meets specifications.

3.35 Suppose it is known from large amounts of historical data that X, the number of cars that arrive at a specific intersection during a 20-second time period, is characterized by the following discrete probability function:

$$f(x) = e^{-6}\frac{6^x}{x!}, \quad \text{for } x = 0, 1, 2, \ldots.$$

(a) Find the probability that in a specific 20-second time period, more than 8 cars arrive at the intersection.

(b) Find the probability that only 2 cars arrive.

3.36 On a laboratory assignment, if the equipment is working, the density function of the observed outcome, X, is

$$f(x) = \begin{cases} 2(1-x), & 0 < x < 1, \\ 0, & \text{otherwise.} \end{cases}$$

(a) Calculate $P(X \leq 1/3)$.

(b) What is the probability that X will exceed 0.5?

(c) Given that $X \geq 0.5$, what is the probability that X will be less than 0.75?

3.4 Joint Probability Distributions

Our study of random variables and their probability distributions in the preceding sections is restricted to one-dimensional sample spaces, in that we recorded outcomes of an experiment as values assumed by a single random variable. There will be situations, however, where we may find it desirable to record the simulta-

3.4 Joint Probability Distributions

neous outcomes of several random variables. For example, we might measure the amount of precipitate P and volume V of gas released from a controlled chemical experiment, giving rise to a two-dimensional sample space consisting of the outcomes (p, v), or we might be interested in the hardness H and tensile strength T of cold-drawn copper, resulting in the outcomes (h, t). In a study to determine the likelihood of success in college based on high school data, we might use a three-dimensional sample space and record for each individual his or her aptitude test score, high school class rank, and grade-point average at the end of freshman year in college.

If X and Y are two discrete random variables, the probability distribution for their simultaneous occurrence can be represented by a function with values $f(x, y)$ for any pair of values (x, y) within the range of the random variables X and Y. It is customary to refer to this function as the **joint probability distribution** of X and Y.

Hence, in the discrete case,

$$f(x, y) = P(X = x, Y = y);$$

that is, the values $f(x, y)$ give the probability that outcomes x and y occur at the same time. For example, if an 18-wheeler is to have its tires serviced and X represents the number of miles these tires have been driven and Y represents the number of tires that need to be replaced, then $f(30000, 5)$ is the probability that the tires are used over 30,000 miles and the truck needs 5 new tires.

Definition 3.8: The function $f(x, y)$ is a **joint probability distribution** or **probability mass function** of the discrete random variables X and Y if

1. $f(x, y) \geq 0$ for all (x, y),

2. $\sum_x \sum_y f(x, y) = 1$,

3. $P(X = x, Y = y) = f(x, y)$.

For any region A in the xy plane, $P[(X, Y) \in A] = \sum\sum_A f(x, y)$.

Example 3.14: Two ballpoint pens are selected at random from a box that contains 3 blue pens, 2 red pens, and 3 green pens. If X is the number of blue pens selected and Y is the number of red pens selected, find

(a) the joint probability function $f(x, y)$,

(b) $P[(X, Y) \in A]$, where A is the region $\{(x, y) | x + y \leq 1\}$.

Solution: The possible pairs of values (x, y) are $(0, 0)$, $(0, 1)$, $(1, 0)$, $(1, 1)$, $(0, 2)$, and $(2, 0)$.

(a) Now, $f(0, 1)$, for example, represents the probability that a red and a green pens are selected. The total number of equally likely ways of selecting any 2 pens from the 8 is $\binom{8}{2} = 28$. The number of ways of selecting 1 red from 2 red pens and 1 green from 3 green pens is $\binom{2}{1}\binom{3}{1} = 6$. Hence, $f(0, 1) = 6/28 = 3/14$. Similar calculations yield the probabilities for the other cases, which are presented in Table 3.1. Note that the probabilities sum to 1. In Chapter

5, it will become clear that the joint probability distribution of Table 3.1 can be represented by the formula

$$f(x,y) = \frac{\binom{3}{x}\binom{2}{y}\binom{3}{2-x-y}}{\binom{8}{2}},$$

for $x = 0, 1, 2$; $y = 0, 1, 2$; and $0 \leq x + y \leq 2$.

(b) The probability that (X, Y) fall in the region A is

$$P[(X,Y) \in A] = P(X + Y \leq 1) = f(0,0) + f(0,1) + f(1,0)$$
$$= \frac{3}{28} + \frac{3}{14} + \frac{9}{28} = \frac{9}{14}.$$

Table 3.1: Joint Probability Distribution for Example 3.14

	$f(x,y)$	x = 0	x = 1	x = 2	Row Totals
y	0	$\frac{3}{28}$	$\frac{9}{28}$	$\frac{3}{28}$	$\frac{15}{28}$
	1	$\frac{3}{14}$	$\frac{3}{14}$	0	$\frac{3}{7}$
	2	$\frac{1}{28}$	0	0	$\frac{1}{28}$
	Column Totals	$\frac{5}{14}$	$\frac{15}{28}$	$\frac{3}{28}$	1

When X and Y are continuous random variables, the **joint density function** $f(x,y)$ is a surface lying above the xy plane, and $P[(X,Y) \in A]$, where A is any region in the xy plane, is equal to the volume of the right cylinder bounded by the base A and the surface.

Definition 3.9: The function $f(x,y)$ is a **joint density function** of the continuous random variables X and Y if

1. $f(x,y) \geq 0$, for all (x,y),
2. $\int_{-\infty}^{\infty} \int_{-\infty}^{\infty} f(x,y)\,dx\,dy = 1$,
3. $P[(X,Y) \in A] = \int \int_A f(x,y)\,dx\,dy$, for any region A in the xy plane.

Example 3.15: A privately owned business operates both a drive-in facility and a walk-in facility. On a randomly selected day, let X and Y, respectively, be the proportions of the time that the drive-in and the walk-in facilities are in use, and suppose that the joint density function of these random variables is

$$f(x,y) = \begin{cases} \frac{2}{5}(2x + 3y), & 0 \leq x \leq 1, 0 \leq y \leq 1, \\ 0, & \text{elsewhere.} \end{cases}$$

(a) Verify condition 2 of Definition 3.9.
(b) Find $P[(X,Y) \in A]$, where $A = \{(x,y) \mid 0 < x < \frac{1}{2}, \frac{1}{4} < y < \frac{1}{2}\}$.

3.4 Joint Probability Distributions

Solution: (a) The integration of $f(x,y)$ over the whole region is

$$\int_{-\infty}^{\infty}\int_{-\infty}^{\infty} f(x,y)\,dx\,dy = \int_0^1 \int_0^1 \frac{2}{5}(2x+3y)\,dx\,dy$$

$$= \int_0^1 \left(\frac{2x^2}{5} + \frac{6xy}{5}\right)\bigg|_{x=0}^{x=1} dy$$

$$= \int_0^1 \left(\frac{2}{5} + \frac{6y}{5}\right) dy = \left(\frac{2y}{5} + \frac{3y^2}{5}\right)\bigg|_0^1 = \frac{2}{5} + \frac{3}{5} = 1.$$

(b) To calculate the probability, we use

$$P[(X,Y) \in A] = P\left(0 < X < \frac{1}{2}, \frac{1}{4} < Y < \frac{1}{2}\right)$$

$$= \int_{1/4}^{1/2} \int_0^{1/2} \frac{2}{5}(2x+3y)\,dx\,dy$$

$$= \int_{1/4}^{1/2} \left(\frac{2x^2}{5} + \frac{6xy}{5}\right)\bigg|_{x=0}^{x=1/2} dy = \int_{1/4}^{1/2} \left(\frac{1}{10} + \frac{3y}{5}\right) dy$$

$$= \left(\frac{y}{10} + \frac{3y^2}{10}\right)\bigg|_{1/4}^{1/2}$$

$$= \frac{1}{10}\left[\left(\frac{1}{2} + \frac{3}{4}\right) - \left(\frac{1}{4} + \frac{3}{16}\right)\right] = \frac{13}{160}.$$

Given the joint probability distribution $f(x,y)$ of the discrete random variables X and Y, the probability distribution $g(x)$ of X alone is obtained by summing $f(x,y)$ over the values of Y. Similarly, the probability distribution $h(y)$ of Y alone is obtained by summing $f(x,y)$ over the values of X. We define $g(x)$ and $h(y)$ to be the **marginal distributions** of X and Y, respectively. When X and Y are continuous random variables, summations are replaced by integrals. We can now make the following general definition.

Definition 3.10: The **marginal distributions** of X alone and of Y alone are

$$g(x) = \sum_y f(x,y) \quad \text{and} \quad h(y) = \sum_x f(x,y)$$

for the discrete case, and

$$g(x) = \int_{-\infty}^{\infty} f(x,y)\,dy \quad \text{and} \quad h(y) = \int_{-\infty}^{\infty} f(x,y)\,dx$$

for the continuous case.

The term *marginal* is used here because, in the discrete case, the values of $g(x)$ and $h(y)$ are just the marginal totals of the respective columns and rows when the values of $f(x,y)$ are displayed in a rectangular table.

Example 3.16: Show that the column and row totals of Table 3.1 give the marginal distribution of X alone and of Y alone.

Solution: For the random variable X, we see that

$$g(0) = f(0,0) + f(0,1) + f(0,2) = \frac{3}{28} + \frac{3}{14} + \frac{1}{28} = \frac{5}{14},$$

$$g(1) = f(1,0) + f(1,1) + f(1,2) = \frac{9}{28} + \frac{3}{14} + 0 = \frac{15}{28},$$

and

$$g(2) = f(2,0) + f(2,1) + f(2,2) = \frac{3}{28} + 0 + 0 = \frac{3}{28},$$

which are just the column totals of Table 3.1. In a similar manner we could show that the values of $h(y)$ are given by the row totals. In tabular form, these marginal distributions may be written as follows:

x	0	1	2
$g(x)$	$\frac{5}{14}$	$\frac{15}{28}$	$\frac{3}{28}$

y	0	1	2
$h(y)$	$\frac{15}{28}$	$\frac{3}{7}$	$\frac{1}{28}$

Example 3.17: Find $g(x)$ and $h(y)$ for the joint density function of Example 3.15.

Solution: By definition,

$$g(x) = \int_{-\infty}^{\infty} f(x,y)\, dy = \int_0^1 \frac{2}{5}(2x + 3y)\, dy = \left(\frac{4xy}{5} + \frac{6y^2}{10}\right)\Big|_{y=0}^{y=1} = \frac{4x + 3}{5},$$

for $0 \leq x \leq 1$, and $g(x) = 0$ elsewhere. Similarly,

$$h(y) = \int_{-\infty}^{\infty} f(x,y)\, dx = \int_0^1 \frac{2}{5}(2x + 3y)\, dx = \frac{2(1 + 3y)}{5},$$

for $0 \leq y \leq 1$, and $h(y) = 0$ elsewhere.

The fact that the marginal distributions $g(x)$ and $h(y)$ are indeed the probability distributions of the individual variables X and Y alone can be verified by showing that the conditions of Definition 3.4 or Definition 3.6 are satisfied. For example, in the continuous case

$$\int_{-\infty}^{\infty} g(x)\, dx = \int_{-\infty}^{\infty} \int_{-\infty}^{\infty} f(x,y)\, dy\, dx = 1,$$

and

$$P(a < X < b) = P(a < X < b, -\infty < Y < \infty)$$
$$= \int_a^b \int_{-\infty}^{\infty} f(x,y)\, dy\, dx = \int_a^b g(x)\, dx.$$

In Section 3.1, we stated that the value x of the random variable X represents an event that is a subset of the sample space. If we use the definition of conditional probability as stated in Chapter 2,

$$P(B|A) = \frac{P(A \cap B)}{P(A)}, \text{ provided } P(A) > 0,$$

3.4 Joint Probability Distributions

where A and B are now the events defined by $X = x$ and $Y = y$, respectively, then

$$P(Y = y \mid X = x) = \frac{P(X = x, Y = y)}{P(X = x)} = \frac{f(x, y)}{g(x)}, \text{ provided } g(x) > 0,$$

where X and Y are discrete random variables.

It is not difficult to show that the function $f(x, y)/g(x)$, which is strictly a function of y with x fixed, satisfies all the conditions of a probability distribution. This is also true when $f(x, y)$ and $g(x)$ are the joint density and marginal distribution, respectively, of continuous random variables. As a result, it is extremely important that we make use of the special type of distribution of the form $f(x, y)/g(x)$ in order to be able to effectively compute conditional probabilities. This type of distribution is called a **conditional probability distribution**; the formal definition follows.

Definition 3.11: Let X and Y be two random variables, discrete or continuous. The **conditional distribution** of the random variable Y given that $X = x$ is

$$f(y|x) = \frac{f(x, y)}{g(x)}, \text{ provided } g(x) > 0.$$

Similarly, the conditional distribution of X given that $Y = y$ is

$$f(x|y) = \frac{f(x, y)}{h(y)}, \text{ provided } h(y) > 0.$$

If we wish to find the probability that the discrete random variable X falls between a and b when it is known that the discrete variable $Y = y$, we evaluate

$$P(a < X < b \mid Y = y) = \sum_{a < x < b} f(x|y),$$

where the summation extends over all values of X between a and b. When X and Y are continuous, we evaluate

$$P(a < X < b \mid Y = y) = \int_a^b f(x|y) \, dx.$$

Example 3.18: Referring to Example 3.14, find the conditional distribution of X, given that $Y = 1$, and use it to determine $P(X = 0 \mid Y = 1)$.

Solution: We need to find $f(x|y)$, where $y = 1$. First, we find that

$$h(1) = \sum_{x=0}^{2} f(x, 1) = \frac{3}{14} + \frac{3}{14} + 0 = \frac{3}{7}.$$

Now

$$f(x|1) = \frac{f(x, 1)}{h(1)} = \left(\frac{7}{3}\right) f(x, 1), \quad x = 0, 1, 2.$$

Therefore,

$$f(0|1) = \left(\frac{7}{3}\right) f(0,1) = \left(\frac{7}{3}\right)\left(\frac{3}{14}\right) = \frac{1}{2}, \quad f(1|1) = \left(\frac{7}{3}\right) f(1,1) = \left(\frac{7}{3}\right)\left(\frac{3}{14}\right) = \frac{1}{2},$$

$$f(2|1) = \left(\frac{7}{3}\right) f(2,1) = \left(\frac{7}{3}\right)(0) = 0,$$

and the conditional distribution of X, given that $Y = 1$, is

x	0	1	2
$f(x|1)$	$\frac{1}{2}$	$\frac{1}{2}$	0

Finally,

$$P(X = 0 \mid Y = 1) = f(0|1) = \frac{1}{2}.$$

Therefore, if it is known that 1 of the 2 pen refills selected is red, we have a probability equal to 1/2 that the other refill is not blue.

Example 3.19: The joint density for the random variables (X, Y), where X is the unit temperature change and Y is the proportion of spectrum shift that a certain atomic particle produces, is

$$f(x, y) = \begin{cases} 10xy^2, & 0 < x < y < 1, \\ 0, & \text{elsewhere.} \end{cases}$$

(a) Find the marginal densities $g(x)$, $h(y)$, and the conditional density $f(y|x)$.

(b) Find the probability that the spectrum shifts more than half of the total observations, given that the temperature is increased by 0.25 unit.

Solution: (a) By definition,

$$g(x) = \int_{-\infty}^{\infty} f(x, y) \, dy = \int_{x}^{1} 10xy^2 \, dy$$

$$= \frac{10}{3} xy^3 \Big|_{y=x}^{y=1} = \frac{10}{3} x(1 - x^3), \ 0 < x < 1,$$

$$h(y) = \int_{-\infty}^{\infty} f(x, y) \, dx = \int_{0}^{y} 10xy^2 \, dx = 5x^2 y^2 \Big|_{x=0}^{x=y} = 5y^4, \ 0 < y < 1.$$

Now

$$f(y|x) = \frac{f(x, y)}{g(x)} = \frac{10xy^2}{\frac{10}{3} x(1 - x^3)} = \frac{3y^2}{1 - x^3}, \ 0 < x < y < 1.$$

(b) Therefore,

$$P\left(Y > \frac{1}{2} \,\Big|\, X = 0.25\right) = \int_{1/2}^{1} f(y \mid x = 0.25) \, dy = \int_{1/2}^{1} \frac{3y^2}{1 - 0.25^3} \, dy = \frac{8}{9}.$$

Example 3.20: Given the joint density function

$$f(x, y) = \begin{cases} \frac{x(1+3y^2)}{4}, & 0 < x < 2, \ 0 < y < 1, \\ 0, & \text{elsewhere,} \end{cases}$$

find $g(x)$, $h(y)$, $f(x|y)$, and evaluate $P(\frac{1}{4} < X < \frac{1}{2} \mid Y = \frac{1}{3})$.

Solution: By definition of the marginal density. for $0 < x < 2$,

$$g(x) = \int_{-\infty}^{\infty} f(x,y)\,dy = \int_{0}^{1} \frac{x(1+3y^2)}{4}\,dy$$

$$= \left(\frac{xy}{4} + \frac{xy^3}{4}\right)\bigg|_{y=0}^{y=1} = \frac{x}{2},$$

and for $0 < y < 1$,

$$h(y) = \int_{-\infty}^{\infty} f(x,y)\,dx = \int_{0}^{2} \frac{x(1+3y^2)}{4}\,dx$$

$$= \left(\frac{x^2}{8} + \frac{3x^2 y^2}{8}\right)\bigg|_{x=0}^{x=2} = \frac{1+3y^2}{2}.$$

Therefore, using the conditional density definition, for $0 < x < 2$,

$$f(x|y) = \frac{f(x,y)}{h(y)} = \frac{x(1+3y^2)/4}{(1+3y^2)/2} = \frac{x}{2},$$

and

$$P\left(\frac{1}{4} < X < \frac{1}{2} \;\bigg|\; Y = \frac{1}{3}\right) = \int_{1/4}^{1/2} \frac{x}{2}\,dx = \frac{3}{64}.$$

Statistical Independence

If $f(x|y)$ does not depend on y, as is the case for Example 3.20, then $f(x|y) = g(x)$ and $f(x,y) = g(x)h(y)$. The proof follows by substituting

$$f(x,y) = f(x|y)h(y)$$

into the marginal distribution of X. That is,

$$g(x) = \int_{-\infty}^{\infty} f(x,y)\,dy = \int_{-\infty}^{\infty} f(x|y)h(y)\,dy.$$

If $f(x|y)$ does not depend on y, we may write

$$g(x) = f(x|y) \int_{-\infty}^{\infty} h(y)\,dy.$$

Now

$$\int_{-\infty}^{\infty} h(y)\,dy = 1,$$

since $h(y)$ is the probability density function of Y. Therefore,

$$g(x) = f(x|y) \quad \text{and then} \quad f(x,y) = g(x)h(y).$$

It should make sense to the reader that if $f(x|y)$ does not depend on y, then of course the outcome of the random variable Y has no impact on the outcome of the random variable X. In other words, we say that X and Y are independent random variables. We now offer the following formal definition of statistical independence.

Definition 3.12: Let X and Y be two random variables, discrete or continuous, with joint probability distribution $f(x,y)$ and marginal distributions $g(x)$ and $h(y)$, respectively. The random variables X and Y are said to be **statistically independent** if and only if

$$f(x,y) = g(x)h(y)$$

for all (x,y) within their range.

The continuous random variables of Example 3.20 are statistically independent, since the product of the two marginal distributions gives the joint density function. This is obviously not the case, however, for the continuous variables of Example 3.19. Checking for statistical independence of discrete random variables requires a more thorough investigation, since it is possible to have the product of the marginal distributions equal to the joint probability distribution for some but not all combinations of (x,y). If you can find any point (x,y) for which $f(x,y)$ is defined such that $f(x,y) \neq g(x)h(y)$, the discrete variables X and Y are not statistically independent.

Example 3.21: Show that the random variables of Example 3.14 are not statistically independent.
Proof: Let us consider the point $(0,1)$. From Table 3.1 we find the three probabilities $f(0,1)$, $g(0)$, and $h(1)$ to be

$$f(0,1) = \frac{3}{14},$$

$$g(0) = \sum_{y=0}^{2} f(0,y) = \frac{3}{28} + \frac{3}{14} + \frac{1}{28} = \frac{5}{14},$$

$$h(1) = \sum_{x=0}^{2} f(x,1) = \frac{3}{14} + \frac{3}{14} + 0 = \frac{3}{7}.$$

Clearly,

$$f(0,1) \neq g(0)h(1),$$

and therefore X and Y are not statistically independent.

All the preceding definitions concerning two random variables can be generalized to the case of n random variables. Let $f(x_1, x_2, \ldots, x_n)$ be the joint probability function of the random variables $X_1, X_2, \ldots, X_n$. The marginal distribution of X_1, for example, is

$$g(x_1) = \sum_{x_2} \cdots \sum_{x_n} f(x_1, x_2, \ldots, x_n)$$

3.4 Joint Probability Distributions

for the discrete case, and

$$g(x_1) = \int_{-\infty}^{\infty} \cdots \int_{-\infty}^{\infty} f(x_1, x_2, \ldots, x_n) \, dx_2 \, dx_3 \cdots dx_n$$

for the continuous case. We can now obtain **joint marginal distributions** such as $g(x_1, x_2)$, where

$$g(x_1, x_2) = \begin{cases} \sum_{x_3} \cdots \sum_{x_n} f(x_1, x_2, \ldots, x_n) & \text{(discrete case)}, \\ \int_{-\infty}^{\infty} \cdots \int_{-\infty}^{\infty} f(x_1, x_2, \ldots, x_n) \, dx_3 \, dx_4 \cdots dx_n & \text{(continuous case)}. \end{cases}$$

We could consider numerous conditional distributions. For example, the **joint conditional distribution** of X_1, X_2, and X_3, given that $X_4 = x_4, X_5 = x_5, \ldots, X_n = x_n$, is written

$$f(x_1, x_2, x_3 \mid x_4, x_5, \ldots, x_n) = \frac{f(x_1, x_2, \ldots, x_n)}{g(x_4, x_5, \ldots, x_n)},$$

where $g(x_4, x_5, \ldots, x_n)$ is the joint marginal distribution of the random variables $X_4, X_5, \ldots, X_n$.

A generalization of Definition 3.12 leads to the following definition for the mutual statistical independence of the variables $X_1, X_2, \ldots, X_n$.

Definition 3.13: Let $X_1, X_2, \ldots, X_n$ be n random variables, discrete or continuous, with joint probability distribution $f(x_1, x_2, \ldots, x_n)$ and marginal distribution $f_1(x_1), f_2(x_2), \ldots, f_n(x_n)$, respectively. The random variables $X_1, X_2, \ldots, X_n$ are said to be mutually **statistically independent** if and only if

$$f(x_1, x_2, \ldots, x_n) = f_1(x_1) f_2(x_2) \cdots f_n(x_n)$$

for all $(x_1, x_2, \ldots, x_n)$ within their range.

Example 3.22: Suppose that the shelf life, in years, of a certain perishable food product packaged in cardboard containers is a random variable whose probability density function is given by

$$f(x) = \begin{cases} e^{-x}, & x > 0, \\ 0, & \text{elsewhere.} \end{cases}$$

Let X_1, X_2, and X_3 represent the shelf lives for three of these containers selected independently and find $P(X_1 < 2, 1 < X_2 < 3, X_3 > 2)$.

Solution: Since the containers were selected independently, we can assume that the random variables X_1, X_2, and X_3 are statistically independent, having the joint probability density

$$f(x_1, x_2, x_3) = f(x_1) f(x_2) f(x_3) = e^{-x_1} e^{-x_2} e^{-x_3} = e^{-x_1 - x_2 - x_3},$$

for $x_1 > 0$, $x_2 > 0$, $x_3 > 0$, and $f(x_1, x_2, x_3) = 0$ elsewhere. Hence

$$P(X_1 < 2, 1 < X_2 < 3, X_3 > 2) = \int_2^{\infty} \int_1^3 \int_0^2 e^{-x_1 - x_2 - x_3} \, dx_1 \, dx_2 \, dx_3$$
$$= (1 - e^{-2})(e^{-1} - e^{-3}) e^{-2} = 0.0372.$$

What Are Important Characteristics of Probability Distributions and Where Do They Come From?

This is an important point in the text to provide the reader with a transition into the next three chapters. We have given illustrations in both examples and exercises of practical scientific and engineering situations in which probability distributions and their properties are used to solve important problems. These probability distributions, either discrete or continuous, were introduced through phrases like "it is known that" or "suppose that" or even in some cases "historical evidence suggests that." These are situations in which the nature of the distribution and even a good estimate of the probability structure can be determined through historical data, data from long-term studies, or even large amounts of planned data. The reader should remember the discussion of the use of histograms in Chapter 1 and from that recall how frequency distributions are estimated from the histograms. However, not all probability functions and probability density functions are derived from large amounts of historical data. There are a substantial number of situations in which the nature of the scientific scenario suggests a distribution type. Indeed, many of these are reflected in exercises in both Chapter 2 and this chapter. When independent repeated observations are binary in nature (e.g., defective or not, survive or not, allergic or not) with value 0 or 1, the distribution covering this situation is called the **binomial distribution** and the probability function is known and will be demonstrated in its generality in Chapter 5. Exercise 3.34 in Section 3.3 and Review Exercise 3.80 are examples, and there are others that the reader should recognize. The scenario of a continuous distribution in time to failure, as in Review Exercise 3.69 or Exercise 3.27 on page 93, often suggests a distribution type called the **exponential distribution**. These types of illustrations are merely two of many so-called standard distributions that are used extensively in real-world problems because the scientific scenario that gives rise to each of them is recognizable and occurs often in practice. Chapters 5 and 6 cover many of these types along with some underlying theory concerning their use.

A second part of this transition to material in future chapters deals with the notion of **population parameters** or **distributional parameters**. Recall in Chapter 1 we discussed the need to use data to provide information about these parameters. We went to some length in discussing the notions of a **mean** and **variance** and provided a vision for the concepts in the context of a population. Indeed, the population mean and variance are easily found from the probability function for the discrete case or probability density function for the continuous case. These parameters and their importance in the solution of many types of real-world problems will provide much of the material in Chapters 8 through 17.

Exercises

3.37 Determine the values of c so that the following functions represent joint probability distributions of the random variables X and Y:
(a) $f(x, y) = cxy$, for $x = 1, 2, 3$; $y = 1, 2, 3$;
(b) $f(x, y) = c|x - y|$, for $x = -2, 0, 2$; $y = -2, 3$.

3.38 If the joint probability distribution of X and Y is given by

$$f(x, y) = \frac{x + y}{30}, \quad \text{for } x = 0, 1, 2, 3;\ y = 0, 1, 2,$$

find

(a) $P(X \le 2, Y = 1)$;
(b) $P(X > 2, Y \le 1)$;
(c) $P(X > Y)$;
(d) $P(X + Y = 4)$.

3.39 From a sack of fruit containing 3 oranges, 2 apples, and 3 bananas, a random sample of 4 pieces of fruit is selected. If X is the number of oranges and Y is the number of apples in the sample, find
(a) the joint probability distribution of X and Y;
(b) $P[(X, Y) \in A]$, where A is the region that is given by $\{(x, y) \mid x + y \le 2\}$.

3.40 A fast-food restaurant operates both a drive-through facility and a walk-in facility. On a randomly selected day, let X and Y, respectively, be the proportions of the time that the drive-through and walk-in facilities are in use, and suppose that the joint density function of these random variables is

$$f(x,y) = \begin{cases} \frac{2}{3}(x + 2y), & 0 \le x \le 1,\ 0 \le y \le 1, \\ 0, & \text{elsewhere.} \end{cases}$$

(a) Find the marginal density of X.
(b) Find the marginal density of Y.
(c) Find the probability that the drive-through facility is busy less than one-half of the time.

3.41 A candy company distributes boxes of chocolates with a mixture of creams, toffees, and cordials. Suppose that the weight of each box is 1 kilogram, but the individual weights of the creams, toffees, and cordials vary from box to box. For a randomly selected box, let X and Y represent the weights of the creams and the toffees, respectively, and suppose that the joint density function of these variables is

$$f(x,y) = \begin{cases} 24xy, & 0 \le x \le 1,\ 0 \le y \le 1,\ x + y \le 1, \\ 0, & \text{elsewhere.} \end{cases}$$

(a) Find the probability that in a given box the cordials account for more than 1/2 of the weight.
(b) Find the marginal density for the weight of the creams.
(c) Find the probability that the weight of the toffees in a box is less than 1/8 of a kilogram if it is known that creams constitute 3/4 of the weight.

3.42 Let X and Y denote the lengths of life, in years, of two components in an electronic system. If the joint density function of these variables is

$$f(x,y) = \begin{cases} e^{-(x+y)}, & x > 0,\ y > 0, \\ 0, & \text{elsewhere,} \end{cases}$$

find $P(0 < X < 1 \mid Y = 2)$.

3.43 Let X denote the reaction time, in seconds, to a certain stimulus and Y denote the temperature (°F) at which a certain reaction starts to take place. Suppose that two random variables X and Y have the joint density

$$f(x,y) = \begin{cases} 4xy, & 0 < x < 1,\ 0 < y < 1, \\ 0, & \text{elsewhere.} \end{cases}$$

Find
(a) $P(0 \le X \le \frac{1}{2}$ and $\frac{1}{4} \le Y \le \frac{1}{2})$;
(b) $P(X < Y)$.

3.44 Each rear tire on an experimental airplane is supposed to be filled to a pressure of 40 pounds per square inch (psi). Let X denote the actual air pressure for the right tire and Y denote the actual air pressure for the left tire. Suppose that X and Y are random variables with the joint density function

$$f(x,y) = \begin{cases} k(x^2 + y^2), & 30 \le x < 50,\ 30 \le y < 50, \\ 0, & \text{elsewhere.} \end{cases}$$

(a) Find k.
(b) Find $P(30 \le X \le 40$ and $40 \le Y < 50)$.
(c) Find the probability that both tires are underfilled.

3.45 Let X denote the diameter of an armored electric cable and Y denote the diameter of the ceramic mold that makes the cable. Both X and Y are scaled so that they range between 0 and 1. Suppose that X and Y have the joint density

$$f(x,y) = \begin{cases} \frac{1}{y}, & 0 < x < y < 1, \\ 0, & \text{elsewhere.} \end{cases}$$

Find $P(X + Y > 1/2)$.

3.46 Referring to Exercise 3.38, find
(a) the marginal distribution of X;
(b) the marginal distribution of Y.

3.47 The amount of kerosene, in thousands of liters, in a tank at the beginning of any day is a random amount Y from which a random amount X is sold during that day. Suppose that the tank is not resupplied during the day so that $x \le y$, and assume that the joint density function of these variables is

$$f(x,y) = \begin{cases} 2, & 0 < x \le y < 1, \\ 0, & \text{elsewhere.} \end{cases}$$

(a) Determine if X and Y are independent.

(b) Find $P(1/4 < X < 1/2 \mid Y = 3/4)$.

3.48 Referring to Exercise 3.39, find
(a) $f(y|2)$ for all values of y;
(b) $P(Y = 0 \mid X = 2)$.

3.49 Let X denote the number of times a certain numerical control machine will malfunction: 1, 2, or 3 times on any given day. Let Y denote the number of times a technician is called on an emergency call. Their joint probability distribution is given as

$f(x,y)$		x	
	1	2	3
y 1	0.05	0.05	0.10
3	0.05	0.10	0.35
5	0.00	0.20	0.10

(a) Evaluate the marginal distribution of X.
(b) Evaluate the marginal distribution of Y.
(c) Find $P(Y = 3 \mid X = 2)$.

3.50 Suppose that X and Y have the following joint probability distribution:

$f(x,y)$		x	
		2	4
y 1		0.10	0.15
3		0.20	0.30
5		0.10	0.15

(a) Find the marginal distribution of X.
(b) Find the marginal distribution of Y.

3.51 Three cards are drawn without replacement from the 12 face cards (jacks, queens, and kings) of an ordinary deck of 52 playing cards. Let X be the number of kings selected and Y the number of jacks. Find
(a) the joint probability distribution of X and Y;
(b) $P[(X, Y) \in A]$, where A is the region given by $\{(x, y) \mid x + y \geq 2\}$.

3.52 A coin is tossed twice. Let Z denote the number of heads on the first toss and W the total number of heads on the 2 tosses. If the coin is unbalanced and a head has a 40% chance of occurring, find
(a) the joint probability distribution of W and Z;
(b) the marginal distribution of W;
(c) the marginal distribution of Z;
(d) the probability that at least 1 head occurs.

3.53 Given the joint density function

$$f(x,y) = \begin{cases} \frac{6-x-y}{8}, & 0 < x < 2,\ 2 < y < 4, \\ 0, & \text{elsewhere}, \end{cases}$$

find $P(1 < Y < 3 \mid X = 1)$.

3.54 Determine whether the two random variables of Exercise 3.49 are dependent or independent.

3.55 Determine whether the two random variables of Exercise 3.50 are dependent or independent.

3.56 The joint density function of the random variables X and Y is

$$f(x,y) = \begin{cases} 6x, & 0 < x < 1,\ 0 < y < 1-x, \\ 0, & \text{elsewhere}. \end{cases}$$

(a) Show that X and Y are not independent.
(b) Find $P(X > 0.3 \mid Y = 0.5)$.

3.57 Let X, Y, and Z have the joint probability density function

$$f(x,y,z) = \begin{cases} kxy^2 z, & 0 < x, y < 1,\ 0 < z < 2, \\ 0, & \text{elsewhere}. \end{cases}$$

(a) Find k.
(b) Find $P(X < \frac{1}{4}, Y > \frac{1}{2}, 1 < Z < 2)$.

3.58 Determine whether the two random variables of Exercise 3.43 are dependent or independent.

3.59 Determine whether the two random variables of Exercise 3.44 are dependent or independent.

3.60 The joint probability density function of the random variables X, Y, and Z is

$$f(x,y,z) = \begin{cases} \frac{4xyz^2}{9}, & 0 < x, y < 1,\ 0 < z < 3, \\ 0, & \text{elsewhere}. \end{cases}$$

Find
(a) the joint marginal density function of Y and Z;
(b) the marginal density of Y;
(c) $P(\frac{1}{4} < X < \frac{1}{2},\ Y > \frac{1}{3},\ 1 < Z < 2)$;
(d) $P(0 < X < \frac{1}{2} \mid Y = \frac{1}{4},\ Z = 2)$.

Review Exercises

3.61 A tobacco company produces blends of tobacco, with each blend containing various proportions of Turkish, domestic, and other tobaccos. The proportions of Turkish and domestic in a blend are random variables with joint density function (X = Turkish and Y = domestic)

$$f(x, y) = \begin{cases} 24xy, & 0 \leq x, y \leq 1, \; x + y \leq 1, \\ 0, & \text{elsewhere.} \end{cases}$$

(a) Find the probability that in a given box the Turkish tobacco accounts for over half the blend.
(b) Find the marginal density function for the proportion of the domestic tobacco.
(c) Find the probability that the proportion of Turkish tobacco is less than 1/8 if it is known that the blend contains 3/4 domestic tobacco.

3.62 An insurance company offers its policyholders a number of different premium payment options. For a randomly selected policyholder, let X be the number of months between successive payments. The cumulative distribution function of X is

$$F(x) = \begin{cases} 0, & \text{if } x < 1, \\ 0.4, & \text{if } 1 \leq x < 3, \\ 0.6, & \text{if } 3 \leq x < 5, \\ 0.8, & \text{if } 5 \leq x < 7, \\ 1.0, & \text{if } x \geq 7. \end{cases}$$

(a) What is the probability mass function of X?
(b) Compute $P(4 < X \leq 7)$.

3.63 Two electronic components of a missile system work in harmony for the success of the total system. Let X and Y denote the life in hours of the two components. The joint density of X and Y is

$$f(x, y) = \begin{cases} ye^{-y(1+x)}, & x, y \geq 0, \\ 0, & \text{elsewhere.} \end{cases}$$

(a) Give the marginal density functions for both random variables.
(b) What is the probability that the lives of both components will exceed 2 hours?

3.64 A service facility operates with two service lines. On a randomly selected day, let X be the proportion of time that the first line is in use whereas Y is the proportion of time that the second line is in use. Suppose that the joint probability density function for (X, Y) is

$$f(x, y) = \begin{cases} \frac{3}{2}(x^2 + y^2), & 0 \leq x, y \leq 1, \\ 0, & \text{elsewhere.} \end{cases}$$

(a) Compute the probability that neither line is busy more than half the time.
(b) Find the probability that the first line is busy more than 75% of the time.

3.65 Let the number of phone calls received by a switchboard during a 5-minute interval be a random variable X with probability function

$$f(x) = \frac{e^{-2}2^x}{x!}, \quad \text{for } x = 0, 1, 2, \ldots.$$

(a) Determine the probability that X equals 0, 1, 2, 3, 4, 5, and 6.
(b) Graph the probability mass function for these values of x.
(c) Determine the cumulative distribution function for these values of X.

3.66 Consider the random variables X and Y with joint density function

$$f(x, y) = \begin{cases} x + y, & 0 \leq x, y \leq 1, \\ 0, & \text{elsewhere.} \end{cases}$$

(a) Find the marginal distributions of X and Y.
(b) Find $P(X > 0.5, Y > 0.5)$.

3.67 An industrial process manufactures items that can be classified as either defective or not defective. The probability that an item is defective is 0.1. An experiment is conducted in which 5 items are drawn randomly from the process. Let the random variable X be the number of defectives in this sample of 5. What is the probability mass function of X?

3.68 Consider the following joint probability density function of the random variables X and Y:

$$f(x, y) = \begin{cases} \frac{3x-y}{9}, & 1 < x < 3, \; 1 < y < 2, \\ 0, & \text{elsewhere.} \end{cases}$$

(a) Find the marginal density functions of X and Y.
(b) Are X and Y independent?
(c) Find $P(X > 2)$.

3.69 The life span in hours of an electrical component is a random variable with cumulative distribution function

$$F(x) = \begin{cases} 1 - e^{-\frac{x}{50}}, & x > 0, \\ 0, & \text{elsewhere.} \end{cases}$$

(a) Determine its probability density function.
(b) Determine the probability that the life span of such a component will exceed 70 hours.

3.70 Pairs of pants are being produced by a particular outlet facility. The pants are checked by a group of 10 workers. The workers inspect pairs of pants taken randomly from the production line. Each inspector is assigned a number from 1 through 10. A buyer selects a pair of pants for purchase. Let the random variable X be the inspector number.

(a) Give a reasonable probability mass function for X.
(b) Plot the cumulative distribution function for X.

3.71 The shelf life of a product is a random variable that is related to consumer acceptance. It turns out that the shelf life Y in days of a certain type of bakery product has a density function

$$f(y) = \begin{cases} \frac{1}{2}e^{-y/2}, & 0 \leq y < \infty, \\ 0, & \text{elsewhere.} \end{cases}$$

What fraction of the loaves of this product stocked today would you expect to be sellable 3 days from now?

3.72 Passenger congestion is a service problem in airports. Trains are installed within the airport to reduce the congestion. With the use of the train, the time X in minutes that it takes to travel from the main terminal to a particular concourse has density function

$$f(x) = \begin{cases} \frac{1}{10}, & 0 \leq x \leq 10, \\ 0, & \text{elsewhere.} \end{cases}$$

(a) Show that the above is a valid probability density function.
(b) Find the probability that the time it takes a passenger to travel from the main terminal to the concourse will not exceed 7 minutes.

3.73 Impurities in a batch of final product of a chemical process often reflect a serious problem. From considerable plant data gathered, it is known that the proportion Y of impurities in a batch has a density function given by

$$f(y) = \begin{cases} 10(1-y)^9, & 0 \leq y \leq 1, \\ 0, & \text{elsewhere.} \end{cases}$$

(a) Verify that the above is a valid density function.
(b) A batch is considered not sellable and then not acceptable if the percentage of impurities exceeds 60%. With the current quality of the process, what is the percentage of batches that are not acceptable?

3.74 The time Z in minutes between calls to an electrical supply system has the probability density function

$$f(z) = \begin{cases} \frac{1}{10}e^{-z/10}, & 0 < z < \infty, \\ 0, & \text{elsewhere.} \end{cases}$$

(a) What is the probability that there are no calls within a 20-minute time interval?
(b) What is the probability that the first call comes within 10 minutes of opening?

3.75 A chemical system that results from a chemical reaction has two important components among others in a blend. The joint distribution describing the proportions X_1 and X_2 of these two components is given by

$$f(x_1, x_2) = \begin{cases} 2, & 0 < x_1 < x_2 < 1, \\ 0, & \text{elsewhere.} \end{cases}$$

(a) Give the marginal distribution of X_1.
(b) Give the marginal distribution of X_2.
(c) What is the probability that component proportions produce the results $X_1 < 0.2$ and $X_2 > 0.5$?
(d) Give the conditional distribution $f_{X_1|X_2}(x_1|x_2)$.

3.76 Consider the situation of Review Exercise 3.75. But suppose the joint distribution of the two proportions is given by

$$f(x_1, x_2) = \begin{cases} 6x_2, & 0 < x_2 < x_1 < 1, \\ 0, & \text{elsewhere.} \end{cases}$$

(a) Give the marginal distribution $f_{X_1}(x_1)$ of the proportion X_1 and verify that it is a valid density function.
(b) What is the probability that proportion X_2 is less than 0.5, given that X_1 is 0.7?

3.77 Consider the random variables X and Y that represent the number of vehicles that arrive at two separate street corners during a certain 2-minute period. These street corners are fairly close together so it is important that traffic engineers deal with them jointly if necessary. The joint distribution of X and Y is known to be

$$f(x, y) = \frac{9}{16} \cdot \frac{1}{4^{(x+y)}},$$

for $x = 0, 1, 2, \ldots$ and $y = 0, 1, 2, \ldots$.

(a) Are the two random variables X and Y independent? Explain why or why not.
(b) What is the probability that during the time period in question less than 4 vehicles arrive at the two street corners?

3.78 The behavior of series of components plays a huge role in scientific and engineering reliability problems. The reliability of the entire system is certainly no better than that of the weakest component in the series. In a series system, the components operate independently of each other. In a particular system containing three components, the probabilities of meeting specifications for components 1, 2, and 3, respectively, are 0.95, 0.99, and 0.92. What is the probability that the entire system works?

3.79 Another type of system that is employed in engineering work is a group of parallel components or a parallel system. In this more conservative approach, the probability that the system operates is larger than the probability that any component operates. The system fails only when all components fail. Consider a situation in which there are 4 independent components in a parallel system with probability of operation given by

Component 1: 0.95;   Component 2: 0.94;
Component 3: 0.90;   Component 4: 0.97.

What is the probability that the system does not fail?

3.80 Consider a system of components in which there are 5 independent components, each of which possesses an operational probability of 0.92. The system does have a redundancy built in such that it does not fail if 3 out of the 5 components are operational. What is the probability that the total system is operational?

3.81 Project: Take 5 class periods to observe the shoe color of individuals in class. Assume the shoe color categories are red, white, black, brown, and other. Complete a frequency table for each color category.

(a) Estimate and interpret the meaning of the probability distribution.

(b) What is the estimated probability that in the next class period a randomly selected student will be wearing a red or a white pair of shoes?

3.5 Potential Misconceptions and Hazards; Relationship to Material in Other Chapters

In future chapters it will become apparent that probability distributions represent the structure through which probabilities that are computed aid in the evaluation and understanding of a process. For example, in Review Exercise 3.65, the probability distribution that quantifies the probability of a heavy load during certain time periods can be very useful in planning for any changes in the system. Review Exercise 3.69 describes a scenario in which the life span of an electronic component is studied. Knowledge of the probability structure for the component will contribute significantly to an understanding of the reliability of a large system of which the component is a part. In addition, an understanding of the general nature of probability distributions will enhance understanding of the concept of a ***P*-value**, which was introduced briefly in Chapter 1 and will play a major role beginning in Chapter 10 and extending throughout the balance of the text.

Chapters 4, 5, and 6 depend heavily on the material in this chapter. In Chapter 4, we discuss the meaning of important **parameters** in probability distributions. These important parameters quantify notions of **central tendency** and **variability** in a system. In fact, knowledge of these quantities themselves, quite apart from the complete distribution, can provide insight into the nature of the system. Chapters 5 and 6 will deal with engineering, biological, or general scientific scenarios that identify special types of distributions. For example, the structure of the probability function in Review Exercise 3.65 will easily be identified under certain assumptions discussed in Chapter 5. The same holds for the scenario of Review Exercise 3.69. This is a special type of **time to failure** problem for which the probability density function will be discussed in Chapter 6.

As far as potential hazards with the use of material in this chapter, the warning to the reader is not to read more into the material than is evident. The general nature of the probability distribution for a specific scientific phenomenon is not obvious from what is learned in this chapter. The purpose of this chapter is for readers to learn how to manipulate a probability distribution, not to learn how to identify a specific type. Chapters 5 and 6 go a long way toward identification according to the general nature of the scientific system.

Chapter 4

Mathematical Expectation

4.1 Mean of a Random Variable

In Chapter 1, we discussed the sample mean, which is the arithmetic mean of the data. Now consider the following. If two coins are tossed 16 times and X is the number of heads that occur per toss, then the values of X are 0, 1, and 2. Suppose that the experiment yields no heads, one head, and two heads a total of 4, 7, and 5 times, respectively. The average number of heads per toss of the two coins is then

$$\frac{(0)(4) + (1)(7) + (2)(5)}{16} = 1.06.$$

This is an average value of the data and yet it is not a possible outcome of $\{0, 1, 2\}$. Hence, an average is not necessarily a possible outcome for the experiment. For instance, a salesman's average monthly income is not likely to be equal to any of his monthly paychecks.

Let us now restructure our computation for the average number of heads so as to have the following equivalent form:

$$(0)\left(\frac{4}{16}\right) + (1)\left(\frac{7}{16}\right) + (2)\left(\frac{5}{16}\right) = 1.06.$$

The numbers 4/16, 7/16, and 5/16 are the fractions of the total tosses resulting in 0, 1, and 2 heads, respectively. These fractions are also the relative frequencies for the different values of X in our experiment. In fact, then, we can calculate the mean, or average, of a set of data by knowing the distinct values that occur and their relative frequencies, without any knowledge of the total number of observations in our set of data. Therefore, if 4/16, or 1/4, of the tosses result in no heads, 7/16 of the tosses result in one head, and 5/16 of the tosses result in two heads, the mean number of heads per toss would be 1.06 no matter whether the total number of tosses were 16, 1000, or even 10,000.

This method of relative frequencies is used to calculate the average number of heads per toss of two coins that we might expect in the long run. We shall refer to this average value as the **mean of the random variable** X or the **mean of the probability distribution of** X and write it as μ_x or simply as μ when it is clear to which random variable we refer. It is also common among statisticians to refer to this mean as the mathematical expectation, or the expected value of the random variable X, and denote it as $E(X)$.

Assuming that 1 fair coin was tossed twice, we find that the sample space for our experiment is

$$S = \{HH, HT, TH, TT\}.$$

Since the 4 sample points are all equally likely, it follows that

$$P(X = 0) = P(TT) = \frac{1}{4}, \quad P(X = 1) = P(TH) + P(HT) = \frac{1}{2},$$

and

$$P(X = 2) = P(HH) = \frac{1}{4},$$

where a typical element, say TH, indicates that the first toss resulted in a tail followed by a head on the second toss. Now, these probabilities are just the relative frequencies for the given events in the long run. Therefore,

$$\mu = E(X) = (0)\left(\frac{1}{4}\right) + (1)\left(\frac{1}{2}\right) + (2)\left(\frac{1}{4}\right) = 1.$$

This result means that a person who tosses 2 coins over and over again will, on the average, get 1 head per toss.

The method described above for calculating the expected number of heads per toss of 2 coins suggests that the mean, or expected value, of any discrete random variable may be obtained by multiplying each of the values $x_1, x_2, \ldots, x_n$ of the random variable X by its corresponding probability $f(x_1), f(x_2), \ldots, f(x_n)$ and summing the products. This is true, however, only if the random variable is discrete. In the case of continuous random variables, the definition of an expected value is essentially the same with summations replaced by integrations.

Definition 4.1: Let X be a random variable with probability distribution $f(x)$. The **mean**, or **expected value**, of X is

$$\mu = E(X) = \sum_x x f(x)$$

if X is discrete, and

$$\mu = E(X) = \int_{-\infty}^{\infty} x f(x)\, dx$$

if X is continuous.

The reader should note that the way to calculate the expected value, or mean, shown here is different from the way to calculate the sample mean described in Chapter 1, where the sample mean is obtained by using data. In mathematical expectation, the expected value is calculated by using the probability distribution.

4.1 Mean of a Random Variable

However, the mean is usually understood as a "center" value of the underlying distribution if we use the expected value, as in Definition 4.1.

Example 4.1: A lot containing 7 components is sampled by a quality inspector; the lot contains 4 good components and 3 defective components. A sample of 3 is taken by the inspector. Find the expected value of the number of good components in this sample.

Solution: Let X represent the number of good components in the sample. The probability distribution of X is

$$f(x) = \frac{\binom{4}{x}\binom{3}{3-x}}{\binom{7}{3}}, \quad x = 0, 1, 2, 3.$$

Simple calculations yield $f(0) = 1/35$, $f(1) = 12/35$, $f(2) = 18/35$, and $f(3) = 4/35$. Therefore,

$$\mu = E(X) = (0)\left(\frac{1}{35}\right) + (1)\left(\frac{12}{35}\right) + (2)\left(\frac{18}{35}\right) + (3)\left(\frac{4}{35}\right) = \frac{12}{7} = 1.7.$$

Thus, if a sample of size 3 is selected at random over and over again from a lot of 4 good components and 3 defective components, it will contain, on average, 1.7 good components.

Example 4.2: A salesperson for a medical device company has two appointments on a given day. At the first appointment, he believes that he has a 70% chance to make the deal, from which he can earn $1000 commission if successful. On the other hand, he thinks he only has a 40% chance to make the deal at the second appointment, from which, if successful, he can make $1500. What is his expected commission based on his own probability belief? Assume that the appointment results are independent of each other.

Solution: First, we know that the salesperson, for the two appointments, can have 4 possible commission totals: $0, $1000, $1500, and $2500. We then need to calculate their associated probabilities. By independence, we obtain

$$f(\$0) = (1-0.7)(1-0.4) = 0.18, \quad f(\$2500) = (0.7)(0.4) = 0.28,$$
$$f(\$1000) = (0.7)(1-0.4) = 0.42, \text{ and } f(\$1500) = (1-0.7)(0.4) = 0.12.$$

Therefore, the expected commission for the salesperson is

$$E(X) = (\$0)(0.18) + (\$1000)(0.42) + (\$1500)(0.12) + (\$2500)(0.28)$$
$$= \$1300.$$

Examples 4.1 and 4.2 are designed to allow the reader to gain some insight into what we mean by the expected value of a random variable. In both cases the random variables are discrete. We follow with an example involving a continuous random variable, where an engineer is interested in the *mean life* of a certain type of electronic device. This is an illustration of a *time to failure* problem that occurs often in practice. The expected value of the life of a device is an important parameter for its evaluation.

Example 4.3: Let X be the random variable that denotes the life in hours of a certain electronic device. The probability density function is

$$f(x) = \begin{cases} \frac{20{,}000}{x^3}, & x > 100, \\ 0, & \text{elsewhere.} \end{cases}$$

Find the expected life of this type of device.

Solution: Using Definition 4.1, we have

$$\mu = E(X) = \int_{100}^{\infty} x \frac{20{,}000}{x^3}\, dx = \int_{100}^{\infty} \frac{20{,}000}{x^2}\, dx = 200.$$

Therefore, we can expect this type of device to last, *on average*, 200 hours.

Now let us consider a new random variable $g(X)$, which depends on X; that is, each value of $g(X)$ is determined by the value of X. For instance, $g(X)$ might be X^2 or $3X - 1$, and whenever X assumes the value 2, $g(X)$ assumes the value $g(2)$. In particular, if X is a discrete random variable with probability distribution $f(x)$, for $x = -1, 0, 1, 2$, and $g(X) = X^2$, then

$$P[g(X) = 0] = P(X = 0) = f(0),$$
$$P[g(X) = 1] = P(X = -1) + P(X = 1) = f(-1) + f(1),$$
$$P[g(X) = 4] = P(X = 2) = f(2),$$

and so the probability distribution of $g(X)$ may be written

$g(x)$	0	1	4
$P[g(X) = g(x)]$	$f(0)$	$f(-1) + f(1)$	$f(2)$

By the definition of the expected value of a random variable, we obtain

$$\mu_{g(X)} = E[g(x)] = 0f(0) + 1[f(-1) + f(1)] + 4f(2)$$
$$= (-1)^2 f(-1) + (0)^2 f(0) + (1)^2 f(1) + (2)^2 f(2) = \sum_x g(x)f(x).$$

This result is generalized in Theorem 4.1 for both discrete and continuous random variables.

Theorem 4.1: Let X be a random variable with probability distribution $f(x)$. The expected value of the random variable $g(X)$ is

$$\mu_{g(X)} = E[g(X)] = \sum_x g(x)f(x)$$

if X is discrete, and

$$\mu_{g(X)} = E[g(X)] = \int_{-\infty}^{\infty} g(x)f(x)\, dx$$

if X is continuous.

Example 4.4: Suppose that the number of cars X that pass through a car wash between 4:00 P.M. and 5:00 P.M. on any sunny Friday has the following probability distribution:

x	4	5	6	7	8	9
$P(X = x)$	$\frac{1}{12}$	$\frac{1}{12}$	$\frac{1}{4}$	$\frac{1}{4}$	$\frac{1}{6}$	$\frac{1}{6}$

Let $g(X) = 2X - 1$ represent the amount of money, in dollars, paid to the attendant by the manager. Find the attendant's expected earnings for this particular time period.

Solution: By Theorem 4.1, the attendant can expect to receive

$$E[g(X)] = E(2X - 1) = \sum_{x=4}^{9} (2x - 1)f(x)$$

$$= (7)\left(\frac{1}{12}\right) + (9)\left(\frac{1}{12}\right) + (11)\left(\frac{1}{4}\right) + (13)\left(\frac{1}{4}\right)$$

$$+ (15)\left(\frac{1}{6}\right) + (17)\left(\frac{1}{6}\right) = \$12.67.$$

Example 4.5: Let X be a random variable with density function

$$f(x) = \begin{cases} \frac{x^2}{3}, & -1 < x < 2, \\ 0, & \text{elsewhere.} \end{cases}$$

Find the expected value of $g(X) = 4X + 3$.

Solution: By Theorem 4.1, we have

$$E(4X + 3) = \int_{-1}^{2} \frac{(4x + 3)x^2}{3}\, dx = \frac{1}{3}\int_{-1}^{2}(4x^3 + 3x^2)\, dx = 8.$$

We shall now extend our concept of mathematical expectation to the case of two random variables X and Y with joint probability distribution $f(x, y)$.

Definition 4.2: Let X and Y be random variables with joint probability distribution $f(x, y)$. The mean, or expected value, of the random variable $g(X, Y)$ is

$$\mu_{g(X,Y)} = E[g(X, Y)] = \sum_x \sum_y g(x, y) f(x, y)$$

if X and Y are discrete, and

$$\mu_{g(X,Y)} = E[g(X, Y)] = \int_{-\infty}^{\infty} \int_{-\infty}^{\infty} g(x, y) f(x, y)\, dx\, dy$$

if X and Y are continuous.

Generalization of Definition 4.2 for the calculation of mathematical expectations of functions of several random variables is straightforward.

Example 4.6: Let X and Y be the random variables with joint probability distribution indicated in Table 3.1 on page 96. Find the expected value of $g(X,Y) = XY$. The table is reprinted here for convenience.

	$f(x,y)$	x = 0	1	2	Row Totals
	0	$\frac{3}{28}$	$\frac{9}{28}$	$\frac{3}{28}$	$\frac{15}{28}$
y	1	$\frac{3}{14}$	$\frac{3}{14}$	0	$\frac{3}{7}$
	2	$\frac{1}{28}$	0	0	$\frac{1}{28}$
	Column Totals	$\frac{5}{14}$	$\frac{15}{28}$	$\frac{3}{28}$	1

Solution: By Definition 4.2, we write

$$E(XY) = \sum_{x=0}^{2}\sum_{y=0}^{2} xy f(x,y)$$
$$= (0)(0)f(0,0) + (0)(1)f(0,1)$$
$$\quad + (1)(0)f(1,0) + (1)(1)f(1,1) + (2)(0)f(2,0)$$
$$= f(1,1) = \frac{3}{14}.$$

Example 4.7: Find $E(Y/X)$ for the density function

$$f(x,y) = \begin{cases} \frac{x(1+3y^2)}{4}, & 0 < x < 2,\ 0 < y < 1, \\ 0, & \text{elsewhere.} \end{cases}$$

Solution: We have

$$E\left(\frac{Y}{X}\right) = \int_0^1 \int_0^2 \frac{y(1+3y^2)}{4}\, dx\, dy = \int_0^1 \frac{y+3y^3}{2}\, dy = \frac{5}{8}.$$

Note that if $g(X,Y) = X$ in Definition 4.2, we have

$$E(X) = \begin{cases} \sum_x \sum_y x f(x,y) = \sum_x x g(x) & \text{(discrete case)}, \\ \int_{-\infty}^{\infty}\int_{-\infty}^{\infty} x f(x,y)\, dy\, dx = \int_{-\infty}^{\infty} x g(x)\, dx & \text{(continuous case)}, \end{cases}$$

where $g(x)$ is the marginal distribution of X. Therefore, in calculating $E(X)$ over a two-dimensional space, one may use either the joint probability distribution of X and Y or the marginal distribution of X. Similarly, we define

$$E(Y) = \begin{cases} \sum_y \sum_x y f(x,y) = \sum_y y h(y) & \text{(discrete case)}, \\ \int_{-\infty}^{\infty}\int_{-\infty}^{\infty} y f(x,y)\, dx\, dy = \int_{-\infty}^{\infty} y h(y)\, dy & \text{(continuous case)}, \end{cases}$$

where $h(y)$ is the marginal distribution of the random variable Y.

Exercises

4.1 The probability distribution of X, the number of imperfections per 10 meters of a synthetic fabric in continuous rolls of uniform width, is given in Exercise 3.13 on page 92 as

x	0	1	2	3	4
$f(x)$	0.41	0.37	0.16	0.05	0.01

Find the average number of imperfections per 10 meters of this fabric.

4.2 The probability distribution of the discrete random variable X is

$$f(x) = \binom{3}{x}\left(\frac{1}{4}\right)^x \left(\frac{3}{4}\right)^{3-x}, \quad x=0,1,2,3.$$

Find the mean of X.

4.3 Find the mean of the random variable T representing the total of the three coins in Exercise 3.25 on page 93.

4.4 A coin is biased such that a head is three times as likely to occur as a tail. Find the expected number of tails when this coin is tossed twice.

4.5 In a gambling game, a woman is paid $3 if she draws a jack or a queen and $5 if she draws a king or an ace from an ordinary deck of 52 playing cards. If she draws any other card, she loses. How much should she pay to play if the game is fair?

4.6 An attendant at a car wash is paid according to the number of cars that pass through. Suppose the probabilities are 1/12, 1/12, 1/4, 1/4, 1/6, and 1/6, respectively, that the attendant receives $7, $9, $11, $13, $15, or $17 between 4:00 P.M. and 5:00 P.M. on any sunny Friday. Find the attendant's expected earnings for this particular period.

4.7 By investing in a particular stock, a person can make a profit in one year of $4000 with probability 0.3 or take a loss of $1000 with probability 0.7. What is this person's expected gain?

4.8 Suppose that an antique jewelry dealer is interested in purchasing a gold necklace for which the probabilities are 0.22, 0.36, 0.28, and 0.14, respectively, that she will be able to sell it for a profit of $250, sell it for a profit of $150, break even, or sell it for a loss of $150. What is her expected profit?

4.9 A private pilot wishes to insure his airplane for $200,000. The insurance company estimates that a total loss will occur with probability 0.002, a 50% loss with probability 0.01, and a 25% loss with probability 0.1. Ignoring all other partial losses, what premium should the insurance company charge each year to realize an average profit of $500?

4.10 Two tire-quality experts examine stacks of tires and assign a quality rating to each tire on a 3-point scale. Let X denote the rating given by expert A and Y denote the rating given by B. The following table gives the joint distribution for X and Y.

		y		
$f(x,y)$		1	2	3
	1	0.10	0.05	0.02
x	2	0.10	0.35	0.05
	3	0.03	0.10	0.20

Find μ_X and μ_Y.

4.11 The density function of coded measurements of the pitch diameter of threads of a fitting is

$$f(x) = \begin{cases} \frac{4}{\pi(1+x^2)}, & 0 < x < 1, \\ 0, & \text{elsewhere.} \end{cases}$$

Find the expected value of X.

4.12 If a dealer's profit, in units of $5000, on a new automobile can be looked upon as a random variable X having the density function

$$f(x) = \begin{cases} 2(1-x), & 0 < x < 1, \\ 0, & \text{elsewhere,} \end{cases}$$

find the average profit per automobile.

4.13 The density function of the continuous random variable X, the total number of hours, in units of 100 hours, that a family runs a vacuum cleaner over a period of one year, is given in Exercise 3.7 on page 92 as

$$f(x) = \begin{cases} x, & 0 < x < 1, \\ 2-x, & 1 \le x < 2, \\ 0, & \text{elsewhere.} \end{cases}$$

Find the average number of hours per year that families run their vacuum cleaners.

4.14 Find the proportion X of individuals who can be expected to respond to a certain mail-order solicitation if X has the density function

$$f(x) = \begin{cases} \frac{2(x+2)}{5}, & 0 < x < 1, \\ 0, & \text{elsewhere.} \end{cases}$$

4.15 Assume that two random variables (X, Y) are uniformly distributed on a circle with radius a. Then the joint probability density function is

$$f(x,y) = \begin{cases} \frac{1}{\pi a^2}, & x^2 + y^2 \leq a^2, \\ 0, & \text{otherwise.} \end{cases}$$

Find μ_X, the expected value of X.

4.16 Suppose that you are inspecting a lot of 1000 light bulbs, among which 20 are defectives. You choose two light bulbs randomly from the lot without replacement. Let

$$X_1 = \begin{cases} 1, & \text{if the 1st light bulb is defective,} \\ 0, & \text{otherwise,} \end{cases}$$

$$X_2 = \begin{cases} 1, & \text{if the 2nd light bulb is defective,} \\ 0, & \text{otherwise.} \end{cases}$$

Find the probability that at least one light bulb chosen is defective. [*Hint*: Compute $P(X_1 + X_2 = 1)$.]

4.17 Let X be a random variable with the following probability distribution:

x	-3	6	9
$f(x)$	$1/6$	$1/2$	$1/3$

Find $\mu_{g(X)}$, where $g(X) = (2X+1)^2$.

4.18 Find the expected value of the random variable $g(X) = X^2$, where X has the probability distribution of Exercise 4.2.

4.19 A large industrial firm purchases several new computers at the end of each year, the exact number depending on the frequency of repairs in the previous year. Suppose that the number of computers, X, purchased each year has the following probability distribution:

x	0	1	2	3
$f(x)$	$1/10$	$3/10$	$2/5$	$1/5$

If the cost of the desired model is \$1200 per unit and at the end of the year a refund of $50X^2$ dollars will be issued, how much can this firm expect to spend on new computers during this year?

4.20 A continuous random variable X has the density function

$$f(x) = \begin{cases} e^{-x}, & x > 0, \\ 0, & \text{elsewhere.} \end{cases}$$

Find the expected value of $g(X) = e^{2X/3}$.

4.21 What is the dealer's average profit per automobile if the profit on each automobile is given by $g(X) = X^2$, where X is a random variable having the density function of Exercise 4.12?

4.22 The hospitalization period, in days, for patients following treatment for a certain type of kidney disorder is a random variable $Y = X + 4$, where X has the density function

$$f(x) = \begin{cases} \frac{32}{(x+4)^3}, & x > 0, \\ 0, & \text{elsewhere.} \end{cases}$$

Find the average number of days that a person is hospitalized following treatment for this disorder.

4.23 Suppose that X and Y have the following joint probability function:

		\multicolumn{2}{c}{x}
$f(x,y)$		2	4
	1	0.10	0.15
y	3	0.20	0.30
	5	0.10	0.15

(a) Find the expected value of $g(X, Y) = XY^2$.
(b) Find μ_X and μ_Y.

4.24 Referring to the random variables whose joint probability distribution is given in Exercise 3.39 on page 105,
(a) find $E(X^2Y - 2XY)$;
(b) find $\mu_X - \mu_Y$.

4.25 Referring to the random variables whose joint probability distribution is given in Exercise 3.51 on page 106, find the mean for the total number of jacks and kings when 3 cards are drawn without replacement from the 12 face cards of an ordinary deck of 52 playing cards.

4.26 Let X and Y be random variables with joint density function

$$f(x,y) = \begin{cases} 4xy, & 0 < x, y < 1, \\ 0, & \text{elsewhere.} \end{cases}$$

Find the expected value of $Z = \sqrt{X^2 + Y^2}$.

4.27 In Exercise 3.27 on page 93, a density function is given for the time to failure of an important component of a DVD player. Find the mean number of hours to failure of the component and thus the DVD player.

4.28 Consider the information in Exercise 3.28 on page 93. The problem deals with the weight in ounces of the product in a cereal box, with

$$f(x) = \begin{cases} \frac{2}{5}, & 23.75 \leq x \leq 26.25, \\ 0, & \text{elsewhere.} \end{cases}$$

(a) Plot the density function.
(b) Compute the expected value, or mean weight, in ounces.
(c) Are you surprised at your answer in (b)? Explain why or why not.

4.29 Exercise 3.29 on page 93 dealt with an important particle size distribution characterized by

$$f(x) = \begin{cases} 3x^{-4}, & x > 1, \\ 0, & \text{elsewhere.} \end{cases}$$

(a) Plot the density function.
(b) Give the mean particle size.

4.30 In Exercise 3.31 on page 94, the distribution of times before a major repair of a washing machine was given as

$$f(y) = \begin{cases} \frac{1}{4}e^{-y/4}, & y \geq 0, \\ 0, & \text{elsewhere.} \end{cases}$$

What is the population mean of the times to repair?

4.31 Consider Exercise 3.32 on page 94.
(a) What is the mean proportion of the budget allocated to environmental and pollution control?
(b) What is the probability that a company selected at random will have allocated to environmental and pollution control a proportion that exceeds the population mean given in (a)?

4.32 In Exercise 3.13 on page 92, the distribution of the number of imperfections per 10 meters of synthetic fabric is given by

x	0	1	2	3	4
$f(x)$	0.41	0.37	0.16	0.05	0.01

(a) Plot the probability function.
(b) Find the expected number of imperfections, $E(X) = \mu$.
(c) Find $E(X^2)$.

4.2 Variance and Covariance of Random Variables

The mean, or expected value, of a random variable X is of special importance in statistics because it describes where the probability distribution is centered. By itself, however, the mean does not give an adequate description of the shape of the distribution. We also need to characterize the variability in the distribution. In Figure 4.1, we have the histograms of two discrete probability distributions that have the same mean, $\mu = 2$, but differ considerably in variability, or the dispersion of their observations about the mean.

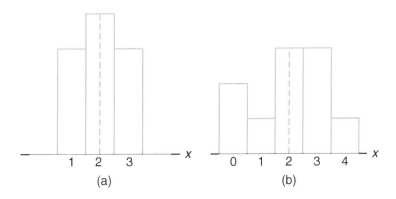

Figure 4.1: Distributions with equal means and unequal dispersions.

The most important measure of variability of a random variable X is obtained by applying Theorem 4.1 with $g(X) = (X - \mu)^2$. The quantity is referred to as the **variance of the random variable** X or the **variance of the probability**

Definition 4.3: Let X be a random variable with probability distribution $f(x)$ and mean μ. The variance of X is

$$\sigma^2 = E[(X-\mu)^2] = \sum_x (x-\mu)^2 f(x), \qquad \text{if } X \text{ is discrete, and}$$

$$\sigma^2 = E[(X-\mu)^2] = \int_{-\infty}^{\infty} (x-\mu)^2 f(x)\, dx, \qquad \text{if } X \text{ is continuous.}$$

The positive square root of the variance, σ, is called the **standard deviation** of X.

The quantity $x - \mu$ in Definition 4.3 is called the **deviation of an observation** from its mean. Since the deviations are squared and then averaged, σ^2 will be much smaller for a set of x values that are close to μ than it will be for a set of values that vary considerably from μ.

Example 4.8: Let the random variable X represent the number of automobiles that are used for official business purposes on any given workday. The probability distribution for company A [Figure 4.1(a)] is

x	1	2	3
$f(x)$	0.3	0.4	0.3

and that for company B [Figure 4.1(b)] is

x	0	1	2	3	4
$f(x)$	0.2	0.1	0.3	0.3	0.1

Show that the variance of the probability distribution for company B is greater than that for company A.

Solution: For company A, we find that

$$\mu_A = E(X) = (1)(0.3) + (2)(0.4) + (3)(0.3) = 2.0,$$

and then

$$\sigma_A^2 = \sum_{x=1}^{3} (x-2)^2 = (1-2)^2(0.3) + (2-2)^2(0.4) + (3-2)^2(0.3) = 0.6.$$

For company B, we have

$$\mu_B = E(X) = (0)(0.2) + (1)(0.1) + (2)(0.3) + (3)(0.3) + (4)(0.1) = 2.0,$$

and then

$$\sigma_B^2 = \sum_{x=0}^{4} (x-2)^2 f(x)$$
$$= (0-2)^2(0.2) + (1-2)^2(0.1) + (2-2)^2(0.3)$$
$$+ (3-2)^2(0.3) + (4-2)^2(0.1) = 1.6.$$

4.2 Variance and Covariance of Random Variables

Clearly, the variance of the number of automobiles that are used for official business purposes is greater for company B than for company A.

An alternative and preferred formula for finding σ^2, which often simplifies the calculations, is stated in the following theorem.

Theorem 4.2: The variance of a random variable X is

$$\sigma^2 = E(X^2) - \mu^2.$$

Proof: For the discrete case, we can write

$$\sigma^2 = \sum_x (x-\mu)^2 f(x) = \sum_x (x^2 - 2\mu x + \mu^2) f(x)$$

$$= \sum_x x^2 f(x) - 2\mu \sum_x x f(x) + \mu^2 \sum_x f(x).$$

Since $\mu = \sum_x x f(x)$ by definition, and $\sum_x f(x) = 1$ for any discrete probability distribution, it follows that

$$\sigma^2 = \sum_x x^2 f(x) - \mu^2 = E(X^2) - \mu^2.$$

For the continuous case the proof is step by step the same, with summations replaced by integrations.

Example 4.9: Let the random variable X represent the number of defective parts for a machine when 3 parts are sampled from a production line and tested. The following is the probability distribution of X.

x	0	1	2	3
$f(x)$	0.51	0.38	0.10	0.01

Using Theorem 4.2, calculate σ^2.

Solution: First, we compute

$$\mu = (0)(0.51) + (1)(0.38) + (2)(0.10) + (3)(0.01) = 0.61.$$

Now,

$$E(X^2) = (0)(0.51) + (1)(0.38) + (4)(0.10) + (9)(0.01) = 0.87.$$

Therefore,

$$\sigma^2 = 0.87 - (0.61)^2 = 0.4979.$$

Example 4.10: The weekly demand for a drinking-water product, in thousands of liters, from a local chain of efficiency stores is a continuous random variable X having the probability density

$$f(x) = \begin{cases} 2(x-1), & 1 < x < 2, \\ 0, & \text{elsewhere.} \end{cases}$$

Find the mean and variance of X.

Solution: Calculating $E(X)$ and $E(X^2)$, we have

$$\mu = E(X) = 2\int_1^2 x(x-1)\,dx = \frac{5}{3}$$

and

$$E(X^2) = 2\int_1^2 x^2(x-1)\,dx = \frac{17}{6}.$$

Therefore,

$$\sigma^2 = \frac{17}{6} - \left(\frac{5}{3}\right)^2 = \frac{1}{18}.$$

At this point, the variance or standard deviation has meaning only when we compare two or more distributions that have the same units of measurement. Therefore, we could compare the variances of the distributions of contents, measured in liters, of bottles of orange juice from two companies, and the larger value would indicate the company whose product was more variable or less uniform. It would not be meaningful to compare the variance of a distribution of heights to the variance of a distribution of aptitude scores. In Section 4.4, we show how the standard deviation can be used to describe a single distribution of observations.

We shall now extend our concept of the variance of a random variable X to include random variables related to X. For the random variable $g(X)$, the variance is denoted by $\sigma^2_{g(X)}$ and is calculated by means of the following theorem.

Theorem 4.3: Let X be a random variable with probability distribution $f(x)$. The variance of the random variable $g(X)$ is

$$\sigma^2_{g(X)} = E\{[g(X) - \mu_{g(X)}]^2\} = \sum_x [g(x) - \mu_{g(X)}]^2 f(x)$$

if X is discrete, and

$$\sigma^2_{g(X)} = E\{[g(X) - \mu_{g(X)}]^2\} = \int_{-\infty}^{\infty} [g(x) - \mu_{g(X)}]^2 f(x)\,dx$$

if X is continuous.

Proof: Since $g(X)$ is itself a random variable with mean $\mu_{g(X)}$ as defined in Theorem 4.1, it follows from Definition 4.3 that

$$\sigma^2_{g(X)} = E\{[g(X) - \mu_{g(X)}]\}.$$

Now, applying Theorem 4.1 again to the random variable $[g(X) - \mu_{g(X)}]^2$ completes the proof.

Example 4.11: Calculate the variance of $g(X) = 2X + 3$, where X is a random variable with probability distribution

x	0	1	2	3
$f(x)$	$\frac{1}{4}$	$\frac{1}{8}$	$\frac{1}{2}$	$\frac{1}{8}$

4.2 Variance and Covariance of Random Variables

Solution: First, we find the mean of the random variable $2X+3$. According to Theorem 4.1,

$$\mu_{2X+3} = E(2X+3) = \sum_{x=0}^{3}(2x+3)f(x) = 6.$$

Now, using Theorem 4.3, we have

$$\sigma^2_{2X+3} = E\{[(2X+3) - \mu_{2x+3}]^2\} = E[(2X+3-6)^2]$$
$$= E(4X^2 - 12X + 9) = \sum_{x=0}^{3}(4x^2 - 12x + 9)f(x) = 4.$$

Example 4.12: Let X be a random variable having the density function given in Example 4.5 on page 115. Find the variance of the random variable $g(X) = 4X + 3$.

Solution: In Example 4.5, we found that $\mu_{4X+3} = 8$. Now, using Theorem 4.3,

$$\sigma^2_{4X+3} = E\{[(4X+3) - 8]^2\} = E[(4X-5)^2]$$
$$= \int_{-1}^{2}(4x-5)^2\frac{x^2}{3}\,dx = \frac{1}{3}\int_{-1}^{2}(16x^4 - 40x^3 + 25x^2)\,dx = \frac{51}{5}.$$

If $g(X,Y) = (X - \mu_X)(Y - \mu_Y)$, where $\mu_X = E(X)$ and $\mu_Y = E(Y)$, Definition 4.2 yields an expected value called the **covariance** of X and Y, which we denote by σ_{XY} or $\text{Cov}(X,Y)$.

Definition 4.4: Let X and Y be random variables with joint probability distribution $f(x,y)$. The covariance of X and Y is

$$\sigma_{XY} = E[(X - \mu_X)(Y - \mu_Y)] = \sum_x \sum_y (x - \mu_X)(y - \mu_y)f(x,y)$$

if X and Y are discrete, and

$$\sigma_{XY} = E[(X - \mu_X)(Y - \mu_Y)] = \int_{-\infty}^{\infty}\int_{-\infty}^{\infty}(x - \mu_X)(y - \mu_y)f(x,y)\,dx\,dy$$

if X and Y are continuous.

The covariance between two random variables is a measure of the nature of the association between the two. If large values of X often result in large values of Y or small values of X result in small values of Y, positive $X - \mu_X$ will often result in positive $Y - \mu_Y$ and negative $X - \mu_X$ will often result in negative $Y - \mu_Y$. Thus, the product $(X - \mu_X)(Y - \mu_Y)$ will tend to be positive. On the other hand, if large X values often result in small Y values, the product $(X - \mu_X)(Y - \mu_Y)$ will tend to be negative. The *sign* of the covariance indicates whether the relationship between two dependent random variables is positive or negative. When X and Y are statistically independent, it can be shown that the covariance is zero (see Corollary 4.5). The converse, however, is not generally true. Two variables may have zero covariance and still not be statistically independent. Note that the covariance only describes the *linear* relationship between two random variables. Therefore, if a covariance between X and Y is zero, X and Y may have a nonlinear relationship, which means that they are not necessarily independent.

The alternative and preferred formula for σ_{XY} is stated by Theorem 4.4.

Theorem 4.4: The covariance of two random variables X and Y with means μ_X and μ_Y, respectively, is given by

$$\sigma_{XY} = E(XY) - \mu_X \mu_Y.$$

Proof: For the discrete case, we can write

$$\sigma_{XY} = \sum_x \sum_y (x - \mu_X)(y - \mu_Y) f(x, y)$$
$$= \sum_x \sum_y xy f(x, y) - \mu_X \sum_x \sum_y y f(x, y)$$
$$- \mu_Y \sum_x \sum_y x f(x, y) + \mu_X \mu_Y \sum_x \sum_y f(x, y).$$

Since

$$\mu_X = \sum_x x f(x, y), \quad \mu_Y = \sum_y y f(x, y), \text{ and } \sum_x \sum_y f(x, y) = 1$$

for any joint discrete distribution, it follows that

$$\sigma_{XY} = E(XY) - \mu_X \mu_Y - \mu_Y \mu_X + \mu_X \mu_Y = E(XY) - \mu_X \mu_Y.$$

For the continuous case, the proof is identical with summations replaced by integrals.

Example 4.13: Example 3.14 on page 95 describes a situation involving the number of blue refills X and the number of red refills Y. Two refills for a ballpoint pen are selected at random from a certain box, and the following is the joint probability distribution:

	$f(x,y)$	x = 0	1	2	$h(y)$
	0	$\frac{3}{28}$	$\frac{9}{28}$	$\frac{3}{28}$	$\frac{15}{28}$
y	1	$\frac{3}{14}$	$\frac{3}{14}$	0	$\frac{3}{7}$
	2	$\frac{1}{28}$	0	0	$\frac{1}{28}$
	$g(x)$	$\frac{5}{14}$	$\frac{15}{28}$	$\frac{3}{28}$	1

Find the covariance of X and Y.

Solution: From Example 4.6, we see that $E(XY) = 3/14$. Now

$$\mu_X = \sum_{x=0}^{2} x g(x) = (0)\left(\frac{5}{14}\right) + (1)\left(\frac{15}{28}\right) + (2)\left(\frac{3}{28}\right) = \frac{3}{4},$$

and

$$\mu_Y = \sum_{y=0}^{2} y h(y) = (0)\left(\frac{15}{28}\right) + (1)\left(\frac{3}{7}\right) + (2)\left(\frac{1}{28}\right) = \frac{1}{2}.$$

4.2 Variance and Covariance of Random Variables

Therefore,

$$\sigma_{XY} = E(XY) - \mu_X\mu_Y = \frac{3}{14} - \left(\frac{3}{4}\right)\left(\frac{1}{2}\right) = -\frac{9}{56}.$$

Example 4.14: The fraction X of male runners and the fraction Y of female runners who compete in marathon races are described by the joint density function

$$f(x,y) = \begin{cases} 8xy, & 0 \le y \le x \le 1, \\ 0, & \text{elsewhere.} \end{cases}$$

Find the covariance of X and Y.

Solution: We first compute the marginal density functions. They are

$$g(x) = \begin{cases} 4x^3, & 0 \le x \le 1, \\ 0, & \text{elsewhere,} \end{cases}$$

and

$$h(y) = \begin{cases} 4y(1-y^2), & 0 \le y \le 1, \\ 0, & \text{elsewhere.} \end{cases}$$

From these marginal density functions, we compute

$$\mu_X = E(X) = \int_0^1 4x^4\,dx = \frac{4}{5} \quad \text{and} \quad \mu_Y = \int_0^1 4y^2(1-y^2)\,dy = \frac{8}{15}.$$

From the joint density function given above, we have

$$E(XY) = \int_0^1 \int_y^1 8x^2y^2\,dx\,dy = \frac{4}{9}.$$

Then

$$\sigma_{XY} = E(XY) - \mu_X\mu_Y = \frac{4}{9} - \left(\frac{4}{5}\right)\left(\frac{8}{15}\right) = \frac{4}{225}.$$

Although the covariance between two random variables does provide information regarding the nature of the relationship, the magnitude of σ_{XY} *does not indicate anything regarding the strength of the relationship, since σ_{XY} is not scale-free.* Its magnitude will depend on the units used to measure both X and Y. There is a scale-free version of the covariance called the **correlation coefficient** that is used widely in statistics.

Definition 4.5: Let X and Y be random variables with covariance σ_{XY} and standard deviations σ_X and σ_Y, respectively. The correlation coefficient of X and Y is

$$\rho_{XY} = \frac{\sigma_{XY}}{\sigma_X\sigma_Y}.$$

It should be clear to the reader that ρ_{XY} is free of the units of X and Y. The correlation coefficient satisfies the inequality $-1 \le \rho_{XY} \le 1$. It assumes a value of zero when $\sigma_{XY} = 0$. Where there is an exact linear dependency, say $Y \equiv a + bX$,

Example 4.15: Find the correlation coefficient between X and Y in Example 4.13.

Solution: Since

$$E(X^2) = (0^2)\left(\frac{5}{14}\right) + (1^2)\left(\frac{15}{28}\right) + (2^2)\left(\frac{3}{28}\right) = \frac{27}{28}$$

and

$$E(Y^2) = (0^2)\left(\frac{15}{28}\right) + (1^2)\left(\frac{3}{7}\right) + (2^2)\left(\frac{1}{28}\right) = \frac{4}{7},$$

we obtain

$$\sigma_X^2 = \frac{27}{28} - \left(\frac{3}{4}\right)^2 = \frac{45}{112} \text{ and } \sigma_Y^2 = \frac{4}{7} - \left(\frac{1}{2}\right)^2 = \frac{9}{28}.$$

Therefore, the correlation coefficient between X and Y is

$$\rho_{XY} = \frac{\sigma_{XY}}{\sigma_X \sigma_Y} = \frac{-9/56}{\sqrt{(45/112)(9/28)}} = -\frac{1}{\sqrt{5}}.$$

Example 4.16: Find the correlation coefficient of X and Y in Example 4.14.

Solution: Because

$$E(X^2) = \int_0^1 4x^5 \, dx = \frac{2}{3} \text{ and } E(Y^2) = \int_0^1 4y^3(1-y^2) \, dy = 1 - \frac{2}{3} = \frac{1}{3},$$

we conclude that

$$\sigma_X^2 = \frac{2}{3} - \left(\frac{4}{5}\right)^2 = \frac{2}{75} \text{ and } \sigma_Y^2 = \frac{1}{3} - \left(\frac{8}{15}\right)^2 = \frac{11}{225}.$$

Hence,

$$\rho_{XY} = \frac{4/225}{\sqrt{(2/75)(11/225)}} = \frac{4}{\sqrt{66}}.$$

Note that although the covariance in Example 4.15 is larger in magnitude (disregarding the sign) than that in Example 4.16, the relationship of the magnitudes of the correlation coefficients in these two examples is just the reverse. This is evidence that we cannot look at the magnitude of the covariance to decide on how strong the relationship is.

Exercises

4.33 Use Definition 4.3 on page 120 to find the variance of the random variable X of Exercise 4.7 on page 117.

4.34 Let X be a random variable with the following probability distribution:

x	-2	3	5
$f(x)$	0.3	0.2	0.5

Find the standard deviation of X.

4.35 The random variable X, representing the number of errors per 100 lines of software code, has the following probability distribution:

x	2	3	4	5	6
$f(x)$	0.01	0.25	0.4	0.3	0.04

Using Theorem 4.2 on page 121, find the variance of X.

4.36 Suppose that the probabilities are 0.4, 0.3, 0.2, and 0.1, respectively, that 0, 1, 2, or 3 power failures will strike a certain subdivision in any given year. Find the mean and variance of the random variable X representing the number of power failures striking this subdivision.

4.37 A dealer's profit, in units of $5000, on a new automobile is a random variable X having the density function given in Exercise 4.12 on page 117. Find the variance of X.

4.38 The proportion of people who respond to a certain mail-order solicitation is a random variable X having the density function given in Exercise 4.14 on page 117. Find the variance of X.

4.39 The total number of hours, in units of 100 hours, that a family runs a vacuum cleaner over a period of one year is a random variable X having the density function given in Exercise 4.13 on page 117. Find the variance of X.

4.40 Referring to Exercise 4.14 on page 117, find $\sigma^2_{g(X)}$ for the function $g(X) = 3X^2 + 4$.

4.41 Find the standard deviation of the random variable $g(X) = (2X+1)^2$ in Exercise 4.17 on page 118.

4.42 Using the results of Exercise 4.21 on page 118, find the variance of $g(X) = X^2$, where X is a random variable having the density function given in Exercise 4.12 on page 117.

4.43 The length of time, in minutes, for an airplane to obtain clearance for takeoff at a certain airport is a random variable $Y = 3X - 2$, where X has the density function

$$f(x) = \begin{cases} \frac{1}{4}e^{-x/4}, & x > 0 \\ 0, & \text{elsewhere.} \end{cases}$$

Find the mean and variance of the random variable Y.

4.44 Find the covariance of the random variables X and Y of Exercise 3.39 on page 105.

4.45 Find the covariance of the random variables X and Y of Exercise 3.49 on page 106.

4.46 Find the covariance of the random variables X and Y of Exercise 3.44 on page 105.

4.47 For the random variables X and Y whose joint density function is given in Exercise 3.40 on page 105, find the covariance.

4.48 Given a random variable X, with standard deviation σ_X, and a random variable $Y = a + bX$, show that if $b < 0$, the correlation coefficient $\rho_{XY} = -1$, and if $b > 0$, $\rho_{XY} = 1$.

4.49 Consider the situation in Exercise 4.32 on page 119. The distribution of the number of imperfections per 10 meters of synthetic failure is given by

x	0	1	2	3	4
$f(x)$	0.41	0.37	0.16	0.05	0.01

Find the variance and standard deviation of the number of imperfections.

4.50 For a laboratory assignment, if the equipment is working, the density function of the observed outcome X is

$$f(x) = \begin{cases} 2(1-x), & 0 < x < 1, \\ 0, & \text{otherwise.} \end{cases}$$

Find the variance and standard deviation of X.

4.51 For the random variables X and Y in Exercise 3.39 on page 105, determine the correlation coefficient between X and Y.

4.52 Random variables X and Y follow a joint distribution

$$f(x,y) = \begin{cases} 2, & 0 < x \leq y < 1, \\ 0, & \text{otherwise.} \end{cases}$$

Determine the correlation coefficient between X and Y.

4.3 Means and Variances of Linear Combinations of Random Variables

We now develop some useful properties that will simplify the calculations of means and variances of random variables that appear in later chapters. These properties will permit us to deal with expectations in terms of other parameters that are either known or easily computed. All the results that we present here are valid for both discrete and continuous random variables. Proofs are given only for the continuous case. We begin with a theorem and two corollaries that should be, intuitively, reasonable to the reader.

Theorem 4.5: If a and b are constants, then
$$E(aX + b) = aE(X) + b.$$

Proof: By the definition of expected value,
$$E(aX + b) = \int_{-\infty}^{\infty} (ax + b)f(x)\, dx = a\int_{-\infty}^{\infty} xf(x)\, dx + b\int_{-\infty}^{\infty} f(x)\, dx.$$

The first integral on the right is $E(X)$ and the second integral equals 1. Therefore, we have
$$E(aX + b) = aE(X) + b.$$

Corollary 4.1: Setting $a = 0$, we see that $E(b) = b$.

Corollary 4.2: Setting $b = 0$, we see that $E(aX) = aE(X)$.

Example 4.17: Applying Theorem 4.5 to the discrete random variable $f(X) = 2X - 1$, rework Example 4.4 on page 115.
Solution: According to Theorem 4.5, we can write
$$E(2X - 1) = 2E(X) - 1.$$
Now
$$\mu = E(X) = \sum_{x=4}^{9} xf(x)$$
$$= (4)\left(\frac{1}{12}\right) + (5)\left(\frac{1}{12}\right) + (6)\left(\frac{1}{4}\right) + (7)\left(\frac{1}{4}\right) + (8)\left(\frac{1}{6}\right) + (9)\left(\frac{1}{6}\right) = \frac{41}{6}.$$
Therefore,
$$\mu_{2X-1} = (2)\left(\frac{41}{6}\right) - 1 = \$12.67,$$
as before.

4.3 Means and Variances of Linear Combinations of Random Variables

Example 4.18: Applying Theorem 4.5 to the continuous random variable $g(X) = 4X + 3$, rework Example 4.5 on page 115.

Solution: For Example 4.5, we may use Theorem 4.5 to write
$$E(4X + 3) = 4E(X) + 3.$$

Now
$$E(X) = \int_{-1}^{2} x \left(\frac{x^2}{3}\right) dx = \int_{-1}^{2} \frac{x^3}{3} dx = \frac{5}{4}.$$

Therefore,
$$E(4X + 3) = (4)\left(\frac{5}{4}\right) + 3 = 8,$$

as before.

Theorem 4.6: The expected value of the sum or difference of two or more functions of a random variable X is the sum or difference of the expected values of the functions. That is,
$$E[g(X) \pm h(X)] = E[g(X)] \pm E[h(X)].$$

Proof: By definition,
$$E[g(X) \pm h(X)] = \int_{-\infty}^{\infty} [g(x) \pm h(x)] f(x)\, dx$$
$$= \int_{-\infty}^{\infty} g(x) f(x)\, dx \pm \int_{-\infty}^{\infty} h(x) f(x)\, dx$$
$$= E[g(X)] \pm E[h(X)].$$

Example 4.19: Let X be a random variable with probability distribution as follows:

x	0	1	2	3
$f(x)$	$\frac{1}{3}$	$\frac{1}{2}$	0	$\frac{1}{6}$

Find the expected value of $Y = (X - 1)^2$.

Solution: Applying Theorem 4.6 to the function $Y = (X - 1)^2$, we can write
$$E[(X-1)^2] = E(X^2 - 2X + 1) = E(X^2) - 2E(X) + E(1).$$

From Corollary 4.1, $E(1) = 1$, and by direct computation,
$$E(X) = (0)\left(\frac{1}{3}\right) + (1)\left(\frac{1}{2}\right) + (2)(0) + (3)\left(\frac{1}{6}\right) = 1 \text{ and}$$
$$E(X^2) = (0)\left(\frac{1}{3}\right) + (1)\left(\frac{1}{2}\right) + (4)(0) + (9)\left(\frac{1}{6}\right) = 2.$$

Hence,
$$E[(X-1)^2] = 2 - (2)(1) + 1 = 1.$$

Example 4.20: The weekly demand for a certain drink, in thousands of liters, at a chain of convenience stores is a continuous random variable $g(X) = X^2 + X - 2$, where X has the density function

$$f(x) = \begin{cases} 2(x-1), & 1 < x < 2, \\ 0, & \text{elsewhere.} \end{cases}$$

Find the expected value of the weekly demand for the drink.

Solution: By Theorem 4.6, we write

$$E(X^2 + X - 2) = E(X^2) + E(X) - E(2).$$

From Corollary 4.1, $E(2) = 2$, and by direct integration,

$$E(X) = \int_1^2 2x(x-1)\, dx = \frac{5}{3} \text{ and } E(X^2) = \int_1^2 2x^2(x-1)\, dx = \frac{17}{6}.$$

Now

$$E(X^2 + X - 2) = \frac{17}{6} + \frac{5}{3} - 2 = \frac{5}{2},$$

so the average weekly demand for the drink from this chain of efficiency stores is 2500 liters.

Suppose that we have two random variables X and Y with joint probability distribution $f(x, y)$. Two additional properties that will be very useful in succeeding chapters involve the expected values of the sum, difference, and product of these two random variables. First, however, let us prove a theorem on the expected value of the sum or difference of functions of the given variables. This, of course, is merely an extension of Theorem 4.6.

Theorem 4.7: The expected value of the sum or difference of two or more functions of the random variables X and Y is the sum or difference of the expected values of the functions. That is,

$$E[g(X,Y) \pm h(X,Y)] = E[g(X,Y)] \pm E[h(X,Y)].$$

Proof: By Definition 4.2,

$$E[g(X,Y) \pm h(X,Y)] = \int_{-\infty}^{\infty} \int_{-\infty}^{\infty} [g(x,y) \pm h(x,y)] f(x,y)\, dx\, dy$$

$$= \int_{-\infty}^{\infty} \int_{-\infty}^{\infty} g(x,y) f(x,y)\, dx\, dy \pm \int_{-\infty}^{\infty} \int_{-\infty}^{\infty} h(x,y) f(x,y)\, dx\, dy$$

$$= E[g(X,Y)] \pm E[h(X,Y)].$$

Corollary 4.3: Setting $g(X,Y) = g(X)$ and $h(X,Y) = h(Y)$, we see that

$$E[g(X) \pm h(Y)] = E[g(X)] \pm E[h(Y)].$$

4.3 Means and Variances of Linear Combinations of Random Variables

Corollary 4.4: Setting $g(X,Y) = X$ and $h(X,Y) = Y$, we see that
$$E[X \pm Y] = E[X] \pm E[Y].$$

If X represents the daily production of some item from machine A and Y the daily production of the same kind of item from machine B, then $X+Y$ represents the total number of items produced daily by both machines. Corollary 4.4 states that the average daily production for both machines is equal to the sum of the average daily production of each machine.

Theorem 4.8: Let X and Y be two independent random variables. Then
$$E(XY) = E(X)E(Y).$$

Proof: By Definition 4.2,
$$E(XY) = \int_{-\infty}^{\infty} \int_{-\infty}^{\infty} xy f(x,y) \, dx \, dy.$$

Since X and Y are independent, we may write
$$f(x,y) = g(x)h(y),$$

where $g(x)$ and $h(y)$ are the marginal distributions of X and Y, respectively. Hence,
$$E(XY) = \int_{-\infty}^{\infty} \int_{-\infty}^{\infty} xy g(x)h(y) \, dx \, dy = \int_{-\infty}^{\infty} xg(x) \, dx \int_{-\infty}^{\infty} yh(y) \, dy$$
$$= E(X)E(Y).$$

Theorem 4.8 can be illustrated for discrete variables by considering the experiment of tossing a green die and a red die. Let the random variable X represent the outcome on the green die and the random variable Y represent the outcome on the red die. Then XY represents the product of the numbers that occur on the pair of dice. In the long run, the average of the products of the numbers is equal to the product of the average number that occurs on the green die and the average number that occurs on the red die.

Corollary 4.5: Let X and Y be two independent random variables. Then $\sigma_{XY} = 0$.

Proof: The proof can be carried out by using Theorems 4.4 and 4.8.

Example 4.21: It is known that the ratio of gallium to arsenide does not affect the functioning of gallium-arsenide wafers, which are the main components of microchips. Let X denote the ratio of gallium to arsenide and Y denote the functional wafers retrieved during a 1-hour period. X and Y are independent random variables with the joint density function

$$f(x,y) = \begin{cases} \frac{x(1+3y^2)}{4}, & 0 < x < 2, \ 0 < y < 1, \\ 0, & \text{elsewhere.} \end{cases}$$

Show that $E(XY) = E(X)E(Y)$, as Theorem 4.8 suggests.

Solution: By definition,

$$E(XY) = \int_0^1 \int_0^2 \frac{x^2 y(1+3y^2)}{4}\, dx\, dy = \frac{5}{6}, \quad E(X) = \frac{4}{3}, \text{ and } E(Y) = \frac{5}{8}.$$

Hence,

$$E(X)E(Y) = \left(\frac{4}{3}\right)\left(\frac{5}{8}\right) = \frac{5}{6} = E(XY).$$

We conclude this section by proving one theorem and presenting several corollaries that are useful for calculating variances or standard deviations.

Theorem 4.9: If X and Y are random variables with joint probability distribution $f(x, y)$ and a, b, and c are constants, then

$$\sigma^2_{aX+bY+c} = a^2\sigma^2_X + b^2\sigma^2_Y + 2ab\sigma_{XY}.$$

Proof: By definition, $\sigma^2_{aX+bY+c} = E\{[(aX+bY+c) - \mu_{aX+bY+c}]^2\}$. Now

$$\mu_{aX+bY+c} = E(aX+bY+c) = aE(X) + bE(Y) + c = a\mu_X + b\mu_Y + c,$$

by using Corollary 4.4 followed by Corollary 4.2. Therefore,

$$\sigma^2_{aX+bY+c} = E\{[a(X-\mu_X) + b(Y-\mu_Y)]^2\}$$
$$= a^2 E[(X-\mu_X)^2] + b^2 E[(Y-\mu_Y)^2] + 2ab E[(X-\mu_X)(Y-\mu_Y)]$$
$$= a^2\sigma^2_X + b^2\sigma^2_Y + 2ab\sigma_{XY}.$$

Using Theorem 4.9, we have the following corollaries.

Corollary 4.6: Setting $b = 0$, we see that

$$\sigma^2_{aX+c} = a^2\sigma^2_X = a^2\sigma^2.$$

Corollary 4.7: Setting $a = 1$ and $b = 0$, we see that

$$\sigma^2_{X+c} = \sigma^2_X = \sigma^2.$$

Corollary 4.8: Setting $b = 0$ and $c = 0$, we see that

$$\sigma^2_{aX} = a^2\sigma^2_X = a^2\sigma^2.$$

Corollaries 4.6 and 4.7 state that the variance is unchanged if a constant is added to or subtracted from a random variable. The addition or subtraction of a constant simply shifts the values of X to the right or to the left but does not change their variability. However, if a random variable is multiplied or divided by a constant, then Corollaries 4.6 and 4.8 state that the variance is multiplied or divided by the square of the constant.

Corollary 4.9: If X and Y are independent random variables, then
$$\sigma^2_{aX+bY} = a^2\sigma^2_X + b^2\sigma^2_Y.$$

The result stated in Corollary 4.9 is obtained from Theorem 4.9 by invoking Corollary 4.5.

Corollary 4.10: If X and Y are independent random variables, then
$$\sigma^2_{aX-bY} = a^2\sigma^2_X + b^2\sigma^2_Y.$$

Corollary 4.10 follows when b in Corollary 4.9 is replaced by $-b$. Generalizing to a linear combination of n independent random variables, we have Corollary 4.11.

Corollary 4.11: If $X_1, X_2, \ldots, X_n$ are independent random variables, then
$$\sigma^2_{a_1X_1+a_2X_2+\cdots+a_nX_n} = a_1^2\sigma^2_{X_1} + a_2^2\sigma^2_{X_2} + \cdots + a_n^2\sigma^2_{X_n}.$$

Example 4.22: If X and Y are random variables with variances $\sigma^2_X = 2$ and $\sigma^2_Y = 4$ and covariance $\sigma_{XY} = -2$, find the variance of the random variable $Z = 3X - 4Y + 8$.

Solution:
$$\begin{aligned}
\sigma^2_Z = \sigma^2_{3X-4Y+8} &= \sigma^2_{3X-4Y} &&\text{(by Corollary 4.6)} \\
&= 9\sigma^2_X + 16\sigma^2_Y - 24\sigma_{XY} &&\text{(by Theorem 4.9)} \\
&= (9)(2) + (16)(4) - (24)(-2) = 130.
\end{aligned}$$

Example 4.23: Let X and Y denote the amounts of two different types of impurities in a batch of a certain chemical product. Suppose that X and Y are independent random variables with variances $\sigma^2_X = 2$ and $\sigma^2_Y = 3$. Find the variance of the random variable $Z = 3X - 2Y + 5$.

Solution:
$$\begin{aligned}
\sigma^2_Z = \sigma^2_{3X-2Y+5} &= \sigma^2_{3X-2Y} &&\text{(by Corollary 4.6)} \\
&= 9\sigma^2_x + 4\sigma^2_y &&\text{(by Corollary 4.10)} \\
&= (9)(2) + (4)(3) = 30.
\end{aligned}$$

What If the Function Is Nonlinear?

In that which has preceded this section, we have dealt with properties of linear functions of random variables for very important reasons. Chapters 8 through 15 will discuss and illustrate practical real-world problems in which the analyst is constructing a **linear model** to describe a data set and thus to describe or explain the behavior of a certain scientific phenomenon. Thus, it is natural that expected values and variances of linear combinations of random variables are encountered. However, there are situations in which properties of **nonlinear** functions of random variables become important. Certainly there are many scientific phenomena that are nonlinear, and certainly statistical modeling using nonlinear functions is very important. In fact, in Chapter 12, we deal with the modeling of what have become standard nonlinear models. Indeed, even a simple function of random variables, such as $Z = X/Y$, occurs quite frequently in practice, and yet unlike in the case of the expected value of linear combinations of random variables, there is no simple general rule. For example,

$$E(Z) = E(X/Y) \neq E(X)/E(Y),$$

except in very special circumstances.

The material provided by Theorems 4.5 through 4.9 and the various corollaries is extremely useful in that there are no restrictions on the form of the density or probability functions, apart from the property of independence when it is required as in the corollaries following Theorems 4.9. To illustrate, consider Example 4.23; the variance of $Z = 3X - 2Y + 5$ does not require restrictions on the distributions of the amounts X and Y of the two types of impurities. Only independence between X and Y is required. Now, we do have at our disposal the capacity to find $\mu_{g(X)}$ and $\sigma^2_{g(X)}$ for any function $g(\cdot)$ from first principles established in Theorems 4.1 and 4.3, where it is assumed that the corresponding distribution $f(x)$ is **known**. Exercises 4.40, 4.41, and 4.42, among others, illustrate the use of these theorems. Thus, if the function $g(x)$ is nonlinear and the density function (or probability function in the discrete case) is known, $\mu_{g(X)}$ and $\sigma^2_{g(X)}$ can be evaluated exactly. But, similar to the rules given for linear combinations, are there rules for nonlinear functions that can be used when the form of the distribution of the pertinent random variables is not known?

In general, suppose X is a random variable and $Y = g(x)$. The general solution for $E(Y)$ or Var(Y) can be difficult to find and depends on the complexity of the function $g(\cdot)$. However, there are approximations available that depend on a linear approximation of the function $g(x)$. For example, suppose we denote $E(X)$ as μ and Var(X) = σ^2_X. Then a Taylor series approximation of $g(x)$ around $X = \mu_X$ gives

$$g(x) = g(\mu_X) + \left.\frac{\partial g(x)}{\partial x}\right|_{x=\mu_X}(x - \mu_X) + \left.\frac{\partial^2 g(x)}{\partial x^2}\right|_{x=\mu_X}\frac{(x-\mu_X)^2}{2} + \cdots.$$

As a result, if we truncate after the linear term and take the expected value of both sides, we obtain $E[g(X)] \approx g(\mu_X)$, which is certainly intuitive and in some cases gives a reasonable approximation. However, if we include the second-order term of the Taylor series, then we have a second-order adjustment for this *first-order approximation* as follows:

Approximation of $E[g(X)]$

$$E[g(X)] \approx g(\mu_X) + \left.\frac{\partial^2 g(x)}{\partial x^2}\right|_{x=\mu_X}\frac{\sigma^2_X}{2}.$$

Example 4.24: Given the random variable X with mean μ_X and variance σ^2_X, give the second-order approximation to $E(e^X)$.

Solution: Since $\frac{\partial e^x}{\partial x} = e^x$ and $\frac{\partial^2 e^x}{\partial x^2} = e^x$, we obtain $E(e^X) \approx e^{\mu_X}(1 + \sigma^2_X/2)$.

Similarly, we can develop an approximation for Var$[g(x)]$ by taking the variance of both sides of the first-order Taylor series expansion of $g(x)$.

Approximation of Var$[g(X)]$

$$\text{Var}[g(X)] \approx \left[\frac{\partial g(x)}{\partial x}\right]^2_{x=\mu_X} \sigma^2_X.$$

Example 4.25: Given the random variable X as in Example 4.24, give an approximate formula for Var$[g(x)]$.

Solution: Again $\frac{\partial e^x}{\partial x} = e^x$; thus, $\text{Var}(X) \approx e^{2\mu_X} \sigma_X^2$.

These approximations can be extended to nonlinear functions of more than one random variable.

Given a set of independent random variables $X_1, X_2, \ldots, X_k$ with means $\mu_1, \mu_2, \ldots, \mu_k$ and variances $\sigma_1^2, \sigma_2^2, \ldots, \sigma_k^2$, respectively, let

$$Y = h(X_1, X_2, \ldots, X_k)$$

be a nonlinear function; then the following are approximations for $E(Y)$ and $\text{Var}(Y)$:

$$E(Y) \approx h(\mu_1, \mu_2, \ldots, \mu_k) + \sum_{i=1}^{k} \frac{\sigma_i^2}{2} \left[\frac{\partial^2 h(x_1, x_2, \ldots, x_k)}{\partial x_i^2} \right]\bigg|_{x_i = \mu_i, \; 1 \leq i \leq k},$$

$$\text{Var}(Y) \approx \sum_{i=1}^{k} \left[\frac{\partial h(x_1, x_2, \ldots, x_k)}{\partial x_i} \right]^2 \bigg|_{x_i = \mu_i, \; 1 \leq i \leq k} \sigma_i^2.$$

Example 4.26: Consider two independent random variables X and Z with means μ_X and μ_Z and variances σ_X^2 and σ_Z^2, respectively. Consider a random variable

$$Y = X/Z.$$

Give approximations for $E(Y)$ and $\text{Var}(Y)$.

Solution: For $E(Y)$, we must use $\frac{\partial y}{\partial x} = \frac{1}{z}$ and $\frac{\partial y}{\partial z} = -\frac{x}{z^2}$. Thus,

$$\frac{\partial^2 y}{\partial x^2} = 0 \text{ and } \frac{\partial^2 y}{\partial z^2} = \frac{2x}{z^3}.$$

As a result,

$$E(Y) \approx \frac{\mu_X}{\mu_Z} + \frac{\mu_X}{\mu_Z^3} \sigma_Z^2 = \frac{\mu_X}{\mu_Z}\left(1 + \frac{\sigma_Z^2}{\mu_Z^2}\right),$$

and the approximation for the variance of Y is given by

$$\text{Var}(Y) \approx \frac{1}{\mu_Z^2} \sigma_X^2 + \frac{\mu_X^2}{\mu_Z^4} \sigma_Z^2 = \frac{1}{\mu_Z^2}\left(\sigma_X^2 + \frac{\mu_X^2}{\mu_Z^2} \sigma_Z^2\right).$$

4.4 Chebyshev's Theorem

In Section 4.2 we stated that the variance of a random variable tells us something about the variability of the observations about the mean. If a random variable has a small variance or standard deviation, we would expect most of the values to be grouped around the mean. Therefore, the probability that the random variable assumes a value within a certain interval about the mean is greater than for a similar random variable with a larger standard deviation. If we think of probability in terms of area, we would expect a continuous distribution with a large value of σ to indicate a greater variability, and therefore we should expect the area to be more spread out, as in Figure 4.2(a). A distribution with a small standard deviation should have most of its area close to μ, as in Figure 4.2(b).

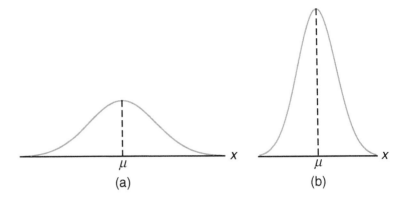

Figure 4.2: Variability of continuous observations about the mean.

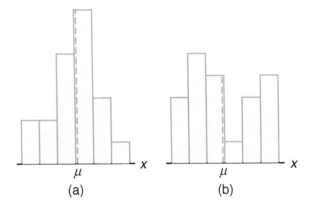

Figure 4.3: Variability of discrete observations about the mean.

We can argue the same way for a discrete distribution. The area in the probability histogram in Figure 4.3(b) is spread out much more than that in Figure 4.3(a) indicating a more variable distribution of measurements or outcomes.

The Russian mathematician P. L. Chebyshev (1821–1894) discovered that the fraction of the area between any two values symmetric about the mean is related to the standard deviation. Since the area under a probability distribution curve or in a probability histogram adds to 1, the area between any two numbers is the probability of the random variable assuming a value between these numbers.

The following theorem, due to Chebyshev, gives a conservative estimate of the probability that a random variable assumes a value within k standard deviations of its mean for any real number k.

Exercises

Theorem 4.10: (**Chebyshev's Theorem**) The probability that any random variable X will assume a value within k standard deviations of the mean is at least $1 - 1/k^2$. That is,

$$P(\mu - k\sigma < X < \mu + k\sigma) \geq 1 - \frac{1}{k^2}.$$

For $k = 2$, the theorem states that the random variable X has a probability of at least $1 - 1/2^2 = 3/4$ of falling within two standard deviations of the mean. That is, three-fourths or more of the observations of any distribution lie in the interval $\mu \pm 2\sigma$. Similarly, the theorem says that at least eight-ninths of the observations of any distribution fall in the interval $\mu \pm 3\sigma$.

Example 4.27: A random variable X has a mean $\mu = 8$, a variance $\sigma^2 = 9$, and an unknown probability distribution. Find

(a) $P(-4 < X < 20)$,

(b) $P(|X - 8| \geq 6)$.

Solution: (a) $P(-4 < X < 20) = P[8 - (4)(3) < X < 8 + (4)(3)] \geq \frac{15}{16}$.

(b) $P(|X - 8| \geq 6) = 1 - P(|X - 8| < 6) = 1 - P(-6 < X - 8 < 6)$

$$= 1 - P[8 - (2)(3) < X < 8 + (2)(3)] \leq \frac{1}{4}.$$

Chebyshev's theorem holds for any distribution of observations, and for this reason the results are usually weak. The value given by the theorem is a lower bound only. That is, we know that the probability of a random variable falling within two standard deviations of the mean can be *no less* than 3/4, but we never know how much more it might actually be. Only when the probability distribution is known can we determine exact probabilities. For this reason we call the theorem a *distribution-free* result. When specific distributions are assumed, as in future chapters, the results will be less conservative. The use of Chebyshev's theorem is relegated to situations where the form of the distribution is unknown.

Exercises

4.53 Referring to Exercise 4.35 on page 127, find the mean and variance of the discrete random variable $Z = 3X - 2$, when X represents the number of errors per 100 lines of code.

4.54 Using Theorem 4.5 and Corollary 4.6, find the mean and variance of the random variable $Z = 5X + 3$, where X has the probability distribution of Exercise 4.36 on page 127.

4.55 Suppose that a grocery store purchases 5 cartons of skim milk at the wholesale price of $1.20 per carton and retails the milk at $1.65 per carton. After the expiration date, the unsold milk is removed from the shelf and the grocer receives a credit from the distributor equal to three-fourths of the wholesale price. If the probability distribution of the random variable X, the number of cartons that are sold from this lot, is

x	0	1	2	3	4	5
$f(x)$	$\frac{1}{15}$	$\frac{2}{15}$	$\frac{2}{15}$	$\frac{3}{15}$	$\frac{4}{15}$	$\frac{3}{15}$

find the expected profit.

4.56 Repeat Exercise 4.43 on page 127 by applying Theorem 4.5 and Corollary 4.6.

4.57 Let X be a random variable with the following probability distribution:

x	-3	6	9
$f(x)$	$\frac{1}{6}$	$\frac{1}{2}$	$\frac{1}{3}$

Find $E(X)$ and $E(X^2)$ and then, using these values, evaluate $E[(2X+1)^2]$.

4.58 The total time, measured in units of 100 hours, that a teenager runs her hair dryer over a period of one year is a continuous random variable X that has the density function

$$f(x) = \begin{cases} x, & 0 < x < 1, \\ 2-x, & 1 \le x < 2, \\ 0, & \text{elsewhere.} \end{cases}$$

Use Theorem 4.6 to evaluate the mean of the random variable $Y = 60X^2 + 39X$, where Y is equal to the number of kilowatt hours expended annually.

4.59 If a random variable X is defined such that

$$E[(X-1)^2] = 10 \text{ and } E[(X-2)^2] = 6,$$

find μ and σ^2.

4.60 Suppose that X and Y are independent random variables having the joint probability distribution

$f(x,y)$		x	
		2	4
	1	0.10	0.15
y	3	0.20	0.30
	5	0.10	0.15

Find
(a) $E(2X - 3Y)$;
(b) $E(XY)$.

4.61 Use Theorem 4.7 to evaluate $E(2XY^2 - X^2Y)$ for the joint probability distribution shown in Table 3.1 on page 96.

4.62 If X and Y are independent random variables with variances $\sigma_X^2 = 5$ and $\sigma_Y^2 = 3$, find the variance of the random variable $Z = -2X + 4Y - 3$.

4.63 Repeat Exercise 4.62 if X and Y are not independent and $\sigma_{XY} = 1$.

4.64 Suppose that X and Y are independent random variables with probability densities and

$$g(x) = \begin{cases} \frac{8}{x^3}, & x > 2, \\ 0, & \text{elsewhere,} \end{cases}$$

and

$$h(y) = \begin{cases} 2y, & 0 < y < 1, \\ 0, & \text{elsewhere.} \end{cases}$$

Find the expected value of $Z = XY$.

4.65 Let X represent the number that occurs when a red die is tossed and Y the number that occurs when a green die is tossed. Find
(a) $E(X + Y)$;
(b) $E(X - Y)$;
(c) $E(XY)$.

4.66 Let X represent the number that occurs when a green die is tossed and Y the number that occurs when a red die is tossed. Find the variance of the random variable
(a) $2X - Y$;
(b) $X + 3Y - 5$.

4.67 If the joint density function of X and Y is given by

$$f(x,y) = \begin{cases} \frac{2}{7}(x+2y), & 0 < x < 1, \ 1 < y < 2, \\ 0, & \text{elsewhere,} \end{cases}$$

find the expected value of $g(X, Y) = \frac{X}{Y^3} + X^2Y$.

4.68 The power P in watts which is dissipated in an electric circuit with resistance R is known to be given by $P = I^2R$, where I is current in amperes and R is a constant fixed at 50 ohms. However, I is a random variable with $\mu_I = 15$ amperes and $\sigma_I^2 = 0.03$ amperes^2. Give numerical approximations to the mean and variance of the power P.

4.69 Consider Review Exercise 3.77 on page 108. The random variables X and Y represent the number of vehicles that arrive at two separate street corners during a certain 2-minute period in the day. The joint distribution is

$$f(x,y) = \left(\frac{1}{4^{(x+y)}}\right)\left(\frac{9}{16}\right),$$

for $x = 0, 1, 2, \ldots$ and $y = 0, 1, 2, \ldots$.
(a) Give $E(X)$, $E(Y)$, Var(X), and Var(Y).
(b) Consider $Z = X + Y$, the sum of the two. Find $E(Z)$ and Var(Z).

4.70 Consider Review Exercise 3.64 on page 107. There are two service lines. The random variables X and Y are the proportions of time that line 1 and line 2 are in use, respectively. The joint probability density function for (X, Y) is given by

$$f(x,y) = \begin{cases} \frac{3}{2}(x^2 + y^2), & 0 \le x, \ y \le 1, \\ 0, & \text{elsewhere.} \end{cases}$$

(a) Determine whether or not X and Y are independent.

(b) It is of interest to know something about the proportion of $Z = X + Y$, the sum of the two proportions. Find $E(X + Y)$. Also find $E(XY)$.

(c) Find $\text{Var}(X)$, $\text{Var}(Y)$, and $\text{Cov}(X, Y)$.

(d) Find $\text{Var}(X + Y)$.

4.71 The length of time Y, in minutes, required to generate a human reflex to tear gas has the density function

$$f(y) = \begin{cases} \frac{1}{4} e^{-y/4}, & 0 \leq y < \infty, \\ 0, & \text{elsewhere.} \end{cases}$$

(a) What is the mean time to reflex?

(b) Find $E(Y^2)$ and $\text{Var}(Y)$.

4.72 A manufacturing company has developed a machine for cleaning carpet that is fuel-efficient because it delivers carpet cleaner so rapidly. Of interest is a random variable Y, the amount in gallons per minute delivered. It is known that the density function is given by

$$f(y) = \begin{cases} 1, & 7 \leq y \leq 8, \\ 0, & \text{elsewhere.} \end{cases}$$

(a) Sketch the density function.

(b) Give $E(Y)$, $E(Y^2)$, and $\text{Var}(Y)$.

4.73 For the situation in Exercise 4.72, compute $E(e^Y)$ using Theorem 4.1, that is, by using

$$E(e^Y) = \int_7^8 e^y f(y) \, dy.$$

Then compute $E(e^Y)$ not by using $f(y)$, but rather by using the second-order adjustment to the first-order approximation of $E(e^Y)$. Comment.

4.74 Consider again the situation of Exercise 4.72. It is required to find $\text{Var}(e^Y)$. Use Theorems 4.2 and 4.3 and define $Z = e^Y$. Thus, use the conditions of Exercise 4.73 to find

$$\text{Var}(Z) = E(Z^2) - [E(Z)]^2.$$

Then do it not by using $f(y)$, but rather by using the first-order Taylor series approximation to $\text{Var}(e^Y)$. Comment!

4.75 An electrical firm manufactures a 100-watt light bulb, which, according to specifications written on the package, has a mean life of 900 hours with a standard deviation of 50 hours. At most, what percentage of the bulbs fail to last even 700 hours? Assume that the distribution is symmetric about the mean.

4.76 Seventy new jobs are opening up at an automobile manufacturing plant, and 1000 applicants show up for the 70 positions. To select the best 70 from among the applicants, the company gives a test that covers mechanical skill, manual dexterity, and mathematical ability. The mean grade on this test turns out to be 60, and the scores have a standard deviation of 6. Can a person who scores 84 count on getting one of the jobs? [*Hint*: Use Chebyshev's theorem.] Assume that the distribution is symmetric about the mean.

4.77 A random variable X has a mean $\mu = 10$ and a variance $\sigma^2 = 4$. Using Chebyshev's theorem, find

(a) $P(|X - 10| \geq 3)$;

(b) $P(|X - 10| < 3)$;

(c) $P(5 < X < 15)$;

(d) the value of the constant c such that

$$P(|X - 10| \geq c) \leq 0.04.$$

4.78 Compute $P(\mu - 2\sigma < X < \mu + 2\sigma)$, where X has the density function

$$f(x) = \begin{cases} 6x(1 - x), & 0 < x < 1, \\ 0, & \text{elsewhere,} \end{cases}$$

and compare with the result given in Chebyshev's theorem.

Review Exercises

4.79 Prove Chebyshev's theorem.

4.80 Find the covariance of random variables X and Y having the joint probability density function

$$f(x, y) = \begin{cases} x + y, & 0 < x < 1, \ 0 < y < 1, \\ 0, & \text{elsewhere.} \end{cases}$$

4.81 Referring to the random variables whose joint probability density function is given in Exercise 3.47 on page 105, find the average amount of kerosene left in the tank at the end of the day.

4.82 Assume the length X, in minutes, of a particular type of telephone conversation is a random variable with probability density function

$$f(x) = \begin{cases} \frac{1}{5}e^{-x/5}, & x > 0, \\ 0, & \text{elsewhere.} \end{cases}$$

(a) Determine the mean length $E(X)$ of this type of telephone conversation.
(b) Find the variance and standard deviation of X.
(c) Find $E[(X+5)^2]$.

4.83 Referring to the random variables whose joint density function is given in Exercise 3.41 on page 105, find the covariance between the weight of the creams and the weight of the toffees in these boxes of chocolates.

4.84 Referring to the random variables whose joint probability density function is given in Exercise 3.41 on page 105, find the expected weight for the sum of the creams and toffees if one purchased a box of these chocolates.

4.85 Suppose it is known that the life X of a particular compressor, in hours, has the density function

$$f(x) = \begin{cases} \frac{1}{900}e^{-x/900}, & x > 0, \\ 0, & \text{elsewhere.} \end{cases}$$

(a) Find the mean life of the compressor.
(b) Find $E(X^2)$.
(c) Find the variance and standard deviation of the random variable X.

4.86 Referring to the random variables whose joint density function is given in Exercise 3.40 on page 105,
(a) find μ_X and μ_Y;
(b) find $E[(X+Y)/2]$.

4.87 Show that $\text{Cov}(aX, bY) = ab\,\text{Cov}(X,Y)$.

4.88 Consider the density function of Review Exercise 4.85. Demonstrate that Chebyshev's theorem holds for $k=2$ and $k=3$.

4.89 Consider the joint density function

$$f(x,y) = \begin{cases} \frac{16y}{x^3}, & x > 2, \ 0 < y < 1, \\ 0, & \text{elsewhere.} \end{cases}$$

Compute the correlation coefficient ρ_{XY}.

4.90 Consider random variables X and Y of Exercise 4.63 on page 138. Compute ρ_{XY}.

4.91 A dealer's profit, in units of $5000, on a new automobile is a random variable X having density function

$$f(x) = \begin{cases} 2(1-x), & 0 \le x \le 1, \\ 0, & \text{elsewhere.} \end{cases}$$

(a) Find the variance of the dealer's profit.
(b) Demonstrate that Chebyshev's theorem holds for $k=2$ with the density function above.
(c) What is the probability that the profit exceeds $500?

4.92 Consider Exercise 4.10 on page 117. Can it be said that the ratings given by the two experts are independent? Explain why or why not.

4.93 A company's marketing and accounting departments have determined that if the company markets its newly developed product, the contribution of the product to the firm's profit during the next 6 months will be described by the following:

Profit Contribution	Probability
−$5,000	0.2
$10,000	0.5
$30,000	0.3

What is the company's expected profit?

4.94 In a support system in the U.S. space program, a single crucial component works only 85% of the time. In order to enhance the reliability of the system, it is decided that 3 components will be installed in parallel such that the system fails only if they all fail. Assume the components act independently and that they are equivalent in the sense that all 3 of them have an 85% success rate. Consider the random variable X as the number of components out of 3 that fail.

(a) Write out a probability function for the random variable X.
(b) What is $E(X)$ (i.e., the mean number of components out of 3 that fail)?
(c) What is $\text{Var}(X)$?
(d) What is the probability that the entire system is successful?
(e) What is the probability that the system fails?
(f) If the desire is to have the system be successful with probability 0.99, are three components sufficient? If not, how many are required?

4.95 In business, it is important to plan and carry out research in order to anticipate what will occur at the end of the year. Research suggests that the profit (loss) spectrum for a certain company, with corresponding probabilities, is as follows:

Review Exercises

Profit	Probability
−$15,000	0.05
$0	0.15
$15,000	0.15
$25,000	0.30
$40,000	0.15
$50,000	0.10
$100,000	0.05
$150,000	0.03
$200,000	0.02

(a) What is the expected profit?

(b) Give the standard deviation of the profit.

4.96 It is known through data collection and considerable research that the amount of time in seconds that a certain employee of a company is late for work is a random variable X with density function

$$f(x) = \begin{cases} \frac{3}{(4)(50^3)}(50^2 - x^2), & -50 \le x \le 50, \\ 0, & \text{elsewhere.} \end{cases}$$

In other words, he not only is slightly late at times, but also can be early to work.

(a) Find the expected value of the time in seconds that he is late.

(b) Find $E(X^2)$.

(c) What is the standard deviation of the amount of time he is late?

4.97 A delivery truck travels from point A to point B and back using the same route each day. There are four traffic lights on the route. Let X_1 denote the number of red lights the truck encounters going from A to B and X_2 denote the number encountered on the return trip. Data collected over a long period suggest that the joint probability distribution for (X_1, X_2) is given by

			x_2		
x_1	0	1	2	3	4
0	0.01	0.01	0.03	0.07	0.01
1	0.03	0.05	0.08	0.03	0.02
2	0.03	0.11	0.15	0.01	0.01
3	0.02	0.07	0.10	0.03	0.01
4	0.01	0.06	0.03	0.01	0.01

(a) Give the marginal density of X_1.

(b) Give the marginal density of X_2.

(c) Give the conditional density distribution of X_1 given $X_2 = 3$.

(d) Give $E(X_1)$.

(e) Give $E(X_2)$.

(f) Give $E(X_1 \mid X_2 = 3)$.

(g) Give the standard deviation of X_1.

4.98 A convenience store has two separate locations where customers can be checked out as they leave. These locations each have two cash registers and two employees who check out customers. Let X be the number of cash registers being used at a particular time for location 1 and Y the number being used at the same time for location 2. The joint probability function is given by

		y	
x	0	1	2
0	0.12	0.04	0.04
1	0.08	0.19	0.05
2	0.06	0.12	0.30

(a) Give the marginal density of both X and Y as well as the probability distribution of X given $Y = 2$.

(b) Give $E(X)$ and $\text{Var}(X)$.

(c) Give $E(X \mid Y = 2)$ and $\text{Var}(X \mid Y = 2)$.

4.99 Consider a ferry that can carry both buses and cars across a waterway. Each trip costs the owner approximately $10. The fee for cars is $3 and the fee for buses is $8. Let X and Y denote the number of buses and cars, respectively, carried on a given trip. The joint distribution of X and Y is given by

		x	
y	0	1	2
0	0.01	0.01	0.03
1	0.03	0.08	0.07
2	0.03	0.06	0.06
3	0.07	0.07	0.13
4	0.12	0.04	0.03
5	0.08	0.06	0.02

Compute the expected profit for the ferry trip.

4.100 As we shall illustrate in Chapter 12, statistical methods associated with linear and nonlinear models are very important. In fact, exponential functions are often used in a wide variety of scientific and engineering problems. Consider a model that is fit to a set of data involving measured values k_1 and k_2 and a certain response Y to the measurements. The model postulated is

$$\hat{Y} = e^{b_0 + b_1 k_1 + b_2 k_2},$$

where $\hat{Y}$ denotes the **estimated value of** Y, k_1 and k_2 are fixed values, and b_0, b_1, and b_2 are **estimates** of constants and hence are random variables. Assume that these random variables are independent and use the approximate formula for the variance of a nonlinear function of more than one variable. Give an expression for $\text{Var}(\hat{Y})$. Assume that the means of b_0, b_1, and b_2 are known and are β_0, β_1, and β_2, and assume that the variances of b_0, b_1, and b_2 are known and are σ_0^2, σ_1^2, and σ_2^2.

4.101 Consider Review Exercise 3.73 on page 108. It involved Y, the proportion of impurities in a batch, and the density function is given by

$$f(y) = \begin{cases} 10(1-y)^9, & 0 \leq y \leq 1, \\ 0, & \text{elsewhere.} \end{cases}$$

(a) Find the expected percentage of impurities.
(b) Find the expected value of the proportion of quality material (i.e., find $E(1-Y)$).
(c) Find the variance of the random variable $Z = 1-Y$.

4.102 Project: Let $X =$ number of hours each student in the class slept the night before. Create a discrete variable by using the following arbitrary intervals: $X < 3$, $3 \leq X < 6$, $6 \leq X < 9$, and $X \geq 9$.

(a) Estimate the probability distribution for X.
(b) Calculate the estimated mean and variance for X.

4.5 Potential Misconceptions and Hazards; Relationship to Material in Other Chapters

The material in this chapter is extremely fundamental in nature, much like that in Chapter 3. Whereas in Chapter 3 we focused on general characteristics of a probability distribution, in this chapter we defined important quantities or *parameters* that characterize the general nature of the system. The **mean** of a distribution reflects *central tendency*, and the **variance** or **standard deviation** reflects *variability* in the system. In addition, covariance reflects the tendency for two random variables to "move together" in a system. These important parameters will remain fundamental to all that follows in this text.

The reader should understand that the distribution type is often dictated by the scientific scenario. However, the parameter values need to be estimated from scientific data. For example, in the case of Review Exercise 4.85, the manufacturer of the compressor may know (material that will be presented in Chapter 6) from experience and knowledge of the type of compressor that the nature of the distribution is as indicated in the exercise. But the mean $\mu = 900$ would be **estimated** from experimentation on the machine. Though the parameter value of 900 is given as known here, it will not be known in real-life situations without the use of experimental data. Chapter 9 is dedicated to **estimation**.

Chapter 5

Some Discrete Probability Distributions

5.1 Introduction and Motivation

No matter whether a discrete probability distribution is represented graphically by a histogram, in tabular form, or by means of a formula, the behavior of a random variable is described. Often, the observations generated by different statistical experiments have the same general type of behavior. Consequently, discrete random variables associated with these experiments can be described by essentially the same probability distribution and therefore can be represented by a single formula. In fact, one needs only a handful of important probability distributions to describe many of the discrete random variables encountered in practice.

Such a handful of distributions describe several real-life random phenomena. For instance, in a study involving testing the effectiveness of a new drug, the number of cured patients among all the patients who use the drug approximately follows a binomial distribution (Section 5.2). In an industrial example, when a sample of items selected from a batch of production is tested, the number of defective items in the sample usually can be modeled as a hypergeometric random variable (Section 5.3). In a statistical quality control problem, the experimenter will signal a shift of the process mean when observational data exceed certain limits. The number of samples required to produce a false alarm follows a geometric distribution which is a special case of the negative binomial distribution (Section 5.4). On the other hand, the number of white cells from a fixed amount of an individual's blood sample is usually random and may be described by a Poisson distribution (Section 5.5). In this chapter, we present these commonly used distributions with various examples.

5.2 Binomial and Multinomial Distributions

An experiment often consists of repeated trials, each with two possible outcomes that may be labeled **success** or **failure**. The most obvious application deals with the testing of items as they come off an assembly line, where each trial may indicate a defective or a nondefective item. We may choose to define either outcome as a success. The process is referred to as a **Bernoulli process**. Each trial is called a **Bernoulli trial**. Observe, for example, if one were drawing cards from a deck, the probabilities for repeated trials change if the cards are not replaced. That is, the probability of selecting a heart on the first draw is 1/4, but on the second draw it is a conditional probability having a value of 13/51 or 12/51, depending on whether a heart appeared on the first draw: this, then, would no longer be considered a set of Bernoulli trials.

The Bernoulli Process

Strictly speaking, the Bernoulli process must possess the following properties:

1. The experiment consists of repeated trials.
2. Each trial results in an outcome that may be classified as a success or a failure.
3. The probability of success, denoted by p, remains constant from trial to trial.
4. The repeated trials are independent.

Consider the set of Bernoulli trials where three items are selected at random from a manufacturing process, inspected, and classified as defective or nondefective. A defective item is designated a success. The number of successes is a random variable X assuming integral values from 0 through 3. The eight possible outcomes and the corresponding values of X are

Outcome	NNN	NDN	NND	DNN	NDD	DND	DDN	DDD
x	0	1	1	1	2	2	2	3

Since the items are selected independently and we assume that the process produces 25% defectives, we have

$$P(NDN) = P(N)P(D)P(N) = \left(\frac{3}{4}\right)\left(\frac{1}{4}\right)\left(\frac{3}{4}\right) = \frac{9}{64}.$$

Similar calculations yield the probabilities for the other possible outcomes. The probability distribution of X is therefore

x	0	1	2	3
$f(x)$	$\frac{27}{64}$	$\frac{27}{64}$	$\frac{9}{64}$	$\frac{1}{64}$

Binomial Distribution

The number X of successes in n Bernoulli trials is called a **binomial random variable**. The probability distribution of this discrete random variable is called the **binomial distribution**, and its values will be denoted by $b(x; n, p)$ since they depend on the number of trials and the probability of a success on a given trial. Thus, for the probability distribution of X, the number of defectives is

$$P(X = 2) = f(2) = b\left(2; 3, \frac{1}{4}\right) = \frac{9}{64}.$$

5.2 Binomial and Multinomial Distributions

Let us now generalize the above illustration to yield a formula for $b(x; n, p)$. That is, we wish to find a formula that gives the probability of x successes in n trials for a binomial experiment. First, consider the probability of x successes and $n - x$ failures in a specified order. Since the trials are independent, we can multiply all the probabilities corresponding to the different outcomes. Each success occurs with probability p and each failure with probability $q = 1 - p$. Therefore, the probability for the specified order is $p^x q^{n-x}$. We must now determine the total number of sample points in the experiment that have x successes and $n-x$ failures. This number is equal to the number of partitions of n outcomes into two groups with x in one group and $n-x$ in the other and is written $\binom{n}{x}$ as introduced in Section 2.3. Because these partitions are mutually exclusive, we add the probabilities of all the different partitions to obtain the general formula, or simply multiply $p^x q^{n-x}$ by $\binom{n}{x}$.

Binomial Distribution A Bernoulli trial can result in a success with probability p and a failure with probability $q = 1 - p$. Then the probability distribution of the binomial random variable X, the number of successes in n independent trials, is

$$b(x; n, p) = \binom{n}{x} p^x q^{n-x}, \quad x = 0, 1, 2, \ldots, n.$$

Note that when $n = 3$ and $p = 1/4$, the probability distribution of X, the number of defectives, may be written as

$$b\left(x; 3, \frac{1}{4}\right) = \binom{3}{x} \left(\frac{1}{4}\right)^x \left(\frac{3}{4}\right)^{3-x}, \quad x = 0, 1, 2, 3,$$

rather than in the tabular form on page 144.

Example 5.1: The probability that a certain kind of component will survive a shock test is $3/4$. Find the probability that exactly 2 of the next 4 components tested survive.

Solution: Assuming that the tests are independent and $p = 3/4$ for each of the 4 tests, we obtain

$$b\left(2; 4, \frac{3}{4}\right) = \binom{4}{2} \left(\frac{3}{4}\right)^2 \left(\frac{1}{4}\right)^2 = \left(\frac{4!}{2!\,2!}\right)\left(\frac{3^2}{4^4}\right) = \frac{27}{128}.$$

Where Does the Name *Binomial* Come From?

The binomial distribution derives its name from the fact that the $n+1$ terms in the binomial expansion of $(q+p)^n$ correspond to the various values of $b(x; n, p)$ for $x = 0, 1, 2, \ldots, n$. That is,

$$(q+p)^n = \binom{n}{0} q^n + \binom{n}{1} p q^{n-1} + \binom{n}{2} p^2 q^{n-2} + \cdots + \binom{n}{n} p^n$$

$$= b(0; n, p) + b(1; n, p) + b(2; n, p) + \cdots + b(n; n, p).$$

Since $p + q = 1$, we see that

$$\sum_{x=0}^{n} b(x; n, p) = 1,$$

a condition that must hold for any probability distribution.

Frequently, we are interested in problems where it is necessary to find $P(X < r)$ or $P(a \leq X \leq b)$. Binomial sums

$$B(r; n, p) = \sum_{x=0}^{r} b(x; n, p)$$

are given in Table A.1 of the Appendix for $n = 1, 2, \ldots, 20$ for selected values of p from 0.1 to 0.9. We illustrate the use of Table A.1 with the following example.

Example 5.2: The probability that a patient recovers from a rare blood disease is 0.4. If 15 people are known to have contracted this disease, what is the probability that (a) at least 10 survive, (b) from 3 to 8 survive, and (c) exactly 5 survive?

Solution: Let X be the number of people who survive.

(a) $P(X \geq 10) = 1 - P(X < 10) = 1 - \sum_{x=0}^{9} b(x; 15, 0.4) = 1 - 0.9662$

$= 0.0338$

(b) $P(3 \leq X \leq 8) = \sum_{x=3}^{8} b(x; 15, 0.4) = \sum_{x=0}^{8} b(x; 15, 0.4) - \sum_{x=0}^{2} b(x; 15, 0.4)$

$= 0.9050 - 0.0271 = 0.8779$

(c) $P(X = 5) = b(5; 15, 0.4) = \sum_{x=0}^{5} b(x; 15, 0.4) - \sum_{x=0}^{4} b(x; 15, 0.4)$

$= 0.4032 - 0.2173 = 0.1859$

Example 5.3: A large chain retailer purchases a certain kind of electronic device from a manufacturer. The manufacturer indicates that the defective rate of the device is 3%.

(a) The inspector randomly picks 20 items from a shipment. What is the probability that there will be at least one defective item among these 20?

(b) Suppose that the retailer receives 10 shipments in a month and the inspector randomly tests 20 devices per shipment. What is the probability that there will be exactly 3 shipments each containing at least one defective device among the 20 that are selected and tested from the shipment?

Solution: (a) Denote by X the number of defective devices among the 20. Then X follows a $b(x; 20, 0.03)$ distribution. Hence,

$$P(X \geq 1) = 1 - P(X = 0) = 1 - b(0; 20, 0.03)$$
$$= 1 - (0.03)^0 (1 - 0.03)^{20-0} = 0.4562.$$

(b) In this case, each shipment can either contain at least one defective item or not. Hence, testing of each shipment can be viewed as a Bernoulli trial with $p = 0.4562$ from part (a). Assuming independence from shipment to shipment and denoting by Y the number of shipments containing at least one defective item, Y follows another binomial distribution $b(y; 10, 0.4562)$. Therefore,

$$P(Y = 3) = \binom{10}{3} 0.4562^3 (1 - 0.4562)^7 = 0.1602.$$

Areas of Application

From Examples 5.1 through 5.3, it should be clear that the binomial distribution finds applications in many scientific fields. An industrial engineer is keenly interested in the "proportion defective" in an industrial process. Often, quality control measures and sampling schemes for processes are based on the binomial distribution. This distribution applies to any industrial situation where an outcome of a process is dichotomous and the results of the process are independent, with the probability of success being constant from trial to trial. The binomial distribution is also used extensively for medical and military applications. In both fields, a success or failure result is important. For example, "cure" or "no cure" is important in pharmaceutical work, and "hit" or "miss" is often the interpretation of the result of firing a guided missile.

Since the probability distribution of any binomial random variable depends only on the values assumed by the parameters n, p, and q, it would seem reasonable to assume that the mean and variance of a binomial random variable also depend on the values assumed by these parameters. Indeed, this is true, and in the proof of Theorem 5.1 we derive general formulas that can be used to compute the mean and variance of any binomial random variable as functions of n, p, and q.

Theorem 5.1: The mean and variance of the binomial distribution $b(x; n, p)$ are
$$\mu = np \text{ and } \sigma^2 = npq.$$

Proof: Let the outcome on the jth trial be represented by a Bernoulli random variable I_j, which assumes the values 0 and 1 with probabilities q and p, respectively. Therefore, in a binomial experiment the number of successes can be written as the sum of the n independent indicator variables. Hence,

$$X = I_1 + I_2 + \cdots + I_n.$$

The mean of any I_j is $E(I_j) = (0)(q) + (1)(p) = p$. Therefore, using Corollary 4.4 on page 131, the mean of the binomial distribution is

$$\mu = E(X) = E(I_1) + E(I_2) + \cdots + E(I_n) = \underbrace{p + p + \cdots + p}_{n \text{ terms}} = np.$$

The variance of any I_j is $\sigma^2_{I_j} = E(I_j^2) - p^2 = (0)^2(q) + (1)^2(p) - p^2 = p(1-p) = pq$. Extending Corollary 4.11 to the case of n independent Bernoulli variables gives the variance of the binomial distribution as

$$\sigma^2_X = \sigma^2_{I_1} + \sigma^2_{I_2} + \cdots + \sigma^2_{I_n} = \underbrace{pq + pq + \cdots + pq}_{n \text{ terms}} = npq.$$

Example 5.4: It is conjectured that an impurity exists in 30% of all drinking wells in a certain rural community. In order to gain some insight into the true extent of the problem, it is determined that some testing is necessary. It is too expensive to test all of the wells in the area, so 10 are randomly selected for testing.

(a) Using the binomial distribution, what is the probability that exactly 3 wells have the impurity, assuming that the conjecture is correct?

(b) What is the probability that more than 3 wells are impure?

Solution: (a) We require
$$b(3; 10, 0.3) = \sum_{x=0}^{3} b(x; 10, 0.3) - \sum_{x=0}^{2} b(x; 10, 0.3) = 0.6496 - 0.3828 = 0.2668.$$

(b) In this case, $P(X > 3) = 1 - 0.6496 = 0.3504$.

Example 5.5: Find the mean and variance of the binomial random variable of Example 5.2, and then use Chebyshev's theorem (on page 137) to interpret the interval $\mu \pm 2\sigma$.

Solution: Since Example 5.2 was a binomial experiment with $n = 15$ and $p = 0.4$, by Theorem 5.1, we have
$$\mu = (15)(0.4) = 6 \text{ and } \sigma^2 = (15)(0.4)(0.6) = 3.6.$$

Taking the square root of 3.6, we find that $\sigma = 1.897$. Hence, the required interval is $6 \pm (2)(1.897)$, or from 2.206 to 9.794. Chebyshev's theorem states that the number of recoveries among 15 patients who contracted the disease has a probability of at least 3/4 of falling between 2.206 and 9.794 or, because the data are discrete, between 2 and 10 inclusive.

There are solutions in which the computation of binomial probabilities may allow us to draw a scientific inference about population after data are collected. An illustration is given in the next example.

Example 5.6: Consider the situation of Example 5.4. The notion that 30% of the wells are impure is merely a conjecture put forth by the area water board. Suppose 10 wells are randomly selected and 6 are found to contain the impurity. What does this imply about the conjecture? Use a probability statement.

Solution: We must first ask: "If the conjecture is correct, is it likely that we would find 6 or more impure wells?"
$$P(X \geq 6) = \sum_{x=0}^{10} b(x; 10, 0.3) - \sum_{x=0}^{5} b(x; 10, 0.3) = 1 - 0.9527 = 0.0473.$$

As a result, it is very unlikely (4.7% chance) that 6 or more wells would be found impure if only 30% of all are impure. This casts considerable doubt on the conjecture and suggests that the impurity problem is much more severe.

As the reader should realize by now, in many applications there are more than two possible outcomes. To borrow an example from the field of genetics, the color of guinea pigs produced as offspring may be red, black, or white. Often the "defective" or "not defective" dichotomy is truly an oversimplification in engineering situations. Indeed, there are often more than two categories that characterize items or parts coming off an assembly line.

Multinomial Experiments and the Multinomial Distribution

The binomial experiment becomes a **multinomial experiment** if we let each trial have more than two possible outcomes. The classification of a manufactured product as being light, heavy, or acceptable and the recording of accidents at a certain intersection according to the day of the week constitute multinomial experiments. The drawing of a card from a deck *with replacement* is also a multinomial experiment if the 4 suits are the outcomes of interest.

In general, if a given trial can result in any one of k possible outcomes $E_1, E_2, \ldots, E_k$ with probabilities $p_1, p_2, \ldots, p_k$, then the **multinomial distribution** will give the probability that E_1 occurs x_1 times, E_2 occurs x_2 times, $\ldots$, and E_k occurs x_k times in n independent trials, where

$$x_1 + x_2 + \cdots + x_k = n.$$

We shall denote this joint probability distribution by

$$f(x_1, x_2, \ldots, x_k; p_1, p_2, \ldots, p_k, n).$$

Clearly, $p_1 + p_2 + \cdots + p_k = 1$, since the result of each trial must be one of the k possible outcomes.

To derive the general formula, we proceed as in the binomial case. Since the trials are independent, any specified order yielding x_1 outcomes for E_1, x_2 for $E_2, \ldots, x_k$ for E_k will occur with probability $p_1^{x_1} p_2^{x_2} \cdots p_k^{x_k}$. The total number of orders yielding similar outcomes for the n trials is equal to the number of partitions of n items into k groups with x_1 in the first group, x_2 in the second group, $\ldots$, and x_k in the kth group. This can be done in

$$\binom{n}{x_1, x_2, \ldots, x_k} = \frac{n!}{x_1! \, x_2! \cdots x_k!}$$

ways. Since all the partitions are mutually exclusive and occur with equal probability, we obtain the multinomial distribution by multiplying the probability for a specified order by the total number of partitions.

Multinomial Distribution If a given trial can result in the k outcomes $E_1, E_2, \ldots, E_k$ with probabilities $p_1, p_2, \ldots, p_k$, then the probability distribution of the random variables $X_1, X_2, \ldots, X_k$, representing the number of occurrences for $E_1, E_2, \ldots, E_k$ in n independent trials, is

$$f(x_1, x_2, \ldots, x_k; p_1, p_2, \ldots, p_k, n) = \binom{n}{x_1, x_2, \ldots, x_k} p_1^{x_1} p_2^{x_2} \cdots p_k^{x_k},$$

with

$$\sum_{i=1}^{k} x_i = n \text{ and } \sum_{i=1}^{k} p_i = 1.$$

The multinomial distribution derives its name from the fact that the terms of the multinomial expansion of $(p_1 + p_2 + \cdots + p_k)^n$ correspond to all the possible values of $f(x_1, x_2, \ldots, x_k; p_1, p_2, \ldots, p_k, n)$.

Example 5.7: The complexity of arrivals and departures of planes at an airport is such that computer simulation is often used to model the "ideal" conditions. For a certain airport with three runways, it is known that in the ideal setting the following are the probabilities that the individual runways are accessed by a randomly arriving commercial jet:

$$\text{Runway 1:} \quad p_1 = 2/9,$$
$$\text{Runway 2:} \quad p_2 = 1/6,$$
$$\text{Runway 3:} \quad p_3 = 11/18.$$

What is the probability that 6 randomly arriving airplanes are distributed in the following fashion?

$$\text{Runway 1:} \quad 2 \text{ airplanes},$$
$$\text{Runway 2:} \quad 1 \text{ airplane},$$
$$\text{Runway 3:} \quad 3 \text{ airplanes}$$

Solution: Using the multinomial distribution, we have

$$f\left(2, 1, 3; \frac{2}{9}, \frac{1}{6}, \frac{11}{18}, 6\right) = \binom{6}{2, 1, 3} \left(\frac{2}{9}\right)^2 \left(\frac{1}{6}\right)^1 \left(\frac{11}{18}\right)^3$$
$$= \frac{6!}{2!\,1!\,3!} \cdot \frac{2^2}{9^2} \cdot \frac{1}{6} \cdot \frac{11^3}{18^3} = 0.1127.$$

Exercises

5.1 A random variable X that assumes the values $x_1, x_2, \ldots, x_k$ is called a discrete uniform random variable if its probability mass function is $f(x) = \frac{1}{k}$ for all of $x_1, x_2, \ldots, x_k$ and 0 otherwise. Find the mean and variance of X.

5.2 Twelve people are given two identical speakers, which they are asked to listen to for differences, if any. Suppose that these people answer simply by guessing. Find the probability that three people claim to have heard a difference between the two speakers.

5.3 An employee is selected from a staff of 10 to supervise a certain project by selecting a tag at random from a box containing 10 tags numbered from 1 to 10. Find the formula for the probability distribution of X representing the number on the tag that is drawn. What is the probability that the number drawn is less than 4?

5.4 In a certain city district, the need for money to buy drugs is stated as the reason for 75% of all thefts. Find the probability that among the next 5 theft cases reported in this district,

(a) exactly 2 resulted from the need for money to buy drugs;

(b) at most 3 resulted from the need for money to buy drugs.

5.5 According to *Chemical Engineering Progress* (November 1990), approximately 30% of all pipework failures in chemical plants are caused by operator error.

(a) What is the probability that out of the next 20 pipework failures at least 10 are due to operator error?

(b) What is the probability that no more than 4 out of 20 such failures are due to operator error?

(c) Suppose, for a particular plant, that out of the random sample of 20 such failures, exactly 5 are due to operator error. Do you feel that the 30% figure stated above applies to this plant? Comment.

5.6 According to a survey by the Administrative Management Society, one-half of U.S. companies give employees 4 weeks of vacation after they have been with the company for 15 years. Find the probability that among 6 companies surveyed at random, the number that give employees 4 weeks of vacation after 15 years of employment is

(a) anywhere from 2 to 5;

(b) fewer than 3.

5.7 One prominent physician claims that 70% of those with lung cancer are chain smokers. If his assertion is correct,

(a) find the probability that of 10 such patients

Exercises

recently admitted to a hospital, fewer than half are chain smokers;

(b) find the probability that of 20 such patients recently admitted to a hospital, fewer than half are chain smokers.

5.8 According to a study published by a group of University of Massachusetts sociologists, approximately 60% of the Valium users in the state of Massachusetts first took Valium for psychological problems. Find the probability that among the next 8 users from this state who are interviewed,

(a) exactly 3 began taking Valium for psychological problems;

(b) at least 5 began taking Valium for problems that were not psychological.

5.9 In testing a certain kind of truck tire over rugged terrain, it is found that 25% of the trucks fail to complete the test run without a blowout. Of the next 15 trucks tested, find the probability that

(a) from 3 to 6 have blowouts;

(b) fewer than 4 have blowouts;

(c) more than 5 have blowouts.

5.10 A nationwide survey of college seniors by the University of Michigan revealed that almost 70% disapprove of daily pot smoking, according to a report in *Parade*. If 12 seniors are selected at random and asked their opinion, find the probability that the number who disapprove of smoking pot daily is

(a) anywhere from 7 to 9;

(b) at most 5;

(c) not less than 8.

5.11 The probability that a patient recovers from a delicate heart operation is 0.9. What is the probability that exactly 5 of the next 7 patients having this operation survive?

5.12 A traffic control engineer reports that 75% of the vehicles passing through a checkpoint are from within the state. What is the probability that fewer than 4 of the next 9 vehicles are from out of state?

5.13 A national study that examined attitudes about antidepressants revealed that approximately 70% of respondents believe "antidepressants do not really cure anything, they just cover up the real trouble." According to this study, what is the probability that at least 3 of the next 5 people selected at random will hold this opinion?

5.14 The percentage of wins for the Chicago Bulls basketball team going into the playoffs for the 1996–97 season was 87.7. Round the 87.7 to 90 in order to use Table A.1.

(a) What is the probability that the Bulls sweep (4-0) the initial best-of-7 playoff series?

(b) What is the probability that the Bulls win the initial best-of-7 playoff series?

(c) What very important assumption is made in answering parts (a) and (b)?

5.15 It is known that 60% of mice inoculated with a serum are protected from a certain disease. If 5 mice are inoculated, find the probability that

(a) none contracts the disease;

(b) fewer than 2 contract the disease;

(c) more than 3 contract the disease.

5.16 Suppose that airplane engines operate independently and fail with probability equal to 0.4. Assuming that a plane makes a safe flight if at least one-half of its engines run, determine whether a 4-engine plane or a 2-engine plane has the higher probability for a successful flight.

5.17 If X represents the number of people in Exercise 5.13 who believe that antidepressants do not cure but only cover up the real problem, find the mean and variance of X when 5 people are selected at random.

5.18 (a) In Exercise 5.9, how many of the 15 trucks would you expect to have blowouts?

(b) What is the variance of the number of blowouts experienced by the 15 trucks? What does that mean?

5.19 As a student drives to school, he encounters a traffic signal. This traffic signal stays green for 35 seconds, yellow for 5 seconds, and red for 60 seconds. Assume that the student goes to school each weekday between 8:00 and 8:30 a.m. Let X_1 be the number of times he encounters a green light, X_2 be the number of times he encounters a yellow light, and X_3 be the number of times he encounters a red light. Find the joint distribution of X_1, X_2, and X_3.

5.20 According to *USA Today* (March 18, 1997), of 4 million workers in the general workforce, 5.8% tested positive for drugs. Of those testing positive, 22.5% were cocaine users and 54.4% marijuana users.

(a) What is the probability that of 10 workers testing positive, 2 are cocaine users, 5 are marijuana users, and 3 are users of other drugs?

(b) What is the probability that of 10 workers testing positive, all are marijuana users?

(c) What is the probability that of 10 workers testing positive, none is a cocaine user?

5.21 The surface of a circular dart board has a small center circle called the bull's-eye and 20 pie-shaped regions numbered from 1 to 20. Each of the pie-shaped regions is further divided into three parts such that a person throwing a dart that lands in a specific region scores the value of the number, double the number, or triple the number, depending on which of the three parts the dart hits. If a person hits the bull's-eye with probability 0.01, hits a double with probability 0.10, hits a triple with probability 0.05, and misses the dart board with probability 0.02, what is the probability that 7 throws will result in no bull's-eyes, no triples, a double twice, and a complete miss once?

5.22 According to a genetics theory, a certain cross of guinea pigs will result in red, black, and white offspring in the ratio 8:4:4. Find the probability that among 8 offspring, 5 will be red, 2 black, and 1 white.

5.23 The probabilities are 0.4, 0.2, 0.3, and 0.1, respectively, that a delegate to a certain convention arrived by air, bus, automobile, or train. What is the probability that among 9 delegates randomly selected at this convention, 3 arrived by air, 3 arrived by bus, 1 arrived by automobile, and 2 arrived by train?

5.24 A safety engineer claims that only 40% of all workers wear safety helmets when they eat lunch at the workplace. Assuming that this claim is right, find the probability that 4 of 6 workers randomly chosen will be wearing their helmets while having lunch at the workplace.

5.25 Suppose that for a very large shipment of integrated-circuit chips, the probability of failure for any one chip is 0.10. Assuming that the assumptions underlying the binomial distributions are met, find the probability that at most 3 chips fail in a random sample of 20.

5.26 Assuming that 6 in 10 automobile accidents are due mainly to a speed violation, find the probability that among 8 automobile accidents, 6 will be due mainly to a speed violation
(a) by using the formula for the binomial distribution;
(b) by using Table A.1.

5.27 If the probability that a fluorescent light has a useful life of at least 800 hours is 0.9, find the probabilities that among 20 such lights
(a) exactly 18 will have a useful life of at least 800 hours;
(b) at least 15 will have a useful life of at least 800 hours;
(c) at least 2 will *not* have a useful life of at least 800 hours.

5.28 A manufacturer knows that on average 20% of the electric toasters produced require repairs within 1 year after they are sold. When 20 toasters are randomly selected, find appropriate numbers x and y such that
(a) the probability that at least x of them will require repairs is less than 0.5;
(b) the probability that at least y of them will *not* require repairs is greater than 0.8.

5.3 Hypergeometric Distribution

The simplest way to view the distinction between the binomial distribution of Section 5.2 and the hypergeometric distribution is to note the way the sampling is done. The types of applications for the hypergeometric are very similar to those for the binomial distribution. We are interested in computing probabilities for the number of observations that fall into a particular category. But in the case of the binomial distribution, independence among trials is required. As a result, if that distribution is applied to, say, sampling from a lot of items (deck of cards, batch of production items), the sampling must be done **with replacement** of each item after it is observed. On the other hand, the hypergeometric distribution does not require independence and is based on sampling done **without replacement**.

Applications for the hypergeometric distribution are found in many areas, with heavy use in acceptance sampling, electronic testing, and quality assurance. Obviously, in many of these fields, testing is done at the expense of the item being tested. That is, the item is destroyed and hence cannot be replaced in the sample. Thus, sampling without replacement is necessary. A simple example with playing cards will serve as our first illustration.

If we wish to find the probability of observing 3 red cards in 5 draws from an ordinary deck of 52 playing cards, the binomial distribution of Section 5.2 does not apply unless each card is replaced and the deck reshuffled before the next draw is made. To solve the problem of sampling without replacement, let us restate the problem. If 5 cards are drawn at random, we are interested in the probability of selecting 3 red cards from the 26 available in the deck and 2 black cards from the 26 available in the deck. There are $\binom{26}{3}$ ways of selecting 3 red cards, and for each of these ways we can choose 2 black cards in $\binom{26}{2}$ ways. Therefore, the total number of ways to select 3 red and 2 black cards in 5 draws is the product $\binom{26}{3}\binom{26}{2}$. The total number of ways to select any 5 cards from the 52 that are available is $\binom{52}{5}$. Hence, the probability of selecting 5 cards without replacement of which 3 are red and 2 are black is given by

$$\frac{\binom{26}{3}\binom{26}{2}}{\binom{52}{5}} = \frac{(26!/3!\,23!)(26!/2!\,24!)}{52!/5!\,47!} = 0.3251.$$

In general, we are interested in the probability of selecting x successes from the k items labeled successes and $n - x$ failures from the $N - k$ items labeled failures when a random sample of size n is selected from N items. This is known as a **hypergeometric experiment**, that is, one that possesses the following two properties:

1. A random sample of size n is selected without replacement from N items.
2. Of the N items, k may be classified as successes and $N - k$ are classified as failures.

The number X of successes of a hypergeometric experiment is called a **hypergeometric random variable**. Accordingly, the probability distribution of the hypergeometric variable is called the **hypergeometric distribution**, and its values are denoted by $h(x; N, n, k)$, since they depend on the number of successes k in the set N from which we select n items.

Hypergeometric Distribution in Acceptance Sampling

Like the binomial distribution, the hypergeometric distribution finds applications in acceptance sampling, where lots of materials or parts are sampled in order to determine whether or not the entire lot is accepted.

Example 5.8: A particular part that is used as an injection device is sold in lots of 10. The producer deems a lot acceptable if no more than one defective is in the lot. A sampling plan involves random sampling and testing 3 of the parts out of 10. If none of the 3 is defective, the lot is accepted. Comment on the utility of this plan.

Solution: Let us assume that the lot is truly **unacceptable** (i.e., that 2 out of 10 parts are defective). The probability that the sampling plan finds the lot acceptable is

$$P(X = 0) = \frac{\binom{2}{0}\binom{8}{3}}{\binom{10}{3}} = 0.467.$$

Thus, if the lot is truly unacceptable, with 2 defective parts, this sampling plan will allow acceptance roughly 47% of the time. As a result, this plan should be considered faulty.

Let us now generalize in order to find a formula for $h(x; N, n, k)$. The total number of samples of size n chosen from N items is $\binom{N}{n}$. These samples are assumed to be equally likely. There are $\binom{k}{x}$ ways of selecting x successes from the k that are available, and for each of these ways we can choose the $n - x$ failures in $\binom{N-k}{n-x}$ ways. Thus, the total number of favorable samples among the $\binom{N}{n}$ possible samples is given by $\binom{k}{x}\binom{N-k}{n-x}$. Hence, we have the following definition.

Hypergeometric Distribution The probability distribution of the hypergeometric random variable X, the number of successes in a random sample of size n selected from N items of which k are labeled **success** and $N - k$ labeled **failure**, is

$$h(x; N, n, k) = \frac{\binom{k}{x}\binom{N-k}{n-x}}{\binom{N}{n}}, \quad \max\{0, n - (N - k)\} \leq x \leq \min\{n, k\}.$$

The range of x can be determined by the three binomial coefficients in the definition, where x and $n - x$ are no more than k and $N - k$, respectively, and both of them cannot be less than 0. Usually, when both k (the number of successes) and $N - k$ (the number of failures) are larger than the sample size n, the range of a hypergeometric random variable will be $x = 0, 1, \ldots, n$.

Example 5.9: Lots of 40 components each are deemed unacceptable if they contain 3 or more defectives. The procedure for sampling a lot is to select 5 components at random and to reject the lot if a defective is found. What is the probability that exactly 1 defective is found in the sample if there are 3 defectives in the entire lot?

Solution: Using the hypergeometric distribution with $n = 5$, $N = 40$, $k = 3$, and $x = 1$, we find the probability of obtaining 1 defective to be

$$h(1; 40, 5, 3) = \frac{\binom{3}{1}\binom{37}{4}}{\binom{40}{5}} = 0.3011.$$

Once again, this plan is not desirable since it detects a bad lot (3 defectives) only about 30% of the time.

Theorem 5.2: The mean and variance of the hypergeometric distribution $h(x; N, n, k)$ are

$$\mu = \frac{nk}{N} \quad \text{and} \quad \sigma^2 = \frac{N-n}{N-1} \cdot n \cdot \frac{k}{N}\left(1 - \frac{k}{N}\right).$$

The proof for the mean is shown in Appendix A.24.

Example 5.10: Let us now reinvestigate Example 3.4 on page 83. The purpose of this example was to illustrate the notion of a random variable and the corresponding sample space. In the example, we have a lot of 100 items of which 12 are defective. What is the probability that in a sample of 10, 3 are defective?

5.3 Hypergeometric Distribution

Solution: Using the hypergeometric probability function, we have

$$h(3; 100, 10, 12) = \frac{\binom{12}{3}\binom{88}{7}}{\binom{100}{10}} = 0.08.$$

Example 5.11: Find the mean and variance of the random variable of Example 5.9 and then use Chebyshev's theorem to interpret the interval $\mu \pm 2\sigma$.

Solution: Since Example 5.9 was a hypergeometric experiment with $N = 40$, $n = 5$, and $k = 3$, by Theorem 5.2, we have

$$\mu = \frac{(5)(3)}{40} = \frac{3}{8} = 0.375,$$

and

$$\sigma^2 = \left(\frac{40-5}{39}\right)(5)\left(\frac{3}{40}\right)\left(1 - \frac{3}{40}\right) = 0.3113.$$

Taking the square root of 0.3113, we find that $\sigma = 0.558$. Hence, the required interval is $0.375 \pm (2)(0.558)$, or from -0.741 to 1.491. Chebyshev's theorem states that the number of defectives obtained when 5 components are selected at random from a lot of 40 components of which 3 are defective has a probability of at least 3/4 of falling between -0.741 and 1.491. That is, at least three-fourths of the time, the 5 components include fewer than 2 defectives.

Relationship to the Binomial Distribution

In this chapter, we discuss several important discrete distributions that have wide applicability. Many of these distributions relate nicely to each other. The beginning student should gain a clear understanding of these relationships. There is an interesting relationship between the hypergeometric and the binomial distribution. As one might expect, if n is small compared to N, the nature of the N items changes very little in each draw. So a binomial distribution can be used to approximate the hypergeometric distribution when n is small compared to N. In fact, as a rule of thumb, the approximation is good when $n/N \leq 0.05$.

Thus, the quantity k/N plays the role of the binomial parameter p. As a result, the binomial distribution may be viewed as a large-population version of the hypergeometric distribution. The mean and variance then come from the formulas

$$\mu = np = \frac{nk}{N} \text{ and } \sigma^2 = npq = n \cdot \frac{k}{N}\left(1 - \frac{k}{N}\right).$$

Comparing these formulas with those of Theorem 5.2, we see that the mean is the same but the variance differs by a correction factor of $(N-n)/(N-1)$, which is negligible when n is small relative to N.

Example 5.12: A manufacturer of automobile tires reports that among a shipment of 5000 sent to a local distributor, 1000 are slightly blemished. If one purchases 10 of these tires at random from the distributor, what is the probability that exactly 3 are blemished?

Solution: Since $N = 5000$ is large relative to the sample size $n = 10$, we shall approximate the desired probability by using the binomial distribution. The probability of obtaining a blemished tire is 0.2. Therefore, the probability of obtaining exactly 3 blemished tires is

$$h(3; 5000, 10, 1000) \approx b(3; 10, 0.2) = 0.8791 - 0.6778 = 0.2013.$$

On the other hand, the exact probability is $h(3; 5000, 10, 1000) = 0.2015.$

The hypergeometric distribution can be extended to treat the case where the N items can be partitioned into k cells $A_1, A_2, \ldots, A_k$ with a_1 elements in the first cell, a_2 elements in the second cell, $\ldots$, a_k elements in the kth cell. We are now interested in the probability that a random sample of size n yields x_1 elements from A_1, x_2 elements from A_2, $\ldots$, and x_k elements from A_k. Let us represent this probability by

$$f(x_1, x_2, \ldots, x_k; a_1, a_2, \ldots, a_k, N, n).$$

To obtain a general formula, we note that the total number of samples of size n that can be chosen from N items is still $\binom{N}{n}$. There are $\binom{a_1}{x_1}$ ways of selecting x_1 items from the items in A_1, and for each of these we can choose x_2 items from the items in A_2 in $\binom{a_2}{x_2}$ ways. Therefore, we can select x_1 items from A_1 and x_2 items from A_2 in $\binom{a_1}{x_1}\binom{a_2}{x_2}$ ways. Continuing in this way, we can select all n items consisting of x_1 from A_1, x_2 from A_2, $\ldots$, and x_k from A_k in

$$\binom{a_1}{x_1}\binom{a_2}{x_2} \cdots \binom{a_k}{x_k} \text{ ways.}$$

The required probability distribution is now defined as follows.

Multivariate Hypergeometric Distribution

If N items can be partitioned into the k cells $A_1, A_2, \ldots, A_k$ with $a_1, a_2, \ldots, a_k$ elements, respectively, then the probability distribution of the random variables $X_1, X_2, \ldots, X_k$, representing the number of elements selected from $A_1, A_2, \ldots, A_k$ in a random sample of size n, is

$$f(x_1, x_2, \ldots, x_k; a_1, a_2, \ldots, a_k, N, n) = \frac{\binom{a_1}{x_1}\binom{a_2}{x_2} \cdots \binom{a_k}{x_k}}{\binom{N}{n}},$$

with $\sum_{i=1}^{k} x_i = n$ and $\sum_{i=1}^{k} a_i = N$.

Example 5.13: A group of 10 individuals is used for a biological case study. The group contains 3 people with blood type O, 4 with blood type A, and 3 with blood type B. What is the probability that a random sample of 5 will contain 1 person with blood type O, 2 people with blood type A, and 2 people with blood type B?

Solution: Using the extension of the hypergeometric distribution with $x_1 = 1$, $x_2 = 2$, $x_3 = 2$, $a_1 = 3$, $a_2 = 4$, $a_3 = 3$, $N = 10$, and $n = 5$, we find that the desired probability is

$$f(1, 2, 2; 3, 4, 3, 10, 5) = \frac{\binom{3}{1}\binom{4}{2}\binom{3}{2}}{\binom{10}{5}} = \frac{3}{14}.$$

Exercises

5.29 A homeowner plants 6 bulbs selected at random from a box containing 5 tulip bulbs and 4 daffodil bulbs. What is the probability that he planted 2 daffodil bulbs and 4 tulip bulbs?

5.30 To avoid detection at customs, a traveler places 6 narcotic tablets in a bottle containing 9 vitamin tablets that are similar in appearance. If the customs official selects 3 of the tablets at random for analysis, what is the probability that the traveler will be arrested for illegal possession of narcotics?

5.31 A random committee of size 3 is selected from 4 doctors and 2 nurses. Write a formula for the probability distribution of the random variable X representing the number of doctors on the committee. Find $P(2 \leq X \leq 3)$.

5.32 From a lot of 10 missiles, 4 are selected at random and fired. If the lot contains 3 defective missiles that will not fire, what is the probability that
(a) all 4 will fire?
(b) at most 2 will not fire?

5.33 If 7 cards are dealt from an ordinary deck of 52 playing cards, what is the probability that
(a) exactly 2 of them will be face cards?
(b) at least 1 of them will be a queen?

5.34 What is the probability that a waitress will refuse to serve alcoholic beverages to only 2 minors if she randomly checks the IDs of 5 among 9 students, 4 of whom are minors?

5.35 A company is interested in evaluating its current inspection procedure for shipments of 50 identical items. The procedure is to take a sample of 5 and pass the shipment if no more than 2 are found to be defective. What proportion of shipments with 20% defectives will be accepted?

5.36 A manufacturing company uses an acceptance scheme on items from a production line before they are shipped. The plan is a two-stage one. Boxes of 25 items are readied for shipment, and a sample of 3 items is tested for defectives. If any defectives are found, the entire box is sent back for 100% screening. If no defectives are found, the box is shipped.
(a) What is the probability that a box containing 3 defectives will be shipped?
(b) What is the probability that a box containing only 1 defective will be sent back for screening?

5.37 Suppose that the manufacturing company of Exercise 5.36 decides to change its acceptance scheme. Under the new scheme, an inspector takes 1 item at random, inspects it, and then replaces it in the box; a second inspector does likewise. Finally, a third inspector goes through the same procedure. The box is not shipped if any of the three inspectors find a defective. Answer the questions in Exercise 5.36 for this new plan.

5.38 Among 150 IRS employees in a large city, only 30 are women. If 10 of the employees are chosen at random to provide free tax assistance for the residents of this city, use the binomial approximation to the hypergeometric distribution to find the probability that at least 3 women are selected.

5.39 An annexation suit against a county subdivision of 1200 residences is being considered by a neighboring city. If the occupants of half the residences object to being annexed, what is the probability that in a random sample of 10 at least 3 favor the annexation suit?

5.40 It is estimated that 4000 of the 10,000 voting residents of a town are against a new sales tax. If 15 eligible voters are selected at random and asked their opinion, what is the probability that at most 7 favor the new tax?

5.41 A nationwide survey of 17,000 college seniors by the University of Michigan revealed that almost 70% disapprove of daily pot smoking. If 18 of these seniors are selected at random and asked their opinion, what is the probability that more than 9 but fewer than 14 disapprove of smoking pot daily?

5.42 Find the probability of being dealt a bridge hand of 13 cards containing 5 spades, 2 hearts, 3 diamonds, and 3 clubs.

5.43 A foreign student club lists as its members 2 Canadians, 3 Japanese, 5 Italians, and 2 Germans. If a committee of 4 is selected at random, find the probability that
(a) all nationalities are represented;
(b) all nationalities except Italian are represented.

5.44 An urn contains 3 green balls, 2 blue balls, and 4 red balls. In a random sample of 5 balls, find the probability that both blue balls and at least 1 red ball are selected.

5.45 Biologists doing studies in a particular environment often tag and release subjects in order to estimate the size of a population or the prevalence of certain features in the population. Ten animals of a certain population thought to be extinct (or near extinction) are caught, tagged, and released in a certain region. After a period of time, a random sample of 15 of this type of animal is selected in the region. What is the probability that 5 of those selected are tagged if there are 25 animals of this type in the region?

5.46 A large company has an inspection system for the batches of small compressors purchased from vendors. A batch typically contains 15 compressors. In the inspection system, a random sample of 5 is selected and all are tested. Suppose there are 2 faulty compressors in the batch of 15.

(a) What is the probability that for a given sample there will be 1 faulty compressor?

(b) What is the probability that inspection will discover both faulty compressors?

5.47 A government task force suspects that some manufacturing companies are in violation of federal pollution regulations with regard to dumping a certain type of product. Twenty firms are under suspicion but not all can be inspected. Suppose that 3 of the firms are in violation.

(a) What is the probability that inspection of 5 firms will find no violations?

(b) What is the probability that the plan above will find two violations?

5.48 Every hour, 10,000 cans of soda are filled by a machine, among which 300 underfilled cans are produced. Each hour, a sample of 30 cans is randomly selected and the number of ounces of soda per can is checked. Denote by X the number of cans selected that are underfilled. Find the probability that at least 1 underfilled can will be among those sampled.

5.4 Negative Binomial and Geometric Distributions

Let us consider an experiment where the properties are the same as those listed for a binomial experiment, with the exception that the trials will be repeated until a *fixed* number of successes occur. Therefore, instead of the probability of x successes in n trials, where n is fixed, we are now interested in the probability that the kth success occurs on the xth trial. Experiments of this kind are called **negative binomial experiments**.

As an illustration, consider the use of a drug that is known to be effective in 60% of the cases where it is used. The drug will be considered a success if it is effective in bringing some degree of relief to the patient. We are interested in finding the probability that the fifth patient to experience relief is the seventh patient to receive the drug during a given week. Designating a success by S and a failure by F, a possible order of achieving the desired result is $SFSSSFS$, which occurs with probability

$$(0.6)(0.4)(0.6)(0.6)(0.6)(0.4)(0.6) = (0.6)^5(0.4)^2.$$

We could list all possible orders by rearranging the F's and S's except for the last outcome, which must be the fifth success. The total number of possible orders is equal to the number of partitions of the first six trials into two groups with 2 failures assigned to the one group and 4 successes assigned to the other group. This can be done in $\binom{6}{4} = 15$ mutually exclusive ways. Hence, if X represents the outcome on which the fifth success occurs, then

$$P(X = 7) = \binom{6}{4}(0.6)^5(0.4)^2 = 0.1866.$$

What Is the Negative Binomial Random Variable?

The number X of trials required to produce k successes in a negative binomial experiment is called a **negative binomial random variable**, and its probability

5.4 Negative Binomial and Geometric Distributions

distribution is called the **negative binomial distribution**. Since its probabilities depend on the number of successes desired and the probability of a success on a given trial, we shall denote them by $b^*(x; k, p)$. To obtain the general formula for $b^*(x; k, p)$, consider the probability of a success on the xth trial preceded by $k - 1$ successes and $x - k$ failures in some specified order. Since the trials are independent, we can multiply all the probabilities corresponding to each desired outcome. Each success occurs with probability p and each failure with probability $q = 1 - p$. Therefore, the probability for the specified order ending in success is

$$p^{k-1} q^{x-k} p = p^k q^{x-k}.$$

The total number of sample points in the experiment ending in a success, after the occurrence of $k - 1$ successes and $x - k$ failures in any order, is equal to the number of partitions of $x - 1$ trials into two groups with $k - 1$ successes corresponding to one group and $x - k$ failures corresponding to the other group. This number is specified by the term $\binom{x-1}{k-1}$, each mutually exclusive and occurring with equal probability $p^k q^{x-k}$. We obtain the general formula by multiplying $p^k q^{x-k}$ by $\binom{x-1}{k-1}$.

Negative Binomial Distribution
If repeated independent trials can result in a success with probability p and a failure with probability $q = 1 - p$, then the probability distribution of the random variable X, the number of the trial on which the kth success occurs, is

$$b^*(x; k, p) = \binom{x-1}{k-1} p^k q^{x-k}, \quad x = k, k+1, k+2, \ldots.$$

Example 5.14: In an NBA (National Basketball Association) championship series, the team that wins four games out of seven is the winner. Suppose that teams A and B face each other in the championship games and that team A has probability 0.55 of winning a game over team B.

(a) What is the probability that team A will win the series in 6 games?

(b) What is the probability that team A will win the series?

(c) If teams A and B were facing each other in a regional playoff series, which is decided by winning three out of five games, what is the probability that team A would win the series?

Solution: (a) $b^*(6; 4, 0.55) = \binom{5}{3} 0.55^4 (1 - 0.55)^{6-4} = 0.1853$

(b) $P(\text{team } A \text{ wins the championship series})$ is

$$b^*(4; 4, 0.55) + b^*(5; 4, 0.55) + b^*(6; 4, 0.55) + b^*(7; 4, 0.55)$$
$$= 0.0915 + 0.1647 + 0.1853 + 0.1668 = 0.6083.$$

(c) $P(\text{team } A \text{ wins the playoff})$ is

$$b^*(3; 3, 0.55) + b^*(4; 3, 0.55) + b^*(5; 3, 0.55)$$
$$= 0.1664 + 0.2246 + 0.2021 = 0.5931.$$

The negative binomial distribution derives its name from the fact that each term in the expansion of $p^k(1-q)^{-k}$ corresponds to the values of $b^*(x;k,p)$ for $x = k,\ k+1,\ k+2,\ \ldots$. If we consider the special case of the negative binomial distribution where $k = 1$, we have a probability distribution for the number of trials required for a single success. An example would be the tossing of a coin until a head occurs. We might be interested in the probability that the first head occurs on the fourth toss. The negative binomial distribution reduces to the form

$$b^*(x;1,p) = pq^{x-1}, \quad x = 1, 2, 3, \ldots.$$

Since the successive terms constitute a geometric progression, it is customary to refer to this special case as the **geometric distribution** and denote its values by $g(x;p)$.

Geometric Distribution If repeated independent trials can result in a success with probability p and a failure with probability $q = 1 - p$, then the probability distribution of the random variable X, the number of the trial on which the first success occurs, is

$$g(x;p) = pq^{x-1}, \quad x = 1, 2, 3, \ldots.$$

Example 5.15: For a certain manufacturing process, it is known that, on the average, 1 in every 100 items is defective. What is the probability that the fifth item inspected is the first defective item found?

Solution: Using the geometric distribution with $x = 5$ and $p = 0.01$, we have

$$g(5; 0.01) = (0.01)(0.99)^4 = 0.0096.$$

Example 5.16: At a "busy time," a telephone exchange is very near capacity, so callers have difficulty placing their calls. It may be of interest to know the number of attempts necessary in order to make a connection. Suppose that we let $p = 0.05$ be the probability of a connection during a busy time. We are interested in knowing the probability that 5 attempts are necessary for a successful call.

Solution: Using the geometric distribution with $x = 5$ and $p = 0.05$ yields

$$P(X = x) = g(5; 0.05) = (0.05)(0.95)^4 = 0.041.$$

Quite often, in applications dealing with the geometric distribution, the mean and variance are important. For example, in Example 5.16, the *expected* number of calls necessary to make a connection is quite important. The following theorem states without proof the mean and variance of the geometric distribution.

Theorem 5.3: The mean and variance of a random variable following the geometric distribution are

$$\mu = \frac{1}{p} \text{ and } \sigma^2 = \frac{1-p}{p^2}.$$

Applications of Negative Binomial and Geometric Distributions

Areas of application for the negative binomial and geometric distributions become obvious when one focuses on the examples in this section and the exercises devoted to these distributions at the end of Section 5.5. In the case of the geometric distribution, Example 5.16 depicts a situation where engineers or managers are attempting to determine how inefficient a telephone exchange system is during busy times. Clearly, in this case, trials occurring prior to a success represent a cost. If there is a high probability of several attempts being required prior to making a connection, then plans should be made to redesign the system.

Applications of the negative binomial distribution are similar in nature. Suppose attempts are costly in some sense and are *occurring in sequence*. A high probability of needing a "large" number of attempts to experience a fixed number of successes is not beneficial to the scientist or engineer. Consider the scenarios of Review Exercises 5.90 and 5.91. In Review Exercise 5.91, the oil driller defines a certain level of success from sequentially drilling locations for oil. If only 6 attempts have been made at the point where the second success is experienced, the profits appear to dominate substantially the investment incurred by the drilling.

5.5 Poisson Distribution and the Poisson Process

Experiments yielding numerical values of a random variable X, the number of outcomes occurring during a given time interval or in a specified region, are called **Poisson experiments**. The given time interval may be of any length, such as a minute, a day, a week, a month, or even a year. For example, a Poisson experiment can generate observations for the random variable X representing the number of telephone calls received per hour by an office, the number of days school is closed due to snow during the winter, or the number of games postponed due to rain during a baseball season. The specified region could be a line segment, an area, a volume, or perhaps a piece of material. In such instances, X might represent the number of field mice per acre, the number of bacteria in a given culture, or the number of typing errors per page. A Poisson experiment is derived from the **Poisson process** and possesses the following properties.

Properties of the Poisson Process

1. The number of outcomes occurring in one time interval or specified region of space is independent of the number that occur in any other disjoint time interval or region. In this sense we say that the Poisson process has no memory.

2. The probability that a single outcome will occur during a very short time interval or in a small region is proportional to the length of the time interval or the size of the region and does not depend on the number of outcomes occurring outside this time interval or region.

3. The probability that more than one outcome will occur in such a short time interval or fall in such a small region is negligible.

The number X of outcomes occurring during a Poisson experiment is called a **Poisson random variable**, and its probability distribution is called the **Poisson**

distribution. The mean number of outcomes is computed from $\mu = \lambda t$, where t is the specific "time," "distance," "area," or "volume" of interest. Since the probabilities depend on λ, the rate of occurrence of outcomes, we shall denote them by $p(x; \lambda t)$. The derivation of the formula for $p(x; \lambda t)$, based on the three properties of a Poisson process listed above, is beyond the scope of this book. The following formula is used for computing Poisson probabilities.

Poisson Distribution The probability distribution of the Poisson random variable X, representing the number of outcomes occurring in a given time interval or specified region denoted by t, is

$$p(x; \lambda t) = \frac{e^{-\lambda t}(\lambda t)^x}{x!}, \quad x = 0, 1, 2, \ldots,$$

where λ is the average number of outcomes per unit time, distance, area, or volume and $e = 2.71828\ldots$.

Table A.2 contains Poisson probability sums,

$$P(r; \lambda t) = \sum_{x=0}^{r} p(x; \lambda t),$$

for selected values of λt ranging from 0.1 to 18.0. We illustrate the use of this table with the following two examples.

Example 5.17: During a laboratory experiment, the average number of radioactive particles passing through a counter in 1 millisecond is 4. What is the probability that 6 particles enter the counter in a given millisecond?

Solution: Using the Poisson distribution with $x = 6$ and $\lambda t = 4$ and referring to Table A.2, we have

$$p(6; 4) = \frac{e^{-4} 4^6}{6!} = \sum_{x=0}^{6} p(x; 4) - \sum_{x=0}^{5} p(x; 4) = 0.8893 - 0.7851 = 0.1042.$$

Example 5.18: Ten is the average number of oil tankers arriving each day at a certain port. The facilities at the port can handle at most 15 tankers per day. What is the probability that on a given day tankers have to be turned away?

Solution: Let X be the number of tankers arriving each day. Then, using Table A.2, we have

$$P(X > 15) = 1 - P(X \leq 15) = 1 - \sum_{x=0}^{15} p(x; 10) = 1 - 0.9513 = 0.0487.$$

Like the binomial distribution, the Poisson distribution is used for quality control, quality assurance, and acceptance sampling. In addition, certain important continuous distributions used in reliability theory and queuing theory depend on the Poisson process. Some of these distributions are discussed and developed in Chapter 6. The following theorem concerning the Poisson random variable is given in Appendix A.25.

Theorem 5.4: Both the mean and the variance of the Poisson distribution $p(x; \lambda t)$ are λt.

Nature of the Poisson Probability Function

Like so many discrete and continuous distributions, the form of the Poisson distribution becomes more and more symmetric, even bell-shaped, as the mean grows large. Figure 5.1 illustrates this, showing plots of the probability function for $\mu = 0.1$, $\mu = 2$, and $\mu = 5$. Note the nearness to symmetry when μ becomes as large as 5. A similar condition exists for the binomial distribution, as will be illustrated later in the text.

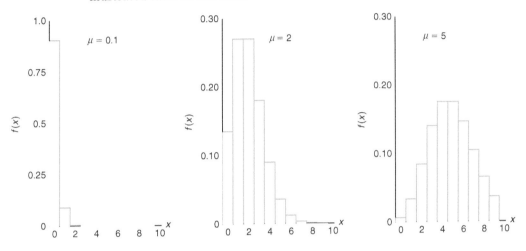

Figure 5.1: Poisson density functions for different means.

Approximation of Binomial Distribution by a Poisson Distribution

It should be evident from the three principles of the Poisson process that the Poisson distribution is related to the binomial distribution. Although the Poisson usually finds applications in space and time problems, as illustrated by Examples 5.17 and 5.18, it can be viewed as a limiting form of the binomial distribution. In the case of the binomial, if n is quite large and p is small, the conditions begin to simulate the *continuous space or time* implications of the Poisson process. The independence among Bernoulli trials in the binomial case is consistent with principle 2 of the Poisson process. Allowing the parameter p to be close to 0 relates to principle 3 of the Poisson process. Indeed, if n is large and p is close to 0, the Poisson distribution can be used, with $\mu = np$, to approximate binomial probabilities. If p is close to 1, we can still use the Poisson distribution to approximate binomial probabilities by interchanging what we have defined to be a success and a failure, thereby changing p to a value close to 0.

Theorem 5.5: Let X be a binomial random variable with probability distribution $b(x; n, p)$. When $n \to \infty$, $p \to 0$, and $np \xrightarrow{n \to \infty} \mu$ remains constant,

$$b(x; n, p) \xrightarrow{n \to \infty} p(x; \mu).$$

Example 5.19: In a certain industrial facility, accidents occur infrequently. It is known that the probability of an accident on any given day is 0.005 and accidents are independent of each other.

(a) What is the probability that in any given period of 400 days there will be an accident on one day?

(b) What is the probability that there are at most three days with an accident?

Solution: Let X be a binomial random variable with $n = 400$ and $p = 0.005$. Thus, $np = 2$. Using the Poisson approximation,

(a) $P(X = 1) = e^{-2}2^1 = 0.271$ and

(b) $P(X \leq 3) = \sum_{x=0}^{3} e^{-2}2^x/x! = 0.857$.

Example 5.20: In a manufacturing process where glass products are made, defects or bubbles occur, occasionally rendering the piece undesirable for marketing. It is known that, on average, 1 in every 1000 of these items produced has one or more bubbles. What is the probability that a random sample of 8000 will yield fewer than 7 items possessing bubbles?

Solution: This is essentially a binomial experiment with $n = 8000$ and $p = 0.001$. Since p is very close to 0 and n is quite large, we shall approximate with the Poisson distribution using

$$\mu = (8000)(0.001) = 8.$$

Hence, if X represents the number of bubbles, we have

$$P(X < 7) = \sum_{x=0}^{6} b(x; 8000, 0.001) \approx p(x; 8) = 0.3134.$$

Exercises

5.49 The probability that a person living in a certain city owns a dog is estimated to be 0.3. Find the probability that the tenth person randomly interviewed in that city is the fifth one to own a dog.

5.50 Find the probability that a person flipping a coin gets

(a) the third head on the seventh flip;
(b) the first head on the fourth flip.

5.51 Three people toss a fair coin and the odd one pays for coffee. If the coins all turn up the same, they are tossed again. Find the probability that fewer than 4 tosses are needed.

5.52 A scientist inoculates mice, one at a time, with a disease germ until he finds 2 that have contracted the disease. If the probability of contracting the disease is 1/6, what is the probability that 8 mice are required?

5.53 An inventory study determines that, on average, demands for a particular item at a warehouse are made 5 times per day. What is the probability that on a given day this item is requested

(a) more than 5 times?
(b) not at all?

5.54 According to a study published by a group of University of Massachusetts sociologists, about two-thirds of the 20 million persons in this country who take Valium are women. Assuming this figure to be a valid estimate, find the probability that on a given day the fifth prescription written by a doctor for Valium is

(a) the first prescribing Valium for a woman;

Exercises

(b) the third prescribing Valium for a woman.

5.55 The probability that a student pilot passes the written test for a private pilot's license is 0.7. Find the probability that a given student will pass the test
(a) on the third try;
(b) before the fourth try.

5.56 On average, 3 traffic accidents per month occur at a certain intersection. What is the probability that in any given month at this intersection
(a) exactly 5 accidents will occur?
(b) fewer than 3 accidents will occur?
(c) at least 2 accidents will occur?

5.57 On average, a textbook author makes two word-processing errors per page on the first draft of her textbook. What is the probability that on the next page she will make
(a) 4 or more errors?
(b) no errors?

5.58 A certain area of the eastern United States is, on average, hit by 6 hurricanes a year. Find the probability that in a given year that area will be hit by
(a) fewer than 4 hurricanes;
(b) anywhere from 6 to 8 hurricanes.

5.59 Suppose the probability that any given person will believe a tale about the transgressions of a famous actress is 0.8. What is the probability that
(a) the sixth person to hear this tale is the fourth one to believe it?
(b) the third person to hear this tale is the first one to believe it?

5.60 The average number of field mice per acre in a 5-acre wheat field is estimated to be 12. Find the probability that fewer than 7 field mice are found
(a) on a given acre;
(b) on 2 of the next 3 acres inspected.

5.61 Suppose that, on average, 1 person in 1000 makes a numerical error in preparing his or her income tax return. If 10,000 returns are selected at random and examined, find the probability that 6, 7, or 8 of them contain an error.

5.62 The probability that a student at a local high school fails the screening test for scoliosis (curvature of the spine) is known to be 0.004. Of the next 1875 students at the school who are screened for scoliosis, find the probability that
(a) fewer than 5 fail the test;
(b) 8, 9, or 10 fail the test.

5.63 Find the mean and variance of the random variable X in Exercise 5.58, representing the number of hurricanes per year to hit a certain area of the eastern United States.

5.64 Find the mean and variance of the random variable X in Exercise 5.61, representing the number of persons among 10,000 who make an error in preparing their income tax returns.

5.65 An automobile manufacturer is concerned about a fault in the braking mechanism of a particular model. The fault can, on rare occasions, cause a catastrophe at high speed. The distribution of the number of cars per year that will experience the catastrophe is a Poisson random variable with $\lambda = 5$.
(a) What is the probability that at most 3 cars per year will experience a catastrophe?
(b) What is the probability that more than 1 car per year will experience a catastrophe?

5.66 Changes in airport procedures require considerable planning. Arrival rates of aircraft are important factors that must be taken into account. Suppose small aircraft arrive at a certain airport, according to a Poisson process, at the rate of 6 per hour. Thus, the Poisson parameter for arrivals over a period of hours is $\mu = 6t$.
(a) What is the probability that exactly 4 small aircraft arrive during a 1-hour period?
(b) What is the probability that at least 4 arrive during a 1-hour period?
(c) If we define a working day as 12 hours, what is the probability that at least 75 small aircraft arrive during a working day?

5.67 The number of customers arriving per hour at a certain automobile service facility is assumed to follow a Poisson distribution with mean $\lambda = 7$.
(a) Compute the probability that more than 10 customers will arrive in a 2-hour period.
(b) What is the mean number of arrivals during a 2-hour period?

5.68 Consider Exercise 5.62. What is the mean number of students who fail the test?

5.69 The probability that a person will die when he or she contracts a virus infection is 0.001. Of the next 4000 people infected, what is the mean number who will die?

5.70 A company purchases large lots of a certain kind of electronic device. A method is used that rejects a lot if 2 or more defective units are found in a random sample of 100 units.

(a) What is the mean number of defective units found in a sample of 100 units if the lot is 1% defective?

(b) What is the variance?

5.71 For a certain type of copper wire, it is known that, on the average, 1.5 flaws occur per millimeter. Assuming that the number of flaws is a Poisson random variable, what is the probability that no flaws occur in a certain portion of wire of length 5 millimeters? What is the mean number of flaws in a portion of length 5 millimeters?

5.72 Potholes on a highway can be a serious problem, and are in constant need of repair. With a particular type of terrain and make of concrete, past experience suggests that there are, on the average, 2 potholes per mile after a certain amount of usage. It is assumed that the Poisson process applies to the random variable "number of potholes."

(a) What is the probability that no more than one pothole will appear in a section of 1 mile?

(b) What is the probability that no more than 4 potholes will occur in a given section of 5 miles?

5.73 Hospital administrators in large cities anguish about traffic in emergency rooms. At a particular hospital in a large city, the staff on hand cannot accommodate the patient traffic if there are more than 10 emergency cases in a given hour. It is assumed that patient arrival follows a Poisson process, and historical data suggest that, on the average, 5 emergencies arrive per hour.

(a) What is the probability that in a given hour the staff cannot accommodate the patient traffic?

(b) What is the probability that more than 20 emergencies arrive during a 3-hour shift?

5.74 It is known that 3% of people whose luggage is screened at an airport have questionable objects in their luggage. What is the probability that a string of 15 people pass through screening successfully before an individual is caught with a questionable object? What is the expected number of people to pass through before an individual is stopped?

5.75 Computer technology has produced an environment in which robots operate with the use of microprocessors. The probability that a robot fails during any 6-hour shift is 0.10. What is the probability that a robot will operate through at most 5 shifts before it fails?

5.76 The refusal rate for telephone polls is known to be approximately 20%. A newspaper report indicates that 50 people were interviewed before the first refusal.

(a) Comment on the validity of the report. Use a probability in your argument.

(b) What is the expected number of people interviewed before a refusal?

Review Exercises

5.77 During a manufacturing process, 15 units are randomly selected each day from the production line to check the percent defective. From historical information it is known that the probability of a defective unit is 0.05. Any time 2 or more defectives are found in the sample of 15, the process is stopped. This procedure is used to provide a signal in case the probability of a defective has increased.

(a) What is the probability that on any given day the production process will be stopped? (Assume 5% defective.)

(b) Suppose that the probability of a defective has increased to 0.07. What is the probability that on any given day the production process will not be stopped?

5.78 An automatic welding machine is being considered for use in a production process. It will be considered for purchase if it is successful on 99% of its welds. Otherwise, it will not be considered efficient. A test is to be conducted with a prototype that is to perform 100 welds. The machine will be accepted for manufacture if it misses no more than 3 welds.

(a) What is the probability that a good machine will be rejected?

(b) What is the probability that an inefficient machine with 95% welding success will be accepted?

5.79 A car rental agency at a local airport has available 5 Fords, 7 Chevrolets, 4 Dodges, 3 Hondas, and 4 Toyotas. If the agency randomly selects 9 of these cars to chauffeur delegates from the airport to the downtown convention center, find the probability that 2 Fords, 3 Chevrolets, 1 Dodge, 1 Honda, and 2 Toyotas are used.

5.80 Service calls come to a maintenance center according to a Poisson process, and on average, 2.7 calls

are received per minute. Find the probability that
(a) no more than 4 calls come in any minute;
(b) fewer than 2 calls come in any minute;
(c) more than 10 calls come in a 5-minute period.

5.81 An electronics firm claims that the proportion of defective units from a certain process is 5%. A buyer has a standard procedure of inspecting 15 units selected randomly from a large lot. On a particular occasion, the buyer found 5 items defective.
(a) What is the probability of this occurrence, given that the claim of 5% defective is correct?
(b) What would be your reaction if you were the buyer?

5.82 An electronic switching device occasionally malfunctions, but the device is considered satisfactory if it makes, on average, no more than 0.20 error per hour. A particular 5-hour period is chosen for testing the device. If no more than 1 error occurs during the time period, the device will be considered satisfactory.
(a) What is the probability that a satisfactory device will be considered unsatisfactory on the basis of the test? Assume a Poisson process.
(b) What is the probability that a device will be accepted as satisfactory when, in fact, the mean number of errors is 0.25? Again, assume a Poisson process.

5.83 A company generally purchases large lots of a certain kind of electronic device. A method is used that rejects a lot if 2 or more defective units are found in a random sample of 100 units.
(a) What is the probability of rejecting a lot that is 1% defective?
(b) What is the probability of accepting a lot that is 5% defective?

5.84 A local drugstore owner knows that, on average, 100 people enter his store each hour.
(a) Find the probability that in a given 3-minute period nobody enters the store.
(b) Find the probability that in a given 3-minute period more than 5 people enter the store.

5.85 (a) Suppose that you throw 4 dice. Find the probability that you get at least one 1.
(b) Suppose that you throw 2 dice 24 times. Find the probability that you get at least one (1, 1), that is, "snake-eyes."

5.86 Suppose that out of 500 lottery tickets sold, 200 pay off at least the cost of the ticket. Now suppose that you buy 5 tickets. Find the probability that you will win back at least the cost of 3 tickets.

5.87 Imperfections in computer circuit boards and computer chips lend themselves to statistical treatment. For a particular type of board, the probability of a diode failure is 0.03 and the board contains 200 diodes.
(a) What is the mean number of failures among the diodes?
(b) What is the variance?
(c) The board will work if there are no defective diodes. What is the probability that a board will work?

5.88 The potential buyer of a particular engine requires (among other things) that the engine start successfully 10 consecutive times. Suppose the probability of a successful start is 0.990. Let us assume that the outcomes of attempted starts are independent.
(a) What is the probability that the engine is accepted after only 10 starts?
(b) What is the probability that 12 attempted starts are made during the acceptance process?

5.89 The acceptance scheme for purchasing lots containing a large number of batteries is to test no more than 75 randomly selected batteries and to reject a lot if a single battery fails. Suppose the probability of a failure is 0.001.
(a) What is the probability that a lot is accepted?
(b) What is the probability that a lot is rejected on the 20th test?
(c) What is the probability that it is rejected in 10 or fewer trials?

5.90 An oil drilling company ventures into various locations, and its success or failure is independent from one location to another. Suppose the probability of a success at any specific location is 0.25.
(a) What is the probability that the driller drills at 10 locations and has 1 success?
(b) The driller will go bankrupt if it drills 10 times before the first success occurs. What are the driller's prospects for bankruptcy?

5.91 Consider the information in Review Exercise 5.90. The drilling company feels that it will "hit it big" if the second success occurs on or before the sixth attempt. What is the probability that the driller will hit it big?

5.92 A couple decides to continue to have children until they have two males. Assuming that $P(\text{male}) = 0.5$, what is the probability that their second male is their fourth child?

5.93 It is known by researchers that 1 in 100 people carries a gene that leads to the inheritance of a certain chronic disease. In a random sample of 1000 individuals, what is the probability that fewer than 7 individuals carry the gene? Use a Poisson approximation. Again, using the approximation, what is the approximate mean number of people out of 1000 carrying the gene?

5.94 A production process produces electronic component parts. It is presumed that the probability of a defective part is 0.01. During a test of this presumption, 500 parts are sampled randomly and 15 defectives are observed.

(a) What is your response to the presumption that the process is 1% defective? Be sure that a computed probability accompanies your comment.

(b) Under the presumption of a 1% defective process, what is the probability that only 3 parts will be found defective?

(c) Do parts (a) and (b) again using the Poisson approximation.

5.95 A production process outputs items in lots of 50. Sampling plans exist in which lots are pulled aside periodically and exposed to a certain type of inspection. It is usually assumed that the proportion defective is very small. It is important to the company that lots containing defectives be a rare event. The current inspection plan is to periodically sample randomly 10 out of the 50 items in a lot and, if none are defective, to perform no intervention.

(a) Suppose in a lot chosen at random, 2 out of 50 are defective. What is the probability that at least 1 in the sample of 10 from the lot is defective?

(b) From your answer to part (a), comment on the quality of this sampling plan.

(c) What is the mean number of defects found out of 10 items sampled?

5.96 Consider the situation of Review Exercise 5.95. It has been determined that the sampling plan should be extensive enough that there is a high probability, say 0.9, that if as many as 2 defectives exist in the lot of 50 being sampled, at least 1 will be found in the sampling. With these restrictions, how many of the 50 items should be sampled?

5.97 National security requires that defense technology be able to detect incoming projectiles or missiles. To make the defense system successful, multiple radar screens are required. Suppose that three independent screens are to be operated and the probability that any one screen will detect an incoming missile is 0.8. Obviously, if no screens detect an incoming projectile, the system is unworthy and must be improved.

(a) What is the probability that an incoming missile will not be detected by any of the three screens?

(b) What is the probability that the missile will be detected by only one screen?

(c) What is the probability that it will be detected by at least two out of three screens?

5.98 Suppose it is important that the overall missile defense system be as near perfect as possible.

(a) Assuming the quality of the screens is as indicated in Review Exercise 5.97, how many are needed to ensure that the probability that a missile gets through undetected is 0.0001?

(b) Suppose it is decided to stay with only 3 screens and attempt to improve the screen detection ability. What must the individual screen effectiveness (i.e., probability of detection) be in order to achieve the effectiveness required in part (a)?

5.99 Go back to Review Exercise 5.95(a). Recompute the probability using the binomial distribution. Comment.

5.100 There are two vacancies in a certain university statistics department. Five individuals apply. Two have expertise in linear models, and one has expertise in applied probability. The search committee is instructed to choose the two applicants randomly.

(a) What is the probability that the two chosen are those with expertise in linear models?

(b) What is the probability that of the two chosen, one has expertise in linear models and one has expertise in applied probability?

5.101 The manufacturer of a tricycle for children has received complaints about defective brakes in the product. According to the design of the product and considerable preliminary testing, it had been determined that the probability of the kind of defect in the complaint was 1 in 10,000 (i.e., 0.0001). After a thorough investigation of the complaints, it was determined that during a certain period of time, 200 products were randomly chosen from production and 5 had defective brakes.

(a) Comment on the "1 in 10,000" claim by the manufacturer. Use a probabilistic argument. Use the binomial distribution for your calculations.

(b) Repeat part (a) using the Poisson approximation.

5.102 Group Project: Divide the class into two groups of approximately equal size. The students in group 1 will each toss a coin 10 times (n_1) and count the number of heads obtained. The students in group 2 will each toss a coin 40 times (n_2) and again count the number of heads. The students in each group should individually compute the proportion of heads observed, which is an estimate of p, the probability of observing a head. Thus, there will be a set of values of p_1 (from group 1) and a set of values p_2 (from group 2). All of the values of p_1 and p_2 are estimates of 0.5, which is the true value of the probability of observing a head for a fair coin.

(a) Which set of values is consistently closer to 0.5, the values of p_1 or p_2? Consider the proof of Theorem 5.1 on page 147 with regard to the estimates of the parameter $p = 0.5$. The values of p_1 were obtained with $n = n_1 = 10$, and the values of p_2 were obtained with $n = n_2 = 40$. Using the notation of the proof, the estimates are given by

$$p_1 = \frac{x_1}{n_1} = \frac{I_1 + \cdots + I_{n_1}}{n_1},$$

where $I_1, \ldots, I_{n_1}$ are 0s and 1s and $n_1 = 10$, and

$$p_2 = \frac{x_2}{n_2} = \frac{I_1 + \cdots + I_{n_2}}{n_2},$$

where $I_1, \ldots, I_{n_2}$, again, are 0s and 1s and $n_2 = 40$.

(b) Referring again to Theorem 5.1, show that

$$E(p_1) = E(p_2) = p = 0.5.$$

(c) Show that $\sigma_{p_1}^2 = \frac{\sigma_{X_1}^2}{n_1}$ is 4 times the value of $\sigma_{p_2}^2 = \frac{\sigma_{X_2}^2}{n_2}$. Then explain further why the values of p_2 from group 2 are more consistently closer to the true value, $p = 0.5$, than the values of p_1 from group 1.

You will continue to learn more and more about parameter estimation beginning in Chapter 9. At that point emphasis will put on the importance of the mean and variance of an estimator of a parameter.

5.6 Potential Misconceptions and Hazards; Relationship to Material in Other Chapters

The discrete distributions discussed in this chapter occur with great frequency in engineering and the biological and physical sciences. The exercises and examples certainly suggest this. Industrial sampling plans and many engineering judgments are based on the binomial and Poisson distributions as well as on the hypergeometric distribution. While the geometric and negative binomial distributions are used to a somewhat lesser extent, they also find applications. In particular, a negative binomial random variable can be viewed as a mixture of Poisson and gamma random variables (the gamma distribution will be discussed in Chapter 6).

Despite the rich heritage that these distributions find in real life, they can be misused unless the scientific practitioner is prudent and cautious. Of course, any probability calculation for the distributions discussed in this chapter is made under the assumption that the parameter value is known. Real-world applications often result in a parameter value that may "move around" due to factors that are difficult to control in the process or because of interventions in the process that have not been taken into account. For example, in Review Exercise 5.77, "historical information" is used. But is the process that exists now the same as that under which the historical data were collected? The use of the Poisson distribution can suffer even more from this kind of difficulty. For example, in Review Exercise 5.80, the questions in parts (a), (b), and (c) are based on the use of $\mu = 2.7$ calls per minute. Based on historical records, this is the number of calls that occur "on average." But in this and many other applications of the Poisson distribution, there are slow times and busy times and so there are times in which the conditions for the Poisson process may appear to hold when in fact they do not. Thus, the probability calculations may be incorrect. In the case of the binomial, the assumption that may fail in certain applications (in addition to nonconstancy of p) is the independence assumption, stating that the Bernoulli trials are independent.

One of the most famous misuses of the binomial distribution occurred in the 1961 baseball season, when Mickey Mantle and Roger Maris were engaged in a friendly battle to break Babe Ruth's all-time record of 60 home runs. A famous magazine article made a prediction, based on probability theory, that Mantle would break the record. The prediction was based on probability calculation with the use of the binomial distribution. The classic error made was to estimate the parameter p (one for each player) based on relative historical frequency of home runs throughout the players' careers. Maris, unlike Mantle, had not been a prodigious home run hitter prior to 1961 so his estimate of p was quite low. As a result, the calculated probability of breaking the record was quite high for Mantle and low for Maris. The end result: Mantle failed to break the record and Maris succeeded.

Chapter 6

Some Continuous Probability Distributions

6.1 Continuous Uniform Distribution

One of the simplest continuous distributions in all of statistics is the **continuous uniform distribution**. This distribution is characterized by a density function that is "flat," and thus the probability is uniform in a closed interval, say $[A, B]$. Although applications of the continuous uniform distribution are not as abundant as those for other distributions discussed in this chapter, it is appropriate for the novice to begin this introduction to continuous distributions with the uniform distribution.

Uniform Distribution The density function of the continuous uniform random variable X on the interval $[A, B]$ is

$$f(x; A, B) = \begin{cases} \frac{1}{B-A}, & A \leq x \leq B, \\ 0, & \text{elsewhere.} \end{cases}$$

The density function forms a rectangle with base $B-A$ and **constant height** $\frac{1}{B-A}$. As a result, the uniform distribution is often called the **rectangular distribution**. Note, however, that the interval may not always be closed: $[A, B]$. It can be (A, B) as well. The density function for a uniform random variable on the interval $[1, 3]$ is shown in Figure 6.1.

Probabilities are simple to calculate for the uniform distribution because of the simple nature of the density function. However, note that the application of this distribution is based on the assumption that the probability of falling in an interval of fixed length within $[A, B]$ is constant.

Example 6.1: Suppose that a large conference room at a certain company can be reserved for no more than 4 hours. Both long and short conferences occur quite often. In fact, it can be assumed that the length X of a conference has a uniform distribution on the interval $[0, 4]$.

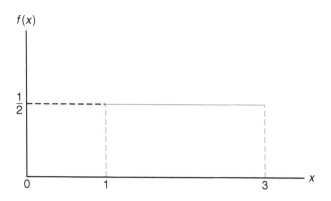

Figure 6.1: The density function for a random variable on the interval $[1, 3]$.

(a) What is the probability density function?

(b) What is the probability that any given conference lasts at least 3 hours?

Solution: (a) The appropriate density function for the uniformly distributed random variable X in this situation is

$$f(x) = \begin{cases} \frac{1}{4}, & 0 \leq x \leq 4, \\ 0, & \text{elsewhere.} \end{cases}$$

(b) $P[X \geq 3] = \int_3^4 \frac{1}{4}\, dx = \frac{1}{4}$.

Theorem 6.1: The mean and variance of the uniform distribution are

$$\mu = \frac{A+B}{2} \text{ and } \sigma^2 = \frac{(B-A)^2}{12}.$$

The proofs of the theorems are left to the reader. See Exercise 6.1 on page 185.

6.2 Normal Distribution

The most important continuous probability distribution in the entire field of statistics is the **normal distribution**. Its graph, called the **normal curve**, is the bell-shaped curve of Figure 6.2, which approximately describes many phenomena that occur in nature, industry, and research. For example, physical measurements in areas such as meteorological experiments, rainfall studies, and measurements of manufactured parts are often more than adequately explained with a normal distribution. In addition, errors in scientific measurements are extremely well approximated by a normal distribution. In 1733, Abraham DeMoivre developed the mathematical equation of the normal curve. It provided a basis from which much of the theory of inductive statistics is founded. The normal distribution is often referred to as the **Gaussian distribution**, in honor of Karl Friedrich Gauss

6.2 Normal Distribution

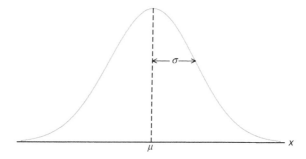

Figure 6.2: The normal curve.

(1777–1855), who also derived its equation from a study of errors in repeated measurements of the same quantity.

A continuous random variable X having the bell-shaped distribution of Figure 6.2 is called a **normal random variable**. The mathematical equation for the probability distribution of the normal variable depends on the two parameters μ and σ, its mean and standard deviation, respectively. Hence, we denote the values of the density of X by $n(x;\mu,\sigma)$.

Normal Distribution The density of the normal random variable X, with mean μ and variance σ^2, is

$$n(x;\mu,\sigma) = \frac{1}{\sqrt{2\pi}\sigma} e^{-\frac{1}{2\sigma^2}(x-\mu)^2}, \quad -\infty < x < \infty,$$

where $\pi = 3.14159\ldots$ and $e = 2.71828\ldots$.

Once μ and σ are specified, the normal curve is completely determined. For example, if $\mu = 50$ and $\sigma = 5$, then the ordinates $n(x;50,5)$ can be computed for various values of x and the curve drawn. In Figure 6.3, we have sketched two normal curves having the same standard deviation but different means. The two curves are identical in form but are centered at different positions along the horizontal axis.

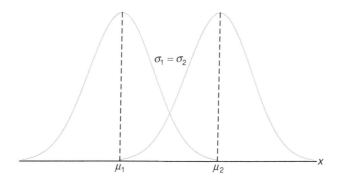

Figure 6.3: Normal curves with $\mu_1 < \mu_2$ and $\sigma_1 = \sigma_2$.

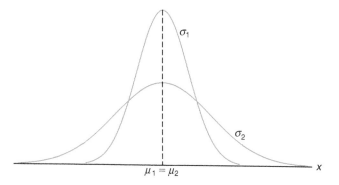

Figure 6.4: Normal curves with $\mu_1 = \mu_2$ and $\sigma_1 < \sigma_2$.

In Figure 6.4, we have sketched two normal curves with the same mean but different standard deviations. This time we see that the two curves are centered at exactly the same position on the horizontal axis, but the curve with the larger standard deviation is lower and spreads out farther. Remember that the area under a probability curve must be equal to 1, and therefore the more variable the set of observations, the lower and wider the corresponding curve will be.

Figure 6.5 shows two normal curves having different means and different standard deviations. Clearly, they are centered at different positions on the horizontal axis and their shapes reflect the two different values of σ.

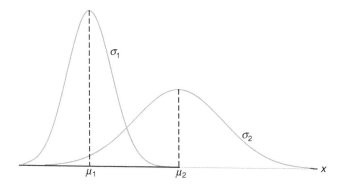

Figure 6.5: Normal curves with $\mu_1 < \mu_2$ and $\sigma_1 < \sigma_2$.

Based on inspection of Figures 6.2 through 6.5 and examination of the first and second derivatives of $n(x; \mu, \sigma)$, we list the following properties of the normal curve:

1. The mode, which is the point on the horizontal axis where the curve is a maximum, occurs at $x = \mu$.

2. The curve is symmetric about a vertical axis through the mean μ.

3. The curve has its points of inflection at $x = \mu \pm \sigma$; it is concave downward if $\mu - \sigma < X < \mu + \sigma$ and is concave upward otherwise.

4. The normal curve approaches the horizontal axis asymptotically as we proceed in either direction away from the mean.
5. The total area under the curve and above the horizontal axis is equal to 1.

Theorem 6.2: The mean and variance of $n(x; \mu, \sigma)$ are μ and σ^2, respectively. Hence, the standard deviation is σ.

Proof: To evaluate the mean, we first calculate

$$E(X - \mu) = \int_{-\infty}^{\infty} \frac{x - \mu}{\sqrt{2\pi}\sigma} e^{-\frac{1}{2}\left(\frac{x-\mu}{\sigma}\right)^2} dx.$$

Setting $z = (x - \mu)/\sigma$ and $dx = \sigma\, dz$, we obtain

$$E(X - \mu) = \frac{1}{\sqrt{2\pi}} \int_{-\infty}^{\infty} z e^{-\frac{1}{2}z^2} dz = 0,$$

since the integrand above is an odd function of z. Using Theorem 4.5 on page 128, we conclude that

$$E(X) = \mu.$$

The variance of the normal distribution is given by

$$E[(X - \mu)^2] = \frac{1}{\sqrt{2\pi}\sigma} \int_{-\infty}^{\infty} (x - \mu)^2 e^{-\frac{1}{2}[(x-\mu)/\sigma]^2} dx.$$

Again setting $z = (x - \mu)/\sigma$ and $dx = \sigma\, dz$, we obtain

$$E[(X - \mu)^2] = \frac{\sigma^2}{\sqrt{2\pi}} \int_{-\infty}^{\infty} z^2 e^{-\frac{z^2}{2}} dz.$$

Integrating by parts with $u = z$ and $dv = ze^{-z^2/2}\, dz$ so that $du = dz$ and $v = -e^{-z^2/2}$, we find that

$$E[(X - \mu)^2] = \frac{\sigma^2}{\sqrt{2\pi}} \left(-ze^{-z^2/2}\Big|_{-\infty}^{\infty} + \int_{-\infty}^{\infty} e^{-z^2/2} dz \right) = \sigma^2(0 + 1) = \sigma^2.$$

Many random variables have probability distributions that can be described adequately by the normal curve once μ and σ^2 are specified. In this chapter, we shall assume that these two parameters are known, perhaps from previous investigations. Later, we shall make statistical inferences when μ and σ^2 are unknown and have been estimated from the available experimental data.

We pointed out earlier the role that the normal distribution plays as a reasonable approximation of scientific variables in real-life experiments. There are other applications of the normal distribution that the reader will appreciate as he or she moves on in the book. The normal distribution finds enormous application as a *limiting distribution*. Under certain conditions, the normal distribution provides a good continuous approximation to the binomial and hypergeometric distributions. The case of the approximation to the binomial is covered in Section 6.5. In Chapter 8, the reader will learn about **sampling distributions**. It turns out that the limiting distribution of sample averages is normal. This provides a broad base for statistical inference that proves very valuable to the data analyst interested in estimation and hypothesis testing. Theory in the important areas such as analysis of variance (Chapters 13, 14, and 15) and quality control (Chapter 17) is based on assumptions that make use of the normal distribution.

In Section 6.3, examples demonstrate the use of tables of the normal distribution. Section 6.4 follows with examples of applications of the normal distribution.

6.3 Areas under the Normal Curve

The curve of any continuous probability distribution or density function is constructed so that the area under the curve bounded by the two ordinates $x = x_1$ and $x = x_2$ equals the probability that the random variable X assumes a value between $x = x_1$ and $x = x_2$. Thus, for the normal curve in Figure 6.6,

$$P(x_1 < X < x_2) = \int_{x_1}^{x_2} n(x; \mu, \sigma)\, dx = \frac{1}{\sqrt{2\pi}\sigma} \int_{x_1}^{x_2} e^{-\frac{1}{2\sigma^2}(x-\mu)^2}\, dx$$

is represented by the area of the shaded region.

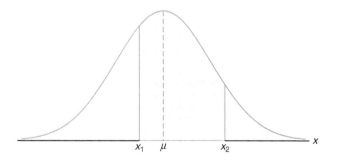

Figure 6.6: $P(x_1 < X < x_2) =$ area of the shaded region.

In Figures 6.3, 6.4, and 6.5 we saw how the normal curve is dependent on the mean and the standard deviation of the distribution under investigation. The area under the curve between any two ordinates must then also depend on the values μ and σ. This is evident in Figure 6.7, where we have shaded regions corresponding to $P(x_1 < X < x_2)$ for two curves with different means and variances. $P(x_1 < X < x_2)$, where X is the random variable describing distribution A, is indicated by the shaded area below the curve of A. If X is the random variable describing distribution B, then $P(x_1 < X < x_2)$ is given by the entire shaded region. Obviously, the two shaded regions are different in size; therefore, the probability associated with each distribution will be different for the two given values of X.

There are many types of statistical software that can be used in calculating areas under the normal curve. The difficulty encountered in solving integrals of normal density functions necessitates the tabulation of normal curve areas for quick reference. However, it would be a hopeless task to attempt to set up separate tables for every conceivable value of μ and σ. Fortunately, we are able to transform all the observations of any normal random variable X into a new set of observations

6.3 Areas under the Normal Curve

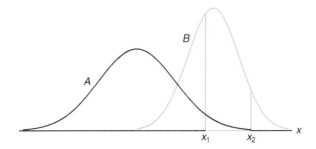

Figure 6.7: $P(x_1 < X < x_2)$ for different normal curves.

of a normal random variable Z with mean 0 and variance 1. This can be done by means of the transformation

$$Z = \frac{X - \mu}{\sigma}.$$

Whenever X assumes a value x, the corresponding value of Z is given by $z = (x - \mu)/\sigma$. Therefore, if X falls between the values $x = x_1$ and $x = x_2$, the random variable Z will fall between the corresponding values $z_1 = (x_1 - \mu)/\sigma$ and $z_2 = (x_2 - \mu)/\sigma$. Consequently, we may write

$$P(x_1 < X < x_2) = \frac{1}{\sqrt{2\pi}\sigma} \int_{x_1}^{x_2} e^{-\frac{1}{2\sigma^2}(x-\mu)^2} dx = \frac{1}{\sqrt{2\pi}} \int_{z_1}^{z_2} e^{-\frac{1}{2}z^2} dz$$

$$= \int_{z_1}^{z_2} n(z; 0, 1) \, dz = P(z_1 < Z < z_2),$$

where Z is seen to be a normal random variable with mean 0 and variance 1.

Definition 6.1: The distribution of a normal random variable with mean 0 and variance 1 is called a **standard normal distribution**.

The original and transformed distributions are illustrated in Figure 6.8. Since all the values of X falling between x_1 and x_2 have corresponding z values between z_1 and z_2, the area under the X-curve between the ordinates $x = x_1$ and $x = x_2$ in Figure 6.8 equals the area under the Z-curve between the transformed ordinates $z = z_1$ and $z = z_2$.

We have now reduced the required number of tables of normal-curve areas to one, that of the standard normal distribution. Table A.3 indicates the area under the standard normal curve corresponding to $P(Z < z)$ for values of z ranging from -3.49 to 3.49. To illustrate the use of this table, let us find the probability that Z is less than 1.74. First, we locate a value of z equal to 1.7 in the left column; then we move across the row to the column under 0.04, where we read 0.9591. Therefore, $P(Z < 1.74) = 0.9591$. To find a z value corresponding to a given probability, the process is reversed. For example, the z value leaving an area of 0.2148 under the curve to the left of z is seen to be -0.79.

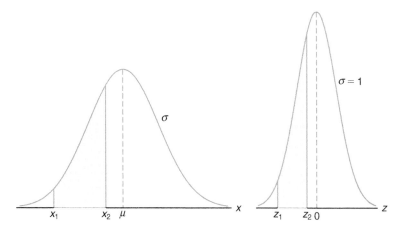

Figure 6.8: The original and transformed normal distributions.

Example 6.2: Given a standard normal distribution, find the area under the curve that lies
(a) to the right of $z = 1.84$ and
(b) between $z = -1.97$ and $z = 0.86$.

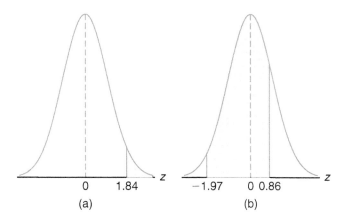

Figure 6.9: Areas for Example 6.2.

Solution: See Figure 6.9 for the specific areas.
(a) The area in Figure 6.9(a) to the right of $z = 1.84$ is equal to 1 minus the area in Table A.3 to the left of $z = 1.84$, namely, $1 - 0.9671 = 0.0329$.
(b) The area in Figure 6.9(b) between $z = -1.97$ and $z = 0.86$ is equal to the area to the left of $z = 0.86$ minus the area to the left of $z = -1.97$. From Table A.3 we find the desired area to be $0.8051 - 0.0244 = 0.7807$.

6.3 Areas under the Normal Curve

Example 6.3: Given a standard normal distribution, find the value of k such that

(a) $P(Z > k) = 0.3015$ and

(b) $P(k < Z < -0.18) = 0.4197$.

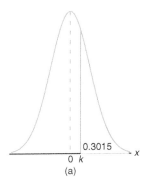 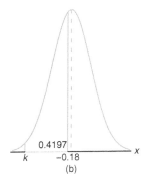

Figure 6.10: Areas for Example 6.3.

Solution: Distributions and the desired areas are shown in Figure 6.10.

(a) In Figure 6.10(a), we see that the k value leaving an area of 0.3015 to the right must then leave an area of 0.6985 to the left. From Table A.3 it follows that $k = 0.52$.

(b) From Table A.3 we note that the total area to the left of -0.18 is equal to 0.4286. In Figure 6.10(b), we see that the area between k and -0.18 is 0.4197, so the area to the left of k must be $0.4286 - 0.4197 = 0.0089$. Hence, from Table A.3, we have $k = -2.37$.

Example 6.4: Given a random variable X having a normal distribution with $\mu = 50$ and $\sigma = 10$, find the probability that X assumes a value between 45 and 62.

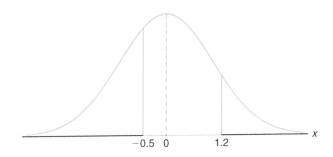

Figure 6.11: Area for Example 6.4.

Solution: The z values corresponding to $x_1 = 45$ and $x_2 = 62$ are
$$z_1 = \frac{45 - 50}{10} = -0.5 \text{ and } z_2 = \frac{62 - 50}{10} = 1.2.$$

Therefore,
$$P(45 < X < 62) = P(-0.5 < Z < 1.2).$$

$P(-0.5 < Z < 1.2)$ is shown by the area of the shaded region in Figure 6.11. This area may be found by subtracting the area to the left of the ordinate $z = -0.5$ from the entire area to the left of $z = 1.2$. Using Table A.3, we have

$$P(45 < X < 62) = P(-0.5 < Z < 1.2) = P(Z < 1.2) - P(Z < -0.5)$$
$$= 0.8849 - 0.3085 = 0.5764.$$

Example 6.5: Given that X has a normal distribution with $\mu = 300$ and $\sigma = 50$, find the probability that X assumes a value greater than 362.

Solution: The normal probability distribution with the desired area shaded is shown in Figure 6.12. To find $P(X > 362)$, we need to evaluate the area under the normal curve to the right of $x = 362$. This can be done by transforming $x = 362$ to the corresponding z value, obtaining the area to the left of z from Table A.3, and then subtracting this area from 1. We find that

$$z = \frac{362 - 300}{50} = 1.24.$$

Hence,

$$P(X > 362) = P(Z > 1.24) = 1 - P(Z < 1.24) = 1 - 0.8925 = 0.1075.$$

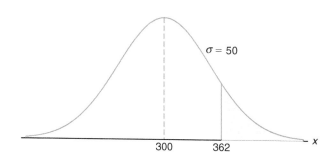

Figure 6.12: Area for Example 6.5.

According to Chebyshev's theorem on page 137, the probability that a random variable assumes a value within 2 standard deviations of the mean is at least $3/4$. If the random variable has a normal distribution, the z values corresponding to $x_1 = \mu - 2\sigma$ and $x_2 = \mu + 2\sigma$ are easily computed to be

$$z_1 = \frac{(\mu - 2\sigma) - \mu}{\sigma} = -2 \text{ and } z_2 = \frac{(\mu + 2\sigma) - \mu}{\sigma} = 2.$$

Hence,

$$P(\mu - 2\sigma < X < \mu + 2\sigma) = P(-2 < Z < 2) = P(Z < 2) - P(Z < -2)$$
$$= 0.9772 - 0.0228 = 0.9544,$$

which is a much stronger statement than that given by Chebyshev's theorem.

6.3 Areas under the Normal Curve

Using the Normal Curve in Reverse

Sometimes, we are required to find the value of z corresponding to a specified probability that falls between values listed in Table A.3 (see Example 6.6). For convenience, we shall always choose the z value corresponding to the tabular probability that comes closest to the specified probability.

The preceding two examples were solved by going first from a value of x to a z value and then computing the desired area. In Example 6.6, we reverse the process and begin with a known area or probability, find the z value, and then determine x by rearranging the formula

$$z = \frac{x - \mu}{\sigma} \quad \text{to give} \quad x = \sigma z + \mu.$$

Example 6.6: Given a normal distribution with $\mu = 40$ and $\sigma = 6$, find the value of x that has

(a) 45% of the area to the left and

(b) 14% of the area to the right.

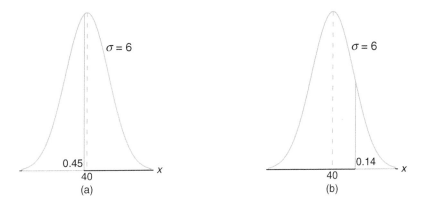

Figure 6.13: Areas for Example 6.6.

Solution: (a) An area of 0.45 to the left of the desired x value is shaded in Figure 6.13(a). We require a z value that leaves an area of 0.45 to the left. From Table A.3 we find $P(Z < -0.13) = 0.45$, so the desired z value is -0.13. Hence,

$$x = (6)(-0.13) + 40 = 39.22.$$

(b) In Figure 6.13(b), we shade an area equal to 0.14 to the right of the desired x value. This time we require a z value that leaves 0.14 of the area to the right and hence an area of 0.86 to the left. Again, from Table A.3, we find $P(Z < 1.08) = 0.86$, so the desired z value is 1.08 and

$$x = (6)(1.08) + 40 = 46.48.$$

6.4 Applications of the Normal Distribution

Some of the many problems for which the normal distribution is applicable are treated in the following examples. The use of the normal curve to approximate binomial probabilities is considered in Section 6.5.

Example 6.7: A certain type of storage battery lasts, on average, 3.0 years with a standard deviation of 0.5 year. Assuming that battery life is normally distributed, find the probability that a given battery will last less than 2.3 years.

Solution: First construct a diagram such as Figure 6.14, showing the given distribution of battery lives and the desired area. To find $P(X < 2.3)$, we need to evaluate the area under the normal curve to the left of 2.3. This is accomplished by finding the area to the left of the corresponding z value. Hence, we find that

$$z = \frac{2.3 - 3}{0.5} = -1.4,$$

and then, using Table A.3, we have

$$P(X < 2.3) = P(Z < -1.4) = 0.0808.$$

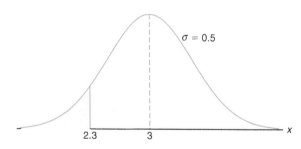

Figure 6.14: Area for Example 6.7.   Figure 6.15: Area for Example 6.8.

Example 6.8: An electrical firm manufactures light bulbs that have a life, before burn-out, that is normally distributed with mean equal to 800 hours and a standard deviation of 40 hours. Find the probability that a bulb burns between 778 and 834 hours.

Solution: The distribution of light bulb life is illustrated in Figure 6.15. The z values corresponding to $x_1 = 778$ and $x_2 = 834$ are

$$z_1 = \frac{778 - 800}{40} = -0.55 \text{ and } z_2 = \frac{834 - 800}{40} = 0.85.$$

Hence,

$$P(778 < X < 834) = P(-0.55 < Z < 0.85) = P(Z < 0.85) - P(Z < -0.55)$$
$$= 0.8023 - 0.2912 = 0.5111.$$

Example 6.9: In an industrial process, the diameter of a ball bearing is an important measurement. The buyer sets specifications for the diameter to be 3.0 ± 0.01 cm. The implication is that no part falling outside these specifications will be accepted. It is known that in the process the diameter of a ball bearing has a normal distribution with mean $\mu = 3.0$ and standard deviation $\sigma = 0.005$. On average, how many manufactured ball bearings will be scrapped?

Solution: The distribution of diameters is illustrated by Figure 6.16. The values corresponding to the specification limits are $x_1 = 2.99$ and $x_2 = 3.01$. The corresponding z values are

$$z_1 = \frac{2.99 - 3.0}{0.005} = -2.0 \text{ and } z_2 = \frac{3.01 - 3.0}{0.005} = +2.0.$$

Hence,

$$P(2.99 < X < 3.01) = P(-2.0 < Z < 2.0).$$

From Table A.3, $P(Z < -2.0) = 0.0228$. Due to symmetry of the normal distribution, we find that

$$P(Z < -2.0) + P(Z > 2.0) = 2(0.0228) = 0.0456.$$

As a result, it is anticipated that, on average, 4.56% of manufactured ball bearings will be scrapped.

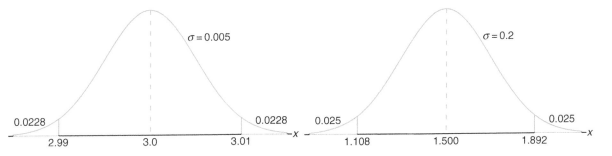

Figure 6.16: Area for Example 6.9.   Figure 6.17: Specifications for Example 6.10.

Example 6.10: Gauges are used to reject all components for which a certain dimension is not within the specification $1.50 \pm d$. It is known that this measurement is normally distributed with mean 1.50 and standard deviation 0.2. Determine the value d such that the specifications "cover" 95% of the measurements.

Solution: From Table A.3 we know that

$$P(-1.96 < Z < 1.96) = 0.95.$$

Therefore,

$$1.96 = \frac{(1.50 + d) - 1.50}{0.2},$$

from which we obtain

$$d = (0.2)(1.96) = 0.392.$$

An illustration of the specifications is shown in Figure 6.17.

Example 6.11: A certain machine makes electrical resistors having a mean resistance of 40 ohms and a standard deviation of 2 ohms. Assuming that the resistance follows a normal distribution and can be measured to any degree of accuracy, what percentage of resistors will have a resistance exceeding 43 ohms?

Solution: A percentage is found by multiplying the relative frequency by 100%. Since the relative frequency for an interval is equal to the probability of a value falling in the interval, we must find the area to the right of $x = 43$ in Figure 6.18. This can be done by transforming $x = 43$ to the corresponding z value, obtaining the area to the left of z from Table A.3, and then subtracting this area from 1. We find

$$z = \frac{43 - 40}{2} = 1.5.$$

Therefore,

$$P(X > 43) = P(Z > 1.5) = 1 - P(Z < 1.5) = 1 - 0.9332 = 0.0668.$$

Hence, 6.68% of the resistors will have a resistance exceeding 43 ohms.

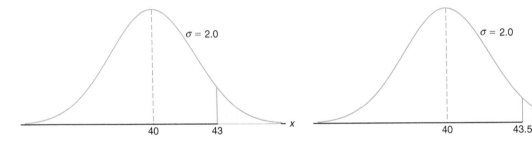

Figure 6.18: Area for Example 6.11.   Figure 6.19: Area for Example 6.12.

Example 6.12: Find the percentage of resistances exceeding 43 ohms for Example 6.11 if resistance is measured to the nearest ohm.

Solution: This problem differs from that in Example 6.11 in that we now assign a measurement of 43 ohms to all resistors whose resistances are greater than 42.5 and less than 43.5. We are actually approximating a discrete distribution by means of a continuous normal distribution. The required area is the region shaded to the right of 43.5 in Figure 6.19. We now find that

$$z = \frac{43.5 - 40}{2} = 1.75.$$

Hence,

$$P(X > 43.5) = P(Z > 1.75) = 1 - P(Z < 1.75) = 1 - 0.9599 = 0.0401.$$

Therefore, 4.01% of the resistances exceed 43 ohms when measured to the nearest ohm. The difference 6.68% − 4.01% = 2.67% between this answer and that of Example 6.11 represents all those resistance values greater than 43 and less than 43.5 that are now being recorded as 43 ohms.

Exercises

Example 6.13: The average grade for an exam is 74, and the standard deviation is 7. If 12% of the class is given As, and the grades are curved to follow a normal distribution, what is the lowest possible A and the highest possible B?

Solution: In this example, we begin with a known area of probability, find the z value, and then determine x from the formula $x = \sigma z + \mu$. An area of 0.12, corresponding to the fraction of students receiving As, is shaded in Figure 6.20. We require a z value that leaves 0.12 of the area to the right and, hence, an area of 0.88 to the left. From Table A.3, $P(Z < 1.18)$ has the closest value to 0.88, so the desired z value is 1.18. Hence,

$$x = (7)(1.18) + 74 = 82.26.$$

Therefore, the lowest A is 83 and the highest B is 82.

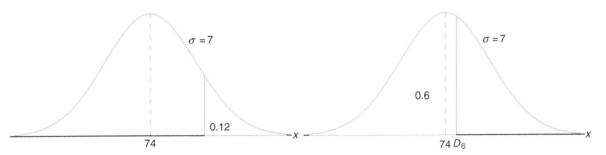

Figure 6.20: Area for Example 6.13. Figure 6.21: Area for Example 6.14.

Example 6.14: Refer to Example 6.13 and find the sixth decile.

Solution: The sixth decile, written D_6, is the x value that leaves 60% of the area to the left, as shown in Figure 6.21. From Table A.3 we find $P(Z < 0.25) \approx 0.6$, so the desired z value is 0.25. Now $x = (7)(0.25) + 74 = 75.75$. Hence, $D_6 = 75.75$. That is, 60% of the grades are 75 or less.

Exercises

6.1 Given a continuous uniform distribution, show that

(a) $\mu = \frac{A+B}{2}$ and

(b) $\sigma^2 = \frac{(B-A)^2}{12}$.

6.2 Suppose X follows a continuous uniform distribution from 1 to 5. Determine the conditional probability $P(X > 2.5 \mid X \leq 4)$.

6.3 The daily amount of coffee, in liters, dispensed by a machine located in an airport lobby is a random variable X having a continuous uniform distribution with $A = 7$ and $B = 10$. Find the probability that on a given day the amount of coffee dispensed by this machine will be

(a) at most 8.8 liters;

(b) more than 7.4 liters but less than 9.5 liters;

(c) at least 8.5 liters.

6.4 A bus arrives every 10 minutes at a bus stop. It is assumed that the waiting time for a particular individual is a random variable with a continuous uniform distribution.

(a) What is the probability that the individual waits more than 7 minutes?

(b) What is the probability that the individual waits between 2 and 7 minutes?

6.5 Given a standard normal distribution, find the area under the curve that lies
(a) to the left of $z = -1.39$;
(b) to the right of $z = 1.96$;
(c) between $z = -2.16$ and $z = -0.65$;
(d) to the left of $z = 1.43$;
(e) to the right of $z = -0.89$;
(f) between $z = -0.48$ and $z = 1.74$.

6.6 Find the value of z if the area under a standard normal curve
(a) to the right of z is 0.3622;
(b) to the left of z is 0.1131;
(c) between 0 and z, with $z > 0$, is 0.4838;
(d) between $-z$ and z, with $z > 0$, is 0.9500.

6.7 Given a standard normal distribution, find the value of k such that
(a) $P(Z > k) = 0.2946$;
(b) $P(Z < k) = 0.0427$;
(c) $P(-0.93 < Z < k) = 0.7235$.

6.8 Given a normal distribution with $\mu = 30$ and $\sigma = 6$, find
(a) the normal curve area to the right of $x = 17$;
(b) the normal curve area to the left of $x = 22$;
(c) the normal curve area between $x = 32$ and $x = 41$;
(d) the value of x that has 80% of the normal curve area to the left;
(e) the two values of x that contain the middle 75% of the normal curve area.

6.9 Given the normally distributed variable X with mean 18 and standard deviation 2.5, find
(a) $P(X < 15)$;
(b) the value of k such that $P(X < k) = 0.2236$;
(c) the value of k such that $P(X > k) = 0.1814$;
(d) $P(17 < X < 21)$.

6.10 According to Chebyshev's theorem, the probability that any random variable assumes a value within 3 standard deviations of the mean is at least 8/9. If it is known that the probability distribution of a random variable X is normal with mean μ and variance σ^2, what is the exact value of $P(\mu - 3\sigma < X < \mu + 3\sigma)$?

6.11 A soft-drink machine is regulated so that it discharges an average of 200 milliliters per cup. If the amount of drink is normally distributed with a standard deviation equal to 15 milliliters,
(a) what fraction of the cups will contain more than 224 milliliters?
(b) what is the probability that a cup contains between 191 and 209 milliliters?
(c) how many cups will probably overflow if 230-milliliter cups are used for the next 1000 drinks?
(d) below what value do we get the smallest 25% of the drinks?

6.12 The loaves of rye bread distributed to local stores by a certain bakery have an average length of 30 centimeters and a standard deviation of 2 centimeters. Assuming that the lengths are normally distributed, what percentage of the loaves are
(a) longer than 31.7 centimeters?
(b) between 29.3 and 33.5 centimeters in length?
(c) shorter than 25.5 centimeters?

6.13 A research scientist reports that mice will live an average of 40 months when their diets are sharply restricted and then enriched with vitamins and proteins. Assuming that the lifetimes of such mice are normally distributed with a standard deviation of 6.3 months, find the probability that a given mouse will live
(a) more than 32 months;
(b) less than 28 months;
(c) between 37 and 49 months.

6.14 The finished inside diameter of a piston ring is normally distributed with a mean of 10 centimeters and a standard deviation of 0.03 centimeter.
(a) What proportion of rings will have inside diameters exceeding 10.075 centimeters?
(b) What is the probability that a piston ring will have an inside diameter between 9.97 and 10.03 centimeters?
(c) Below what value of inside diameter will 15% of the piston rings fall?

6.15 A lawyer commutes daily from his suburban home to his midtown office. The average time for a one-way trip is 24 minutes, with a standard deviation of 3.8 minutes. Assume the distribution of trip times to be normally distributed.
(a) What is the probability that a trip will take at least 1/2 hour?
(b) If the office opens at 9:00 A.M. and the lawyer leaves his house at 8:45 A.M. daily, what percentage of the time is he late for work?

(c) If he leaves the house at 8:35 A.M. and coffee is served at the office from 8:50 A.M. until 9:00 A.M., what is the probability that he misses coffee?

(d) Find the length of time above which we find the slowest 15% of the trips.

(e) Find the probability that 2 of the next 3 trips will take at least 1/2 hour.

6.16 In the November 1990 issue of *Chemical Engineering Progress*, a study discussed the percent purity of oxygen from a certain supplier. Assume that the mean was 99.61 with a standard deviation of 0.08. Assume that the distribution of percent purity was approximately normal.

(a) What percentage of the purity values would you expect to be between 99.5 and 99.7?

(b) What purity value would you expect to exceed exactly 5% of the population?

6.17 The average life of a certain type of small motor is 10 years with a standard deviation of 2 years. The manufacturer replaces free all motors that fail while under guarantee. If she is willing to replace only 3% of the motors that fail, how long a guarantee should be offered? Assume that the lifetime of a motor follows a normal distribution.

6.18 The heights of 1000 students are normally distributed with a mean of 174.5 centimeters and a standard deviation of 6.9 centimeters. Assuming that the heights are recorded to the nearest half-centimeter, how many of these students would you expect to have heights

(a) less than 160.0 centimeters?

(b) between 171.5 and 182.0 centimeters inclusive?

(c) equal to 175.0 centimeters?

(d) greater than or equal to 188.0 centimeters?

6.19 A company pays its employees an average wage of $15.90 an hour with a standard deviation of $1.50. If the wages are approximately normally distributed and paid to the nearest cent,

(a) what percentage of the workers receive wages between $13.75 and $16.22 an hour inclusive?

(b) the highest 5% of the employee hourly wages is greater than what amount?

6.20 The weights of a large number of miniature poodles are approximately normally distributed with a mean of 8 kilograms and a standard deviation of 0.9 kilogram. If measurements are recorded to the nearest tenth of a kilogram, find the fraction of these poodles with weights

(a) over 9.5 kilograms;

(b) of at most 8.6 kilograms;

(c) between 7.3 and 9.1 kilograms inclusive.

6.21 The tensile strength of a certain metal component is normally distributed with a mean of 10,000 kilograms per square centimeter and a standard deviation of 100 kilograms per square centimeter. Measurements are recorded to the nearest 50 kilograms per square centimeter.

(a) What proportion of these components exceed 10,150 kilograms per square centimeter in tensile strength?

(b) If specifications require that all components have tensile strength between 9800 and 10,200 kilograms per square centimeter inclusive, what proportion of pieces would we expect to scrap?

6.22 If a set of observations is normally distributed, what percent of these differ from the mean by

(a) more than 1.3σ?

(b) less than 0.52σ?

6.23 The IQs of 600 applicants to a certain college are approximately normally distributed with a mean of 115 and a standard deviation of 12. If the college requires an IQ of at least 95, how many of these students will be rejected on this basis of IQ, regardless of their other qualifications? Note that IQs are recorded to the nearest integers.

6.5 Normal Approximation to the Binomial

Probabilities associated with binomial experiments are readily obtainable from the formula $b(x; n, p)$ of the binomial distribution or from Table A.1 when n is small. In addition, binomial probabilities are readily available in many computer software packages. However, it is instructive to learn the relationship between the binomial and the normal distribution. In Section 5.5, we illustrated how the Poisson distribution can be used to approximate binomial probabilities when n is quite large and p is very close to 0 or 1. Both the binomial and the Poisson distributions are discrete. The first application of a continuous probability distribution to approximate probabilities over a discrete sample space was demonstrated in Example 6.12, where the normal curve was used. The normal distribution is often a good approximation to a discrete distribution when the latter takes on a symmetric bell shape. From a theoretical point of view, some distributions converge to the normal as their parameters approach certain limits. The normal distribution is a convenient approximating distribution because the cumulative distribution function is so easily tabled. The binomial distribution is nicely approximated by the normal in practical problems when one works with the cumulative distribution function. We now state a theorem that allows us to use areas under the normal curve to approximate binomial properties when n is sufficiently large.

Theorem 6.3: If X is a binomial random variable with mean $\mu = np$ and variance $\sigma^2 = npq$, then the limiting form of the distribution of

$$Z = \frac{X - np}{\sqrt{npq}},$$

as $n \to \infty$, is the standard normal distribution $n(z; 0, 1)$.

It turns out that the normal distribution with $\mu = np$ and $\sigma^2 = np(1-p)$ not only provides a very accurate approximation to the binomial distribution when n is large and p is not extremely close to 0 or 1 but also provides a fairly good approximation even when n is small and p is reasonably close to $1/2$.

To illustrate the normal approximation to the binomial distribution, we first draw the histogram for $b(x; 15, 0.4)$ and then superimpose the particular normal curve having the same mean and variance as the binomial variable X. Hence, we draw a normal curve with

$$\mu = np = (15)(0.4) = 6 \text{ and } \sigma^2 = npq = (15)(0.4)(0.6) = 3.6.$$

The histogram of $b(x; 15, 0.4)$ and the corresponding superimposed normal curve, which is completely determined by its mean and variance, are illustrated in Figure 6.22.

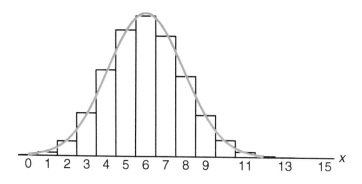

Figure 6.22: Normal approximation of $b(x; 15, 0.4)$.

6.5 Normal Approximation to the Binomial

The exact probability that the binomial random variable X assumes a given value x is equal to the area of the bar whose base is centered at x. For example, the exact probability that X assumes the value 4 is equal to the area of the rectangle with base centered at $x = 4$. Using Table A.1, we find this area to be

$$P(X = 4) = b(4; 15, 0.4) = 0.1268,$$

which is approximately equal to the area of the shaded region under the normal curve between the two ordinates $x_1 = 3.5$ and $x_2 = 4.5$ in Figure 6.23. Converting to z values, we have

$$z_1 = \frac{3.5 - 6}{1.897} = -1.32 \quad \text{and} \quad z_2 = \frac{4.5 - 6}{1.897} = -0.79.$$

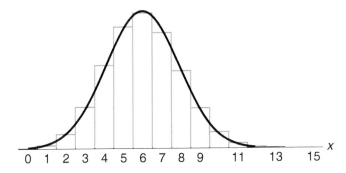

Figure 6.23: Normal approximation of $b(x; 15, 0.4)$ and $\sum_{x=7}^{9} b(x; 15, 0.4)$.

If X is a binomial random variable and Z a standard normal variable, then

$$P(X = 4) = b(4; 15, 0.4) \approx P(-1.32 < Z < -0.79)$$
$$= P(Z < -0.79) - P(Z < -1.32) = 0.2148 - 0.0934 = 0.1214.$$

This agrees very closely with the exact value of 0.1268.

The normal approximation is most useful in calculating binomial sums for large values of n. Referring to Figure 6.23, we might be interested in the probability that X assumes a value from 7 to 9 inclusive. The exact probability is given by

$$P(7 \leq X \leq 9) = \sum_{x=0}^{9} b(x; 15, 0.4) - \sum_{x=0}^{6} b(x; 15, 0.4)$$
$$= 0.9662 - 0.6098 = 0.3564,$$

which is equal to the sum of the areas of the rectangles with bases centered at $x = 7$, 8, and 9. For the normal approximation, we find the area of the shaded region under the curve between the ordinates $x_1 = 6.5$ and $x_2 = 9.5$ in Figure 6.23. The corresponding z values are

$$z_1 = \frac{6.5 - 6}{1.897} = 0.26 \quad \text{and} \quad z_2 = \frac{9.5 - 6}{1.897} = 1.85.$$

Now,
$$P(7 \leq X \leq 9) \approx P(0.26 < Z < 1.85) = P(Z < 1.85) - P(Z < 0.26)$$
$$= 0.9678 - 0.6026 = 0.3652.$$

Once again, the normal curve approximation provides a value that agrees very closely with the exact value of 0.3564. The degree of accuracy, which depends on how well the curve fits the histogram, will increase as n increases. This is particularly true when p is not very close to $1/2$ and the histogram is no longer symmetric. Figures 6.24 and 6.25 show the histograms for $b(x; 6, 0.2)$ and $b(x; 15, 0.2)$, respectively. It is evident that a normal curve would fit the histogram considerably better when $n = 15$ than when $n = 6$.

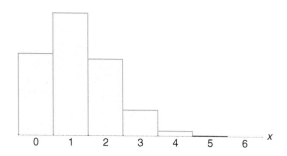

Figure 6.24: Histogram for $b(x; 6, 0.2)$.

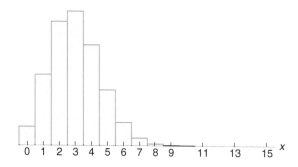

Figure 6.25: Histogram for $b(x; 15, 0.2)$.

In our illustrations of the normal approximation to the binomial, it became apparent that if we seek the area under the normal curve to the left of, say, x, it is more accurate to use $x + 0.5$. This is a correction to accommodate the fact that a discrete distribution is being approximated by a continuous distribution. The correction $+0.5$ is called a **continuity correction**. The foregoing discussion leads to the following formal normal approximation to the binomial.

Normal Approximation to the Binomial Distribution

Let X be a binomial random variable with parameters n and p. For large n, X has approximately a normal distribution with $\mu = np$ and $\sigma^2 = npq = np(1-p)$ and

$$P(X \leq x) = \sum_{k=0}^{x} b(k; n, p)$$
$$\approx \text{ area under normal curve to the left of } x + 0.5$$
$$= P\left(Z \leq \frac{x + 0.5 - np}{\sqrt{npq}}\right),$$

and the approximation will be good if np and $n(1-p)$ are greater than or equal to 5.

As we indicated earlier, the quality of the approximation is quite good for large n. If p is close to $1/2$, a moderate or small sample size will be sufficient for a reasonable approximation. We offer Table 6.1 as an indication of the quality of the

6.5 Normal Approximation to the Binomial

approximation. Both the normal approximation and the true binomial cumulative probabilities are given. Notice that at $p = 0.05$ and $p = 0.10$, the approximation is fairly crude for $n = 10$. However, even for $n = 10$, note the improvement for $p = 0.50$. On the other hand, when p is fixed at $p = 0.05$, note the improvement of the approximation as we go from $n = 20$ to $n = 100$.

Table 6.1: Normal Approximation and True Cumulative Binomial Probabilities

r	$p = 0.05, n = 10$		$p = 0.10, n = 10$		$p = 0.50, n = 10$	
	Binomial	Normal	Binomial	Normal	Binomial	Normal
0	0.5987	0.5000	0.3487	0.2981	0.0010	0.0022
1	0.9139	0.9265	0.7361	0.7019	0.0107	0.0136
2	0.9885	0.9981	0.9298	0.9429	0.0547	0.0571
3	0.9990	1.0000	0.9872	0.9959	0.1719	0.1711
4	1.0000	1.0000	0.9984	0.9999	0.3770	0.3745
5			1.0000	1.0000	0.6230	0.6255
6					0.8281	0.8289
7					0.9453	0.9429
8					0.9893	0.9864
9					0.9990	0.9978
10					1.0000	0.9997

	$p = 0.05$					
r	$n = 20$		$n = 50$		$n = 100$	
	Binomial	Normal	Binomial	Normal	Binomial	Normal
0	0.3585	0.3015	0.0769	0.0968	0.0059	0.0197
1	0.7358	0.6985	0.2794	0.2578	0.0371	0.0537
2	0.9245	0.9382	0.5405	0.5000	0.1183	0.1251
3	0.9841	0.9948	0.7604	0.7422	0.2578	0.2451
4	0.9974	0.9998	0.8964	0.9032	0.4360	0.4090
5	0.9997	1.0000	0.9622	0.9744	0.6160	0.5910
6	1.0000	1.0000	0.9882	0.9953	0.7660	0.7549
7			0.9968	0.9994	0.8720	0.8749
8			0.9992	0.9999	0.9369	0.9463
9			0.9998	1.0000	0.9718	0.9803
10			1.0000	1.0000	0.9885	0.9941

Example 6.15: The probability that a patient recovers from a rare blood disease is 0.4. If 100 people are known to have contracted this disease, what is the probability that fewer than 30 survive?

Solution: Let the binomial variable X represent the number of patients who survive. Since $n = 100$, we should obtain fairly accurate results using the normal-curve approximation with

$$\mu = np = (100)(0.4) = 40 \text{ and } \sigma = \sqrt{npq} = \sqrt{(100)(0.4)(0.6)} = 4.899.$$

To obtain the desired probability, we have to find the area to the left of $x = 29.5$.

The z value corresponding to 29.5 is

$$z = \frac{29.5 - 40}{4.899} = -2.14,$$

and the probability of fewer than 30 of the 100 patients surviving is given by the shaded region in Figure 6.26. Hence,

$$P(X < 30) \approx P(Z < -2.14) = 0.0162.$$

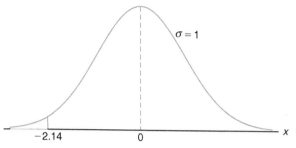

Figure 6.26: Area for Example 6.15.

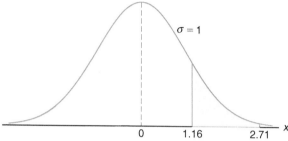

Figure 6.27: Area for Example 6.16.

Example 6.16: A multiple-choice quiz has 200 questions, each with 4 possible answers of which only 1 is correct. What is the probability that sheer guesswork yields from 25 to 30 correct answers for the 80 of the 200 problems about which the student has no knowledge?

Solution: The probability of guessing a correct answer for each of the 80 questions is $p = 1/4$. If X represents the number of correct answers resulting from guesswork, then

$$P(25 \leq X \leq 30) = \sum_{x=25}^{30} b(x; 80, 1/4).$$

Using the normal curve approximation with

$$\mu = np = (80)\left(\frac{1}{4}\right) = 20$$

and

$$\sigma = \sqrt{npq} = \sqrt{(80)(1/4)(3/4)} = 3.873,$$

we need the area between $x_1 = 24.5$ and $x_2 = 30.5$. The corresponding z values are

$$z_1 = \frac{24.5 - 20}{3.873} = 1.16 \text{ and } z_2 = \frac{30.5 - 20}{3.873} = 2.71.$$

The probability of correctly guessing from 25 to 30 questions is given by the shaded region in Figure 6.27. From Table A.3 we find that

$$P(25 \leq X \leq 30) = \sum_{x=25}^{30} b(x; 80, 0.25) \approx P(1.16 < Z < 2.71)$$
$$= P(Z < 2.71) - P(Z < 1.16) = 0.9966 - 0.8770 = 0.1196.$$

Exercises

6.24 A coin is tossed 400 times. Use the normal curve approximation to find the probability of obtaining
(a) between 185 and 210 heads inclusive;
(b) exactly 205 heads;
(c) fewer than 176 or more than 227 heads.

6.25 A process for manufacturing an electronic component yields items of which 1% are defective. A quality control plan is to select 100 items from the process, and if none are defective, the process continues. Use the normal approximation to the binomial to find
(a) the probability that the process continues given the sampling plan described;
(b) the probability that the process continues even if the process has gone bad (i.e., if the frequency of defective components has shifted to 5.0% defective).

6.26 A process yields 10% defective items. If 100 items are randomly selected from the process, what is the probability that the number of defectives
(a) exceeds 13?
(b) is less than 8?

6.27 The probability that a patient recovers from a delicate heart operation is 0.9. Of the next 100 patients having this operation, what is the probability that
(a) between 84 and 95 inclusive survive?
(b) fewer than 86 survive?

6.28 Researchers at George Washington University and the National Institutes of Health claim that approximately 75% of people believe "tranquilizers work very well to make a person more calm and relaxed." Of the next 80 people interviewed, what is the probability that
(a) at least 50 are of this opinion?
(b) at most 56 are of this opinion?

6.29 If 20% of the residents in a U.S. city prefer a white cell phone over any other color available, what is the probability that among the next 1000 cell phone purchased in that city
(a) between 170 and 185 inclusive will be white?
(b) at least 210 but not more than 225 will be white?

6.30 A drug manufacturer claims that a certain drug cures a blood disease, on the average, 80% of the time. To check the claim, government testers use the drug on a sample of 100 individuals and decide to accept the claim if 75 or more are cured.
(a) What is the probability that the claim will be rejected when the cure probability is, in fact, 0.8?
(b) What is the probability that the claim will be accepted by the government when the cure probability is as low as 0.7?

6.31 One-sixth of the male freshmen entering a large state school are out-of-state students. If the students are assigned at random to dormitories, 180 to a building, what is the probability that in a given dormitory at least one-fifth of the students are from out of state?

6.32 A pharmaceutical company knows that approximately 5% of its birth-control pills have an ingredient that is below the minimum strength, thus rendering the pill ineffective. What is the probability that fewer than 10 in a sample of 200 pills will be ineffective?

6.33 Statistics released by the National Highway Traffic Safety Administration and the National Safety Council show that on an average weekend night, 1 out of every 10 drivers on the road is drunk. If 400 drivers are randomly checked next Saturday night, what is the probability that the number of drunk drivers will be
(a) less than 32?
(b) more than 49?
(c) at least 35 but less than 47?

6.34 A pair of dice is rolled 180 times. What is the probability that a total of 7 occurs
(a) at least 25 times?
(b) between 33 and 41 times inclusive?
(c) exactly 30 times?

6.35 A company produces component parts for an engine. Parts specifications suggest that 95% of items meet specifications. The parts are shipped to customers in lots of 100.
(a) What is the probability that more than 2 items in a given lot will be defective?
(b) What is the probability that more than 10 items in a lot will be defective?

6.36 A common practice of airline companies is to sell more tickets for a particular flight than there are seats on the plane, because customers who buy tickets do not always show up for the flight. Suppose that the percentage of no-shows at flight time is 2%. For a particular flight with 197 seats, a total of 200 tick-cts were sold. What is the probability that the airline overbooked this flight?

6.37 The serum cholesterol level X in 14-year-old boys has approximately a normal distribution with mean 170 and standard deviation 30.

(a) Find the probability that the serum cholesterol level of a randomly chosen 14-year-old boy exceeds 230.

(b) In a middle school there are 300 14-year-old boys. Find the probability that at least 8 boys have a serum cholesterol level that exceeds 230.

6.38 A telemarketing company has a special letter-opening machine that opens and removes the contents of an envelope. If the envelope is fed improperly into the machine, the contents of the envelope may not be removed or may be damaged. In this case, the machine is said to have "failed."

(a) If the machine has a probability of failure of 0.01, what is the probability of more than 1 failure occurring in a batch of 20 envelopes?

(b) If the probability of failure of the machine is 0.01 and a batch of 500 envelopes is to be opened, what is the probability that more than 8 failures will occur?

6.6 Gamma and Exponential Distributions

Although the normal distribution can be used to solve many problems in engineering and science, there are still numerous situations that require different types of density functions. Two such density functions, the **gamma** and **exponential distributions**, are discussed in this section.

It turns out that the exponential distribution is a special case of the gamma distribution. Both find a large number of applications. The exponential and gamma distributions play an important role in both queuing theory and reliability problems. Time between arrivals at service facilities and time to failure of component parts and electrical systems often are nicely modeled by the exponential distribution. The relationship between the gamma and the exponential allows the gamma to be used in similar types of problems. More details and illustrations will be supplied later in the section.

The gamma distribution derives its name from the well-known **gamma function**, studied in many areas of mathematics. Before we proceed to the gamma distribution, let us review this function and some of its important properties.

Definition 6.2: The **gamma function** is defined by

$$\Gamma(\alpha) = \int_0^\infty x^{\alpha-1} e^{-x}\, dx, \quad \text{for } \alpha > 0.$$

The following are a few simple properties of the gamma function.

(a) $\Gamma(n) = (n-1)(n-2)\cdots(1)\Gamma(1)$, for a positive integer n.

To see the proof, integrating by parts with $u = x^{\alpha-1}$ and $dv = e^{-x}\, dx$, we obtain

$$\Gamma(\alpha) = -e^{-x} x^{\alpha-1} \Big|_0^\infty + \int_0^\infty e^{-x}(\alpha-1)x^{\alpha-2}\, dx = (\alpha-1)\int_0^\infty x^{\alpha-2} e^{-x}\, dx,$$

for $\alpha > 1$, which yields the recursion formula

$$\Gamma(\alpha) = (\alpha-1)\Gamma(\alpha-1).$$

The result follows after repeated application of the recursion formula. Using this result, we can easily show the following two properties.

6.6 Gamma and Exponential Distributions

(b) $\Gamma(n) = (n-1)!$ for a positive integer n.

(c) $\Gamma(1) = 1$.

Furthermore, we have the following property of $\Gamma(\alpha)$, which is left for the reader to verify (see Exercise 6.39 on page 206).

(d) $\Gamma(1/2) = \sqrt{\pi}$.

The following is the definition of the **gamma distribution**.

Gamma Distribution The continuous random variable X has a **gamma distribution**, with parameters α and β, if its density function is given by

$$f(x;\alpha,\beta) = \begin{cases} \frac{1}{\beta^\alpha \Gamma(\alpha)} x^{\alpha-1} e^{-x/\beta}, & x > 0, \\ 0, & \text{elsewhere,} \end{cases}$$

where $\alpha > 0$ and $\beta > 0$.

Graphs of several gamma distributions are shown in Figure 6.28 for certain specified values of the parameters α and β. The special gamma distribution for which $\alpha = 1$ is called the **exponential distribution**.

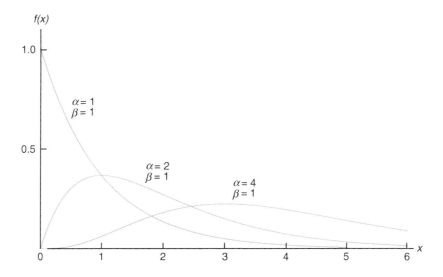

Figure 6.28: Gamma distributions.

Exponential Distribution The continuous random variable X has an **exponential distribution**, with parameter β, if its density function is given by

$$f(x;\beta) = \begin{cases} \frac{1}{\beta} e^{-x/\beta}, & x > 0, \\ 0, & \text{elsewhere,} \end{cases}$$

where $\beta > 0$.

The following theorem and corollary give the mean and variance of the gamma and exponential distributions.

Theorem 6.4: The mean and variance of the gamma distribution are
$$\mu = \alpha\beta \text{ and } \sigma^2 = \alpha\beta^2.$$

The proof of this theorem is found in Appendix A.26.

Corollary 6.1: The mean and variance of the exponential distribution are
$$\mu = \beta \text{ and } \sigma^2 = \beta^2.$$

Relationship to the Poisson Process

We shall pursue applications of the exponential distribution and then return to the gamma distribution. The most important applications of the exponential distribution are situations where the Poisson process applies (see Section 5.5). The reader should recall that the Poisson process allows for the use of the discrete distribution called the Poisson distribution. Recall that the Poisson distribution is used to compute the probability of specific numbers of "events" during a particular *period of time or span of space*. In many applications, the time period or span of space is the random variable. For example, an industrial engineer may be interested in modeling the time T between arrivals at a congested intersection during rush hour in a large city. An arrival represents the Poisson event.

The relationship between the exponential distribution (often called the negative exponential) and the Poisson process is quite simple. In Chapter 5, the Poisson distribution was developed as a single-parameter distribution with parameter λ, where λ may be interpreted as the mean number of events *per unit "time."* Consider now the *random variable* described by the time required for the first event to occur. Using the Poisson distribution, we find that the probability of no events occurring in the span up to time t is given by

$$p(0; \lambda t) = \frac{e^{-\lambda t}(\lambda t)^0}{0!} = e^{-\lambda t}.$$

We can now make use of the above and let X be the time to the first Poisson event. The probability that the length of time until the first event will exceed x is the same as the probability that no Poisson events will occur in x. The latter, of course, is given by $e^{-\lambda x}$. As a result,

$$P(X > x) = e^{-\lambda x}.$$

Thus, the cumulative distribution function for X is given by

$$P(0 \leq X \leq x) = 1 - e^{-\lambda x}.$$

Now, in order that we may recognize the presence of the exponential distribution, we differentiate the cumulative distribution function above to obtain the density

function
$$f(x) = \lambda e^{-\lambda x},$$
which is the density function of the exponential distribution with $\lambda = 1/\beta$.

Applications of the Exponential and Gamma Distributions

In the foregoing, we provided the foundation for the application of the exponential distribution in "time to arrival" or time to Poisson event problems. We will illustrate some applications here and then proceed to discuss the role of the gamma distribution in these modeling applications. Notice that the mean of the exponential distribution is the parameter β, the reciprocal of the parameter in the Poisson distribution. The reader should recall that it is often said that the Poisson distribution has no memory, implying that occurrences in successive time periods are independent. The important parameter β is the mean time between events. In reliability theory, where equipment failure often conforms to this Poisson process, β is called **mean time between failures**. Many equipment breakdowns do follow the Poisson process, and thus the exponential distribution does apply. Other applications include survival times in biomedical experiments and computer response time.

In the following example, we show a simple application of the exponential distribution to a problem in reliability. The binomial distribution also plays a role in the solution.

Example 6.17: Suppose that a system contains a certain type of component whose time, in years, to failure is given by T. The random variable T is modeled nicely by the exponential distribution with mean time to failure $\beta = 5$. If 5 of these components are installed in different systems, what is the probability that at least 2 are still functioning at the end of 8 years?

Solution: The probability that a given component is still functioning after 8 years is given by
$$P(T > 8) = \frac{1}{5} \int_8^\infty e^{-t/5} \, dt = e^{-8/5} \approx 0.2.$$
Let X represent the number of components functioning after 8 years. Then using the binomial distribution, we have
$$P(X \geq 2) = \sum_{x=2}^{5} b(x; 5, 0.2) = 1 - \sum_{x=0}^{1} b(x; 5, 0.2) = 1 - 0.7373 = 0.2627.$$

There are exercises and examples in Chapter 3 where the reader has already encountered the exponential distribution. Others involving waiting time and reliability include Example 6.24 and some of the exercises and review exercises at the end of this chapter.

The Memoryless Property and Its Effect on the Exponential Distribution

The types of applications of the exponential distribution in reliability and component or machine lifetime problems are influenced by the **memoryless** (or lack-of-memory) **property** of the exponential distribution. For example, in the case of, say, an electronic component where lifetime has an exponential distribution, the probability that the component lasts, say, t hours, that is, $P(X \geq t)$, is the same as the conditional probability

$$P(X \geq t_0 + t \mid X \geq t_0).$$

So if the component "makes it" to t_0 hours, the probability of lasting an additional t hours is the same as the probability of lasting t hours. There is no "punishment" through wear that may have ensued for lasting the first t_0 hours. Thus, the exponential distribution is more appropriate when the memoryless property is justified. But if the failure of the component is a result of gradual or slow wear (as in mechanical wear), then the exponential does not apply and either the gamma or the Weibull distribution (Section 6.10) may be more appropriate.

The importance of the gamma distribution lies in the fact that it defines a family of which other distributions are special cases. But the gamma itself has important applications in waiting time and reliability theory. Whereas the exponential distribution describes the time until the occurrence of a Poisson event (or the time between Poisson events), the time (or space) occurring until a *specified number of Poisson events occur* is a random variable whose density function is described by the gamma distribution. This specific number of events is the parameter α in the gamma density function. Thus, it becomes easy to understand that when $\alpha = 1$, the special case of the exponential distribution occurs. The gamma density can be developed from its relationship to the Poisson process in much the same manner as we developed the exponential density. The details are left to the reader. The following is a numerical example of the use of the gamma distribution in a waiting-time application.

Example 6.18: Suppose that telephone calls arriving at a particular switchboard follow a Poisson process with an average of 5 calls coming per minute. What is the probability that up to a minute will elapse by the time 2 calls have come in to the switchboard?

Solution: The Poisson process applies, with time until 2 Poisson events following a gamma distribution with $\beta = 1/5$ and $\alpha = 2$. Denote by X the time in minutes that transpires before 2 calls come. The required probability is given by

$$P(X \leq 1) = \int_0^1 \frac{1}{\beta^2} x e^{-x/\beta} \, dx = 25 \int_0^1 x e^{-5x} \, dx = 1 - e^{-5}(1+5) = 0.96.$$

While the origin of the gamma distribution deals in time (or space) until the occurrence of α Poisson events, there are many instances where a gamma distribution works very well even though there is no clear Poisson structure. This is particularly true for **survival time** problems in both engineering and biomedical applications.

Example 6.19: In a biomedical study with rats, a dose-response investigation is used to determine the effect of the dose of a toxicant on their survival time. The toxicant is one that is frequently discharged into the atmosphere from jet fuel. For a certain dose of the toxicant, the study determines that the survival time, in weeks, has a gamma distribution with $\alpha = 5$ and $\beta = 10$. What is the probability that a rat survives no longer than 60 weeks?

6.6 Gamma and Exponential Distributions

Solution: Let the random variable X be the survival time (time to death). The required probability is

$$P(X \leq 60) = \frac{1}{\beta^5} \int_0^{60} \frac{x^{\alpha-1} e^{-x/\beta}}{\Gamma(5)} \, dx.$$

The integral above can be solved through the use of the **incomplete gamma function**, which becomes the cumulative distribution function for the gamma distribution. This function is written as

$$F(x; \alpha) = \int_0^x \frac{y^{\alpha-1} e^{-y}}{\Gamma(\alpha)} \, dy.$$

If we let $y = x/\beta$, so $x = \beta y$, we have

$$P(X \leq 60) = \int_0^6 \frac{y^4 e^{-y}}{\Gamma(5)} \, dy,$$

which is denoted as $F(6; 5)$ in the table of the incomplete gamma function in Appendix A.23. Note that this allows a quick computation of probabilities for the gamma distribution. Indeed, for this problem, the probability that the rat survives no longer than 60 days is given by

$$P(X \leq 60) = F(6; 5) = 0.715.$$

Example 6.20: It is known, from previous data, that the length of time in months between customer complaints about a certain product is a gamma distribution with $\alpha = 2$ and $\beta = 4$. Changes were made to tighten quality control requirements. Following these changes, 20 months passed before the first complaint. Does it appear as if the quality control tightening was effective?

Solution: Let X be the time to the first complaint, which, under conditions prior to the changes, followed a gamma distribution with $\alpha = 2$ and $\beta = 4$. The question centers around how rare $X \geq 20$ is, given that α and β remain at values 2 and 4, respectively. In other words, under the prior conditions is a "time to complaint" as large as 20 months reasonable? Thus, following the solution to Example 6.19,

$$P(X \geq 20) = 1 - \frac{1}{\beta^\alpha} \int_0^{20} \frac{x^{\alpha-1} e^{-x/\beta}}{\Gamma(\alpha)} \, dx.$$

Again, using $y = x/\beta$, we have

$$P(X \geq 20) = 1 - \int_0^5 \frac{y e^{-y}}{\Gamma(2)} \, dy = 1 - F(5; 2) = 1 - 0.96 = 0.04,$$

where $F(5; 2) = 0.96$ is found from Table A.23.

As a result, we could conclude that the conditions of the gamma distribution with $\alpha = 2$ and $\beta = 4$ are not supported by the data that an observed time to complaint is as large as 20 months. Thus, it is reasonable to conclude that the quality control work was effective.

Example 6.21: Consider Exercise 3.31 on page 94. Based on extensive testing, it is determined that the time Y in years before a major repair is required for a certain washing machine is characterized by the density function

$$f(y) = \begin{cases} \frac{1}{4} e^{-y/4}, & y \geq 0, \\ 0, & \text{elsewhere.} \end{cases}$$

Note that Y is an exponential random variable with $\mu = 4$ years. The machine is considered a bargain if it is unlikely to require a major repair before the sixth year. What is the probability $P(Y > 6)$? What is the probability that a major repair is required in the first year?

Solution: Consider the cumulative distribution function $F(y)$ for the exponential distribution,

$$F(y) = \frac{1}{\beta}\int_0^y e^{-t/\beta}\,dt = 1 - e^{-y/\beta}.$$

Then

$$P(Y > 6) = 1 - F(6) = e^{-3/2} = 0.2231.$$

Thus, the probability that the washing machine will require major repair after year six is 0.223. Of course, it will require repair before year six with probability 0.777. Thus, one might conclude the machine is not really a bargain. The probability that a major repair is necessary in the first year is

$$P(Y < 1) = 1 - e^{-1/4} = 1 - 0.779 = 0.221.$$

6.7 Chi-Squared Distribution

Another very important special case of the gamma distribution is obtained by letting $\alpha = v/2$ and $\beta = 2$, where v is a positive integer. The result is called **the chi-squared distribution**. The distribution has a single parameter, v, called the **degrees of freedom**.

Chi-Squared Distribution The continuous random variable X has a **chi-squared distribution**, with v **degrees of freedom**, if its density function is given by

$$f(x;v) = \begin{cases} \frac{1}{2^{v/2}\Gamma(v/2)} x^{v/2-1} e^{-x/2}, & x > 0, \\ 0, & \text{elsewhere}, \end{cases}$$

where v is a positive integer.

The chi-squared distribution plays a vital role in statistical inference. It has considerable applications in both methodology and theory. While we do not discuss applications in detail in this chapter, it is important to understand that Chapters 8, 9, and 16 contain important applications. The chi-squared distribution is an important component of statistical hypothesis testing and estimation.

Topics dealing with sampling distributions, analysis of variance, and nonparametric statistics involve extensive use of the chi-squared distribution.

Theorem 6.5: The mean and variance of the chi-squared distribution are

$$\mu = v \text{ and } \sigma^2 = 2v.$$

6.8 Beta Distribution

An extension to the uniform distribution is a beta distribution. Let us start by defining a **beta function**.

Definition 6.3: A **beta function** is defined by

$$B(\alpha, \beta) = \int_0^1 x^{\alpha-1}(1-x)^{\beta-1}dx = \frac{\Gamma(\alpha)\Gamma(\beta)}{\Gamma(\alpha+\beta)}, \text{ for } \alpha, \beta > 0,$$

where $\Gamma(\alpha)$ is the gamma function.

Beta Distribution — The continuous random variable X has a **beta distribution** with parameters $\alpha > 0$ and $\beta > 0$ if its density function is given by

$$f(x) = \begin{cases} \frac{1}{B(\alpha,\beta)} x^{\alpha-1}(1-x)^{\beta-1}, & 0 < x < 1, \\ 0, & \text{elsewhere.} \end{cases}$$

Note that the uniform distribution on $(0,1)$ is a beta distribution with parameters $\alpha = 1$ and $\beta = 1$.

Theorem 6.6: The mean and variance of a beta distribution with parameters α and β are

$$\mu = \frac{\alpha}{\alpha+\beta} \text{ and } \sigma^2 = \frac{\alpha\beta}{(\alpha+\beta)^2(\alpha+\beta+1)},$$

respectively.

For the uniform distribution on $(0,1)$, the mean and variance are

$$\mu = \frac{1}{1+1} = \frac{1}{2} \text{ and } \sigma^2 = \frac{(1)(1)}{(1+1)^2(1+1+1)} = \frac{1}{12},$$

respectively.

6.9 Lognormal Distribution

The lognormal distribution is used for a wide variety of applications. The distribution applies in cases where a natural log transformation results in a normal distribution.

Lognormal Distribution — The continuous random variable X has a **lognormal distribution** if the random variable $Y = \ln(X)$ has a normal distribution with mean μ and standard deviation σ. The resulting density function of X is

$$f(x; \mu, \sigma) = \begin{cases} \frac{1}{\sqrt{2\pi}\sigma x} e^{-\frac{1}{2\sigma^2}[\ln(x)-\mu]^2}, & x \geq 0, \\ 0, & x < 0. \end{cases}$$

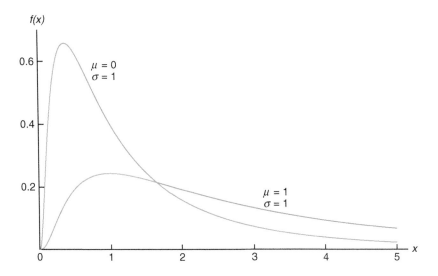

Figure 6.29: Lognormal distributions.

The graphs of the lognormal distributions are illustrated in Figure 6.29.

Theorem 6.7: The mean and variance of the lognormal distribution are

$$\mu = e^{\mu+\sigma^2/2} \text{ and } \sigma^2 = e^{2\mu+\sigma^2}(e^{\sigma^2} - 1).$$

The cumulative distribution function is quite simple due to its relationship to the normal distribution. The use of the distribution function is illustrated by the following example.

Example 6.22: Concentrations of pollutants produced by chemical plants historically are known to exhibit behavior that resembles a lognormal distribution. This is important when one considers issues regarding compliance with government regulations. Suppose it is assumed that the concentration of a certain pollutant, in parts per million, has a lognormal distribution with parameters $\mu = 3.2$ and $\sigma = 1$. What is the probability that the concentration exceeds 8 parts per million?

Solution: Let the random variable X be pollutant concentration. Then

$$P(X > 8) = 1 - P(X \leq 8).$$

Since $\ln(X)$ has a normal distribution with mean $\mu = 3.2$ and standard deviation $\sigma = 1$,

$$P(X \leq 8) = \Phi\left[\frac{\ln(8) - 3.2}{1}\right] = \Phi(-1.12) = 0.1314.$$

Here, we use Φ to denote the cumulative distribution function of the standard normal distribution. As a result, the probability that the pollutant concentration exceeds 8 parts per million is 0.1314.

Example 6.23: The life, in thousands of miles, of a certain type of electronic control for locomotives has an approximately lognormal distribution with $\mu = 5.149$ and $\sigma = 0.737$. Find the 5th percentile of the life of such an electronic control.

Solution: From Table A.3, we know that $P(Z < -1.645) = 0.05$. Denote by X the life of such an electronic control. Since $\ln(X)$ has a normal distribution with mean $\mu = 5.149$ and $\sigma = 0.737$, the 5th percentile of X can be calculated as

$$\ln(x) = 5.149 + (0.737)(-1.645) = 3.937.$$

Hence, $x = 51.265$. This means that only 5% of the controls will have lifetimes less than 51,265 miles.

6.10 Weibull Distribution (Optional)

Modern technology has enabled engineers to design many complicated systems whose operation and safety depend on the reliability of the various components making up the systems. For example, a fuse may burn out, a steel column may buckle, or a heat-sensing device may fail. Identical components subjected to identical environmental conditions will fail at different and unpredictable times. We have seen the role that the gamma and exponential distributions play in these types of problems. Another distribution that has been used extensively in recent years to deal with such problems is the **Weibull distribution**, introduced by the Swedish physicist Waloddi Weibull in 1939.

Weibull Distribution The continuous random variable X has a **Weibull distribution**, with parameters α and β, if its density function is given by

$$f(x; \alpha, \beta) = \begin{cases} \alpha \beta x^{\beta-1} e^{-\alpha x^{\beta}}, & x > 0, \\ 0, & \text{elsewhere,} \end{cases}$$

where $\alpha > 0$ and $\beta > 0$.

The graphs of the Weibull distribution for $\alpha = 1$ and various values of the parameter β are illustrated in Figure 6.30. We see that the curves change considerably in shape for different values of the parameter β. If we let $\beta = 1$, the Weibull distribution reduces to the exponential distribution. For values of $\beta > 1$, the curves become somewhat bell shaped and resemble the normal curve but display some skewness.

The mean and variance of the Weibull distribution are stated in the following theorem. The reader is asked to provide the proof in Exercise 6.52 on page 206.

Theorem 6.8: The mean and variance of the Weibull distribution are

$$\mu = \alpha^{-1/\beta} \Gamma\left(1 + \frac{1}{\beta}\right) \quad \text{and} \quad \sigma^2 = \alpha^{-2/\beta} \left\{ \Gamma\left(1 + \frac{2}{\beta}\right) - \left[\Gamma\left(1 + \frac{1}{\beta}\right)\right]^2 \right\}.$$

Like the gamma and exponential distributions, the Weibull distribution is also applied to reliability and life-testing problems such as the **time to failure** or

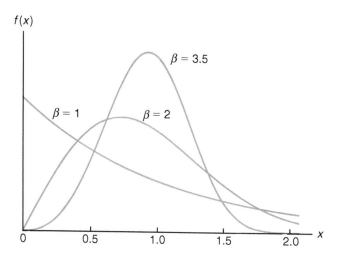

Figure 6.30: Weibull distributions ($\alpha = 1$).

life length of a component, measured from some specified time until it fails. Let us represent this time to failure by the continuous random variable T, with probability density function $f(t)$, where $f(t)$ is the Weibull distribution. The Weibull distribution has inherent flexibility in that it does not require the lack of memory property of the exponential distribution. The cumulative distribution function (cdf) for the Weibull can be written in closed form and certainly is useful in computing probabilities.

cdf for Weibull Distribution The **cumulative distribution function for the Weibull distribution** is given by

$$F(x) = 1 - e^{-\alpha x^\beta}, \qquad \text{for } x \geq 0,$$

for $\alpha > 0$ and $\beta > 0$.

Example 6.24: The length of life X, in hours, of an item in a machine shop has a Weibull distribution with $\alpha = 0.01$ and $\beta = 2$. What is the probability that it fails before eight hours of usage?

Solution: $P(X < 8) = F(8) = 1 - e^{-(0.01)8^2} = 1 - 0.527 = 0.473$.

The Failure Rate for the Weibull Distribution

When the Weibull distribution applies, it is often helpful to determine the **failure rate** (sometimes called the hazard rate) in order to get a sense of wear or deterioration of the component. Let us first define the reliability of a component or product as the *probability that it will function properly for at least a specified time under specified experimental conditions*. Therefore, if $R(t)$ is defined to be

6.10 Weibull Distribution (Optional)

the reliability of the given component at time t, we may write

$$R(t) = P(T > t) = \int_t^\infty f(t)\, dt = 1 - F(t),$$

where $F(t)$ is the cumulative distribution function of T. The conditional probability that a component will fail in the interval from $T = t$ to $T = t + \Delta t$, given that it survived to time t, is

$$\frac{F(t + \Delta t) - F(t)}{R(t)}.$$

Dividing this ratio by Δt and taking the limit as $\Delta t \to 0$, we get the **failure rate**, denoted by $Z(t)$. Hence,

$$Z(t) = \lim_{\Delta t \to 0} \frac{F(t + \Delta t) - F(t)}{\Delta t} \frac{1}{R(t)} = \frac{F'(t)}{R(t)} = \frac{f(t)}{R(t)} = \frac{f(t)}{1 - F(t)},$$

which expresses the failure rate in terms of the distribution of the time to failure. Since $Z(t) = f(t)/[1 - F(t)]$, the failure rate is given as follows:

Failure Rate for Weibull Distribution

The **failure rate** at time t for the Weibull distribution is given by

$$Z(t) = \alpha \beta t^{\beta - 1}, \qquad t > 0.$$

Interpretation of the Failure Rate

The quantity $Z(t)$ is aptly named as a failure rate since it does quantify the rate of change over time of the conditional probability that the component lasts an additional Δt *given that it has lasted to time t*. The rate of decrease (or increase) with time is important. The following are crucial points.

(a) If $\beta = 1$, the failure rate $= \alpha$, a constant. This, as indicated earlier, is the special case of the exponential distribution in which lack of memory prevails.

(b) If $\beta > 1$, $Z(t)$ is an increasing function of time t, which indicates that the component wears over time.

(c) If $\beta < 1$, $Z(t)$ is a decreasing function of time t and hence the component strengthens or hardens over time.

For example, the item in the machine shop in Example 6.24 has $\beta = 2$, and hence it wears over time. In fact, the failure rate function is given by $Z(t) = 0.02t$. On the other hand, suppose the parameters were $\beta = 3/4$ and $\alpha = 2$. In that case, $Z(t) = 1.5/t^{1/4}$ and hence the component gets stronger over time.

Exercises

6.39 Use the gamma function with $y = \sqrt{2x}$ to show that $\Gamma(1/2) = \sqrt{\pi}$.

6.40 In a certain city, the daily consumption of water (in millions of liters) follows approximately a gamma distribution with $\alpha = 2$ and $\beta = 3$. If the daily capacity of that city is 9 million liters of water, what is the probability that on any given day the water supply is inadequate?

6.41 If a random variable X has the gamma distribution with $\alpha = 2$ and $\beta = 1$, find $P(1.8 < X < 2.4)$.

6.42 Suppose that the time, in hours, required to repair a heat pump is a random variable X having a gamma distribution with parameters $\alpha = 2$ and $\beta = 1/2$. What is the probability that on the next service call

(a) at most 1 hour will be required to repair the heat pump?

(b) at least 2 hours will be required to repair the heat pump?

6.43 (a) Find the mean and variance of the daily water consumption in Exercise 6.40.

(b) According to Chebyshev's theorem, there is a probability of at least $3/4$ that the water consumption on any given day will fall within what interval?

6.44 In a certain city, the daily consumption of electric power, in millions of kilowatt-hours, is a random variable X having a gamma distribution with mean $\mu = 6$ and variance $\sigma^2 = 12$.

(a) Find the values of α and β.

(b) Find the probability that on any given day the daily power consumption will exceed 12 million kilowatt-hours.

6.45 The length of time for one individual to be served at a cafeteria is a random variable having an exponential distribution with a mean of 4 minutes. What is the probability that a person is served in less than 3 minutes on at least 4 of the next 6 days?

6.46 The life, in years, of a certain type of electrical switch has an exponential distribution with an average life $\beta = 2$. If 100 of these switches are installed in different systems, what is the probability that at most 30 fail during the first year?

6.47 Suppose that the service life, in years, of a hearing aid battery is a random variable having a Weibull distribution with $\alpha = 1/2$ and $\beta = 2$.

(a) How long can such a battery be expected to last?

(b) What is the probability that such a battery will be operating after 2 years?

6.48 Derive the mean and variance of the beta distribution.

6.49 Suppose the random variable X follows a beta distribution with $\alpha = 1$ and $\beta = 3$.

(a) Determine the mean and median of X.

(b) Determine the variance of X.

(c) Find the probability that $X > 1/3$.

6.50 If the proportion of a brand of television requiring service during the first year of operation is a random variable having a beta distribution with $\alpha = 3$ and $\beta = 2$, what is the probability that at least 80% of the new models of this brand sold this year will require service during their first year of operation?

6.51 The lives of a certain automobile seal have the Weibull distribution with failure rate $Z(t) = 1/\sqrt{t}$. Find the probability that such a seal is still intact after 4 years.

6.52 Derive the mean and variance of the Weibull distribution.

6.53 In a biomedical research study, it was determined that the survival time, in weeks, of an animal subjected to a certain exposure of gamma radiation has a gamma distribution with $\alpha = 5$ and $\beta = 10$.

(a) What is the mean survival time of a randomly selected animal of the type used in the experiment?

(b) What is the standard deviation of survival time?

(c) What is the probability that an animal survives more than 30 weeks?

6.54 The lifetime, in weeks, of a certain type of transistor is known to follow a gamma distribution with mean 10 weeks and standard deviation $\sqrt{50}$ weeks.

(a) What is the probability that a transistor of this type will last at most 50 weeks?

(b) What is the probability that a transistor of this type will not survive the first 10 weeks?

6.55 Computer response time is an important application of the gamma and exponential distributions. Suppose that a study of a certain computer system reveals that the response time, in seconds, has an exponential distribution with a mean of 3 seconds.

(a) What is the probability that response time exceeds 5 seconds?

(b) What is the probability that response time exceeds 10 seconds?

6.56 Rate data often follow a lognormal distribution. Average power usage (dB per hour) for a particular company is studied and is known to have a lognormal distribution with parameters $\mu = 4$ and $\sigma = 2$. What is the probability that the company uses more than 270 dB during any particular hour?

6.57 For Exercise 6.56, what is the mean power usage (average dB per hour)? What is the variance?

6.58 The number of automobiles that arrive at a certain intersection per minute has a Poisson distribution with a mean of 5. Interest centers around the time that elapses before 10 automobiles appear at the intersection.

(a) What is the probability that more than 10 automobiles appear at the intersection during any given minute of time?

(b) What is the probability that more than 2 minutes elapse before 10 cars arrive?

6.59 Consider the information in Exercise 6.58.

(a) What is the probability that more than 1 minute elapses between arrivals?

(b) What is the mean number of minutes that elapse between arrivals?

6.60 Show that the failure-rate function is given by

$$Z(t) = \alpha\beta t^{\beta-1}, \quad t > 0,$$

if and only if the time to failure distribution is the Weibull distribution

$$f(t) = \alpha\beta t^{\beta-1} e^{-\alpha t^\beta}, \quad t > 0.$$

Review Exercises

6.61 According to a study published by a group of sociologists at the University of Massachusetts, approximately 49% of the Valium users in the state of Massachusetts are white-collar workers. What is the probability that between 482 and 510, inclusive, of the next 1000 randomly selected Valium users from this state are white-collar workers?

6.62 The exponential distribution is frequently applied to the waiting times between successes in a Poisson process. If the number of calls received per hour by a telephone answering service is a Poisson random variable with parameter $\lambda = 6$, we know that the time, in hours, between successive calls has an exponential distribution with parameter $\beta = 1/6$. What is the probability of waiting more than 15 minutes between any two successive calls?

6.63 When α is a positive integer n, the gamma distribution is also known as the **Erlang distribution**. Setting $\alpha = n$ in the gamma distribution on page 195, the Erlang distribution is

$$f(x) = \begin{cases} \frac{x^{n-1} e^{-x/\beta}}{\beta^n (n-1)!}, & x > 0, \\ 0, & \text{elsewhere}. \end{cases}$$

It can be shown that if the times between successive events are independent, each having an exponential distribution with parameter β, then the total elapsed waiting time X until all n events occur has the Erlang distribution. Referring to Review Exercise 6.62, what is the probability that the next 3 calls will be received within the next 30 minutes?

6.64 A manufacturer of a certain type of large machine wishes to buy rivets from one of two manufacturers. It is important that the breaking strength of each rivet exceed 10,000 psi. Two manufacturers (A and B) offer this type of rivet and both have rivets whose breaking strength is normally distributed. The mean breaking strengths for manufacturers A and B are 14,000 psi and 13,000 psi, respectively. The standard deviations are 2000 psi and 1000 psi, respectively. Which manufacturer will produce, on the average, the fewest number of defective rivets?

6.65 According to a recent census, almost 65% of all households in the United States were composed of only one or two persons. Assuming that this percentage is still valid today, what is the probability that between 590 and 625, inclusive, of the next 1000 randomly selected households in America consist of either one or two persons?

6.66 A certain type of device has an advertised failure rate of 0.01 per hour. The failure rate is constant and the exponential distribution applies.

(a) What is the mean time to failure?

(b) What is the probability that 200 hours will pass before a failure is observed?

6.67 In a chemical processing plant, it is important that the yield of a certain type of batch product stay above 80%. If it stays below 80% for an extended period of time, the company loses money. Occasional defective batches are of little concern. But if several batches per day are defective, the plant shuts down and adjustments are made. It is known that the yield is normally distributed with standard deviation 4%.

(a) What is the probability of a "false alarm" (yield below 80%) when the mean yield is 85%?

(b) What is the probability that a batch will have a yield that exceeds 80% when in fact the mean yield is 79%?

6.68 For an electrical component with a failure rate of once every 5 hours, it is important to consider the time that it takes for 2 components to fail.

(a) Assuming that the gamma distribution applies, what is the mean time that it takes for 2 components to fail?

(b) What is the probability that 12 hours will elapse before 2 components fail?

6.69 The elongation of a steel bar under a particular load has been established to be normally distributed with a mean of 0.05 inch and $\sigma = 0.01$ inch. Find the probability that the elongation is

(a) above 0.1 inch;
(b) below 0.04 inch;
(c) between 0.025 and 0.065 inch.

6.70 A controlled satellite is known to have an error (distance from target) that is normally distributed with mean zero and standard deviation 4 feet. The manufacturer of the satellite defines a success as a firing in which the satellite comes within 10 feet of the target. Compute the probability that the satellite fails.

6.71 A technician plans to test a certain type of resin developed in the laboratory to determine the nature of the time required before bonding takes place. It is known that the mean time to bonding is 3 hours and the standard deviation is 0.5 hour. It will be considered an undesirable product if the bonding time is either less than 1 hour or more than 4 hours. Comment on the utility of the resin. How often would its performance be considered undesirable? Assume that time to bonding is normally distributed.

6.72 Consider the information in Review Exercise 6.66. What is the probability that less than 200 hours will elapse before 2 failures occur?

6.73 For Review Exercise 6.72, what are the mean and variance of the time that elapses before 2 failures occur?

6.74 The average rate of water usage (thousands of gallons per hour) by a certain community is known to involve the lognormal distribution with parameters $\mu = 5$ and $\sigma = 2$. It is important for planning purposes to get a sense of periods of high usage. What is the probability that, for any given hour, 50,000 gallons of water are used?

6.75 For Review Exercise 6.74, what is the mean of the average water usage per hour in thousands of gallons?

6.76 In Exercise 6.54 on page 206, the lifetime of a transistor is assumed to have a gamma distribution with mean 10 weeks and standard deviation $\sqrt{50}$ weeks. Suppose that the gamma distribution assumption is incorrect. Assume that the distribution is normal.

(a) What is the probability that a transistor will last at most 50 weeks?

(b) What is the probability that a transistor will not survive for the first 10 weeks?

(c) Comment on the difference between your results here and those found in Exercise 6.54 on page 206.

6.77 The beta distribution has considerable application in reliability problems in which the basic random variable is a proportion, as in the practical scenario illustrated in Exercise 6.50 on page 206. In that regard, consider Review Exercise 3.73 on page 108. Impurities in batches of product of a chemical process reflect a serious problem. It is known that the proportion of impurities Y in a batch has the density function

$$f(y) = \begin{cases} 10(1-y)^9, & 0 \le y \le 1, \\ 0, & \text{elsewhere.} \end{cases}$$

(a) Verify that the above is a valid density function.

(b) What is the probability that a batch is considered not acceptable (i.e., $Y > 0.6$)?

(c) What are the parameters α and β of the beta distribution illustrated here?

(d) The mean of the beta distribution is $\frac{\alpha}{\alpha+\beta}$. What is the mean proportion of impurities in the batch?

(e) The variance of a beta distributed random variable is

$$\sigma^2 = \frac{\alpha\beta}{(\alpha+\beta)^2(\alpha+\beta+1)}.$$

What is the variance of Y in this problem?

6.78 Consider now Review Exercise 3.74 on page 108. The density function of the time Z in minutes between calls to an electrical supply store is given by

$$f(z) = \begin{cases} \frac{1}{10}e^{-z/10}, & 0 < z < \infty, \\ 0, & \text{elsewhere.} \end{cases}$$

(a) What is the mean time between calls?

(b) What is the variance in the time between calls?

(c) What is the probability that the time between calls exceeds the mean?

6.79 Consider Review Exercise 6.78. Given the assumption of the exponential distribution, what is the mean number of calls per hour? What is the variance in the number of calls per hour?

6.80 In a human factor experimental project, it has been determined that the reaction time of a pilot to a visual stimulus is normally distributed with a mean of 1/2 second and standard deviation of 2/5 second.

(a) What is the probability that a reaction from the pilot takes more than 0.3 second?

(b) What reaction time is that which is exceeded 95% of the time?

6.81 The length of time between breakdowns of an essential piece of equipment is important in the decision of the use of auxiliary equipment. An engineer thinks that the best model for time between breakdowns of a generator is the exponential distribution with a mean of 15 days.

(a) If the generator has just broken down, what is the probability that it will break down in the next 21 days?

(b) What is the probability that the generator will operate for 30 days without a breakdown?

6.82 The length of life, in hours, of a drill bit in a mechanical operation has a Weibull distribution with $\alpha = 2$ and $\beta = 50$. Find the probability that the bit will fail before 10 hours of usage.

6.83 Derive the cdf for the Weibull distribution. [*Hint*: In the definition of a cdf, make the transformation $z = y^\beta$.]

6.84 Explain why the nature of the scenario in Review Exercise 6.82 would likely not lend itself to the exponential distribution.

6.85 From the relationship between the chi-squared random variable and the gamma random variable, prove that the mean of the chi-squared random variable is v and the variance is $2v$.

6.86 The length of time, in seconds, that a computer user takes to read his or her e-mail is distributed as a lognormal random variable with $\mu = 1.8$ and $\sigma^2 = 4.0$.

(a) What is the probability that a user reads e-mail for more than 20 seconds? More than a minute?

(b) What is the probability that a user reads e-mail for a length of time that is equal to the mean of the underlying lognormal distribution?

6.87 Group Project: Have groups of students observe the number of people who enter a specific coffee shop or fast food restaurant over the course of an hour, beginning at the same time every day, for two weeks. The hour should be a time of peak traffic at the shop or restaurant. The data collected will be the number of customers who enter the shop in each half hour of time. Thus, two data points will be collected each day. Let us assume that the random variable X, the number of people entering each half hour, follows a Poisson distribution. The students should calculate the sample mean and variance of X using the 28 data points collected.

(a) What evidence indicates that the Poisson distribution assumption may or may not be correct?

(b) Given that X is Poisson, what is the distribution of T, the time between arrivals into the shop during a half hour period? Give a numerical estimate of the parameter of that distribution.

(c) Give an estimate of the probability that the time between two arrivals is less than 15 minutes.

(d) What is the estimated probability that the time between two arrivals is more than 10 minutes?

(e) What is the estimated probability that 20 minutes after the start of data collection not one customer has appeared?

6.11 Potential Misconceptions and Hazards; Relationship to Material in Other Chapters

Many of the hazards in the use of material in this chapter are quite similar to those of Chapter 5. One of the biggest misuses of statistics is the assumption of an underlying normal distribution in carrying out a type of statistical inference when indeed it is not normal. The reader will be exposed to tests of hypotheses in Chapters 10 through 15 in which the normality assumption is made. In addition, however, the reader will be reminded that there are **tests of goodness of fit** as well as graphical routines discussed in Chapters 8 and 10 that allow for checks on data to determine if the normality assumption is reasonable.

Similar warnings should be conveyed regarding assumptions that are often made concerning other distributions, apart from the normal. This chapter has presented examples in which one is required to calculate probabilities to failure of a certain item or the probability that one observes a complaint during a certain time period. Assumptions are made concerning a certain distribution type as well as values of parameters of the distributions. Note that parameter values (for example, the value of β for the exponential distribution) were given in the example problems. However, in real-life problems, parameter values must be estimates from real-life experience or data. Note the emphasis placed on estimation in the projects that appear in Chapters 1, 5, and 6. Note also the reference in Chapter 5 to parameter estimation, which will be discussed extensively beginning in Chapter 9.

Chapter 7

Functions of Random Variables (Optional)

7.1 Introduction

This chapter contains a broad spectrum of material. Chapters 5 and 6 deal with specific types of distributions, both discrete and continuous. These are distributions that find use in many subject matter applications, including reliability, quality control, and acceptance sampling. In the present chapter, we begin with a more general topic, that of distributions of functions of random variables. General techniques are introduced and illustrated by examples. This discussion is followed by coverage of a related concept, *moment-generating functions*, which can be helpful in learning about distributions of linear functions of random variables.

In standard statistical methods, the result of statistical hypothesis testing, estimation, or even statistical graphics does not involve a single random variable but, rather, *functions of one or more random variables*. As a result, statistical inference requires the distributions of these functions. For example, the use of **averages of random variables** is common. In addition, sums and more general linear combinations are important. We are often interested in the distribution of sums of squares of random variables, particularly in the use of analysis of variance techniques discussed in Chapters 11–14.

7.2 Transformations of Variables

Frequently in statistics, one encounters the need to derive the probability distribution of a function of one or more random variables. For example, suppose that X is a discrete random variable with probability distribution $f(x)$, and suppose further that $Y = u(X)$ defines a one-to-one transformation between the values of X and Y. We wish to find the probability distribution of Y. It is important to note that the one-to-one transformation implies that each value x is related to one, and only one, value $y = u(x)$ and that each value y is related to one, and only one, value $x = w(y)$, where $w(y)$ is obtained by solving $y = u(x)$ for x in terms of y.

From our discussion of discrete probability distributions in Chapter 3, it is clear that the random variable Y assumes the value y when X assumes the value $w(y)$. Consequently, the probability distribution of Y is given by

$$g(y) = P(Y = y) = P[X = w(y)] = f[w(y)].$$

Theorem 7.1: Suppose that X is a **discrete** random variable with probability distribution $f(x)$. Let $Y = u(X)$ define a one-to-one transformation between the values of X and Y so that the equation $y = u(x)$ can be uniquely solved for x in terms of y, say $x = w(y)$. Then the probability distribution of Y is

$$g(y) = f[w(y)].$$

Example 7.1: Let X be a geometric random variable with probability distribution

$$f(x) = \frac{3}{4}\left(\frac{1}{4}\right)^{x-1}, \quad x = 1, 2, 3, \ldots.$$

Find the probability distribution of the random variable $Y = X^2$.

Solution: Since the values of X are all positive, the transformation defines a one-to-one correspondence between the x and y values, $y = x^2$ and $x = \sqrt{y}$. Hence

$$g(y) = \begin{cases} f(\sqrt{y}) = \frac{3}{4}\left(\frac{1}{4}\right)^{\sqrt{y}-1}, & y = 1, 4, 9, \ldots, \\ 0, & \text{elsewhere}. \end{cases}$$

Similarly, for a two-dimension transformation, we have the result in Theorem 7.2.

Theorem 7.2: Suppose that X_1 and X_2 are **discrete** random variables with joint probability distribution $f(x_1, x_2)$. Let $Y_1 = u_1(X_1, X_2)$ and $Y_2 = u_2(X_1, X_2)$ define a one-to-one transformation between the points (x_1, x_2) and (y_1, y_2) so that the equations

$$y_1 = u_1(x_1, x_2) \quad \text{and} \quad y_2 = u_2(x_1, x_2)$$

may be uniquely solved for x_1 and x_2 in terms of y_1 and y_2, say $x_1 = w_1(y_1, y_2)$ and $x_2 = w_2(y_1, y_2)$. Then the joint probability distribution of Y_1 and Y_2 is

$$g(y_1, y_2) = f[w_1(y_1, y_2), w_2(y_1, y_2)].$$

Theorem 7.2 is extremely useful for finding the distribution of some random variable $Y_1 = u_1(X_1, X_2)$, where X_1 and X_2 are discrete random variables with joint probability distribution $f(x_1, x_2)$. We simply define a second function, say $Y_2 = u_2(X_1, X_2)$, maintaining a one-to-one correspondence between the points (x_1, x_2) and (y_1, y_2), and obtain the joint probability distribution $g(y_1, y_2)$. The distribution of Y_1 is just the marginal distribution of $g(y_1, y_2)$, found by summing over the y_2 values. Denoting the distribution of Y_1 by $h(y_1)$, we can then write

$$h(y_1) = \sum_{y_2} g(y_1, y_2).$$

7.2 Transformations of Variables

Example 7.2: Let X_1 and X_2 be two independent random variables having Poisson distributions with parameters μ_1 and μ_2, respectively. Find the distribution of the random variable $Y_1 = X_1 + X_2$.

Solution: Since X_1 and X_2 are independent, we can write

$$f(x_1, x_2) = f(x_1)f(x_2) = \frac{e^{-\mu_1}\mu_1^{x_1}}{x_1!} \frac{e^{-\mu_2}\mu_2^{x_2}}{x_2!} = \frac{e^{-(\mu_1+\mu_2)}\mu_1^{x_1}\mu_2^{x_2}}{x_1! x_2!},$$

where $x_1 = 0, 1, 2, \ldots$ and $x_2 = 0, 1, 2, \ldots$. Let us now define a second random variable, say $Y_2 = X_2$. The inverse functions are given by $x_1 = y_1 - y_2$ and $x_2 = y_2$. Using Theorem 7.2, we find the joint probability distribution of Y_1 and Y_2 to be

$$g(y_1, y_2) = \frac{e^{-(\mu_1+\mu_2)}\mu_1^{y_1-y_2}\mu_2^{y_2}}{(y_1 - y_2)! y_2!},$$

where $y_1 = 0, 1, 2, \ldots$ and $y_2 = 0, 1, 2, \ldots, y_1$. Note that since $x_1 > 0$, the transformation $x_1 = y_1 - x_2$ implies that y_2 and hence x_2 must always be less than or equal to y_1. Consequently, the marginal probability distribution of Y_1 is

$$h(y_1) = \sum_{y_2=0}^{y_1} g(y_1, y_2) = e^{-(\mu_1+\mu_2)} \sum_{y_2=0}^{y_1} \frac{\mu_1^{y_1-y_2}\mu_2^{y_2}}{(y_1 - y_2)! y_2!}$$

$$= \frac{e^{-(\mu_1+\mu_2)}}{y_1!} \sum_{y_2=0}^{y_1} \frac{y_1!}{y_2!(y_1 - y_2)!} \mu_1^{y_1-y_2}\mu_2^{y_2}$$

$$= \frac{e^{-(\mu_1+\mu_2)}}{y_1!} \sum_{y_2=0}^{y_1} \binom{y_1}{y_2} \mu_1^{y_1-y_2}\mu_2^{y_2}.$$

Recognizing this sum as the binomial expansion of $(\mu_1 + \mu_2)^{y_1}$ we obtain

$$h(y_1) = \frac{e^{-(\mu_1+\mu_2)}(\mu_1 + \mu_2)^{y_1}}{y_1!}, \qquad y_1 = 0, 1, 2, \ldots,$$

from which we conclude that the sum of the two independent random variables having Poisson distributions, with parameters μ_1 and μ_2, has a Poisson distribution with parameter $\mu_1 + \mu_2$.

To find the probability distribution of the random variable $Y = u(X)$ when X is a continuous random variable and the transformation is one-to-one, we shall need Theorem 7.3. The proof of the theorem is left to the reader.

Theorem 7.3: Suppose that X is a **continuous** random variable with probability distribution $f(x)$. Let $Y = u(X)$ define a one-to-one correspondence between the values of X and Y so that the equation $y = u(x)$ can be uniquely solved for x in terms of y, say $x = w(y)$. Then the probability distribution of Y is

$$g(y) = f[w(y)]|J|,$$

where $J = w'(y)$ and is called the **Jacobian** of the transformation.

Example 7.3: Let X be a continuous random variable with probability distribution

$$f(x) = \begin{cases} \frac{x}{12}, & 1 < x < 5, \\ 0, & \text{elsewhere.} \end{cases}$$

Find the probability distribution of the random variable $Y = 2X - 3$.

Solution: The inverse solution of $y = 2x - 3$ yields $x = (y+3)/2$, from which we obtain $J = w'(y) = dx/dy = 1/2$. Therefore, using Theorem 7.3, we find the density function of Y to be

$$g(y) = \begin{cases} \frac{(y+3)/2}{12}\left(\frac{1}{2}\right) = \frac{y+3}{48}, & -1 < y < 7, \\ 0, & \text{elsewhere.} \end{cases}$$

To find the joint probability distribution of the random variables $Y_1 = u_1(X_1, X_2)$ and $Y_2 = u_2(X_1, X_2)$ when X_1 and X_2 are continuous and the transformation is one-to-one, we need an additional theorem, analogous to Theorem 7.2, which we state without proof.

Theorem 7.4: Suppose that X_1 and X_2 are **continuous** random variables with joint probability distribution $f(x_1, x_2)$. Let $Y_1 = u_1(X_1, X_2)$ and $Y_2 = u_2(X_1, X_2)$ define a one-to-one transformation between the points (x_1, x_2) and (y_1, y_2) so that the equations $y_1 = u_1(x_1, x_2)$ and $y_2 = u_2(x_1, x_2)$ may be uniquely solved for x_1 and x_2 in terms of y_1 and y_2, say $x_1 = w_1(y_l, y_2)$ and $x_2 = w_2(y_1, y_2)$. Then the joint probability distribution of Y_1 and Y_2 is

$$g(y_1, y_2) = f[w_1(y_1, y_2), w_2(y_1, y_2)]|J|,$$

where the Jacobian is the 2×2 determinant

$$J = \begin{vmatrix} \frac{\partial x_1}{\partial y_1} & \frac{\partial x_1}{\partial y_2} \\ \frac{\partial x_2}{\partial y_1} & \frac{\partial x_2}{\partial y_2} \end{vmatrix}$$

and $\frac{\partial x_1}{\partial y_1}$ is simply the derivative of $x_1 = w_1(y_1, y_2)$ with respect to y_1 with y_2 held constant, referred to in calculus as the partial derivative of x_1 with respect to y_1. The other partial derivatives are defined in a similar manner.

Example 7.4: Let X_1 and X_2 be two continuous random variables with joint probability distribution

$$f(x_1, x_2) = \begin{cases} 4x_1 x_2, & 0 < x_1 < 1, \ 0 < x_2 < 1, \\ 0, & \text{elsewhere.} \end{cases}$$

Find the joint probability distribution of $Y_1 = X_1^2$ and $Y_2 = X_1 X_2$.

Solution: The inverse solutions of $y_1 = x_1^2$ and $y_2 = x_1 x_2$ are $x_1 = \sqrt{y_1}$ and $x_2 = y_2/\sqrt{y_1}$, from which we obtain

$$J = \begin{vmatrix} 1/(2\sqrt{y_1}) & 0 \\ -y_2/2y_1^{3/2} & 1/\sqrt{y_1} \end{vmatrix} = \frac{1}{2y_1}.$$

7.2 Transformations of Variables

To determine the set B of points in the $y_1 y_2$ plane into which the set A of points in the $x_1 x_2$ plane is mapped, we write

$$x_1 = \sqrt{y_1} \quad \text{and} \quad x_2 = y_2/\sqrt{y_1}.$$

Then setting $x_1 = 0$, $x_2 = 0$, $x_1 = 1$, and $x_2 = 1$, the boundaries of set A are transformed to $y_1 = 0$, $y_2 = 0$, $y_1 = 1$, and $y_2 = \sqrt{y_1}$, or $y_2^2 = y_1$. The two regions are illustrated in Figure 7.1. Clearly, the transformation is one-to-one, mapping the set $A = \{(x_1, x_2) \mid 0 < x_1 < 1, \ 0 < x_2 < 1\}$ into the set $B = \{(y_1, y_2) \mid y_2^2 < y_1 < 1, \ 0 < y_2 < 1\}$. From Theorem 7.4 the joint probability distribution of Y_1 and Y_2 is

$$g(y_1, y_2) = 4(\sqrt{y_1}) \frac{y_2}{\sqrt{y_1}} \frac{1}{2 y_1} = \begin{cases} \frac{2 y_2}{y_1}, & y_2^2 < y_1 < 1, \ 0 < y_2 < 1, \\ 0, & \text{elsewhere.} \end{cases}$$

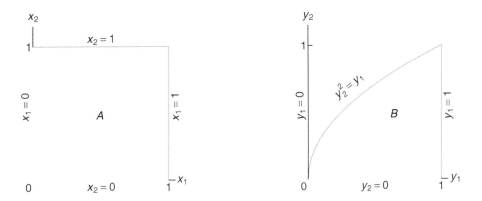

Figure 7.1: Mapping set A into set B.

Problems frequently arise when we wish to find the probability distribution of the random variable $Y = u(X)$ when X is a continuous random variable and the transformation is not one-to-one. That is, to each value x there corresponds exactly one value y, but to each y value there corresponds more than one x value. For example, suppose that $f(x)$ is positive over the interval $-1 < x < 2$ and zero elsewhere. Consider the transformation $y = x^2$. In this case, $x = \pm\sqrt{y}$ for $0 < y < 1$ and $x = \sqrt{y}$ for $1 < y < 4$. For the interval $1 < y < 4$, the probability distribution of Y is found as before, using Theorem 7.3. That is,

$$g(y) = f[w(y)]|J| = \frac{f(\sqrt{y})}{2\sqrt{y}}, \quad 1 < y < 4.$$

However, when $0 < y < 1$, we may partition the interval $-1 < x < 1$ to obtain the two inverse functions

$$x = -\sqrt{y}, \quad -1 < x < 0, \quad \text{and} \quad x = \sqrt{y}, \quad 0 < x < 1.$$

Then to every y value there corresponds a single x value for each partition. From Figure 7.2 we see that

$$P(a < Y < b) = P(-\sqrt{b} < X < -\sqrt{a}) + P(\sqrt{a} < X < \sqrt{b})$$
$$= \int_{-\sqrt{b}}^{-\sqrt{a}} f(x)\, dx + \int_{\sqrt{a}}^{\sqrt{b}} f(x)\, dx.$$

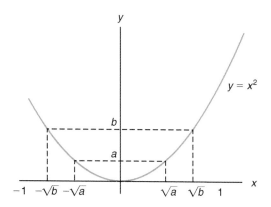

Figure 7.2: Decreasing and increasing function.

Changing the variable of integration from x to y, we obtain

$$P(a < Y < b) = \int_b^a f(-\sqrt{y}) J_1 \, dy + \int_a^b f(\sqrt{y}) J_2 \, dy$$
$$= -\int_a^b f(-\sqrt{y}) J_1 \, dy + \int_a^b f(\sqrt{y}) J_2 \, dy,$$

where

$$J_1 = \frac{d(-\sqrt{y})}{dy} = \frac{-1}{2\sqrt{y}} = -|J_1|$$

and

$$J_2 = \frac{d(\sqrt{y})}{dy} = \frac{1}{2\sqrt{y}} = |J_2|.$$

Hence, we can write

$$P(a < Y < b) = \int_a^b [f(-\sqrt{y})|J_1| + f(\sqrt{y})|J_2|] \, dy,$$

and then

$$g(y) = f(-\sqrt{y})|J_1| + f(\sqrt{y})|J_2| = \frac{f(-\sqrt{y}) + f(\sqrt{y})}{2\sqrt{y}}, \qquad 0 < y < 1.$$

7.2 Transformations of Variables

The probability distribution of Y for $0 < y < 4$ may now be written

$$g(y) = \begin{cases} \frac{f(-\sqrt{y})+f(\sqrt{y})}{2\sqrt{y}}, & 0 < y < 1, \\ \frac{f(\sqrt{y})}{2\sqrt{y}}, & 1 < y < 4, \\ 0, & \text{elsewhere.} \end{cases}$$

This procedure for finding $g(y)$ when $0 < y < 1$ is generalized in Theorem 7.5 for k inverse functions. For transformations not one-to-one of functions of several variables, the reader is referred to *Introduction to Mathematical Statistics* by Hogg, McKean, and Craig (2005; see the Bibliography).

Theorem 7.5: Suppose that X is a **continuous** random variable with probability distribution $f(x)$. Let $Y = u(X)$ define a transformation between the values of X and Y that is not one-to-one. If the interval over which X is defined can be partitioned into k mutually disjoint sets such that each of the inverse functions

$$x_1 = w_1(y), \quad x_2 = w_2(y), \quad \ldots, \quad x_k = w_k(y)$$

of $y = u(x)$ defines a one-to-one correspondence, then the probability distribution of Y is

$$g(y) = \sum_{i=1}^{k} f[w_i(y)]|J_i|,$$

where $J_i = w_i'(y)$, $i = 1, 2, \ldots, k$.

Example 7.5: Show that $Y = (X-\mu)^2/\sigma^2$ has a chi-squared distribution with 1 degree of freedom when X has a normal distribution with mean μ and variance σ^2.

Solution: Let $Z = (X - \mu)/\sigma$, where the random variable Z has the standard normal distribution

$$f(z) = \frac{1}{\sqrt{2\pi}} e^{-z^2/2}, \quad -\infty < z < \infty.$$

We shall now find the distribution of the random variable $Y = Z^2$. The inverse solutions of $y = z^2$ are $z = \pm\sqrt{y}$. If we designate $z_1 = -\sqrt{y}$ and $z_2 = \sqrt{y}$, then $J_1 = -1/2\sqrt{y}$ and $J_2 = 1/2\sqrt{y}$. Hence, by Theorem 7.5, we have

$$g(y) = \frac{1}{\sqrt{2\pi}} e^{-y/2} \left|\frac{-1}{2\sqrt{y}}\right| + \frac{1}{\sqrt{2\pi}} e^{-y/2} \left|\frac{1}{2\sqrt{y}}\right| = \frac{1}{\sqrt{2\pi}} y^{1/2-1} e^{-y/2}, \quad y > 0.$$

Since $g(y)$ is a density function, it follows that

$$1 = \frac{1}{\sqrt{2\pi}} \int_0^\infty y^{1/2-1} e^{-y/2} \, dy = \frac{\Gamma(1/2)}{\sqrt{\pi}} \int_0^\infty \frac{y^{1/2-1} e^{-y/2}}{\sqrt{2}\Gamma(1/2)} \, dy = \frac{\Gamma(1/2)}{\sqrt{\pi}},$$

the integral being the area under a gamma probability curve with parameters $\alpha = 1/2$ and $\beta = 2$. Hence, $\sqrt{\pi} = \Gamma(1/2)$ and the density of Y is given by

$$g(y) = \begin{cases} \frac{1}{\sqrt{2}\Gamma(1/2)} y^{1/2-1} e^{-y/2}, & y > 0, \\ 0, & \text{elsewhere,} \end{cases}$$

which is seen to be a chi-squared distribution with 1 degree of freedom.

7.3 Moments and Moment-Generating Functions

In this section, we concentrate on applications of moment-generating functions. The obvious purpose of the moment-generating function is in determining moments of random variables. However, the most important contribution is to establish distributions of functions of random variables.

If $g(X) = X^r$ for $r = 0, 1, 2, 3, \ldots$, Definition 7.1 yields an expected value called the rth **moment about the origin** of the random variable X, which we denote by μ'_r.

Definition 7.1: The rth **moment about the origin** of the random variable X is given by

$$\mu'_r = E(X^r) = \begin{cases} \sum_x x^r f(x), & \text{if } X \text{ is discrete,} \\ \int_{-\infty}^{\infty} x^r f(x)\, dx, & \text{if } X \text{ is continuous.} \end{cases}$$

Since the first and second moments about the origin are given by $\mu'_1 = E(X)$ and $\mu'_2 = E(X^2)$, we can write the mean and variance of a random variable as

$$\mu = \mu'_1 \quad \text{and} \quad \sigma^2 = \mu'_2 - \mu^2.$$

Although the moments of a random variable can be determined directly from Definition 7.1, an alternative procedure exists. This procedure requires us to utilize a **moment-generating function**.

Definition 7.2: The **moment-generating function** of the random variable X is given by $E(e^{tX})$ and is denoted by $M_X(t)$. Hence,

$$M_X(t) = E(e^{tX}) = \begin{cases} \sum_x e^{tx} f(x), & \text{if } X \text{ is discrete,} \\ \int_{-\infty}^{\infty} e^{tx} f(x)\, dx, & \text{if } X \text{ is continuous.} \end{cases}$$

Moment-generating functions will exist only if the sum or integral of Definition 7.2 converges. If a moment-generating function of a random variable X does exist, it can be used to generate all the moments of that variable. The method is described in Theorem 7.6 without proof.

Theorem 7.6: Let X be a random variable with moment-generating function $M_X(t)$. Then

$$\left. \frac{d^r M_X(t)}{dt^r} \right|_{t=0} = \mu'_r.$$

Example 7.6: Find the moment-generating function of the binomial random variable X and then use it to verify that $\mu = np$ and $\sigma^2 = npq$.

Solution: From Definition 7.2 we have

$$M_X(t) = \sum_{x=0}^{n} e^{tx} \binom{n}{x} p^x q^{n-x} = \sum_{x=0}^{n} \binom{n}{x} (pe^t)^x q^{n-x}.$$

7.3 Moments and Moment-Generating Functions

Recognizing this last sum as the binomial expansion of $(pe^t + q)^n$, we obtain

$$M_X(t) = (pe^t + q)^n.$$

Now

$$\frac{dM_X(t)}{dt} = n(pe^t + q)^{n-1} pe^t$$

and

$$\frac{d^2 M_X(t)}{dt^2} = np[e^t(n-1)(pe^t + q)^{n-2} pe^t + (pe^t + q)^{n-1} e^t].$$

Setting $t = 0$, we get

$$\mu'_1 = np \text{ and } \mu'_2 = np[(n-1)p + 1].$$

Therefore,

$$\mu = \mu'_1 = np \text{ and } \sigma^2 = \mu'_2 - \mu^2 = np(1-p) = npq,$$

which agrees with the results obtained in Chapter 5.

Example 7.7: Show that the moment-generating function of the random variable X having a normal probability distribution with mean μ and variance σ^2 is given by

$$M_X(t) = \exp\left(\mu t + \frac{1}{2}\sigma^2 t^2\right).$$

Solution: From Definition 7.2 the moment-generating function of the normal random variable X is

$$M_X(t) = \int_{-\infty}^{\infty} e^{tx} \frac{1}{\sqrt{2\pi}\sigma} \exp\left[-\frac{1}{2}\left(\frac{x-\mu}{\sigma}\right)^2\right] dx$$

$$= \int_{-\infty}^{\infty} \frac{1}{\sqrt{2\pi}\sigma} \exp\left[-\frac{x^2 - 2(\mu + t\sigma^2)x + \mu^2}{2\sigma^2}\right] dx.$$

Completing the square in the exponent, we can write

$$x^2 - 2(\mu + t\sigma^2)x + \mu^2 = [x - (\mu + t\sigma^2)]^2 - 2\mu t\sigma^2 - t^2\sigma^4$$

and then

$$M_X(t) = \int_{-\infty}^{\infty} \frac{1}{\sqrt{2\pi}\sigma} \exp\left\{-\frac{[x - (\mu + t\sigma^2)]^2 - 2\mu t\sigma^2 - t^2\sigma^4}{2\sigma^2}\right\} dx$$

$$= \exp\left(\frac{2\mu t + \sigma^2 t^2}{2}\right) \int_{-\infty}^{\infty} \frac{1}{\sqrt{2\pi}\sigma} \exp\left\{-\frac{[x - (\mu + t\sigma^2)]^2}{2\sigma^2}\right\} dx.$$

Let $w = [x - (\mu + t\sigma^2)]/\sigma$; then $dx = \sigma\, dw$ and

$$M_X(t) = \exp\left(\mu t + \frac{1}{2}\sigma^2 t^2\right) \int_{-\infty}^{\infty} \frac{1}{\sqrt{2\pi}} e^{-w^2/2}\, dw = \exp\left(\mu t + \frac{1}{2}\sigma^2 t^2\right),$$

since the last integral represents the area under a standard normal density curve and hence equals 1.

Although the method of transforming variables provides an effective way of finding the distribution of a function of several variables, there is an alternative and often preferred procedure when the function in question is a linear combination of independent random variables. This procedure utilizes the properties of moment-generating functions discussed in the following four theorems. In keeping with the mathematical scope of this book, we state Theorem 7.7 without proof.

Theorem 7.7: (**Uniqueness Theorem**) Let X and Y be two random variables with moment-generating functions $M_X(t)$ and $M_Y(t)$, respectively. If $M_X(t) = M_Y(t)$ for all values of t, then X and Y have the same probability distribution.

Theorem 7.8: $M_{X+a}(t) = e^{at} M_X(t).$

Proof: $M_{X+a}(t) = E[e^{t(X+a)}] = e^{at} E(e^{tX}) = e^{at} M_X(t).$

Theorem 7.9: $M_{aX}(t) = M_X(at).$

Proof: $M_{aX}(t) = E[e^{t(aX)}] = E[e^{(at)X}] = M_X(at).$

Theorem 7.10: If $X_1, X_2, \ldots, X_n$ are independent random variables with moment-generating functions $M_{X_1}(t), M_{X_2}(t), \ldots, M_{X_n}(t)$, respectively, and $Y = X_1 + X_2 + \cdots + X_n$, then
$$M_Y(t) = M_{X_1}(t) M_{X_2}(t) \cdots M_{X_n}(t).$$

The proof of Theorem 7.10 is left for the reader.

Theorems 7.7 through 7.10 are vital for understanding moment-generating functions. An example follows to illustrate. There are many situations in which we need to know the distribution of the sum of random variables. We may use Theorems 7.7 and 7.10 and the result of Exercise 7.19 on page 224 to find the distribution of a sum of two independent Poisson random variables with moment-generating functions given by

$$M_{X_1}(t) = e^{\mu_1(e^t - 1)} \text{ and } M_{X_2}(t) = e^{\mu_2(e^t - 1)},$$

respectively. According to Theorem 7.10, the moment-generating function of the random variable $Y_1 = X_1 + X_2$ is

$$M_{Y_1}(t) = M_{X_1}(t) M_{X_2}(t) = e^{\mu_1(e^t - 1)} e^{\mu_2(e^t - 1)} = e^{(\mu_1 + \mu_2)(e^t - 1)},$$

which we immediately identify as the moment-generating function of a random variable having a Poisson distribution with the parameter $\mu_1 + \mu_2$. Hence, according to Theorem 7.7, we again conclude that the sum of two independent random variables having Poisson distributions, with parameters μ_1 and μ_2, has a Poisson distribution with parameter $\mu_1 + \mu_2$.

7.3 Moments and Moment-Generating Functions

Linear Combinations of Random Variables

In applied statistics one frequently needs to know the probability distribution of a linear combination of independent normal random variables. Let us obtain the distribution of the random variable $Y = a_1 X_1 + a_2 X_2$ when X_1 is a normal variable with mean μ_1 and variance σ_1^2 and X_2 is also a normal variable but independent of X_1 with mean μ_2 and variance σ_2^2. First, by Theorem 7.10, we find

$$M_Y(t) = M_{a_1 X_1}(t) M_{a_2 X_2}(t),$$

and then, using Theorem 7.9, we find

$$M_Y(t) = M_{X_1}(a_1 t) M_{X_2}(a_2 t).$$

Substituting $a_1 t$ for t and then $a_2 t$ for t in a moment-generating function of the normal distribution derived in Example 7.7, we have

$$\begin{aligned} M_Y(t) &= \exp(a_1 \mu_1 t + a_1^2 \sigma_1^2 t^2 / 2 + a_2 \mu_2 t + a_2^2 \sigma_2^2 t^2 / 2) \\ &= \exp[(a_1 \mu_1 + a_2 \mu_2) t + (a_1^2 \sigma_1^2 + a_2^2 \sigma_2^2) t^2 / 2], \end{aligned}$$

which we recognize as the moment-generating function of a distribution that is normal with mean $a_1 \mu_1 + a_2 \mu_2$ and variance $a_1^2 \sigma_1^2 + a_2^2 \sigma_2^2$.

Generalizing to the case of n independent normal variables, we state the following result.

Theorem 7.11: If $X_1, X_2, \ldots, X_n$ are independent random variables having normal distributions with means $\mu_1, \mu_2, \ldots, \mu_n$ and variances $\sigma_1^2, \sigma_2^2, \ldots, \sigma_n^2$, respectively, then the random variable

$$Y = a_1 X_1 + a_2 X_2 + \cdots + a_n X_n$$

has a normal distribution with mean

$$\mu_Y = a_1 \mu_1 + a_2 \mu_2 + \cdots + a_n \mu_n$$

and variance

$$\sigma_Y^2 = a_1^2 \sigma_1^2 + a_2^2 \sigma_2^2 + \cdots + a_n^2 \sigma_n^2.$$

It is now evident that the Poisson distribution and the normal distribution possess a reproductive property in that the sum of independent random variables having either of these distributions is a random variable that also has the same type of distribution. The chi-squared distribution also has this reproductive property.

Theorem 7.12: If $X_1, X_2, \ldots, X_n$ are mutually independent random variables that have, respectively, chi-squared distributions with $v_1, v_2, \ldots, v_n$ degrees of freedom, then the random variable

$$Y = X_1 + X_2 + \cdots + X_n$$

has a chi-squared distribution with $v = v_1 + v_2 + \cdots + v_n$ degrees of freedom.

Proof: By Theorem 7.10 and Exercise 7.21,

$$M_Y(t) = M_{X_1}(t) M_{X_2}(t) \cdots M_{X_n}(t) \text{ and } M_{X_i}(t) = (1 - 2t)^{-v_i/2}, \; i = 1, 2, \ldots, n.$$

Therefore,
$$M_Y(t) = (1-2t)^{-v_1/2}(1-2t)^{-v_2/2}\cdots(1-2t)^{-v_n/2} = (1-2t)^{-(v_1+v_2+\cdots+v_n)/2},$$
which we recognize as the moment-generating function of a chi-squared distribution with $v = v_1 + v_2 + \cdots + v_n$ degrees of freedom.

Corollary 7.1: If $X_1, X_2, \ldots, X_n$ are independent random variables having identical normal distributions with mean μ and variance σ^2, then the random variable
$$Y = \sum_{i=1}^{n}\left(\frac{X_i - \mu}{\sigma}\right)^2$$
has a chi-squared distribution with $v = n$ degrees of freedom.

This corollary is an immediate consequence of Example 7.5. It establishes a relationship between the very important chi-squared distribution and the normal distribution. It also should provide the reader with a clear idea of what we mean by the parameter that we call degrees of freedom. In future chapters, the notion of degrees of freedom will play an increasingly important role.

Corollary 7.2: If $X_1, X_2, \ldots, X_n$ are independent random variables and X_i follows a normal distribution with mean μ_i and variance σ_i^2 for $i = 1, 2, \ldots, n$, then the random variable
$$Y = \sum_{i=1}^{n}\left(\frac{X_i - \mu_i}{\sigma_i}\right)^2$$
has a chi-squared distribution with $v = n$ degrees of freedom.

Exercises

7.1 Let X be a random variable with probability
$$f(x) = \begin{cases} \frac{1}{3}, & x = 1, 2, 3, \\ 0, & \text{elsewhere.} \end{cases}$$
Find the probability distribution of the random variable $Y = 2X - 1$.

7.2 Let X be a binomial random variable with probability distribution
$$f(x) = \begin{cases} \binom{3}{x}\left(\frac{2}{5}\right)^x\left(\frac{3}{5}\right)^{3-x}, & x = 0, 1, 2, 3, \\ 0, & \text{elsewhere.} \end{cases}$$
Find the probability distribution of the random variable $Y = X^2$.

7.3 Let X_1 and X_2 be discrete random variables with the joint multinomial distribution
$$f(x_1, x_2) = \binom{2}{x_1, x_2, 2-x_1-x_2}\left(\frac{1}{4}\right)^{x_1}\left(\frac{1}{3}\right)^{x_2}\left(\frac{5}{12}\right)^{2-x_1-x_2}$$
for $x_1 = 0, 1, 2$; $x_2 = 0, 1, 2$; $x_1 + x_2 \leq 2$; and zero elsewhere. Find the joint probability distribution of $Y_1 = X_1 + X_2$ and $Y_2 = X_1 - X_2$.

7.4 Let X_1 and X_2 be discrete random variables with joint probability distribution
$$f(x_1, x_2) = \begin{cases} \frac{x_1 x_2}{18}, & x_1 = 1, 2; x_2 = 1, 2, 3, \\ 0, & \text{elsewhere.} \end{cases}$$
Find the probability distribution of the random variable $Y = X_1 X_2$.

Exercises

7.5 Let X have the probability distribution

$$f(x) = \begin{cases} 1, & 0 < x < 1, \\ 0, & \text{elsewhere.} \end{cases}$$

Show that the random variable $Y = -2\ln X$ has a chi-squared distribution with 2 degrees of freedom.

7.6 Given the random variable X with probability distribution

$$f(x) = \begin{cases} 2x, & 0 < x < 1, \\ 0, & \text{elsewhere,} \end{cases}$$

find the probability distribution of $Y = 8X^3$.

7.7 The speed of a molecule in a uniform gas at equilibrium is a random variable V whose probability distribution is given by

$$f(v) = \begin{cases} kv^2 e^{-bv^2}, & v > 0, \\ 0, & \text{elsewhere,} \end{cases}$$

where k is an appropriate constant and b depends on the absolute temperature and mass of the molecule. Find the probability distribution of the kinetic energy of the molecule W, where $W = mV^2/2$.

7.8 A dealer's profit, in units of $5000, on a new automobile is given by $Y = X^2$, where X is a random variable having the density function

$$f(x) = \begin{cases} 2(1-x), & 0 < x < 1, \\ 0, & \text{elsewhere.} \end{cases}$$

(a) Find the probability density function of the random variable Y.

(b) Using the density function of Y, find the probability that the profit on the next new automobile sold by this dealership will be less than $500.

7.9 The hospital period, in days, for patients following treatment for a certain type of kidney disorder is a random variable $Y = X + 4$, where X has the density function

$$f(x) = \begin{cases} \frac{32}{(x+4)^3}, & x > 0, \\ 0, & \text{elsewhere.} \end{cases}$$

(a) Find the probability density function of the random variable Y.

(b) Using the density function of Y, find the probability that the hospital period for a patient following this treatment will exceed 8 days.

7.10 The random variables X and Y, representing the weights of creams and toffees, respectively, in 1-kilogram boxes of chocolates containing a mixture of creams, toffees, and cordials, have the joint density function

$$f(x,y) = \begin{cases} 24xy, & 0 \le x \le 1, 0 \le y \le 1, x+y \le 1, \\ 0, & \text{elsewhere.} \end{cases}$$

(a) Find the probability density function of the random variable $Z = X + Y$.

(b) Using the density function of Z, find the probability that, in a given box, the sum of the weights of creams and toffees accounts for at least 1/2 but less than 3/4 of the total weight.

7.11 The amount of kerosene, in thousands of liters, in a tank at the beginning of any day is a random amount Y from which a random amount X is sold during that day. Assume that the joint density function of these variables is given by

$$f(x,y) = \begin{cases} 2, & 0 < x < y, 0 < y < 1, \\ 0, & \text{elsewhere.} \end{cases}$$

Find the probability density function for the amount of kerosene left in the tank at the end of the day.

7.12 Let X_1 and X_2 be independent random variables each having the probability distribution

$$f(x) = \begin{cases} e^{-x}, & x > 0, \\ 0, & \text{elsewhere.} \end{cases}$$

Show that the random variables Y_1 and Y_2 are independent when $Y_1 = X_1 + X_2$ and $Y_2 = X_1/(X_1 + X_2)$.

7.13 A current of I amperes flowing through a resistance of R ohms varies according to the probability distribution

$$f(i) = \begin{cases} 6i(1-i), & 0 < i < 1, \\ 0, & \text{elsewhere.} \end{cases}$$

If the resistance varies independently of the current according to the probability distribution

$$g(r) = \begin{cases} 2r, & 0 < r < 1, \\ 0, & \text{elsewhere,} \end{cases}$$

find the probability distribution for the power $W = I^2 R$ watts.

7.14 Let X be a random variable with probability distribution

$$f(x) = \begin{cases} \frac{1+x}{2}, & -1 < x < 1, \\ 0, & \text{elsewhere.} \end{cases}$$

Find the probability distribution of the random variable $Y = X^2$.

7.15 Let X have the probability distribution

$$f(x) = \begin{cases} \frac{2(x+1)}{9}, & -1 < x < 2, \\ 0, & \text{elsewhere.} \end{cases}$$

Find the probability distribution of the random variable $Y = X^2$.

7.16 Show that the rth moment about the origin of the gamma distribution is

$$\mu'_r = \frac{\beta^r \Gamma(\alpha + r)}{\Gamma(\alpha)}.$$

[*Hint*: Substitute $y = x/\beta$ in the integral defining μ'_r and then use the gamma function to evaluate the integral.]

7.17 A random variable X has the discrete uniform distribution

$$f(x;k) = \begin{cases} \frac{1}{k}, & x = 1, 2, \ldots, k, \\ 0, & \text{elsewhere.} \end{cases}$$

Show that the moment-generating function of X is

$$M_X(t) = \frac{e^t(1 - e^{kt})}{k(1 - e^t)}.$$

7.18 A random variable X has the geometric distribution $g(x;p) = pq^{x-1}$ for $x = 1, 2, 3, \ldots$. Show that the moment-generating function of X is

$$M_X(t) = \frac{pe^t}{1 - qe^t}, \quad t < \ln q,$$

and then use $M_X(t)$ to find the mean and variance of the geometric distribution.

7.19 A random variable X has the Poisson distribution $p(x;\mu) = e^{-\mu}\mu^x/x!$ for $x = 0, 1, 2, \ldots$. Show that the moment-generating function of X is

$$M_X(t) = e^{\mu(e^t - 1)}.$$

Using $M_X(t)$, find the mean and variance of the Poisson distribution.

7.20 The moment-generating function of a certain Poisson random variable X is given by

$$M_X(t) = e^{4(e^t - 1)}.$$

Find $P(\mu - 2\sigma < X < \mu + 2\sigma)$.

7.21 Show that the moment-generating function of the random variable X having a chi-squared distribution with v degrees of freedom is

$$M_X(t) = (1 - 2t)^{-v/2}.$$

7.22 Using the moment-generating function of Exercise 7.21, show that the mean and variance of the chi-squared distribution with v degrees of freedom are, respectively, v and $2v$.

7.23 If both X and Y, distributed independently, follow exponential distributions with mean parameter 1, find the distributions of
(a) $U = X + Y$;
(b) $V = X/(X + Y)$.

7.24 By expanding e^{tx} in a Maclaurin series and integrating term by term, show that

$$M_X(t) = \int_{-\infty}^{\infty} e^{tx} f(x)\, dx$$

$$= 1 + \mu t + \mu'_2 \frac{t^2}{2!} + \cdots + \mu'_r \frac{t^r}{r!} + \cdots.$$

Chapter 8

Fundamental Sampling Distributions and Data Descriptions

8.1 Random Sampling

The outcome of a statistical experiment may be recorded either as a numerical value or as a descriptive representation. When a pair of dice is tossed and the total is the outcome of interest, we record a numerical value. However, if the students of a certain school are given blood tests and the type of blood is of interest, then a descriptive representation might be more useful. A person's blood can be classified in 8 ways: AB, A, B, or O, each with a plus or minus sign, depending on the presence or absence of the Rh antigen.

In this chapter, we focus on sampling from distributions or populations and study such important quantities as the *sample mean* and *sample variance*, which will be of vital importance in future chapters. In addition, we attempt to give the reader an introduction to the role that the sample mean and variance will play in statistical inference in later chapters. The use of modern high-speed computers allows the scientist or engineer to greatly enhance his or her use of formal statistical inference with graphical techniques. Much of the time, formal inference appears quite dry and perhaps even abstract to the practitioner or to the manager who wishes to let statistical analysis be a guide to decision-making.

Populations and Samples

We begin this section by discussing the notions of *populations* and *samples*. Both are mentioned in a broad fashion in Chapter 1. However, much more needs to be presented about them here, particularly in the context of the concept of random variables. The totality of observations with which we are concerned, whether their number be finite or infinite, constitutes what we call a **population**. There was a time when the word *population* referred to observations obtained from statistical studies about people. Today, statisticians use the term to refer to observations relevant to anything of interest, whether it be groups of people, animals, or all possible outcomes from some complicated biological or engineering system.

Definition 8.1: A **population** consists of the totality of the observations with which we are concerned.

The number of observations in the population is defined to be the size of the population. If there are 600 students in the school whom we classified according to blood type, we say that we have a population of size 600. The numbers on the cards in a deck, the heights of residents in a certain city, and the lengths of fish in a particular lake are examples of populations with finite size. In each case, the total number of observations is a finite number. The observations obtained by measuring the atmospheric pressure every day, from the past on into the future, or all measurements of the depth of a lake, from any conceivable position, are examples of populations whose sizes are infinite. Some finite populations are so large that in theory we assume them to be infinite. This is true in the case of the population of lifetimes of a certain type of storage battery being manufactured for mass distribution throughout the country.

Each observation in a population is a value of a random variable X having some probability distribution $f(x)$. If one is inspecting items coming off an assembly line for defects, then each observation in the population might be a value 0 or 1 of the Bernoulli random variable X with probability distribution

$$b(x; 1, p) = p^x q^{1-x}, \quad x = 0, 1$$

where 0 indicates a nondefective item and 1 indicates a defective item. Of course, it is assumed that p, the probability of any item being defective, remains constant from trial to trial. In the blood-type experiment, the random variable X represents the type of blood and is assumed to take on values from 1 to 8. Each student is given one of the values of the discrete random variable. The lives of the storage batteries are values assumed by a continuous random variable having perhaps a normal distribution. When we refer hereafter to a "binomial population," a "normal population," or, in general, the "population $f(x)$," we shall mean a population whose observations are values of a random variable having a binomial distribution, a normal distribution, or the probability distribution $f(x)$. Hence, the mean and variance of a random variable or probability distribution are also referred to as the mean and variance of the corresponding population.

In the field of statistical inference, statisticians are interested in arriving at conclusions concerning a population when it is impossible or impractical to observe the entire set of observations that make up the population. For example, in attempting to determine the average length of life of a certain brand of light bulb, it would be impossible to test all such bulbs if we are to have any left to sell. Exorbitant costs can also be a prohibitive factor in studying an entire population. Therefore, we must depend on a subset of observations from the population to help us make inferences concerning that same population. This brings us to consider the notion of sampling.

Definition 8.2: A **sample** is a subset of a population.

If our inferences from the sample to the population are to be valid, we must obtain samples that are representative of the population. All too often we are tempted to choose a sample by selecting the most convenient members of the population. Such a procedure may lead to erroneous inferences concerning the population. Any sampling procedure that produces inferences that consistently overestimate or consistently underestimate some characteristic of the population is said to be **biased**. To eliminate any possibility of bias in the sampling procedure, it is desirable to choose a **random sample** in the sense that the observations are made independently and at random.

In selecting a random sample of size n from a population $f(x)$, let us define the random variable X_i, $i = 1, 2, \ldots, n$, to represent the ith measurement or sample value that we observe. The random variables $X_1, X_2, \ldots, X_n$ will then constitute a random sample from the population $f(x)$ with numerical values $x_1, x_2, \ldots, x_n$ if the measurements are obtained by repeating the experiment n independent times under essentially the same conditions. Because of the identical conditions under which the elements of the sample are selected, it is reasonable to assume that the n random variables $X_1, X_2, \ldots, X_n$ are independent and that each has the same probability distribution $f(x)$. That is, the probability distributions of $X_1, X_2, \ldots, X_n$ are, respectively, $f(x_1), f(x_2), \ldots, f(x_n)$, and their joint probability distribution is $f(x_1, x_2, \ldots, x_n) = f(x_1)f(x_2) \cdots f(x_n)$. The concept of a random sample is described formally by the following definition.

Definition 8.3: Let $X_1, X_2, \ldots, X_n$ be n independent random variables, each having the same probability distribution $f(x)$. Define $X_1, X_2, \ldots, X_n$ to be a **random sample** of size n from the population $f(x)$ and write its joint probability distribution as

$$f(x_1, x_2, \ldots, x_n) = f(x_1)f(x_2) \cdots f(x_n).$$

If one makes a random selection of $n = 8$ storage batteries from a manufacturing process that has maintained the same specification throughout and records the length of life for each battery, with the first measurement x_1 being a value of X_1, the second measurement x_2 a value of X_2, and so forth, then $x_1, x_2, \ldots, x_8$ are the values of the random sample $X_1, X_2, \ldots, X_8$. If we assume the population of battery lives to be normal, the possible values of any X_i, $i = 1, 2, \ldots, 8$, will be precisely the same as those in the original population, and hence X_i has the same identical normal distribution as X.

8.2 Some Important Statistics

Our main purpose in selecting random samples is to elicit information about the unknown population parameters. Suppose, for example, that we wish to arrive at a conclusion concerning the proportion of coffee-drinkers in the United States who prefer a certain brand of coffee. It would be impossible to question every coffee-drinking American in order to compute the value of the parameter p representing the population proportion. Instead, a large random sample is selected and the proportion $\hat{p}$ of people in this sample favoring the brand of coffee in question is calculated. The value $\hat{p}$ is now used to make an inference concerning the true proportion p.

Now, $\hat{p}$ is a function of the observed values in the random sample; since many random samples are possible from the same population, we would expect $\hat{p}$ to vary somewhat from sample to sample. That is, $\hat{p}$ is a value of a random variable that we represent by P. Such a random variable is called a **statistic**.

Definition 8.4: Any function of the random variables constituting a random sample is called a **statistic**.

Location Measures of a Sample: The Sample Mean, Median, and Mode

In Chapter 4 we introduced the two parameters μ and σ^2, which measure the center of location and the variability of a probability distribution. These are constant population parameters and are in no way affected or influenced by the observations of a random sample. We shall, however, define some important statistics that describe corresponding measures of a random sample. The most commonly used statistics for measuring the center of a set of data, arranged in order of magnitude, are the **mean**, **median**, and **mode**. Although the first two of these statistics were defined in Chapter 1, we repeat the definitions here. Let $X_1, X_2, \ldots, X_n$ represent n random variables.

(a) Sample mean:
$$\bar{X} = \frac{1}{n} \sum_{i=1}^{n} X_i.$$

Note that the statistic $\bar{X}$ assumes the value $\bar{x} = \frac{1}{n} \sum_{i=1}^{n} x_i$ when X_1 assumes the value x_1, X_2 assumes the value x_2, and so forth. The term *sample mean* is applied to both the statistic $\bar{X}$ and its computed value $\bar{x}$.

(b) Sample median:
$$\tilde{x} = \begin{cases} x_{(n+1)/2}, & \text{if } n \text{ is odd,} \\ \frac{1}{2}(x_{n/2} + x_{n/2+1}), & \text{if } n \text{ is even.} \end{cases}$$

The sample median is also a location measure that shows the middle value of the sample. Examples for both the sample mean and the sample median can be found in Section 1.3. The sample mode is defined as follows.

(c) The sample mode is the value of the sample that occurs most often.

Example 8.1: Suppose a data set consists of the following observations:

$$0.32 \quad 0.53 \quad 0.28 \quad 0.37 \quad 0.47 \quad 0.43 \quad 0.36 \quad 0.42 \quad 0.38 \quad 0.43.$$

The sample mode is 0.43, since this value occurs more than any other value. ⌋

As we suggested in Chapter 1, a measure of location or central tendency in a sample does not by itself give a clear indication of the nature of the sample. Thus, a measure of variability in the sample must also be considered.

Variability Measures of a Sample: The Sample Variance, Standard Deviation, and Range

The variability in a sample displays how the observations spread out from the average. The reader is referred to Chapter 1 for more discussion. It is possible to have two sets of observations with the same mean or median that differ considerably in the variability of their measurements about the average.

Consider the following measurements, in liters, for two samples of orange juice bottled by companies A and B:

Sample A	0.97	1.00	0.94	1.03	1.06
Sample B	1.06	1.01	0.88	0.91	1.14

Both samples have the same mean, 1.00 liter. It is obvious that company A bottles orange juice with a more uniform content than company B. We say that the **variability**, or the **dispersion**, of the observations from the average is less for sample A than for sample B. Therefore, in buying orange juice, we would feel more confident that the bottle we select will be close to the advertised average if we buy from company A.

In Chapter 1 we introduced several measures of sample variability, including the **sample variance**, **sample standard deviation**, and **sample range**. In this chapter, we will focus mainly on the sample variance. Again, let $X_1, \ldots, X_n$ represent n random variables.

(a) Sample variance:
$$S^2 = \frac{1}{n-1} \sum_{i=1}^{n} (X_i - \bar{X})^2. \tag{8.2.1}$$

The computed value of S^2 for a given sample is denoted by s^2. Note that S^2 is essentially defined to be the average of the squares of the deviations of the observations from their mean. The reason for using $n-1$ as a divisor rather than the more obvious choice n will become apparent in Chapter 9.

Example 8.2: A comparison of coffee prices at 4 randomly selected grocery stores in San Diego showed increases from the previous month of 12, 15, 17, and 20 cents for a 1-pound bag. Find the variance of this random sample of price increases.

Solution: Calculating the sample mean, we get
$$\bar{x} = \frac{12 + 15 + 17 + 20}{4} = 16 \text{ cents}.$$

Therefore,
$$s^2 = \frac{1}{3} \sum_{i=1}^{4}(x_i - 16)^2 = \frac{(12-16)^2 + (15-16)^2 + (17-16)^2 + (20-16)^2}{3}$$
$$= \frac{(-4)^2 + (-1)^2 + (1)^2 + (4)^2}{3} = \frac{34}{3}.$$

Whereas the expression for the sample variance best illustrates that S^2 is a measure of variability, an alternative expression does have some merit and thus the reader should be aware of it. The following theorem contains this expression.

Theorem 8.1: If S^2 is the variance of a random sample of size n, we may write

$$S^2 = \frac{1}{n(n-1)}\left[n\sum_{i=1}^{n}X_i^2 - \left(\sum_{i=1}^{n}X_i\right)^2\right].$$

Proof: By definition,

$$S^2 = \frac{1}{n-1}\sum_{i=1}^{n}(X_i - \bar{X})^2 = \frac{1}{n-1}\sum_{i=1}^{n}(X_i^2 - 2\bar{X}X_i + \bar{X}^2)$$

$$= \frac{1}{n-1}\left[\sum_{i=1}^{n}X_i^2 - 2\bar{X}\sum_{i=1}^{n}X_i + n\bar{X}^2\right].$$

As in Chapter 1, the **sample standard deviation** and the **sample range** are defined below.

(b) Sample standard deviation:

$$S = \sqrt{S^2},$$

where S^2 is the sample variance.

Let $X_{\max}$ denote the largest of the X_i values and $X_{\min}$ the smallest.

(c) Sample range:

$$R = X_{\max} - X_{\min}.$$

Example 8.3: Find the variance of the data 3, 4, 5, 6, 6, and 7, representing the number of trout caught by a random sample of 6 fishermen on June 19, 1996, at Lake Muskoka.

Solution: We find that $\sum_{i=1}^{6}x_i^2 = 171$, $\sum_{i=1}^{6}x_i = 31$, and $n = 6$. Hence,

$$s^2 = \frac{1}{(6)(5)}[(6)(171) - (31)^2] = \frac{13}{6}.$$

Thus, the sample standard deviation $s = \sqrt{13/6} = 1.47$ and the sample range is $7 - 3 = 4$.

Exercises

8.1 Define suitable populations from which the following samples are selected:

(a) Persons in 200 homes in the city of Richmond are called on the phone and asked to name the candidate they favor for election to the school board.

(b) A coin is tossed 100 times and 34 tails are recorded.

(c) Two hundred pairs of a new type of tennis shoe were tested on the professional tour and, on average, lasted 4 months.

(d) On five different occasions it took a lawyer 21, 26, 24, 22, and 21 minutes to drive from her suburban home to her midtown office.

Exercises

8.2 The lengths of time, in minutes, that 10 patients waited in a doctor's office before receiving treatment were recorded as follows: 5, 11, 9, 5, 10, 15, 6, 10, 5, and 10. Treating the data as a random sample, find

(a) the mean;

(b) the median;

(c) the mode.

8.3 The reaction times for a random sample of 9 subjects to a stimulant were recorded as 2.5, 3.6, 3.1, 4.3, 2.9, 2.3, 2.6, 4.1, and 3.4 seconds. Calculate

(a) the mean;

(b) the median.

8.4 The number of tickets issued for traffic violations by 8 state troopers during the Memorial Day weekend are 5, 4, 7, 7, 6, 3, 8, and 6.

(a) If these values represent the number of tickets issued by a random sample of 8 state troopers from Montgomery County in Virginia, define a suitable population.

(b) If the values represent the number of tickets issued by a random sample of 8 state troopers from South Carolina, define a suitable population.

8.5 The numbers of incorrect answers on a true-false competency test for a random sample of 15 students were recorded as follows: 2, 1, 3, 0, 1, 3, 6, 0, 3, 3, 5, 2, 1, 4, and 2. Find

(a) the mean;

(b) the median;

(c) the mode.

8.6 Find the mean, median, and mode for the sample whose observations, 15, 7, 8, 95, 19, 12, 8, 22, and 14, represent the number of sick days claimed on 9 federal income tax returns. Which value appears to be the best measure of the center of these data? State reasons for your preference.

8.7 A random sample of employees from a local manufacturing plant pledged the following donations, in dollars, to the United Fund: 100, 40, 75, 15, 20, 100, 75, 50, 30, 10, 55, 75, 25, 50, 90, 80, 15, 25, 45, and 100. Calculate

(a) the mean;

(b) the mode.

8.8 According to ecology writer Jacqueline Killeen, phosphates contained in household detergents pass right through our sewer systems, causing lakes to turn into swamps that eventually dry up into deserts. The following data show the amount of phosphates per load of laundry, in grams, for a random sample of various types of detergents used according to the prescribed directions:

Laundry Detergent	Phosphates per Load (grams)
A & P Blue Sail	48
Dash	47
Concentrated All	42
Cold Water All	42
Breeze	41
Oxydol	34
Ajax	31
Sears	30
Fab	29
Cold Power	29
Bold	29
Rinso	26

For the given phosphate data, find

(a) the mean;

(b) the median;

(c) the mode.

8.9 Consider the data in Exercise 8.2, find

(a) the range;

(b) the standard deviation.

8.10 For the sample of reaction times in Exercise 8.3, calculate

(a) the range;

(b) the variance, using the formula of form (8.2.1).

8.11 For the data of Exercise 8.5, calculate the variance using the formula

(a) of form (8.2.1);

(b) in Theorem 8.1.

8.12 The tar contents of 8 brands of cigarettes selected at random from the latest list released by the Federal Trade Commission are as follows: 7.3, 8.6, 10.4, 16.1, 12.2, 15.1, 14.5, and 9.3 milligrams. Calculate

(a) the mean;

(b) the variance.

8.13 The grade-point averages of 20 college seniors selected at random from a graduating class are as follows:

3.2	1.9	2.7	2.4	2.8
2.9	3.8	3.0	2.5	3.3
1.8	2.5	3.7	2.8	2.0
3.2	2.3	2.1	2.5	1.9

Calculate the standard deviation.

8.14 (a) Show that the sample variance is unchanged if a constant c is added to or subtracted from each

value in the sample.
(b) Show that the sample variance becomes c^2 times its original value if each observation in the sample is multiplied by c.

8.15 Verify that the variance of the sample 4, 9, 3, 6, 4, and 7 is 5.1, and using this fact, along with the results of Exercise 8.14, find
(a) the variance of the sample 12, 27, 9, 18, 12, and 21;
(b) the variance of the sample 9, 14, 8, 11, 9, and 12.

8.16 In the 2004-05 football season, University of Southern California had the following score differences for the 13 games it played.

11 49 32 3 6 38 38 30 8 40 31 5 36

Find
(a) the mean score difference;
(b) the median score difference.

8.3 Sampling Distributions

The field of statistical inference is basically concerned with generalizations and predictions. For example, we might claim, based on the opinions of several people interviewed on the street, that in a forthcoming election 60% of the eligible voters in the city of Detroit favor a certain candidate. In this case, we are dealing with a random sample of opinions from a very large finite population. As a second illustration we might state that the average cost to build a residence in Charleston, South Carolina, is between $330,000 and $335,000, based on the estimates of 3 contractors selected at random from the 30 now building in this city. The population being sampled here is again finite but very small. Finally, let us consider a soft-drink machine designed to dispense, on average, 240 milliliters per drink. A company official who computes the mean of 40 drinks obtains $\bar{x} = 236$ milliliters and, on the basis of this value, decides that the machine is still dispensing drinks with an average content of $\mu = 240$ milliliters. The 40 drinks represent a sample from the infinite population of possible drinks that will be dispensed by this machine.

Inference about the Population from Sample Information

In each of the examples above, we computed a statistic from a sample selected from the population, and from this statistic we made various statements concerning the values of population parameters that may or may not be true. The company official made the decision that the soft-drink machine dispenses drinks with an average content of 240 milliliters, even though the sample mean was 236 milliliters, because he knows from sampling theory that, if $\mu = 240$ milliliters, such a sample value could easily occur. In fact, if he ran similar tests, say every hour, he would expect the values of the statistic $\bar{x}$ to fluctuate above and below $\mu = 240$ milliliters. Only when the value of $\bar{x}$ is substantially different from 240 milliliters will the company official initiate action to adjust the machine.

Since a statistic is a random variable that depends only on the observed sample, it must have a probability distribution.

Definition 8.5: The probability distribution of a statistic is called a **sampling distribution**.

The sampling distribution of a statistic depends on the distribution of the population, the size of the samples, and the method of choosing the samples. In the remainder of this chapter we study several of the important sampling distributions of frequently used statistics. Applications of these sampling distributions to problems of statistical inference are considered throughout most of the remaining chapters. The probability distribution of $\bar{X}$ is called the **sampling distribution of the mean**.

What Is the Sampling Distribution of $\bar{X}$?

We should view the sampling distributions of $\bar{X}$ and S^2 as the mechanisms from which we will be able to make inferences on the parameters μ and σ^2. The sampling distribution of $\bar{X}$ with sample size n is the distribution that results when an **experiment is conducted over and over** (always with sample size n) **and the many values of $\bar{X}$ result**. This sampling distribution, then, describes the variability of sample averages around the population mean μ. In the case of the soft-drink machine, knowledge of the sampling distribution of $\bar{X}$ arms the analyst with the knowledge of a "typical" discrepancy between an observed $\bar{x}$ value and true μ. The same principle applies in the case of the distribution of S^2. The sampling distribution produces information about the variability of s^2 values around σ^2 in repeated experiments.

8.4 Sampling Distribution of Means and the Central Limit Theorem

The first important sampling distribution to be considered is that of the mean $\bar{X}$. Suppose that a random sample of n observations is taken from a normal population with mean μ and variance σ^2. Each observation X_i, $i = 1, 2, \ldots, n$, of the random sample will then have the same normal distribution as the population being sampled. Hence, by the reproductive property of the normal distribution established in Theorem 7.11, we conclude that

$$\bar{X} = \frac{1}{n}(X_1 + X_2 + \cdots + X_n)$$

has a normal distribution with mean

$$\mu_{\bar{X}} = \frac{1}{n}\underbrace{(\mu + \mu + \cdots + \mu)}_{n \text{ terms}} = \mu \text{ and variance } \sigma^2_{\bar{X}} = \frac{1}{n^2}\underbrace{(\sigma^2 + \sigma^2 + \cdots + \sigma^2)}_{n \text{ terms}} = \frac{\sigma^2}{n}.$$

If we are sampling from a population with unknown distribution, either finite or infinite, the sampling distribution of $\bar{X}$ will still be approximately normal with mean μ and variance σ^2/n, provided that the sample size is large. This amazing result is an immediate consequence of the following theorem, called the Central Limit Theorem.

The Central Limit Theorem

Theorem 8.2: **Central Limit Theorem:** If $\bar{X}$ is the mean of a random sample of size n taken from a population with mean μ and finite variance σ^2, then the limiting form of the distribution of

$$Z = \frac{\bar{X} - \mu}{\sigma/\sqrt{n}},$$

as $n \to \infty$, is the standard normal distribution $n(z; 0, 1)$.

The normal approximation for $\bar{X}$ will generally be good if $n \geq 30$, provided the population distribution is not terribly skewed. If $n < 30$, the approximation is good only if the population is not too different from a normal distribution and, as stated above, if the population is known to be normal, the sampling distribution of $\bar{X}$ will follow a normal distribution exactly, no matter how small the size of the samples.

The sample size $n = 30$ is a guideline to use for the Central Limit Theorem. However, as the statement of the theorem implies, the presumption of normality on the distribution of $\bar{X}$ becomes more accurate as n grows larger. In fact, Figure 8.1 illustrates how the theorem works. It shows how the distribution of $\bar{X}$ becomes closer to normal as n grows larger, beginning with the clearly nonsymmetric distribution of an individual observation ($n = 1$). It also illustrates that the mean of $\bar{X}$ remains μ for any sample size and the variance of $\bar{X}$ gets smaller as n increases.

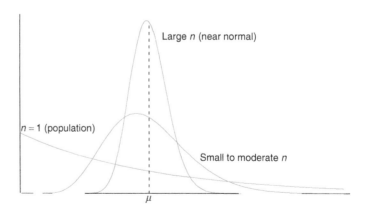

Figure 8.1: Illustration of the Central Limit Theorem (distribution of $\bar{X}$ for $n = 1$, moderate n, and large n).

Example 8.4: An electrical firm manufactures light bulbs that have a length of life that is approximately normally distributed, with mean equal to 800 hours and a standard deviation of 40 hours. Find the probability that a random sample of 16 bulbs will have an average life of less than 775 hours.

Solution: The sampling distribution of $\bar{X}$ will be approximately normal, with $\mu_{\bar{X}} = 800$ and $\sigma_{\bar{X}} = 40/\sqrt{16} = 10$. The desired probability is given by the area of the shaded

8.4 Sampling Distribution of Means and the Central Limit Theorem

region in Figure 8.2.

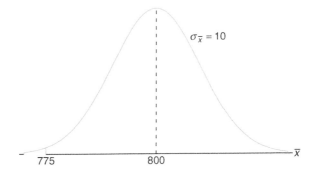

Figure 8.2: Area for Example 8.4.

Corresponding to $\bar{x} = 775$, we find that

$$z = \frac{775 - 800}{10} = -2.5,$$

and therefore

$$P(\bar{X} < 775) = P(Z < -2.5) = 0.0062.$$

Inferences on the Population Mean

One very important application of the Central Limit Theorem is the determination of reasonable values of the population mean μ. Topics such as hypothesis testing, estimation, quality control, and many others make use of the Central Limit Theorem. The following example illustrates the use of the Central Limit Theorem with regard to its relationship with μ, the mean of the population, although the formal application to the foregoing topics is relegated to future chapters.

In the following case study, an illustration is given which draws an inference that makes use of the sampling distribution of $\bar{X}$. In this simple illustration, μ and σ are both known. The Central Limit Theorem and the general notion of sampling distributions are often used to produce evidence about some important aspect of a distribution such as a parameter of the distribution. In the case of the Central Limit Theorem, the parameter of interest is the mean μ. The inference made concerning μ may take one of many forms. Often there is a desire on the part of the analyst that the data (in the form of $\bar{x}$) support (or not) some predetermined conjecture concerning the value of μ. The use of what we know about the sampling distribution can contribute to answering this type of question. In the following case study, the concept of hypothesis testing leads to a formal objective that we will highlight in future chapters.

Case Study 8.1: **Automobile Parts:** An important manufacturing process produces cylindrical component parts for the automotive industry. It is important that the process produce parts having a mean diameter of 5.0 millimeters. The engineer involved conjectures that the population mean is 5.0 millimeters. An experiment is conducted in which 100 parts produced by the process are selected randomly and the diameter measured on each. It is known that the population standard deviation is $\sigma = 0.1$ millimeter. The experiment indicates a sample average diameter of $\bar{x} = 5.027$ millimeters. Does this sample information appear to support or refute the engineer's conjecture?

Solution: This example reflects the kind of problem often posed and solved with hypothesis testing machinery introduced in future chapters. We will not use the formality associated with hypothesis testing here, but we will illustrate the principles and logic used.

Whether the data support or refute the conjecture depends on the probability that data similar to those obtained in this experiment ($\bar{x} = 5.027$) can readily occur when in fact $\mu = 5.0$ (Figure 8.3). In other words, how likely is it that one can obtain $\bar{x} \geq 5.027$ with $n = 100$ if the population mean is $\mu = 5.0$? If this probability suggests that $\bar{x} = 5.027$ is not unreasonable, the conjecture is not refuted. If the probability is quite low, one can certainly argue that the data do not support the conjecture that $\mu = 5.0$. The probability that we choose to compute is given by $P(|\bar{X} - 5| \geq 0.027)$.

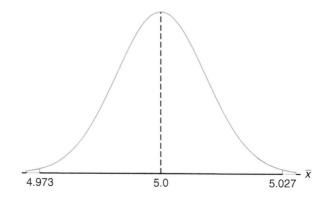

Figure 8.3: Area for Case Study 8.1.

In other words, if the mean μ is 5, what is the chance that $\bar{X}$ will deviate by as much as 0.027 millimeter?

$$P(|\bar{X} - 5| \geq 0.027) = P(\bar{X} - 5 \geq 0.027) + P(\bar{X} - 5 \leq -0.027)$$
$$= 2P\left(\frac{\bar{X} - 5}{0.1/\sqrt{100}} \geq 2.7\right).$$

Here we are simply standardizing $\bar{X}$ according to the Central Limit Theorem. If the conjecture $\mu = 5.0$ is true, $\frac{\bar{X}-5}{0.1/\sqrt{100}}$ should follow $N(0,1)$. Thus,

$$2P\left(\frac{\bar{X} - 5}{0.1/\sqrt{100}} \geq 2.7\right) = 2P(Z \geq 2.7) = 2(0.0035) = 0.007.$$

8.4 Sampling Distribution of Means and the Central Limit Theorem

Therefore, one would experience by chance that an $\bar{x}$ would be 0.027 millimeter from the mean in only 7 in 1000 experiments. As a result, this experiment with $\bar{x} = 5.027$ certainly does not give supporting evidence to the conjecture that $\mu = 5.0$. In fact, it strongly refutes the conjecture!

Example 8.5: Traveling between two campuses of a university in a city via shuttle bus takes, on average, 28 minutes with a standard deviation of 5 minutes. In a given week, a bus transported passengers 40 times. What is the probability that the average transport time was more than 30 minutes? Assume the mean time is measured to the nearest minute.

Solution: In this case, $\mu = 28$ and $\sigma = 3$. We need to calculate the probability $P(\bar{X} > 30)$ with $n = 40$. Since the time is measured on a continuous scale to the nearest minute, an $\bar{x}$ greater than 30 is equivalent to $\bar{x} \geq 30.5$. Hence,

$$P(\bar{X} > 30) = P\left(\frac{\bar{X} - 28}{5/\sqrt{40}} \geq \frac{30.5 - 28}{5/\sqrt{40}}\right) = P(Z \geq 3.16) = 0.0008.$$

There is only a slight chance that the average time of one bus trip will exceed 30 minutes. An illustrative graph is shown in Figure 8.4.

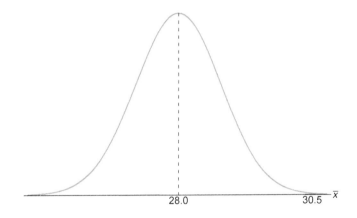

Figure 8.4: Area for Example 8.5.

Sampling Distribution of the Difference between Two Means

The illustration in Case Study 8.1 deals with notions of statistical inference on a single mean μ. The engineer was interested in supporting a conjecture regarding a single population mean. A far more important application involves two populations. A scientist or engineer may be interested in a comparative experiment in which two manufacturing methods, 1 and 2, are to be compared. The basis for that comparison is $\mu_1 - \mu_2$, the difference in the population means.

Suppose that we have two populations, the first with mean μ_1 and variance σ_1^2, and the second with mean μ_2 and variance σ_2^2. Let the statistic $\bar{X}_1$ represent the mean of a random sample of size n_1 selected from the first population, and the statistic $\bar{X}_2$ represent the mean of a random sample of size n_2 selected from the second population, independent of the sample from the first population. What can we say about the sampling distribution of the difference $\bar{X}_1 - \bar{X}_2$ for repeated samples of size n_1 and n_2? According to Theorem 8.2, the variables $\bar{X}_1$ and $\bar{X}_2$ are both approximately normally distributed with means μ_1 and μ_2 and variances σ_1^2/n_1 and σ_2^2/n_2, respectively. This approximation improves as n_1 and n_2 increase. By choosing independent samples from the two populations we ensure that the variables $\bar{X}_1$ and $\bar{X}_2$ will be independent, and then using Theorem 7.11, with $a_1 = 1$ and $a_2 = -1$, we can conclude that $\bar{X}_1 - \bar{X}_2$ is approximately normally distributed with mean

$$\mu_{\bar{X}_1-\bar{X}_2} = \mu_{\bar{X}_1} - \mu_{\bar{X}_2} = \mu_1 - \mu_2$$

and variance

$$\sigma^2_{\bar{X}_1-\bar{X}_2} = \sigma^2_{\bar{X}_1} + \sigma^2_{\bar{X}_2} = \frac{\sigma_1^2}{n_1} + \frac{\sigma_2^2}{n_2}.$$

The Central Limit Theorem can be easily extended to the two-sample, two-population case.

Theorem 8.3: If independent samples of size n_1 and n_2 are drawn at random from two populations, discrete or continuous, with means μ_1 and μ_2 and variances σ_1^2 and σ_2^2, respectively, then the sampling distribution of the differences of means, $\bar{X}_1 - \bar{X}_2$, is approximately normally distributed with mean and variance given by

$$\mu_{\bar{X}_1-\bar{X}_2} = \mu_1 - \mu_2 \text{ and } \sigma^2_{\bar{X}_1-\bar{X}_2} = \frac{\sigma_1^2}{n_1} + \frac{\sigma_2^2}{n_2}.$$

Hence,

$$Z = \frac{(\bar{X}_1 - \bar{X}_2) - (\mu_1 - \mu_2)}{\sqrt{(\sigma_1^2/n_1) + (\sigma_2^2/n_2)}}$$

is approximately a standard normal variable.

If both n_1 and n_2 are greater than or equal to 30, the normal approximation for the distribution of $\bar{X}_1 - \bar{X}_2$ is very good when the underlying distributions are not too far away from normal. However, even when n_1 and n_2 are less than 30, the normal approximation is reasonably good except when the populations are decidedly nonnormal. Of course, if both populations are normal, then $\bar{X}_1 - \bar{X}_2$ has a normal distribution no matter what the sizes of n_1 and n_2 are.

The utility of the sampling distribution of the difference between two sample averages is very similar to that described in Case Study 8.1 on page 235 for the case of a single mean. Case Study 8.2 that follows focuses on the use of the difference between two sample means to support (or not) the conjecture that two population means are the same.

Case Study 8.2: Paint Drying Time: Two independent experiments are run in which two different types of paint are compared. Eighteen specimens are painted using type A, and the drying time, in hours, is recorded for each. The same is done with type B. The population standard deviations are both known to be 1.0.

Assuming that the mean drying time is equal for the two types of paint, find $P(\bar{X}_A - \bar{X}_B > 1.0)$, where $\bar{X}_A$ and $\bar{X}_B$ are average drying times for samples of size $n_A = n_B = 18$.

Solution: From the sampling distribution of $\bar{X}_A - \bar{X}_B$, we know that the distribution is approximately normal with mean

$$\mu_{\bar{X}_A - \bar{X}_B} = \mu_A - \mu_B = 0$$

and variance

$$\sigma^2_{\bar{X}_A - \bar{X}_B} = \frac{\sigma_A^2}{n_A} + \frac{\sigma_B^2}{n_B} = \frac{1}{18} + \frac{1}{18} = \frac{1}{9}.$$

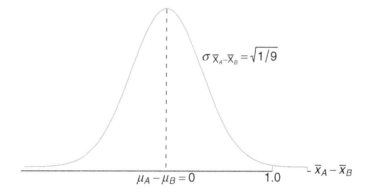

Figure 8.5: Area for Case Study 8.2.

The desired probability is given by the shaded region in Figure 8.5. Corresponding to the value $\bar{X}_A - \bar{X}_B = 1.0$, we have

$$z = \frac{1 - (\mu_A - \mu_B)}{\sqrt{1/9}} = \frac{1 - 0}{\sqrt{1/9}} = 3.0;$$

so

$$P(Z > 3.0) = 1 - P(Z < 3.0) = 1 - 0.9987 = 0.0013.$$

What Do We Learn from Case Study 8.2?

The machinery in the calculation is based on the presumption that $\mu_A = \mu_B$. Suppose, however, that the experiment is actually conducted for the purpose of drawing an inference regarding the equality of μ_A and μ_B, the two population mean drying times. If the two averages differ by as much as 1 hour (or more), this clearly is evidence that would lead one to conclude that the population mean drying time is not equal for the two types of paint. On the other hand, suppose that the difference in the two sample averages is as small as, say, 15 minutes. If $\mu_A = \mu_B$,

$$P[(\bar{X}_A - \bar{X}_B) > 0.25 \text{ hour}] = P\left(\frac{\bar{X}_A - \bar{X}_B - 0}{\sqrt{1/9}} > \frac{3}{4}\right)$$

$$= P\left(Z > \frac{3}{4}\right) = 1 - P(Z < 0.75) = 1 - 0.7734 = 0.2266.$$

Since this probability is not low, one would conclude that a difference in sample means of 15 minutes can happen by chance (i.e., it happens frequently even though $\mu_A = \mu_B$). As a result, that type of difference in average drying times certainly *is not a clear signal* that $\mu_A \neq \mu_B$.

As we indicated earlier, a more detailed formalism regarding this and other types of statistical inference (e.g., hypothesis testing) will be supplied in future chapters. The Central Limit Theorem and sampling distributions discussed in the next three sections will also play a vital role.

Example 8.6: The television picture tubes of manufacturer A have a mean lifetime of 6.5 years and a standard deviation of 0.9 year, while those of manufacturer B have a mean lifetime of 6.0 years and a standard deviation of 0.8 year. What is the probability that a random sample of 36 tubes from manufacturer A will have a mean lifetime that is at least 1 year more than the mean lifetime of a sample of 49 tubes from manufacturer B?

Solution: We are given the following information:

Population 1	Population 2
$\mu_1 = 6.5$	$\mu_2 = 6.0$
$\sigma_1 = 0.9$	$\sigma_2 = 0.8$
$n_1 = 36$	$n_2 = 49$

If we use Theorem 8.3, the sampling distribution of $\bar{X}_1 - \bar{X}_2$ will be approximately normal and will have a mean and standard deviation

$$\mu_{\bar{X}_1 - \bar{X}_2} = 6.5 - 6.0 = 0.5 \quad \text{and} \quad \sigma_{\bar{X}_1 - \bar{X}_2} = \sqrt{\frac{0.81}{36} + \frac{0.64}{49}} = 0.189.$$

The probability that the mean lifetime for 36 tubes from manufacturer A will be at least 1 year longer than the mean lifetime for 49 tubes from manufacturer B is given by the area of the shaded region in Figure 8.6. Corresponding to the value $\bar{x}_1 - \bar{x}_2 = 1.0$, we find that

$$z = \frac{1.0 - 0.5}{0.189} = 2.65,$$

and hence

$$P(\bar{X}_1 - \bar{X}_2 \geq 1.0) = P(Z > 2.65) = 1 - P(Z < 2.65)$$
$$= 1 - 0.9960 = 0.0040.$$

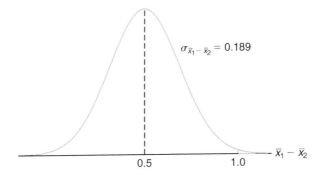

Figure 8.6: Area for Example 8.6.

More on Sampling Distribution of Means—Normal Approximation to the Binomial Distribution

Section 6.5 presented the normal approximation to the binomial distribution at length. Conditions were given on the parameters n and p for which the distribution of a binomial random variable can be approximated by the normal distribution. Examples and exercises reflected the importance of the concept of the "normal approximation." It turns out that the Central Limit Theorem sheds even more light on how and why this approximation works. We certainly know that a binomial random variable is the number X of successes in n independent trials, where the outcome of each trial is binary. We also illustrated in Chapter 1 that the proportion computed in such an experiment is an average of a set of 0s and 1s. Indeed, while the proportion X/n is an average, X is the sum of this set of 0s and 1s, and both X and X/n are approximately normal if n is sufficiently large. Of course, from what we learned in Chapter 6, we know that there are conditions on n and p that affect the quality of the approximation, namely $np \geq 5$ and $nq \geq 5$.

Exercises

8.17 If all possible samples of size 16 are drawn from a normal population with mean equal to 50 and standard deviation equal to 5, what is the probability that a sample mean $\bar{X}$ will fall in the interval from $\mu_{\bar{X}} - 1.9\sigma_{\bar{X}}$ to $\mu_{\bar{X}} - 0.4\sigma_{\bar{X}}$? Assume that the sample means can be measured to any degree of accuracy.

8.18 If the standard deviation of the mean for the sampling distribution of random samples of size 36 from a large or infinite population is 2, how large must the sample size become if the standard deviation is to be reduced to 1.2?

8.19 A certain type of thread is manufactured with a mean tensile strength of 78.3 kilograms and a standard deviation of 5.6 kilograms. How is the variance of the sample mean changed when the sample size is

(a) increased from 64 to 196?

(b) decreased from 784 to 49?

8.20 Given the discrete uniform population

$$f(x) = \begin{cases} \frac{1}{3}, & x = 2, 4, 6, \\ 0, & \text{elsewhere}, \end{cases}$$

find the probability that a random sample of size 54, selected with replacement, will yield a sample mean greater than 4.1 but less than 4.4. Assume the means are measured to the nearest tenth.

8.21 A soft-drink machine is regulated so that the amount of drink dispensed averages 240 milliliters with a standard deviation of 15 milliliters. Periodically, the machine is checked by taking a sample of 40 drinks and computing the average content. If the mean of the 40 drinks is a value within the interval $\mu_{\bar{X}} \pm 2\sigma_{\bar{X}}$, the machine is thought to be operating satisfactorily; otherwise, adjustments are made. In Section 8.3, the company official found the mean of 40 drinks to be $\bar{x} = 236$ milliliters and concluded that the machine needed no adjustment. Was this a reasonable decision?

8.22 The heights of 1000 students are approximately normally distributed with a mean of 174.5 centimeters and a standard deviation of 6.9 centimeters. Suppose 200 random samples of size 25 are drawn from this population and the means recorded to the nearest tenth of a centimeter. Determine

(a) the mean and standard deviation of the sampling distribution of $\bar{X}$;

(b) the number of sample means that fall between 172.5 and 175.8 centimeters inclusive;

(c) the number of sample means falling below 172.0 centimeters.

8.23 The random variable X, representing the number of cherries in a cherry puff, has the following probability distribution:

x	4	5	6	7
$P(X=x)$	0.2	0.4	0.3	0.1

(a) Find the mean μ and the variance σ^2 of X.

(b) Find the mean $\mu_{\bar{X}}$ and the variance $\sigma_{\bar{X}}^2$ of the mean $\bar{X}$ for random samples of 36 cherry puffs.

(c) Find the probability that the average number of cherries in 36 cherry puffs will be less than 5.5.

8.24 If a certain machine makes electrical resistors having a mean resistance of 40 ohms and a standard deviation of 2 ohms, what is the probability that a random sample of 36 of these resistors will have a combined resistance of more than 1458 ohms?

8.25 The average life of a bread-making machine is 7 years, with a standard deviation of 1 year. Assuming that the lives of these machines follow approximately a normal distribution, find

(a) the probability that the mean life of a random sample of 9 such machines falls between 6.4 and 7.2 years;

(b) the value of x to the right of which 15% of the means computed from random samples of size 9 would fall.

8.26 The amount of time that a drive-through bank teller spends on a customer is a random variable with a mean $\mu = 3.2$ minutes and a standard deviation $\sigma = 1.6$ minutes. If a random sample of 64 customers is observed, find the probability that their mean time at the teller's window is

(a) at most 2.7 minutes;

(b) more than 3.5 minutes;

(c) at least 3.2 minutes but less than 3.4 minutes.

8.27 In a chemical process, the amount of a certain type of impurity in the output is difficult to control and is thus a random variable. Speculation is that the population mean amount of the impurity is 0.20 gram per gram of output. It is known that the standard deviation is 0.1 gram per gram. An experiment is conducted to gain more insight regarding the speculation that $\mu = 0.2$. The process is run on a lab scale 50 times and the sample average $\bar{x}$ turns out to be 0.23 gram per gram. Comment on the speculation that the mean amount of impurity is 0.20 gram per gram. Make use of the Central Limit Theorem in your work.

8.28 A random sample of size 25 is taken from a normal population having a mean of 80 and a standard deviation of 5. A second random sample of size 36 is taken from a different normal population having a mean of 75 and a standard deviation of 3. Find the probability that the sample mean computed from the 25 measurements will exceed the sample mean computed from the 36 measurements by at least 3.4 but less than 5.9. Assume the difference of the means to be measured to the nearest tenth.

8.29 The distribution of heights of a certain breed of terrier has a mean of 72 centimeters and a standard deviation of 10 centimeters, whereas the distribution of heights of a certain breed of poodle has a mean of 28 centimeters with a standard deviation of 5 centimeters. Assuming that the sample means can be measured to any degree of accuracy, find the probability that the sample mean for a random sample of heights of 64 terriers exceeds the sample mean for a random sample of heights of 100 poodles by at most 44.2 centimeters.

8.30 The mean score for freshmen on an aptitude test at a certain college is 540, with a standard deviation of 50. Assume the means to be measured to any degree of accuracy. What is the probability that two groups selected at random, consisting of 32 and 50 students, respectively, will differ in their mean scores by

(a) more than 20 points?

(b) an amount between 5 and 10 points?

8.31 Consider Case Study 8.2 on page 238. Suppose 18 specimens were used for each type of paint in an experiment and $\bar{x}_A - \bar{x}_B$, the actual difference in mean drying time, turned out to be 1.0.

(a) Does this seem to be a reasonable result if the

two population mean drying times truly are equal? Make use of the result in the solution to Case Study 8.2.

(b) If someone did the experiment 10,000 times under the condition that $\mu_A = \mu_B$, in how many of those 10,000 experiments would there be a difference $\bar{x}_A - \bar{x}_B$ that was as large as (or larger than) 1.0?

8.32 Two different box-filling machines are used to fill cereal boxes on an assembly line. The critical measurement influenced by these machines is the weight of the product in the boxes. Engineers are quite certain that the variance of the weight of product is $\sigma^2 = 1$ ounce. Experiments are conducted using both machines with sample sizes of 36 each. The sample averages for machines A and B are $\bar{x}_A = 4.5$ ounces and $\bar{x}_B = 4.7$ ounces. Engineers are surprised that the two sample averages for the filling machines are so different.

(a) Use the Central Limit Theorem to determine
$$P(\bar{X}_B - \bar{X}_A \geq 0.2)$$
under the condition that $\mu_A = \mu_B$.

(b) Do the aforementioned experiments seem to, in any way, strongly support a conjecture that the population means for the two machines are different? Explain using your answer in (a).

8.33 The chemical benzene is highly toxic to humans. However, it is used in the manufacture of many medicine dyes, leather, and coverings. Government regulations dictate that for any production process involving benzene, the water in the output of the process must not exceed 7950 parts per million (ppm) of benzene. For a particular process of concern, the water sample was collected by a manufacturer 25 times randomly and the sample average $\bar{x}$ was 7960 ppm. It is known from historical data that the standard deviation σ is 100 ppm.

(a) What is the probability that the sample average in this experiment would exceed the government limit if the population mean is equal to the limit? Use the Central Limit Theorem.

(b) Is an observed $\bar{x} = 7960$ in this experiment firm evidence that the population mean for the process exceeds the government limit? Answer your question by computing
$$P(\bar{X} \geq 7960 \mid \mu = 7950).$$
Assume that the distribution of benzene concentration is normal.

8.34 Two alloys A and B are being used to manufacture a certain steel product. An experiment needs to be designed to compare the two in terms of maximum load capacity in tons (the maximum weight that can be tolerated without breaking). It is known that the two standard deviations in load capacity are equal at 5 tons each. An experiment is conducted in which 30 specimens of each alloy (A and B) are tested and the results recorded as follows:
$$\bar{x}_A = 49.5, \quad \bar{x}_B = 45.5; \quad \bar{x}_A - \bar{x}_B = 4.$$

The manufacturers of alloy A are convinced that this evidence shows conclusively that $\mu_A > \mu_B$ and strongly supports the claim that their alloy is superior. Manufacturers of alloy B claim that the experiment could easily have given $\bar{x}_A - \bar{x}_B = 4$ *even if* the two population means are equal. In other words, "the results are inconclusive!"

(a) Make an argument that manufacturers of alloy B are wrong. Do it by computing
$$P(\bar{X}_A - \bar{X}_B > 4 \mid \mu_A = \mu_B).$$

(b) Do you think these data strongly support alloy A?

8.35 Consider the situation described in Example 8.4 on page 234. Do these results prompt you to question the premise that $\mu = 800$ hours? Give a probabilistic result that indicates how *rare* an event $\bar{X} \leq 775$ is when $\mu = 800$. On the other hand, how rare would it be if μ truly were, say, 760 hours?

8.36 Let $X_1, X_2, \ldots, X_n$ be a random sample from a distribution that can take on only positive values. Use the Central Limit Theorem to produce an argument that if n is sufficiently large, then $Y = X_1 X_2 \cdots X_n$ has approximately a lognormal distribution.

8.5 Sampling Distribution of S^2

In the preceding section we learned about the sampling distribution of $\bar{X}$. The Central Limit Theorem allowed us to make use of the fact that
$$\frac{\bar{X} - \mu}{\sigma/\sqrt{n}}$$

tends toward $N(0,1)$ as the sample size grows large. *Sampling distributions of important statistics* allow us to learn information about parameters. Usually, the parameters are the counterpart to the statistics in question. For example, if an engineer is interested in the population mean resistance of a certain type of resistor, the sampling distribution of $\bar{X}$ will be exploited once the sample information is gathered. On the other hand, if the variability in resistance is to be studied, clearly the sampling distribution of S^2 will be used in learning about the parametric counterpart, the population variance σ^2.

If a random sample of size n is drawn from a normal population with mean μ and variance σ^2, and the sample variance is computed, we obtain a value of the statistic S^2. We shall proceed to consider the distribution of the statistic $(n-1)S^2/\sigma^2$.

By the addition and subtraction of the sample mean $\bar{X}$, it is easy to see that

$$\sum_{i=1}^{n}(X_i - \mu)^2 = \sum_{i=1}^{n}[(X_i - \bar{X}) + (\bar{X} - \mu)]^2$$

$$= \sum_{i=1}^{n}(X_i - \bar{X})^2 + \sum_{i=1}^{n}(\bar{X} - \mu)^2 + 2(\bar{X} - \mu)\sum_{i=1}^{n}(X_i - \bar{X})$$

$$= \sum_{i=1}^{n}(X_i - \bar{X})^2 + n(\bar{X} - \mu)^2.$$

Dividing each term of the equality by σ^2 and substituting $(n-1)S^2$ for $\sum_{i=1}^{n}(X_i - \bar{X})^2$, we obtain

$$\frac{1}{\sigma^2}\sum_{i=1}^{n}(X_i - \mu)^2 = \frac{(n-1)S^2}{\sigma^2} + \frac{(\bar{X} - \mu)^2}{\sigma^2/n}.$$

Now, according to Corollary 7.1 on page 222, we know that

$$\sum_{i=1}^{n}\frac{(X_i - \mu)^2}{\sigma^2}$$

is a chi-squared random variable with n degrees of freedom. We have a chi-squared random variable with n degrees of freedom partitioned into two components. Note that in Section 6.7 we showed that a chi-squared distribution is a special case of a gamma distribution. The second term on the right-hand side is Z^2, which is a chi-squared random variable with 1 degree of freedom, and it turns out that $(n-1)S^2/\sigma^2$ is a chi-squared random variable with $n-1$ degree of freedom. We formalize this in the following theorem.

Theorem 8.4: If S^2 is the variance of a random sample of size n taken from a normal population having the variance σ^2, then the statistic

$$\chi^2 = \frac{(n-1)S^2}{\sigma^2} = \sum_{i=1}^{n}\frac{(X_i - \bar{X})^2}{\sigma^2}$$

has a chi-squared distribution with $v = n-1$ degrees of freedom.

The values of the random variable χ^2 are calculated from each sample by the

8.5 Sampling Distribution of S^2

formula

$$\chi^2 = \frac{(n-1)s^2}{\sigma^2}.$$

The probability that a random sample produces a χ^2 value greater than some specified value is equal to the area under the curve to the right of this value. It is customary to let χ^2_α represent the χ^2 value above which we find an area of α. This is illustrated by the shaded region in Figure 8.7.

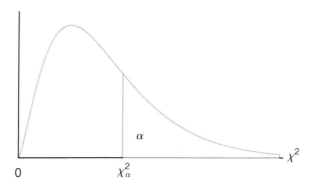

Figure 8.7: The chi-squared distribution.

Table A.5 gives values of χ^2_α for various values of α and v. The areas, α, are the column headings; the degrees of freedom, v, are given in the left column; and the table entries are the χ^2 values. Hence, the χ^2 value with 7 degrees of freedom, leaving an area of 0.05 to the right, is $\chi^2_{0.05} = 14.067$. Owing to lack of symmetry, we must also use the tables to find $\chi^2_{0.95} = 2.167$ for $v = 7$.

Exactly 95% of a chi-squared distribution lies between $\chi^2_{0.975}$ and $\chi^2_{0.025}$. A χ^2 value falling to the right of $\chi^2_{0.025}$ is not likely to occur unless our assumed value of σ^2 is too small. Similarly, a χ^2 value falling to the left of $\chi^2_{0.975}$ is unlikely unless our assumed value of σ^2 is too large. In other words, it is possible to have a χ^2 value to the left of $\chi^2_{0.975}$ or to the right of $\chi^2_{0.025}$ when σ^2 is correct, but if this should occur, it is more probable that the assumed value of σ^2 is in error.

Example 8.7: A manufacturer of car batteries guarantees that the batteries will last, on average, 3 years with a standard deviation of 1 year. If five of these batteries have lifetimes of 1.9, 2.4, 3.0, 3.5, and 4.2 years, should the manufacturer still be convinced that the batteries have a standard deviation of 1 year? Assume that the battery lifetime follows a normal distribution.

Solution: We first find the sample variance using Theorem 8.1,

$$s^2 = \frac{(5)(48.26) - (15)^2}{(5)(4)} = 0.815.$$

Then

$$\chi^2 = \frac{(4)(0.815)}{1} = 3.26$$

is a value from a chi-squared distribution with 4 degrees of freedom. Since 95% of the χ^2 values with 4 degrees of freedom fall between 0.484 and 11.143, the computed value with $\sigma^2 = 1$ is reasonable, and therefore the manufacturer has no reason to suspect that the standard deviation is other than 1 year.

Degrees of Freedom as a Measure of Sample Information

Recall from Corollary 7.1 in Section 7.3 that

$$\sum_{i=1}^{n} \frac{(X_i - \mu)^2}{\sigma^2}$$

has a χ^2-distribution with n *degrees of freedom*. Note also Theorem 8.4, which indicates that the random variable

$$\frac{(n-1)S^2}{\sigma^2} = \sum_{i=1}^{n} \frac{(X_i - \bar{X})^2}{\sigma^2}$$

has a χ^2-distribution with $n-1$ *degrees of freedom*. The reader may also recall that the term *degrees of freedom*, used in this identical context, is discussed in Chapter 1.

As we indicated earlier, the proof of Theorem 8.4 will not be given. However, the reader can view Theorem 8.4 as indicating that when μ is not known and one considers the distribution of

$$\sum_{i=1}^{n} \frac{(X_i - \bar{X})^2}{\sigma^2},$$

there is **1 less degree of freedom**, or a degree of freedom is lost in the estimation of μ (i.e., when μ is replaced by $\bar{x}$). In other words, there are n degrees of freedom, or independent *pieces of information*, in the random sample from the normal distribution. When the data (the values in the sample) are used to compute the mean, there is 1 less degree of freedom in the information used to estimate σ^2.

8.6 *t*-Distribution

In Section 8.4, we discussed the utility of the Central Limit Theorem. Its applications revolve around inferences on a population mean or the difference between two population means. Use of the Central Limit Theorem and the normal distribution is certainly helpful in this context. However, it was assumed that the population standard deviation is known. This assumption may not be unreasonable in situations where the engineer is quite familiar with the system or process. However, in many experimental scenarios, knowledge of σ is certainly no more reasonable than knowledge of the population mean μ. Often, in fact, an estimate of σ must be supplied by the same sample information that produced the sample average $\bar{x}$. As a result, a natural statistic to consider to deal with inferences on μ is

$$T = \frac{\bar{X} - \mu}{S/\sqrt{n}},$$

8.6 t-Distribution

since S is the sample analog to σ. If the sample size is small, the values of S^2 fluctuate considerably from sample to sample (see Exercise 8.43 on page 260) and the distribution of T deviates appreciably from that of a standard normal distribution.

If the sample size is large enough, say $n \geq 30$, the distribution of T does not differ considerably from the standard normal. However, for $n < 30$, it is useful to deal with the exact distribution of T. In developing the sampling distribution of T, we shall assume that our random sample was selected from a normal population. We can then write

$$T = \frac{(\bar{X} - \mu)/(\sigma/\sqrt{n})}{\sqrt{S^2/\sigma^2}} = \frac{Z}{\sqrt{V/(n-1)}},$$

where

$$Z = \frac{\bar{X} - \mu}{\sigma/\sqrt{n}}$$

has the standard normal distribution and

$$V = \frac{(n-1)S^2}{\sigma^2}$$

has a chi-squared distribution with $v = n - 1$ degrees of freedom. In sampling from normal populations, we can show that $\bar{X}$ and S^2 are independent, and consequently so are Z and V. The following theorem gives the definition of a random variable T as a function of Z (standard normal) and χ^2. For completeness, the density function of the t-distribution is given.

Theorem 8.5: Let Z be a standard normal random variable and V a chi-squared random variable with v degrees of freedom. If Z and V are independent, then the distribution of the random variable T, where

$$T = \frac{Z}{\sqrt{V/v}},$$

is given by the density function

$$h(t) = \frac{\Gamma[(v+1)/2]}{\Gamma(v/2)\sqrt{\pi v}} \left(1 + \frac{t^2}{v}\right)^{-(v+1)/2}, \quad -\infty < t < \infty.$$

This is known as the **t-distribution** with v degrees of freedom.

From the foregoing and the theorem above we have the following corollary.

Corollary 8.1: Let $X_1, X_2, \ldots, X_n$ be independent random variables that are all normal with mean μ and standard deviation σ. Let

$$\bar{X} = \frac{1}{n}\sum_{i=1}^{n} X_i \quad \text{and} \quad S^2 = \frac{1}{n-1}\sum_{i=1}^{n}(X_i - \bar{X})^2.$$

Then the random variable $T = \frac{\bar{X}-\mu}{S/\sqrt{n}}$ has a t-distribution with $v = n-1$ degrees of freedom.

The probability distribution of T was first published in 1908 in a paper written by W. S. Gosset. At the time, Gosset was employed by an Irish brewery that prohibited publication of research by members of its staff. To circumvent this restriction, he published his work secretly under the name "Student." Consequently, the distribution of T is usually called the Student t-distribution or simply the t-distribution. In deriving the equation of this distribution, Gosset assumed that the samples were selected from a normal population. Although this would seem to be a very restrictive assumption, it can be shown that nonnormal populations possessing nearly bell-shaped distributions will still provide values of T that approximate the t-distribution very closely.

What Does the *t*-Distribution Look Like?

The distribution of T is similar to the distribution of Z in that they both are symmetric about a mean of zero. Both distributions are bell shaped, but the t-distribution is more variable, owing to the fact that the T-values depend on the fluctuations of two quantities, $\bar{X}$ and S^2, whereas the Z-values depend only on the changes in $\bar{X}$ from sample to sample. The distribution of T differs from that of Z in that the variance of T depends on the sample size n and is always greater than 1. Only when the sample size $n \to \infty$ will the two distributions become the same. In Figure 8.8, we show the relationship between a standard normal distribution ($v = \infty$) and t-distributions with 2 and 5 degrees of freedom. The percentage points of the t-distribution are given in Table A.4.

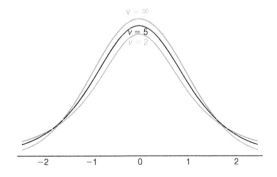

Figure 8.8: The t-distribution curves for $v = 2, 5,$ and ∞.

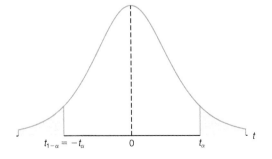

Figure 8.9: Symmetry property (about 0) of the t-distribution.

8.6 t-Distribution

It is customary to let t_α represent the t-value above which we find an area equal to α. Hence, the t-value with 10 degrees of freedom leaving an area of 0.025 to the right is $t = 2.228$. Since the t-distribution is symmetric about a mean of zero, we have $t_{1-\alpha} = -t_\alpha$; that is, the t-value leaving an area of $1 - \alpha$ to the right and therefore an area of α to the left is equal to the negative t-value that leaves an area of α in the right tail of the distribution (see Figure 8.9). That is, $t_{0.95} = -t_{0.05}$, $t_{0.99} = -t_{0.01}$, and so forth.

Example 8.8: The t-value with $v = 14$ degrees of freedom that leaves an area of 0.025 to the left, and therefore an area of 0.975 to the right, is

$$t_{0.975} = -t_{0.025} = -2.145.$$

Example 8.9: Find $P(-t_{0.025} < T < t_{0.05})$.

Solution: Since $t_{0.05}$ leaves an area of 0.05 to the right, and $-t_{0.025}$ leaves an area of 0.025 to the left, we find a total area of

$$1 - 0.05 - 0.025 = 0.925$$

between $-t_{0.025}$ and $t_{0.05}$. Hence

$$P(-t_{0.025} < T < t_{0.05}) = 0.925.$$

Example 8.10: Find k such that $P(k < T < -1.761) = 0.045$ for a random sample of size 15 selected from a normal distribution and $\frac{\overline{X}-\mu}{s/\sqrt{n}}$.

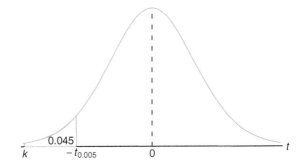

Figure 8.10: The t-values for Example 8.10.

Solution: From Table A.4 we note that 1.761 corresponds to $t_{0.05}$ when $v = 14$. Therefore, $-t_{0.05} = -1.761$. Since k in the original probability statement is to the left of $-t_{0.05} = -1.761$, let $k = -t_\alpha$. Then, from Figure 8.10, we have

$$0.045 = 0.05 - \alpha, \text{ or } \alpha = 0.005.$$

Hence, from Table A.4 with $v = 14$,

$$k = -t_{0.005} = -2.977 \text{ and } P(-2.977 < T < -1.761) = 0.045.$$

Exactly 95% of the values of a t-distribution with $v = n-1$ degrees of freedom lie between $-t_{0.025}$ and $t_{0.025}$. Of course, there are other t-values that contain 95% of the distribution, such as $-t_{0.02}$ and $t_{0.03}$, but these values do not appear in Table A.4, and furthermore, the shortest possible interval is obtained by choosing t-values that leave exactly the same area in the two tails of our distribution. A t-value that falls below $-t_{0.025}$ or above $t_{0.025}$ would tend to make us believe either that a very rare event has taken place or that our assumption about μ is in error. Should this happen, we shall make the the decision that our assumed value of μ is in error. In fact, a t-value falling below $-t_{0.01}$ or above $t_{0.01}$ would provide even stronger evidence that our assumed value of μ is quite unlikely. General procedures for testing claims concerning the value of the parameter μ will be treated in Chapter 10. A preliminary look into the foundation of these procedure is illustrated by the following example.

Example 8.11: A chemical engineer claims that the population mean yield of a certain batch process is 500 grams per milliliter of raw material. To check this claim he samples 25 batches each month. If the computed t-value falls between $-t_{0.05}$ and $t_{0.05}$, he is satisfied with this claim. What conclusion should he draw from a sample that has a mean $\bar{x} = 518$ grams per milliliter and a sample standard deviation $s = 40$ grams? Assume the distribution of yields to be approximately normal.

Solution: From Table A.4 we find that $t_{0.05} = 1.711$ for 24 degrees of freedom. Therefore, the engineer can be satisfied with his claim if a sample of 25 batches yields a t-value between -1.711 and 1.711. If $\mu = 500$, then

$$t = \frac{518 - 500}{40/\sqrt{25}} = 2.25,$$

a value well above 1.711. The probability of obtaining a t-value, with $v = 24$, equal to or greater than 2.25 is approximately 0.02. If $\mu > 500$, the value of t computed from the sample is more reasonable. Hence, the engineer is likely to conclude that the process produces a better product than he thought.

What Is the t-Distribution Used For?

The t-distribution is used extensively in problems that deal with inference about the population mean (as illustrated in Example 8.11) or in problems that involve comparative samples (i.e., in cases where one is trying to determine if means from two samples are significantly different). The use of the distribution will be extended in Chapters 9, 10, 11, and 12. The reader should note that use of the t-distribution for the statistic

$$T = \frac{\bar{X} - \mu}{S/\sqrt{n}}$$

requires that $X_1, X_2, \ldots, X_n$ be normal. The use of the t-distribution and the sample size consideration do not relate to the Central Limit Theorem. The use of the standard normal distribution rather than T for $n \geq 30$ merely implies that S is a sufficiently good estimator of σ in this case. In chapters that follow the t-distribution finds extensive usage.

8.7 F-Distribution

We have motivated the t-distribution in part by its application to problems in which there is comparative sampling (i.e., a comparison between two sample means). For example, some of our examples in future chapters will take a more formal approach, chemical engineer collects data on two catalysts, biologist collects data on two growth media, or chemist gathers data on two methods of coating material to inhibit corrosion. While it is of interest to let sample information shed light on two population means, it is often the case that a comparison of variability is equally important, if not more so. The F-distribution finds enormous application in comparing sample variances. Applications of the F-distribution are found in problems involving two or more samples.

The statistic F is defined to be the ratio of two independent chi-squared random variables, each divided by its number of degrees of freedom. Hence, we can write

$$F = \frac{U/v_1}{V/v_2},$$

where U and V are independent random variables having chi-squared distributions with v_1 and v_2 degrees of freedom, respectively. We shall now state the sampling distribution of F.

Theorem 8.6: Let U and V be two independent random variables having chi-squared distributions with v_1 and v_2 degrees of freedom, respectively. Then the distribution of the random variable $F = \frac{U/v_1}{V/v_2}$ is given by the density function

$$h(f) = \begin{cases} \frac{\Gamma[(v_1+v_2)/2](v_1/v_2)^{v_1/2}}{\Gamma(v_1/2)\Gamma(v_2/2)} \frac{f^{(v_1/2)-1}}{(1+v_1 f/v_2)^{(v_1+v_2)/2}}, & f > 0, \\ 0, & f \leq 0. \end{cases}$$

This is known as the **F-distribution** with v_1 and v_2 degrees of freedom (d.f.).

We will make considerable use of the random variable F in future chapters. However, the density function will not be used and is given only for completeness. The curve of the F-distribution depends not only on the two parameters v_1 and v_2 but also on the order in which we state them. Once these two values are given, we can identify the curve. Typical F-distributions are shown in Figure 8.11.

Let f_α be the f-value above which we find an area equal to α. This is illustrated by the shaded region in Figure 8.12. Table A.6 gives values of f_α only for $\alpha = 0.05$ and $\alpha = 0.01$ for various combinations of the degrees of freedom v_1 and v_2. Hence, the f-value with 6 and 10 degrees of freedom, leaving an area of 0.05 to the right, is $f_{0.05} = 3.22$. By means of the following theorem, Table A.6 can also be used to find values of $f_{0.95}$ and $f_{0.99}$. The proof is left for the reader.

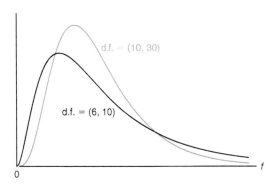
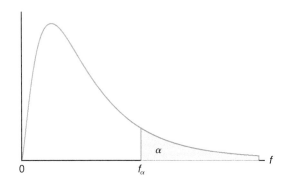

Figure 8.11: Typical F-distributions.

Figure 8.12: Illustration of the f_α for the F-distribution.

Theorem 8.7: Writing $f_\alpha(v_1, v_2)$ for f_α with v_1 and v_2 degrees of freedom, we obtain

$$f_{1-\alpha}(v_1, v_2) = \frac{1}{f_\alpha(v_2, v_1)}.$$

Thus, the f-value with 6 and 10 degrees of freedom, leaving an area of 0.95 to the right, is

$$f_{0.95}(6, 10) = \frac{1}{f_{0.05}(10, 6)} = \frac{1}{4.06} = 0.246.$$

The F-Distribution with Two Sample Variances

Suppose that random samples of size n_1 and n_2 are selected from two normal populations with variances σ_1^2 and σ_2^2, respectively. From Theorem 8.4, we know that

$$\chi_1^2 = \frac{(n_1 - 1)S_1^2}{\sigma_1^2} \text{ and } \chi_2^2 = \frac{(n_2 - 1)S_2^2}{\sigma_2^2}$$

are random variables having chi-squared distributions with $v_1 = n_1 - 1$ and $v_2 = n_2 - 1$ degrees of freedom. Furthermore, since the samples are selected at random, we are dealing with independent random variables. Then, using Theorem 8.6 with $\chi_1^2 = U$ and $\chi_2^2 = V$, we obtain the following result.

Theorem 8.8: If S_1^2 and S_2^2 are the variances of independent random samples of size n_1 and n_2 taken from normal populations with variances σ_1^2 and σ_2^2, respectively, then

$$F = \frac{S_1^2/\sigma_1^2}{S_2^2/\sigma_2^2} = \frac{\sigma_2^2 S_1^2}{\sigma_1^2 S_2^2}$$

has an F-distribution with $v_1 = n_1 - 1$ and $v_2 = n_2 - 1$ degrees of freedom.

8.7 F-Distribution

What Is the F-Distribution Used For?

We answered this question, in part, at the beginning of this section. The F-distribution is used in two-sample situations to draw inferences about the population variances. This involves the application of Theorem 8.8. However, the F-distribution can also be applied to many other types of problems involving sample variances. In fact, the F-distribution is called the *variance ratio distribution*. As an illustration, consider Case Study 8.2, in which two paints, A and B, were compared with regard to mean drying time. The normal distribution applies nicely (assuming that σ_A and σ_B are known). However, suppose that there are three types of paints to compare, say A, B, and C. We wish to determine if the population means are equivalent. Suppose that important summary information from the experiment is as follows:

Paint	Sample Mean	Sample Variance	Sample Size
A	$\bar{X}_A = 4.5$	$s_A^2 = 0.20$	10
B	$\bar{X}_B = 5.5$	$s_B^2 = 0.14$	10
C	$\bar{X}_C = 6.5$	$s_C^2 = 0.11$	10

The problem centers around whether or not the sample averages ($\bar{x}_A, \bar{x}_B, \bar{x}_C$) are far enough apart. The implication of "far enough apart" is very important. It would seem reasonable that if the variability between sample averages is larger than what one would expect by chance, the data do not support the conclusion that $\mu_A = \mu_B = \mu_C$. Whether these sample averages could have occurred by chance depends on the *variability within samples*, as quantified by s_A^2, s_B^2, and s_C^2. The notion of the important components of variability is best seen through some simple graphics. Consider the plot of raw data from samples A, B, and C, shown in Figure 8.13. These data could easily have generated the above summary information.

```
A        A A A A A     A B A AB    A B B B B B  BBCCB       C C CC    C C C C
                       4.5                      5.5                   6.5
                        ↑                        ↑                     ↑
                       X̄_A                      X̄_B                   X̄_C
```

Figure 8.13: Data from three distinct samples.

It appears evident that the data came from distributions with different population means, although there is some overlap between the samples. An analysis that involves all of the data would attempt to determine if the variability between the sample averages *and* the variability within the samples could have occurred jointly *if in fact the populations have a common mean*. Notice that the key to this analysis centers around the two following sources of variability.

(1) Variability within samples (between observations in distinct samples)

(2) Variability between samples (between sample averages)

Clearly, if the variability in (1) is considerably larger than that in (2), there will be considerable overlap in the sample data, a signal that the data could all have come from a common distribution. An example is found in the data set shown in Figure 8.14. On the other hand, it is very unlikely that data from distributions with a common mean could have variability between sample averages that is considerably larger than the variability within samples.

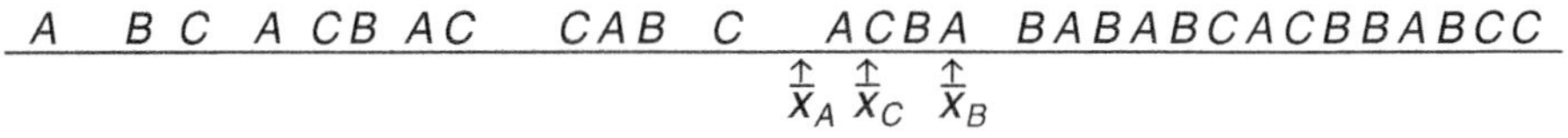

$$\underline{A \quad BC \quad A \quad CB \quad AC \quad \underset{\underset{\bar{x}_A}{\uparrow}}{} CAB \quad \underset{\underset{\bar{x}_C}{\uparrow}}{C} \quad ACBA \quad \underset{\underset{\bar{x}_B}{\uparrow}}{} BABABCACBBABCC}$$

Figure 8.14: Data that easily could have come from the same population.

The sources of variability in (1) and (2) above generate important ratios of *sample variances*, and ratios are used in conjunction with the F-distribution. The general procedure involved is called **analysis of variance**. It is interesting that in the paint example described here, we are dealing with inferences on three population means, but two sources of variability are used. We will not supply details here, but in Chapters 13 through 15 we make extensive use of analysis of variance, and, of course, the F-distribution plays an important role.

8.8 Quantile and Probability Plots

In Chapter 1 we introduced the reader to empirical distributions. The motivation is to use creative displays to extract information about properties of a set of data. For example, stem-and-leaf plots provide the viewer with a look at symmetry and other properties of the data. In this chapter we deal with samples, which, of course, are collections of experimental data from which we draw conclusions about populations. Often the appearance of the sample provides information about the distribution from which the data are taken. For example, in Chapter 1 we illustrated the general nature of pairs of samples with point plots that displayed a relative comparison between central tendency and variability in two samples.

In chapters that follow, we often make the assumption that a distribution is normal. Graphical information regarding the validity of this assumption can be retrieved from displays like stem-and-leaf plots and frequency histograms. In addition, we will introduce the notion of *normal probability plots* and *quantile plots* in this section. These plots are used in studies that have varying degrees of complexity, with the main objective of the plots being to provide a diagnostic check on the assumption that the data came from a normal distribution.

We can characterize statistical analysis as the process of drawing conclusions about systems in the presence of system variability. For example, an engineer's attempt to learn about a chemical process is often clouded by *process variability*. A study involving the number of defective items in a production process is often made more difficult by variability in the method of manufacture of the items. In what has preceded, we have learned about samples and statistics that express center of location and variability in the sample. These statistics provide single measures, whereas a graphical display adds additional information through a picture.

One type of plot that can be particularly useful in characterizing the nature of a data set is the *quantile plot*. As in the case of the box-and-whisker plot (Section

Quantile Plot

The purpose of the quantile plot is to depict, in sample form, the cumulative distribution function discussed in Chapter 3.

Definition 8.6: A **quantile** of a sample, $q(f)$, is a value for which a specified fraction f of the data values is less than or equal to $q(f)$.

Obviously, a quantile represents an estimate of a characteristic of a population, or rather, the theoretical distribution. The sample median is $q(0.5)$. The 75th percentile (upper quartile) is $q(0.75)$ and the lower quartile is $q(0.25)$.

A **quantile plot** simply *plots the data values on the vertical axis against an empirical assessment of the fraction of observations exceeded by the data value.* For theoretical purposes, this fraction is computed as

$$f_i = \frac{i - \frac{3}{8}}{n + \frac{1}{4}},$$

where i is the order of the observations when they are ranked from low to high. In other words, if we denote the ranked observations as

$$y_{(1)} \leq y_{(2)} \leq y_{(3)} \leq \cdots \leq y_{(n-1)} \leq y_{(n)},$$

then the quantile plot depicts a plot of $y_{(i)}$ against f_i. In Figure 8.15, the quantile plot is given for the paint can ear data discussed previously.

Unlike the box-and-whisker plot, the quantile plot actually shows all observations. All quantiles, including the median and the upper and lower quantile, can be approximated visually. For example, we readily observe a median of 35 and an upper quartile of about 36. Relatively large clusters around specific values are indicated by slopes near zero, while sparse data in certain areas produce steeper slopes. Figure 8.15 depicts sparsity of data from the values 28 through 30 but relatively high density at 36 through 38. In Chapters 9 and 10 we pursue quantile plotting further by illustrating useful ways of comparing distinct samples.

It should be somewhat evident to the reader that detection of whether or not a data set came from a normal distribution can be an important tool for the data analyst. As we indicated earlier in this section, we often make the assumption that all or subsets of observations in a data set are realizations of independent identically distributed normal random variables. Once again, the diagnostic plot can often nicely augment (for display purposes) a formal *goodness-of-fit test* on the data. Goodness-of-fit tests are discussed in Chapter 10. Readers of a scientific paper or report tend to find diagnostic information much clearer, less dry, and perhaps less boring than a formal analysis. In later chapters (Chapters 9 through 13), we focus

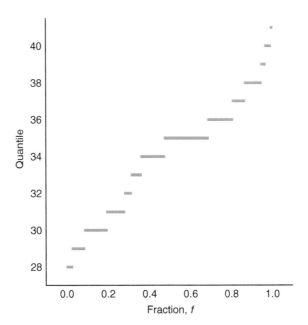

Figure 8.15: Quantile plot for paint data.

again on methods of detecting deviations from normality as an augmentation of formal statistical inference. Quantile plots are useful in detection of distribution types. There are also situations in both model building and design of experiments in which the plots are used to detect important **model terms** or **effects** that are active. In other situations, they are used to determine whether or not the underlying assumptions made by the scientist or engineer in building the model are reasonable. Many examples with illustrations will be encountered in Chapters 11, 12, and 13. The following subsection provides a discussion and illustration of a diagnostic plot called the *normal quantile-quantile plot*.

Normal Quantile-Quantile Plot

The normal quantile-quantile plot takes advantage of what is known about the quantiles of the normal distribution. The methodology involves a plot of the empirical quantiles recently discussed against the corresponding quantile of the normal distribution. Now, the expression for a quantile of an $N(\mu, \sigma)$ random variable is very complicated. However, a good approximation is given by

$$q_{\mu,\sigma}(f) = \mu + \sigma\{4.91[f^{0.14} - (1-f)^{0.14}]\}.$$

The expression in braces (the multiple of σ) is the approximation for the corresponding quantile for the $N(0, 1)$ random variable, that is,

$$q_{0,1}(f) = 4.91[f^{0.14} - (1-f)^{0.14}].$$

8.8 Quantile and Probability Plots

Definition 8.7: The **normal quantile-quantile plot** is a plot of $y_{(i)}$ (ordered observations) against $q_{0,1}(f_i)$, where $f_i = \frac{i - \frac{3}{8}}{n + \frac{1}{4}}$.

A nearly straight-line relationship suggests that the data came from a normal distribution. The intercept on the vertical axis is an estimate of the population mean μ and the slope is an estimate of the standard deviation σ. Figure 8.16 shows a normal quantile-quantile plot for the paint can data.

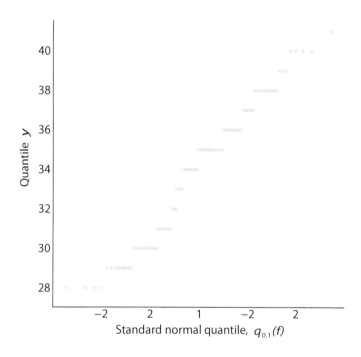

Figure 8.16: Normal quantile-quantile plot for paint data.

Normal Probability Plotting

Notice how the deviation from normality becomes clear from the appearance of the plot. The asymmetry exhibited in the data results in changes in the slope.

The ideas of probability plotting are manifested in plots other than the normal quantile-quantile plot discussed here. For example, much attention is given to the so-called **normal probability plot**, in which f is plotted against the ordered data values on special paper and the scale used results in a straight line. In addition, an alternative plot makes use of the expected values of the ranked observations for the normal distribution and plots the ranked observations against their expected value, under the assumption of data from $N(\mu, \sigma)$. Once again, the straight line is the graphical yardstick used. We continue to suggest that the foundation in graphical analytical methods developed in this section will aid in understanding formal methods of distinguishing between distinct samples of data.

8.8 Quantile and Probability Plots

Example 8.12: Consider the data in Exercise 10.41 on page 358 in Chapter 10. In a study "Nutrient Retention and Macro Invertebrate Community Response to Sewage Stress in a Stream Ecosystem," conducted in the Department of Zoology at the Virginia Polytechnic Institute and State University, data were collected on density measurements (number of organisms per square meter) at two different collecting stations. Details are given in Chapter 10 regarding analytical methods of comparing samples to determine if both are from the same $N(\mu, \sigma)$ distribution. The data are given in Table 8.1.

Table 8.1: Data for Example 8.12

Number of Organisms per Square Meter			
Station 1		Station 2	
5,030	4,980	2,800	2,810
13,700	11,910	4,670	1,330
10,730	8,130	6,890	3,320
11,400	26,850	7,720	1,230
860	17,660	7,030	2,130
2,200	22,800	7,330	2,190
4,250	1,130		
15,040	1,690		

Construct a normal quantile-quantile plot and draw conclusions regarding whether or not it is reasonable to assume that the two samples are from the same $n(x; \mu, \sigma)$ distribution.

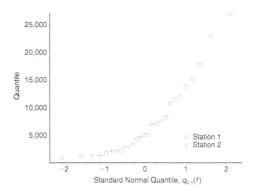

Figure 8.17: Normal quantile-quantile plot for density data of Example 8.12.

Solution: Figure 8.17 shows the normal quantile-quantile plot for the density measurements. The plot is far from a single straight line. In fact, the data from station 1 reflect a few values in the lower tail of the distribution and several in the upper tail. The "clustering" of observations would make it seem unlikely that the two samples came from a common $N(\mu, \sigma)$ distribution.

Exercises

8.37 For a chi-squared distribution, find
(a) $\chi^2_{0.025}$ when $v = 15$;
(b) $\chi^2_{0.01}$ when $v = 7$;
(c) $\chi^2_{0.05}$ when $v = 24$.

8.38 For a chi-squared distribution, find
(a) $\chi^2_{0.005}$ when $v = 5$;
(b) $\chi^2_{0.05}$ when $v = 19$;
(c) $\chi^2_{0.01}$ when $v = 12$.

8.39 For a chi-squared distribution, find χ^2_α such that
(a) $P(X^2 > \chi^2_\alpha) = 0.99$ when $v = 4$;
(b) $P(X^2 > \chi^2_\alpha) = 0.025$ when $v = 19$;
(c) $P(37.652 < X^2 < \chi^2_\alpha) = 0.045$ when $v = 25$.

8.40 For a chi-squared distribution, find χ^2_α such that
(a) $P(X^2 > \chi^2_\alpha) = 0.01$ when $v = 21$;
(b) $P(X^2 < \chi^2_\alpha) = 0.95$ when $v = 6$;
(c) $P(\chi^2_\alpha < X^2 < 23.209) = 0.015$ when $v = 10$.

8.41 Assume the sample variances to be continuous measurements. Find the probability that a random sample of 25 observations, from a normal population with variance $\sigma^2 = 6$, will have a sample variance S^2
(a) greater than 9.1;
(b) between 3.462 and 10.745.

8.42 The scores on a placement test given to college freshmen for the past five years are approximately normally distributed with a mean $\mu = 74$ and a variance $\sigma^2 = 8$. Would you still consider $\sigma^2 = 8$ to be a valid value of the variance if a random sample of 20 students who take the placement test this year obtain a value of $s^2 = 20$?

8.43 Show that the variance of S^2 for random samples of size n from a normal population decreases as n becomes large. [*Hint*: First find the variance of $(n-1)S^2/\sigma^2$.]

8.44 (a) Find $t_{0.025}$ when $v = 14$.
(b) Find $-t_{0.10}$ when $v = 10$.
(c) Find $t_{0.995}$ when $v = 7$.

8.45 (a) Find $P(T < 2.365)$ when $v = 7$.
(b) Find $P(T > 1.318)$ when $v = 24$.
(c) Find $P(-1.356 < T < 2.179)$ when $v = 12$.
(d) Find $P(T > -2.567)$ when $v = 17$.

8.46 (a) Find $P(-t_{0.005} < T < t_{0.01})$ for $v = 20$.
(b) Find $P(T > -t_{0.025})$.

8.47 Given a random sample of size 24 from a normal distribution, find k such that
(a) $P(-2.069 < T < k) = 0.965$;
(b) $P(k < T < 2.807) = 0.095$;
(c) $P(-k < T < k) = 0.90$.

8.48 A manufacturing firm claims that the batteries used in their electronic games will last an average of 30 hours. To maintain this average, 16 batteries are tested each month. If the computed t-value falls between $-t_{0.025}$ and $t_{0.025}$, the firm is satisfied with its claim. What conclusion should the firm draw from a sample that has a mean of $\bar{x} = 27.5$ hours and a standard deviation of $s = 5$ hours? Assume the distribution of battery lives to be approximately normal.

8.49 A normal population with unknown variance has a mean of 20. Is one likely to obtain a random sample of size 9 from this population with a mean of 24 and a standard deviation of 4.1? If not, what conclusion would you draw?

8.50 A maker of a certain brand of low-fat cereal bars claims that the average saturated fat content is 0.5 gram. In a random sample of 8 cereal bars of this brand, the saturated fat content was 0.6, 0.7, 0.7, 0.3, 0.4, 0.5, 0.4, and 0.2. Would you agree with the claim? Assume a normal distribution.

8.51 For an F-distribution, find
(a) $f_{0.05}$ with $v_1 = 7$ and $v_2 = 15$;
(b) $f_{0.05}$ with $v_1 = 15$ and $v_2 = 7$;
(c) $f_{0.01}$ with $v_1 = 24$ and $v_2 = 19$;

(d) $f_{0.95}$ with $v_1 = 19$ and $v_2 = 24$;
(e) $f_{0.99}$ with $v_1 = 28$ and $v_2 = 12$.

8.52 Pull-strength tests on 10 soldered leads for a semiconductor device yield the following results, in pounds of force required to rupture the bond:

    19.8 12.7 13.2 16.9 10.6
    18.8 11.1 14.3 17.0 12.5

Another set of 8 leads was tested after encapsulation to determine whether the pull strength had been increased by encapsulation of the device, with the following results:

    24.9 22.8 23.6 22.1 20.4 21.6 21.8 22.5

Comment on the evidence available concerning equality of the two population variances.

8.53 Consider the following measurements of the heat-producing capacity of the coal produced by two mines (in millions of calories per ton):

    Mine 1: 8260 8130 8350 8070 8340
    Mine 2: 7950 7890 7900 8140 7920 7840

Can it be concluded that the two population variances are equal?

8.54 Construct a quantile plot of these data, which represent the lifetimes, in hours, of fifty 40-watt, 110-volt internally frosted incandescent lamps taken from forced life tests:

919	1196	785	1126	936	918
1156	920	948	1067	1092	1162
1170	929	950	905	972	1035
1045	855	1195	1195	1340	1122
938	970	1237	956	1102	1157
978	832	1009	1157	1151	1009
765	958	902	1022	1333	811
1217	1085	896	958	1311	1037
702	923				

8.55 Construct a normal quantile-quantile plot of these data, which represent the diameters of 36 rivet heads in 1/100 of an inch:

6.72	6.77	6.82	6.70	6.78	6.70	6.62
6.75	6.66	6.66	6.64	6.76	6.73	6.80
6.72	6.76	6.76	6.68	6.66	6.62	6.72
6.76	6.70	6.78	6.76	6.67	6.70	6.72
6.74	6.81	6.79	6.78	6.66	6.76	6.76
6.72						

Review Exercises

8.56 Consider the data displayed in Exercise 1.20 on page 31. Construct a box-and-whisker plot and comment on the nature of the sample. Compute the sample mean and sample standard deviation.

8.57 If $X_1, X_2, \ldots, X_n$ are independent random variables having identical exponential distributions with parameter θ, show that the density function of the random variable $Y = X_1 + X_2 + \cdots + X_n$ is that of a gamma distribution with parameters $\alpha = n$ and $\beta = \theta$.

8.58 In testing for carbon monoxide in a certain brand of cigarette, the data, in milligrams per cigarette, were coded by subtracting 12 from each observation. Use the results of Exercise 8.14 on page 231 to find the standard deviation for the carbon monoxide content of a random sample of 15 cigarettes of this brand if the coded measurements are 3.8, −0.9, 5.4, 4.5, 5.2, 5.6, 2.7, −0.1, −0.3, −1.7, 5.7, 3.3, 4.4, −0.5, and 1.9.

8.59 If S_1^2 and S_2^2 represent the variances of independent random samples of size $n_1 = 8$ and $n_2 = 12$, taken from normal populations with equal variances, find $P(S_1^2/S_2^2 < 4.89)$.

8.60 A random sample of 5 bank presidents indicated annual salaries of $395,000, $521,000, $483,000, $479,000, and $510,000. Find the variance of this set.

8.61 If the number of hurricanes that hit a certain area of the eastern United States per year is a random variable having a Poisson distribution with $\mu = 6$, find the probability that this area will be hit by

(a) exactly 15 hurricanes in 2 years;
(b) at most 9 hurricanes in 2 years.

8.62 A taxi company tests a random sample of 10 steel-belted radial tires of a certain brand and records the following tread wear: 48,000, 53,000, 45,000, 61,000, 59,000, 56,000, 63,000, 49,000, 53,000, and 54,000 kilometers. Use the results of Exercise 8.14 on page 231 to find the standard deviation of this set of data by first dividing each observation by 1000 and then subtracting 55.

8.63 Consider the data of Exercise 1.19 on page 31. Construct a box-and-whisker plot. Comment. Compute the sample mean and sample standard deviation.

8.64 If S_1^2 and S_2^2 represent the variances of independent random samples of size $n_1 = 25$ and $n_2 = 31$, taken from normal populations with variances $\sigma_1^2 = 10$

and $\sigma_2^2 = 15$, respectively, find
$$P(S_1^2/S_2^2 > 1.26).$$

8.65 Consider Example 1.5 on page 25. Comment on any outliers.

8.66 Consider Review Exercise 8.56. Comment on any outliers in the data.

8.67 The breaking strength X of a certain rivet used in a machine engine has a mean 5000 psi and standard deviation 400 psi. A random sample of 36 rivets is taken. Consider the distribution of $\bar{X}$, the sample mean breaking strength.

(a) What is the probability that the sample mean falls between 4800 psi and 5200 psi?

(b) What sample n would be necessary in order to have
$$P(4900 < \bar{X} < 5100) = 0.99?$$

8.68 Consider the situation of Review Exercise 8.62. If the population from which the sample was taken has population mean $\mu = 53{,}000$ kilometers, does the sample information here seem to support that claim? In your answer, compute
$$t = \frac{\bar{x} - 53{,}000}{s/\sqrt{10}}$$
and determine from Table A.4 (with 9 d.f.) whether the computed t-value is reasonable or appears to be a rare event.

8.69 Two distinct solid fuel propellants, type A and type B, are being considered for a space program activity. Burning rates of the propellant are crucial. Random samples of 20 specimens of the two propellants are taken with sample means 20.5 cm/sec for propellant A and 24.50 cm/sec for propellant B. It is generally assumed that the variability in burning rate is roughly the same for the two propellants and is given by a population standard deviation of 5 cm/sec. Assume that the burning rates for each propellant are approximately normal and hence make use of the Central Limit Theorem. Nothing is known about the two population mean burning rates, and it is hoped that this experiment might shed some light on them.

(a) If, indeed, $\mu_A = \mu_B$, what is $P(\bar{X}_B - \bar{X}_A \geq 4.0)$?

(b) Use your answer in (a) to shed some light on the proposition that $\mu_A = \mu_B$.

8.70 The concentration of an active ingredient in the output of a chemical reaction is strongly influenced by the catalyst that is used in the reaction. It is felt that when catalyst A is used, the population mean concentration exceeds 65%. The standard deviation is known to be $\sigma = 5\%$. A sample of outputs from 30 independent experiments gives the average concentration of $\bar{x}_A = 64.5\%$.

(a) Does this sample information with an average concentration of $\bar{x}_A = 64.5\%$ provide disturbing information that perhaps μ_A is not 65%, but less than 65%? Support your answer with a probability statement.

(b) Suppose a similar experiment is done with the use of another catalyst, catalyst B. The standard deviation σ is still assumed to be 5% and $\bar{x}_B$ turns out to be 70%. Comment on whether or not the sample information on catalyst B strongly suggests that μ_B is truly greater than μ_A. Support your answer by computing
$$P(\bar{X}_B - \bar{X}_A \geq 5.5 \mid \mu_B = \mu_A).$$

(c) Under the condition that $\mu_A = \mu_B = 65\%$, give the approximate distribution of the following quantities (with mean and variance of each). Make use of the Central Limit Theorem.
i) $\bar{X}_B$;
ii) $\bar{X}_A - \bar{X}_B$;
iii) $\frac{\bar{X}_A - \bar{X}_B}{\sigma\sqrt{2/30}}$.

8.71 From the information in Review Exercise 8.70, compute (assuming $\mu_B = 65\%$) $P(\bar{X}_B \geq 70)$.

8.72 Given a normal random variable X with mean 20 and variance 9, and a random sample of size n taken from the distribution, what sample size n is necessary in order that
$$P(19.9 \leq \bar{X} \leq 20.1) = 0.95?$$

8.73 In Chapter 9, the concept of **parameter estimation** will be discussed at length. Suppose X is a random variable with mean μ and variance $\sigma^2 = 1.0$. Suppose also that a random sample of size n is to be taken and $\bar{x}$ is to be used as an *estimate* of μ. When the data are taken and the sample mean is measured, we wish it to be within 0.05 unit of the true mean with probability 0.99. That is, we want there to be a good chance that the computed $\bar{x}$ from the sample is "very close" to the population mean (wherever it is!), so we wish
$$P(|\bar{X} - \mu| > 0.05) = 0.99.$$
What sample size is required?

8.74 Suppose a filling machine is used to fill cartons with a liquid product. The specification that is strictly enforced for the filling machine is 9 ± 1.5 oz. If any carton is produced with weight outside these bounds, it is considered by the supplier to be defective. It is hoped that at least 99% of cartons will meet these specifications. With the conditions $\mu = 9$ and $\sigma = 1$, what proportion of cartons from the process are defective? If changes are made to reduce variability, what must σ be reduced to in order to meet specifications with probability 0.99? Assume a normal distribution for the weight.

8.75 Consider the situation in Review Exercise 8.74. Suppose a considerable effort is conducted to "tighten" the variability in the system. Following the effort, a random sample of size 40 is taken from the new assembly line and the sample variance is $s^2 = 0.188$ ounces^2. Do we have strong numerical evidence that σ^2 has been reduced below 1.0? Consider the probability
$$P(S^2 \leq 0.188 \mid \sigma^2 = 1.0),$$
and give your conclusion.

8.76 Group Project: The class should be divided into groups of four people. The four students in each group should go to the college gym or a local fitness center. The students should ask each person who comes through the door his or her height in inches. Each group will then divide the height data by gender and work together to answer the following questions.

(a) Construct a normal quantile-quantile plot of the data. Based on the plot, do the data appear to follow a normal distribution?

(b) Use the estimated sample variance as the true variance for each gender. Assume that the population mean height for male students is actually three inches larger than that of female students. What is the probability that the average height of the male students will be 4 inches larger than that of the female students in your sample?

(c) What factors could render these results misleading?

8.9 Potential Misconceptions and Hazards; Relationship to Material in Other Chapters

The Central Limit Theorem is one of the most powerful tools in all of statistics, and even though this chapter is relatively short, it contains a wealth of fundamental information about tools that will be used throughout the balance of the text.

The notion of a sampling distribution is one of the most important fundamental concepts in all of statistics, and the student at this point in his or her training should gain a clear understanding of it before proceeding beyond this chapter. All chapters that follow will make considerable use of sampling distributions. Suppose one wants to use the statistic $\bar{X}$ to draw inferences about the population mean μ. This will be done by using the observed value $\bar{x}$ from a single sample of size n. Then any inference made must be accomplished by taking into account not just the single value but rather the theoretical structure, or **distribution of all $\bar{x}$ values that could be observed from samples of size n**. Thus, the concept of a *sampling distribution* comes to the surface. This distribution is the basis for the Central Limit Theorem. The t, χ^2, and F-distributions are also used in the context of sampling distributions. For example, the t-distribution, pictured in Figure 8.8, represents the structure that occurs if all of the values of $\frac{\bar{x}-\mu}{s/\sqrt{n}}$ are formed, where $\bar{x}$ and s are taken from samples of size n from a $n(x; \mu, \sigma)$ distribution. Similar remarks can be made about χ^2 and F, and the reader should not forget that the sample information forming the statistics for all of these distributions is the normal. So it can be said that **where there is a t, F, or χ^2, the source was a sample from a normal distribution**.

The three distributions described above may appear to have been introduced in a rather self-contained fashion with no indication of what they are about. However, they will appear in practical problem-solving throughout the balance of the text.

Now, there are three things that one must bear in mind, lest confusion set in regarding these fundamental sampling distributions:

(i) One cannot use the Central Limit Theorem unless σ is known. When σ is not known, it should be replaced by s, the sample standard deviation, in order to use the Central Limit Theorem.

(ii) The T statistic is **not** a result of the Central Limit Theorem and $x_1, x_2, \ldots, x_n$ must come from a $n(x; \mu, \sigma)$ distribution in order for $\frac{\bar{x}-\mu}{s/\sqrt{n}}$ to be a t-distribution; s is, of course, merely an estimate of σ.

(iii) While the notion of **degrees of freedom** is new at this point, the concept should be very intuitive, since it is reasonable that the nature of the distribution of S and also t should depend on the amount of information in the sample $x_1, x_2, \ldots, x_n$.

Chapter 9

One- and Two-Sample Estimation Problems

9.1 Introduction

In previous chapters, we emphasized sampling properties of the sample mean and variance. We also emphasized displays of data in various forms. The purpose of these presentations is to build a foundation that allows us to draw conclusions about the population parameters from experimental data. For example, the Central Limit Theorem provides information about the distribution of the sample mean $\bar{X}$. The distribution involves the population mean μ. Thus, any conclusions concerning μ drawn from an observed sample average must depend on knowledge of this sampling distribution. Similar comments apply to S^2 and σ^2. Clearly, any conclusions we draw about the variance of a normal distribution will likely involve the sampling distribution of S^2.

In this chapter, we begin by formally outlining the purpose of statistical inference. We follow this by discussing the problem of **estimation of population parameters**. We confine our formal developments of specific estimation procedures to problems involving one and two samples.

9.2 Statistical Inference

In Chapter 1, we discussed the general philosophy of formal statistical inference. **Statistical inference** consists of those methods by which one makes inferences or generalizations about a population. The trend today is to distinguish between the **classical method** of estimating a population parameter, whereby inferences are based strictly on information obtained from a random sample selected from the population, and the **Bayesian method**, which utilizes prior subjective knowledge about the probability distribution of the unknown parameters in conjunction with the information provided by the sample data. Throughout most of this chapter, we shall use classical methods to estimate unknown population parameters such as the mean, the proportion, and the variance by computing statistics from random samples and applying the theory of sampling distributions, much of which was covered in Chapter 8. Bayesian estimation will be discussed in Chapter 18.

Statistical inference may be divided into two major areas: **estimation** and **tests of hypotheses**. We treat these two areas separately, dealing with theory and applications of estimation in this chapter and hypothesis testing in Chapter 10. To distinguish clearly between the two areas, consider the following examples. A candidate for public office may wish to estimate the true proportion of voters favoring him by obtaining opinions from a random sample of 100 eligible voters. The fraction of voters in the sample favoring the candidate could be used as an estimate of the true proportion in the population of voters. A knowledge of the sampling distribution of a proportion enables one to establish the degree of accuracy of such an estimate. This problem falls in the area of estimation.

Now consider the case in which one is interested in finding out whether brand A floor wax is more scuff-resistant than brand B floor wax. He or she might hypothesize that brand A is better than brand B and, after proper testing, accept or reject this hypothesis. In this example, we do not attempt to estimate a parameter, but instead we try to arrive at a correct decision about a prestated hypothesis. Once again we are dependent on sampling theory and the use of data to provide us with some measure of accuracy for our decision.

9.3 Classical Methods of Estimation

A **point estimate** of some population parameter θ is a single value $\hat{\theta}$ of a statistic $\hat{\Theta}$. For example, the value $\bar{x}$ of the statistic $\bar{X}$, computed from a sample of size n, is a point estimate of the population parameter μ. Similarly, $\hat{p} = x/n$ is a point estimate of the true proportion p for a binomial experiment.

An estimator is not expected to estimate the population parameter without error. We do not expect $\bar{X}$ to estimate μ exactly, but we certainly hope that it is not far off. For a particular sample, it is possible to obtain a closer estimate of μ by using the sample median $\tilde{X}$ as an estimator. Consider, for instance, a sample consisting of the values 2, 5, and 11 from a population whose mean is 4 but is supposedly unknown. We would estimate μ to be $\bar{x} = 6$, using the sample mean as our estimate, or $\tilde{x} = 5$, using the sample median as our estimate. In this case, the estimator $\tilde{X}$ produces an estimate closer to the true parameter than does the estimator $\bar{X}$. On the other hand, if our random sample contains the values 2, 6, and 7, then $\bar{x} = 5$ and $\tilde{x} = 6$, so $\bar{X}$ is the better estimator. Not knowing the true value of μ, we must decide in advance whether to use $\bar{X}$ or $\tilde{X}$ as our estimator.

Unbiased Estimator

What are the desirable properties of a "good" decision function that would influence us to choose one estimator rather than another? Let $\hat{\Theta}$ be an estimator whose value $\hat{\theta}$ is a point estimate of some unknown population parameter θ. Certainly, we would like the sampling distribution of $\hat{\Theta}$ to have a mean equal to the parameter estimated. An estimator possessing this property is said to be **unbiased**.

9.3 Classical Methods of Estimation

Definition 9.1: A statistic $\hat{\Theta}$ is said to be an **unbiased estimator** of the parameter θ if
$$\mu_{\hat{\Theta}} = E(\hat{\Theta}) = \theta.$$

Example 9.1: Show that S^2 is an unbiased estimator of the parameter σ^2.

Solution: In Section 8.5 on page 244, we showed that
$$\sum_{i=1}^{n}(X_i - \bar{X})^2 = \sum_{i=1}^{n}(X_i - \mu)^2 - n(\bar{X} - \mu)^2.$$

Now
$$E(S^2) = E\left[\frac{1}{n-1}\sum_{i=1}^{n}(X_i - \bar{X})^2\right]$$
$$= \frac{1}{n-1}\left[\sum_{i=1}^{n}E(X_i - \mu)^2 - nE(\bar{X} - \mu)^2\right] = \frac{1}{n-1}\left(\sum_{i=1}^{n}\sigma_{X_i}^2 - n\sigma_{\bar{X}}^2\right).$$

However,
$$\sigma_{X_i}^2 = \sigma^2, \text{ for } i = 1, 2, \ldots, n, \text{ and } \sigma_{\bar{X}}^2 = \frac{\sigma^2}{n}.$$

Therefore,
$$E(S^2) = \frac{1}{n-1}\left(n\sigma^2 - n\frac{\sigma^2}{n}\right) = \sigma^2.$$

Although S^2 is an unbiased estimator of σ^2, S, on the other hand, is usually a biased estimator of σ, with the bias becoming insignificant for large samples. This example illustrates **why we divide by $n-1$** rather than n when the variance is estimated.

Variance of a Point Estimator

If $\hat{\Theta}_1$ and $\hat{\Theta}_2$ are two unbiased estimators of the same population parameter θ, we want to choose the estimator whose sampling distribution has the smaller variance. Hence, if $\sigma_{\hat{\theta}_1}^2 < \sigma_{\hat{\theta}_2}^2$, we say that $\hat{\Theta}_1$ is a **more efficient estimator** of θ than $\hat{\Theta}_2$.

Definition 9.2: If we consider all possible unbiased estimators of some parameter θ, the one with the smallest variance is called the **most efficient estimator** of θ.

Figure 9.1 illustrates the sampling distributions of three different estimators, $\hat{\Theta}_1$, $\hat{\Theta}_2$, and $\hat{\Theta}_3$, all estimating θ. It is clear that only $\hat{\Theta}_1$ and $\hat{\Theta}_2$ are unbiased, since their distributions are centered at θ. The estimator $\hat{\Theta}_1$ has a smaller variance than $\hat{\Theta}_2$ and is therefore more efficient. Hence, our choice for an estimator of θ, among the three considered, would be $\hat{\Theta}_1$.

For normal populations, one can show that both $\bar{X}$ and $\tilde{X}$ are unbiased estimators of the population mean μ, but the variance of $\bar{X}$ is smaller than the variance

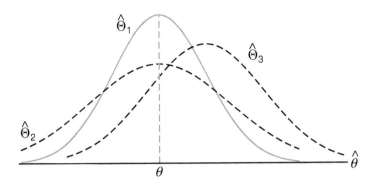

Figure 9.1: Sampling distributions of different estimators of θ.

of $\tilde{X}$. Thus, both estimates $\bar{x}$ and $\tilde{x}$ will, on average, equal the population mean μ, but $\bar{x}$ is likely to be closer to μ for a given sample, and thus $\bar{X}$ is more efficient than $\tilde{X}$.

Interval Estimation

Even the most efficient unbiased estimator is unlikely to estimate the population parameter exactly. It is true that estimation accuracy increases with large samples, but there is still no reason we should expect a **point estimate** from a given sample to be exactly equal to the population parameter it is supposed to estimate. There are many situations in which it is preferable to determine an interval within which we would expect to find the value of the parameter. Such an interval is called an **interval estimate**.

An interval estimate of a population parameter θ is an interval of the form $\hat{\theta}_L < \theta < \hat{\theta}_U$, where $\hat{\theta}_L$ and $\hat{\theta}_U$ depend on the value of the statistic $\hat{\Theta}$ for a particular sample and also on the sampling distribution of $\hat{\Theta}$. For example, a random sample of SAT verbal scores for students in the entering freshman class might produce an interval from 530 to 550, within which we expect to find the true average of all SAT verbal scores for the freshman class. The values of the endpoints, 530 and 550, will depend on the computed sample mean $\bar{x}$ and the sampling distribution of $\bar{X}$. As the sample size increases, we know that $\sigma_{\bar{X}}^2 = \sigma^2/n$ decreases, and consequently our estimate is likely to be closer to the parameter μ, resulting in a shorter interval. Thus, the interval estimate indicates, by its length, the accuracy of the point estimate. An engineer will gain some insight into the population proportion defective by taking a sample and computing the *sample proportion defective*. But an interval estimate might be more informative.

Interpretation of Interval Estimates

Since different samples will generally yield different values of $\hat{\Theta}$ and, therefore, different values for $\hat{\theta}_L$ and $\hat{\theta}_U$, these endpoints of the interval are values of corresponding random variables $\hat{\Theta}_L$ and $\hat{\Theta}_U$. From the sampling distribution of $\hat{\Theta}$ we shall be able to determine $\hat{\Theta}_L$ and $\hat{\Theta}_U$ such that $P(\hat{\Theta}_L < \theta < \hat{\Theta}_U)$ is equal to any positive fractional value we care to specify. If, for instance, we find $\hat{\Theta}_L$ and $\hat{\Theta}_U$ such that

$$P(\hat{\Theta}_L < \theta < \hat{\Theta}_U) = 1 - \alpha,$$

for $0 < \alpha < 1$, then we have a probability of $1 - \alpha$ of selecting a random sample that will produce an interval containing θ. The interval $\hat{\theta}_L < \theta < \hat{\theta}_U$, computed from the selected sample, is called a $100(1 - \alpha)\%$ **confidence interval**, the fraction $1 - \alpha$ is called the **confidence coefficient** or the **degree of confidence**, and the endpoints, $\hat{\theta}_L$ and $\hat{\theta}_U$, are called the lower and upper **confidence limits**. Thus, when $\alpha = 0.05$, we have a 95% confidence interval, and when $\alpha = 0.01$, we obtain a wider 99% confidence interval. The wider the confidence interval is, the more confident we can be that the interval contains the unknown parameter. Of course, it is better to be 95% confident that the average life of a certain television transistor is between 6 and 7 years than to be 99% confident that it is between 3 and 10 years. Ideally, we prefer a short interval with a high degree of confidence. Sometimes, restrictions on the size of our sample prevent us from achieving short intervals without sacrificing some degree of confidence.

In the sections that follow, we pursue the notions of point and interval estimation, with each section presenting a different special case. The reader should notice that while point and interval estimation represent different approaches to gaining information regarding a parameter, they are related in the sense that confidence interval estimators are based on point estimators. In the following section, for example, we will see that $\bar{X}$ is a very reasonable point estimator of μ. As a result, the important confidence interval estimator of μ depends on knowledge of the sampling distribution of $\bar{X}$.

We begin the following section with the simplest case of a confidence interval. The scenario is simple and yet unrealistic. We are interested in estimating a population mean μ and yet σ is known. Clearly, if μ is unknown, it is quite unlikely that σ is known. Any historical results that produced enough information to allow the assumption that σ is known would likely have produced similar information about μ. Despite this argument, we begin with this case because the concepts and indeed the resulting mechanics associated with confidence interval estimation remain the same for the more realistic situations presented later in Section 9.4 and beyond.

9.4 Single Sample: Estimating the Mean

The sampling distribution of $\bar{X}$ is centered at μ, and in most applications the variance is smaller than that of any other estimators of μ. Thus, the sample mean $\bar{x}$ will be used as a point estimate for the population mean μ. Recall that $\sigma_{\bar{X}}^2 = \sigma^2/n$, so a large sample will yield a value of $\bar{X}$ that comes from a sampling distribution with a small variance. Hence, $\bar{x}$ is likely to be a very accurate estimate of μ when n is large.

Let us now consider the interval estimate of μ. If our sample is selected from a normal population or, failing this, if n is sufficiently large, we can establish a confidence interval for μ by considering the sampling distribution of $\bar{X}$.

According to the Central Limit Theorem, we can expect the sampling distribution of $\bar{X}$ to be approximately normally distributed with mean $\mu_{\bar{X}} = \mu$ and standard deviation $\sigma_{\bar{X}} = \sigma/\sqrt{n}$. Writing $z_{\alpha/2}$ for the z-value above which we find an area of $\alpha/2$ under the normal curve, we can see from Figure 9.2 that

$$P(-z_{\alpha/2} < Z < z_{\alpha/2}) = 1 - \alpha,$$

where

$$Z = \frac{\bar{X} - \mu}{\sigma/\sqrt{n}}.$$

Hence,

$$P\left(-z_{\alpha/2} < \frac{\bar{X} - \mu}{\sigma/\sqrt{n}} < z_{\alpha/2}\right) = 1 - \alpha.$$

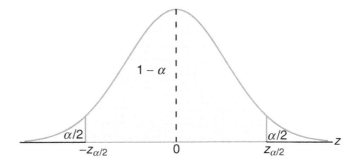

Figure 9.2: $P(-z_{\alpha/2} < Z < z_{\alpha/2}) = 1 - \alpha$.

Multiplying each term in the inequality by $\sigma/\sqrt{n}$ and then subtracting $\bar{X}$ from each term and multiplying by -1 (reversing the sense of the inequalities), we obtain

$$P\left(\bar{X} - z_{\alpha/2}\frac{\sigma}{\sqrt{n}} < \mu < \bar{X} + z_{\alpha/2}\frac{\sigma}{\sqrt{n}}\right) = 1 - \alpha.$$

A random sample of size n is selected from a population whose variance σ^2 is known, and the mean $\bar{x}$ is computed to give the $100(1 - \alpha)\%$ confidence interval below. It is important to emphasize that we have invoked the Central Limit Theorem above. As a result, it is important to note the conditions for applications that follow.

Confidence Interval on μ, σ^2 Known

If $\bar{x}$ is the mean of a random sample of size n from a population with known variance σ^2, a $100(1 - \alpha)\%$ confidence interval for μ is given by

$$\bar{x} - z_{\alpha/2}\frac{\sigma}{\sqrt{n}} < \mu < \bar{x} + z_{\alpha/2}\frac{\sigma}{\sqrt{n}},$$

where $z_{\alpha/2}$ is the z-value leaving an area of $\alpha/2$ to the right.

For small samples selected from nonnormal populations, we cannot expect our degree of confidence to be accurate. However, for samples of size $n \geq 30$, with

9.4 Single Sample: Estimating the Mean

the shape of the distributions not too skewed, sampling theory guarantees good results.

Clearly, the values of the random variables $\hat{\Theta}_L$ and $\hat{\Theta}_U$, defined in Section 9.3, are the confidence limits

$$\hat{\theta}_L = \bar{x} - z_{\alpha/2}\frac{\sigma}{\sqrt{n}} \quad \text{and} \quad \hat{\theta}_U = \bar{x} + z_{\alpha/2}\frac{\sigma}{\sqrt{n}}.$$

Different samples will yield different values of $\bar{x}$ and therefore produce different interval estimates of the parameter μ, as shown in Figure 9.3. The dot at the center of each interval indicates the position of the point estimate $\bar{x}$ for that random sample. Note that all of these intervals are of the same width, since their widths depend only on the choice of $z_{\alpha/2}$ once $\bar{x}$ is determined. The larger the value we choose for $z_{\alpha/2}$, the wider we make all the intervals and the more confident we can be that the particular sample selected will produce an interval that contains the unknown parameter μ. In general, for a selection of $z_{\alpha/2}$, $100(1-\alpha)\%$ of the intervals will cover μ.

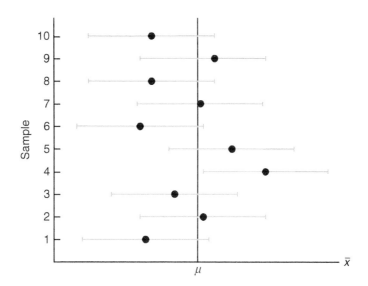

Figure 9.3: Interval estimates of μ for different samples.

Example 9.2: The average zinc concentration recovered from a sample of measurements taken in 36 different locations in a river is found to be 2.6 grams per milliliter. Find the 95% and 99% confidence intervals for the mean zinc concentration in the river. Assume that the population standard deviation is 0.3 gram per milliliter.

Solution: The point estimate of μ is $\bar{x} = 2.6$. The z-value leaving an area of 0.025 to the right, and therefore an area of 0.975 to the left, is $z_{0.025} = 1.96$ (Table A.3). Hence, the 95% confidence interval is

$$2.6 - (1.96)\left(\frac{0.3}{\sqrt{36}}\right) < \mu < 2.6 + (1.96)\left(\frac{0.3}{\sqrt{36}}\right),$$

which reduces to $2.50 < \mu < 2.70$. To find a 99% confidence interval, we find the z-value leaving an area of 0.005 to the right and 0.995 to the left. From Table A.3 again, $z_{0.005} = 2.575$, and the 99% confidence interval is

$$2.6 - (2.575)\left(\frac{0.3}{\sqrt{36}}\right) < \mu < 2.6 + (2.575)\left(\frac{0.3}{\sqrt{36}}\right),$$

or simply

$$2.47 < \mu < 2.73.$$

We now see that a longer interval is required to estimate μ with a higher degree of confidence.

The $100(1-\alpha)\%$ confidence interval provides an estimate of the accuracy of our point estimate. If μ is actually the center value of the interval, then $\bar{x}$ estimates μ without error. Most of the time, however, $\bar{x}$ will not be exactly equal to μ and the point estimate will be in error. The size of this error will be the absolute value of the difference between μ and $\bar{x}$, and we can be $100(1-\alpha)\%$ confident that this difference will not exceed $z_{\alpha/2}\frac{\sigma}{\sqrt{n}}$. We can readily see this if we draw a diagram of a hypothetical confidence interval, as in Figure 9.4.

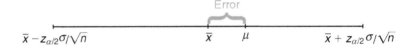

Figure 9.4: Error in estimating μ by $\bar{x}$.

Theorem 9.1: If $\bar{x}$ is used as an estimate of μ, we can be $100(1-\alpha)\%$ confident that the error will not exceed $z_{\alpha/2}\frac{\sigma}{\sqrt{n}}$.

In Example 9.2, we are 95% confident that the sample mean $\bar{x} = 2.6$ differs from the true mean μ by an amount less than $(1.96)(0.3)/\sqrt{36} = 0.1$ and 99% confident that the difference is less than $(2.575)(0.3)/\sqrt{36} = 0.13$.

Frequently, we wish to know how large a sample is necessary to ensure that the error in estimating μ will be less than a specified amount e. By Theorem 9.1, we must choose n such that $z_{\alpha/2}\frac{\sigma}{\sqrt{n}} = e$. Solving this equation gives the following formula for n.

Theorem 9.2: If $\bar{x}$ is used as an estimate of μ, we can be $100(1-\alpha)\%$ confident that the error will not exceed a specified amount e when the sample size is

$$n = \left(\frac{z_{\alpha/2}\sigma}{e}\right)^2.$$

When solving for the sample size, n, we round all fractional values up to the next whole number. By adhering to this principle, we can be sure that our degree of confidence never falls below $100(1-\alpha)\%$.

9.4 Single Sample: Estimating the Mean

Strictly speaking, the formula in Theorem 9.2 is applicable only if we know the variance of the population from which we select our sample. Lacking this information, we could take a preliminary sample of size $n \geq 30$ to provide an estimate of σ. Then, using s as an approximation for σ in Theorem 9.2, we could determine approximately how many observations are needed to provide the desired degree of accuracy.

Example 9.3: How large a sample is required if we want to be 95% confident that our estimate of μ in Example 9.2 is off by less than 0.05?

Solution: The population standard deviation is $\sigma = 0.3$. Then, by Theorem 9.2,

$$n = \left[\frac{(1.96)(0.3)}{0.05}\right]^2 = 138.3.$$

Therefore, we can be 95% confident that a random sample of size 139 will provide an estimate $\bar{x}$ differing from μ by an amount less than 0.05.

One-Sided Confidence Bounds

The confidence intervals and resulting confidence bounds discussed thus far are *two-sided* (i.e., both upper and lower bounds are given). However, there are many applications in which only one bound is sought. For example, if the measurement of interest is tensile strength, the engineer receives better information from a lower bound only. This bound communicates the worst-case scenario. On the other hand, if the measurement is something for which a relatively large value of μ is not profitable or desirable, then an upper confidence bound is of interest. An example would be a case in which inferences need to be made concerning the mean mercury composition in a river. An upper bound is very informative in this case.

One-sided confidence bounds are developed in the same fashion as two-sided intervals. However, the source is a one-sided probability statement that makes use of the Central Limit Theorem:

$$P\left(\frac{\bar{X} - \mu}{\sigma/\sqrt{n}} < z_\alpha\right) = 1 - \alpha.$$

One can then manipulate the probability statement much as before and obtain

$$P(\mu > \bar{X} - z_\alpha \sigma/\sqrt{n}) = 1 - \alpha.$$

Similar manipulation of $P\left(\frac{\bar{X}-\mu}{\sigma/\sqrt{n}} > -z_\alpha\right) = 1 - \alpha$ gives

$$P(\mu < \bar{X} + z_\alpha \sigma/\sqrt{n}) = 1 - \alpha.$$

As a result, the upper and lower one-sided bounds follow.

One-Sided Confidence Bounds on μ, σ^2 Known

If $\bar{X}$ is the mean of a random sample of size n from a population with variance σ^2, the one-sided $100(1-\alpha)\%$ confidence bounds for μ are given by

upper one-sided bound: $\bar{x} + z_\alpha \sigma/\sqrt{n}$;
lower one-sided bound: $\bar{x} - z_\alpha \sigma/\sqrt{n}$.

Example 9.4: In a psychological testing experiment, 25 subjects are selected randomly and their reaction time, in seconds, to a particular stimulus is measured. Past experience suggests that the variance in reaction times to these types of stimuli is 4 sec^2 and that the distribution of reaction times is approximately normal. The average time for the subjects is 6.2 seconds. Give an upper 95% bound for the mean reaction time.

Solution: The upper 95% bound is given by

$$\bar{x} + z_\alpha \sigma/\sqrt{n} = 6.2 + (1.645)\sqrt{4/25} = 6.2 + 0.658$$
$$= 6.858 \text{ seconds.}$$

Hence, we are 95% confident that the mean reaction time is less than 6.858 seconds.

The Case of σ Unknown

Frequently, we must attempt to estimate the mean of a population when the variance is unknown. The reader should recall learning in Chapter 8 that if we have a random sample from a *normal distribution*, then the random variable

$$T = \frac{\bar{X} - \mu}{S/\sqrt{n}}$$

has a Student t-distribution with $n-1$ degrees of freedom. Here S is the sample standard deviation. In this situation, with σ unknown, T can be used to construct a confidence interval on μ. The procedure is the same as that with σ known except that σ is replaced by S and the standard normal distribution is replaced by the t-distribution. Referring to Figure 9.5, we can assert that

$$P(-t_{\alpha/2} < T < t_{\alpha/2}) = 1 - \alpha,$$

where $t_{\alpha/2}$ is the t-value with $n-1$ degrees of freedom, above which we find an area of $\alpha/2$. Because of symmetry, an equal area of $\alpha/2$ will fall to the left of $-t_{\alpha/2}$. Substituting for T, we write

$$P\left(-t_{\alpha/2} < \frac{\bar{X} - \mu}{S/\sqrt{n}} < t_{\alpha/2}\right) = 1 - \alpha.$$

Multiplying each term in the inequality by $S/\sqrt{n}$, and then subtracting $\bar{X}$ from each term and multiplying by -1, we obtain

$$P\left(\bar{X} - t_{\alpha/2}\frac{S}{\sqrt{n}} < \mu < \bar{X} + t_{\alpha/2}\frac{S}{\sqrt{n}}\right) = 1 - \alpha.$$

For a particular random sample of size n, the mean $\bar{x}$ and standard deviation s are computed and the following $100(1-\alpha)\%$ confidence interval for μ is obtained.

9.4 Single Sample: Estimating the Mean

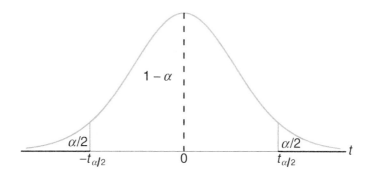

Figure 9.5: $P(-t_{\alpha/2} < T < t_{\alpha/2}) = 1 - \alpha$.

Confidence Interval on μ, σ^2 Unknown

If $\bar{x}$ and s are the mean and standard deviation of a random sample from a normal population with unknown variance σ^2, a $100(1-\alpha)\%$ confidence interval for μ is

$$\bar{x} - t_{\alpha/2}\frac{s}{\sqrt{n}} < \mu < \bar{x} + t_{\alpha/2}\frac{s}{\sqrt{n}},$$

where $t_{\alpha/2}$ is the t-value with $v = n - 1$ degrees of freedom, leaving an area of $\alpha/2$ to the right.

We have made a distinction between the cases of σ known and σ unknown in computing confidence interval estimates. We should emphasize that for σ known we exploited the Central Limit Theorem, whereas for σ unknown we made use of the sampling distribution of the random variable T. However, the use of the t-distribution is based on the premise that the sampling is from a normal distribution. As long as the distribution is approximately bell shaped, confidence intervals can be computed when σ^2 is unknown by using the t-distribution and we may expect very good results.

Computed one-sided confidence bounds for μ with σ unknown are as the reader would expect, namely

$$\bar{x} + t_\alpha \frac{s}{\sqrt{n}} \quad \text{and} \quad \bar{x} - t_\alpha \frac{s}{\sqrt{n}}.$$

They are the upper and lower $100(1 - \alpha)\%$ bounds, respectively. Here t_α is the t-value having an area of α to the right.

Example 9.5: The contents of seven similar containers of sulfuric acid are 9.8, 10.2, 10.4, 9.8, 10.0, 10.2, and 9.6 liters. Find a 95% confidence interval for the mean contents of all such containers, assuming an approximately normal distribution.

Solution: The sample mean and standard deviation for the given data are

$$\bar{x} = 10.0 \quad \text{and} \quad s = 0.283.$$

Using Table A.4, we find $t_{0.025} = 2.447$ for $v = 6$ degrees of freedom. Hence, the

95% confidence interval for μ is

$$10.0 - (2.447)\left(\frac{0.283}{\sqrt{7}}\right) < \mu < 10.0 + (2.447)\left(\frac{0.283}{\sqrt{7}}\right),$$

which reduces to $9.74 < \mu < 10.26$.

Concept of a Large-Sample Confidence Interval

Often statisticians recommend that even when normality cannot be assumed, σ is unknown, and $n \geq 30$, s can replace σ and the confidence interval

$$\bar{x} \pm z_{\alpha/2} \frac{s}{\sqrt{n}}$$

may be used. This is often referred to as a *large-sample confidence interval*. The justification lies only in the presumption that with a sample as large as 30 and the population distribution not too skewed, s will be very close to the true σ and thus the Central Limit Theorem prevails. It should be emphasized that this is only an approximation and the quality of the result becomes better as the sample size grows larger.

Example 9.6: Scholastic Aptitude Test (SAT) mathematics scores of a random sample of 500 high school seniors in the state of Texas are collected, and the sample mean and standard deviation are found to be 501 and 112, respectively. Find a 99% confidence interval on the mean SAT mathematics score for seniors in the state of Texas.

Solution: Since the sample size is large, it is reasonable to use the normal approximation. Using Table A.3, we find $z_{0.005} = 2.575$. Hence, a 99% confidence interval for μ is

$$501 \pm (2.575)\left(\frac{112}{\sqrt{500}}\right) = 501 \pm 12.9,$$

which yields $488.1 < \mu < 513.9$.

9.5 Standard Error of a Point Estimate

We have made a rather sharp distinction between the goal of a point estimate and that of a confidence interval estimate. The former supplies a single number extracted from a set of experimental data, and the latter provides an interval that is reasonable for the parameter, *given the experimental data*; that is, $100(1-\alpha)\%$ of such computed intervals "cover" the parameter.

These two approaches to estimation are related to each other. The common thread is the sampling distribution of the point estimator. Consider, for example, the estimator $\bar{X}$ of μ with σ known. We indicated earlier that a measure of the quality of an unbiased estimator is its variance. The variance of $\bar{X}$ is

$$\sigma_{\bar{X}}^2 = \frac{\sigma^2}{n}.$$

Thus, the standard deviation of $\bar{X}$, or *standard error* of $\bar{X}$, is $\sigma/\sqrt{n}$. Simply put, the standard error of an estimator is its standard deviation. For $\bar{X}$, the computed confidence limit

$$\bar{x} \pm z_{\alpha/2}\frac{\sigma}{\sqrt{n}} \text{ is written as } \bar{x} \pm z_{\alpha/2} \text{ s.e.}(\bar{x}),$$

where "s.e." is the "standard error." The important point is that the width of the confidence interval on μ is dependent on the quality of the point estimator through its standard error. In the case where σ is unknown and sampling is from a normal distribution, s replaces σ and the *estimated standard error* $s/\sqrt{n}$ is involved. Thus, the confidence limits on μ are

Confidence Limits on μ, σ^2 Unknown

$$\bar{x} \pm t_{\alpha/2}\frac{s}{\sqrt{n}} = \bar{x} \pm t_{\alpha/2} \text{ s.e.}(\bar{x})$$

Again, the confidence interval is *no better* (in terms of width) *than the quality of the point estimate*, in this case through its estimated standard error. Computer packages often refer to estimated standard errors simply as "standard errors."

As we move to more complex confidence intervals, there is a prevailing notion that widths of confidence intervals become shorter as the quality of the corresponding point estimate becomes better, although it is not always quite as simple as we have illustrated here. It can be argued that a confidence interval is merely an augmentation of the point estimate to take into account the precision of the point estimate.

9.6 Prediction Intervals

The point and interval estimations of the mean in Sections 9.4 and 9.5 provide good information about the unknown parameter μ of a normal distribution or a nonnormal distribution from which a large sample is drawn. Sometimes, other than the population mean, the experimenter may also be interested in predicting the possible **value of a future observation**. For instance, in quality control, the experimenter may need to use the observed data to predict a new observation. A process that produces a metal part may be evaluated on the basis of whether the part meets specifications on tensile strength. On certain occasions, a customer may be interested in purchasing a **single part**. In this case, a confidence interval on the mean tensile strength does not capture the required information. The customer requires a statement regarding the uncertainty of a **single observation**. This type of requirement is nicely fulfilled by the construction of a **prediction interval**.

It is quite simple to obtain a prediction interval for the situations we have considered so far. Assume that the random sample comes from a normal population with unknown mean μ and known variance σ^2. A natural point estimator of a new observation is $\bar{X}$. It is known, from Section 8.4, that the variance of $\bar{X}$ is σ^2/n. However, to predict a new observation, not only do we need to account for the variation due to estimating the mean, but also we should account for the **variation of a future observation**. From the assumption, we know that the variance of the random error in a new observation is σ^2. The development of a prediction interval is best illustrated by beginning with a normal random variable $x_0 - \bar{x}$, where x_0 is the new observation and $\bar{x}$ comes from the sample. Since x_0 and $\bar{x}$ are independent, we know that

$$z = \frac{x_0 - \bar{x}}{\sqrt{\sigma^2 + \sigma^2/n}} = \frac{x_0 - \bar{x}}{\sigma\sqrt{1 + 1/n}}$$

is $n(z; 0, 1)$. As a result, if we use the probability statement

$$P(-z_{\alpha/2} < Z < z_{\alpha/2}) = 1 - \alpha$$

with the z-statistic above and place x_0 in the center of the probability statement, we have the following event occurring with probability $1 - \alpha$:

$$\bar{x} - z_{\alpha/2}\sigma\sqrt{1 + 1/n} < x_0 < \bar{x} + z_{\alpha/2}\sigma\sqrt{1 + 1/n}.$$

As a result, computation of the prediction interval is formalized as follows.

Prediction Interval of a Future Observation, σ^2 Known

For a normal distribution of measurements with unknown mean μ and known variance σ^2, a $100(1 - \alpha)\%$ **prediction interval** of a future observation x_0 is

$$\bar{x} - z_{\alpha/2}\sigma\sqrt{1 + 1/n} < x_0 < \bar{x} + z_{\alpha/2}\sigma\sqrt{1 + 1/n},$$

where $z_{\alpha/2}$ is the z-value leaving an area of $\alpha/2$ to the right.

Example 9.7: Due to the decrease in interest rates, the First Citizens Bank received a lot of mortgage applications. A recent sample of 50 mortgage loans resulted in an average loan amount of \$257,300. Assume a population standard deviation of \$25,000. For the next customer who fills out a mortgage application, find a 95% prediction interval for the loan amount.

Solution: The point prediction of the next customer's loan amount is $\bar{x} = \$257,300$. The z-value here is $z_{0.025} = 1.96$. Hence, a 95% prediction interval for the future loan amount is

$$257,300 - (1.96)(25,000)\sqrt{1 + 1/50} < x_0 < 257,300 + (1.96)(25,000)\sqrt{1 + 1/50},$$

which gives the interval (\$207,812.43, \$306,787.57).

The prediction interval provides a good estimate of the location of a future observation, which is quite different from the estimate of the sample mean value. It should be noted that the variation of this prediction is the sum of the variation due to an estimation of the mean and the variation of a single observation. However, as in the past, we first consider the case with known variance. It is also important to deal with the prediction interval of a future observation in the situation where the variance is unknown. Indeed a Student t-distribution may be used in this case, as described in the following result. The normal distribution is merely replaced by the t-distribution.

9.6 Prediction Intervals

Prediction Interval of a Future Observation, σ^2 Unknown

For a normal distribution of measurements with unknown mean μ and unknown variance σ^2, a $100(1-\alpha)\%$ **prediction interval** of a future observation x_0 is

$$\bar{x} - t_{\alpha/2} s \sqrt{1 + 1/n} < x_0 < \bar{x} + t_{\alpha/2} s \sqrt{1 + 1/n},$$

where $t_{\alpha/2}$ is the t-value with $v = n - 1$ degrees of freedom, leaving an area of $\alpha/2$ to the right.

One-sided prediction intervals can also be constructed. Upper prediction bounds apply in cases where focus must be placed on future large observations. Concern over future small observations calls for the use of lower prediction bounds. The upper bound is given by

$$\bar{x} + t_\alpha s \sqrt{1 + 1/n}$$

and the lower bound by

$$\bar{x} - t_\alpha s \sqrt{1 + 1/n}.$$

Example 9.8: A meat inspector has randomly selected 30 packs of 95% lean beef. The sample resulted in a mean of 96.2% with a sample standard deviation of 0.8%. Find a 99% prediction interval for the leanness of a new pack. Assume normality.

Solution: For $v = 29$ degrees of freedom, $t_{0.005} = 2.756$. Hence, a 99% prediction interval for a new observation x_0 is

$$96.2 - (2.756)(0.8)\sqrt{1 + \frac{1}{30}} < x_0 < 96.2 + (2.756)(0.8)\sqrt{1 + \frac{1}{30}},$$

which reduces to $(93.96, 98.44)$.

Use of Prediction Limits for Outlier Detection

To this point in the text very little attention has been paid to the concept of **outliers**, or aberrant observations. The majority of scientific investigators are keenly sensitive to the existence of outlying observations or so-called faulty or "bad data." We deal with the concept of outlier detection extensively in Chapter 12. However, it is certainly of interest here since there is an important relationship between outlier detection and prediction intervals.

It is convenient for our purposes to view an outlying observation as one that comes from a population with a mean that is different from the mean that governs the rest of the sample of size n being studied. The prediction interval produces a bound that "covers" a future single observation with probability $1 - \alpha$ if it comes from the population from which the sample was drawn. As a result, a methodology for outlier detection involves the rule that **an observation is an outlier if it falls outside the prediction interval computed without including the questionable observation in the sample.** As a result, for the prediction interval of Example 9.8, if a new pack of beef is measured and its leanness is outside the interval $(93.96, 98.44)$, that observation can be viewed as an outlier.

9.7 Tolerance Limits

As discussed in Section 9.6, the scientist or engineer may be less interested in estimating parameters than in gaining a notion about where an individual *observation* or measurement might fall. Such situations call for the use of prediction intervals. However, there is yet a third type of interval that is of interest in many applications. Once again, suppose that interest centers around the manufacturing of a component part and specifications exist on a dimension of that part. In addition, there is little concern about the mean of the dimension. But unlike in the scenario in Section 9.6, one may be less interested in a single observation and more interested in where the majority of the population falls. If process specifications are important, the manager of the process is concerned about long-range performance, **not the next observation**. One must attempt to determine bounds that, in some probabilistic sense, "cover" values in the population (i.e., the measured values of the dimension).

One method of establishing the desired bounds is to determine a confidence interval on a *fixed proportion* of the measurements. This is best motivated by visualizing a situation in which we are doing random sampling from a normal distribution with known mean μ and variance σ^2. Clearly, a bound that covers the middle 95% of the population of observations is

$$\mu \pm 1.96\sigma.$$

This is called a **tolerance interval**, and indeed its coverage of 95% of measured observations is exact. However, in practice, μ and σ are seldom known; thus, the user must apply

$$\bar{x} \pm ks.$$

Now, of course, the interval is a random variable, and hence the *coverage* of a proportion of the population by the interval is not exact. As a result, a $100(1-\gamma)\%$ confidence interval must be used since $\bar{x} \pm ks$ cannot be expected to cover any specified proportion all the time. As a result, we have the following definition.

Tolerance Limits For a normal distribution of measurements with unknown mean μ and unknown standard deviation σ, **tolerance limits** are given by $\bar{x} \pm ks$, where k is determined such that one can assert with $100(1-\gamma)\%$ confidence that the given limits contain at least the proportion $1-\alpha$ of the measurements.

Table A.7 gives values of k for $1-\alpha = 0.90, 0.95, 0.99$; $\gamma = 0.05, 0.01$; and selected values of n from 2 to 300.

Example 9.9: Consider Example 9.8. With the information given, find a tolerance interval that gives two-sided 95% bounds on 90% of the distribution of packages of 95% lean beef. Assume the data came from an approximately normal distribution.

***Solution*:** Recall from Example 9.8 that $n = 30$, the sample mean is 96.2%, and the sample standard deviation is 0.8%. From Table A.7, $k = 2.14$. Using

$$\bar{x} \pm ks = 96.2 \pm (2.14)(0.8),$$

9.7 Tolerance Limits

we find that the lower and upper bounds are 94.5 and 97.9.

We are 95% confident that the above range covers the central 90% of the distribution of 95% lean beef packages.

Distinction among Confidence Intervals, Prediction Intervals, and Tolerance Intervals

It is important to reemphasize the difference among the three types of intervals discussed and illustrated in the preceding sections. The computations are straightforward, but interpretation can be confusing. In real-life applications, these intervals are not interchangeable because their interpretations are quite distinct.

In the case of confidence intervals, one is attentive only to the **population mean**. For example, Exercise 9.13 on page 283 deals with an engineering process that produces shearing pins. A specification will be set on Rockwell hardness, below which a customer will not accept any pins. Here, a population parameter must take a backseat. It is important that the engineer know where the *majority of the values of Rockwell hardness are going to be*. Thus, tolerance limits should be used. Surely, when tolerance limits on any process output are tighter than process specifications, that is good news for the process manager.

It is true that the tolerance limit interpretation is somewhat related to the confidence interval. The $100(1-\alpha)\%$ tolerance interval on, say, the proportion 0.95 can be viewed as a confidence interval **on the middle 95%** of the corresponding normal distribution. One-sided tolerance limits are also relevant. In the case of the Rockwell hardness problem, it is desirable to have a lower bound of the form $\bar{x} - ks$ such that there is 99% confidence that at least 99% of Rockwell hardness values will exceed the computed value.

Prediction intervals are applicable when it is important to determine a bound on a **single value**. The mean is not the issue here, nor is the location of the majority of the population. Rather, the location of a single new observation is required.

Case Study 9.1: Machine Quality: A machine produces metal pieces that are cylindrical in shape. A sample of these pieces is taken and the diameters are found to be 1.01, 0.97, 1.03, 1.04, 0.99, 0.98, 0.99, 1.01, and 1.03 centimeters. Use these data to calculate three interval types and draw interpretations that illustrate the distinction between them in the context of the system. For all computations, assume an approximately normal distribution. The sample mean and standard deviation for the given data are $\bar{x} = 1.0056$ and $s = 0.0246$.

(a) Find a 99% confidence interval on the mean diameter.

(b) Compute a 99% prediction interval on a measured diameter of a single metal piece taken from the machine.

(c) Find the 99% tolerance limits that will contain 95% of the metal pieces produced by this machine.

Solution: (a) The 99% confidence interval for the mean diameter is given by

$$\bar{x} \pm t_{0.005} s/\sqrt{n} = 1.0056 \pm (3.355)(0.0246/3) = 1.0056 \pm 0.0275.$$

Thus, the 99% confidence bounds are 0.9781 and 1.0331.

(b) The 99% prediction interval for a future observation is given by

$$\bar{x} \pm t_{0.005} s \sqrt{1 + 1/n} = 1.0056 \pm (3.355)(0.0246)\sqrt{1 + 1/9},$$

with the bounds being 0.9186 and 1.0926.

(c) From Table A.7, for $n = 9$, $1 - \gamma = 0.99$, and $1 - \alpha = 0.95$, we find $k = 4.550$ for two-sided limits. Hence, the 99% tolerance limits are given by

$$\bar{x} + ks = 1.0056 \pm (4.550)(0.0246),$$

with the bounds being 0.8937 and 1.1175. We are 99% confident that the tolerance interval from 0.8937 to 1.1175 will contain the central 95% of the distribution of diameters produced.

This case study illustrates that the three types of limits can give appreciably different results even though they are all 99% bounds. In the case of the confidence interval on the mean, 99% of such intervals cover the population mean diameter. Thus, we say that we are 99% confident that the mean diameter produced by the process is between 0.9781 and 1.0331 centimeters. Emphasis is placed on the mean, with less concern about a single reading or the general nature of the distribution of diameters in the population. In the case of the prediction limits, the bounds 0.9186 and 1.0926 are based on the distribution of a single "new" metal piece taken from the process, and again 99% of such limits will cover the diameter of a new measured piece. On the other hand, the tolerance limits, as suggested in the previous section, give the engineer a sense of where the "majority," say the central 95%, of the diameters of measured pieces in the population reside. The 99% tolerance limits, 0.8937 and 1.1175, are numerically quite different from the other two bounds. If these bounds appear alarmingly wide to the engineer, it reflects negatively on process quality. On the other hand, if the bounds represent a desirable result, the engineer may conclude that a majority (95% in here) of the diameters are in a desirable range. Again, a confidence interval interpretation may be used: namely, 99% of such calculated bounds will cover the middle 95% of the population of diameters.

Exercises

9.1 A UCLA researcher claims that the life span of mice can be extended by as much as 25% when the calories in their diet are reduced by approximately 40% from the time they are weaned. The restricted diet is enriched to normal levels by vitamins and protein. Assuming that it is known from previous studies that $\sigma = 5.8$ months, how many mice should be included in our sample if we wish to be 99% confident that the mean life span of the sample will be within 2 months of the population mean for all mice subjected to this reduced diet?

9.2 An electrical firm manufactures light bulbs that have a length of life that is approximately normally distributed with a standard deviation of 40 hours. If a sample of 30 bulbs has an average life of 780 hours, find a 96% confidence interval for the population mean of all bulbs produced by this firm.

9.3 Many cardiac patients wear an implanted pacemaker to control their heartbeat. A plastic connector module mounts on the top of the pacemaker. Assuming a standard deviation of 0.0015 inch and an approximately normal distribution, find a 95% confidence interval for the mean of the depths of all connector modules made by a certain manufacturing company. A random sample of 75 modules has an average depth of 0.310 inch.

9.4 The heights of a random sample of 50 college students showed a mean of 174.5 centimeters and a standard deviation of 6.9 centimeters.

(a) Construct a 98% confidence interval for the mean height of all college students.

(b) What can we assert with 98% confidence about the possible size of our error if we estimate the mean height of all college students to be 174.5 centimeters?

9.5 A random sample of 100 automobile owners in the state of Virginia shows that an automobile is driven on average 23,500 kilometers per year with a standard deviation of 3900 kilometers. Assume the distribution of measurements to be approximately normal.

(a) Construct a 99% confidence interval for the average number of kilometers an automobile is driven annually in Virginia.

(b) What can we assert with 99% confidence about the possible size of our error if we estimate the average number of kilometers driven by car owners in Virginia to be 23,500 kilometers per year?

9.6 How large a sample is needed in Exercise 9.2 if we wish to be 96% confident that our sample mean will be within 10 hours of the true mean?

9.7 How large a sample is needed in Exercise 9.3 if we wish to be 95% confident that our sample mean will be within 0.0005 inch of the true mean?

9.8 An efficiency expert wishes to determine the average time that it takes to drill three holes in a certain metal clamp. How large a sample will she need to be 95% confident that her sample mean will be within 15 seconds of the true mean? Assume that it is known from previous studies that $\sigma = 40$ seconds.

9.9 Regular consumption of presweetened cereals contributes to tooth decay, heart disease, and other degenerative diseases, according to studies conducted by Dr. W. H. Bowen of the National Institute of Health and Dr. J. Yudben, Professor of Nutrition and Dietetics at the University of London. In a random sample consisting of 20 similar single servings of Alpha-Bits, the average sugar content was 11.3 grams with a standard deviation of 2.45 grams. Assuming that the sugar contents are normally distributed, construct a 95% confidence interval for the mean sugar content for single servings of Alpha-Bits.

9.10 A random sample of 12 graduates of a certain secretarial school typed an average of 79.3 words per minute with a standard deviation of 7.8 words per minute. Assuming a normal distribution for the number of words typed per minute, find a 95% confidence interval for the average number of words typed by all graduates of this school.

9.11 A machine produces metal pieces that are cylindrical in shape. A sample of pieces is taken, and the diameters are found to be 1.01, 0.97, 1.03, 1.04, 0.99, 0.98, 0.99, 1.01, and 1.03 centimeters. Find a 99% confidence interval for the mean diameter of pieces from this machine, assuming an approximately normal distribution.

9.12 A random sample of 10 chocolate energy bars of a certain brand has, on average, 230 calories per bar, with a standard deviation of 15 calories. Construct a 99% confidence interval for the true mean calorie content of this brand of energy bar. Assume that the distribution of the calorie content is approximately normal.

9.13 A random sample of 12 shearing pins is taken in a study of the Rockwell hardness of the pin head. Measurements on the Rockwell hardness are made for each of the 12, yielding an average value of 48.50 with a sample standard deviation of 1.5. Assuming the measurements to be normally distributed, construct a 90% confidence interval for the mean Rockwell hardness.

9.14 The following measurements were recorded for the drying time, in hours, of a certain brand of latex paint:

3.4	2.5	4.8	2.9	3.6
2.8	3.3	5.6	3.7	2.8
4.4	4.0	5.2	3.0	4.8

Assuming that the measurements represent a random sample from a normal population, find a 95% prediction interval for the drying time for the next trial of the paint.

9.15 Referring to Exercise 9.5, construct a 99% prediction interval for the kilometers traveled annually by an automobile owner in Virginia.

9.16 Consider Exercise 9.10. Compute the 95% prediction interval for the next observed number of words per minute typed by a graduate of the secretarial school.

9.17 Consider Exercise 9.9. Compute a 95% prediction interval for the sugar content of the next single serving of Alpha-Bits.

9.18 Referring to Exercise 9.13, construct a 95% tolerance interval containing 90% of the measurements.

9.19 A random sample of 25 tablets of buffered aspirin contains, on average, 325.05 mg of aspirin per tablet, with a standard deviation of 0.5 mg. Find the 95% tolerance limits that will contain 90% of the tablet contents for this brand of buffered aspirin. Assume that the aspirin content is normally distributed.

9.20 Consider the situation of Exercise 9.11. Estimation of the mean diameter, while important, is not nearly as important as trying to pin down the location of the majority of the distribution of diameters. Find the 95% tolerance limits that contain 95% of the diameters.

9.21 In a study conducted by the Department of Biological Sciences at Virginia Tech, fifteen samples of water were collected from a certain station in the James River in order to gain some insight regarding the amount of orthophosphorus in the river. The concentration of the chemical is measured in milligrams per liter. Let us suppose that the mean at the station is not as important as the upper extreme of the distribution of the concentration of the chemical at the station. Concern centers around whether the concentration at the extreme is too large. Readings for the fifteen water samples gave a sample mean of 3.84 milligrams per liter and a sample standard deviation of 3.07 milligrams per liter. Assume that the readings are a random sample from a normal distribution. Calculate a prediction interval (upper 95% prediction limit) and a tolerance limit (95% upper tolerance limit that exceeds 95% of the population of values). Interpret both; that is, tell what each communicates about the upper extreme of the distribution of orthophosphorus at the sampling station.

9.22 A type of thread is being studied for its tensile strength properties. Fifty pieces were tested under similar conditions, and the results showed an average tensile strength of 78.3 kilograms and a standard deviation of 5.6 kilograms. Assuming a normal distribution of tensile strengths, give a lower 95% prediction limit on a single observed tensile strength value. In addition, give a lower 95% tolerance limit that is exceeded by 99% of the tensile strength values.

9.23 Refer to Exercise 9.22. Why are the quantities requested in the exercise likely to be more important to the manufacturer of the thread than, say, a confidence interval on the mean tensile strength?

9.24 Refer to Exercise 9.22 again. Suppose that specifications by a buyer of the thread are that the tensile strength of the material must be at least 62 kilograms. The manufacturer is satisfied if at most 5% of the manufactured pieces have tensile strength less than 62 kilograms. Is there cause for concern? Use a one-sided 99% tolerance limit that is exceeded by 95% of the tensile strength values.

9.25 Consider the drying time measurements in Exercise 9.14. Suppose the 15 observations in the data set are supplemented by a 16th value of 6.9 hours. In the context of the original 15 observations, is the 16th value an outlier? Show work.

9.26 Consider the data in Exercise 9.13. Suppose the manufacturer of the shearing pins insists that the Rockwell hardness of the product be less than or equal to 44.0 only 5% of the time. What is your reaction? Use a tolerance limit calculation as the basis for your judgment.

9.27 Consider the situation of Case Study 9.1 on page 281 with a larger sample of metal pieces. The diameters are as follows: 1.01, 0.97, 1.03, 1.04, 0.99, 0.98, 1.01, 1.03, 0.99, 1.00, 1.00, 0.99, 0.98, 1.01, 1.02, 0.99 centimeters. Once again the normality assumption may be made. Do the following and compare your results to those of the case study. Discuss how they are different and why.

(a) Compute a 99% confidence interval on the mean diameter.

(b) Compute a 99% prediction interval on the next diameter to be measured.

(c) Compute a 99% tolerance interval for coverage of the central 95% of the distribution of diameters.

9.28 In Section 9.3, we emphasized the notion of "most efficient estimator" by comparing the variance of two unbiased estimators $\hat{\Theta}_1$ and $\hat{\Theta}_2$. However, this does not take into account bias in case one or both estimators are not unbiased. Consider the quantity

$$MSE = E(\hat{\Theta} - \theta),$$

where MSE denotes **mean squared error**. The MSE is often used to compare two estimators $\hat{\Theta}_1$ and $\hat{\Theta}_2$ of θ when either or both is unbiased because (i) it is intuitively reasonable and (ii) it accounts for bias. Show that MSE can be written

$$MSE = E[\hat{\Theta} - E(\hat{\Theta})]^2 + [E(\hat{\Theta} - \theta)]^2$$
$$= \text{Var}(\hat{\Theta}) + [\text{Bias}(\hat{\Theta})]^2.$$

9.29 Let us define $S'^2 = \sum_{i=1}^{n}(X_i - \bar{X})^2/n$. Show that

$$E(S'^2) = [(n-1)/n]\sigma^2,$$

and hence S'^2 is a biased estimator for σ^2.

9.30 Consider S'^2, the estimator of σ^2, from Exercise 9.29. Analysts often use S'^2 rather than dividing $\sum_{i=1}^{n}(X_i - \bar{X})^2$ by $n-1$, the degrees of freedom in the sample.

(a) What is the bias of S'^2?
(b) Show that the bias of S'^2 approaches zero as $n \to \infty$.

9.31 If X is a binomial random variable, show that
(a) $\hat{P} = X/n$ is an unbiased estimator of p;
(b) $P' = \frac{X+\sqrt{n}/2}{n+\sqrt{n}}$ is a biased estimator of p.

9.32 Show that the estimator P' of Exercise 9.31(b) becomes unbiased as $n \to \infty$.

9.33 Compare S^2 and S'^2 (see Exercise 9.29), the two estimators of σ^2, to determine which is more efficient. Assume these estimators are found using $X_1, X_2, \ldots, X_n$, independent random variables from $n(x; \mu, \sigma)$. Which estimator is more efficient considering only the variance of the estimators? [*Hint*: Make use of Theorem 8.4 and the fact that the variance of χ_v^2 is $2v$, from Section 6.7.]

9.34 Consider Exercise 9.33. Use the MSE discussed in Exercise 9.28 to determine which estimator is more efficient. Write out

$$\frac{MSE(S^2)}{MSE(S'^2)}.$$

9.8 Two Samples: Estimating the Difference between Two Means

If we have two populations with means μ_1 and μ_2 and variances σ_1^2 and σ_2^2, respectively, a point estimator of the difference between μ_1 and μ_2 is given by the statistic $\bar{X}_1 - \bar{X}_2$. Therefore, to obtain a point estimate of $\mu_1 - \mu_2$, we shall select two independent random samples, one from each population, of sizes n_1 and n_2, and compute $\bar{x}_1 - \bar{x}_2$, the difference of the sample means. Clearly, we must consider the sampling distribution of $\bar{X}_1 - \bar{X}_2$.

According to Theorem 8.3, we can expect the sampling distribution of $\bar{X}_1 - \bar{X}_2$ to be approximately normally distributed with mean $\mu_{\bar{X}_1 - \bar{X}_2} = \mu_1 - \mu_2$ and standard deviation $\sigma_{\bar{X}_1 - \bar{X}_2} = \sqrt{\sigma_1^2/n_1 + \sigma_2^2/n_2}$. Therefore, we can assert with a probability of $1 - \alpha$ that the standard normal variable

$$Z = \frac{(\bar{X}_1 - \bar{X}_2) - (\mu_1 - \mu_2)}{\sqrt{\sigma_1^2/n_1 + \sigma_2^2/n_2}}$$

will fall between $-z_{\alpha/2}$ and $z_{\alpha/2}$. Referring once again to Figure 9.2, we write

$$P(-z_{\alpha/2} < Z < z_{\alpha/2}) = 1 - \alpha.$$

Substituting for Z, we state equivalently that

$$P\left(-z_{\alpha/2} < \frac{(\bar{X}_1 - \bar{X}_2) - (\mu_1 - \mu_2)}{\sqrt{\sigma_1^2/n_1 + \sigma_2^2/n_2}} < z_{\alpha/2}\right) = 1 - \alpha,$$

which leads to the following $100(1-\alpha)\%$ confidence interval for $\mu_1 - \mu_2$.

Confidence Interval for $\mu_1 - \mu_2$, σ_1^2 and σ_2^2 Known

If $\bar{x}_1$ and $\bar{x}_2$ are means of independent random samples of sizes n_1 and n_2 from populations with known variances σ_1^2 and σ_2^2, respectively, a $100(l-\alpha)\%$ confidence interval for $\mu_1 - \mu_2$ is given by

$$(\bar{x}_1 - \bar{x}_2) - z_{\alpha/2}\sqrt{\frac{\sigma_1^2}{n_1} + \frac{\sigma_2^2}{n_2}} < \mu_1 - \mu_2 < (\bar{x}_1 - \bar{x}_2) + z_{\alpha/2}\sqrt{\frac{\sigma_1^2}{n_1} + \frac{\sigma_2^2}{n_2}},$$

where $z_{\alpha/2}$ is the z-value leaving an area of $\alpha/2$ to the right.

The Experimental Conditions and the Experimental Unit

For the case of confidence interval estimation on the difference between two means, we need to consider the experimental conditions in the data-taking process. It is assumed that we have two independent random samples from distributions with means μ_1 and μ_2, respectively. It is important that experimental conditions emulate this ideal described by these assumptions as closely as possible. Quite often, the experimenter should plan the strategy of the experiment accordingly. For almost any study of this type, there is a so-called *experimental unit*, which is that part of the experiment that produces experimental error and is responsible for the population variance we refer to as σ^2. In a drug study, the experimental unit is the patient or subject. In an agricultural experiment, it may be a plot of ground. In a chemical experiment, it may be a quantity of raw materials. It is important that differences between the experimental units have minimal impact on the results. The experimenter will have a degree of insurance that experimental units will not bias results if the conditions that define the two populations are *randomly assigned* to the experimental units. We shall again focus on randomization in future chapters that deal with hypothesis testing.

Example 9.10: A study was conducted in which two types of engines, A and B, were compared. Gas mileage, in miles per gallon, was measured. Fifty experiments were conducted using engine type A and 75 experiments were done with engine type B. The gasoline used and other conditions were held constant. The average gas mileage was 36 miles per gallon for engine A and 42 miles per gallon for engine B. Find a 96% confidence interval on $\mu_B - \mu_A$, where μ_A and μ_B are population mean gas mileages for engines A and B, respectively. Assume that the population standard deviations are 6 and 8 for engines A and B, respectively.

Solution: The point estimate of $\mu_B - \mu_A$ is $\bar{x}_B - \bar{x}_A = 42 - 36 = 6$. Using $\alpha = 0.04$, we find $z_{0.02} = 2.05$ from Table A.3. Hence, with substitution in the formula above, the 96% confidence interval is

$$6 - 2.05\sqrt{\frac{64}{75} + \frac{36}{50}} < \mu_B - \mu_A < 6 + 2.05\sqrt{\frac{64}{75} + \frac{36}{50}},$$

or simply $3.43 < \mu_B - \mu_A < 8.57$.

This procedure for estimating the difference between two means is applicable if σ_1^2 and σ_2^2 are known. If the variances are not known and the two distributions involved are approximately normal, the t-distribution becomes involved, as in the case of a single sample. If one is not willing to assume normality, large samples (say greater than 30) will allow the use of s_1 and s_2 in place of σ_1 and σ_2, respectively, with the rationale that $s_1 \approx \sigma_1$ and $s_2 \approx \sigma_2$. Again, of course, the confidence interval is an approximate one.

9.8 Two Samples: Estimating the Difference between Two Means

Variances Unknown but Equal

Consider the case where σ_1^2 and σ_2^2 are unknown. If $\sigma_1^2 = \sigma_2^2 = \sigma^2$, we obtain a standard normal variable of the form

$$Z = \frac{(\bar{X}_1 - \bar{X}_2) - (\mu_1 - \mu_2)}{\sqrt{\sigma^2[(1/n_1) + (1/n_2)]}}.$$

According to Theorem 8.4, the two random variables

$$\frac{(n_1 - 1)S_1^2}{\sigma^2} \quad \text{and} \quad \frac{(n_2 - 1)S_2^2}{\sigma^2}$$

have chi-squared distributions with $n_1 - 1$ and $n_2 - 1$ degrees of freedom, respectively. Furthermore, they are independent chi-squared variables, since the random samples were selected independently. Consequently, their sum

$$V = \frac{(n_1 - 1)S_1^2}{\sigma^2} + \frac{(n_2 - 1)S_2^2}{\sigma^2} = \frac{(n_1 - 1)S_1^2 + (n_2 - 1)S_2^2}{\sigma^2}$$

has a chi-squared distribution with $v = n_1 + n_2 - 2$ degrees of freedom.

Since the preceding expressions for Z and V can be shown to be independent, it follows from Theorem 8.5 that the statistic

$$T = \frac{(\bar{X}_1 - \bar{X}_2) - (\mu_1 - \mu_2)}{\sqrt{\sigma^2[(1/n_1) + (1/n_2)]}} \bigg/ \sqrt{\frac{(n_1 - 1)S_1^2 + (n_2 - 1)S_2^2}{\sigma^2(n_1 + n_2 - 2)}}$$

has the t-distribution with $v = n_1 + n_2 - 2$ degrees of freedom.

A point estimate of the unknown common variance σ^2 can be obtained by pooling the sample variances. Denoting the pooled estimator by S_p^2, we have the following.

Pooled Estimate of Variance

$$S_p^2 = \frac{(n_1 - 1)S_1^2 + (n_2 - 1)S_2^2}{n_1 + n_2 - 2}.$$

Substituting S_p^2 in the T statistic, we obtain the less cumbersome form

$$T = \frac{(\bar{X}_1 - \bar{X}_2) - (\mu_1 - \mu_2)}{S_p\sqrt{(1/n_1) + (1/n_2)}}.$$

Using the T statistic, we have

$$P(-t_{\alpha/2} < T < t_{\alpha/2}) = 1 - \alpha,$$

where $t_{\alpha/2}$ is the t-value with $n_1 + n_2 - 2$ degrees of freedom, above which we find an area of $\alpha/2$. Substituting for T in the inequality, we write

$$P\left[-t_{\alpha/2} < \frac{(\bar{X}_1 - \bar{X}_2) - (\mu_1 - \mu_2)}{S_p\sqrt{(1/n_1) + (1/n_2)}} < t_{\alpha/2}\right] = 1 - \alpha.$$

After the usual mathematical manipulations, the difference of the sample means $\bar{x}_1 - \bar{x}_2$ and the pooled variance are computed and then the following $100(1-\alpha)\%$ confidence interval for $\mu_1 - \mu_2$ is obtained.

The value of s_p^2 is easily seen to be a weighted average of the two sample variances s_1^2 and s_2^2, where the weights are the degrees of freedom.

Confidence Interval for $\mu_1 - \mu_2$, $\sigma_1^2 = \sigma_2^2$ but Both Unknown

If $\bar{x}_1$ and $\bar{x}_2$ are the means of independent random samples of sizes n_1 and n_2, respectively, from approximately normal populations with unknown but equal variances, a $100(1-\alpha)\%$ confidence interval for $\mu_1 - \mu_2$ is given by

$$(\bar{x}_1 - \bar{x}_2) - t_{\alpha/2} s_p \sqrt{\frac{1}{n_1} + \frac{1}{n_2}} < \mu_1 - \mu_2 < (\bar{x}_1 - \bar{x}_2) + t_{\alpha/2} s_p \sqrt{\frac{1}{n_1} + \frac{1}{n_2}},$$

where s_p is the pooled estimate of the population standard deviation and $t_{\alpha/2}$ is the t-value with $v = n_1 + n_2 - 2$ degrees of freedom, leaving an area of $\alpha/2$ to the right.

Example 9.11: The article "Macroinvertebrate Community Structure as an Indicator of Acid Mine Pollution," published in the *Journal of Environmental Pollution*, reports on an investigation undertaken in Cane Creek, Alabama, to determine the relationship between selected physiochemical parameters and different measures of macroinvertebrate community structure. One facet of the investigation was an evaluation of the effectiveness of a numerical species diversity index to indicate aquatic degradation due to acid mine drainage. Conceptually, a high index of macroinvertebrate species diversity should indicate an unstressed aquatic system, while a low diversity index should indicate a stressed aquatic system.

Two independent sampling stations were chosen for this study, one located downstream from the acid mine discharge point and the other located upstream. For 12 monthly samples collected at the downstream station, the species diversity index had a mean value $\bar{x}_1 = 3.11$ and a standard deviation $s_1 = 0.771$, while 10 monthly samples collected at the upstream station had a mean index value $\bar{x}_2 = 2.04$ and a standard deviation $s_2 = 0.448$. Find a 90% confidence interval for the difference between the population means for the two locations, assuming that the populations are approximately normally distributed with equal variances.

Solution: Let μ_1 and μ_2 represent the population means, respectively, for the species diversity indices at the downstream and upstream stations. We wish to find a 90% confidence interval for $\mu_1 - \mu_2$. Our point estimate of $\mu_1 - \mu_2$ is

$$\bar{x}_1 - \bar{x}_2 = 3.11 - 2.04 = 1.07.$$

The pooled estimate, s_p^2, of the common variance, σ^2, is

$$s_p^2 = \frac{(n_1 - 1)s_1^2 + (n_2 - 1)s_2^2}{n_1 + n_2 - 2} = \frac{(11)(0.771^2) + (9)(0.448^2)}{12 + 10 - 2} = 0.417.$$

Taking the square root, we obtain $s_p = 0.646$. Using $\alpha = 0.1$, we find in Table A.4 that $t_{0.05} = 1.725$ for $v = n_1 + n_2 - 2 = 20$ degrees of freedom. Therefore, the 90% confidence interval for $\mu_1 - \mu_2$ is

$$1.07 - (1.725)(0.646)\sqrt{\frac{1}{12} + \frac{1}{10}} < \mu_1 - \mu_2 < 1.07 + (1.725)(0.646)\sqrt{\frac{1}{12} + \frac{1}{10}},$$

which simplifies to $0.593 < \mu_1 - \mu_2 < 1.547$.

Interpretation of the Confidence Interval

For the case of a single parameter, the confidence interval simply provides error bounds on the parameter. Values contained in the interval should be viewed as reasonable values given the experimental data. In the case of a difference between two means, the interpretation can be extended to one of comparing the two means. For example, if we have high confidence that a difference $\mu_1 - \mu_2$ is positive, we would certainly infer that $\mu_1 > \mu_2$ with little risk of being in error. For example, in Example 9.11, we are 90% confident that the interval from 0.593 to 1.547 contains the difference of the population means for values of the species diversity index at the two stations. The fact that both confidence limits are positive indicates that, on the average, the index for the station located downstream from the discharge point is greater than the index for the station located upstream.

Equal Sample Sizes

The procedure for constructing confidence intervals for $\mu_1 - \mu_2$ with $\sigma_1 = \sigma_2 = \sigma$ unknown requires the assumption that the populations are normal. Slight departures from either the equal variance or the normality assumption do not seriously alter the degree of confidence for our interval. (A procedure is presented in Chapter 10 for testing the equality of two unknown population variances based on the information provided by the sample variances.) If the population variances are considerably different, we still obtain reasonable results when the populations are normal, provided that $n_1 = n_2$. Therefore, in planning an experiment, one should make every effort to equalize the size of the samples.

Unknown and Unequal Variances

Let us now consider the problem of finding an interval estimate of $\mu_1 - \mu_2$ when the unknown population variances are not likely to be equal. The statistic most often used in this case is

$$T' = \frac{(\bar{X}_1 - \bar{X}_2) - (\mu_1 - \mu_2)}{\sqrt{(S_1^2/n_1) + (S_2^2/n_2)}},$$

which has approximately a t-distribution with v degrees of freedom, where

$$v = \frac{(s_1^2/n_1 + s_2^2/n_2)^2}{[(s_1^2/n_1)^2/(n_1 - 1)] + [(s_2^2/n_2)^2/(n_2 - 1)]}.$$

Since v is seldom an integer, we *round it down* to the nearest whole number. The above estimate of the degrees of freedom is called the Satterthwaite approximation (Satterthwaite, 1946, in the Bibliography).

Using the statistic T', we write

$$P(-t_{\alpha/2} < T' < t_{\alpha/2}) \approx 1 - \alpha,$$

where $t_{\alpha/2}$ is the value of the t-distribution with v degrees of freedom, above which we find an area of $\alpha/2$. Substituting for T' in the inequality and following the same steps as before, we state the final result.

Confidence Interval for $\mu_1 - \mu_2$, $\sigma_1^2 \neq \sigma_2^2$ and Both Unknown

If $\bar{x}_1$ and s_1^2 and $\bar{x}_2$ and s_2^2 are the means and variances of independent random samples of sizes n_1 and n_2, respectively, from approximately normal populations with unknown and unequal variances, an approximate $100(1-\alpha)\%$ confidence interval for $\mu_1 - \mu_2$ is given by

$$(\bar{x}_1 - \bar{x}_2) - t_{\alpha/2}\sqrt{\frac{s_1^2}{n_1} + \frac{s_2^2}{n_2}} < \mu_1 - \mu_2 < (\bar{x}_1 - \bar{x}_2) + t_{\alpha/2}\sqrt{\frac{s_1^2}{n_1} + \frac{s_2^2}{n_2}},$$

where $t_{\alpha/2}$ is the t-value with

$$v = \frac{(s_1^2/n_1 + s_2^2/n_2)^2}{[(s_1^2/n_1)^2/(n_1-1)] + [(s_2^2/n_2)^2/(n_2-1)]}$$

degrees of freedom, leaving an area of $\alpha/2$ to the right.

Note that the expression for v above involves random variables, and thus v is an *estimate* of the degrees of freedom. In applications, this estimate will not result in a whole number, and thus the analyst must round down to the nearest integer to achieve the desired confidence.

Before we illustrate the above confidence interval with an example, we should point out that all the confidence intervals on $\mu_1 - \mu_2$ are of the same general form as those on a single mean; namely, they can be written as

$$\text{point estimate } \pm t_{\alpha/2} \, \widehat{\text{s.e.}}\text{(point estimate)}$$

or

$$\text{point estimate } \pm z_{\alpha/2} \, \text{s.e.(point estimate)}.$$

For example, in the case where $\sigma_1 = \sigma_2 = \sigma$, the estimated standard error of $\bar{x}_1 - \bar{x}_2$ is $s_p\sqrt{1/n_1 + 1/n_2}$. For the case where $\sigma_1^2 \neq \sigma_2^2$,

$$\widehat{\text{s.e.}}(\bar{x}_1 - \bar{x}_2) = \sqrt{\frac{s_1^2}{n_1} + \frac{s_2^2}{n_2}}.$$

Example 9.12: A study was conducted by the Department of Biological Sciences at the Virginia Tech to estimate the difference in the amounts of the chemical orthophosphorus measured at two different stations on the James River. Orthophosphorus was measured in milligrams per liter. Fifteen samples were collected from station 1, and 12 samples were obtained from station 2. The 15 samples from station 1 had an average orthophosphorus content of 3.84 milligrams per liter and a standard deviation of 3.07 milligrams per liter, while the 12 samples from station 2 had an average content of 1.49 milligrams per liter and a standard deviation of 0.80 milligram per liter. Find a 95% confidence interval for the difference in the true average orthophosphorus contents at these two stations, assuming that the observations came from normal populations with different variances.

Solution: For station 1, we have $\bar{x}_1 = 3.84$, $s_1 = 3.07$, and $n_1 = 15$. For station 2, $\bar{x}_2 = 1.49$, $s_2 = 0.80$, and $n_2 = 12$. We wish to find a 95% confidence interval for $\mu_1 - \mu_2$.

Since the population variances are assumed to be unequal, we can only find an approximate 95% confidence interval based on the t-distribution with v degrees of freedom, where

$$v = \frac{(3.07^2/15 + 0.80^2/12)^2}{[(3.07^2/15)^2/14] + [(0.80^2/12)^2/11]} = 16.3 \approx 16.$$

Our point estimate of $\mu_1 - \mu_2$ is

$$\bar{x}_1 - \bar{x}_2 = 3.84 - 1.49 = 2.35.$$

Using $\alpha = 0.05$, we find in Table A.4 that $t_{0.025} = 2.120$ for $v = 16$ degrees of freedom. Therefore, the 95% confidence interval for $\mu_1 - \mu_2$ is

$$2.35 - 2.120\sqrt{\frac{3.07^2}{15} + \frac{0.80^2}{12}} < \mu_1 - \mu_2 < 2.35 + 2.120\sqrt{\frac{3.07^2}{15} + \frac{0.80^2}{12}},$$

which simplifies to $0.60 < \mu_1 - \mu_2 < 4.10$. Hence, we are 95% confident that the interval from 0.60 to 4.10 milligrams per liter contains the difference of the true average orthophosphorus contents for these two locations.

When two population variances are unknown, the assumption of equal variances or unequal variances may be precarious. In Section 10.10, a procedure will be introduced that will aid in discriminating between the equal variance and the unequal variance situation.

9.9 Paired Observations

At this point, we shall consider estimation procedures for the difference of two means when the samples are not independent and the variances of the two populations are not necessarily equal. The situation considered here deals with a very special experimental condition, namely that of *paired observations*. Unlike in the situation described earlier, the conditions of the two populations are not assigned randomly to experimental units. Rather, each homogeneous experimental unit receives both population conditions; as a result, each experimental unit has a pair of observations, one for each population. For example, if we run a test on a new diet using 15 individuals, the weights before and after going on the diet form the information for our two samples. The two populations are "before" and "after," and the experimental unit is the individual. Obviously, the observations in a pair have something in common. To determine if the diet is effective, we consider the differences $d_1, d_2, \ldots, d_n$ in the paired observations. These differences are the values of a random sample $D_1, D_2, \ldots, D_n$ from a population of differences that we shall assume to be normally distributed with mean $\mu_D = \mu_1 - \mu_2$ and variance σ_D^2. We estimate σ_D^2 by s_d^2, the variance of the differences that constitute our sample. The point estimator of μ_D is given by $\bar{D}$.

When Should Pairing Be Done?

Pairing observations in an experiment is a strategy that can be employed in many fields of application. The reader will be exposed to this concept in material related to hypothesis testing in Chapter 10 and experimental design issues in Chapters 13 and 15. Selecting experimental units that are relatively homogeneous (within the units) and allowing each unit to experience both population conditions reduces the effective experimental error variance (in this case, σ_D^2). The reader may visualize the ith pair difference as

$$D_i = X_{1i} - X_{2i}.$$

Since the two observations are taken on the sample experimental unit, they are not independent and, in fact,

$$\text{Var}(D_i) = \text{Var}(X_{1i} - X_{2i}) = \sigma_1^2 + \sigma_2^2 - 2\,\text{Cov}(X_{1i}, X_{2i}).$$

Now, intuitively, we expect that σ_D^2 should be reduced because of the similarity in nature of the "errors" of the two observations within a given experimental unit, and this comes through in the expression above. One certainly expects that if the unit is homogeneous, the covariance is positive. As a result, the gain in quality of the confidence interval over that obtained without pairing will be greatest when there is homogeneity within units and large differences as one goes from unit to unit. One should keep in mind that the performance of the confidence interval will depend on the standard error of $\bar{D}$, which is, of course, $\sigma_D/\sqrt{n}$, where n is the number of pairs. As we indicated earlier, the intent of pairing is to reduce σ_D.

Tradeoff between Reducing Variance and Losing Degrees of Freedom

Comparing the confidence intervals obtained with and without pairing makes apparent that there is a tradeoff involved. Although pairing should indeed reduce variance and hence reduce the standard error of the point estimate, the degrees of freedom are reduced by reducing the problem to a one-sample problem. As a result, the $t_{\alpha/2}$ point attached to the standard error is adjusted accordingly. Thus, pairing may be counterproductive. This would certainly be the case if one experienced only a modest reduction in variance (through σ_D^2) by pairing.

Another illustration of pairing involves choosing n pairs of subjects, with each pair having a similar characteristic such as IQ, age, or breed, and then selecting one member of each pair at random to yield a value of X_1, leaving the other member to provide the value of X_2. In this case, X_1 and X_2 might represent the grades obtained by two individuals of equal IQ when one of the individuals is assigned at random to a class using the conventional lecture approach while the other individual is assigned to a class using programmed materials.

A $100(1-\alpha)\%$ confidence interval for μ_D can be established by writing

$$P(-t_{\alpha/2} < T < t_{\alpha/2}) = 1 - \alpha,$$

where $T = \frac{\bar{D}-\mu_D}{S_d/\sqrt{n}}$ and $t_{\alpha/2}$, as before, is a value of the t-distribution with $n-1$ degrees of freedom.

It is now a routine procedure to replace T by its definition in the inequality above and carry out the mathematical steps that lead to the following $100(1-\alpha)\%$ confidence interval for $\mu_1 - \mu_2 = \mu_D$.

Confidence Interval for $\mu_D = \mu_1 - \mu_2$ for Paired Observations

If $\bar{d}$ and s_d are the mean and standard deviation, respectively, of the normally distributed differences of n random pairs of measurements, a $100(1-\alpha)\%$ confidence interval for $\mu_D = \mu_1 - \mu_2$ is

$$\bar{d} - t_{\alpha/2}\frac{s_d}{\sqrt{n}} < \mu_D < \bar{d} + t_{\alpha/2}\frac{s_d}{\sqrt{n}},$$

where $t_{\alpha/2}$ is the t-value with $v = n - 1$ degrees of freedom, leaving an area of $\alpha/2$ to the right.

Example 9.13: A study published in *Chemosphere* reported the levels of the dioxin TCDD of 20 Massachusetts Vietnam veterans who were possibly exposed to Agent Orange. The TCDD levels in plasma and in fat tissue are listed in Table 9.1.

Find a 95% confidence interval for $\mu_1 - \mu_2$, where μ_1 and μ_2 represent the true mean TCDD levels in plasma and in fat tissue, respectively. Assume the distribution of the differences to be approximately normal.

Table 9.1: Data for Example 9.13

Veteran	TCDD Levels in Plasma	TCDD Levels in Fat Tissue	d_i	Veteran	TCDD Levels in Plasma	TCDD Levels in Fat Tissue	d_i
1	2.5	4.9	−2.4	11	6.9	7.0	−0.1
2	3.1	5.9	−2.8	12	3.3	2.9	0.4
3	2.1	4.4	−2.3	13	4.6	4.6	0.0
4	3.5	6.9	−3.4	14	1.6	1.4	0.2
5	3.1	7.0	−3.9	15	7.2	7.7	−0.5
6	1.8	4.2	−2.4	16	1.8	1.1	0.7
7	6.0	10.0	−4.0	17	20.0	11.0	9.0
8	3.0	5.5	−2.5	18	2.0	2.5	−0.5
9	36.0	41.0	−5.0	19	2.5	2.3	0.2
10	4.7	4.4	0.3	20	4.1	2.5	1.6

Reprinted from *Chemosphere*, Vol. 20, Nos. 7-9 (Tables I and II), Schecter et al., "Partitioning 2,3,7,8-chlorinated dibenzo-p-dioxins and dibenzofurans between adipose tissue and plasma lipid of 20 Massachusetts Vietnam veterans," pp. 954–955, Copyright ©1990, with permission from Elsevier.

Solution: We wish to find a 95% confidence interval for $\mu_1 - \mu_2$. Since the observations are paired, $\mu_1 - \mu_2 = \mu_D$. The point estimate of μ_D is $\bar{d} = -0.87$. The standard deviation, s_d, of the sample differences is

$$s_d = \sqrt{\frac{1}{n-1}\sum_{i=1}^{n}(d_i - \bar{d})^2} = \sqrt{\frac{168.4220}{19}} = 2.9773.$$

Using $\alpha = 0.05$, we find in Table A.4 that $t_{0.025} = 2.093$ for $v = n - 1 = 19$ degrees of freedom. Therefore, the 95% confidence interval is

$$-0.8700 - (2.093)\left(\frac{2.9773}{\sqrt{20}}\right) < \mu_D < -0.8700 + (2.093)\left(\frac{2.9773}{\sqrt{20}}\right),$$

or simply $-2.2634 < \mu_D < 0.5234$, from which we can conclude that there is no significant difference between the mean TCDD level in plasma and the mean TCDD level in fat tissue.

Exercises

9.35 A random sample of size $n_1 = 25$, taken from a normal population with a standard deviation $\sigma_1 = 5$, has a mean $\bar{x}_1 = 80$. A second random sample of size $n_2 = 36$, taken from a different normal population with a standard deviation $\sigma_2 = 3$, has a mean $\bar{x}_2 = 75$. Find a 94% confidence interval for $\mu_1 - \mu_2$.

9.36 Two kinds of thread are being compared for strength. Fifty pieces of each type of thread are tested under similar conditions. Brand A has an average tensile strength of 78.3 kilograms with a standard deviation of 5.6 kilograms, while brand B has an average tensile strength of 87.2 kilograms with a standard deviation of 6.3 kilograms. Construct a 95% confidence interval for the difference of the population means.

9.37 A study was conducted to determine if a certain treatment has any effect on the amount of metal removed in a pickling operation. A random sample of 100 pieces was immersed in a bath for 24 hours without the treatment, yielding an average of 12.2 millimeters of metal removed and a sample standard deviation of 1.1 millimeters. A second sample of 200 pieces was exposed to the treatment, followed by the 24-hour immersion in the bath, resulting in an average removal of 9.1 millimeters of metal with a sample standard deviation of 0.9 millimeter. Compute a 98% confidence interval estimate for the difference between the population means. Does the treatment appear to reduce the mean amount of metal removed?

9.38 Two catalysts in a batch chemical process, are being compared for their effect on the output of the process reaction. A sample of 12 batches was prepared using catalyst 1, and a sample of 10 batches was prepared using catalyst 2. The 12 batches for which catalyst 1 was used in the reaction gave an average yield of 85 with a sample standard deviation of 4, and the 10 batches for which catalyst 2 was used gave an average yield of 81 and a sample standard deviation of 5. Find a 90% confidence interval for the difference between the population means, assuming that the populations are approximately normally distributed with equal variances.

9.39 Students may choose between a 3-semester-hour physics course without labs and a 4-semester-hour course with labs. The final written examination is the same for each section. If 12 students in the section with labs made an average grade of 84 with a standard deviation of 4, and 18 students in the section without labs made an average grade of 77 with a standard deviation of 6, find a 99% confidence interval for the difference between the average grades for the two courses. Assume the populations to be approximately normally distributed with equal variances.

9.40 In a study conducted at Virginia Tech on the development of ectomycorrhizal, a symbiotic relationship between the roots of trees and a fungus, in which minerals are transferred from the fungus to the trees and sugars from the trees to the fungus, 20 northern red oak seedlings exposed to the fungus *Pisolithus tinctorus* were grown in a greenhouse. All seedlings were planted in the same type of soil and received the same amount of sunshine and water. Half received no nitrogen at planting time, to serve as a control, and the other half received 368 ppm of nitrogen in the form $NaNO_3$. The stem weights, in grams, at the end of 140 days were recorded as follows:

No Nitrogen	Nitrogen
0.32	0.26
0.53	0.43
0.28	0.47
0.37	0.49
0.47	0.52
0.43	0.75
0.36	0.79
0.42	0.86
0.38	0.62
0.43	0.46

Construct a 95% confidence interval for the difference in the mean stem weight between seedlings that receive no nitrogen and those that receive 368 ppm of nitrogen. Assume the populations to be normally distributed with equal variances.

9.41 The following data represent the length of time, in days, to recovery for patients randomly treated with one of two medications to clear up severe bladder infections:

Medication 1	Medication 2
$n_1 = 14$	$n_2 = 16$
$\bar{x}_1 = 17$	$\bar{x}_2 = 19$
$s_1^2 = 1.5$	$s_2^2 = 1.8$

Find a 99% confidence interval for the difference $\mu_2 - \mu_1$

in the mean recovery times for the two medications, assuming normal populations with equal variances.

9.42 An experiment reported in *Popular Science* compared fuel economies for two types of similarly equipped diesel mini-trucks. Let us suppose that 12 Volkswagen and 10 Toyota trucks were tested in 90-kilometer-per-hour steady-paced trials. If the 12 Volkswagen trucks averaged 16 kilometers per liter with a standard deviation of 1.0 kilometer per liter and the 10 Toyota trucks averaged 11 kilometers per liter with a standard deviation of 0.8 kilometer per liter, construct a 90% confidence interval for the difference between the average kilometers per liter for these two mini-trucks. Assume that the distances per liter for the truck models are approximately normally distributed with equal variances.

9.43 A taxi company is trying to decide whether to purchase brand A or brand B tires for its fleet of taxis. To estimate the difference in the two brands, an experiment is conducted using 12 of each brand. The tires are run until they wear out. The results are

Brand A: $\bar{x}_1 = 36,300$ kilometers,
               $s_1 = 5000$ kilometers.
Brand B: $\bar{x}_2 = 38,100$ kilometers,
               $s_2 = 6100$ kilometers.

Compute a 95% confidence interval for $\mu_A - \mu_B$ assuming the populations to be approximately normally distributed. You may not assume that the variances are equal.

9.44 Referring to Exercise 9.43, find a 99% confidence interval for $\mu_1 - \mu_2$ if tires of the two brands are assigned at random to the left and right rear wheels of 8 taxis and the following distances, in kilometers, are recorded:

Taxi	Brand A	Brand B
1	34,400	36,700
2	45,500	46,800
3	36,700	37,700
4	32,000	31,100
5	48,400	47,800
6	32,800	36,400
7	38,100	38,900
8	30,100	31,500

Assume that the differences of the distances are approximately normally distributed.

9.45 The federal government awarded grants to the agricultural departments of 9 universities to test the yield capabilities of two new varieties of wheat. Each variety was planted on a plot of equal area at each university, and the yields, in kilograms per plot, were recorded as follows:

	University								
Variety	1	2	3	4	5	6	7	8	9
1	38	23	35	41	44	29	37	31	38
2	45	25	31	38	50	33	36	40	43

Find a 95% confidence interval for the mean difference between the yields of the two varieties, assuming the differences of yields to be approximately normally distributed. Explain why pairing is necessary in this problem.

9.46 The following data represent the running times of films produced by two motion-picture companies.

Company	Time (minutes)
I	103 94 110 87 98
II	97 82 123 92 175 88 118

Compute a 90% confidence interval for the difference between the average running times of films produced by the two companies. Assume that the running-time differences are approximately normally distributed with unequal variances.

9.47 *Fortune* magazine (March 1997) reported the total returns to investors for the 10 years prior to 1996 and also for 1996 for 431 companies. The total returns for 10 of the companies are listed below. Find a 95% confidence interval for the mean change in percent return to investors.

	Total Return to Investors	
Company	1986–96	1996
Coca-Cola	29.8%	43.3%
Mirage Resorts	27.9%	25.4%
Merck	22.1%	24.0%
Microsoft	44.5%	88.3%
Johnson & Johnson	22.2%	18.1%
Intel	43.8%	131.2%
Pfizer	21.7%	34.0%
Procter & Gamble	21.9%	32.1%
Berkshire Hathaway	28.3%	6.2%
S&P 500	11.8%	20.3%

9.48 An automotive company is considering two types of batteries for its automobile. Sample information on battery life is collected for 20 batteries of type A and 20 batteries of type B. The summary statistics are $\bar{x}_A = 32.91$, $\bar{x}_B = 30.47$, $s_A = 1.57$, and $s_B = 1.74$. Assume the data on each battery are normally distributed and assume $\sigma_A = \sigma_B$.

(a) Find a 95% confidence interval on $\mu_A - \mu_B$.

(b) Draw a conclusion from (a) that provides insight into whether A or B should be adopted.

9.49 Two different brands of latex paint are being considered for use. Fifteen specimens of each type of paint were selected, and the drying times, in hours, were as follows:

Paint A	Paint B
3.5 2.7 3.9 4.2 3.6	4.7 3.9 4.5 5.5 4.0
2.7 3.3 5.2 4.2 2.9	5.3 4.3 6.0 5.2 3.7
4.4 5.2 4.0 4.1 3.4	5.5 6.2 5.1 5.4 4.8

Assume the drying time is normally distributed with $\sigma_A = \sigma_B$. Find a 95% confidence interval on $\mu_B - \mu_A$, where μ_A and μ_B are the mean drying times.

9.50 Two levels (low and high) of insulin doses are given to two groups of diabetic rats to check the insulin-binding capacity, yielding the following data:

Low dose: $n_1 = 8$ $\bar{x}_1 = 1.98$ $s_1 = 0.51$
High dose: $n_2 = 13$ $\bar{x}_2 = 1.30$ $s_2 = 0.35$

Assume that the variances are equal. Give a 95% confidence interval for the difference in the true average insulin-binding capacity between the two samples.

9.10 Single Sample: Estimating a Proportion

A point estimator of the proportion p in a binomial experiment is given by the statistic $\widehat{P} = X/n$, where X represents the number of successes in n trials. Therefore, the sample proportion $\hat{p} = x/n$ will be used as the point estimate of the parameter p.

If the unknown proportion p is not expected to be too close to 0 or 1, we can establish a confidence interval for p by considering the sampling distribution of $\widehat{P}$. Designating a failure in each binomial trial by the value 0 and a success by the value 1, the number of successes, x, can be interpreted as the sum of n values consisting only of 0 and 1s, and $\hat{p}$ is just the sample mean of these n values. Hence, by the Central Limit Theorem, for n sufficiently large, $\widehat{P}$ is approximately normally distributed with mean

$$\mu_{\widehat{P}} = E(\widehat{P}) = E\left(\frac{X}{n}\right) = \frac{np}{n} = p$$

and variance

$$\sigma_{\widehat{P}}^2 = \sigma_{X/n}^2 = \frac{\sigma_X^2}{n^2} = \frac{npq}{n^2} = \frac{pq}{n}.$$

Therefore, we can assert that

$$P(-z_{\alpha/2} < Z < z_{\alpha/2}) = 1 - \alpha, \text{ with } Z = \frac{\widehat{P} - p}{\sqrt{pq/n}},$$

and $z_{\alpha/2}$ is the value above which we find an area of $\alpha/2$ under the standard normal curve. Substituting for Z, we write

$$P\left(-z_{\alpha/2} < \frac{\widehat{P} - p}{\sqrt{pq/n}} < z_{\alpha/2}\right) = 1 - \alpha.$$

When n is large, very little error is introduced by substituting the point estimate $\hat{p} = x/n$ for the p under the radical sign. Then we can write

$$P\left(\widehat{P} - z_{\alpha/2}\sqrt{\frac{\hat{p}\hat{q}}{n}} < p < \widehat{P} + z_{\alpha/2}\sqrt{\frac{\hat{p}\hat{q}}{n}}\right) \approx 1 - \alpha.$$

9.10 Single Sample: Estimating a Proportion

On the other hand, by solving for p in the quadratic inequality above,

$$-z_{\alpha/2} < \frac{\hat{P} - p}{\sqrt{pq/n}} < z_{\alpha/2},$$

we obtain another form of the confidence interval for p with limits

$$\frac{\hat{p} + \frac{z_{\alpha/2}^2}{2n}}{1 + \frac{z_{\alpha/2}^2}{n}} \pm \frac{z_{\alpha/2}}{1 + \frac{z_{\alpha/2}^2}{n}} \sqrt{\frac{\hat{p}\hat{q}}{n} + \frac{z_{\alpha/2}^2}{4n^2}}.$$

For a random sample of size n, the sample proportion $\hat{p} = x/n$ is computed, and the following approximate $100(1-\alpha)\%$ confidence intervals for p can be obtained.

Large-Sample Confidence Intervals for p
If $\hat{p}$ is the proportion of successes in a random sample of size n and $\hat{q} = 1 - \hat{p}$, an approximate $100(1-\alpha)\%$ confidence interval, for the binomial parameter p is given by (method 1)

$$\hat{p} - z_{\alpha/2}\sqrt{\frac{\hat{p}\hat{q}}{n}} < p < \hat{p} + z_{\alpha/2}\sqrt{\frac{\hat{p}\hat{q}}{n}}$$

or by (method 2)

$$\frac{\hat{p} + \frac{z_{\alpha/2}^2}{2n}}{1 + \frac{z_{\alpha/2}^2}{n}} - \frac{z_{\alpha/2}}{1 + \frac{z_{\alpha/2}^2}{n}}\sqrt{\frac{\hat{p}\hat{q}}{n} + \frac{z_{\alpha/2}^2}{4n^2}} < p < \frac{\hat{p} + \frac{z_{\alpha/2}^2}{2n}}{1 + \frac{z_{\alpha/2}^2}{n}} + \frac{z_{\alpha/2}}{1 + \frac{z_{\alpha/2}^2}{n}}\sqrt{\frac{\hat{p}\hat{q}}{n} + \frac{z_{\alpha/2}^2}{4n^2}},$$

where $z_{\alpha/2}$ is the z-value leaving an area of $\alpha/2$ to the right.

When n is small and the unknown proportion p is believed to be close to 0 or to 1, the confidence-interval procedure established here is unreliable and, therefore, should not be used. To be on the safe side, one should require both $n\hat{p}$ and $n\hat{q}$ to be greater than or equal to 5. The methods for finding a confidence interval for the binomial parameter p are also applicable when the binomial distribution is being used to approximate the hypergeometric distribution, that is, when n is small relative to N, as illustrated by Example 9.14.

Note that although method 2 yields more accurate results, it is more complicated to calculate, and the gain in accuracy that it provides diminishes when the sample size is large enough. Hence, method 1 is commonly used in practice.

Example 9.14: In a random sample of $n = 500$ families owning televisions in the city of Hamilton, Canada, it is found that $x = 340$ subscribe to HBO. Find a 95% confidence interval for the actual proportion of families with televisions in this city that subscribe to HBO.

Solution: The point estimate of p is $\hat{p} = 340/500 = 0.68$. Using Table A.3, we find that $z_{0.025} = 1.96$. Therefore, using method 1, the 95% confidence interval for p is

$$0.68 - 1.96\sqrt{\frac{(0.68)(0.32)}{500}} < p < 0.68 + 1.96\sqrt{\frac{(0.68)(0.32)}{500}},$$

which simplifies to $0.6391 < p < 0.7209$.

If we use method 2, we can obtain

$$\frac{0.68 + \frac{1.96^2}{(2)(500)}}{1 + \frac{1.96^2}{500}} \pm \frac{1.96}{1 + \frac{1.96^2}{500}} \sqrt{\frac{(0.68)(0.32)}{500} + \frac{1.96^2}{(4)(500^2)}} = 0.6786 \pm 0.0408,$$

which simplifies to $0.6378 < p < 0.7194$. Apparently, when n is large (500 here), both methods yield very similar results.

If p is the center value of a $100(1-\alpha)\%$ confidence interval, then $\hat{p}$ estimates p without error. Most of the time, however, $\hat{p}$ will not be exactly equal to p and the point estimate will be in error. The size of this error will be the positive difference that separates p and $\hat{p}$, and we can be $100(1-\alpha)\%$ confident that this difference will not exceed $z_{\alpha/2}\sqrt{\hat{p}\hat{q}/n}$. We can readily see this if we draw a diagram of a typical confidence interval, as in Figure 9.6. Here we use method 1 to estimate the error.

Figure 9.6: Error in estimating p by $\hat{p}$.

Theorem 9.3: If $\hat{p}$ is used as an estimate of p, we can be $100(1-\alpha)\%$ confident that the error will not exceed $z_{\alpha/2}\sqrt{\hat{p}\hat{q}/n}$.

In Example 9.14, we are 95% confident that the sample proportion $\hat{p} = 0.68$ differs from the true proportion p by an amount not exceeding 0.04.

Choice of Sample Size

Let us now determine how large a sample is necessary to ensure that the error in estimating p will be less than a specified amount e. By Theorem 9.3, we must choose n such that $z_{\alpha/2}\sqrt{\hat{p}\hat{q}/n} = e$.

Theorem 9.4: If $\hat{p}$ is used as an estimate of p, we can be $100(1-\alpha)\%$ confident that the error will be less than a specified amount e when the sample size is approximately

$$n = \frac{z_{\alpha/2}^2 \hat{p}\hat{q}}{e^2}.$$

Theorem 9.4 is somewhat misleading in that we must use $\hat{p}$ to determine the sample size n, but $\hat{p}$ is computed from the sample. If a crude estimate of p can be made without taking a sample, this value can be used to determine n. Lacking such an estimate, we could take a preliminary sample of size $n \geq 30$ to provide an estimate of p. Using Theorem 9.4, we could determine approximately how many observations are needed to provide the desired degree of accuracy. Note that fractional values of n are rounded up to the next whole number.

9.10 Single Sample: Estimating a Proportion

Example 9.15: How large a sample is required if we want to be 95% confident that our estimate of p in Example 9.14 is within 0.02 of the true value?

Solution: Let us treat the 500 families as a preliminary sample, providing an estimate $\hat{p} = 0.68$. Then, by Theorem 9.4,

$$n = \frac{(1.96)^2(0.68)(0.32)}{(0.02)^2} = 2089.8 \approx 2090.$$

Therefore, if we base our estimate of p on a random sample of size 2090, we can be 95% confident that our sample proportion will not differ from the true proportion by more than 0.02.

Occasionally, it will be impractical to obtain an estimate of p to be used for determining the sample size for a specified degree of confidence. If this happens, an upper bound for n is established by noting that $\hat{p}\hat{q} = \hat{p}(1-\hat{p})$, which must be at most $1/4$, since $\hat{p}$ must lie between 0 and 1. This fact may be verified by completing the square. Hence

$$\hat{p}(1-\hat{p}) = -(\hat{p}^2 - \hat{p}) = \frac{1}{4} - \left(\hat{p}^2 - \hat{p} + \frac{1}{4}\right) = \frac{1}{4} - \left(\hat{p} - \frac{1}{2}\right)^2,$$

which is always less than $1/4$ except when $\hat{p} = 1/2$, and then $\hat{p}\hat{q} = 1/4$. Therefore, if we substitute $\hat{p} = 1/2$ into the formula for n in Theorem 9.4 when, in fact, p actually differs from $1/2$, n will turn out to be larger than necessary for the specified degree of confidence; as a result, our degree of confidence will increase.

Theorem 9.5: If $\hat{p}$ is used as an estimate of p, we can be **at least** $100(1-\alpha)\%$ confident that the error will not exceed a specified amount e when the sample size is

$$n = \frac{z_{\alpha/2}^2}{4e^2}.$$

Example 9.16: How large a sample is required if we want to be at least 95% confident that our estimate of p in Example 9.14 is within 0.02 of the true value?

Solution: Unlike in Example 9.15, we shall now assume that no preliminary sample has been taken to provide an estimate of p. Consequently, we can be at least 95% confident that our sample proportion will not differ from the true proportion by more than 0.02 if we choose a sample of size

$$n = \frac{(1.96)^2}{(4)(0.02)^2} = 2401.$$

Comparing the results of Examples 9.15 and 9.16, we see that information concerning p, provided by a preliminary sample or from experience, enables us to choose a smaller sample while maintaining our required degree of accuracy.

9.11 Two Samples: Estimating the Difference between Two Proportions

Consider the problem where we wish to estimate the difference between two binomial parameters p_1 and p_2. For example, p_1 might be the proportion of smokers with lung cancer and p_2 the proportion of nonsmokers with lung cancer, and the problem is to estimate the difference between these two proportions. First, we select independent random samples of sizes n_1 and n_2 from the two binomial populations with means $n_1 p_1$ and $n_2 p_2$ and variances $n_1 p_1 q_1$ and $n_2 p_2 q_2$, respectively; then we determine the numbers x_1 and x_2 of people in each sample with lung cancer and form the proportions $\hat{p}_1 = x_1/n$ and $\hat{p}_2 = x_2/n$. A point estimator of the difference between the two proportions, $p_1 - p_2$, is given by the statistic $\widehat{P}_1 - \widehat{P}_2$. Therefore, the difference of the sample proportions, $\hat{p}_1 - \hat{p}_2$, will be used as the point estimate of $p_1 - p_2$.

A confidence interval for $p_1 - p_2$ can be established by considering the sampling distribution of $\widehat{P}_1 - \widehat{P}_2$. From Section 9.10 we know that $\widehat{P}_1$ and $\widehat{P}_2$ are each approximately normally distributed, with means p_1 and p_2 and variances $p_1 q_1/n_1$ and $p_2 q_2/n_2$, respectively. Choosing independent samples from the two populations ensures that the variables $\widehat{P}_1$ and $\widehat{P}_2$ will be independent, and then by the reproductive property of the normal distribution established in Theorem 7.11, we conclude that $\widehat{P}_1 - \widehat{P}_2$ is approximately normally distributed with mean

$$\mu_{\widehat{P}_1 - \widehat{P}_2} = p_1 - p_2$$

and variance

$$\sigma^2_{\widehat{P}_1 - \widehat{P}_2} = \frac{p_1 q_1}{n_1} + \frac{p_2 q_2}{n_2}.$$

Therefore, we can assert that

$$P(-z_{\alpha/2} < Z < z_{\alpha/2}) = 1 - \alpha,$$

where

$$Z = \frac{(\widehat{P}_1 - \widehat{P}_2) - (p_1 - p_2)}{\sqrt{p_1 q_1/n_1 + p_2 q_2/n_2}}$$

and $z_{\alpha/2}$ is the value above which we find an area of $\alpha/2$ under the standard normal curve. Substituting for Z, we write

$$P\left[-z_{\alpha/2} < \frac{(\widehat{P}_1 - \widehat{P}_2) - (p_1 - p_2)}{\sqrt{p_1 q_1/n_1 + p_2 q_2/n_2}} < z_{\alpha/2}\right] = 1 - \alpha.$$

After performing the usual mathematical manipulations, we replace p_1, p_2, q_1, and q_2 under the radical sign by their estimates $\hat{p}_1 = x_1/n_1$, $\hat{p}_2 = x_2/n_2$, $\hat{q}_1 = 1 - \hat{p}_1$, and $\hat{q}_2 = 1 - \hat{p}_2$, provided that $n_1 \hat{p}_1$, $n_1 \hat{q}_1$, $n_2 \hat{p}_2$, and $n_2 \hat{q}_2$ are all greater than or equal to 5, and the following approximate $100(1-\alpha)\%$ confidence interval for $p_1 - p_2$ is obtained.

9.11 Two Samples: Estimating the Difference between Two Proportions

Large-Sample Confidence Interval for $p_1 - p_2$

If $\hat{p}_1$ and $\hat{p}_2$ are the proportions of successes in random samples of sizes n_1 and n_2, respectively, $\hat{q}_1 = 1 - \hat{p}_1$, and $\hat{q}_2 = 1 - \hat{p}_2$, an approximate $100(1-\alpha)\%$ confidence interval for the difference of two binomial parameters, $p_1 - p_2$, is given by

$$(\hat{p}_1 - \hat{p}_2) - z_{\alpha/2}\sqrt{\frac{\hat{p}_1 \hat{q}_1}{n_1} + \frac{\hat{p}_2 \hat{q}_2}{n_2}} < p_1 - p_2 < (\hat{p}_1 - \hat{p}_2) + z_{\alpha/2}\sqrt{\frac{\hat{p}_1 \hat{q}_1}{n_1} + \frac{\hat{p}_2 \hat{q}_2}{n_2}},$$

where $z_{\alpha/2}$ is the z-value leaving an area of $\alpha/2$ to the right.

Example 9.17: A certain change in a process for manufacturing component parts is being considered. Samples are taken under both the existing and the new process so as to determine if the new process results in an improvement. If 75 of 1500 items from the existing process are found to be defective and 80 of 2000 items from the new process are found to be defective, find a 90% confidence interval for the true difference in the proportion of defectives between the existing and the new process.

Solution: Let p_1 and p_2 be the true proportions of defectives for the existing and new processes, respectively. Hence, $\hat{p}_1 = 75/1500 = 0.05$ and $\hat{p}_2 = 80/2000 = 0.04$, and the point estimate of $p_1 - p_2$ is

$$\hat{p}_1 - \hat{p}_2 = 0.05 - 0.04 = 0.01.$$

Using Table A.3, we find $z_{0.05} = 1.645$. Therefore, substituting into the formula, with

$$1.645\sqrt{\frac{(0.05)(0.95)}{1500} + \frac{(0.04)(0.96)}{2000}} = 0.0117,$$

we find the 90% confidence interval to be $-0.0017 < p_1 - p_2 < 0.0217$. Since the interval contains the value 0, there is no reason to believe that the new process produces a significant decrease in the proportion of defectives over the existing method.

Up to this point, all confidence intervals presented were of the form

$$\text{point estimate} \pm K \text{ s.e.(point estimate)},$$

where K is a constant (either t or normal percent point). This form is valid when the parameter is a mean, a difference between means, a proportion, or a difference between proportions, due to the symmetry of the t- and Z-distributions. However, it does not extend to variances and ratios of variances, which will be discussed in Sections 9.12 and 9.13.

Exercises

In this set of exercises, for estimation concerning one proportion, use only method 1 to obtain the confidence intervals, unless instructed otherwise.

9.51 In a random sample of 1000 homes in a certain city, it is found that 228 are heated by oil. Find 99% confidence intervals for the proportion of homes in this city that are heated by oil using both methods presented on page 297.

9.52 Compute 95% confidence intervals, using both methods on page 297, for the proportion of defective items in a process when it is found that a sample of size 100 yields 8 defectives.

9.53 (a) A random sample of 200 voters in a town is selected, and 114 are found to support an annexation suit. Find the 96% confidence interval for the fraction of the voting population favoring the suit.

(b) What can we assert with 96% confidence about the possible size of our error if we estimate the fraction of voters favoring the annexation suit to be 0.57?

9.54 A manufacturer of MP3 players conducts a set of comprehensive tests on the electrical functions of its product. All MP3 players must pass all tests prior to being sold. Of a random sample of 500 MP3 players, 15 failed one or more tests. Find a 90% confidence interval for the proportion of MP3 players from the population that pass all tests.

9.55 A new rocket-launching system is being considered for deployment of small, short-range rockets. The existing system has $p = 0.8$ as the probability of a successful launch. A sample of 40 experimental launches is made with the new system, and 34 are successful.

(a) Construct a 95% confidence interval for p.

(b) Would you conclude that the new system is better?

9.56 A geneticist is interested in the proportion of African males who have a certain minor blood disorder. In a random sample of 100 African males, 24 are found to be afflicted.

(a) Compute a 99% confidence interval for the proportion of African males who have this blood disorder.

(b) What can we assert with 99% confidence about the possible size of our error if we estimate the proportion of African males with this blood disorder to be 0.24?

9.57 (a) According to a report in the *Roanoke Times*, approximately 2/3 of 1600 adults polled by telephone said they think the space shuttle program is a good investment for the country. Find a 95% confidence interval for the proportion of American adults who think the space shuttle program is a good investment for the country.

(b) What can we assert with 95% confidence about the possible size of our error if we estimate the proportion of American adults who think the space shuttle program is a good investment to be 2/3?

9.58 In the newspaper article referred to in Exercise 9.57, 32% of the 1600 adults polled said the U.S. space program should emphasize scientific exploration. How large a sample of adults is needed for the poll if one wishes to be 95% confident that the estimated percentage will be within 2% of the true percentage?

9.59 How large a sample is needed if we wish to be 96% confident that our sample proportion in Exercise 9.53 will be within 0.02 of the true fraction of the voting population?

9.60 How large a sample is needed if we wish to be 99% confident that our sample proportion in Exercise 9.51 will be within 0.05 of the true proportion of homes in the city that are heated by oil?

9.61 How large a sample is needed in Exercise 9.52 if we wish to be 98% confident that our sample proportion will be within 0.05 of the true proportion defective?

9.62 A conjecture by a faculty member in the microbiology department at Washington University School of Dental Medicine in St. Louis, Missouri, states that a couple of cups of either green or oolong tea each day will provide sufficient fluoride to protect your teeth from decay. How large a sample is needed to estimate the percentage of citizens in a certain town who favor having their water fluoridated if one wishes to be at least 99% confident that the estimate is within 1% of the true percentage?

9.63 A study is to be made to estimate the percentage of citizens in a town who favor having their water fluoridated. How large a sample is needed if one wishes to be at least 95% confident that the estimate is within 1% of the true percentage?

9.64 A study is to be made to estimate the proportion of residents of a certain city and its suburbs who favor the construction of a nuclear power plant near the city. How large a sample is needed if one wishes to be at least 95% confident that the estimate is within 0.04 of the true proportion of residents who favor the construction of the nuclear power plant?

9.65 A certain geneticist is interested in the proportion of males and females in the population who have a minor blood disorder. In a random sample of 1000 males, 250 are found to be afflicted, whereas 275 of 1000 females tested appear to have the disorder. Compute a 95% confidence interval for the difference between the proportions of males and females who have the blood disorder.

9.66 Ten engineering schools in the United States were surveyed. The sample contained 250 electrical engineers, 80 being women; 175 chemical engineers, 40 being women. Compute a 90% confidence interval for the difference between the proportions of women in these two fields of engineering. Is there a significant difference between the two proportions?

9.67 A clinical trial was conducted to determine if a certain type of inoculation has an effect on the incidence of a certain disease. A sample of 1000 rats was kept in a controlled environment for a period of 1 year, and 500 of the rats were given the inoculation. In the group not inoculated, there were 120 incidences of the disease, while 98 of the rats in the inoculated group contracted it. If p_1 is the probability of incidence of the disease in uninoculated rats and p_2 the probability of incidence in inoculated rats, compute a 90% confidence interval for $p_1 - p_2$.

9.68 In the study *Germination and Emergence of Broccoli*, conducted by the Department of Horticulture at Virginia Tech, a researcher found that at 5°C, 10 broccoli seeds out of 20 germinated, while at 15°C, 15 out of 20 germinated. Compute a 95% confidence interval for the difference between the proportions of germination at the two different temperatures and decide if there is a significant difference.

9.69 A survey of 1000 students found that 274 chose professional baseball team A as their favorite team. In a similar survey involving 760 students, 240 of them chose team A as their favorite. Compute a 95% confidence interval for the difference between the proportions of students favoring team A in the two surveys. Is there a significant difference?

9.70 According to *USA Today* (March 17, 1997), women made up 33.7% of the editorial staff at local TV stations in the United States in 1990 and 36.2% in 1994. Assume 20 new employees were hired as editorial staff.

(a) Estimate the number that would have been women in 1990 and 1994, respectively.

(b) Compute a 95% confidence interval to see if there is evidence that the proportion of women hired as editorial staff was higher in 1994 than in 1990.

9.12 Single Sample: Estimating the Variance

If a sample of size n is drawn from a normal population with variance σ^2 and the sample variance s^2 is computed, we obtain a value of the statistic S^2. This computed sample variance is used as a point estimate of σ^2. Hence, the statistic S^2 is called an estimator of σ^2.

An interval estimate of σ^2 can be established by using the statistic

$$X^2 = \frac{(n-1)S^2}{\sigma^2}.$$

According to Theorem 8.4, the statistic X^2 has a chi-squared distribution with $n-1$ degrees of freedom when samples are chosen from a normal population. We may write (see Figure 9.7)

$$P(\chi^2_{1-\alpha/2} < X^2 < \chi^2_{\alpha/2}) = 1 - \alpha,$$

where $\chi^2_{1-\alpha/2}$ and $\chi^2_{\alpha/2}$ are values of the chi-squared distribution with $n-1$ degrees of freedom, leaving areas of $1-\alpha/2$ and $\alpha/2$, respectively, to the right. Substituting for X^2, we write

$$P\left[\chi^2_{1-\alpha/2} < \frac{(n-1)S^2}{\sigma^2} < \chi^2_{\alpha/2}\right] = 1 - \alpha.$$

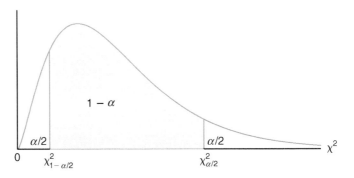

Figure 9.7: $P(\chi^2_{1-\alpha/2} < X^2 < \chi^2_{\alpha/2}) = 1 - \alpha$.

Dividing each term in the inequality by $(n-1)S^2$ and then inverting each term (thereby changing the sense of the inequalities), we obtain

$$P\left[\frac{(n-1)S^2}{\chi^2_{\alpha/2}} < \sigma^2 < \frac{(n-1)S^2}{\chi^2_{1-\alpha/2}}\right] = 1 - \alpha.$$

For a random sample of size n from a normal population, the sample variance s^2 is computed, and the following $100(1-\alpha)\%$ confidence interval for σ^2 is obtained.

Confidence Interval for σ^2   If s^2 is the variance of a random sample of size n from a normal population, a $100(1-\alpha)\%$ confidence interval for σ^2 is

$$\frac{(n-1)s^2}{\chi^2_{\alpha/2}} < \sigma^2 < \frac{(n-1)s^2}{\chi^2_{1-\alpha/2}},$$

where $\chi^2_{\alpha/2}$ and $\chi^2_{1-\alpha/2}$ are χ^2-values with $v = n-1$ degrees of freedom, leaving areas of $\alpha/2$ and $1 - \alpha/2$, respectively, to the right.

An approximate $100(1-\alpha)\%$ confidence interval for σ is obtained by taking the square root of each endpoint of the interval for σ^2.

Example 9.18: The following are the weights, in decagrams, of 10 packages of grass seed distributed by a certain company: 46.4, 46.1, 45.8, 47.0, 46.1, 45.9, 45.8, 46.9, 45.2, and 46.0. Find a 95% confidence interval for the variance of the weights of all such packages of grass seed distributed by this company, assuming a normal population.

Solution: First we find

$$s^2 = \frac{n\sum_{i=1}^{n}x_i^2 - \left(\sum_{i=1}^{n}x_i\right)^2}{n(n-1)}$$

$$= \frac{(10)(21,273.12) - (461.2)^2}{(10)(9)} = 0.286.$$

To obtain a 95% confidence interval, we choose $\alpha = 0.05$. Then, using Table A.5 with $v = 9$ degrees of freedom, we find $\chi^2_{0.025} = 19.023$ and $\chi^2_{0.975} = 2.700$. Therefore, the 95% confidence interval for σ^2 is

$$\frac{(9)(0.286)}{19.023} < \sigma^2 < \frac{(9)(0.286)}{2.700},$$

or simply $0.135 < \sigma^2 < 0.953$.

9.13 Two Samples: Estimating the Ratio of Two Variances

A point estimate of the ratio of two population variances σ_1^2/σ_2^2 is given by the ratio s_1^2/s_2^2 of the sample variances. Hence, the statistic S_1^2/S_2^2 is called an estimator of σ_1^2/σ_2^2.

If σ_1^2 and σ_2^2 are the variances of normal populations, we can establish an interval estimate of σ_1^2/σ_2^2 by using the statistic

$$F = \frac{\sigma_2^2 S_1^2}{\sigma_1^2 S_2^2}.$$

According to Theorem 8.8, the random variable F has an F-distribution with $v_1 = n_1 - 1$ and $v_2 = n_2 - 1$ degrees of freedom. Therefore, we may write (see Figure 9.8)

$$P[f_{1-\alpha/2}(v_1, v_2) < F < f_{\alpha/2}(v_1, v_2)] = 1 - \alpha,$$

where $f_{1-\alpha/2}(v_1, v_2)$ and $f_{\alpha/2}(v_1, v_2)$ are the values of the F-distribution with v_1 and v_2 degrees of freedom, leaving areas of $1 - \alpha/2$ and $\alpha/2$, respectively, to the right.

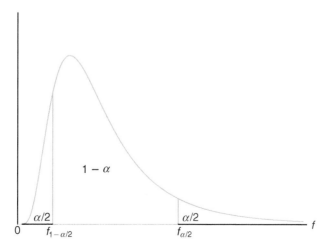

Figure 9.8: $P[f_{1-\alpha/2}(v_1, v_2) < F < f_{\alpha/2}(v_1, v_2)] = 1 - \alpha$.

Substituting for F, we write

$$P\left[f_{1-\alpha/2}(v_1, v_2) < \frac{\sigma_2^2 S_1^2}{\sigma_1^2 S_2^2} < f_{\alpha/2}(v_1, v_2)\right] = 1 - \alpha.$$

Multiplying each term in the inequality by S_2^2/S_1^2 and then inverting each term, we obtain

$$P\left[\frac{S_1^2}{S_2^2}\frac{1}{f_{\alpha/2}(v_1, v_2)} < \frac{\sigma_1^2}{\sigma_2^2} < \frac{S_1^2}{S_2^2}\frac{1}{f_{1-\alpha/2}(v_1, v_2)}\right] = 1 - \alpha.$$

The results of Theorem 8.7 enable us to replace the quantity $f_{1-\alpha/2}(v_1, v_2)$ by $1/f_{\alpha/2}(v_2, v_1)$. Therefore,

$$P\left[\frac{S_1^2}{S_2^2}\frac{1}{f_{\alpha/2}(v_1, v_2)} < \frac{\sigma_1^2}{\sigma_2^2} < \frac{S_1^2}{S_2^2}f_{\alpha/2}(v_2, v_1)\right] = 1 - \alpha.$$

For any two independent random samples of sizes n_1 and n_2 selected from two normal populations, the ratio of the sample variances s_1^2/s_2^2 is computed, and the following $100(1-\alpha)\%$ confidence interval for σ_1^2/σ_2^2 is obtained.

Confidence Interval for σ_1^2/σ_2^2   If s_1^2 and s_2^2 are the variances of independent samples of sizes n_1 and n_2, respectively, from normal populations, then a $100(1-\alpha)\%$ confidence interval for σ_1^2/σ_2^2 is

$$\frac{s_1^2}{s_2^2}\frac{1}{f_{\alpha/2}(v_1, v_2)} < \frac{\sigma_1^2}{\sigma_2^2} < \frac{s_1^2}{s_2^2}f_{\alpha/2}(v_2, v_1),$$

where $f_{\alpha/2}(v_1, v_2)$ is an f-value with $v_1 = n_1 - 1$ and $v_2 = n_2 - 1$ degrees of freedom, leaving an area of $\alpha/2$ to the right, and $f_{\alpha/2}(v_2, v_1)$ is a similar f-value with $v_2 = n_2 - 1$ and $v_1 = n_1 - 1$ degrees of freedom.

As in Section 9.12, an approximate $100(1-\alpha)\%$ confidence interval for σ_1/σ_2 is obtained by taking the square root of each endpoint of the interval for σ_1^2/σ_2^2.

Example 9.19: A confidence interval for the difference in the mean orthophosphorus contents, measured in milligrams per liter, at two stations on the James River was constructed in Example 9.12 on page 290 by assuming the normal population variance to be unequal. Justify this assumption by constructing 98% confidence intervals for σ_1^2/σ_2^2 and for σ_1/σ_2, where σ_1^2 and σ_2^2 are the variances of the populations of orthophosphorus contents at station 1 and station 2, respectively.

Solution: From Example 9.12, we have $n_1 = 15$, $n_2 = 12$, $s_1 = 3.07$, and $s_2 = 0.80$. For a 98% confidence interval, $\alpha = 0.02$. Interpolating in Table A.6, we find $f_{0.01}(14, 11) \approx 4.30$ and $f_{0.01}(11, 14) \approx 3.87$. Therefore, the 98% confidence interval for σ_1^2/σ_2^2 is

$$\left(\frac{3.07^2}{0.80^2}\right)\left(\frac{1}{4.30}\right) < \frac{\sigma_1^2}{\sigma_2^2} < \left(\frac{3.07^2}{0.80^2}\right)(3.87),$$

which simplifies to $3.425 < \frac{\sigma_1^2}{\sigma_2^2} < 56.991$. Taking square roots of the confidence limits, we find that a 98% confidence interval for σ_1/σ_2 is

$$1.851 < \frac{\sigma_1}{\sigma_2} < 7.549.$$

Since this interval does not allow for the possibility of σ_1/σ_2 being equal to 1, we were correct in assuming that $\sigma_1 \neq \sigma_2$ or $\sigma_1^2 \neq \sigma_2^2$ in Example 9.12.

Exercises

9.71 A manufacturer of car batteries claims that the batteries will last, on average, 3 years with a variance of 1 year. If 5 of these batteries have lifetimes of 1.9, 2.4, 3.0, 3.5, and 4.2 years, construct a 95% confidence interval for σ^2 and decide if the manufacturer's claim that $\sigma^2 = 1$ is valid. Assume the population of battery lives to be approximately normally distributed.

9.72 A random sample of 20 students yielded a mean of $\bar{x} = 72$ and a variance of $s^2 = 16$ for scores on a college placement test in mathematics. Assuming the scores to be normally distributed, construct a 98% confidence interval for σ^2.

9.73 Construct a 95% confidence interval for σ^2 in Exercise 9.9 on page 283.

9.74 Construct a 99% confidence interval for σ^2 in Exercise 9.11 on page 283.

9.75 Construct a 99% confidence interval for σ in Exercise 9.12 on page 283.

9.76 Construct a 90% confidence interval for σ in Exercise 9.13 on page 283.

9.77 Construct a 98% confidence interval for σ_1/σ_2 in Exercise 9.42 on page 295, where σ_1 and σ_2 are, respectively, the standard deviations for the distances traveled per liter of fuel by the Volkswagen and Toyota mini-trucks.

9.78 Construct a 90% confidence interval for σ_1^2/σ_2^2 in Exercise 9.43 on page 295. Were we justified in assuming that $\sigma_1^2 \neq \sigma_2^2$ when we constructed the confidence interval for $\mu_1 - \mu_2$?

9.79 Construct a 90% confidence interval for σ_1^2/σ_2^2 in Exercise 9.46 on page 295. Should we have assumed $\sigma_1^2 = \sigma_2^2$ in constructing our confidence interval for $\mu_I - \mu_{II}$?

9.80 Construct a 95% confidence interval for σ_A^2/σ_B^2 in Exercise 9.49 on page 295. Should the equal-variance assumption be used?

9.14 Maximum Likelihood Estimation (Optional)

Often the estimators of parameters have been those that appeal to intuition. The estimator $\bar{X}$ certainly seems reasonable as an estimator of a population mean μ. The virtue of S^2 as an estimator of σ^2 is underscored through the discussion of unbiasedness in Section 9.3. The estimator for a binomial parameter p is merely a sample proportion, which, of course, is an *average* and appeals to common sense. But there are many situations in which it is not at all obvious what the proper estimator should be. As a result, there is much to be learned by the student of statistics concerning different philosophies that produce different methods of estimation. In this section, we deal with the **method of maximum likelihood**.

Maximum likelihood estimation is one of the most important approaches to estimation in all of statistical inference. We will not give a thorough development of the method. Rather, we will attempt to communicate the philosophy of maximum likelihood and illustrate with examples that relate to other estimation problems discussed in this chapter.

The Likelihood Function

As the name implies, the method of maximum likelihood is that for which the *likelihood function* is maximized. The likelihood function is best illustrated through the use of an example with a discrete distribution and a single parameter. Denote by $X_1, X_2, \ldots, X_n$ the independent random variables taken from a discrete probability distribution represented by $f(\mathbf{x}, \theta)$, where θ is a single parameter of the distribution. Now

$$L(x_1, x_2, \ldots, x_n; \theta) = f(x_1, x_2, \ldots, x_n; \theta)$$
$$= f(x_1, \theta) f(x_2, \theta) \cdots f(x_n, \theta)$$

is the *joint distribution of the random variables*, often referred to as the likelihood function. Note that the variable of the likelihood function is θ, not $\mathbf{x}$. Denote by $x_1, x_2, \ldots, x_n$ the observed values in a sample. In the case of a discrete random variable, the interpretation is very clear. The quantity $L(x_1, x_2, \ldots, x_n; \theta)$, the *likelihood of the sample*, is the following joint probability:

$$P(X_1 = x_1, X_2 = x_2, \ldots, X_n = x_n \mid \theta),$$

which is the probability of obtaining the sample values $x_1, x_2, \ldots, x_n$. For the discrete case, the maximum likelihood estimator is one that results in a maximum value for this joint probability or maximizes the likelihood of the sample.

Consider a fictitious example where three items from an assembly line are inspected. The items are ruled either defective or nondefective, and thus the Bernoulli process applies. Testing the three items results in two nondefective items followed by a defective item. It is of interest to estimate p, the proportion nondefective in the process. The likelihood of the sample for this illustration is given by

$$p \cdot p \cdot q = p^2 q = p^2 - p^3,$$

where $q = 1 - p$. Maximum likelihood estimation would give an estimate of p for which the likelihood is maximized. It is clear that if we differentiate the likelihood with respect to p, set the derivative to zero, and solve, we obtain the value

$$\hat{p} = \frac{2}{3}.$$

Now, of course, in this situation $\hat{p} = 2/3$ is the sample proportion defective and is thus a reasonable estimator of the probability of a defective. The reader should attempt to understand that the philosophy of maximum likelihood estimation evolves from the notion that the reasonable estimator of a parameter based on sample information *is that parameter value that produces the largest probability of obtaining the sample*. This is, indeed, the interpretation for the discrete case, since the likelihood is the probability of jointly observing the values in the sample.

Now, while the interpretation of the likelihood function as a joint probability is confined to the discrete case, the notion of maximum likelihood extends to the estimation of parameters of a continuous distribution. We now present a formal definition of maximum likelihood estimation.

9.14 Maximum Likelihood Estimation (Optional)

Definition 9.3: Given independent observations $x_1, x_2, \ldots, x_n$ from a probability density function (continuous case) or probability mass function (discrete case) $f(\mathbf{x}; \theta)$, the maximum likelihood estimator $\hat{\theta}$ is that which maximizes the likelihood function

$$L(x_1, x_2, \ldots, x_n; \theta) = f(\mathbf{x}; \theta) = f(x_1, \theta) f(x_2, \theta) \cdots f(x_n, \theta).$$

Quite often it is convenient to work with the natural log of the likelihood function in finding the maximum of that function. Consider the following example dealing with the parameter μ of a Poisson distribution.

Example 9.20: Consider a Poisson distribution with probability mass function

$$f(x|\mu) = \frac{e^{-\mu}\mu^x}{x!}, \quad x = 0, 1, 2, \ldots.$$

Suppose that a random sample $x_1, x_2, \ldots, x_n$ is taken from the distribution. What is the maximum likelihood estimate of μ?

Solution: The likelihood function is

$$L(x_1, x_2, \ldots, x_n; \mu) = \prod_{i=1}^{n} f(x_i|\mu) = \frac{e^{-n\mu}\mu^{\sum_{i=1}^{n} x_i}}{\prod_{i=1}^{n} x_i!}.$$

Now consider

$$\ln L(x_1, x_2, \ldots, x_n; \mu) = -n\mu + \sum_{i=1}^{n} x_i \ln \mu - \ln \prod_{i=1}^{n} x_i!$$

$$\frac{\partial \ln L(x_1, x_2, \ldots, x_n; \mu)}{\partial \mu} = -n + \sum_{i=1}^{n} \frac{x_i}{\mu}.$$

Solving for $\hat{\mu}$, the maximum likelihood estimator, involves setting the derivative to zero and solving for the parameter. Thus,

$$\hat{\mu} = \sum_{i=1}^{n} \frac{x_i}{n} = \bar{x}.$$

The second derivative of the log-likelihood function is negative, which implies that the solution above indeed is a maximum. Since μ is the mean of the Poisson distribution (Chapter 5), the sample average would certainly seem like a reasonable estimator.

The following example shows the use of the method of maximum likelihood for finding estimates of two parameters. We simply find the values of the parameters that maximize (jointly) the likelihood function.

Example 9.21: Consider a random sample $x_1, x_2, \ldots, x_n$ from a normal distribution $N(\mu, \sigma)$. Find the maximum likelihood estimators for μ and σ^2.

Solution: The likelihood function for the normal distribution is

$$L(x_1, x_2, \ldots, x_n; \mu, \sigma^2) = \frac{1}{(2\pi)^{n/2}(\sigma^2)^{n/2}} \exp\left[-\frac{1}{2}\sum_{i=1}^n \left(\frac{x_i - \mu}{\sigma}\right)^2\right].$$

Taking logarithms gives us

$$\ln L(x_1, x_2, \ldots, x_n; \mu, \sigma^2) = -\frac{n}{2}\ln(2\pi) - \frac{n}{2}\ln\sigma^2 - \frac{1}{2}\sum_{i=1}^n \left(\frac{x_i - \mu}{\sigma}\right)^2.$$

Hence,

$$\frac{\partial \ln L}{\partial \mu} = \sum_{i=1}^n \left(\frac{x_i - \mu}{\sigma^2}\right)$$

and

$$\frac{\partial \ln L}{\partial \sigma^2} = -\frac{n}{2\sigma^2} + \frac{1}{2(\sigma^2)^2}\sum_{i=1}^n (x_i - \mu)^2.$$

Setting both derivatives equal to 0, we obtain

$$\sum_{i=1}^n x_i - n\mu = 0 \quad \text{and} \quad n\sigma^2 = \sum_{i=1}^n (x_i - \mu)^2.$$

Thus, the maximum likelihood estimator of μ is given by

$$\hat{\mu} = \frac{1}{n}\sum_{i=1}^n x_i = \bar{x},$$

which is a pleasing result since $\bar{x}$ has played such an important role in this chapter as a point estimate of μ. On the other hand, the maximum likelihood estimator of σ^2 is

$$\hat{\sigma}^2 = \frac{1}{n}\sum_{i=1}^n (x_i - \bar{x})^2.$$

Checking the second-order partial derivative matrix confirms that the solution results in a maximum of the likelihood function.

It is interesting to note the distinction between the maximum likelihood estimator of σ^2 and the unbiased estimator S^2 developed earlier in this chapter. The numerators are identical, of course, and the denominator is the degrees of freedom $n-1$ for the unbiased estimator and n for the maximum likelihood estimator. Maximum likelihood estimators do not necessarily enjoy the property of unbiasedness. However, they do have very important asymptotic properties.

Example 9.22: Suppose 10 rats are used in a biomedical study where they are injected with cancer cells and then given a cancer drug that is designed to increase their survival rate. The survival times, in months, are 14, 17, 27, 18, 12, 8, 22, 13, 19, and 12. Assume

9.14 Maximum Likelihood Estimation (Optional)

that the exponential distribution applies. Give a maximum likelihood estimate of the mean survival time.

Solution: From Chapter 6, we know that the probability density function for the exponential random variable X is

$$f(x, \beta) = \begin{cases} \frac{1}{\beta} e^{-x/\beta}, & x > 0, \\ 0, & \text{elsewhere.} \end{cases}$$

Thus, the log-likelihood function for the data, given $n = 10$, is

$$\ln L(x_1, x_2, \ldots, x_{10}; \beta) = -10 \ln \beta - \frac{1}{\beta} \sum_{i=1}^{10} x_i.$$

Setting

$$\frac{\partial \ln L}{\partial \beta} = -\frac{10}{\beta} + \frac{1}{\beta^2} \sum_{i=1}^{10} x_i = 0$$

implies that

$$\hat{\beta} = \frac{1}{10} \sum_{i=1}^{10} x_i = \bar{x} = 16.2.$$

Evaluating the second derivative of the log-likelihood function at the value $\hat{\beta}$ above yields a negative value. As a result, the estimator of the parameter β, the population mean, is the sample average $\bar{x}$.

The following example shows the maximum likelihood estimator for a distribution that does not appear in previous chapters.

Example 9.23: It is known that a sample consisting of the values 12, 11.2, 13.5, 12.3, 13.8, and 11.9 comes from a population with the density function

$$f(x; \theta) = \begin{cases} \frac{\theta}{x^{\theta+1}}, & x > 1, \\ 0, & \text{elsewhere,} \end{cases}$$

where $\theta > 0$. Find the maximum likelihood estimate of θ.

Solution: The likelihood function of n observations from this population can be written as

$$L(x_1, x_2, \ldots, x_{10}; \theta) = \prod_{i=1}^{n} \frac{\theta}{x_i^{\theta+1}} = \frac{\theta^n}{(\prod_{i=1}^{n} x_i)^{\theta+1}},$$

which implies that

$$\ln L(x_1, x_2, \ldots, x_{10}; \theta) = n \ln(\theta) - (\theta + 1) \sum_{i=1}^{n} \ln(x_i).$$

Setting $0 = \frac{\partial \ln L}{\partial \theta} = \frac{n}{\theta} - \sum_{i=1}^{n} \ln(x_i)$ results in

$$\hat{\theta} = \frac{n}{\sum_{i=1}^{n} \ln(x_i)}$$

$$= \frac{6}{\ln(12) + \ln(11.2) + \ln(13.5) + \ln(12.3) + \ln(13.8) + \ln(11.9)} = 0.3970.$$

Since the second derivative of L is $-n/\theta^2$, which is always negative, the likelihood function does achieve its maximum value at $\hat{\theta}$.

Additional Comments Concerning Maximum Likelihood Estimation

A thorough discussion of the properties of maximum likelihood estimation is beyond the scope of this book and is usually a major topic of a course in the theory of statistical inference. The method of maximum likelihood allows the analyst to make use of knowledge of the distribution in determining an appropriate estimator. *The method of maximum likelihood cannot be applied without knowledge of the underlying distribution.* We learned in Example 9.21 that the maximum likelihood estimator is not necessarily unbiased. The maximum likelihood estimator is unbiased *asymptotically* or *in the limit*; that is, the amount of bias approaches zero as the sample size becomes large. Earlier in this chapter the notion of efficiency was discussed, efficiency being linked to the variance property of an estimator. Maximum likelihood estimators possess desirable variance properties in the limit. The reader should consult Lehmann and D'Abrera (1998) for details.

Exercises

9.81 Suppose that there are n trials $x_1, x_2, \ldots, x_n$ from a Bernoulli process with parameter p, the probability of a success. That is, the probability of r successes is given by $\binom{n}{r} p^r (1-p)^{n-r}$. Work out the maximum likelihood estimator for the parameter p.

9.82 Consider the lognormal distribution with the density function given in Section 6.9. Suppose we have a random sample $x_1, x_2, \ldots, x_n$ from a lognormal distribution.
(a) Write out the likelihood function.
(b) Develop the maximum likelihood estimators of μ and σ^2.

9.83 Consider a random sample of $x_1, \ldots, x_n$ coming from the gamma distribution discussed in Section 6.6. Suppose the parameter α is known, say 5, and determine the maximum likelihood estimation for parameter β.

9.84 Consider a random sample of $x_1, x_2, \ldots, x_n$ observations from a Weibull distribution with parameters α and β and density function

$$f(x) = \begin{cases} \alpha \beta x^{\beta-1} e^{-\alpha x^\beta}, & x > 0, \\ 0, & \text{elsewhere}, \end{cases}$$

for $\alpha, \beta > 0$.
(a) Write out the likelihood function.
(b) Write out the equations that, when solved, give the maximum likelihood estimators of α and β.

9.85 Consider a random sample of $x_1, \ldots, x_n$ from a uniform distribution $U(0, \theta)$ with unknown parameter θ, where $\theta > 0$. Determine the maximum likelihood estimator of θ.

9.86 Consider the independent observations $x_1, x_2, \ldots, x_n$ from the gamma distribution discussed in Section 6.6.
(a) Write out the likelihood function.

(b) Write out a set of equations that, when solved, give the maximum likelihood estimators of α and β.

9.87 Consider a hypothetical experiment where a man with a fungus uses an antifungal drug and is cured. Consider this, then, a sample of one from a Bernoulli distribution with probability function

$$f(x) = p^x q^{1-x}, \quad x = 0, 1,$$

where p is the probability of a success (cure) and $q = 1 - p$. Now, of course, the sample information gives $x = 1$. Write out a development that shows that $\hat{p} = 1.0$ is the maximum likelihood estimator of the probability of a cure.

9.88 Consider the observation X from the negative binomial distribution given in Section 5.4. Find the maximum likelihood estimator for p, assuming k is known.

Review Exercises

9.89 Consider two estimators of σ^2 for a sample $x_1, x_2, \ldots, x_n$, which is drawn from a normal distribution with mean μ and variance σ^2. The estimators are the unbiased estimator $s^2 = \frac{1}{n-1} \sum_{i=1}^{n} (x_i - \bar{x})^2$ and the maximum likelihood estimator $\hat{\sigma}^2 = \frac{1}{n} \sum_{i=1}^{n} (x_i - \bar{x})^2$. Discuss the variance properties of these two estimators.

9.90 According to the *Roanoke Times*, McDonald's sold 42.1% of the market share of hamburgers. A random sample of 75 burgers sold resulted in 28 of them being from McDonald's. Use material in Section 9.10 to determine if this information supports the claim in the *Roanoke Times*.

9.91 It is claimed that a new diet will reduce a person's weight by 4.5 kilograms on average in a period of 2 weeks. The weights of 7 women who followed this diet were recorded before and after the 2-week period.

Woman	Weight Before	Weight After
1	58.5	60.0
2	60.3	54.9
3	61.7	58.1
4	69.0	62.1
5	64.0	58.5
6	62.6	59.9
7	56.7	54.4

Test the claim about the diet by computing a 95% confidence interval for the mean difference in weights. Assume the differences of weights to be approximately normally distributed.

9.92 A study was undertaken at Virginia Tech to determine if fire can be used as a viable management tool to increase the amount of forage available to deer during the critical months in late winter and early spring. Calcium is a required element for plants and animals. The amount taken up and stored in plants is closely correlated to the amount present in the soil. It was hypothesized that a fire may change the calcium levels present in the soil and thus affect the amount available to deer. A large tract of land in the Fishburn Forest was selected for a prescribed burn. Soil samples were taken from 12 plots of equal area just prior to the burn and analyzed for calcium. Postburn calcium levels were analyzed from the same plots. These values, in kilograms per plot, are presented in the following table:

Plot	Calcium Level (kg/plot) Preburn	Postburn
1	50	9
2	50	18
3	82	45
4	64	18
5	82	18
6	73	9
7	77	32
8	54	9
9	23	18
10	45	9
11	36	9
12	54	9

Construct a 95% confidence interval for the mean difference in calcium levels in the soil prior to and after the prescribed burn. Assume the distribution of differences in calcium levels to be approximately normal.

9.93 A health spa claims that a new exercise program will reduce a person's waist size by 2 centimeters on average over a 5-day period. The waist sizes, in centimeters, of 6 men who participated in this exercise program are recorded before and after the 5-day period in the following table:

Man	Waist Size Before	Waist Size After
1	90.4	91.7
2	95.5	93.9
3	98.7	97.4
4	115.9	112.8
5	104.0	101.3
6	85.6	84.0

By computing a 95% confidence interval for the mean reduction in waist size, determine whether the health spa's claim is valid. Assume the distribution of differences in waist sizes before and after the program to be approximately normal.

9.94 The Department of Civil and Environmental Engineering at Virginia Tech compared a modified (M-5 hr) assay technique for recovering fecal coliforms in stormwater runoff from an urban area to a most probable number (MPN) technique. A total of 12 runoff samples were collected and analyzed by the two techniques. Fecal coliform counts per 100 milliliters are recorded in the following table.

Sample	MPN Count	M-5 hr Count
1	2300	2010
2	1200	930
3	450	400
4	210	436
5	270	4100
6	450	2090
7	154	219
8	179	169
9	192	194
10	230	174
11	340	274
12	194	183

Construct a 90% confidence interval for the difference in the mean fecal coliform counts between the M-5 hr and the MPN techniques. Assume that the count differences are approximately normally distributed.

9.95 An experiment was conducted to determine whether surface finish has an effect on the endurance limit of steel. There is a theory that polishing increases the average endurance limit (for reverse bending). From a practical point of view, polishing should not have any effect on the standard deviation of the endurance limit, which is known from numerous endurance limit experiments to be 4000 psi. An experiment was performed on 0.4% carbon steel using both unpolished and polished smooth-turned specimens. The data are as follows:

Endurance Limit (psi)	
Polished 0.4% Carbon	Unpolished 0.4% Carbon
85,500	82,600
91,900	82,400
89,400	81,700
84,000	79,500
89,900	79,400
78,700	69,800
87,500	79,900
83,100	83,400

Find a 95% confidence interval for the difference between the population means for the two methods, assuming that the populations are approximately normally distributed.

9.96 An anthropologist is interested in the proportion of individuals in two Indian tribes with double occipital hair whorls. Suppose that independent samples are taken from each of the two tribes, and it is found that 24 of 100 Indians from tribe A and 36 of 120 Indians from tribe B possess this characteristic. Construct a 95% confidence interval for the difference $p_B - p_A$ between the proportions of these two tribes with occipital hair whorls.

9.97 A manufacturer of electric irons produces these items in two plants. Both plants have the same suppliers of small parts. A saving can be made by purchasing thermostats for plant B from a local supplier. A single lot was purchased from the local supplier, and a test was conducted to see whether or not these new thermostats were as accurate as the old. The thermostats were tested on tile irons on the 550°F setting, and the actual temperatures were read to the nearest 0.1°F with a thermocouple. The data are as follows:

New Supplier (°F)					
530.3	559.3	549.4	544.0	551.7	566.3
549.9	556.9	536.7	558.8	538.8	543.3
559.1	555.0	538.6	551.1	565.4	554.9
550.0	554.9	554.7	536.1	569.1	
Old Supplier (°F)					
559.7	534.7	554.8	545.0	544.6	538.0
550.7	563.1	551.1	553.8	538.8	564.6
554.5	553.0	538.4	548.3	552.9	535.1
555.0	544.8	558.4	548.7	560.3	

Find 95% confidence intervals for σ_1^2/σ_2^2 and for σ_1/σ_2, where σ_1^2 and σ_2^2 are the population variances of the thermostat readings for the new and old suppliers, respectively.

9.98 It is argued that the resistance of wire A is greater than the resistance of wire B. An experiment on the wires shows the following results (in ohms):

Wire A	Wire B
0.140	0.135
0.138	0.140
0.143	0.136
0.142	0.142
0.144	0.138
0.137	0.140

Assuming equal variances, what conclusions do you draw? Justify your answer.

9.99 An alternative form of estimation is accomplished through the method of moments. This method involves equating the population mean and variance to the corresponding sample mean $\bar{x}$ and sample variance

Review Exercises

s^2 and solving for the parameters, the results being the **moment estimators**. In the case of a single parameter, only the means are used. Give an argument that in the case of the Poisson distribution the maximum likelihood estimator and moment estimators are the same.

9.100 Specify the moment estimators for μ and σ^2 for the normal distribution.

9.101 Specify the moment estimators for μ and σ^2 for the lognormal distribution.

9.102 Specify the moment estimators for α and β for the gamma distribution.

9.103 A survey was done with the hope of comparing salaries of chemical plant managers employed in two areas of the country, the northern and west central regions. An independent random sample of 300 plant managers was selected from each of the two regions. These managers were asked their annual salaries. The results are as follows

North	West Central
$\bar{x}_1 = \$102,300$	$\bar{x}_2 = \$98,500$
$s_1 = \$5700$	$s_2 = \$3800$

(a) Construct a 99% confidence interval for $\mu_1 - \mu_2$, the difference in the mean salaries.

(b) What assumption did you make in (a) about the distribution of annual salaries for the two regions? Is the assumption of normality necessary? Why or why not?

(c) What assumption did you make about the two variances? Is the assumption of equality of variances reasonable? Explain!

9.104 Consider Review Exercise 9.103. Let us assume that the data have not been collected yet and that previous statistics suggest that $\sigma_1 = \sigma_2 = \$4000$. Are the sample sizes in Review Exercise 9.103 sufficient to produce a 95% confidence interval on $\mu_1 - \mu_2$ having a width of only $1000? Show all work.

9.105 A labor union is becoming defensive about gross absenteeism by its members. The union leaders had always claimed that, in a typical month, 95% of its members were absent less than 10 hours. The union decided to check this by monitoring a random sample of 300 of its members. The number of hours absent was recorded for each of the 300 members. The results were $\bar{x} = 6.5$ hours and $s = 2.5$ hours. Use the data to respond to this claim, using a one-sided tolerance limit and choosing the confidence level to be 99%. Be sure to interpret what you learn from the tolerance limit calculation.

9.106 A random sample of 30 firms dealing in wireless products was selected to determine the proportion of such firms that have implemented new software to improve productivity. It turned out that 8 of the 30 had implemented such software. Find a 95% confidence interval on p, the true proportion of such firms that have implemented new software.

9.107 Refer to Review Exercise 9.106. Suppose there is concern about whether the point estimate $\hat{p} = 8/30$ is accurate enough because the confidence interval around p is not sufficiently narrow. Using $\hat{p}$ as the estimate of p, how many companies would need to be sampled in order to have a 95% confidence interval with a width of only 0.05?

9.108 A manufacturer turns out a product item that is labeled either "defective" or "not defective." In order to estimate the proportion defective, a random sample of 100 items is taken from production, and 10 are found to be defective. Following implementation of a quality improvement program, the experiment is conducted again. A new sample of 100 is taken, and this time only 6 are found to be defective.

(a) Give a 95% confidence interval on $p_1 - p_2$, where p_1 is the population proportion defective before improvement and p_2 is the proportion defective after improvement.

(b) Is there information in the confidence interval found in (a) that would suggest that $p_1 > p_2$? Explain.

9.109 A machine is used to fill boxes with product in an assembly line operation. Much concern centers around the variability in the number of ounces of product in a box. The standard deviation in weight of product is known to be 0.3 ounce. An improvement is implemented, after which a random sample of 20 boxes is selected and the sample variance is found to be 0.045 ounce^2. Find a 95% confidence interval on the variance in the weight of the product. Does it appear from the range of the confidence interval that the improvement of the process enhanced quality as far as variability is concerned? Assume normality on the distribution of weights of product.

9.110 A consumer group is interested in comparing operating costs for two different types of automobile engines. The group is able to find 15 owners whose cars have engine type A and 15 whose cars have engine type B. All 30 owners bought their cars at roughly the same time, and all have kept good records for a certain 12-month period. In addition, these owners drove roughly the same number of miles. The cost statistics are $\bar{y}_A = \$87.00/1000$ miles, $\bar{y}_B = \$75.00/1000$ miles, $s_A = \$5.99$, and $s_B = \$4.85$. Compute a 95% confidence interval to estimate $\mu_A - \mu_B$, the difference in the mean operating costs. Assume normality and equal variances.

9.111 Consider the statistic S_p^2, the pooled estimate of σ^2 discussed in Section 9.8. It is used when one is willing to assume that $\sigma_1^2 = \sigma_2^2 = \sigma^2$. Show that the estimator is unbiased for σ^2 [i.e., show that $E(S_p^2) = \sigma^2$]. You may make use of results from any theorem or example in this chapter.

9.112 A group of human factor researchers are concerned about reaction to a stimulus by airplane pilots in a certain cockpit arrangement. An experiment was conducted in a simulation laboratory, and 15 pilots were used with average reaction time of 3.2 seconds with a sample standard deviation of 0.6 second. It is of interest to characterize the extreme (i.e., worst case scenario). To that end, do the following:

(a) Give a particular important one-sided 99% confidence bound on the mean reaction time. What assumption, if any, must you make on the distribution of reaction times?

(b) Give a 99% one-sided prediction interval and give an interpretation of what it means. Must you make an assumption about the distribution of reaction times to compute this bound?

(c) Compute a one-sided tolerance bound with 99% confidence that involves 95% of reaction times. Again, give an interpretation and assumptions about the distribution, if any. (Note: The one-sided tolerance limit values are also included in Table A.7.)

9.113 A certain supplier manufactures a type of rubber mat that is sold to automotive companies. The material used to produce the mats must have certain hardness characteristics. Defective mats are occasionally discovered and rejected. The supplier claims that the proportion defective is 0.05. A challenge was made by one of the clients who purchased the mats, so an experiment was conducted in which 400 mats are tested and 17 were found defective.

(a) Compute a 95% two-sided confidence interval on the proportion defective.

(b) Compute an appropriate 95% one-sided confidence interval on the proportion defective.

(c) Interpret both intervals from (a) and (b) and comment on the claim made by the supplier.

9.15 Potential Misconceptions and Hazards; Relationship to Material in Other Chapters

The concept of a *large-sample confidence interval* on a population is often confusing to the beginning student. It is based on the notion that even when σ is unknown and one is not convinced that the distribution being sampled is normal, a confidence interval on μ can be computed from

$$\bar{x} \pm z_{\alpha/2} \frac{s}{\sqrt{n}}.$$

In practice, this formula is often used when the sample is too small. The genesis of this large sample interval is, of course, the Central Limit Theorem (CLT), under which normality is not necessary. Here the CLT requires a known σ, of which s is only an estimate. Thus, n must be at least as large as 30 and the underlying distribution must be close to symmetric, in which case the interval is still an approximation.

There are instances in which the appropriateness of the practical application of material in this chapter depends very much on the specific context. One very important illustration is the use of the t-distribution for the confidence interval on μ when σ is unknown. Strictly speaking, the use of the t-distribution requires that the distribution sampled from be normal. However, it is well known that any application of the t-distribution is reasonably insensitive (i.e., **robust**) to the normality assumption. This represents one of those fortunate situations which occur often in the field of statistics in which a basic assumption does not hold and yet "everything turns out all right!" However, one population from which the sample is drawn cannot deviate substantially from normal. Thus, the normal probability plots discussed in Chapter 8 and the goodness-of-fit tests introduced in Chapter 10 often need be called upon to ascertain some sense of "nearness to normality." This idea of "robustness to normality" will reappear in Chapter 10.

It is our experience that one of the most serious "misuses of statistics" in practice evolves from confusion about distinctions in the interpretation of the types of statistical intervals. Thus, the subsection in this chapter where differences among the three types of intervals are discussed is important. It is very likely that in practice the **confidence interval is heavily overused**. That is, it is used when there is really no interest in the mean; rather, the question is "Where is the next observation going to fall?" or often, more importantly, "Where is the large bulk of the distribution?" These are crucial questions that are not answered by computing an interval on the mean. The interpretation of a confidence interval is often misunderstood. It is tempting to conclude that the parameter falls inside the interval with probability 0.95. While this is a correct interpretation of a **Bayesian posterior interval** (readers are referred to Chapter 18 for more information on Bayesian inference), it is not the proper frequency interpretation.

A confidence interval merely suggests that if the experiment is conducted and data are observed again and again, about 95% of such intervals will contain the true parameter. Any beginning student of practical statistics should be very clear on the difference among these statistical intervals.

Another potential serious misuse of statistics centers around the use of the χ^2-distribution for a confidence interval on a single variance. Again, normality of the distribution from which the sample is drawn is assumed. Unlike the use of the t-distribution, the use of the χ^2 test for this application is **not robust to the normality assumption** (i.e., the sampling distribution of $\frac{(n-1)S^2}{\sigma^2}$ deviates far from χ^2 if the underlying distribution is not normal). Thus, strict use of goodness-of-fit (Chapter 10) tests and/or normal probability plotting can be extremely important in such contexts. More information about this general issue will be given in future chapters.

Chapter 10

One- and Two-Sample Tests of Hypotheses

10.1 Statistical Hypotheses: General Concepts

Often, the problem confronting the scientist or engineer is not so much the estimation of a population parameter, as discussed in Chapter 9, but rather the formation of a data-based decision procedure that can produce a conclusion about some scientific system. For example, a medical researcher may decide on the basis of experimental evidence whether coffee drinking increases the risk of cancer in humans; an engineer might have to decide on the basis of sample data whether there is a difference between the accuracy of two kinds of gauges; or a sociologist might wish to collect appropriate data to enable him or her to decide whether a person's blood type and eye color are independent variables. In each of these cases, the scientist or engineer *postulates* or *conjectures* something about a system. In addition, each must make use of experimental data and make a decision based on the data. In each case, the conjecture can be put in the form of a statistical hypothesis. Procedures that lead to the acceptance or rejection of statistical hypotheses such as these comprise a major area of statistical inference. First, let us define precisely what we mean by a **statistical hypothesis**.

Definition 10.1: A **statistical hypothesis** is an assertion or conjecture concerning one or more populations.

The truth or falsity of a statistical hypothesis is never known with absolute certainty unless we examine the entire population. This, of course, would be impractical in most situations. Instead, we take a random sample from the population of interest and use the data contained in this sample to provide evidence that either supports or does not support the hypothesis. Evidence from the sample that is inconsistent with the stated hypothesis leads to a rejection of the hypothesis.

The Role of Probability in Hypothesis Testing

It should be made clear to the reader that the decision procedure must include an awareness of the *probability of a wrong conclusion*. For example, suppose that the hypothesis postulated by the engineer is that the fraction defective p in a certain process is 0.10. The experiment is to observe a random sample of the product in question. Suppose that 100 items are tested and 12 items are found defective. It is reasonable to conclude that this evidence does not refute the condition that the binomial parameter $p = 0.10$, and thus it may lead one not to reject the hypothesis. However, it also does not refute $p = 0.12$ or perhaps even $p = 0.15$. As a result, the reader must be accustomed to understanding that **rejection of a hypothesis implies that the sample evidence refutes it**. Put another way, **rejection means that there is a small probability of obtaining the sample information observed when, in fact, the hypothesis is true**. For example, for our proportion-defective hypothesis, a sample of 100 revealing 20 defective items is certainly evidence for rejection. Why? If, indeed, $p = 0.10$, the probability of obtaining 20 or more defectives is approximately 0.002. With the resulting small risk of a wrong conclusion, it would seem safe to **reject the hypothesis** that $p = 0.10$. In other words, rejection of a hypothesis tends to all but "rule out" the hypothesis. On the other hand, it is very important to emphasize that acceptance or, rather, failure to reject does not rule out other possibilities. As a result, the *firm conclusion is established by the data analyst when a hypothesis is rejected.*

The formal statement of a hypothesis is often influenced by the structure of the probability of a wrong conclusion. If the scientist is interested in *strongly supporting* a contention, he or she hopes to arrive at the contention in the form of rejection of a hypothesis. If the medical researcher wishes to show strong evidence in favor of the contention that coffee drinking increases the risk of cancer, the hypothesis tested should be of the form "there is no increase in cancer risk produced by drinking coffee." As a result, the contention is reached via a rejection. Similarly, to support the claim that one kind of gauge is more accurate than another, the engineer tests the hypothesis that there is no difference in the accuracy of the two kinds of gauges.

The foregoing implies that when the data analyst formalizes experimental evidence on the basis of hypothesis testing, the formal **statement of the hypothesis** is very important.

The Null and Alternative Hypotheses

The structure of hypothesis testing will be formulated with the use of the term **null hypothesis**, which refers to any hypothesis we wish to test and is denoted by H_0. The rejection of H_0 leads to the acceptance of an **alternative hypothesis**, denoted by H_1. An understanding of the different roles played by the null hypothesis (H_0) and the alternative hypothesis (H_1) is crucial to one's understanding of the rudiments of hypothesis testing. The alternative hypothesis H_1 usually represents the *question to be answered or the theory to be tested,* and thus its specification is crucial. The null hypothesis H_0 *nullifies or opposes* H_1 and is often the logical complement to H_1. As the reader gains more understanding of hypothesis testing, he or she should note that the analyst arrives at one of the two following conclusions:

> ***reject*** H_0 in favor of H_1 because of sufficient evidence in the data or
> ***fail to reject*** H_0 because of insufficient evidence in the data.

Note that the *conclusions do not involve a formal and literal "accept* H_0*."* The statement of H_0 often represents the "status quo" in opposition to the new idea, conjecture, and so on, stated in H_1, while failure to reject H_0 represents the proper conclusion. In our binomial example, the practical issue may be a concern that the historical defective probability of 0.10 no longer is true. Indeed, the conjecture may be that p exceeds 0.10. We may then state

$$H_0: p = 0.10,$$
$$H_1: p > 0.10.$$

Now 12 defective items out of 100 does not refute $p = 0.10$, so the conclusion is "fail to reject H_0." However, if the data produce 20 out of 100 defective items, then the conclusion is "reject H_0" in favor of H_1: $p > 0.10$.

Though the applications of hypothesis testing are quite abundant in scientific and engineering work, perhaps the best illustration for a novice lies in the predicament encountered in a jury trial. The null and alternative hypotheses are

$$H_0: \text{defendant is innocent},$$
$$H_1: \text{defendant is guilty}.$$

The indictment comes because of suspicion of guilt. The hypothesis H_0 (the status quo) stands in opposition to H_1 and is maintained unless H_1 is supported by evidence "beyond a reasonable doubt." However, "failure to reject H_0" in this case does not imply innocence, but merely that the evidence was insufficient to convict. So the jury does not necessarily *accept* H_0 but *fails to reject* H_0.

10.2 Testing a Statistical Hypothesis

To illustrate the concepts used in testing a statistical hypothesis about a population, we present the following example. A certain type of cold vaccine is known to be only 25% effective after a period of 2 years. To determine if a new and somewhat more expensive vaccine is superior in providing protection against the same virus for a longer period of time, suppose that 20 people are chosen at random and inoculated. (In an actual study of this type, the participants receiving the new vaccine might number several thousand. The number 20 is being used here only to demonstrate the basic steps in carrying out a statistical test.) If more than 8 of those receiving the new vaccine surpass the 2-year period without contracting the virus, the new vaccine will be considered superior to the one presently in use. The requirement that the number exceed 8 is somewhat arbitrary but appears reasonable in that it represents a modest gain over the 5 people who could be expected to receive protection if the 20 people had been inoculated with the vaccine already in use. We are essentially testing the null hypothesis that the new vaccine is equally effective after a period of 2 years as the one now commonly used. The alternative hypothesis is that the new vaccine is in fact superior. This is equivalent to testing the hypothesis that the binomial parameter for the probability of a success on a given trial is $p = 1/4$ against the alternative that $p > 1/4$. This is usually written as follows:

$$H_0: p = 0.25,$$
$$H_1: p > 0.25.$$

The Test Statistic

The **test statistic** on which we base our decision is X, the number of individuals in our test group who receive protection from the new vaccine for a period of at least 2 years. The possible values of X, from 0 to 20, are divided into two groups: those numbers less than or equal to 8 and those greater than 8. All possible scores greater than 8 constitute the **critical region**. The last number that we observe in passing into the critical region is called the **critical value**. In our illustration, the critical value is the number 8. Therefore, if $x > 8$, we reject H_0 in favor of the alternative hypothesis H_1. If $x \leq 8$, we fail to reject H_0. This decision criterion is illustrated in Figure 10.1.

Figure 10.1: Decision criterion for testing $p = 0.25$ versus $p > 0.25$.

The Probability of a Type I Error

The decision procedure just described could lead to either of two wrong conclusions. For instance, the new vaccine may be no better than the one now in use (H_0 true) and yet, in this particular randomly selected group of individuals, more than 8 surpass the 2-year period without contracting the virus. We would be committing an error by rejecting H_0 in favor of H_1 when, in fact, H_0 is true. Such an error is called a **type I error**.

Definition 10.2: Rejection of the null hypothesis when it is true is called a **type I error**.

A second kind of error is committed if 8 or fewer of the group surpass the 2-year period successfully and we are unable to conclude that the vaccine is better when it actually is better (H_1 true). Thus, in this case, we fail to reject H_0 when in fact H_0 is false. This is called a **type II error**.

Definition 10.3: Nonrejection of the null hypothesis when it is false is called a **type II error**.

In testing any statistical hypothesis, there are four possible situations that determine whether our decision is correct or in error. These four situations are

10.2 Testing a Statistical Hypothesis

summarized in Table 10.1.

Table 10.1: Possible Situations for Testing a Statistical Hypothesis

	H_0 is true	H_0 is false
Do not reject H_0	Correct decision	Type II error
Reject H_0	Type I error	Correct decision

The probability of committing a type I error, also called the **level of significance**, is denoted by the Greek letter α. In our illustration, a type I error will occur when more than 8 individuals inoculated with the new vaccine surpass the 2-year period without contracting the virus and researchers conclude that the new vaccine is better when it is actually equivalent to the one in use. Hence, if X is the number of individuals who remain free of the virus for at least 2 years,

$$\alpha = P(\text{type I error}) = P\left(X > 8 \text{ when } p = \frac{1}{4}\right) = \sum_{x=9}^{20} b\left(x; 20, \frac{1}{4}\right)$$

$$= 1 - \sum_{x=0}^{8} b\left(x; 20, \frac{1}{4}\right) = 1 - 0.9591 = 0.0409.$$

We say that the null hypothesis, $p = 1/4$, is being tested at the $\alpha = 0.0409$ level of significance. Sometimes the level of significance is called the **size of the test**. A critical region of size 0.0409 is very small, and therefore it is unlikely that a type I error will be committed. Consequently, it would be most unusual for more than 8 individuals to remain immune to a virus for a 2-year period using a new vaccine that is essentially equivalent to the one now on the market.

The Probability of a Type II Error

The probability of committing a type II error, denoted by β, is impossible to compute unless we have a specific alternative hypothesis. If we test the null hypothesis that $p = 1/4$ against the alternative hypothesis that $p = 1/2$, then we are able to compute the probability of not rejecting H_0 when it is false. We simply find the probability of obtaining 8 or fewer in the group that surpass the 2-year period when $p = 1/2$. In this case,

$$\beta = P(\text{type II error}) = P\left(X \leq 8 \text{ when } p = \frac{1}{2}\right)$$

$$= \sum_{x=0}^{8} b\left(x; 20, \frac{1}{2}\right) = 0.2517.$$

This is a rather high probability, indicating a test procedure in which it is quite likely that we shall reject the new vaccine when, in fact, it is superior to what is now in use. Ideally, we like to use a test procedure for which the type I and type II error probabilities are both small.

It is possible that the director of the testing program is willing to make a type II error if the more expensive vaccine is not significantly superior. In fact, the only time he wishes to guard against the type II error is when the true value of p is at least 0.7. If $p = 0.7$, this test procedure gives

$$\beta = P(\text{type II error}) = P(X \leq 8 \text{ when } p = 0.7)$$

$$= \sum_{x=0}^{8} b(x; 20, 0.7) = 0.0051.$$

With such a small probability of committing a type II error, it is extremely unlikely that the new vaccine would be rejected when it was 70% effective after a period of 2 years. As the alternative hypothesis approaches unity, the value of β diminishes to zero.

The Role of α, β, and Sample Size

Let us assume that the director of the testing program is unwilling to commit a type II error when the alternative hypothesis $p = 1/2$ is true, even though we have found the probability of such an error to be $\beta = 0.2517$. It is always possible to reduce β by increasing the size of the critical region. For example, consider what happens to the values of α and β when we change our critical value to 7 so that all scores greater than 7 fall in the critical region and those less than or equal to 7 fall in the nonrejection region. Now, in testing $p = 1/4$ against the alternative hypothesis that $p = 1/2$, we find that

$$\alpha = \sum_{x=8}^{20} b\left(x; 20, \frac{1}{4}\right) = 1 - \sum_{x=0}^{7} b\left(x; 20, \frac{1}{4}\right) = 1 - 0.8982 = 0.1018$$

and

$$\beta = \sum_{x=0}^{7} b\left(x; 20, \frac{1}{2}\right) = 0.1316.$$

By adopting a new decision procedure, we have reduced the probability of committing a type II error at the expense of increasing the probability of committing a type I error. For a fixed sample size, a decrease in the probability of one error will usually result in an increase in the probability of the other error. Fortunately, **the probability of committing both types of error can be reduced by increasing the sample size**. Consider the same problem using a random sample of 100 individuals. If more than 36 of the group surpass the 2-year period, we reject the null hypothesis that $p = 1/4$ and accept the alternative hypothesis that $p > 1/4$. The critical value is now 36. All possible scores above 36 constitute the critical region, and all possible scores less than or equal to 36 fall in the acceptance region.

To determine the probability of committing a type I error, we shall use the normal curve approximation with

$$\mu = np = (100)\left(\frac{1}{4}\right) = 25 \quad \text{and} \quad \sigma = \sqrt{npq} = \sqrt{(100)(1/4)(3/4)} = 4.33.$$

Referring to Figure 10.2, we need the area under the normal curve to the right of $x = 36.5$. The corresponding z-value is

$$z = \frac{36.5 - 25}{4.33} = 2.66.$$

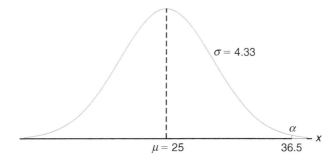

Figure 10.2: Probability of a type I error.

From Table A.3 we find that

$$\alpha = P(\text{type I error}) = P\left(X > 36 \text{ when } p = \frac{1}{4}\right) \approx P(Z > 2.66)$$
$$= 1 - P(Z < 2.66) = 1 - 0.9961 = 0.0039.$$

If H_0 is false and the true value of H_1 is $p = 1/2$, we can determine the probability of a type II error using the normal curve approximation with

$$\mu = np = (100)(1/2) = 50 \quad \text{and} \quad \sigma = \sqrt{npq} = \sqrt{(100)(1/2)(1/2)} = 5.$$

The probability of a value falling in the nonrejection region when H_0 is true is given by the area of the shaded region to the left of $x = 36.5$ in Figure 10.3. The z-value corresponding to $x = 36.5$ is

$$z = \frac{36.5 - 50}{5} = -2.7.$$

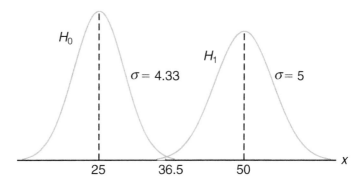

Figure 10.3: Probability of a type II error.

Therefore,

$$\beta = P(\text{type II error}) = P\left(X \leq 36 \text{ when } p = \frac{1}{2}\right) \approx P(Z < -2.7) = 0.0035.$$

Obviously, the type I and type II errors will rarely occur if the experiment consists of 100 individuals.

The illustration above underscores the strategy of the scientist in hypothesis testing. After the null and alternative hypotheses are stated, it is important to consider the sensitivity of the test procedure. By this we mean that there should be a determination, for a fixed α, of a reasonable value for the probability of wrongly accepting H_0 (i.e., the value of β) when the true situation represents some *important deviation from* H_0. A value for the sample size can usually be determined for which there is a reasonable balance between the values of α and β computed in this fashion. The vaccine problem provides an illustration.

Illustration with a Continuous Random Variable

The concepts discussed here for a discrete population can be applied equally well to continuous random variables. Consider the null hypothesis that the average weight of male students in a certain college is 68 kilograms against the alternative hypothesis that it is unequal to 68. That is, we wish to test

$$H_0: \mu = 68,$$
$$H_1: \mu \neq 68.$$

The alternative hypothesis allows for the possibility that $\mu < 68$ or $\mu > 68$.

A sample mean that falls close to the hypothesized value of 68 would be considered evidence in favor of H_0. On the other hand, a sample mean that is considerably less than or more than 68 would be evidence inconsistent with H_0 and therefore favoring H_1. The sample mean is the test statistic in this case. A critical region for the test statistic might arbitrarily be chosen to be the two intervals $\bar{x} < 67$ and $\bar{x} > 69$. The nonrejection region will then be the interval $67 \leq \bar{x} \leq 69$. This decision criterion is illustrated in Figure 10.4.

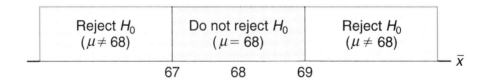

Figure 10.4: Critical region (in blue).

Let us now use the decision criterion of Figure 10.4 to calculate the probabilities of committing type I and type II errors when testing the null hypothesis that $\mu = 68$ kilograms against the alternative that $\mu \neq 68$ kilograms.

Assume the standard deviation of the population of weights to be $\sigma = 3.6$. For large samples, we may substitute s for σ if no other estimate of σ is available. Our decision statistic, based on a random sample of size $n = 36$, will be $\bar{X}$, the most efficient estimator of μ. From the Central Limit Theorem, we know that the sampling distribution of $\bar{X}$ is approximately normal with standard deviation $\sigma_{\bar{X}} = \sigma/\sqrt{n} = 3.6/6 = 0.6$.

10.2 Testing a Statistical Hypothesis

The probability of committing a type I error, or the level of significance of our test, is equal to the sum of the areas that have been shaded in each tail of the distribution in Figure 10.5. Therefore,

$$\alpha = P(\bar{X} < 67 \text{ when } \mu = 68) + P(\bar{X} > 69 \text{ when } \mu = 68).$$

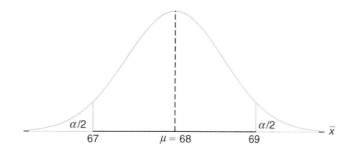

Figure 10.5: Critical region for testing $\mu = 68$ versus $\mu \neq 68$.

The z-values corresponding to $\bar{x}_1 = 67$ and $\bar{x}_2 = 69$ when H_0 is true are

$$z_1 = \frac{67 - 68}{0.6} = -1.67 \quad \text{and} \quad z_2 = \frac{69 - 68}{0.6} = 1.67.$$

Therefore,

$$\alpha = P(Z < -1.67) + P(Z > 1.67) = 2P(Z < -1.67) = 0.0950.$$

Thus, 9.5% of all samples of size 36 would lead us to reject $\mu = 68$ kilograms when, in fact, it is true. To reduce α, we have a choice of increasing the sample size or widening the fail-to-reject region. Suppose that we increase the sample size to $n = 64$. Then $\sigma_{\bar{X}} = 3.6/8 = 0.45$. Now

$$z_1 = \frac{67 - 68}{0.45} = -2.22 \quad \text{and} \quad z_2 = \frac{69 - 68}{0.45} = 2.22.$$

Hence,

$$\alpha = P(Z < -2.22) + P(Z > 2.22) = 2P(Z < -2.22) = 0.0264.$$

The reduction in α is not sufficient by itself to guarantee a good testing procedure. We must also evaluate β for various alternative hypotheses. If it is important to reject H_0 when the true mean is some value $\mu \geq 70$ or $\mu \leq 66$, then the probability of committing a type II error should be computed and examined for the alternatives $\mu = 66$ and $\mu = 70$. Because of symmetry, it is only necessary to consider the probability of not rejecting the null hypothesis that $\mu = 68$ when the alternative $\mu = 70$ is true. A type II error will result when the sample mean $\bar{x}$ falls between 67 and 69 when H_1 is true. Therefore, referring to Figure 10.6, we find that

$$\beta = P(67 \leq \bar{X} \leq 69 \text{ when } \mu = 70).$$

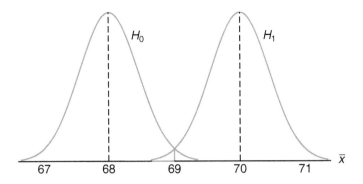

Figure 10.6: Probability of type II error for testing $\mu = 68$ versus $\mu = 70$.

The z-values corresponding to $\bar{x}_1 = 67$ and $\bar{x}_2 = 69$ when H_1 is true are

$$z_1 = \frac{67 - 70}{0.45} = -6.67 \quad \text{and} \quad z_2 = \frac{69 - 70}{0.45} = -2.22.$$

Therefore,

$$\beta = P(-6.67 < Z < -2.22) = P(Z < -2.22) - P(Z < -6.67)$$
$$= 0.0132 - 0.0000 = 0.0132.$$

If the true value of μ is the alternative $\mu = 66$, the value of β will again be 0.0132. For all possible values of $\mu < 66$ or $\mu > 70$, the value of β will be even smaller when $n = 64$, and consequently there would be little chance of not rejecting H_0 when it is false.

The probability of committing a type II error increases rapidly when the true value of μ approaches, but is not equal to, the hypothesized value. Of course, this is usually the situation where we do not mind making a type II error. For example, if the alternative hypothesis $\mu = 68.5$ is true, we do not mind committing a type II error by concluding that the true answer is $\mu = 68$. The probability of making such an error will be high when $n = 64$. Referring to Figure 10.7, we have

$$\beta = P(67 \leq \bar{X} \leq 69 \text{ when } \mu = 68.5).$$

The z-values corresponding to $\bar{x}_1 = 67$ and $\bar{x}_2 = 69$ when $\mu = 68.5$ are

$$z_1 = \frac{67 - 68.5}{0.45} = -3.33 \quad \text{and} \quad z_2 = \frac{69 - 68.5}{0.45} = 1.11.$$

Therefore,

$$\beta = P(-3.33 < Z < 1.11) = P(Z < 1.11) - P(Z < -3.33)$$
$$= 0.8665 - 0.0004 = 0.8661.$$

The preceding examples illustrate the following important properties:

10.2 Testing a Statistical Hypothesis

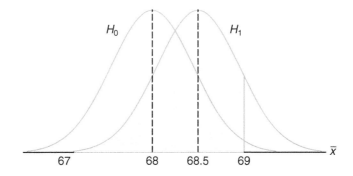

Figure 10.7: Type II error for testing $\mu = 68$ versus $\mu = 68.5$.

Important Properties of a Test of Hypothesis

1. The type I error and type II error are related. A decrease in the probability of one generally results in an increase in the probability of the other.

2. The size of the critical region, and therefore the probability of committing a type I error, can always be reduced by adjusting the critical value(s).

3. An increase in the sample size n will reduce α and β simultaneously.

4. If the null hypothesis is false, β is a maximum when the true value of a parameter approaches the hypothesized value. The greater the distance between the true value and the hypothesized value, the smaller β will be.

One very important concept that relates to error probabilities is the notion of the **power** of a test.

Definition 10.4: The **power** of a test is the probability of rejecting H_0 given that a specific alternative is true.

The power of a test can be computed as $1 - \beta$. Often **different types of tests are compared by contrasting power properties**. Consider the previous illustration, in which we were testing $H_0\colon \mu = 68$ and $H_1\colon \mu \neq 68$. As before, suppose we are interested in assessing the sensitivity of the test. The test is governed by the rule that we do not reject H_0 if $67 \leq \bar{x} \leq 69$. We seek the capability of the test to properly reject H_0 when indeed $\mu = 68.5$. We have seen that the probability of a type II error is given by $\beta = 0.8661$. Thus, the **power** of the test is $1 - 0.8661 = 0.1339$. In a sense, the power is a more succinct measure of how sensitive the test is for detecting differences between a mean of 68 and a mean of 68.5. In this case, if μ is truly 68.5, the test as described will *properly reject H_0 only 13.39% of the time*. As a result, the test would not be a good one if it was important that the analyst have a reasonable chance of truly distinguishing between a mean of 68.0 (specified by H_0) and a mean of 68.5. From the foregoing, it is clear that to produce a desirable power (say, greater than 0.8), one must either increase α or increase the sample size.

So far in this chapter, much of the discussion of hypothesis testing has focused on foundations and definitions. In the sections that follow, we get more specific and put hypotheses in categories as well as discuss tests of hypotheses on various parameters of interest. We begin by drawing the distinction between a one-sided and a two-sided hypothesis.

One- and Two-Tailed Tests

A test of any statistical hypothesis where the alternative is **one sided**, such as

$$H_0: \theta = \theta_0,$$
$$H_1: \theta > \theta_0$$

or perhaps

$$H_0: \theta = \theta_0,$$
$$H_1: \theta < \theta_0,$$

is called a **one-tailed test**. Earlier in this section, we referred to the **test statistic** for a hypothesis. Generally, the critical region for the alternative hypothesis $\theta > \theta_0$ lies in the right tail of the distribution of the test statistic, while the critical region for the alternative hypothesis $\theta < \theta_0$ lies entirely in the left tail. (In a sense, the inequality symbol points in the direction of the critical region.) A one-tailed test was used in the vaccine experiment to test the hypothesis $p = 1/4$ against the one-sided alternative $p > 1/4$ for the binomial distribution. The one-tailed critical region is usually obvious; the reader should visualize the behavior of the test statistic and notice the obvious *signal* that would produce evidence supporting the alternative hypothesis.

A test of any statistical hypothesis where the alternative is **two sided**, such as

$$H_0: \theta = \theta_0,$$
$$H_1: \theta \neq \theta_0,$$

is called a **two-tailed test**, since the critical region is split into two parts, often having equal probabilities, in each tail of the distribution of the test statistic. The alternative hypothesis $\theta \neq \theta_0$ states that either $\theta < \theta_0$ or $\theta > \theta_0$. A two-tailed test was used to test the null hypothesis that $\mu = 68$ kilograms against the two-sided alternative $\mu \neq 68$ kilograms in the example of the continuous population of student weights.

How Are the Null and Alternative Hypotheses Chosen?

The null hypothesis H_0 will often be stated using the *equality sign*. With this approach, it is clear how the probability of type I error is controlled. However, there are situations in which "do not reject H_0" implies that the parameter θ might be any value defined by the natural complement to the alternative hypothesis. For example, in the vaccine example, where the alternative hypothesis is $H_1: p > 1/4$, it is quite possible that nonrejection of H_0 cannot rule out a value of p less than $1/4$. Clearly though, in the case of one-tailed tests, the statement of the alternative is the most important consideration.

Whether one sets up a one-tailed or a two-tailed test will depend on the conclusion to be drawn if H_0 is rejected. The location of the critical region can be determined only after H_1 has been stated. For example, in testing a new drug, one sets up the hypothesis that it is no better than similar drugs now on the market and tests this against the alternative hypothesis that the new drug is superior. Such an alternative hypothesis will result in a one-tailed test with the critical region in the right tail. However, if we wish to compare a new teaching technique with the conventional classroom procedure, the alternative hypothesis should allow for the new approach to be either inferior or superior to the conventional procedure. Hence, the test is two-tailed with the critical region divided equally so as to fall in the extreme left and right tails of the distribution of our statistic.

Example 10.1: A manufacturer of a certain brand of rice cereal claims that the average saturated fat content does not exceed 1.5 grams per serving. State the null and alternative hypotheses to be used in testing this claim and determine where the critical region is located.

Solution: The manufacturer's claim should be rejected only if μ is greater than 1.5 milligrams and should not be rejected if μ is less than or equal to 1.5 milligrams. We test

$$H_0: \mu = 1.5,$$
$$H_1: \mu > 1.5.$$

Nonrejection of H_0 does not rule out values less than 1.5 milligrams. Since we have a one-tailed test, the greater than symbol indicates that the critical region lies entirely in the right tail of the distribution of our test statistic $\bar{X}$.

Example 10.2: A real estate agent claims that 60% of all private residences being built today are 3-bedroom homes. To test this claim, a large sample of new residences is inspected; the proportion of these homes with 3 bedrooms is recorded and used as the test statistic. State the null and alternative hypotheses to be used in this test and determine the location of the critical region.

Solution: If the test statistic were substantially higher or lower than $p = 0.6$, we would reject the agent's claim. Hence, we should make the hypothesis

$$H_0: p = 0.6,$$
$$H_1: p \neq 0.6.$$

The alternative hypothesis implies a two-tailed test with the critical region divided equally in both tails of the distribution of $\widehat{P}$, our test statistic.

10.3 The Use of P-Values for Decision Making in Testing Hypotheses

In testing hypotheses in which the test statistic is discrete, the critical region may be chosen arbitrarily and its size determined. If α is too large, it can be reduced by making an adjustment in the critical value. It may be necessary to increase the sample size to offset the decrease that occurs automatically in the power of the test.

Over a number of generations of statistical analysis, it had become customary to choose an α of 0.05 or 0.01 and select the critical region accordingly. Then, of course, strict rejection or nonrejection of H_0 would depend on that critical region. For example, if the test is two tailed and α is set at the 0.05 level of significance and the test statistic involves, say, the standard normal distribution, then a z-value is observed from the data and the critical region is

$$z > 1.96 \quad \text{or} \quad z < -1.96,$$

where the value 1.96 is found as $z_{0.025}$ in Table A.3. A value of z in the critical region prompts the statement "The value of the test statistic is significant," which we can then translate into the user's language. For example, if the hypothesis is given by

$$H_0: \mu = 10,$$
$$H_1: \mu \neq 10,$$

one might say, "The mean differs significantly from the value 10."

Preselection of a Significance Level

This preselection of a significance level α has its roots in the philosophy that the maximum risk of making a type I error should be controlled. However, this approach does not account for values of test statistics that are "close" to the critical region. Suppose, for example, in the illustration with $H_0: \mu = 10$ versus $H_1: \mu \neq 10$, a value of $z = 1.87$ is observed; strictly speaking, with $\alpha = 0.05$, the value is not significant. But the risk of committing a type I error if one rejects H_0 in this case could hardly be considered severe. In fact, in a two-tailed scenario, one can quantify this risk as

$$P = 2P(Z > 1.87 \text{ when } \mu = 10) = 2(0.0307) = 0.0614.$$

As a result, 0.0614 is the probability of obtaining a value of z as large as or larger (in magnitude) than 1.87 when in fact $\mu = 10$. Although this evidence against H_0 is not as strong as that which would result from rejection at an $\alpha = 0.05$ level, it is important information to the user. Indeed, continued use of $\alpha = 0.05$ or 0.01 is only a result of what standards have been passed down through the generations. **The P-value approach has been adopted extensively by users of applied statistics.** The approach is designed to give the user an alternative (in terms of a probability) to a mere "reject" or "do not reject" conclusion. The P-value computation also gives the user important information when the z-value falls well *into the ordinary critical region*. For example, if z is 2.73, it is informative for the user to observe that

$$P = 2(0.0032) = 0.0064,$$

and thus the z-value is significant at a level considerably less than 0.05. It is important to know that under the condition of H_0, a value of $z = 2.73$ is an extremely rare event. That is, a value at least that large in magnitude would only occur 64 times in 10,000 experiments.

10.3 The Use of P-Values for Decision Making in Testing Hypotheses

A Graphical Demonstration of a P-Value

One very simple way of explaining a P-value graphically is to consider two distinct samples. Suppose that two materials are being considered for coating a particular type of metal in order to inhibit corrosion. Specimens are obtained, and one collection is coated with material 1 and one collection coated with material 2. The sample sizes are $n_1 = n_2 = 10$, and corrosion is measured in percent of surface area affected. The hypothesis is that the samples came from common distributions with mean $\mu = 10$. Let us assume that the population variance is 1.0. Then we are testing

$$H_0: \mu_1 = \mu_2 = 10.$$

Let Figure 10.8 represent a point plot of the data; the data are placed on the distribution stated by the null hypothesis. Let us assume that the "×" data refer to material 1 and the "∘" data refer to material 2. Now it seems clear that the data do refute the null hypothesis. But how can this be summarized in one number? **The P-value can be viewed as simply the probability of obtaining these data given that both samples come from the same distribution.** Clearly, this probability is quite small, say 0.00000001! Thus, the small P-value clearly refutes H_0, and the conclusion is that the population means are significantly different.

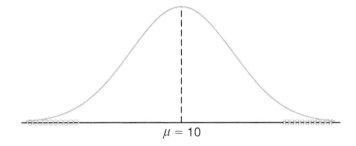

Figure 10.8: Data that are likely generated from populations having two different means.

Use of the P-value approach as an aid in decision-making is quite natural, and nearly all computer packages that provide hypothesis-testing computation print out P-values along with values of the appropriate test statistic. The following is a formal definition of a P-value.

Definition 10.5: A **P-value** is the lowest level (of significance) at which the observed value of the test statistic is significant.

How Does the Use of P-Values Differ from Classic Hypothesis Testing?

It is tempting at this point to summarize the procedures associated with testing, say, $H_0: \theta = \theta_0$. However, the student who is a novice in this area should understand that there are differences in approach and philosophy between the classic fixed α approach that is climaxed with either a "reject H_0" or a "do not reject H_0" conclusion and the P-value approach. In the latter, no fixed α is determined and conclusions are drawn on the basis of the size of the P-value in harmony with the subjective judgment of the engineer or scientist. While modern computer software will output P-values, nevertheless it is important that readers understand both approaches in order to appreciate the totality of the concepts. Thus, we offer a brief list of procedural steps for both the classical and the P-value approach.

Approach to Hypothesis Testing with Fixed Probability of Type I Error
1. State the null and alternative hypotheses.
2. Choose a fixed significance level α.
3. Choose an appropriate test statistic and establish the critical region based on α.
4. Reject H_0 if the computed test statistic is in the critical region. Otherwise, do not reject.
5. Draw scientific or engineering conclusions.

Significance Testing (P-Value Approach)
1. State null and alternative hypotheses.
2. Choose an appropriate test statistic.
3. Compute the P-value based on the computed value of the test statistic.
4. Use judgment based on the P-value and knowledge of the scientific system.

In later sections of this chapter and chapters that follow, many examples and exercises emphasize the P-value approach to drawing scientific conclusions.

Exercises

10.1 Suppose that an allergist wishes to test the hypothesis that at least 30% of the public is allergic to some cheese products. Explain how the allergist could commit

(a) a type I error;

(b) a type II error.

10.2 A sociologist is concerned about the effectiveness of a training course designed to get more drivers to use seat belts in automobiles.

(a) What hypothesis is she testing if she commits a type I error by erroneously concluding that the training course is ineffective?

(b) What hypothesis is she testing if she commits a type II error by erroneously concluding that the training course is effective?

10.3 A large manufacturing firm is being charged with discrimination in its hiring practices.

(a) What hypothesis is being tested if a jury commits a type I error by finding the firm guilty?

(b) What hypothesis is being tested if a jury commits a type II error by finding the firm guilty?

10.4 A fabric manufacturer believes that the proportion of orders for raw material arriving late is $p = 0.6$. If a random sample of 10 orders shows that 3 or fewer arrived late, the hypothesis that $p = 0.6$ should be rejected in favor of the alternative $p < 0.6$. Use the binomial distribution.

(a) Find the probability of committing a type I error if the true proportion is $p = 0.6$.

(b) Find the probability of committing a type II error for the alternatives $p = 0.3$, $p = 0.4$, and $p = 0.5$.

10.5 Repeat Exercise 10.4 but assume that 50 orders are selected and the critical region is defined to be $x \leq 24$, where x is the number of orders in the sample that arrived late. Use the normal approximation.

10.6 The proportion of adults living in a small town who are college graduates is estimated to be $p = 0.6$. To test this hypothesis, a random sample of 15 adults is selected. If the number of college graduates in the sample is anywhere from 6 to 12, we shall not reject the null hypothesis that $p = 0.6$; otherwise, we shall conclude that $p \neq 0.6$.

(a) Evaluate α assuming that $p = 0.6$. Use the binomial distribution.

(b) Evaluate β for the alternatives $p = 0.5$ and $p = 0.7$.
(c) Is this a good test procedure?

10.7 Repeat Exercise 10.6 but assume that 200 adults are selected and the fail-to-reject region is defined to be $110 \leq x \leq 130$, where x is the number of college graduates in our sample. Use the normal approximation.

10.8 In *Relief from Arthritis* published by Thorsons Publishers, Ltd., John E. Croft claims that over 40% of those who suffer from osteoarthritis receive measurable relief from an ingredient produced by a particular species of mussel found off the coast of New Zealand. To test this claim, the mussel extract is to be given to a group of 7 osteoarthritic patients. If 3 or more of the patients receive relief, we shall not reject the null hypothesis that $p = 0.4$; otherwise, we conclude that $p < 0.4$.
(a) Evaluate α, assuming that $p = 0.4$.
(b) Evaluate β for the alternative $p = 0.3$.

10.9 A dry cleaning establishment claims that a new spot remover will remove more than 70% of the spots to which it is applied. To check this claim, the spot remover will be used on 12 spots chosen at random. If fewer than 11 of the spots are removed, we shall not reject the null hypothesis that $p = 0.7$; otherwise, we conclude that $p > 0.7$.
(a) Evaluate α, assuming that $p = 0.7$.
(b) Evaluate β for the alternative $p = 0.9$.

10.10 Repeat Exercise 10.9 but assume that 100 spots are treated and the critical region is defined to be $x > 82$, where x is the number of spots removed.

10.11 Repeat Exercise 10.8 but assume that 70 patients are given the mussel extract and the critical region is defined to be $x < 24$, where x is the number of osteoarthritic patients who receive relief.

10.12 A random sample of 400 voters in a certain city are asked if they favor an additional 4% gasoline sales tax to provide badly needed revenues for street repairs. If more than 220 but fewer than 260 favor the sales tax, we shall conclude that 60% of the voters are for it.
(a) Find the probability of committing a type I error if 60% of the voters favor the increased tax.
(b) What is the probability of committing a type II error using this test procedure if actually only 48% of the voters are in favor of the additional gasoline tax?

10.13 Suppose, in Exercise 10.12, we conclude that 60% of the voters favor the gasoline sales tax if more than 214 but fewer than 266 voters in our sample favor it. Show that this new critical region results in a smaller value for α at the expense of increasing β.

10.14 A manufacturer has developed a new fishing line, which the company claims has a mean breaking strength of 15 kilograms with a standard deviation of 0.5 kilogram. To test the hypothesis that $\mu = 15$ kilograms against the alternative that $\mu < 15$ kilograms, a random sample of 50 lines will be tested. The critical region is defined to be $\bar{x} < 14.9$.
(a) Find the probability of committing a type I error when H_0 is true.
(b) Evaluate β for the alternatives $\mu = 14.8$ and $\mu = 14.9$ kilograms.

10.15 A soft-drink machine at a steak house is regulated so that the amount of drink dispensed is approximately normally distributed with a mean of 200 milliliters and a standard deviation of 15 milliliters. The machine is checked periodically by taking a sample of 9 drinks and computing the average content. If $\bar{x}$ falls in the interval $191 < \bar{x} < 209$, the machine is thought to be operating satisfactorily; otherwise, we conclude that $\mu \neq 200$ milliliters.
(a) Find the probability of committing a type I error when $\mu = 200$ milliliters.
(b) Find the probability of committing a type II error when $\mu = 215$ milliliters.

10.16 Repeat Exercise 10.15 for samples of size $n = 25$. Use the same critical region.

10.17 A new curing process developed for a certain type of cement results in a mean compressive strength of 5000 kilograms per square centimeter with a standard deviation of 120 kilograms. To test the hypothesis that $\mu = 5000$ against the alternative that $\mu < 5000$, a random sample of 50 pieces of cement is tested. The critical region is defined to be $\bar{x} < 4970$.
(a) Find the probability of committing a type I error when H_0 is true.
(b) Evaluate β for the alternatives $\mu = 4970$ and $\mu = 4960$.

10.18 If we plot the probabilities of failing to reject H_0 corresponding to various alternatives for μ (including the value specified by H_0) and connect all the points by a smooth curve, we obtain the **operating characteristic curve** of the test criterion, or simply the OC curve. Note that the probability of failing to reject H_0 when it is true is simply $1 - \alpha$. Operating characteristic curves are widely used in industrial applications to provide a visual display of the merits of the test criterion. With reference to Exercise 10.15, find the probabilities of failing to reject H_0 for the following 9 values of μ and plot the OC curve: 184, 188, 192, 196, 200, 204, 208, 212, and 216.

10.4 Single Sample: Tests Concerning a Single Mean

In this section, we formally consider tests of hypotheses on a single population mean. Many of the illustrations from previous sections involved tests on the mean, so the reader should already have insight into some of the details that are outlined here.

Tests on a Single Mean (Variance Known)

We should first describe the assumptions on which the experiment is based. The model for the underlying situation centers around an experiment with $X_1, X_2, \ldots, X_n$ representing a random sample from a distribution with mean μ and variance $\sigma^2 > 0$. Consider first the hypothesis

$$H_0: \mu = \mu_0,$$
$$H_1: \mu \neq \mu_0.$$

The appropriate test statistic should be based on the random variable $\bar{X}$. In Chapter 8, the Central Limit Theorem was introduced, which essentially states that despite the distribution of X, the random variable $\bar{X}$ has approximately a normal distribution with mean μ and variance σ^2/n for reasonably large sample sizes. So, $\mu_{\bar{X}} = \mu$ and $\sigma^2_{\bar{X}} = \sigma^2/n$. We can then determine a critical region based on the computed sample average, $\bar{x}$. It should be clear to the reader by now that there will be a two-tailed critical region for the test.

Standardization of $\bar{X}$

It is convenient to standardize $\bar{X}$ and formally involve the **standard normal** random variable Z, where

$$Z = \frac{\bar{X} - \mu}{\sigma/\sqrt{n}}.$$

We know that *under H_0*, that is, if $\mu = \mu_0$, $\sqrt{n}(\bar{X} - \mu_0)/\sigma$ follows an $n(x; 0, 1)$ distribution, and hence the expression

$$P\left(-z_{\alpha/2} < \frac{\bar{X} - \mu_0}{\sigma/\sqrt{n}} < z_{\alpha/2}\right) = 1 - \alpha$$

can be used to write an appropriate nonrejection region. The reader should keep in mind that, formally, the critical region is designed to control α, the probability of type I error. It should be obvious that a *two-tailed signal* of evidence is needed to support H_1. Thus, given a computed value $\bar{x}$, the formal test involves rejecting H_0 if the computed *test statistic z* falls in the critical region described next.

10.4 Single Sample: Tests Concerning a Single Mean

Test Procedure for a Single Mean (Variance Known)

$$z = \frac{\bar{x} - \mu_0}{\sigma/\sqrt{n}} > z_{\alpha/2} \quad \text{or} \quad z = \frac{\bar{x} - \mu_0}{\sigma/\sqrt{n}} < -z_{\alpha/2}$$

If $-z_{\alpha/2} < z < z_{\alpha/2}$, do not reject H_0. Rejection of H_0, of course, implies acceptance of the alternative hypothesis $\mu \neq \mu_0$. With this definition of the critical region, it should be clear that there will be probability α of rejecting H_0 (falling into the critical region) when, indeed, $\mu = \mu_0$.

Although it is easier to understand the critical region written in terms of z, we can write the same critical region in terms of the computed average $\bar{x}$. The following can be written as an identical decision procedure:

$$\text{reject } H_0 \text{ if } \bar{x} < a \text{ or } \bar{x} > b,$$

where

$$a = \mu_0 - z_{\alpha/2}\frac{\sigma}{\sqrt{n}}, \quad b = \mu_0 + z_{\alpha/2}\frac{\sigma}{\sqrt{n}}.$$

Hence, for a significance level α, the critical values of the random variable z and $\bar{x}$ are both depicted in Figure 10.9.

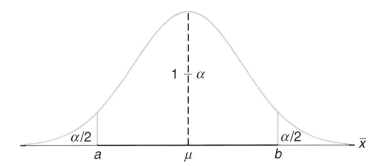

Figure 10.9: Critical region for the alternative hypothesis $\mu \neq \mu_0$.

Tests of one-sided hypotheses on the mean involve the same statistic described in the two-sided case. The difference, of course, is that the critical region is only in one tail of the standard normal distribution. For example, suppose that we seek to test

$$H_0: \mu = \mu_0,$$
$$H_1: \mu > \mu_0.$$

The signal that favors H_1 comes from *large values* of z. Thus, rejection of H_0 results when the computed $z > z_\alpha$. Obviously, if the alternative is $H_1: \mu < \mu_0$, the critical region is entirely in the lower tail and thus rejection results from $z < -z_\alpha$. Although in a one-sided testing case the null hypothesis can be written as $H_0: \mu \leq \mu_0$ or $H_0: \mu \geq \mu_0$, it is usually written as $H_0: \mu = \mu_0$.

The following two examples illustrate tests on means for the case in which σ is known.

Example 10.3: A random sample of 100 recorded deaths in the United States during the past year showed an average life span of 71.8 years. Assuming a population standard deviation of 8.9 years, does this seem to indicate that the mean life span today is greater than 70 years? Use a 0.05 level of significance.

Solution:
1. H_0: $\mu = 70$ years.
2. H_1: $\mu > 70$ years.
3. $\alpha = 0.05$.
4. Critical region: $z > 1.645$, where $z = \frac{\bar{x} - \mu_0}{\sigma/\sqrt{n}}$.
5. Computations: $\bar{x} = 71.8$ years, $\sigma = 8.9$ years, and hence $z = \frac{71.8 - 70}{8.9/\sqrt{100}} = 2.02$.
6. Decision: Reject H_0 and conclude that the mean life span today is greater than 70 years.

The P-value corresponding to $z = 2.02$ is given by the area of the shaded region in Figure 10.10.
Using Table A.3, we have

$$P = P(Z > 2.02) = 0.0217.$$

As a result, the evidence in favor of H_1 is even stronger than that suggested by a 0.05 level of significance.

Example 10.4: A manufacturer of sports equipment has developed a new synthetic fishing line that the company claims has a mean breaking strength of 8 kilograms with a standard deviation of 0.5 kilogram. Test the hypothesis that $\mu = 8$ kilograms against the alternative that $\mu \neq 8$ kilograms if a random sample of 50 lines is tested and found to have a mean breaking strength of 7.8 kilograms. Use a 0.01 level of significance.

Solution:
1. H_0: $\mu = 8$ kilograms.
2. H_1: $\mu \neq 8$ kilograms.
3. $\alpha = 0.01$.
4. Critical region: $z < -2.575$ and $z > 2.575$, where $z = \frac{\bar{x} - \mu_0}{\sigma/\sqrt{n}}$.
5. Computations: $\bar{x} = 7.8$ kilograms, $n = 50$, and hence $z = \frac{7.8 - 8}{0.5/\sqrt{50}} = -2.83$.
6. Decision: Reject H_0 and conclude that the average breaking strength is not equal to 8 but is, in fact, less than 8 kilograms.

Since the test in this example is two tailed, the desired P-value is twice the area of the shaded region in Figure 10.11 to the left of $z = -2.83$. Therefore, using Table A.3, we have

$$P = P(|Z| > 2.83) = 2P(Z < -2.83) = 0.0046,$$

which allows us to reject the null hypothesis that $\mu = 8$ kilograms at a level of significance smaller than 0.01.

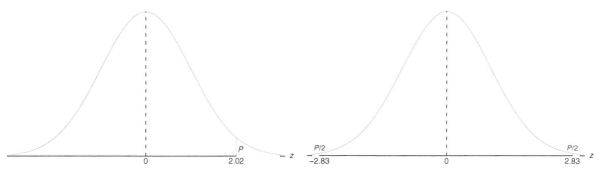

Figure 10.10: *P*-value for Example 10.3. Figure 10.11: *P*-value for Example 10.4.

Relationship to Confidence Interval Estimation

The reader should realize by now that the hypothesis-testing approach to statistical inference in this chapter is very closely related to the confidence interval approach in Chapter 9. Confidence interval estimation involves computation of bounds within which it is "reasonable" for the parameter in question to lie. For the case of a single population mean μ with σ^2 known, the structure of both hypothesis testing and confidence interval estimation is based on the random variable

$$Z = \frac{\bar{X} - \mu}{\sigma/\sqrt{n}}.$$

It turns out that the testing of H_0: $\mu = \mu_0$ against H_1: $\mu \neq \mu_0$ at a significance level α is equivalent to computing a $100(1-\alpha)\%$ confidence interval on μ and rejecting H_0 if μ_0 is outside the confidence interval. If μ_0 is inside the confidence interval, the hypothesis is not rejected. The equivalence is very intuitive and quite simple to illustrate. Recall that with an observed value $\bar{x}$, failure to reject H_0 at significance level α implies that

$$-z_{\alpha/2} \leq \frac{\bar{x} - \mu_0}{\sigma/\sqrt{n}} \leq z_{\alpha/2},$$

which is equivalent to

$$\bar{x} - z_{\alpha/2}\frac{\sigma}{\sqrt{n}} \leq \mu_0 \leq \bar{x} + z_{\alpha/2}\frac{\sigma}{\sqrt{n}}.$$

The equivalence of confidence interval estimation to hypothesis testing extends to differences between two means, variances, ratios of variances, and so on. As a result, the student of statistics should not consider confidence interval estimation and hypothesis testing as separate forms of statistical inference. For example, consider Example 9.2 on page 271. The 95% confidence interval on the mean is given by the bounds (2.50, 2.70). Thus, with the same sample information, a two-sided hypothesis on μ involving any hypothesized value between 2.50 and 2.70 will not be rejected. As we turn to different areas of hypothesis testing, the equivalence to the confidence interval estimation will continue to be exploited.

Tests on a Single Sample (Variance Unknown)

One would certainly suspect that tests on a population mean μ with σ^2 unknown, like confidence interval estimation, should involve the use of Student t-distribution. Strictly speaking, the application of Student t for both confidence intervals and hypothesis testing is developed under the following assumptions. The random variables $X_1, X_2, \ldots, X_n$ represent a random sample from a normal distribution with unknown μ and σ^2. Then the random variable $\sqrt{n}(\bar{X} - \mu)/S$ has a Student t-distribution with $n-1$ degrees of freedom. The structure of the test is identical to that for the case of σ known, with the exception that the value σ in the test statistic is replaced by the computed estimate S and the standard normal distribution is replaced by a t-distribution.

The t-Statistic for a Test on a Single Mean (Variance Unknown)

For the two-sided hypothesis

$$H_0: \mu = \mu_0,$$
$$H_1: \mu \neq \mu_0,$$

we reject H_0 at significance level α when the computed t-statistic

$$t = \frac{\bar{x} - \mu_0}{s/\sqrt{n}}$$

exceeds $t_{\alpha/2, n-1}$ or is less than $-t_{\alpha/2, n-1}$.

The reader should recall from Chapters 8 and 9 that the t-distribution is symmetric around the value zero. Thus, this two-tailed critical region applies in a fashion similar to that for the case of known σ. For the two-sided hypothesis at significance level α, the two-tailed critical regions apply. For $H_1: \mu > \mu_0$, rejection results when $t > t_{\alpha, n-1}$. For $H_1: \mu < \mu_0$, the critical region is given by $t < -t_{\alpha, n-1}$.

Example 10.5: The Edison Electric Institute has published figures on the number of kilowatt hours used annually by various home appliances. It is claimed that a vacuum cleaner uses an average of 46 kilowatt hours per year. If a random sample of 12 homes included in a planned study indicates that vacuum cleaners use an average of 42 kilowatt hours per year with a standard deviation of 11.9 kilowatt hours, does this suggest at the 0.05 level of significance that vacuum cleaners use, on average, less than 46 kilowatt hours annually? Assume the population of kilowatt hours to be normal.

Solution: 1. $H_0: \mu = 46$ kilowatt hours.

2. $H_1: \mu < 46$ kilowatt hours.

3. $\alpha = 0.05$.

4. Critical region: $t < -1.796$, where $t = \frac{\bar{x} - \mu_0}{s/\sqrt{n}}$ with 11 degrees of freedom.

5. Computations: $\bar{x} = 42$ kilowatt hours, $s = 11.9$ kilowatt hours, and $n = 12$. Hence,

$$t = \frac{42 - 46}{11.9/\sqrt{12}} = -1.16, \qquad P = P(T < -1.16) \approx 0.135.$$

6. **Decision:** Do not reject H_0 and conclude that the average number of kilowatt hours used annually by home vacuum cleaners is not significantly less than 46.

Comment on the Single-Sample t-Test

The reader has probably noticed that the equivalence of the two-tailed t-test for a single mean and the computation of a confidence interval on μ with σ replaced by s is maintained. For example, consider Example 9.5 on page 275. Essentially, we can view that computation as one in which we have found all values of μ_0, the hypothesized mean volume of containers of sulfuric acid, for which the hypothesis $H_0: \mu = \mu_0$ *will not be rejected* at $\alpha = 0.05$. Again, this is consistent with the statement "Based on the sample information, values of the population mean volume between 9.74 and 10.26 liters are not unreasonable."

Comments regarding the normality assumption are worth emphasizing at this point. We have indicated that when σ is known, the Central Limit Theorem allows for the use of a test statistic or a confidence interval which is based on Z, the standard normal random variable. Strictly speaking, of course, the Central Limit Theorem, and thus the use of the standard normal distribution, does not apply unless σ is known. In Chapter 8, the development of the t-distribution was given. There we pointed out that normality on $X_1, X_2, \ldots, X_n$ was an underlying assumption. Thus, *strictly speaking*, the Student's t-tables of percentage points for tests or confidence intervals should not be used unless it is known that the sample comes from a normal population. In practice, σ can rarely be assumed to be known. However, a very good estimate may be available from previous experiments. Many statistics textbooks suggest that one can safely replace σ by s in the test statistic

$$z = \frac{\bar{x} - \mu_0}{\sigma/\sqrt{n}}$$

when $n \geq 30$ with a bell-shaped population and still use the Z-tables for the appropriate critical region. The implication here is that the Central Limit Theorem is indeed being invoked and one is relying on the fact that $s \approx \sigma$. Obviously, when this is done, the results must be viewed as approximate. Thus, a computed P-value (from the Z-distribution) of 0.15 may be 0.12 or perhaps 0.17, or a computed confidence interval may be a 93% confidence interval rather than a 95% interval as desired. Now what about situations where $n \leq 30$? The user cannot rely on s being close to σ, and in order to take into account the inaccuracy of the estimate, the confidence interval should be wider or the critical value larger in magnitude. The t-distribution percentage points accomplish this but are correct only when the sample is from a normal distribution. Of course, normal probability plots can be used to ascertain some sense of the deviation of normality in a data set.

For small samples, it is often difficult to detect deviations from a normal distribution. (Goodness-of-fit tests are discussed in a later section of this chapter.) For bell-shaped distributions of the random variables $X_1, X_2, \ldots, X_n$, the use of the t-distribution for tests or confidence intervals is likely to produce quite good results. When in doubt, the user should resort to nonparametric procedures, which are presented in Chapter 16.

Annotated Computer Printout for Single-Sample t-Test

It should be of interest for the reader to see an annotated computer printout showing the result of a single-sample t-test. Suppose that an engineer is interested in testing the bias in a pH meter. Data are collected on a neutral substance (pH = 7.0). A sample of the measurements were taken with the data as follows:

$$7.07 \quad 7.00 \quad 7.10 \quad 6.97 \quad 7.00 \quad 7.03 \quad 7.01 \quad 7.01 \quad 6.98 \quad 7.08$$

It is, then, of interest to test

$$H_0: \mu = 7.0,$$
$$H_1: \mu \neq 7.0.$$

In this illustration, we use the computer package *MINITAB* to illustrate the analysis of the data set above. Notice the key components of the printout shown in Figure 10.12. Of course, the mean $\bar{y}$ is 7.0250, StDev is simply the sample standard deviation $s = 0.044$, and SE Mean is the estimated standard error of the mean and is computed as $s/\sqrt{n} = 0.0139$. The t-value is the ratio

$$(7.0250 - 7)/0.0139 = 1.80.$$

```
pH-meter
   7.07    7.00    7.10    6.97    7.00    7.03    7.01    7.01    6.98    7.08
MTB > Onet 'pH-meter'; SUBC>    Test 7.

One-Sample T: pH-meter Test of mu = 7 vs not = 7
Variable   N    Mean    StDev   SE Mean       95% CI            T     P
pH-meter  10  7.02500  0.04403  0.01392  (6.99350, 7.05650)  1.80  0.106
```

Figure 10.12: *MINITAB* printout for one sample t-test for pH meter.

The P-value of 0.106 suggests results that are inconclusive. There is no evidence suggesting a strong rejection of H_0 (based on an α of 0.05 or 0.10), **yet one certainly cannot truly conclude that the pH meter is unbiased**. Notice that the sample size of 10 is rather small. An increase in sample size (perhaps another experiment) may sort things out. A discussion regarding appropriate sample size appears in Section 10.6.

10.5 Two Samples: Tests on Two Means

The reader should now understand the relationship between tests and confidence intervals, and can only heavily rely on details supplied by the confidence interval material in Chapter 9. Tests concerning two means represent a set of very important analytical tools for the scientist or engineer. The experimental setting is very much like that described in Section 9.8. Two independent random samples of sizes

10.5 Two Samples: Tests on Two Means

n_1 and n_2, respectively, are drawn from two populations with means μ_1 and μ_2 and variances σ_1^2 and σ_2^2. We know that the random variable

$$Z = \frac{(\bar{X}_1 - \bar{X}_2) - (\mu_1 - \mu_2)}{\sqrt{\sigma_1^2/n_1 + \sigma_2^2/n_2}}$$

has a standard normal distribution. Here we are assuming that n_1 and n_2 are sufficiently large that the Central Limit Theorem applies. Of course, if the two populations are normal, the statistic above has a standard normal distribution even for small n_1 and n_2. Obviously, if we can assume that $\sigma_1 = \sigma_2 = \sigma$, the statistic above reduces to

$$Z = \frac{(\bar{X}_1 - \bar{X}_2) - (\mu_1 - \mu_2)}{\sigma\sqrt{1/n_1 + 1/n_2}}.$$

The two statistics above serve as a basis for the development of the test procedures involving two means. The equivalence between tests and confidence intervals, along with the technical detail involving tests on one mean, allow a simple transition to tests on two means.

The two-sided hypothesis on two means can be written generally as

$$H_0: \mu_1 - \mu_2 = d_0.$$

Obviously, the alternative can be two sided or one sided. Again, the distribution used is the distribution of the test statistic under H_0. Values $\bar{x}_1$ and $\bar{x}_2$ are computed and, for σ_1 and σ_2 known, the test statistic is given by

$$z = \frac{(\bar{x}_1 - \bar{x}_2) - d_0}{\sqrt{\sigma_1^2/n_1 + \sigma_2^2/n_2}},$$

with a two-tailed critical region in the case of a two-sided alternative. That is, reject H_0 in favor of H_1: $\mu_1 - \mu_2 \neq d_0$ if $z > z_{\alpha/2}$ or $z < -z_{\alpha/2}$. One-tailed critical regions are used in the case of the one-sided alternatives. The reader should, as before, study the test statistic and be satisfied that for, say, H_1: $\mu_1 - \mu_2 > d_0$, the signal favoring H_1 comes from large values of z. Thus, the upper-tailed critical region applies.

Unknown But Equal Variances

The more prevalent situations involving tests on two means are those in which variances are unknown. If the scientist involved is willing to assume that both distributions are normal and that $\sigma_1 = \sigma_2 = \sigma$, the *pooled t-test* (often called the two-sample *t*-test) may be used. The test statistic (see Section 9.8) is given by the following test procedure.

Two-Sample Pooled t-Test

For the two-sided hypothesis

$$H_0: \mu_1 = \mu_2,$$
$$H_1: \mu_1 \neq \mu_2,$$

we reject H_0 at significance level α when the computed t-statistic

$$t = \frac{(\bar{x}_1 - \bar{x}_2) - d_0}{s_p\sqrt{1/n_1 + 1/n_2}},$$

where

$$s_p^2 = \frac{s_1^2(n_1 - 1) + s_2^2(n_2 - 1)}{n_1 + n_2 - 2}$$

exceeds $t_{\alpha/2, n_1+n_2-2}$ or is less than $-t_{\alpha/2, n_1+n_2-2}$.

Recall from Chapter 9 that the degrees of freedom for the t-distribution are a result of pooling of information from the two samples to estimate σ^2. One-sided alternatives suggest one-sided critical regions, as one might expect. For example, for H_1: $\mu_1 - \mu_2 > d_0$, reject H_1: $\mu_1 - \mu_2 = d_0$ when $t > t_{\alpha, n_1+n_2-2}$.

Example 10.6: An experiment was performed to compare the abrasive wear of two different laminated materials. Twelve pieces of material 1 were tested by exposing each piece to a machine measuring wear. Ten pieces of material 2 were similarly tested. In each case, the depth of wear was observed. The samples of material 1 gave an average (coded) wear of 85 units with a sample standard deviation of 4, while the samples of material 2 gave an average of 81 with a sample standard deviation of 5. Can we conclude at the 0.05 level of significance that the abrasive wear of material 1 exceeds that of material 2 by more than 2 units? Assume the populations to be approximately normal with equal variances.

Solution: Let μ_1 and μ_2 represent the population means of the abrasive wear for material 1 and material 2, respectively.

1. H_0: $\mu_1 - \mu_2 = 2$.
2. H_1: $\mu_1 - \mu_2 > 2$.
3. $\alpha = 0.05$.
4. Critical region: $t > 1.725$, where $t = \frac{(\bar{x}_1 - \bar{x}_2) - d_0}{s_p\sqrt{1/n_1 + 1/n_2}}$ with $v = 20$ degrees of freedom.
5. Computations:

$$\bar{x}_1 = 85, \quad s_1 = 4, \quad n_1 = 12,$$
$$\bar{x}_2 = 81, \quad s_2 = 5, \quad n_2 = 10.$$

Hence

$$s_p = \sqrt{\frac{(11)(16)+(9)(25)}{12+10-2}} = 4.478,$$

$$t = \frac{(85-81)-2}{4.478\sqrt{1/12+1/10}} = 1.04,$$

$$P = P(T > 1.04) \approx 0.16. \quad \text{(See Table A.4.)}$$

6. Decision: Do not reject H_0. We are unable to conclude that the abrasive wear of material 1 exceeds that of material 2 by more than 2 units.

Unknown But Unequal Variances

There are situations where the analyst is **not** able to assume that $\sigma_1 = \sigma_2$. Recall from Section 9.8 that, if the populations are normal, the statistic

$$T' = \frac{(\bar{X}_1 - \bar{X}_2) - d_0}{\sqrt{s_1^2/n_1 + s_2^2/n_2}}$$

has an approximate t-distribution with approximate degrees of freedom

$$v = \frac{(s_1^2/n_1 + s_2^2/n_2)^2}{(s_1^2/n_1)^2/(n_1-1) + (s_2^2/n_2)^2/(n_2-1)}.$$

As a result, the test procedure is to *not reject* H_0 when

$$-t_{\alpha/2,v} < t' < t_{\alpha/2,v},$$

with v given as above. Again, as in the case of the pooled t-test, one-sided alternatives suggest one-sided critical regions.

Paired Observations

A study of the two-sample t-test or confidence interval on the difference between means should suggest the need for experimental design. Recall the discussion of experimental units in Chapter 9, where it was suggested that the conditions of the two populations (often referred to as the two treatments) should be assigned randomly to the experimental units. This is done to avoid biased results due to systematic differences between experimental units. In other words, in hypothesis-testing jargon, it is important that any significant difference found between means be due to the different conditions of the populations and not due to the experimental units in the study. For example, consider Exercise 9.40 in Section 9.9. The 20 seedlings play the role of the experimental units. Ten of them are to be treated with nitrogen and 10 with no nitrogen. It may be very important that this assignment to the "nitrogen" and "no-nitrogen" treatments be random to ensure that systematic differences between the seedlings do not interfere with a valid comparison between the means.

In Example 10.6, time of measurement is the most likely choice for the experimental unit. The 22 pieces of material should be measured in random order. We need to guard against the possibility that wear measurements made close together in time might tend to give similar results. **Systematic** (nonrandom) **differences in experimental units** *are not expected*. However, random assignments guard against the problem.

References to planning of experiments, randomization, choice of sample size, and so on, will continue to influence much of the development in Chapters 13, 14, and 15. Any scientist or engineer whose interest lies in analysis of real data should study this material. The pooled *t*-test is extended in Chapter 13 to cover more than two means.

Testing of two means can be accomplished when data are in the form of paired observations, as discussed in Chapter 9. In this pairing structure, the conditions of the two populations (treatments) are assigned randomly within homogeneous units. Computation of the confidence interval for $\mu_1 - \mu_2$ in the situation with paired observations is based on the random variable

$$T = \frac{\bar{D} - \mu_D}{S_d/\sqrt{n}},$$

where $\bar{D}$ and S_d are random variables representing the sample mean and standard deviation of the differences of the observations in the experimental units. As in the case of the *pooled t-test*, the assumption is that the observations from each population are normal. This two-sample problem is essentially reduced to a one-sample problem by using the computed differences $d_1, d_2, \ldots, d_n$. Thus, the hypothesis reduces to

$$H_0\!:\ \mu_D = d_0.$$

The computed test statistic is then given by

$$t = \frac{\bar{d} - d_0}{s_d/\sqrt{n}}.$$

Critical regions are constructed using the *t*-distribution with $n-1$ degrees of freedom.

Problem of Interaction in a Paired *t*-Test

Not only will the case study that follows illustrate the use of the paired *t*-test but the discussion will shed considerable light on the difficulties that arise when there is an interaction between the treatments and the experimental units in the paired *t* structure. Recall that interaction between factors was introduced in Section 1.7 in a discussion of general types of statistical studies. The concept of interaction will be an important issue from Chapter 13 through Chapter 15.

There are some types of statistical tests in which the existence of interaction results in difficulty. The paired *t*-test is one such example. In Section 9.9, the paired structure was used in the computation of a confidence interval on the difference between two means, and the advantage in pairing was revealed for situations in which the experimental units are homogeneous. The pairing results in a reduction in σ_D, the standard deviation of a difference $D_i = X_{1i} - X_{2i}$, as discussed in

Section 9.9. If interaction exists between treatments and experimental units, the advantage gained in pairing may be substantially reduced. Thus, in Example 9.13 on page 293, the no interaction assumption allowed the difference in mean TCDD levels (plasma vs. fat tissue) to be the same across veterans. A quick glance at the data would suggest that there is no significant violation of the assumption of no interaction.

In order to demonstrate how interaction influences $\text{Var}(D)$ and hence the quality of the paired t-test, it is instructive to revisit the ith difference given by $D_i = X_{1i} - X_{2i} = (\mu_1 - \mu_2) + (\epsilon_1 - \epsilon_2)$, where X_{1i} and X_{2i} are taken on the ith experimental unit. If the pairing unit is homogeneous, the errors in X_{1i} and in X_{2i} should be similar and not independent. We noted in Chapter 9 that the positive covariance between the errors results in a reduced $\text{Var}(D)$. Thus, the size of the difference in the treatments and the relationship between the errors in X_{1i} and X_{2i} contributed by the experimental unit will tend to allow a significant difference to be detected.

What Conditions Result in Interaction?

Let us consider a situation in which the experimental units are not homogeneous. Rather, consider the ith experimental unit with random variables X_{1i} and X_{2i} that are not similar. Let ϵ_{1i} and ϵ_{2i} be random variables representing the errors in the values X_{1i} and X_{2i}, respectively, at the ith unit. Thus, we may write

$$X_{1i} = \mu_1 + \epsilon_{1i} \text{ and } X_{2i} = \mu_2 + \epsilon_{2i}.$$

The errors with expectation zero may tend to cause the response values X_{1i} and X_{2i} to move in opposite directions, resulting in a negative value for $\text{Cov}(\epsilon_{1i}, \epsilon_{2i})$ and hence negative $\text{Cov}(X_{1i}, X_{2i})$. In fact, the model may be complicated even more by the fact that $\sigma_1^2 = \text{Var}(\epsilon_{1i}) \neq \sigma_2^2 = \text{Var}(\epsilon_{2i})$. The variance and covariance parameters may vary among the n experimental units. Thus, unlike in the homogeneous case, D_i will tend to be quite different across experimental units due to the heterogeneous nature of the difference in $\epsilon_1 - \epsilon_2$ among the units. This produces the interaction between treatments and units. In addition, for a specific experimental unit (see Theorem 4.9),

$$\sigma_D^2 = \text{Var}(D) = \text{Var}(\epsilon_1) + \text{Var}(\epsilon_2) - 2\,\text{Cov}(\epsilon_1, \epsilon_2)$$

is inflated by the negative covariance term, and thus the advantage gained in pairing in the homogeneous unit case is lost in the case described here. While the inflation in $\text{Var}(D)$ will vary from case to case, there is a danger in some cases that the increase in variance may neutralize any difference that exists between μ_1 and μ_2. Of course, a large value of $\bar{d}$ in the t-statistic may reflect a treatment difference that overcomes the inflated variance estimate, s_d^2.

Case Study 10.1: **Blood Sample Data**: In a study conducted in the Forestry and Wildlife Department at Virginia Tech, J. A. Wesson examined the influence of the drug succinylcholine on the circulation levels of androgens in the blood. Blood samples were taken from wild, free-ranging deer immediately after they had received an intramuscular injection of succinylcholine administered using darts and a capture gun. A second blood sample was obtained from each deer 30 minutes after the first sample, after which the deer was released. The levels of androgens at time of capture and 30 minutes later, measured in nanograms per milliliter (ng/mL), for 15 deer are given in Table 10.2.

Assuming that the populations of androgen levels at time of injection and 30 minutes later are normally distributed, test at the 0.05 level of significance whether the androgen concentrations are altered after 30 minutes.

Table 10.2: Data for Case Study 10.1

Deer	Androgen (ng/mL)		d_i
	At Time of Injection	30 Minutes after Injection	
1	2.76	7.02	4.26
2	5.18	3.10	−2.08
3	2.68	5.44	2.76
4	3.05	3.99	0.94
5	4.10	5.21	1.11
6	7.05	10.26	3.21
7	6.60	13.91	7.31
8	4.79	18.53	13.74
9	7.39	7.91	0.52
10	7.30	4.85	−2.45
11	11.78	11.10	−0.68
12	3.90	3.74	−0.16
13	26.00	94.03	68.03
14	67.48	94.03	26.55
15	17.04	41.70	24.66

Solution: Let μ_1 and μ_2 be the average androgen concentration at the time of injection and 30 minutes later, respectively. We proceed as follows:

1. H_0: $\mu_1 = \mu_2$ or $\mu_D = \mu_1 - \mu_2 = 0$.
2. H_1: $\mu_1 \neq \mu_2$ or $\mu_D = \mu_1 - \mu_2 \neq 0$.
3. $\alpha = 0.05$.
4. Critical region: $t < -2.145$ and $t > 2.145$, where $t = \frac{\bar{d}-d_0}{s_D/\sqrt{n}}$ with $v = 14$ degrees of freedom.
5. Computations: The sample mean and standard deviation for the d_i are

$$\bar{d} = 9.848 \quad \text{and} \quad s_d = 18.474.$$

Therefore,

$$t = \frac{9.848 - 0}{18.474/\sqrt{15}} = 2.06.$$

6. Though the t-statistic is not significant at the 0.05 level, from Table A.4,

$$P = P(|T| > 2.06) \approx 0.06.$$

As a result, there is some evidence that there is a difference in mean circulating levels of androgen.

The assumption of no interaction would imply that the effect on androgen levels of the deer is roughly the same in the data for both treatments, i.e., at the time of injection of succinylcholine and 30 minutes following injection. This can be expressed with the two factors switching roles; for example, the difference in treatments is roughly the same across the units (i.e., the deer). There certainly are some deer/treatment combinations for which the no interaction assumption seems to hold, but there is hardly any strong evidence that the experimental units are homogeneous. However, the nature of the interaction and the resulting increase in $\text{Var}(\bar{D})$ appear to be dominated by a substantial difference in the treatments. This is further demonstrated by the fact that 11 of the 15 deer exhibited positive signs for the computed d_i and the negative d_i (for deer 2, 10, 11, and 12) are small in magnitude compared to the 12 positive ones. Thus, it appears that the mean level of androgen is significantly higher 30 minutes following injection than at injection, and the conclusions may be stronger than $p = 0.06$ would suggest.

Annotated Computer Printout for Paired *t*-Test

Figure 10.13 displays a *SAS* computer printout for a paired *t*-test using the data of Case Study 10.1. Notice that the printout looks like that for a single sample *t*-test and, of course, that is exactly what is accomplished, since the test seeks to determine if $\bar{d}$ is significantly different from zero.

```
                    Analysis Variable : Diff

       N             Mean        Std Error    t Value    Pr > |t|
       ----------------------------------------------------------
      15          9.8480000      4.7698699     2.06       0.0580
       ----------------------------------------------------------
```

Figure 10.13: *SAS* printout of paired *t*-test for data of Case Study 10.1.

Summary of Test Procedures

As we complete the formal development of tests on population means, we offer Table 10.3, which summarizes the test procedure for the cases of a single mean and two means. Notice the approximate procedure when distributions are normal and variances are unknown but not assumed to be equal. This statistic was introduced in Chapter 9.

10.6 Choice of Sample Size for Testing Means

In Section 10.2, we demonstrated how the analyst can exploit relationships among the sample size, the significance level α, and the power of the test to achieve a certain standard of quality. In most practical circumstances, the experiment should be planned, with a choice of sample size made prior to the data-taking process if possible. The sample size is usually determined to achieve good power for a fixed α and fixed specific alternative. This fixed alternative may be in the

Table 10.3: Tests Concerning Means

H_0	Value of Test Statistic	H_1	Critical Region
$\mu = \mu_0$	$z = \dfrac{\bar{x} - \mu_0}{\sigma/\sqrt{n}};\ \ \sigma$ known	$\mu < \mu_0$ $\mu > \mu_0$ $\mu \neq \mu_0$	$z < -z_\alpha$ $z > z_\alpha$ $z < -z_{\alpha/2}$ or $z > z_{\alpha/2}$
$\mu = \mu_0$	$t = \dfrac{\bar{x} - \mu_0}{s/\sqrt{n}};\ \ v = n-1,$ σ unknown	$\mu < \mu_0$ $\mu > \mu_0$ $\mu \neq \mu_0$	$t < -t_\alpha$ $t > t_\alpha$ $t < -t_{\alpha/2}$ or $t > t_{\alpha/2}$
$\mu_1 - \mu_2 = d_0$	$z = \dfrac{(\bar{x}_1 - \bar{x}_2) - d_0}{\sqrt{\sigma_1^2/n_1 + \sigma_2^2/n_2}};$ σ_1 and σ_2 known	$\mu_1 - \mu_2 < d_0$ $\mu_1 - \mu_2 > d_0$ $\mu_1 - \mu_2 \neq d_0$	$z < -z_\alpha$ $z > z_\alpha$ $z < -z_{\alpha/2}$ or $z > z_{\alpha/2}$
$\mu_1 - \mu_2 = d_0$	$t = \dfrac{(\bar{x}_1 - \bar{x}_2) - d_0}{s_p\sqrt{1/n_1 + 1/n_2}};$ $v = n_1 + n_2 - 2,$ $\sigma_1 = \sigma_2$ but unknown, $s_p^2 = \dfrac{(n_1-1)s_1^2 + (n_2-1)s_2^2}{n_1 + n_2 - 2}$	$\mu_1 - \mu_2 < d_0$ $\mu_1 - \mu_2 > d_0$ $\mu_1 - \mu_2 \neq d_0$	$t < -t_\alpha$ $t > t_\alpha$ $t < -t_{\alpha/2}$ or $t > t_{\alpha/2}$
$\mu_1 - \mu_2 = d_0$	$t' = \dfrac{(\bar{x}_1 - \bar{x}_2) - d_0}{\sqrt{s_1^2/n_1 + s_2^2/n_2}};$ $v = \dfrac{(s_1^2/n_1 + s_2^2/n_2)^2}{\frac{(s_1^2/n_1)^2}{n_1-1} + \frac{(s_2^2/n_2)^2}{n_2-1}},$ $\sigma_1 \neq \sigma_2$ and unknown	$\mu_1 - \mu_2 < d_0$ $\mu_1 - \mu_2 > d_0$ $\mu_1 - \mu_2 \neq d_0$	$t' < -t_\alpha$ $t' > t_\alpha$ $t' < -t_{\alpha/2}$ or $t' > t_{\alpha/2}$
$\mu_D = d_0$ paired observations	$t = \dfrac{\bar{d} - d_0}{s_d/\sqrt{n}};$ $v = n - 1$	$\mu_D < d_0$ $\mu_D > d_0$ $\mu_D \neq d_0$	$t < -t_\alpha$ $t > t_\alpha$ $t < -t_{\alpha/2}$ or $t > t_{\alpha/2}$

form of $\mu - \mu_0$ in the case of a hypothesis involving a single mean or $\mu_1 - \mu_2$ in the case of a problem involving two means. Specific cases will provide illustrations.

Suppose that we wish to test the hypothesis

$$H_0:\ \mu = \mu_0,$$
$$H_1:\ \mu > \mu_0,$$

with a significance level α, when the variance σ^2 is known. For a specific alternative, say $\mu = \mu_0 + \delta$, the power of our test is shown in Figure 10.14 to be

$$1 - \beta = P(\bar{X} > a \text{ when } \mu = \mu_0 + \delta).$$

Therefore,

$$\beta = P(\bar{X} < a \text{ when } \mu = \mu_0 + \delta)$$
$$= P\left[\frac{\bar{X} - (\mu_0 + \delta)}{\sigma/\sqrt{n}} < \frac{a - (\mu_0 + \delta)}{\sigma/\sqrt{n}} \text{ when } \mu = \mu_0 + \delta\right].$$

10.6 Choice of Sample Size for Testing Means

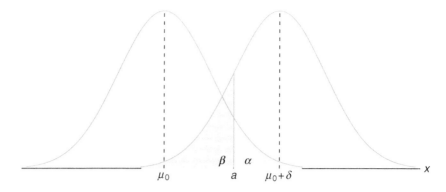

Figure 10.14: Testing $\mu = \mu_0$ versus $\mu = \mu_0 + \delta$.

Under the alternative hypothesis $\mu = \mu_0 + \delta$, the statistic

$$\frac{\bar{X} - (\mu_0 + \delta)}{\sigma/\sqrt{n}}$$

is the standard normal variable Z. So

$$\beta = P\left(Z < \frac{a - \mu_0}{\sigma/\sqrt{n}} - \frac{\delta}{\sigma/\sqrt{n}}\right) = P\left(Z < z_\alpha - \frac{\delta}{\sigma/\sqrt{n}}\right),$$

from which we conclude that

$$-z_\beta = z_\alpha - \frac{\delta\sqrt{n}}{\sigma},$$

and hence

$$\text{Choice of sample size:} \quad n = \frac{(z_\alpha + z_\beta)^2 \sigma^2}{\delta^2},$$

a result that is also true when the alternative hypothesis is $\mu < \mu_0$.

In the case of a two-tailed test, we obtain the power $1 - \beta$ for a specified alternative when

$$n \approx \frac{(z_{\alpha/2} + z_\beta)^2 \sigma^2}{\delta^2}.$$

Example 10.7: Suppose that we wish to test the hypothesis

$$H_0: \mu = 68 \text{ kilograms},$$
$$H_1: \mu > 68 \text{ kilograms}$$

for the weights of male students at a certain college, using an $\alpha = 0.05$ level of significance, when it is known that $\sigma = 5$. Find the sample size required if the power of our test is to be 0.95 when the true mean is 69 kilograms.

Solution: Since $\alpha = \beta = 0.05$, we have $z_\alpha = z_\beta = 1.645$. For the alternative $\beta = 69$, we take $\delta = 1$ and then

$$n = \frac{(1.645 + 1.645)^2(25)}{1} = 270.6.$$

Therefore, 271 observations are required if the test is to reject the null hypothesis 95% of the time when, in fact, μ is as large as 69 kilograms.

Two-Sample Case

A similar procedure can be used to determine the sample size $n = n_1 = n_2$ required for a specific power of the test in which two population means are being compared. For example, suppose that we wish to test the hypothesis

$$H_0: \mu_1 - \mu_2 = d_0,$$
$$H_1: \mu_1 - \mu_2 \neq d_0,$$

when σ_1 and σ_2 are known. For a specific alternative, say $\mu_1 - \mu_2 = d_0 + \delta$, the power of our test is shown in Figure 10.15 to be

$$1 - \beta = P(|\bar{X}_1 - \bar{X}_2| > a \text{ when } \mu_1 - \mu_2 = d_0 + \delta).$$

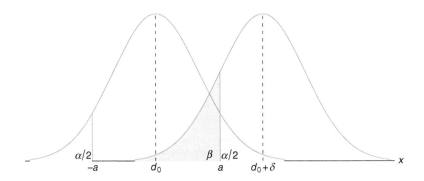

Figure 10.15: Testing $\mu_1 - \mu_2 = d_0$ versus $\mu_1 - \mu_2 = d_0 + \delta$.

Therefore,

$$\beta = P(-a < \bar{X}_1 - \bar{X}_2 < a \text{ when } \mu_1 - \mu_2 = d_0 + \delta)$$

$$= P\left[\frac{-a - (d_0 + \delta)}{\sqrt{(\sigma_1^2 + \sigma_2^2)/n}} < \frac{(\bar{X}_1 - \bar{X}_2) - (d_0 + \delta)}{\sqrt{(\sigma_1^2 + \sigma_2^2)/n}}\right.$$

$$\left.< \frac{a - (d_0 + \delta)}{\sqrt{(\sigma_1^2 + \sigma_2^2)/n}} \text{ when } \mu_1 - \mu_2 = d_0 + \delta\right].$$

Under the alternative hypothesis $\mu_1 - \mu_2 = d_0 + \delta$, the statistic

$$\frac{\bar{X}_1 - \bar{X}_2 - (d_0 + \delta)}{\sqrt{(\sigma_1^2 + \sigma_2^2)/n}}$$

10.6 Choice of Sample Size for Testing Means

is the standard normal variable Z. Now, writing

$$-z_{\alpha/2} = \frac{-a - d_0}{\sqrt{(\sigma_1^2 + \sigma_2^2)/n}} \quad \text{and} \quad z_{\alpha/2} = \frac{a - d_0}{\sqrt{(\sigma_1^2 + \sigma_2^2)/n}},$$

we have

$$\beta = P\left[-z_{\alpha/2} - \frac{\delta}{\sqrt{(\sigma_1^2 + \sigma_2^2)/n}} < Z < z_{\alpha/2} - \frac{\delta}{\sqrt{(\sigma_1^2 + \sigma_2^2)/n}}\right],$$

from which we conclude that

$$-z_\beta \approx z_{\alpha/2} - \frac{\delta}{\sqrt{(\sigma_1^2 + \sigma_2^2)/n}},$$

and hence

$$n \approx \frac{(z_{\alpha/2} + z_\beta)^2(\sigma_1^2 + \sigma_2^2)}{\delta^2}.$$

For the one-tailed test, the expression for the required sample size when $n = n_1 = n_2$ is

$$\text{Choice of sample size:} \quad n = \frac{(z_\alpha + z_\beta)^2(\sigma_1^2 + \sigma_2^2)}{\delta^2}.$$

When the population variance (or variances, in the two-sample situation) is unknown, the choice of sample size is not straightforward. In testing the hypothesis $\mu = \mu_0$ when the true value is $\mu = \mu_0 + \delta$, the statistic

$$\frac{\bar{X} - (\mu_0 + \delta)}{S/\sqrt{n}}$$

does not follow the t-distribution, as one might expect, but instead follows the **noncentral t-distribution**. However, tables or charts based on the noncentral t-distribution do exist for determining the appropriate sample size if some estimate of σ is available or if δ is a multiple of σ. Table A.8 gives the sample sizes needed to control the values of α and β for various values of

$$\Delta = \frac{|\delta|}{\sigma} = \frac{|\mu - \mu_0|}{\sigma}$$

for both one- and two-tailed tests. In the case of the two-sample t-test in which the variances are unknown but assumed equal, we obtain the sample sizes $n = n_1 = n_2$ needed to control the values of α and β for various values of

$$\Delta = \frac{|\delta|}{\sigma} = \frac{|\mu_1 - \mu_2 - d_0|}{\sigma}$$

from Table A.9.

Example 10.8: In comparing the performance of two catalysts on the effect of a reaction yield, a two-sample t-test is to be conducted with $\alpha = 0.05$. The variances in the yields are considered to be the same for the two catalysts. How large a sample for each catalyst is needed to test the hypothesis

$$H_0: \mu_1 = \mu_2,$$
$$H_1: \mu_1 \neq \mu_2$$

if it is essential to detect a difference of 0.8σ between the catalysts with probability 0.9?

Solution: From Table A.9, with $\alpha = 0.05$ for a two-tailed test, $\beta = 0.1$, and

$$\Delta = \frac{|0.8\sigma|}{\sigma} = 0.8,$$

we find the required sample size to be $n = 34$.

In practical situations, it might be difficult to force a scientist or engineer to make a commitment on information from which a value of Δ can be found. The reader is reminded that the Δ-value quantifies the kind of difference between the means that the scientist considers important, that is, a difference considered *significant* from a scientific, not a statistical, point of view. Example 10.8 illustrates how this choice is often made, namely, by selecting a fraction of σ. Obviously, if the sample size is based on a choice of $|\delta|$ that is a small fraction of σ, the resulting sample size may be quite large compared to what the study allows.

10.7 Graphical Methods for Comparing Means

In Chapter 1, considerable attention was directed to displaying data in graphical form, such as stem-and-leaf plots and box-and-whisker plots. In Section 8.8, quantile plots and quantile-quantile normal plots were used to provide a "picture" to summarize a set of experimental data. Many computer software packages produce graphical displays. As we proceed to other forms of data analysis (e.g., regression analysis and analysis of variance), graphical methods become even more informative.

Graphical aids cannot be used as a replacement for the test procedure itself. Certainly, the value of the test statistic indicates the proper type of evidence in support of H_0 or H_1. However, a pictorial display provides a good illustration and is often a better communicator of evidence to the beneficiary of the analysis. Also, a picture will often clarify why a significant difference was found. Failure of an important assumption may be exposed by a summary type of graphical tool.

For the comparison of means, side-by-side box-and-whisker plots provide a telling display. The reader should recall that these plots display the 25th percentile, 75th percentile, and the median in a data set. In addition, the whiskers display the extremes in a data set. Consider Exercise 10.40 at the end of this section. Plasma ascorbic acid levels were measured in two groups of pregnant women, smokers and nonsmokers. Figure 10.16 shows the box-and-whisker plots for both groups of women. Two things are very apparent. Taking into account variability, there appears to be a negligible difference in the sample means. In addition, the variability in the two groups appears to be somewhat different. Of course, the analyst must keep in mind the rather sizable differences between the sample sizes in this case.

10.7 Graphical Methods for Comparing Means

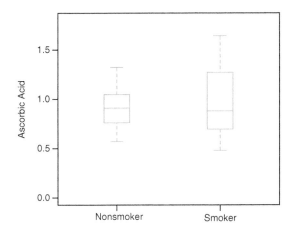

Figure 10.16: Two box-and-whisker plots of plasma ascorbic acid in smokers and nonsmokers.

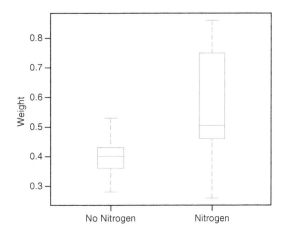

Figure 10.17: Two box-and-whisker plots of seedling data.

Consider Exercise 9.40 in Section 9.9. Figure 10.17 shows the multiple box-and-whisker plot for the data on 10 seedlings, half given nitrogen and half given no nitrogen. The display reveals a smaller variability for the group containing no nitrogen. In addition, the lack of overlap of the box plots suggests a significant difference between the mean stem weights for the two groups. It would appear that the presence of nitrogen increases the stem weights and perhaps increases the variability in the weights.

There are no certain rules of thumb regarding when two box-and-whisker plots give evidence of significant difference between the means. However, a rough guideline is that if the 25th percentile line for one sample exceeds the median line for the other sample, there is strong evidence of a difference between means.

More emphasis is placed on graphical methods in a real-life case study presented later in this chapter.

Annotated Computer Printout for Two-Sample t-Test

Consider once again Exercise 9.40 on page 294, where seedling data under conditions of nitrogen and no nitrogen were collected. Test

$$H_0: \mu_{\text{NIT}} = \mu_{\text{NON}},$$
$$H_1: \mu_{\text{NIT}} > \mu_{\text{NON}},$$

where the population means indicate mean weights. Figure 10.18 is an annotated computer printout generated using the *SAS* package. Notice that sample standard deviation and standard error are shown for both samples. The t-statistics under the assumption of equal variance and unequal variance are both given. From the box-and-whisker plot of Figure 10.17 it would certainly appear that the equal variance assumption is violated. A P-value of 0.0229 suggests a conclusion of unequal means. This concurs with the diagnostic information given in Figure 10.18. Incidentally, notice that t and t' are equal in this case, since $n_1 = n_2$.

```
                    TTEST Procedure
      Variable Weight
         Mineral       N    Mean    Std Dev  Std Err
      No nitrogen     10   0.3990   0.0728   0.0230
         Nitrogen     10   0.5650   0.1867   0.0591

         Variances     DF    t Value    Pr > |t|
            Equal      18      2.62      0.0174
          Unequal     11.7     2.62      0.0229

               Test the Equality of Variances
      Variable   Num DF    Den DF    F Value    Pr > F
       Weight      9         9        6.58      0.0098
```

Figure 10.18: *SAS* printout for two-sample *t*-test.

Exercises

10.19 In a research report, Richard H. Weindruch of the UCLA Medical School claims that mice with an average life span of 32 months will live to be about 40 months old when 40% of the calories in their diet are replaced by vitamins and protein. Is there any reason to believe that $\mu < 40$ if 64 mice that are placed on this diet have an average life of 38 months with a standard deviation of 5.8 months? Use a *P*-value in your conclusion.

10.20 A random sample of 64 bags of white cheddar popcorn weighed, on average, 5.23 ounces with a standard deviation of 0.24 ounce. Test the hypothesis that $\mu = 5.5$ ounces against the alternative hypothesis, $\mu < 5.5$ ounces, at the 0.05 level of significance.

10.21 An electrical firm manufactures light bulbs that have a lifetime that is approximately normally distributed with a mean of 800 hours and a standard deviation of 40 hours. Test the hypothesis that $\mu = 800$ hours against the alternative, $\mu \neq 800$ hours, if a random sample of 30 bulbs has an average life of 788 hours. Use a *P*-value in your answer.

10.22 In the American Heart Association journal *Hypertension*, researchers report that individuals who practice Transcendental Meditation (TM) lower their blood pressure significantly. If a random sample of 225 male TM practitioners meditate for 8.5 hours per week with a standard deviation of 2.25 hours, does that suggest that, on average, men who use TM meditate more than 8 hours per week? Quote a *P*-value in your conclusion.

10.23 Test the hypothesis that the average content of containers of a particular lubricant is 10 liters if the contents of a random sample of 10 containers are 10.2, 9.7, 10.1, 10.3, 10.1, 9.8, 9.9, 10.4, 10.3, and 9.8 liters. Use a 0.01 level of significance and assume that the distribution of contents is normal.

10.24 The average height of females in the freshman class of a certain college has historically been 162.5 centimeters with a standard deviation of 6.9 centimeters. Is there reason to believe that there has been a change in the average height if a random sample of 50 females in the present freshman class has an average height of 165.2 centimeters? Use a *P*-value in your conclusion. Assume the standard deviation remains the same.

10.25 It is claimed that automobiles are driven on average more than 20,000 kilometers per year. To test this claim, 100 randomly selected automobile owners are asked to keep a record of the kilometers they travel. Would you agree with this claim if the random sample showed an average of 23,500 kilometers and a standard deviation of 3900 kilometers? Use a *P*-value in your conclusion.

10.26 According to a dietary study, high sodium intake may be related to ulcers, stomach cancer, and migraine headaches. The human requirement for salt is only 220 milligrams per day, which is surpassed in most single servings of ready-to-eat cereals. If a random sample of 20 similar servings of a certain cereal has a mean sodium content of 244 milligrams and a standard deviation of 24.5 milligrams, does this suggest at the 0.05 level of significance that the average sodium content for a single serving of such cereal is greater than 220 milligrams? Assume the distribution of sodium contents to be normal.

10.27 A study at the University of Colorado at Boulder shows that running increases the percent resting metabolic rate (RMR) in older women. The average RMR of 30 elderly women runners was 34.0% higher than the average RMR of 30 sedentary elderly women, and the standard deviations were reported to be 10.5 and 10.2%, respectively. Was there a significant increase in RMR of the women runners over the sedentary women? Assume the populations to be approximately normally distributed with equal variances. Use a P-value in your conclusions.

10.28 According to *Chemical Engineering*, an important property of fiber is its water absorbency. The average percent absorbency of 25 randomly selected pieces of cotton fiber was found to be 20 with a standard deviation of 1.5. A random sample of 25 pieces of acetate yielded an average percent of 12 with a standard deviation of 1.25. Is there strong evidence that the population mean percent absorbency is significantly higher for cotton fiber than for acetate? Assume that the percent absorbency is approximately normally distributed and that the population variances in percent absorbency for the two fibers are the same. Use a significance level of 0.05.

10.29 Past experience indicates that the time required for high school seniors to complete a standardized test is a normal random variable with a mean of 35 minutes. If a random sample of 20 high school seniors took an average of 33.1 minutes to complete this test with a standard deviation of 4.3 minutes, test the hypothesis, at the 0.05 level of significance, that $\mu = 35$ minutes against the alternative that $\mu < 35$ minutes.

10.30 A random sample of size $n_1 = 25$, taken from a normal population with a standard deviation $\sigma_1 = 5.2$, has a mean $\bar{x}_1 = 81$. A second random sample of size $n_2 = 36$, taken from a different normal population with a standard deviation $\sigma_2 = 3.4$, has a mean $\bar{x}_2 = 76$. Test the hypothesis that $\mu_1 = \mu_2$ against the alternative, $\mu_1 \neq \mu_2$. Quote a P-value in your conclusion.

10.31 A manufacturer claims that the average tensile strength of thread A exceeds the average tensile strength of thread B by at least 12 kilograms. To test this claim, 50 pieces of each type of thread were tested under similar conditions. Type A thread had an average tensile strength of 86.7 kilograms with a standard deviation of 6.28 kilograms, while type B thread had an average tensile strength of 77.8 kilograms with a standard deviation of 5.61 kilograms. Test the manufacturer's claim using a 0.05 level of significance.

10.32 *Amstat News* (December 2004) lists median salaries for associate professors of statistics at research institutions and at liberal arts and other institutions in the United States. Assume that a sample of 200 associate professors from research institutions has an average salary of $70,750 per year with a standard deviation of $6000. Assume also that a sample of 200 associate professors from other types of institutions has an average salary of $65,200 with a standard deviation of $5000. Test the hypothesis that the mean salary for associate professors in research institutions is $2000 higher than for those in other institutions. Use a 0.01 level of significance.

10.33 A study was conducted to see if increasing the substrate concentration has an appreciable effect on the velocity of a chemical reaction. With a substrate concentration of 1.5 moles per liter, the reaction was run 15 times, with an average velocity of 7.5 micromoles per 30 minutes and a standard deviation of 1.5. With a substrate concentration of 2.0 moles per liter, 12 runs were made, yielding an average velocity of 8.8 micromoles per 30 minutes and a sample standard deviation of 1.2. Is there any reason to believe that this increase in substrate concentration causes an increase in the mean velocity of the reaction of more than 0.5 micromole per 30 minutes? Use a 0.01 level of significance and assume the populations to be approximately normally distributed with equal variances.

10.34 A study was made to determine if the subject matter in a physics course is better understood when a lab constitutes part of the course. Students were randomly selected to participate in either a 3-semester-hour course without labs or a 4-semester-hour course with labs. In the section with labs, 11 students made an average grade of 85 with a standard deviation of 4.7, and in the section without labs, 17 students made an average grade of 79 with a standard deviation of 6.1. Would you say that the laboratory course increases the average grade by as much as 8 points? Use a P-value in your conclusion and assume the populations to be approximately normally distributed with equal variances.

10.35 To find out whether a new serum will arrest leukemia, 9 mice, all with an advanced stage of the disease, are selected. Five mice receive the treatment and 4 do not. Survival times, in years, from the time the experiment commenced are as follows:

Treatment	2.1	5.3	1.4	4.6	0.9
No Treatment	1.9	0.5	2.8	3.1	

At the 0.05 level of significance, can the serum be said to be effective? Assume the two populations to be normally distributed with equal variances.

10.36 Engineers at a large automobile manufacturing company are trying to decide whether to purchase brand A or brand B tires for the company's new models. To help them arrive at a decision, an experiment is conducted using 12 of each brand. The tires are run until they wear out. The results are as follows:

Brand A: $\bar{x}_1 = 37{,}900$ kilometers,
$s_1 = 5100$ kilometers.

Brand B: $\bar{x}_1 = 39{,}800$ kilometers,
$s_2 = 5900$ kilometers.

Test the hypothesis that there is no difference in the average wear of the two brands of tires. Assume the populations to be approximately normally distributed with equal variances. Use a P-value.

10.37 In Exercise 9.42 on page 295, test the hypothesis that the fuel economy of Volkswagen mini-trucks, on average, exceeds that of similarly equipped Toyota mini-trucks by 4 kilometers per liter. Use a 0.10 level of significance.

10.38 A UCLA researcher claims that the average life span of mice can be extended by as much as 8 months when the calories in their diet are reduced by approximately 40% from the time they are weaned. The restricted diets are enriched to normal levels by vitamins and protein. Suppose that a random sample of 10 mice is fed a normal diet and has an average life span of 32.1 months with a standard deviation of 3.2 months, while a random sample of 15 mice is fed the restricted diet and has an average life span of 37.6 months with a standard deviation of 2.8 months. Test the hypothesis, at the 0.05 level of significance, that the average life span of mice on this restricted diet is increased by 8 months against the alternative that the increase is less than 8 months. Assume the distributions of life spans for the regular and restricted diets are approximately normal with equal variances.

10.39 The following data represent the running times of films produced by two motion-picture companies:

Company	Time (minutes)
1	102 86 98 109 92
2	81 165 97 134 92 87 114

Test the hypothesis that the average running time of films produced by company 2 exceeds the average running time of films produced by company 1 by 10 minutes against the one-sided alternative that the difference is less than 10 minutes. Use a 0.1 level of significance and assume the distributions of times to be approximately normal with unequal variances.

10.40 In a study conducted at Virginia Tech, the plasma ascorbic acid levels of pregnant women were compared for smokers versus nonsmokers. Thirty-two women in the last three months of pregnancy, free of major health disorders and ranging in age from 15 to 32 years, were selected for the study. Prior to the collection of 20 ml of blood, the participants were told to avoid breakfast, forgo their vitamin supplements, and avoid foods high in ascorbic acid content. From the blood samples, the following plasma ascorbic acid values were determined, in milligrams per 100 milliliters:

Plasma Ascorbic Acid Values

Nonsmokers		Smokers
0.97	1.16	0.48
0.72	0.86	0.71
1.00	0.85	0.98
0.81	0.58	0.68
0.62	0.57	1.18
1.32	0.64	1.36
1.24	0.98	0.78
0.99	1.09	1.64
0.90	0.92	
0.74	0.78	
0.88	1.24	
0.94	1.18	

Is there sufficient evidence to conclude that there is a difference between plasma ascorbic acid levels of smokers and nonsmokers? Assume that the two sets of data came from normal populations with unequal variances. Use a P-value.

10.41 A study was conducted by the Department of Biological Sciences at Virginia Tech to determine if there is a significant difference in the density of organisms at two different stations located on Cedar Run, a secondary stream in the Roanoke River drainage basin. Sewage from a sewage treatment plant and overflow from the Federal Mogul Corporation settling pond enter the stream near its headwaters. The following data give the density measurements, in number of organisms per square meter, at the two collecting stations:

Number of Organisms per Square Meter

Station 1		Station 2	
5030	4980	2800	2810
13,700	11,910	4670	1330
10,730	8130	6890	3320
11,400	26,850	7720	1230
860	17,660	7030	2130
2200	22,800	7330	2190
4250	1130		
15,040	1690		

Can we conclude, at the 0.05 level of significance, that the average densities at the two stations are equal? Assume that the observations come from normal populations with different variances.

10.42 Five samples of a ferrous-type substance were used to determine if there is a difference between a laboratory chemical analysis and an X-ray fluorescence analysis of the iron content. Each sample was split into two subsamples and the two types of analysis were applied. Following are the coded data showing the iron content analysis:

Analysis	Sample 1	2	3	4	5
X-ray	2.0	2.0	2.3	2.1	2.4
Chemical	2.2	1.9	2.5	2.3	2.4

Assuming that the populations are normal, test at the 0.05 level of significance whether the two methods of analysis give, on the average, the same result.

10.43 According to published reports, practice under fatigued conditions distorts mechanisms that govern performance. An experiment was conducted using 15 college males, who were trained to make a continuous horizontal right-to-left arm movement from a microswitch to a barrier, knocking over the barrier coincident with the arrival of a clock sweephand to the 6 o'clock position. The absolute value of the difference between the time, in milliseconds, that it took to knock over the barrier and the time for the sweephand to reach the 6 o'clock position (500 msec) was recorded. Each participant performed the task five times under prefatigue and postfatigue conditions, and the sums of the absolute differences for the five performances were recorded.

Subject	Absolute Time Differences Prefatigue	Postfatigue
1	158	91
2	92	59
3	65	215
4	98	226
5	33	223
6	89	91
7	148	92
8	58	177
9	142	134
10	117	116
11	74	153
12	66	219
13	109	143
14	57	164
15	85	100

An increase in the mean absolute time difference when the task is performed under postfatigue conditions would support the claim that practice under fatigued conditions distorts mechanisms that govern performance. Assuming the populations to be normally distributed, test this claim.

10.44 In a study conducted by the Department of Human Nutrition, Foods and Exercise at Virginia Tech, the following data were recorded on sorbic acid residuals, in parts per million, in ham immediately after dipping in a sorbate solution and after 60 days of storage:

	Sorbic Acid Residuals in Ham	
Slice	Before Storage	After Storage
1	224	116
2	270	96
3	400	239
4	444	329
5	590	437
6	660	597
7	1400	689
8	680	576

Assuming the populations to be normally distributed, is there sufficient evidence, at the 0.05 level of significance, to say that the length of storage influences sorbic acid residual concentrations?

10.45 A taxi company manager is trying to decide whether the use of radial tires instead of regular belted tires improves fuel economy. Twelve cars were equipped with radial tires and driven over a prescribed test course. Without changing drivers, the same cars were then equipped with regular belted tires and driven once again over the test course. The gasoline consumption, in kilometers per liter, was recorded as follows:

	Kilometers per Liter	
Car	Radial Tires	Belted Tires
1	4.2	4.1
2	4.7	4.9
3	6.6	6.2
4	7.0	6.9
5	6.7	6.8
6	4.5	4.4
7	5.7	5.7
8	6.0	5.8
9	7.4	6.9
10	4.9	4.7
11	6.1	6.0
12	5.2	4.9

Can we conclude that cars equipped with radial tires give better fuel economy than those equipped with belted tires? Assume the populations to be normally distributed. Use a P-value in your conclusion.

10.46 In Review Exercise 9.91 on page 313, use the t-distribution to test the hypothesis that the diet reduces a woman's weight by 4.5 kilograms on average against the alternative hypothesis that the mean difference in weight is less than 4.5 kilograms. Use a P-value.

10.47 How large a sample is required in Exercise 10.20 if the power of the test is to be 0.90 when the true mean is 5.20? Assume that $\sigma = 0.24$.

10.48 If the distribution of life spans in Exercise 10.19 is approximately normal, how large a sample is required in order that the probability of committing a type II error be 0.1 when the true mean is 35.9 months? Assume that $\sigma = 5.8$ months.

10.49 How large a sample is required in Exercise 10.24 if the power of the test is to be 0.95 when the true average height differs from 162.5 by 3.1 centimeters? Use $\alpha = 0.02$.

10.50 How large should the samples be in Exercise 10.31 if the power of the test is to be 0.95 when the true difference between thread types A and B is 8 kilograms?

10.51 How large a sample is required in Exercise 10.22 if the power of the test is to be 0.8 when the true mean meditation time exceeds the hypothesized value by 1.2σ? Use $\alpha = 0.05$.

10.52 For testing

$$H_0: \mu = 14,$$
$$H_1: \mu \neq 14,$$

an $\alpha = 0.05$ level t-test is being considered. What sample size is necessary in order for the probability to be 0.1 of falsely failing to reject H_0 when the true population mean differs from 14 by 0.5? From a preliminary sample we estimate σ to be 1.25.

10.53 A study was conducted at Virginia Tech to determine if the "strength" of a wound from surgical incision is affected by the temperature of the knife. Eight dogs were used in the experiment. "Hot" and "cold" incisions were made on the abdomen of each dog, and the strength was measured. The resulting data appear below.

Dog	Knife	Strength
1	Hot	5120
1	Cold	8200
2	Hot	10,000
2	Cold	8600
3	Hot	10,000
3	Cold	9200
4	Hot	10,000
4	Cold	6200
5	Hot	10,000
5	Cold	10,000
6	Hot	7900
6	Cold	5200
7	Hot	510
7	Cold	885
8	Hot	1020
8	Cold	460

(a) Write an appropriate hypothesis to determine if there is a significant difference in strength between the hot and cold incisions.

(b) Test the hypothesis using a paired t-test. Use a P-value in your conclusion.

10.54 Nine subjects were used in an experiment to determine if exposure to carbon monoxide has an impact on breathing capability. The data were collected by personnel in the Department of Human Nutrition, Foods and Exercise at Virginia Tech and were analyzed in the Statistics Consulting Center at Hokie Land. The subjects were exposed to breathing chambers, one of which contained a high concentration of CO. Breathing frequency measures were made for each subject for each chamber. The subjects were exposed to the breathing chambers in random sequence. The data give the breathing frequency, in number of breaths taken per minute. Make a one-sided test of the hypothesis that mean breathing frequency is the same for the two environments. Use $\alpha = 0.05$. Assume that breathing frequency is approximately normal.

Subject	With CO	Without CO
1	30	30
2	45	40
3	26	25
4	25	23
5	34	30
6	51	49
7	46	41
8	32	35
9	30	28

10.8 One Sample: Test on a Single Proportion

Tests of hypotheses concerning proportions are required in many areas. Politicians are certainly interested in knowing what fraction of the voters will favor them in the next election. All manufacturing firms are concerned about the proportion of defective items when a shipment is made. Gamblers depend on a knowledge of the proportion of outcomes that they consider favorable.

We shall consider the problem of testing the hypothesis that the proportion of successes in a binomial experiment equals some specified value. That is, we are testing the null hypothesis H_0 that $p = p_0$, where p is the parameter of the binomial distribution. The alternative hypothesis may be one of the usual one-sided

10.8 One Sample: Test on a Single Proportion

or two-sided alternatives:

$$p < p_0, \quad p > p_0, \quad \text{or} \quad p \neq p_0.$$

The appropriate random variable on which we base our decision criterion is the binomial random variable X, although we could just as well use the statistic $\hat{p} = X/n$. Values of X that are far from the mean $\mu = np_0$ will lead to the rejection of the null hypothesis. Because X is a discrete binomial variable, it is unlikely that a critical region can be established whose size is *exactly* equal to a prespecified value of α. For this reason it is preferable, in dealing with small samples, to base our decisions on P-values. To test the hypothesis

$$H_0: p = p_0,$$
$$H_1: p < p_0,$$

we use the binomial distribution to compute the P-value

$$P = P(X \leq x \text{ when } p = p_0).$$

The value x is the number of successes in our sample of size n. If this P-value is less than or equal to α, our test is significant at the α level and we reject H_0 in favor of H_1. Similarly, to test the hypothesis

$$H_0: p = p_0,$$
$$H_1: p > p_0,$$

at the α-level of significance, we compute

$$P = P(X \geq x \text{ when } p = p_0)$$

and reject H_0 in favor of H_1 if this P-value is less than or equal to α. Finally, to test the hypothesis

$$H_0: p = p_0,$$
$$H_1: p \neq p_0,$$

at the α-level of significance, we compute

$$P = 2P(X \leq x \text{ when } p = p_0) \quad \text{if } x < np_0$$

or

$$P = 2P(X \geq x \text{ when } p = p_0) \quad \text{if } x > np_0$$

and reject H_0 in favor of H_1 if the computed P-value is less than or equal to α.

The steps for testing a null hypothesis about a proportion against various alternatives using the binomial probabilities of Table A.1 are as follows:

Testing a Proportion (Small Samples)

1. H_0: $p = p_0$.
2. One of the alternatives H_1: $p < p_0$, $p > p_0$, or $p \neq p_0$.
3. Choose a level of significance equal to α.
4. Test statistic: Binomial variable X with $p = p_0$.
5. Computations: Find x, the number of successes, and compute the appropriate P-value.
6. Decision: Draw appropriate conclusions based on the P-value.

Example 10.9: A builder claims that heat pumps are installed in 70% of all homes being constructed today in the city of Richmond, Virginia. Would you agree with this claim if a random survey of new homes in this city showed that 8 out of 15 had heat pumps installed? Use a 0.10 level of significance.

Solution: 1. H_0: $p = 0.7$.
2. H_1: $p \neq 0.7$.
3. $\alpha = 0.10$.
4. Test statistic: Binomial variable X with $p = 0.7$ and $n = 15$.
5. Computations: $x = 8$ and $np_0 = (15)(0.7) = 10.5$. Therefore, from Table A.1, the computed P-value is

$$P = 2P(X \leq 8 \text{ when } p = 0.7) = 2 \sum_{x=0}^{8} b(x; 15, 0.7) = 0.2622 > 0.10.$$

6. Decision: Do not reject H_0. Conclude that there is insufficient reason to doubt the builder's claim.

In Section 5.2, we saw that binomial probabilities can be obtained from the actual binomial formula or from Table A.1 when n is small. For large n, approximation procedures are required. When the hypothesized value p_0 is very close to 0 or 1, the Poisson distribution, with parameter $\mu = np_0$, may be used. However, the normal curve approximation, with parameters $\mu = np_0$ and $\sigma^2 = np_0q_0$, is usually preferred for large n and is very accurate as long as p_0 is not extremely close to 0 or to 1. If we use the normal approximation, the **z-value for testing $p = p_0$** is given by

$$z = \frac{x - np_0}{\sqrt{np_0q_0}} = \frac{\hat{p} - p_0}{\sqrt{p_0q_0/n}},$$

which is a value of the standard normal variable Z. Hence, for a two-tailed test at the α-level of significance, the critical region is $z < -z_{\alpha/2}$ or $z > z_{\alpha/2}$. For the one-sided alternative $p < p_0$, the critical region is $z < -z_\alpha$, and for the alternative $p > p_0$, the critical region is $z > z_\alpha$.

Example 10.10: A commonly prescribed drug for relieving nervous tension is believed to be only 60% effective. Experimental results with a new drug administered to a random sample of 100 adults who were suffering from nervous tension show that 70 received relief. Is this sufficient evidence to conclude that the new drug is superior to the one commonly prescribed? Use a 0.05 level of significance.

Solution: 1. H_0: $p = 0.6$.
2. H_1: $p > 0.6$.
3. $\alpha = 0.05$.
4. Critical region: $z > 1.645$.

5. Computations: $x = 70$, $n = 100$, $\hat{p} = 70/100 = 0.7$, and

$$z = \frac{0.7 - 0.6}{\sqrt{(0.6)(0.4)/100}} = 2.04, \quad P = P(Z > 2.04) < 0.0207.$$

6. Decision: Reject H_0 and conclude that the new drug is superior.

10.9 Two Samples: Tests on Two Proportions

Situations often arise where we wish to test the hypothesis that two proportions are equal. For example, we might want to show evidence that the proportion of doctors who are pediatricians in one state is equal to the proportion in another state. A person may decide to give up smoking only if he or she is convinced that the proportion of smokers with lung cancer exceeds the proportion of nonsmokers with lung cancer.

In general, we wish to test the null hypothesis that two proportions, or binomial parameters, are equal. That is, we are testing $p_1 = p_2$ against one of the alternatives $p_1 < p_2$, $p_1 > p_2$, or $p_1 \neq p_2$. Of course, this is equivalent to testing the null hypothesis that $p_1 - p_2 = 0$ against one of the alternatives $p_1 - p_2 < 0$, $p_1 - p_2 > 0$, or $p_1 - p_2 \neq 0$. The statistic on which we base our decision is the random variable $\hat{P}_1 - \hat{P}_2$. Independent samples of sizes n_1 and n_2 are selected at random from two binomial populations and the proportions of successes $\hat{P}_1$ and $\hat{P}_2$ for the two samples are computed.

In our construction of confidence intervals for p_1 and p_2 we noted, for n_1 and n_2 sufficiently large, that the point estimator $\hat{P}_1$ minus $\hat{P}_2$ was approximately normally distributed with mean

$$\mu_{\hat{P}_1 - \hat{P}_2} = p_1 - p_2$$

and variance

$$\sigma^2_{\hat{P}_1 - \hat{P}_2} = \frac{p_1 q_1}{n_1} + \frac{p_2 q_2}{n_2}.$$

Therefore, our critical region(s) can be established by using the standard normal variable

$$Z = \frac{(\hat{P}_1 - \hat{P}_2) - (p_1 - p_2)}{\sqrt{p_1 q_1/n_1 + p_2 q_2/n_2}}.$$

When H_0 is true, we can substitute $p_1 = p_2 = p$ and $q_1 = q_2 = q$ (where p and q are the common values) in the preceding formula for Z to give the form

$$Z = \frac{\hat{P}_1 - \hat{P}_2}{\sqrt{pq(1/n_1 + 1/n_2)}}.$$

To compute a value of Z, however, we must estimate the parameters p and q that appear in the radical. Upon pooling the data from both samples, the **pooled estimate of the proportion** p is

$$\hat{p} = \frac{x_1 + x_2}{n_1 + n_2},$$

where x_1 and x_2 are the numbers of successes in each of the two samples. Substituting $\hat{p}$ for p and $\hat{q} = 1 - \hat{p}$ for q, the **z-value for testing $p_1 = p_2$** is determined from the formula

$$z = \frac{\hat{p}_1 - \hat{p}_2}{\sqrt{\hat{p}\hat{q}(1/n_1 + 1/n_2)}}.$$

The critical regions for the appropriate alternative hypotheses are set up as before, using critical points of the standard normal curve. Hence, for the alternative $p_1 \neq p_2$ at the α-level of significance, the critical region is $z < -z_{\alpha/2}$ or $z > z_{\alpha/2}$. For a test where the alternative is $p_1 < p_2$, the critical region is $z < -z_\alpha$, and when the alternative is $p_1 > p_2$, the critical region is $z > z_\alpha$.

Example 10.11: A vote is to be taken among the residents of a town and the surrounding county to determine whether a proposed chemical plant should be constructed. The construction site is within the town limits, and for this reason many voters in the county believe that the proposal will pass because of the large proportion of town voters who favor the construction. To determine if there is a significant difference in the proportions of town voters and county voters favoring the proposal, a poll is taken. If 120 of 200 town voters favor the proposal and 240 of 500 county residents favor it, would you agree that the proportion of town voters favoring the proposal is higher than the proportion of county voters? Use an $\alpha = 0.05$ level of significance.

Solution: Let p_1 and p_2 be the true proportions of voters in the town and county, respectively, favoring the proposal.

1. H_0: $p_1 = p_2$.
2. H_1: $p_1 > p_2$.
3. $\alpha = 0.05$.
4. Critical region: $z > 1.645$.
5. Computations:

$$\hat{p}_1 = \frac{x_1}{n_1} = \frac{120}{200} = 0.60, \quad \hat{p}_2 = \frac{x_2}{n_2} = \frac{240}{500} = 0.48, \quad \text{and}$$

$$\hat{p} = \frac{x_1 + x_2}{n_1 + n_2} = \frac{120 + 240}{200 + 500} = 0.51.$$

Therefore,

$$z = \frac{0.60 - 0.48}{\sqrt{(0.51)(0.49)(1/200 + 1/500)}} = 2.9,$$

$$P = P(Z > 2.9) = 0.0019.$$

6. Decision: Reject H_0 and agree that the proportion of town voters favoring the proposal is higher than the proportion of county voters.

Exercises

10.55 A marketing expert for a pasta-making company believes that 40% of pasta lovers prefer lasagna. If 9 out of 20 pasta lovers choose lasagna over other pastas, what can be concluded about the expert's claim? Use a 0.05 level of significance.

10.56 Suppose that, in the past, 40% of all adults favored capital punishment. Do we have reason to believe that the proportion of adults favoring capital punishment has increased if, in a random sample of 15 adults, 8 favor capital punishment? Use a 0.05 level of significance.

10.57 A new radar device is being considered for a certain missile defense system. The system is checked by experimenting with aircraft in which a kill or a no kill is simulated. If, in 300 trials, 250 kills occur, accept or reject, at the 0.04 level of significance, the claim that the probability of a kill with the new system does not exceed the 0.8 probability of the existing device.

10.58 It is believed that at least 60% of the residents in a certain area favor an annexation suit by a neighboring city. What conclusion would you draw if only 110 in a sample of 200 voters favored the suit? Use a 0.05 level of significance.

10.59 A fuel oil company claims that one-fifth of the homes in a certain city are heated by oil. Do we have reason to believe that fewer than one-fifth are heated by oil if, in a random sample of 1000 homes in this city, 136 are heated by oil? Use a P-value in your conclusion.

10.60 At a certain college, it is estimated that at most 25% of the students ride bicycles to class. Does this seem to be a valid estimate if, in a random sample of 90 college students, 28 are found to ride bicycles to class? Use a 0.05 level of significance.

10.61 In a winter of an epidemic flu, the parents of 2000 babies were surveyed by researchers at a well-known pharmaceutical company to determine if the company's new medicine was effective after two days. Among 120 babies who had the flu and were given the medicine, 29 were cured within two days. Among 280 babies who had the flu but were not given the medicine, 56 recovered within two days. Is there any significant indication that supports the company's claim of the effectiveness of the medicine?

10.62 In a controlled laboratory experiment, scientists at the University of Minnesota discovered that 25% of a certain strain of rats subjected to a 20% coffee bean diet and then force-fed a powerful cancer-causing chemical later developed cancerous tumors. Would we have reason to believe that the proportion of rats developing tumors when subjected to this diet has increased if the experiment were repeated and 16 of 48 rats developed tumors? Use a 0.05 level of significance.

10.63 In a study to estimate the proportion of residents in a certain city and its suburbs who favor the construction of a nuclear power plant, it is found that 63 of 100 urban residents favor the construction while only 59 of 125 suburban residents are in favor. Is there a significant difference between the proportions of urban and suburban residents who favor construction of the nuclear plant? Make use of a P-value.

10.64 In a study on the fertility of married women conducted by Martin O'Connell and Carolyn C. Rogers for the Census Bureau in 1979, two groups of childless wives aged 25 to 29 were selected at random, and each was asked if she eventually planned to have a child. One group was selected from among wives married less than two years and the other from among wives married five years. Suppose that 240 of the 300 wives married less than two years planned to have children some day compared to 288 of the 400 wives married five years. Can we conclude that the proportion of wives married less than two years who planned to have children is significantly higher than the proportion of wives married five years? Make use of a P-value.

10.65 An urban community would like to show that the incidence of breast cancer is higher in their area than in a nearby rural area. (PCB levels were found to be higher in the soil of the urban community.) If it is found that 20 of 200 adult women in the urban community have breast cancer and 10 of 150 adult women in the rural community have breast cancer, can we conclude at the 0.05 level of significance that breast cancer is more prevalent in the urban community?

10.66 Group Project: The class should be divided into pairs of students for this project. Suppose it is conjectured that at least 25% of students at your university exercise for more than two hours a week. Collect data from a random sample of 50 students. Ask each student if he or she works out for at least two hours per week. Then do the computations that allow either rejection or nonrejection of the above conjecture. Show all work and quote a P-value in your conclusion.

10.10 One- and Two-Sample Tests Concerning Variances

In this section, we are concerned with testing hypotheses concerning population variances or standard deviations. Applications of one- and two-sample tests on variances are certainly not difficult to motivate. Engineers and scientists are confronted with studies in which they are required to demonstrate that measurements involving products or processes adhere to specifications set by consumers. The specifications are often met if the process variance is sufficiently small. Attention is also focused on comparative experiments between methods or processes, where inherent reproducibility or variability must formally be compared. In addition, to determine if the equal variance assumption is violated, a test comparing two variances is often applied prior to conducting a t-test on two means.

Let us first consider the problem of testing the null hypothesis H_0 that the population variance σ^2 equals a specified value σ_0^2 against one of the usual alternatives $\sigma^2 < \sigma_0^2$, $\sigma^2 > \sigma_0^2$, or $\sigma^2 \neq \sigma_0^2$. The appropriate statistic on which to base our decision is the chi-squared statistic of Theorem 8.4, which was used in Chapter 9 to construct a confidence interval for σ^2. Therefore, if we assume that the distribution of the population being sampled is normal, the chi-squared value for testing $\sigma^2 = \sigma_0^2$ is given by

$$\chi^2 = \frac{(n-1)s^2}{\sigma_0^2},$$

where n is the sample size, s^2 is the sample variance, and σ_0^2 is the value of σ^2 given by the null hypothesis. If H_0 is true, χ^2 is a value of the chi-squared distribution with $v = n - 1$ degrees of freedom. Hence, for a two-tailed test at the α-level of significance, the critical region is $\chi^2 < \chi^2_{1-\alpha/2}$ or $\chi^2 > \chi^2_{\alpha/2}$. For the one-sided alternative $\sigma^2 < \sigma_0^2$, the critical region is $\chi^2 < \chi^2_{1-\alpha}$, and for the one-sided alternative $\sigma^2 > \sigma_0^2$, the critical region is $\chi^2 > \chi^2_{\alpha}$.

Robustness of χ^2-Test to Assumption of Normality

The reader may have discerned that various tests depend, at least theoretically, on the assumption of normality. In general, many procedures in applied statistics have theoretical underpinnings that depend on the normal distribution. These procedures vary in the degree of their dependency on the assumption of normality. A procedure that is reasonably insensitive to the assumption is called a **robust procedure** (i.e., robust to normality). The χ^2-test on a single variance is very nonrobust to normality (i.e., the practical success of the procedure depends on normality). As a result, the P-value computed may be appreciably different from the actual P-value if the population sampled is not normal. Indeed, it is quite feasible that a statistically significant P-value may not truly signal H_1: $\sigma \neq \sigma_0$; rather, a significant value may be a result of the violation of the normality assumptions. Therefore, the analyst should approach the use of this particular χ^2-test with caution.

Example 10.12: A manufacturer of car batteries claims that the life of the company's batteries is approximately normally distributed with a standard deviation equal to 0.9 year.

10.10 One- and Two-Sample Tests Concerning Variances

If a random sample of 10 of these batteries has a standard deviation of 1.2 years, do you think that $\sigma > 0.9$ year? Use a 0.05 level of significance.

Solution:
1. H_0: $\sigma^2 = 0.81$.
2. H_1: $\sigma^2 > 0.81$.
3. $\alpha = 0.05$.
4. Critical region: From Figure 10.19 we see that the null hypothesis is rejected when $\chi^2 > 16.919$, where $\chi^2 = \frac{(n-1)s^2}{\sigma_0^2}$, with $v = 9$ degrees of freedom.

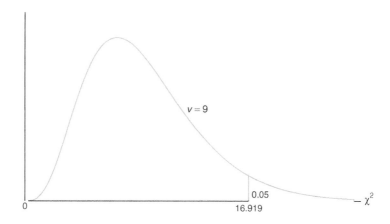

Figure 10.19: Critical region for the alternative hypothesis $\sigma > 0.9$.

5. Computations: $s^2 = 1.44$, $n = 10$, and

$$\chi^2 = \frac{(9)(1.44)}{0.81} = 16.0, \qquad P \approx 0.07.$$

6. Decision: The χ^2-statistic is not significant at the 0.05 level. However, based on the P-value 0.07, there is evidence that $\sigma > 0.9$.

Now let us consider the problem of testing the equality of the variances σ_1^2 and σ_2^2 of two populations. That is, we shall test the null hypothesis H_0 that $\sigma_1^2 = \sigma_2^2$ against one of the usual alternatives

$$\sigma_1^2 < \sigma_2^2, \quad \sigma_1^2 > \sigma_2^2, \quad \text{or} \quad \sigma_1^2 \neq \sigma_2^2.$$

For independent random samples of sizes n_1 and n_2, respectively, from the two populations, the **f-value for testing $\sigma_1^2 = \sigma_2^2$** is the ratio

$$f = \frac{s_1^2}{s_2^2},$$

where s_1^2 and s_2^2 are the variances computed from the two samples. If the two populations are approximately normally distributed and the null hypothesis is true, according to Theorem 8.8 the ratio $f = s_1^2/s_2^2$ is a value of the F-distribution with $v_1 = n_1 - 1$ and $v_2 = n_2 - 1$ degrees of freedom. Therefore, the critical regions of size α corresponding to the one-sided alternatives $\sigma_1^2 < \sigma_2^2$ and $\sigma_1^2 > \sigma_2^2$ are, respectively, $f < f_{1-\alpha}(v_1, v_2)$ and $f > f_\alpha(v_1, v_2)$. For the two-sided alternative $\sigma_1^2 \neq \sigma_2^2$, the critical region is $f < f_{1-\alpha/2}(v_1, v_2)$ or $f > f_{\alpha/2}(v_1, v_2)$.

Example 10.13: In testing for the difference in the abrasive wear of the two materials in Example 10.6, we assumed that the two unknown population variances were equal. Were we justified in making this assumption? Use a 0.10 level of significance.

Solution: Let σ_1^2 and σ_2^2 be the population variances for the abrasive wear of material 1 and material 2, respectively.

1. H_0: $\sigma_1^2 = \sigma_2^2$.
2. H_1: $\sigma_1^2 \neq \sigma_2^2$.
3. $\alpha = 0.10$.
4. Critical region: From Figure 10.20, we see that $f_{0.05}(11, 9) = 3.11$, and, by using Theorem 8.7, we find

$$f_{0.95}(11, 9) = \frac{1}{f_{0.05}(9, 11)} = 0.34.$$

Therefore, the null hypothesis is rejected when $f < 0.34$ or $f > 3.11$, where $f = s_1^2/s_2^2$ with $v_1 = 11$ and $v_2 = 9$ degrees of freedom.

5. Computations: $s_1^2 = 16$, $s_2^2 = 25$, and hence $f = \frac{16}{25} = 0.64$.
6. Decision: Do not reject H_0. Conclude that there is insufficient evidence that the variances differ.

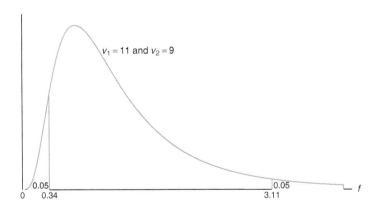

Figure 10.20: Critical region for the alternative hypothesis $\sigma_1^2 \neq \sigma_2^2$.

F-Test for Testing Variances in *SAS*

Figure 10.18 on page 356 displays the printout of a two-sample t-test where two means from the seedling data in Exercise 9.40 were compared. Box-and-whisker plots in Figure 10.17 on page 355 suggest that variances are not homogeneous, and thus the t'-statistic and its corresponding P-value are relevant. Note also that the printout displays the F-statistic for H_0: $\sigma_1 = \sigma_2$ with a P-value of 0.0098, additional evidence that more variability is to be expected when nitrogen is used than under the no-nitrogen condition.

Exercises

10.67 The content of containers of a particular lubricant is known to be normally distributed with a variance of 0.03 liter. Test the hypothesis that $\sigma^2 = 0.03$ against the alternative that $\sigma^2 \neq 0.03$ for the random sample of 10 containers in Exercise 10.23 on page 356. Use a P-value in your conclusion.

10.68 Past experience indicates that the time required for high school seniors to complete a standardized test is a normal random variable with a standard deviation of 6 minutes. Test the hypothesis that $\sigma = 6$ against the alternative that $\sigma < 6$ if a random sample of the test times of 20 high school seniors has a standard deviation $s = 4.51$. Use a 0.05 level of significance.

10.69 Aflatoxins produced by mold on peanut crops in Virginia must be monitored. A sample of 64 batches of peanuts reveals levels of 24.17 ppm, on average, with a variance of 4.25 ppm. Test the hypothesis that $\sigma^2 = 4.2$ ppm against the alternative that $\sigma^2 \neq 4.2$ ppm. Use a P-value in your conclusion.

10.70 Past data indicate that the amount of money contributed by the working residents of a large city to a volunteer rescue squad is a normal random variable with a standard deviation of $1.40. It has been suggested that the contributions to the rescue squad from just the employees of the sanitation department are much more variable. If the contributions of a random sample of 12 employees from the sanitation department have a standard deviation of $1.75, can we conclude at the 0.01 level of significance that the standard deviation of the contributions of all sanitation workers is greater than that of all workers living in the city?

10.71 A soft-drink dispensing machine is said to be out of control if the variance of the contents exceeds 1.15 deciliters. If a random sample of 25 drinks from this machine has a variance of 2.03 deciliters, does this indicate at the 0.05 level of significance that the machine is out of control? Assume that the contents are approximately normally distributed.

10.72 Large-Sample Test of $\sigma^2 = \sigma_0^2$: When $n \geq 30$, we can test the null hypothesis that $\sigma^2 = \sigma_0^2$, or $\sigma = \sigma_0$, by computing

$$z = \frac{s - \sigma_0}{\sigma_0/\sqrt{2n}},$$

which is a value of a random variable whose sampling distribution is approximately the standard normal distribution.

(a) With reference to Example 10.4, test at the 0.05 level of significance whether $\sigma = 10.0$ years against the alternative that $\sigma \neq 10.0$ years.

(b) It is suspected that the variance of the distribution of distances in kilometers traveled on 5 liters of fuel by a new automobile model equipped with a diesel engine is less than the variance of the distribution of distances traveled by the same model equipped with a six-cylinder gasoline engine, which is known to be $\sigma^2 = 6.25$. If 72 test runs of the diesel model have a variance of 4.41, can we conclude at the 0.05 level of significance that the variance of the distances traveled by the diesel model is less than that of the gasoline model?

10.73 A study is conducted to compare the lengths of time required by men and women to assemble a certain product. Past experience indicates that the distribution of times for both men and women is approximately normal but the variance of the times for women is less than that for men. A random sample of times for 11 men and 14 women produced the following data:

Men	Women
$n_1 = 11$	$n_2 = 14$
$s_1 = 6.1$	$s_2 = 5.3$

Test the hypothesis that $\sigma_1^2 = \sigma_2^2$ against the alternative that $\sigma_1^2 > \sigma_2^2$. Use a P-value in your conclusion.

10.74 For Exercise 10.41 on page 358, test the hypothesis at the 0.05 level of significance that $\sigma_1^2 = \sigma_2^2$ against the alternative that $\sigma_1^2 \neq \sigma_2^2$, where σ_1^2 and σ_2^2 are the variances of the number of organisms per square meter of water at the two different locations on Cedar Run.

10.75 With reference to Exercise 10.39 on page 358, test the hypothesis that $\sigma_1^2 = \sigma_2^2$ against the alternative that $\sigma_1^2 \neq \sigma_2^2$, where σ_1^2 and σ_2^2 are the variances for the running times of films produced by company 1 and company 2, respectively. Use a P-value.

10.76 Two types of instruments for measuring the amount of sulfur monoxide in the atmosphere are being compared in an air-pollution experiment. Researchers wish to determine whether the two types of instruments yield measurements having the same variability. The readings in the following table were recorded for the two instruments.

Sulfur Monoxide	
Instrument A	Instrument B
0.86	0.87
0.82	0.74
0.75	0.63
0.61	0.55
0.89	0.76
0.64	0.70
0.81	0.69
0.68	0.57
0.65	0.53

Assuming the populations of measurements to be approximately normally distributed, test the hypothesis that $\sigma_A = \sigma_B$ against the alternative that $\sigma_A \neq \sigma_B$. Use a P-value.

10.77 An experiment was conducted to compare the alcohol content of soy sauce on two different production lines. Production was monitored eight times a day. The data are shown here.

Production line 1:
  0.48 0.39 0.42 0.52 0.40 0.48 0.52 0.52
Production line 2:
  0.38 0.37 0.39 0.41 0.38 0.39 0.40 0.39

Assume both populations are normal. It is suspected that production line 1 is not producing as consistently as production line 2 in terms of alcohol content. Test the hypothesis that $\sigma_1 = \sigma_2$ against the alternative that $\sigma_1 \neq \sigma_2$. Use a P-value.

10.78 Hydrocarbon emissions from cars are known to have decreased dramatically during the 1980s. A study was conducted to compare the hydrocarbon emissions at idling speed, in parts per million (ppm), for automobiles from 1980 and 1990. Twenty cars of each model year were randomly selected, and their hydrocarbon emission levels were recorded. The data are as follows:

1980 models:
141 359 247 940 882 494 306 210 105 880
200 223 188 940 241 190 300 435 241 380
1990 models:
140 160 20 20 223 60 20 95 360 70
220 400 217 58 235 380 200 175 85 65

Test the hypothesis that $\sigma_1 = \sigma_2$ against the alternative that $\sigma_1 \neq \sigma_2$. Assume both populations are normal. Use a P-value.

10.11 Goodness-of-Fit Test

Throughout this chapter, we have been concerned with the testing of statistical hypotheses about single population parameters such as μ, σ^2, and p. Now we shall consider a test to determine if a population has a specified theoretical distribution. The test is based on how good a fit we have between the frequency of occurrence of observations in an observed sample and the expected frequencies obtained from the hypothesized distribution.

To illustrate, we consider the tossing of a die. We hypothesize that the die is honest, which is equivalent to testing the hypothesis that the distribution of outcomes is the discrete uniform distribution

$$f(x) = \frac{1}{6}, \quad x = 1, 2, \ldots, 6.$$

Suppose that the die is tossed 120 times and each outcome is recorded. Theoretically, if the die is balanced, we would expect each face to occur 20 times. The results are given in Table 10.4.

Table 10.4: Observed and Expected Frequencies of 120 Tosses of a Die

Face:	1	2	3	4	5	6
Observed	20	22	17	18	19	24
Expected	20	20	20	20	20	20

10.11 Goodness-of-Fit Test

By comparing the observed frequencies with the corresponding expected frequencies, we must decide whether these discrepancies are likely to occur as a result of sampling fluctuations and the die is balanced or whether the die is not honest and the distribution of outcomes is not uniform. It is common practice to refer to each possible outcome of an experiment as a cell. In our illustration, we have 6 cells. The appropriate statistic on which we base our decision criterion for an experiment involving k cells is defined by the following.

A **goodness-of-fit test** between observed and expected frequencies is based on the quantity

Goodness-of-Fit Test

$$\chi^2 = \sum_{i=1}^{k} \frac{(o_i - e_i)^2}{e_i},$$

where χ^2 is a value of a random variable whose sampling distribution is approximated very closely by the chi-squared distribution with $v = k - 1$ degrees of freedom. The symbols o_i and e_i represent the observed and expected frequencies, respectively, for the ith cell.

The number of degrees of freedom associated with the chi-squared distribution used here is equal to $k - 1$, since there are only $k - 1$ freely determined cell frequencies. That is, once $k - 1$ cell frequencies are determined, so is the frequency for the kth cell.

If the observed frequencies are close to the corresponding expected frequencies, the χ^2-value will be small, indicating a good fit. If the observed frequencies differ considerably from the expected frequencies, the χ^2-value will be large and the fit is poor. A good fit leads to the acceptance of H_0, whereas a poor fit leads to its rejection. The critical region will, therefore, fall in the right tail of the chi-squared distribution. For a level of significance equal to α, we find the critical value χ^2_α from Table A.5, and then $\chi^2 > \chi^2_\alpha$ constitutes the critical region. **The decision criterion described here should not be used unless each of the expected frequencies is at least equal to 5**. This restriction may require the combining of adjacent cells, resulting in a reduction in the number of degrees of freedom.

From Table 10.4, we find the χ^2-value to be

$$\chi^2 = \frac{(20-20)^2}{20} + \frac{(22-20)^2}{20} + \frac{(17-20)^2}{20}$$
$$+ \frac{(18-20)^2}{20} + \frac{(19-20)^2}{20} + \frac{(24-20)^2}{20} = 1.7.$$

Using Table A.5, we find $\chi^2_{0.05} = 11.070$ for $v = 5$ degrees of freedom. Since 1.7 is less than the critical value, we fail to reject H_0. We conclude that there is insufficient evidence that the die is not balanced.

As a second illustration, let us test the hypothesis that the frequency distribution of battery lives given in Table 1.7 on page 23 may be approximated by a normal distribution with mean $\mu = 3.5$ and standard deviation $\sigma = 0.7$. The expected frequencies for the 7 classes (cells), listed in Table 10.5, are obtained by computing the areas under the hypothesized normal curve that fall between the various class boundaries.

Table 10.5: Observed and Expected Frequencies of Battery Lives, Assuming Normality

Class Boundaries	o_i		e_i	
1.45–1.95	2		0.5	
1.95–2.45	1	7	2.1	8.5
2.45–2.95	4		5.9	
2.95–3.45	15		10.3	
3.45–3.95	10		10.7	
3.95–4.45	5	8	7.0	10.5
4.45–4.95	3		3.5	

For example, the z-values corresponding to the boundaries of the fourth class are

$$z_1 = \frac{2.95 - 3.5}{0.7} = -0.79 \quad \text{and} \quad z_2 = \frac{3.45 - 3.5}{0.7} = -0.07.$$

From Table A.3 we find the area between $z_1 = -0.79$ and $z_2 = -0.07$ to be

$$\text{area} = P(-0.79 < Z < -0.07) = P(Z < -0.07) - P(Z < -0.79)$$
$$= 0.4721 - 0.2148 = 0.2573.$$

Hence, the expected frequency for the fourth class is

$$e_4 = (0.2573)(40) = 10.3.$$

It is customary to round these frequencies to one decimal.

The expected frequency for the first class interval is obtained by using the total area under the normal curve to the left of the boundary 1.95. For the last class interval, we use the total area to the right of the boundary 4.45. All other expected frequencies are determined by the method described for the fourth class. Note that we have combined adjacent classes in Table 10.5 where the expected frequencies are less than 5 (a rule of thumb in the goodness-of-fit test). Consequently, the total number of intervals is reduced from 7 to 4, resulting in $v = 3$ degrees of freedom. The χ^2-value is then given by

$$\chi^2 = \frac{(7-8.5)^2}{8.5} + \frac{(15-10.3)^2}{10.3} + \frac{(10-10.7)^2}{10.7} + \frac{(8-10.5)^2}{10.5} = 3.05.$$

Since the computed χ^2-value is less than $\chi^2_{0.05} = 7.815$ for 3 degrees of freedom, we have no reason to reject the null hypothesis and conclude that the normal distribution with $\mu = 3.5$ and $\sigma = 0.7$ provides a good fit for the distribution of battery lives.

The chi-squared goodness-of-fit test is an important resource, particularly since so many statistical procedures in practice depend, in a theoretical sense, on the assumption that the data gathered come from a specific type of distribution. As we have already seen, the normality assumption is often made. In the chapters that follow, we shall continue to make normality assumptions in order to provide a theoretical basis for certain tests and confidence intervals.

There are tests in the literature that are more powerful than the chi-squared test for testing normality. One such test is called **Geary's test**. This test is based on a very simple statistic which is a ratio of two estimators of the population standard deviation σ. Suppose that a random sample $X_1, X_2, \ldots, X_n$ is taken from a normal distribution, $N(\mu, \sigma)$. Consider the ratio

$$U = \frac{\sqrt{\pi/2} \sum_{i=1}^{n} |X_i - \bar{X}|/n}{\sqrt{\sum_{i=1}^{n}(X_i - \bar{X})^2/n}}.$$

The reader should recognize that the denominator is a reasonable estimator of σ whether the distribution is normal or not. The numerator is a good estimator of σ if the distribution is normal but may overestimate or underestimate σ when there are departures from normality. Thus, values of U differing considerably from 1.0 represent the signal that the hypothesis of normality should be rejected.

For large samples, a reasonable test is based on approximate normality of U. The test statistic is then a standardization of U, given by

$$Z = \frac{U-1}{0.2661/\sqrt{n}}.$$

Of course, the test procedure involves the two-sided critical region. We compute a value of z from the data and do not reject the hypothesis of normality when

$$-z_{\alpha/2} < Z < z_{\alpha/2}.$$

A paper dealing with Geary's test is cited in the Bibliography (Geary, 1947).

10.12 Test for Independence (Categorical Data)

The chi-squared test procedure discussed in Section 10.11 can also be used to test the hypothesis of independence of two variables of classification. Suppose that we wish to determine whether the opinions of the voting residents of the state of Illinois concerning a new tax reform are independent of their levels of income. Members of a random sample of 1000 registered voters from the state of Illinois are classified as to whether they are in a low, medium, or high income bracket and whether or not they favor the tax reform. The observed frequencies are presented in Table 10.6, which is known as a **contingency table**.

Table 10.6: 2 × 3 Contingency Table

Tax Reform	Low	Medium	High	Total
For	182	213	203	598
Against	154	138	110	402
Total	336	351	313	1000

(Income Level spans Low, Medium, High)

A contingency table with r rows and c columns is referred to as an $r \times c$ table ("$r \times c$" is read "r by c"). The row and column totals in Table 10.6 are called **marginal frequencies**. Our decision to accept or reject the null hypothesis, H_0, of independence between a voter's opinion concerning the tax reform and his or her level of income is based upon how good a fit we have between the observed frequencies in each of the 6 cells of Table 10.6 and the frequencies that we would expect for each cell under the assumption that H_0 is true. To find these expected frequencies, let us define the following events:

L: A person selected is in the low-income level.
M: A person selected is in the medium-income level.
H: A person selected is in the high-income level.
F: A person selected is for the tax reform.
A: A person selected is against the tax reform.

By using the marginal frequencies, we can list the following probability estimates:

$$P(L) = \frac{336}{1000}, \quad P(M) = \frac{351}{1000}, \quad P(H) = \frac{313}{1000},$$
$$P(F) = \frac{598}{1000}, \quad P(A) = \frac{402}{1000}.$$

Now, if H_0 is true and the two variables are independent, we should have

$$P(L \cap F) = P(L)P(F) = \left(\frac{336}{1000}\right)\left(\frac{598}{1000}\right),$$
$$P(L \cap A) = P(L)P(A) = \left(\frac{336}{1000}\right)\left(\frac{402}{1000}\right),$$
$$P(M \cap F) = P(M)P(F) = \left(\frac{351}{1000}\right)\left(\frac{598}{1000}\right),$$
$$P(M \cap A) = P(M)P(A) = \left(\frac{351}{1000}\right)\left(\frac{402}{1000}\right),$$
$$P(H \cap F) = P(H)P(F) = \left(\frac{313}{1000}\right)\left(\frac{598}{1000}\right),$$
$$P(H \cap A) = P(H)P(A) = \left(\frac{313}{1000}\right)\left(\frac{402}{1000}\right).$$

The expected frequencies are obtained by multiplying each cell probability by the total number of observations. As before, we round these frequencies to one decimal. Thus, the expected number of low-income voters in our sample who favor the tax reform is estimated to be

$$\left(\frac{336}{1000}\right)\left(\frac{598}{1000}\right)(1000) = \frac{(336)(598)}{1000} = 200.9.$$

when H_0 is true. The general rule for obtaining the expected frequency of any cell is given by the following formula:

$$\text{expected frequency} = \frac{(\text{column total}) \times (\text{row total})}{\text{grand total}}.$$

The expected frequency for each cell is recorded in parentheses beside the actual observed value in Table 10.7. Note that the expected frequencies in any row or column add up to the appropriate marginal total. In our example, we need to compute only two expected frequencies in the top row of Table 10.7 and then find the others by subtraction. The number of degrees of freedom associated with the chi-squared test used here is equal to the number of cell frequencies that may be filled in freely when we are given the marginal totals and the grand total, and in this illustration that number is 2. A simple formula providing the correct number of degrees of freedom is

$$v = (r-1)(c-1).$$

Table 10.7: Observed and Expected Frequencies

Tax Reform	Income Level			Total
	Low	Medium	High	
For	182 (200.9)	213 (209.9)	203 (187.2)	598
Against	154 (135.1)	138 (141.1)	110 (125.8)	402
Total	336	351	313	1000

Hence, for our example, $v = (2-1)(3-1) = 2$ degrees of freedom. To test the null hypothesis of independence, we use the following decision criterion.

Test for Independence Calculate

$$\chi^2 = \sum_i \frac{(o_i - e_i)^2}{e_i},$$

where the summation extends over all rc cells in the $r \times c$ contingency table. If $\chi^2 > \chi_\alpha^2$ with $v = (r-1)(c-1)$ degrees of freedom, reject the null hypothesis of independence at the α-level of significance; otherwise, fail to reject the null hypothesis.

Applying this criterion to our example, we find that

$$\chi^2 = \frac{(182-200.9)^2}{200.9} + \frac{(213-209.9)^2}{209.9} + \frac{(203-187.2)^2}{187.2}$$
$$+ \frac{(154-135.1)^2}{135.1} + \frac{(138-141.1)^2}{141.1} + \frac{(110-125.8)^2}{125.8} = 7.85,$$

$P \approx 0.02$.

From Table A.5 we find that $\chi_{0.05}^2 = 5.991$ for $v = (2-1)(3-1) = 2$ degrees of freedom. The null hypothesis is rejected and we conclude that a voter's opinion concerning the tax reform and his or her level of income are not independent.

It is important to remember that the statistic on which we base our decision has a distribution that is only approximated by the chi-squared distribution. The computed χ^2-values depend on the cell frequencies and consequently are discrete. The continuous chi-squared distribution seems to approximate the discrete sampling distribution of χ^2 very well, provided that the number of degrees of freedom is greater than 1. In a 2 × 2 contingency table, where we have only 1 degree of freedom, a correction called **Yates' correction for continuity** is applied. The corrected formula then becomes

$$\chi^2(\text{corrected}) = \sum_i \frac{(|o_i - e_i| - 0.5)^2}{e_i}.$$

If the expected cell frequencies are large, the corrected and uncorrected results are almost the same. When the expected frequencies are between 5 and 10, Yates' correction should be applied. For expected frequencies less than 5, the Fisher-Irwin exact test should be used. A discussion of this test may be found in *Basic Concepts of Probability and Statistics* by Hodges and Lehmann (2005; see the Bibliography). The Fisher-Irwin test may be avoided, however, by choosing a larger sample.

10.13 Test for Homogeneity

When we tested for independence in Section 10.12, a random sample of 1000 voters was selected and the row and column totals for our contingency table were determined by chance. Another type of problem for which the method of Section 10.12 applies is one in which either the row or column totals are predetermined. Suppose, for example, that we decide in advance to select 200 Democrats, 150 Republicans, and 150 Independents from the voters of the state of North Carolina and record whether they are for a proposed abortion law, against it, or undecided. The observed responses are given in Table 10.8.

Table 10.8: Observed Frequencies

Abortion Law	Political Affiliation			Total
	Democrat	Republican	Independent	
For	82	70	62	214
Against	93	62	67	222
Undecided	25	18	21	64
Total	200	150	150	500

Now, rather than test for independence, we test the hypothesis that the population proportions within each row are the same. That is, we test the hypothesis that the proportions of Democrats, Republicans, and Independents favoring the abortion law are the same; the proportions of each political affiliation against the law are the same; and the proportions of each political affiliation that are undecided are the same. We are basically interested in determining whether the three categories of voters are **homogeneous** with respect to their opinions concerning the proposed abortion law. Such a test is called a test for homogeneity.

Assuming homogeneity, we again find the expected cell frequencies by multiplying the corresponding row and column totals and then dividing by the grand

10.13 Test for Homogeneity

total. The analysis then proceeds using the same chi-squared statistic as before. We illustrate this process for the data of Table 10.8 in the following example.

Example 10.14: Referring to the data of Table 10.8, test the hypothesis that opinions concerning the proposed abortion law are the same within each political affiliation. Use a 0.05 level of significance.

Solution:
1. H_0: For each opinion, the proportions of Democrats, Republicans, and Independents are the same.
2. H_1: For at least one opinion, the proportions of Democrats, Republicans, and Independents are not the same.
3. $\alpha = 0.05$.
4. Critical region: $\chi^2 > 9.488$ with $v = 4$ degrees of freedom.
5. Computations: Using the expected cell frequency formula on page 375, we need to compute 4 cell frequencies. All other frequencies are found by subtraction. The observed and expected cell frequencies are displayed in Table 10.9.

Table 10.9: Observed and Expected Frequencies

Abortion Law	Political Affiliation			Total
	Democrat	Republican	Independent	
For	82 (85.6)	70 (64.2)	62 (64.2)	214
Against	93 (88.8)	62 (66.6)	67 (66.6)	222
Undecided	25 (25.6)	18 (19.2)	21 (19.2)	64
Total	200	150	150	500

Now,

$$\chi^2 = \frac{(82-85.6)^2}{85.6} + \frac{(70-64.2)^2}{64.2} + \frac{(62-64.2)^2}{64.2}$$
$$+ \frac{(93-88.8)^2}{88.8} + \frac{(62-66.6)^2}{66.6} + \frac{(67-66.6)^2}{66.6}$$
$$+ \frac{(25-25.6)^2}{25.6} + \frac{(18-19.2)^2}{19.2} + \frac{(21-19.2)^2}{19.2}$$
$$= 1.53.$$

6. Decision: Do not reject H_0. There is insufficient evidence to conclude that the proportions of Democrats, Republicans, and Independents differ for each stated opinion.

Testing for Several Proportions

The chi-squared statistic for testing for homogeneity is also applicable when testing the hypothesis that k binomial parameters have the same value. This is, therefore, an extension of the test presented in Section 10.9 for determining differences between two proportions to a test for determining differences among k proportions. Hence, we are interested in testing the null hypothesis

$$H_0: p_1 = p_2 = \cdots = p_k$$

against the alternative hypothesis, H_1, that the population proportions are *not all equal*. To perform this test, we first observe independent random samples of size $n_1, n_2, \ldots, n_k$ from the k populations and arrange the data in a $2 \times k$ contingency table, Table 10.10.

Table 10.10: k Independent Binomial Samples

Sample:	1	2	$\cdots$	k
Successes	x_1	x_2	$\cdots$	x_k
Failures	n_1-x_1	n_2-x_2	$\cdots$	n_k-x_k

Depending on whether the sizes of the random samples were predetermined or occurred at random, the test procedure is identical to the test for homogeneity or the test for independence. Therefore, the expected cell frequencies are calculated as before and substituted, together with the observed frequencies, into the chi-squared statistic

$$\chi^2 = \sum_i \frac{(o_i - e_i)^2}{e_i},$$

with

$$v = (2-1)(k-1) = k-1$$

degrees of freedom.

By selecting the appropriate upper-tail critical region of the form $\chi^2 > \chi_\alpha^2$, we can now reach a decision concerning H_0.

Example 10.15: In a shop study, a set of data was collected to determine whether or not the proportion of defectives produced was the same for workers on the day, evening, and night shifts. The data collected are shown in Table 10.11.

Table 10.11: Data for Example 10.15

Shift:	Day	Evening	Night
Defectives	45	55	70
Nondefectives	905	890	870

Use a 0.025 level of significance to determine if the proportion of defectives is the same for all three shifts.

Solution: Let p_1, p_2, and p_3 represent the true proportions of defectives for the day, evening, and night shifts, respectively.

1. H_0: $p_1 = p_2 = p_3$.
2. H_1: p_1, p_2, and p_3 are not all equal.
3. $\alpha = 0.025$.
4. Critical region: $\chi^2 > 7.378$ for $v = 2$ degrees of freedom.

5. Computations: Corresponding to the observed frequencies $o_1 = 45$ and $o_2 = 55$, we find

$$e_1 = \frac{(950)(170)}{2835} = 57.0 \quad \text{and} \quad e_2 = \frac{(945)(170)}{2835} = 56.7.$$

All other expected frequencies are found by subtraction and are displayed in Table 10.12.

Table 10.12: Observed and Expected Frequencies

Shift:	Day	Evening	Night	Total
Defectives	45 (57.0)	55 (56.7)	70 (56.3)	170
Nondefectives	905 (893.0)	890 (888.3)	870 (883.7)	2665
Total	950	945	940	2835

Now

$$\chi^2 = \frac{(45-57.0)^2}{57.0} + \frac{(55-56.7)^2}{56.7} + \frac{(70-56.3)^2}{56.3}$$
$$+ \frac{(905-893.0)^2}{893.0} + \frac{(890-888.3)^2}{888.3} + \frac{(870-883.7)^2}{883.7} = 6.29,$$

$P \approx 0.04$.

6. Decision: We do not reject H_0 at $\alpha = 0.025$. Nevertheless, with the above P-value computed, it would certainly be dangerous to conclude that the proportion of defectives produced is the same for all shifts.

Often a complete study involving the use of statistical methods in hypothesis testing can be illustrated for the scientist or engineer using both test statistics, complete with P-values and statistical graphics. The graphics supplement the numerical diagnostics with pictures that show intuitively why the P-values appear as they do, as well as how reasonable (or not) the operative assumptions are.

10.14 Two-Sample Case Study

In this section, we consider a study involving a thorough graphical and formal analysis, along with annotated computer printout and conclusions. In a data analysis study conducted by personnel at the Laboratory for Interdisciplinary Statistical Analysis at Virginia Tech, two different materials, alloy A and alloy B, were compared in terms of breaking strength. Alloy B is more expensive, but it should certainly be adopted if it can be shown to be stronger than alloy A. The consistency of performance of the two alloys should also be taken into account.

Random samples of beams made from each alloy were selected, and strength was measured in units of 0.001-inch deflection as a fixed force was applied at both ends of the beam. Twenty specimens were used for each of the two alloys. The data are given in Table 10.13.

It is important that the engineer compare the two alloys. Of concern is average strength and reproducibility. It is of interest to determine if there is a severe

Table 10.13: Data for Two-Sample Case Study

Alloy A			Alloy B		
88	82	87	75	81	80
79	85	90	77	78	81
84	88	83	86	78	77
89	80	81	84	82	78
81	85		80	80	
83	87		78	76	
82	80		83	85	
79	78		76	79	

violation of the normality assumption required of both the t- and F-tests. Figures 10.21 and 10.22 are normal quantile-quantile plots of the samples of the two alloys.

There does not appear to be any serious violation of the normality assumption. In addition, Figure 10.23 shows two box-and-whisker plots on the same graph. The box-and-whisker plots suggest that there is no appreciable difference in the variability of deflection for the two alloys. However, it seems that the mean deflection for alloy B is significantly smaller, suggesting, at least graphically, that alloy B is stronger. The sample means and standard deviations are

$$\bar{y}_A = 83.55, \quad s_A = 3.663; \quad \bar{y}_B = 79.70, \quad s_B = 3.097.$$

The *SAS* printout for the PROC TTEST is shown in Figure 10.24. The F-test suggests no significant difference in variances ($P = 0.4709$), and the two-sample t-statistic for testing

$$H_0: \mu_A = \mu_B,$$
$$H_1: \mu_A > \mu_B$$

($t = 3.59$, $P = 0.0009$) rejects H_0 in favor of H_1 and thus confirms what the graphical information suggests. Here we use the t-test that pools the two-sample variances together in light of the results of the F-test. On the basis of this analysis, the adoption of alloy B would seem to be in order.

Statistical Significance and Engineering or Scientific Significance

While the statistician may feel quite comfortable with the results of the comparison between the two alloys in the case study above, a dilemma remains for the engineer. The analysis demonstrated a statistically significant improvement with the use of alloy B. However, is the difference found really worth pursuing, since alloy B is more expensive? This illustration highlights a very important issue often overlooked by statisticians and data analysts—the *distinction between statistical significance and engineering or scientific significance*. Here the average difference in deflection is $\bar{y}_A - \bar{y}_B = 0.00385$ inch. In a complete analysis, the engineer must determine if the difference is sufficient to justify the extra cost in the long run. This is an economic and engineering issue. The reader should understand that a statistically significant difference merely implies that the difference in the sample

10.14 Two-Sample Case Study

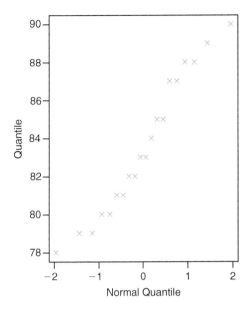

Figure 10.21: Normal quantile-quantile plot of data for alloy A.

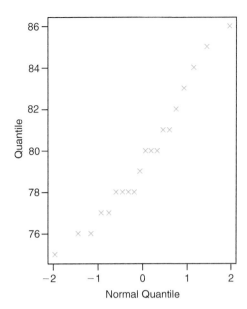

Figure 10.22: Normal quantile-quantile plot of data for alloy B.

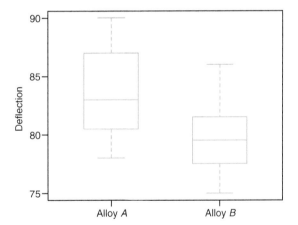

Figure 10.23: Box-and-whisker plots for both alloys.

means found in the data could hardly have occurred by chance. It does not imply that the difference in the population means is profound or particularly significant in the context of the problem. For example, in Section 10.4, an annotated computer printout was used to show evidence that a pH meter was, in fact, biased. That is, it does not demonstrate a mean pH of 7.00 for the material on which it was tested. But the variability among the observations in the sample is very small. The engineer may decide that the small deviations from 7.0 render the pH meter adequate.

```
                    The TTEST Procedure
        Alloy        N    Mean    Std Dev   Std Err
        Alloy A      20   83.55   3.6631    0.8191
        Alloy B      20   79.7    3.0967    0.6924

        Variances        DF      t Value    Pr > |t|
        Equal            38      3.59       0.0009
        Unequal          37      3.59       0.0010
                    Equality of Variances
        Num DF       Den DF      F Value    Pr > F
          19           19         1.40      0.4709
```

Figure 10.24: Annotated *SAS* printout for alloy data.

Exercises

10.79 A machine is supposed to mix peanuts, hazelnuts, cashews, and pecans in the ratio 5:2:2:1. A can containing 500 of these mixed nuts was found to have 269 peanuts, 112 hazelnuts, 74 cashews, and 45 pecans. At the 0.05 level of significance, test the hypothesis that the machine is mixing the nuts in the ratio 5:2:2:1.

10.80 The grades in a statistics course for a particular semester were as follows:

Grade	A	B	C	D	F
f	14	18	32	20	16

Test the hypothesis, at the 0.05 level of significance, that the distribution of grades is uniform.

10.81 A die is tossed 180 times with the following results:

x	1	2	3	4	5	6
f	28	36	36	30	27	23

Is this a balanced die? Use a 0.01 level of significance.

10.82 Three marbles are selected from an urn containing 5 red marbles and 3 green marbles. After the number X of red marbles is recorded, the marbles are replaced in the urn and the experiment repeated 112 times. The results obtained are as follows:

x	0	1	2	3
f	1	31	55	25

Test the hypothesis, at the 0.05 level of significance, that the recorded data may be fitted by the hypergeometric distribution $h(x; 8, 3, 5)$, $x = 0, 1, 2, 3$.

10.83 A coin is thrown until a head occurs and the number X of tosses recorded. After repeating the experiment 256 times, we obtained the following results:

x	1	2	3	4	5	6	7	8
f	136	60	34	12	9	1	3	1

Test the hypothesis, at the 0.05 level of significance, that the observed distribution of X may be fitted by the geometric distribution $g(x; 1/2)$, $x = 1, 2, 3, \ldots$.

10.84 For Exercise 1.18 on page 31, test the goodness of fit between the observed class frequencies and the corresponding expected frequencies of a normal distribution with $\mu = 65$ and $\sigma = 21$, using a 0.05 level of significance.

10.85 For Exercise 1.19 on page 31, test the goodness of fit between the observed class frequencies and the corresponding expected frequencies of a normal distribution with $\mu = 1.8$ and $\sigma = 0.4$, using a 0.01 level of significance.

10.86 In an experiment to study the dependence of hypertension on smoking habits, the following data were taken on 180 individuals:

	Non-smokers	Moderate Smokers	Heavy Smokers
Hypertension	21	36	30
No hypertension	48	26	19

Test the hypothesis that the presence or absence of hypertension is independent of smoking habits. Use a 0.05 level of significance.

10.87 A random sample of 90 adults is classified according to gender and the number of hours of television watched during a week:

	Gender	
	Male	Female
Over 25 hours	15	29
Under 25 hours	27	19

Use a 0.01 level of significance and test the hypothesis that the time spent watching television is independent of whether the viewer is male or female.

10.88 A random sample of 200 married men, all retired, was classified according to education and number of children:

Education	Number of Children		
	0–1	2–3	Over 3
Elementary	14	37	32
Secondary	19	42	17
College	12	17	10

Test the hypothesis, at the 0.05 level of significance, that the size of a family is independent of the level of education attained by the father.

10.89 A criminologist conducted a survey to determine whether the incidence of certain types of crime varied from one part of a large city to another. The particular crimes of interest were assault, burglary, larceny, and homicide. The following table shows the numbers of crimes committed in four areas of the city during the past year.

District	Type of Crime			
	Assault	Burglary	Larceny	Homicide
1	162	118	451	18
2	310	196	996	25
3	258	193	458	10
4	280	175	390	19

Can we conclude from these data at the 0.01 level of significance that the occurrence of these types of crime is dependent on the city district?

10.90 According to a Johns Hopkins University study published in the *American Journal of Public Health*, widows live longer than widowers. Consider the following survival data collected on 100 widows and 100 widowers following the death of a spouse:

Years Lived	Widow	Widower
Less than 5	25	39
5 to 10	42	40
More than 10	33	21

Can we conclude at the 0.05 level of significance that the proportions of widows and widowers are equal with respect to the different time periods that a spouse survives after the death of his or her mate?

10.91 The following responses concerning the standard of living at the time of an independent opinion poll of 1000 households versus one year earlier seem to be in agreement with the results of a study published in *Across the Board* (June 1981):

Period	Standard of Living			Total
	Somewhat Better	Same	Not as Good	
1980: Jan.	72	144	84	300
May	63	135	102	300
Sept.	47	100	53	200
1981: Jan.	40	105	55	200

Test the hypothesis that the proportions of households within each standard of living category are the same for each of the four time periods. Use a *P*-value.

10.92 A college infirmary conducted an experiment to determine the degree of relief provided by three cough remedies. Each cough remedy was tried on 50 students and the following data recorded:

	Cough Remedy		
	NyQuil	Robitussin	Triaminic
No relief	11	13	9
Some relief	32	28	27
Total relief	7	9	14

Test the hypothesis that the three cough remedies are equally effective. Use a *P*-value in your conclusion.

10.93 To determine current attitudes about prayer in public schools, a survey was conducted in four Virginia counties. The following table gives the attitudes of 200 parents from Craig County, 150 parents from Giles County, 100 parents from Franklin County, and 100 parents from Montgomery County:

Attitude	County			
	Craig	Giles	Franklin	Mont.
Favor	65	66	40	34
Oppose	42	30	33	42
No opinion	93	54	27	24

Test for homogeneity of attitudes among the four counties concerning prayer in the public schools. Use a *P*-value in your conclusion.

10.94 A survey was conducted in Indiana, Kentucky, and Ohio to determine the attitude of voters concerning school busing. A poll of 200 voters from each of these states yielded the following results:

State	Voter Attitude		
	Support	Do Not Support	Undecided
Indiana	82	97	21
Kentucky	107	66	27
Ohio	93	74	33

At the 0.05 level of significance, test the null hypothesis that the proportions of voters within each attitude category are the same for each of the three states.

10.95 A survey was conducted in two Virginia cities to determine voter sentiment about two gubernatorial candidates in an upcoming election. Five hundred voters were randomly selected from each city and the following data were recorded:

Voter Sentiment	City	
	Richmond	Norfolk
Favor A	204	225
Favor B	211	198
Undecided	85	77

At the 0.05 level of significance, test the null hypothesis that proportions of voters favoring candidate A, favoring candidate B, and undecided are the same for each city.

10.96 In a study to estimate the proportion of wives who regularly watch soap operas, it is found that 52 of 200 wives in Denver, 31 of 150 wives in Phoenix, and 37 of 150 wives in Rochester watch at least one soap opera. Use a 0.05 level of significance to test the hypothesis that there is no difference among the true proportions of wives who watch soap operas in these three cities.

Review Exercises

10.97 State the null and alternative hypotheses to be used in testing the following claims and determine generally where the critical region is located:

(a) The mean snowfall at Lake George during the month of February is 21.8 centimeters.

(b) No more than 20% of the faculty at the local university contributed to the annual giving fund.

(c) On the average, children attend schools within 6.2 kilometers of their homes in suburban St. Louis.

(d) At least 70% of next year's new cars will be in the compact and subcompact category.

(e) The proportion of voters favoring the incumbent in the upcoming election is 0.58.

(f) The average rib-eye steak at the Longhorn Steak house weighs at least 340 grams.

10.98 A geneticist is interested in the proportions of males and females in a population who have a certain minor blood disorder. In a random sample of 100 males, 31 are found to be afflicted, whereas only 24 of 100 females tested have the disorder. Can we conclude at the 0.01 level of significance that the proportion of men in the population afflicted with this blood disorder is significantly greater than the proportion of women afflicted?

10.99 A study was made to determine whether more Italians than Americans prefer white champagne to pink champagne at weddings. Of the 300 Italians selected at random, 72 preferred white champagne, and of the 400 Americans selected, 70 preferred white champagne. Can we conclude that a higher proportion of Italians than Americans prefer white champagne at weddings? Use a 0.05 level of significance.

10.100 Consider the situation of Exercise 10.54 on page 360. Oxygen consumption in mL/kg/min, was also measured.

Subject	With CO	Without CO
1	26.46	25.41
2	17.46	22.53
3	16.32	16.32
4	20.19	27.48
5	19.84	24.97
6	20.65	21.77
7	28.21	28.17
8	33.94	32.02
9	29.32	28.96

It is conjectured that oxygen consumption should be higher in an environment relatively free of CO. Do a significance test and discuss the conjecture.

10.101 In a study analyzed by the Laboratory for Interdisciplinary Statistical Analysis at Virginia Tech, a group of subjects was asked to complete a certain task on the computer. The response measured was the time to completion. The purpose of the experiment was to test a set of facilitation tools developed by the Department of Computer Science at the university. There were 10 subjects involved. With a random assignment, five were given a standard procedure using Fortran language for completion of the task. The other five were asked to do the task with the use of the facilitation tools. The data on the completion times for the task are given here.

Group 1 (Standard Procedure)	Group 2 (Facilitation Tool)
161	132
169	162
174	134
158	138
163	133

Assuming that the population distributions are normal and variances are the same for the two groups, support or refute the conjecture that the facilitation tools increase the speed with which the task can be accomplished.

10.102 State the null and alternative hypotheses to be used in testing the following claims, and determine

generally where the critical region is located:
(a) At most, 20% of next year's wheat crop will be exported to the Soviet Union.
(b) On the average, American homemakers drink 3 cups of coffee per day.
(c) The proportion of college graduates in Virginia this year who majored in the social sciences is at least 0.15.
(d) The average donation to the American Lung Association is no more than $10.
(e) Residents in suburban Richmond commute, on the average, 15 kilometers to their place of employment.

10.103 If one can containing 500 nuts is selected at random from each of three different distributors of mixed nuts and there are, respectively, 345, 313, and 359 peanuts in each of the cans, can we conclude at the 0.01 level of significance that the mixed nuts of the three distributors contain equal proportions of peanuts?

10.104 A study was made to determine whether there is a difference between the proportions of parents in the states of Maryland (MD), Virginia (VA), Georgia (GA), and Alabama (AL) who favor placing Bibles in the elementary schools. The responses of 100 parents selected at random in each of these states are recorded in the following table:

Preference	State			
	MD	VA	GA	AL
Yes	65	71	78	82
No	35	29	22	18

Can we conclude that the proportions of parents who favor placing Bibles in the schools are the same for these four states? Use a 0.01 level of significance.

10.105 A study was conducted at the Virginia-Maryland Regional College of Veterinary Medicine Equine Center to determine if the performance of a certain type of surgery on young horses had any effect on certain kinds of blood cell types in the animal. Fluid samples were taken from each of six foals before and after surgery. The samples were analyzed for the number of postoperative white blood cell (WBC) leukocytes. A preoperative measure of WBC leukocytes was also measured. The data are given as follows:

Foal	Presurgery*	Postsurgery*
1	10.80	10.60
2	12.90	16.60
3	9.59	17.20
4	8.81	14.00
5	12.00	10.60
6	6.07	8.60

*All values $\times 10^{-3}$.

Use a paired sample t-test to determine if there is a significant change in WBC leukocytes with the surgery.

10.106 A study was conducted at the Department of Human Nutrition, Foods and Exercise at Virginia Tech to determine if 8 weeks of training truly reduces the cholesterol levels of the participants. A treatment group consisting of 15 people was given lectures twice a week on how to reduce cholesterol level. Another group of 18 people of similar age was randomly selected as a control group. All participants' cholesterol levels were recorded at the end of the 8-week program and are listed below.

Treatment:
    129 131 154 172 115 126 175 191
    122 238 159 156 176 175 126
Control:
    151 132 196 195 188 198 187 168 115
    165 137 208 133 217 191 193 140 146

Can we conclude, at the 5% level of significance, that the average cholesterol level has been reduced due to the program? Make the appropriate test on means.

10.107 In a study conducted by the Department of Mechanical Engineering and analyzed by the Laboratory for Interdisciplinary Statistical Analysis at Virginia Tech, steel rods supplied by two different companies were compared. Ten sample springs were made out of the steel rods supplied by each company, and the "bounciness" was studied. The data are as follows:

Company A:
    9.3 8.8 6.8 8.7 8.5 6.7 8.0 6.5 9.2 7.0
Company B:
    11.0 9.8 9.9 10.2 10.1 9.7 11.0 11.1 10.2 9.6

Can you conclude that there is virtually no difference in means between the steel rods supplied by the two companies? Use a P-value to reach your conclusion. Should variances be pooled here?

10.108 In a study conducted by the Water Resources Research Center and analyzed by the Laboratory for Interdisciplinary Statistical Analysis at Virginia Tech, two different wastewater treatment plants are compared. Plant A is located where the median household income is below $22,000 a year, and plant B is located where the median household income is above $60,000 a year. The amount of waste-water treated at each plant (thousands of gallons/day) was randomly sampled for 10 days. The data are as follows:

Plant A:
21 19 20 23 22 28 32 19 13 18

Plant B:
20 39 24 33 30 28 30 22 33 24

Can we conclude, at the 5% level of significance, that

the average amount of wastewater treated at the plant in the high-income neighborhood is more than that treated at the plant in the low-income area? Assume normality.

10.109 The following data show the numbers of defects in 100,000 lines of code in a particular type of software program developed in the United States and Japan. Is there enough evidence to claim that there is a significant difference between the programs developed in the two countries? Test on means. Should variances be pooled?

U.S.	48	39	42	52	40	48	52	52
	54	48	52	55	43	46	48	52
Japan	50	48	42	40	43	48	50	46
	38	38	36	40	40	48	48	45

10.110 Studies show that the concentration of PCBs is much higher in malignant breast tissue than in normal breast tissue. If a study of 50 women with breast cancer reveals an average PCB concentration of 22.8×10^{-4} gram, with a standard deviation of 4.8×10^{-4} gram, is the mean concentration of PCBs less than 24×10^{-4} gram?

10.111 z-Value for Testing $p_1 - p_2 = d_0$: To test the null hypothesis H_0 that $p_1 - p_2 = d_0$, where $d_0 \neq 0$, we base our decision on

$$z = \frac{\hat{p}_1 - \hat{p}_2 - d_0}{\sqrt{\hat{p}_1\hat{q}_1/n_1 + \hat{p}_2\hat{q}_2/n_2}},$$

which is a value of a random variable whose distribution approximates the standard normal distribution as long as n_1 and n_2 are both large. With reference to Example 10.11 on page 364, test the hypothesis that the percentage of town voters favoring the construction of the chemical plant will not exceed the percentage of county voters by more than 3%. Use a P-value in your conclusion.

10.15 Potential Misconceptions and Hazards; Relationship to Material in Other Chapters

One of the easiest ways to misuse statistics relates to the final scientific conclusion drawn when the analyst does not reject the null hypothesis H_0. In this text, we have attempted to make clear what the null hypothesis means and what the alternative means, and to stress that, in a large sense, the alternative hypothesis is much more important. Put in the form of an example, if an engineer is attempting to compare two gauges using a two-sample t-test, and H_0 is "the gauges are equivalent" while H_1 is "the gauges are not equivalent," not rejecting H_0 does not lead to the conclusion of equivalent gauges. In fact, a case can be made for never writing or saying "accept H_0"! Not rejecting H_0 merely implies insufficient evidence. Depending on the nature of the hypothesis, a lot of possibilities are still not ruled out.

In Chapter 9, we considered the case of the large-sample confidence interval using

$$z = \frac{\bar{x} - \mu}{s/\sqrt{n}}.$$

In hypothesis testing, replacing σ by s for $n < 30$ is risky. If $n \geq 30$ and the distribution is not normal but somehow close to normal, the Central Limit Theorem is being called upon and one is relying on the fact that with $n \geq 30$, $s \approx \sigma$. Of course, any t-test is accompanied by the concomitant assumption of normality. As in the case of confidence intervals, the t-test is relatively robust to normality. However, one should still use normal probability plotting, goodness-of-fit tests, or other graphical procedures when the sample is not too small.

Most of the chapters in this text include discussions whose purpose is to relate the chapter in question to other material that will follow. The topics of estimation and hypothesis testing are both used in a major way in nearly all of the techniques that fall under the umbrella of "statistical methods." This will be readily noted by students who advance to Chapters 11 through 16. It will be obvious that these chapters depend heavily on statistical modeling. Students will be exposed to the use of modeling in a wide variety of applications in many scientific and engineering fields. It will become obvious quite quickly that the framework of a statistical model is useless unless data are available with which to estimate parameters in the formulated model. This will become particularly apparent in Chapters 11 and 12 as we introduce the notion of regression models. The concepts and theory associated with Chapter 9 will carry over. As far as material in the present chapter is concerned, the framework of hypothesis testing, P-values, power of tests, and choice of sample size will collectively play a major role. Since initial model formulation quite often must be supplemented by model editing before the analyst is sufficiently comfortable to use the model for either process understanding or prediction, Chapters 11, 12, and 15 make major use of hypothesis testing to supplement diagnostic measures that are used to assess model quality.

Chapter 11

Simple Linear Regression and Correlation

11.1 Introduction to Linear Regression

Often, in practice, one is called upon to solve problems involving sets of variables when it is known that there exists some inherent relationship among the variables. For example, in an industrial situation it may be known that the tar content in the outlet stream in a chemical process is related to the inlet temperature. It may be of interest to develop a method of prediction, that is, a procedure for estimating the tar content for various levels of the inlet temperature from experimental information. Now, of course, it is highly likely that for many example runs in which the inlet temperature is the same, say 130°C, the outlet tar content will not be the same. This is much like what happens when we study several automobiles with the same engine volume. They will not all have the same gas mileage. Houses in the same part of the country that have the same square footage of living space will not all be sold for the same price. Tar content, gas mileage (mpg), and the price of houses (in thousands of dollars) are natural **dependent variables**, or responses, in these three scenarios. Inlet temperature, engine volume (cubic feet), and square feet of living space are, respectively, natural **independent variables**, or **regressors**. A reasonable form of a relationship between the **response Y** and the regressor x is the linear relationship

$$Y = \beta_0 + \beta_1 x,$$

where, of course, β_0 is the **intercept** and β_1 is the **slope**. The relationship is illustrated in Figure 11.1.

If the relationship is exact, then it is a **deterministic** relationship between two scientific variables and there is no random or probabilistic component to it. However, in the examples listed above, as well as in countless other scientific and engineering phenomena, the relationship is not deterministic (i.e., a given x does not always give the same value for Y). As a result, important problems here are probabilistic in nature since the relationship above cannot be viewed as being exact. The concept of **regression analysis** deals with finding the best relationship

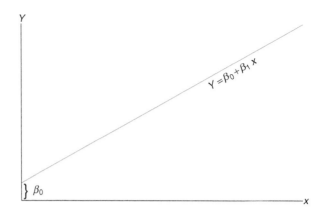

Figure 11.1: A linear relationship; β_0: intercept; β_1: slope.

between Y and x, quantifying the strength of that relationship, and using methods that allow for prediction of the response values given values of the regressor x.

In many applications, there will be more than one regressor (i.e., more than one independent variable **that helps to explain Y**). For example, in the case where the response is the price of a house, one would expect the age of the house to contribute to the explanation of the price, so in this case the multiple regression structure might be written

$$Y = \beta_0 + \beta_1 x_1 + \beta_2 x_2,$$

where Y is price, x_1 is square footage, and x_2 is age in years. In the next chapter, we will consider problems with multiple regressors. The resulting analysis is termed **multiple regression**, while the analysis of the single regressor case is called **simple regression**. As a second illustration of multiple regression, a chemical engineer may be concerned with the amount of hydrogen lost from samples of a particular metal when the material is placed in storage. In this case, there may be two inputs, storage time x_1 in hours and storage temperature x_2 in degrees centigrade. The response would then be hydrogen loss Y in parts per million.

In this chapter, we deal with the topic of **simple linear regression**, treating only the case of a single regressor variable in which the relationship between y and x is linear. For the case of more than one regressor variable, the reader is referred to Chapter 12. Denote a random sample of size n by the set $\{(x_i, y_i); \ i = 1, 2, \ldots, n\}$. If additional samples were taken using exactly the same values of x, we should expect the y values to vary. Hence, the value y_i in the ordered pair (x_i, y_i) is a value of some random variable Y_i.

11.2 The Simple Linear Regression (SLR) Model

We have already confined the terminology *regression analysis* to situations in which relationships among variables are not deterministic (i.e., not exact). In other words, there must be a **random component** to the equation that relates the variables.

11.2 The Simple Linear Regression Model

This random component takes into account considerations that are not being measured or, in fact, are not understood by the scientists or engineers. Indeed, in most applications of regression, the linear equation, say $Y = \beta_0 + \beta_1 x$, is an approximation that is a simplification of something unknown and much more complicated. For example, in our illustration involving the response $Y =$ tar content and $x =$ inlet temperature, $Y = \beta_0 + \beta_1 x$ is likely a reasonable approximation that may be operative within a confined range on x. More often than not, the models that are simplifications of more complicated and unknown structures are linear in nature (i.e., linear in the **parameters** β_0 and β_1 or, in the case of the model involving the price, size, and age of the house, linear in the **parameters** β_0, β_1, and β_2). These linear structures are simple and empirical in nature and are thus called **empirical models**.

An analysis of the relationship between Y and x requires the statement of a **statistical model**. A model is often used by a statistician as a representation of an **ideal** that essentially defines how we perceive that the data were generated by the system in question. The model must include the set $\{(x_i, y_i); \ i = 1, 2, \ldots, n\}$ of data involving n pairs of (x, y) values. One must bear in mind that the value y_i depends on x_i via a linear structure that also has the random component involved. The basis for the use of a statistical model relates to how the random variable Y moves with x and the random component. The model also includes what is assumed about the statistical properties of the random component. The statistical model for simple linear regression is given below. The response Y is related to the independent variable x through the equation

Simple Linear Regression Model

$$Y = \beta_0 + \beta_1 x + \epsilon.$$

In the above, β_0 and β_1 are unknown intercept and slope parameters, respectively, and ϵ is a random variable that is assumed to be distributed with $E(\epsilon) = 0$ and $\text{Var}(\epsilon) = \sigma^2$. The quantity σ^2 is often called the error variance or residual variance.

From the model above, several things become apparent. The quantity Y is a random variable since ϵ is random. The value x of the regressor variable is not random and, in fact, is measured with negligible error. The quantity ϵ, often called a **random error** or **random disturbance**, has constant variance. This portion of the assumptions is often called the **homogeneous variance assumption**. The presence of this random error, ϵ, keeps the model from becoming simply a deterministic equation. Now, the fact that $E(\epsilon) = 0$ implies that at a specific x the y-values are distributed around the **true**, or population, **regression line** $y = \beta_0 + \beta_1 x$. If the model is well chosen (i.e., there are no additional important regressors and the linear approximation is good within the ranges of the data), then positive and negative errors around the true regression are reasonable. We must keep in mind that in practice β_0 and β_1 are not known and must be estimated from data. In addition, the model described above is conceptual in nature. As a result, we never observe the actual ϵ values in practice and thus we can never draw the true regression line (but we assume it is there). We can only draw an estimated line. Figure 11.2 depicts the nature of hypothetical (x, y) data scattered around a true regression line for a case in which only $n = 5$ observations are available. Let us emphasize that what we see in Figure 11.2 is not the line that is used by the scientist or engineer. Rather, the picture merely describes what the assumptions mean! The regression that the user has at his or her disposal will now be described.

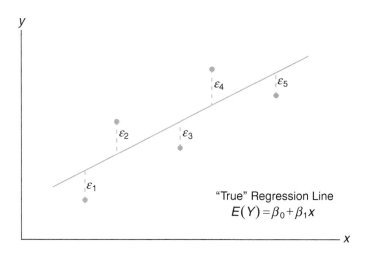

Figure 11.2: Hypothetical (x, y) data scattered around the true regression line for $n = 5$.

The Fitted Regression Line

An important aspect of regression analysis is, very simply, to estimate the parameters β_0 and β_1 (i.e., estimate the so-called **regression coefficients**). The method of estimation will be discussed in the next section. Suppose we denote the estimates b_0 for β_0 and b_1 for β_1. Then the estimated or **fitted regression** line is given by

$$\hat{y} = b_0 + b_1 x,$$

where $\hat{y}$ is the predicted or fitted value. Obviously, the fitted line is an estimate of the true regression line. We expect that the fitted line should be closer to the true regression line when a large amount of data are available. In the following example, we illustrate the fitted line for a real-life pollution study.

One of the more challenging problems confronting the water pollution control field is presented by the tanning industry. Tannery wastes are chemically complex. They are characterized by high values of chemical oxygen demand, volatile solids, and other pollution measures. Consider the experimental data in Table 11.1, which were obtained from 33 samples of chemically treated waste in a study conducted at Virginia Tech. Readings on x, the percent reduction in total solids, and y, the percent reduction in chemical oxygen demand, were recorded.

The data of Table 11.1 are plotted in a **scatter diagram** in Figure 11.3. From an inspection of this scatter diagram, it can be seen that the points closely follow a straight line, indicating that the assumption of linearity between the two variables appears to be reasonable.

11.2 The Simple Linear Regression Model

Table 11.1: Measures of Reduction in Solids and Oxygen Demand

Solids Reduction, x (%)	Oxygen Demand Reduction, y (%)	Solids Reduction, x (%)	Oxygen Demand Reduction, y (%)
3	5	36	34
7	11	37	36
11	21	38	38
15	16	39	37
18	16	39	36
27	28	39	45
29	27	40	39
30	25	41	41
30	35	42	40
31	30	42	44
31	40	43	37
32	32	44	44
33	34	45	46
33	32	46	46
34	34	47	49
36	37	50	51
36	38		

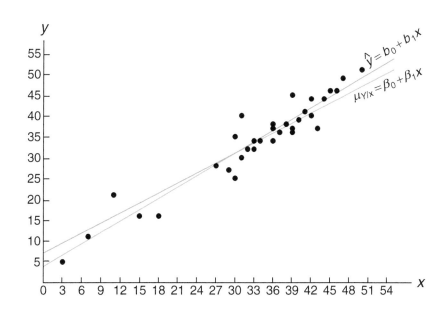

Figure 11.3: Scatter diagram with regression lines.

The fitted regression line and a *hypothetical true regression line* are shown on the scatter diagram of Figure 11.3. This example will be revisited as we move on to the method of estimation, discussed in Section 11.3.

Another Look at the Model Assumptions

It may be instructive to revisit the simple linear regression model presented previously and discuss in a graphical sense how it relates to the so-called true regression. Let us expand on Figure 11.2 by illustrating not merely where the ϵ_i fall on a graph but also what the implication is of the normality assumption on the ϵ_i.

Suppose we have a simple linear regression with $n = 6$ evenly spaced values of x and a single y-value at each x. Consider the graph in Figure 11.4. This illustration should give the reader a clear representation of the model and the assumptions involved. The line in the graph is the true regression line. The points plotted are actual (y, x) points which are scattered about the line. Each point is on its own normal distribution with the center of the distribution (i.e., the mean of y) falling on the line. This is certainly expected since $E(Y) = \beta_0 + \beta_1 x$. As a result, the true regression line **goes through the means of the response**, and the actual observations are on the distribution around the means. Note also that all distributions have the same variance, which we referred to as σ^2. Of course, the deviation between an individual y and the point on the line will be its individual ϵ value. This is clear since

$$y_i - E(Y_i) = y_i - (\beta_0 + \beta_1 x_i) = \epsilon_i.$$

Thus, at a given x, Y and the corresponding ϵ both have variance σ^2.

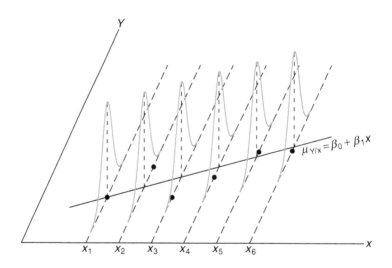

Figure 11.4: Individual observations around true regression line.

Note also that we have written the true regression line here as $\mu_{Y|x} = \beta_0 + \beta_1 x$ in order to reaffirm that the line goes through the mean of the Y random variable.

11.3 Least Squares and the Fitted Model

In this section, we discuss the method of fitting an estimated regression line to the data. This is tantamount to the determination of estimates b_0 for β_0 and b_1

11.3 Least Squares and the Fitted Model

for β_1. This of course allows for the computation of predicted values from the fitted line $\hat{y} = b_0 + b_1 x$ and other types of analyses and diagnostic information that will ascertain the strength of the relationship and the adequacy of the fitted model. Before we discuss the method of least squares estimation, it is important to introduce the concept of a **residual**. A residual is essentially an error in the fit of the model $\hat{y} = b_0 + b_1 x$.

Residual: Error in Fit Given a set of regression data $\{(x_i, y_i); i = 1, 2, \ldots, n\}$ and a fitted model, $\hat{y}_i = b_0 + b_1 x_i$, the ith residual e_i is given by

$$e_i = y_i - \hat{y}_i, \quad i = 1, 2, \ldots, n.$$

Obviously, if a set of n residuals is large, then the fit of the model is not good. Small residuals are a sign of a good fit. Another interesting relationship which is useful at times is the following:

$$y_i = b_0 + b_1 x_i + e_i.$$

The use of the above equation should result in clarification of the distinction between the residuals, e_i, and the conceptual model errors, ϵ_i. One must bear in mind that whereas the ϵ_i are not observed, the e_i not only are observed but also play an important role in the total analysis.

Figure 11.5 depicts the line fit to this set of data, namely $\hat{y} = b_0 + b_1 x$, and the line reflecting the model $\mu_{Y|x} = \beta_0 + \beta_1 x$. Now, of course, β_0 and β_1 are unknown parameters. The fitted line is an estimate of the line produced by the statistical model. Keep in mind that the line $\mu_{Y|x} = \beta_0 + \beta_1 x$ is not known.

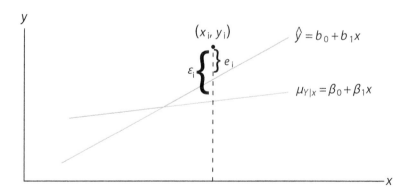

Figure 11.5: Comparing ϵ_i with the residual, e_i.

The Method of Least Squares

We shall find b_0 and b_1, the estimates of β_0 and β_1, so that the sum of the squares of the residuals is a minimum. The residual sum of squares is often called the sum of squares of the errors about the regression line and is denoted by SSE. This minimization procedure for estimating the parameters is called the **method of least squares**. Hence, we shall find a and b so as to minimize

$$SSE = \sum_{i=1}^{n} e_i^2 = \sum_{i=1}^{n}(y_i - \hat{y}_i)^2 = \sum_{i=1}^{n}(y_i - b_0 - b_1 x_i)^2.$$

Differentiating SSE with respect to b_0 and b_1, we have

$$\frac{\partial(SSE)}{\partial b_0} = -2\sum_{i=1}^{n}(y_i - b_0 - b_1 x_i), \quad \frac{\partial(SSE)}{\partial b_1} = -2\sum_{i=1}^{n}(y_i - b_0 - b_1 x_i)x_i.$$

Setting the partial derivatives equal to zero and rearranging the terms, we obtain the equations (called the **normal equations**)

$$nb_0 + b_1 \sum_{i=1}^{n} x_i = \sum_{i=1}^{n} y_i, \quad b_0 \sum_{i=1}^{n} x_i + b_1 \sum_{i=1}^{n} x_i^2 = \sum_{i=1}^{n} x_i y_i,$$

which may be solved simultaneously to yield computing formulas for b_0 and b_1.

Estimating the Regression Coefficients Given the sample $\{(x_i, y_i);\ i = 1, 2, \ldots, n\}$, the least squares estimates b_0 and b_1 of the regression coefficients β_0 and β_1 are computed from the formulas

$$b_1 = \frac{n\sum_{i=1}^{n} x_i y_i - \left(\sum_{i=1}^{n} x_i\right)\left(\sum_{i=1}^{n} y_i\right)}{n\sum_{i=1}^{n} x_i^2 - \left(\sum_{i=1}^{n} x_i\right)^2} = \frac{\sum_{i=1}^{n}(x_i - \bar{x})(y_i - \bar{y})}{\sum_{i=1}^{n}(x_i - \bar{x})^2} \quad \text{and}$$

$$b_0 = \frac{\sum_{i=1}^{n} y_i - b_1 \sum_{i=1}^{n} x_i}{n} = \bar{y} - b_1 \bar{x}.$$

The calculations of b_0 and b_1, using the data of Table 11.1, are illustrated by the following example.

Example 11.1: Estimate the regression line for the pollution data of Table 11.1.

Solution:
$$\sum_{i=1}^{33} x_i = 1104,\quad \sum_{i=1}^{33} y_i = 1124,\quad \sum_{i=1}^{33} x_i y_i = 41{,}355,\quad \sum_{i=1}^{33} x_i^2 = 41{,}086$$

Therefore,

$$b_1 = \frac{(33)(41{,}355) - (1104)(1124)}{(33)(41{,}086) - (1104)^2} = 0.903643 \text{ and}$$

$$b_0 = \frac{1124 - (0.903643)(1104)}{33} = 3.829633.$$

Thus, the estimated regression line is given by

$$\hat{y} = 3.8296 + 0.9036x.$$

Using the regression line of Example 11.1, we would predict a 31% reduction in the chemical oxygen demand when the reduction in the total solids is 30%. The

31% reduction in the chemical oxygen demand may be interpreted as an estimate of the population mean $\mu_{Y|30}$ or as an estimate of a new observation when the reduction in total solids is 30%. Such estimates, however, are subject to error. Even if the experiment were controlled so that the reduction in total solids was 30%, it is unlikely that we would measure a reduction in the chemical oxygen demand exactly equal to 31%. In fact, the original data recorded in Table 11.1 show that measurements of 25% and 35% were recorded for the reduction in oxygen demand when the reduction in total solids was kept at 30%.

What Is Good about Least Squares?

It should be noted that the least squares criterion is designed to provide a fitted line that results in a "closeness" between the line and the plotted points. There are many ways of measuring closeness. For example, one may wish to determine b_0 and b_1 for which $\sum_{i=1}^{n} |y_i - \hat{y}_i|$ is minimized or for which $\sum_{i=1}^{n} |y_i - \hat{y}_i|^{1.5}$ is minimized. These are both viable and reasonable methods. Note that both of these, as well as the least squares procedure, result in forcing residuals to be "small" in some sense. One should remember that the residuals are the empirical counterpart to the ϵ values. Figure 11.6 illustrates a set of residuals. One should note that the fitted line has predicted values as points on the line and hence the residuals are vertical deviations from points to the line. As a result, the least squares procedure produces a line that **minimizes the sum of squares of vertical deviations from the points to the line.**

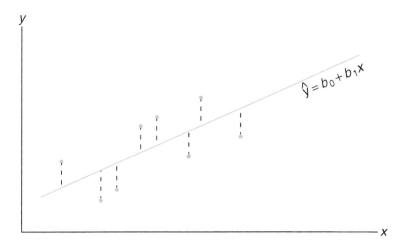

Figure 11.6: Residuals as vertical deviations.

Exercises

11.1 A study was conducted at Virginia Tech to determine if certain static arm-strength measures have an influence on the "dynamic lift" characteristics of an individual. Twenty-five individuals were subjected to strength tests and then were asked to perform a weight-lifting test in which weight was dynamically lifted overhead. The data are given here.

Individual	Arm Strength, x	Dynamic Lift, y
1	17.3	71.7
2	19.3	48.3
3	19.5	88.3
4	19.7	75.0
5	22.9	91.7
6	23.1	100.0
7	26.4	73.3
8	26.8	65.0
9	27.6	75.0
10	28.1	88.3
11	28.2	68.3
12	28.7	96.7
13	29.0	76.7
14	29.6	78.3
15	29.9	60.0
16	29.9	71.7
17	30.3	85.0
18	31.3	85.0
19	36.0	88.3
20	39.5	100.0
21	40.4	100.0
22	44.3	100.0
23	44.6	91.7
24	50.4	100.0
25	55.9	71.7

(a) Estimate β_0 and β_1 for the linear regression curve $\mu_{Y|x} = \beta_0 + \beta_1 x$.
(b) Find a point estimate of $\mu_{Y|30}$.
(c) Plot the residuals versus the x's (arm strength). Comment.

11.2 The grades of a class of 9 students on a midterm report (x) and on the final examination (y) are as follows:

x	77	50	71	72	81	94	96	99	67
y	82	66	78	34	47	85	99	99	68

(a) Estimate the linear regression line.
(b) Estimate the final examination grade of a student who received a grade of 85 on the midterm report.

11.3 The amounts of a chemical compound y that dissolved in 100 grams of water at various temperatures x were recorded as follows:

x (°C)	y (grams)		
0	8	6	8
15	12	10	14
30	25	21	24
45	31	33	28
60	44	39	42
75	48	51	44

(a) Find the equation of the regression line.
(b) Graph the line on a scatter diagram.
(c) Estimate the amount of chemical that will dissolve in 100 grams of water at 50°C.

11.4 The following data were collected to determine the relationship between pressure and the corresponding scale reading for the purpose of calibration.

Pressure, x (lb/sq in.)	Scale Reading, y
10	13
10	18
10	16
10	15
10	20
50	86
50	90
50	88
50	88
50	92

(a) Find the equation of the regression line.
(b) The purpose of calibration in this application is to estimate pressure from an observed scale reading. Estimate the pressure for a scale reading of 54 using $\hat{x} = (54 - b_0)/b_1$.

11.5 A study was made on the amount of converted sugar in a certain process at various temperatures. The data were coded and recorded as follows:

Temperature, x	Converted Sugar, y
1.0	8.1
1.1	7.8
1.2	8.5
1.3	9.8
1.4	9.5
1.5	8.9
1.6	8.6
1.7	10.2
1.8	9.3
1.9	9.2
2.0	10.5

(a) Estimate the linear regression line.
(b) Estimate the mean amount of converted sugar produced when the coded temperature is 1.75.
(c) Plot the residuals versus temperature. Comment.

Exercises

11.6 In a certain type of metal test specimen, the normal stress on a specimen is known to be functionally related to the shear resistance. The following is a set of coded experimental data on the two variables:

Normal Stress, x	Shear Resistance, y
26.8	26.5
25.4	27.3
28.9	24.2
23.6	27.1
27.7	23.6
23.9	25.9
24.7	26.3
28.1	22.5
26.9	21.7
27.4	21.4
22.6	25.8
25.6	24.9

(a) Estimate the regression line $\mu_{Y|x} = \beta_0 + \beta_1 x$.

(b) Estimate the shear resistance for a normal stress of 24.5.

11.7 The following is a portion of a classic data set called the "pilot plot data" in *Fitting Equations to Data* by Daniel and Wood, published in 1971. The response y is the acid content of material produced by titration, whereas the regressor x is the organic acid content produced by extraction and weighing.

y	x	y	x
76	123	70	109
62	55	37	48
66	100	82	138
58	75	88	164
88	159	43	28

(a) Plot the data; does it appear that a simple linear regression will be a suitable model?

(b) Fit a simple linear regression; estimate a slope and intercept.

(c) Graph the regression line on the plot in (a).

11.8 A mathematics placement test is given to all entering freshmen at a small college. A student who receives a grade below 35 is denied admission to the regular mathematics course and placed in a remedial class. The placement test scores and the final grades for 20 students who took the regular course were recorded.

(a) Plot a scatter diagram.

(b) Find the equation of the regression line to predict course grades from placement test scores.

(c) Graph the line on the scatter diagram.

(d) If 60 is the minimum passing grade, below which placement test score should students in the future be denied admission to this course?

Placement Test	Course Grade
50	53
35	41
35	61
40	56
55	68
65	36
35	11
60	70
90	79
35	59
90	54
80	91
60	48
60	71
60	71
40	47
55	53
50	68
65	57
50	79

11.9 A study was made by a retail merchant to determine the relation between weekly advertising expenditures and sales.

Advertising Costs (\$)	Sales (\$)
40	385
20	400
25	395
20	365
30	475
50	440
40	490
20	420
50	560
40	525
25	480
50	510

(a) Plot a scatter diagram.

(b) Find the equation of the regression line to predict weekly sales from advertising expenditures.

(c) Estimate the weekly sales when advertising costs are \$35.

(d) Plot the residuals versus advertising costs. Comment.

11.10 The following data are the selling prices z of a certain make and model of used car w years old. Fit a curve of the form $\mu_{z|w} = \gamma \delta^w$ by means of the nonlinear sample regression equation $\hat{z} = cd^w$. [*Hint*: Write $\ln \hat{z} = \ln c + (\ln d)w = b_0 + b_1 w$.]

w (years)	z (dollars)	w (years)	z (dollars)
1	6350	3	5395
2	5695	5	4985
2	5750	5	4895

11.11 The thrust of an engine (y) is a function of exhaust temperature (x) in °F when other important variables are held constant. Consider the following data.

y	x	y	x
4300	1760	4010	1665
4650	1652	3810	1550
3200	1485	4500	1700
3150	1390	3008	1270
4950	1820		

(a) Plot the data.
(b) Fit a simple linear regression to the data and plot the line through the data.

11.12 A study was done to study the effect of ambient temperature x on the electric power consumed by a chemical plant y. Other factors were held constant, and the data were collected from an experimental pilot plant.

y (BTU)	x (°F)	y (BTU)	x (°F)
250	27	265	31
285	45	298	60
320	72	267	34
295	58	321	74

(a) Plot the data.
(b) Estimate the slope and intercept in a simple linear regression model.
(c) Predict power consumption for an ambient temperature of 65°F.

11.13 A study of the amount of rainfall and the quantity of air pollution removed produced the following data:

Daily Rainfall, x (0.01 cm)	Particulate Removed, y (μg/m^3)
4.3	126
4.5	121
5.9	116
5.6	118
6.1	114
5.2	118
3.8	132
2.1	141
7.5	108

(a) Find the equation of the regression line to predict the particulate removed from the amount of daily rainfall.
(b) Estimate the amount of particulate removed when the daily rainfall is $x = 4.8$ units.

11.14 A professor in the School of Business in a university polled a dozen colleagues about the number of professional meetings they attended in the past five years (x) and the number of papers they submitted to refereed journals (y) during the same period. The summary data are given as follows:

$$n = 12, \quad \bar{x} = 4, \quad \bar{y} = 12,$$

$$\sum_{i=1}^{n} x_i^2 = 232, \quad \sum_{i=1}^{n} x_i y_i = 318.$$

Fit a simple linear regression model between x and y by finding out the estimates of intercept and slope. Comment on whether attending more professional meetings would result in publishing more papers.

11.4 Properties of the Least Squares Estimators

In addition to the assumptions that the error term in the model

$$Y_i = \beta_0 + \beta_1 x_i + \epsilon_i$$

is a random variable with mean 0 and constant variance σ^2, suppose that we make the further assumption that $\epsilon_1, \epsilon_2, \ldots, \epsilon_n$ are independent from run to run in the experiment. This provides a foundation for finding the means and variances for the estimators of β_0 and β_1.

It is important to remember that our values of b_0 and b_1, based on a given sample of n observations, are only estimates of true parameters β_0 and β_1. If the experiment is repeated over and over again, each time using the same fixed values of x, the resulting estimates of β_0 and β_1 will most likely differ from experiment to experiment. These different estimates may be viewed as values assumed by the random variables B_0 and B_1, while b_0 and b_1 are specific realizations.

Since the values of x remain fixed, the values of B_0 and B_1 depend on the variations in the values of y or, more precisely, on the values of the random variables,

11.4 Properties of the Least Squares Estimators

$Y_1, Y_2, \ldots, Y_n$. The distributional assumptions imply that the Y_i, $i = 1, 2, \ldots, n$, are also independently distributed, with mean $\mu_{Y|x_i} = \beta_0 + \beta_1 x_i$ and equal variances σ^2; that is,

$$\sigma^2_{Y|x_i} = \sigma^2 \quad \text{for} \quad i = 1, 2, \ldots, n.$$

Mean and Variance of Estimators

In what follows, we show that the estimator B_1 is unbiased for β_1 and demonstrate the variances of both B_0 and B_1. This will begin a series of developments that lead to hypothesis testing and confidence interval estimation on the intercept and slope.

Since the estimator

$$B_1 = \frac{\sum_{i=1}^{n}(x_i - \bar{x})(Y_i - \bar{Y})}{\sum_{i=1}^{n}(x_i - \bar{x})^2} = \frac{\sum_{i=1}^{n}(x_i - \bar{x})Y_i}{\sum_{i=1}^{n}(x_i - \bar{x})^2}$$

is of the form $\sum_{i=1}^{n} c_i Y_i$, where

$$c_i = \frac{x_i - \bar{x}}{\sum_{i=1}^{n}(x_i - \bar{x})^2}, \quad i = 1, 2, \ldots, n,$$

we may conclude from Theorem 7.11 that B_1 has a $n(\mu_{B_1}, \sigma_{B_1})$ distribution with

$$\mu_{B_1} = \frac{\sum_{i=1}^{n}(x_i - \bar{x})(\beta_0 + \beta_1 x_i)}{\sum_{i=1}^{n}(x_i - \bar{x})^2} = \beta_1 \quad \text{and} \quad \sigma^2_{B_1} = \frac{\sum_{i=1}^{n}(x_i - \bar{x})^2 \sigma^2_{Y_i}}{\left[\sum_{i=1}^{n}(x_i - \bar{x})^2\right]^2} = \frac{\sigma^2}{\sum_{i=1}^{n}(x_i - \bar{x})^2}.$$

It can also be shown (Review Exercise 11.60 on page 438) that the random variable B_0 is normally distributed with

$$\text{mean } \mu_{B_0} = \beta_0 \quad \text{and variance } \sigma^2_{B_0} = \frac{\sum_{i=1}^{n} x_i^2}{n \sum_{i=1}^{n}(x_i - \bar{x})^2} \sigma^2.$$

From the foregoing results, it is apparent that the **least squares estimators for β_0 and β_1 are both unbiased estimators.**

Partition of Total Variability and Estimation of σ^2

To draw inferences on β_0 and β_1, it becomes necessary to arrive at an estimate of the parameter σ^2 appearing in the two preceding variance formulas for B_0 and B_1. The parameter σ^2, the model error variance, reflects random variation or experimental error variation around the regression line. In much of what follows, it is advantageous to use the notation

$$S_{xx} = \sum_{i=1}^{n}(x_i - \bar{x})^2, \quad S_{yy} = \sum_{i=1}^{n}(y_i - \bar{y})^2, \quad S_{xy} = \sum_{i=1}^{n}(x_i - \bar{x})(y_i - \bar{y}).$$

Now we may write the error sum of squares as follows:

$$\begin{aligned} SSE &= \sum_{i=1}^{n}(y_i - b_0 - b_1 x_i)^2 = \sum_{i=1}^{n}[(y_i - \bar{y}) - b_1(x_i - \bar{x})]^2 \\ &= \sum_{i=1}^{n}(y_i - \bar{y})^2 - 2b_1\sum_{i=1}^{n}(x_i - \bar{x})(y_i - \bar{y}) + b_1^2\sum_{i=1}^{n}(x_i - \bar{x})^2 \\ &= S_{yy} - 2b_1 S_{xy} + b_1^2 S_{xx} = S_{yy} - b_1 S_{xy}, \end{aligned}$$

the final step following from the fact that $b_1 = S_{xy}/S_{xx}$.

Theorem 11.1: An unbiased estimate of σ^2 is

$$s^2 = \frac{SSE}{n-2} = \sum_{i=1}^{n}\frac{(y_i - \hat{y}_i)^2}{n-2} = \frac{S_{yy} - b_1 S_{xy}}{n-2}.$$

The proof of Theorem 11.1 is left as an exercise (see Review Exercise 11.59).

The Estimator of σ^2 as a Mean Squared Error

One should observe the result of Theorem 11.1 in order to gain some intuition about the estimator of σ^2. The parameter σ^2 measures variance or squared deviations between Y values and their mean given by $\mu_{Y|x}$ (i.e., squared deviations between Y and $\beta_0 + \beta_1 x$). Of course, $\beta_0 + \beta_1 x$ is estimated by $\hat{y} = b_0 + b_1 x$. Thus, it would make sense that the variance σ^2 is best depicted as a squared deviation of the typical observation y_i from the estimated mean, $\hat{y}_i$, which is the corresponding point on the fitted line. Thus, $(y_i - \hat{y}_i)^2$ values reveal the appropriate variance, much like the way $(y_i - \bar{y})^2$ values measure variance when one is sampling in a nonregression scenario. In other words, $\bar{y}$ estimates the mean in the latter simple situation, whereas $\hat{y}_i$ estimates the mean of y_i in a regression structure. Now, what about the divisor $n-2$? In future sections, we shall note that these are the degrees of freedom associated with the estimator s^2 of σ^2. Whereas in the standard normal i.i.d. scenario, one degree of freedom is subtracted from n in the denominator and a reasonable explanation is that one parameter is estimated, namely the mean μ by, say, $\bar{y}$, but in the regression problem, **two parameters are estimated**, namely β_0 and β_1 by b_0 and b_1. Thus, the important parameter σ^2, estimated by

$$s^2 = \sum_{i=1}^{n}(y_i - \hat{y}_i)^2/(n-2),$$

is called a **mean squared error**, depicting a type of mean (division by $n-2$) of the squared residuals.

11.5 Inferences Concerning the Regression Coefficients

Aside from merely estimating the linear relationship between x and Y for purposes of prediction, the experimenter may also be interested in drawing certain inferences about the slope and intercept. In order to allow for the testing of hypotheses and the construction of confidence intervals on β_0 and β_1, one must be willing to make the further assumption that each ϵ_i, $i = 1, 2, \ldots, n$, is normally distributed. This assumption implies that $Y_1, Y_2, \ldots, Y_n$ are also normally distributed, each with probability distribution $n(y_i; \beta_0 + \beta_1 x_i, \sigma)$.

From Section 11.4 we know that B_1 follows a normal distribution. It turns out that under the normality assumption, a result very much analogous to that given in Theorem 8.4 allows us to conclude that $(n-2)S^2/\sigma^2$ is a chi-squared variable with $n-2$ degrees of freedom, independent of the random variable B_1. Theorem 8.5 then assures us that the statistic

$$T = \frac{(B_1 - \beta_1)/(\sigma/\sqrt{S_{xx}})}{S/\sigma} = \frac{B_1 - \beta_1}{S/\sqrt{S_{xx}}}$$

has a t-distribution with $n-2$ degrees of freedom. The statistic T can be used to construct a $100(1-\alpha)\%$ confidence interval for the coefficient β_1.

Confidence Interval for β_1 A $100(1-\alpha)\%$ confidence interval for the parameter β_1 in the regression line $\mu_{Y|x} = \beta_0 + \beta_1 x$ is

$$b_1 - t_{\alpha/2}\frac{s}{\sqrt{S_{xx}}} < \beta_1 < b_1 + t_{\alpha/2}\frac{s}{\sqrt{S_{xx}}},$$

where $t_{\alpha/2}$ is a value of the t-distribution with $n-2$ degrees of freedom.

Example 11.2: Find a 95% confidence interval for β_1 in the regression line $\mu_{Y|x} = \beta_0 + \beta_1 x$, based on the pollution data of Table 11.1.

Solution: From the results given in Example 11.1 we find that $S_{xx} = 4152.18$ and $S_{xy} = 3752.09$. In addition, we find that $S_{yy} = 3713.88$. Recall that $b_1 = 0.903643$. Hence,

$$s^2 = \frac{S_{yy} - b_1 S_{xy}}{n-2} = \frac{3713.88 - (0.903643)(3752.09)}{31} = 10.4299.$$

Therefore, taking the square root, we obtain $s = 3.2295$. Using Table A.4, we find $t_{0.025} \approx 2.045$ for 31 degrees of freedom. Therefore, a 95% confidence interval for β_1 is

$$0.903643 - \frac{(2.045)(3.2295)}{\sqrt{4152.18}} < \beta < 0.903643 + \frac{(2.045)(3.2295)}{\sqrt{4152.18}},$$

which simplifies to

$$0.8012 < \beta_1 < 1.0061.$$

Hypothesis Testing on the Slope

To test the null hypothesis H_0 that $\beta_1 = \beta_{10}$ against a suitable alternative, we again use the t-distribution with $n-2$ degrees of freedom to establish a critical region and then base our decision on the value of

$$t = \frac{b_1 - \beta_{10}}{s/\sqrt{S_{xx}}}.$$

The method is illustrated by the following example.

Example 11.3: Using the estimated value $b_1 = 0.903643$ of Example 11.1, test the hypothesis that $\beta_1 = 1.0$ against the alternative that $\beta_1 < 1.0$.

Solution: The hypotheses are H_0: $\beta_1 = 1.0$ and H_1: $\beta_1 < 1.0$. So

$$t = \frac{0.903643 - 1.0}{3.2295/\sqrt{4152.18}} = -1.92,$$

with $n - 2 = 31$ degrees of freedom ($P \approx 0.03$).

Decision: The t-value is significant at the 0.03 level, suggesting strong evidence that $\beta_1 < 1.0$.

One important t-test on the slope is the test of the hypothesis

$$H_0\colon \beta_1 = 0 \text{ versus } H_1\colon \beta_1 \neq 0.$$

When the null hypothesis is not rejected, the conclusion is that there is no significant linear relationship between $E(y)$ and the independent variable x. The plot of the data for Example 11.1 would suggest that a linear relationship exists. However, in some applications in which σ^2 is large and thus considerable "noise" is present in the data, a plot, while useful, may not produce clear information for the researcher. Rejection of H_0 above implies that a significant linear regression exists.

Figure 11.7 displays a *MINITAB* printout showing the t-test for

$$H_0\colon \beta_1 = 0 \text{ versus } H_1\colon \beta_1 \neq 0,$$

for the data of Example 11.1. Note the regression coefficient (Coef), standard error (SE Coef), t-value (T), and P-value (P). The null hypothesis is rejected. Clearly, there is a significant linear relationship between mean chemical oxygen demand reduction and solids reduction. Note that the t-statistic is computed as

$$t = \frac{\text{coefficient}}{\text{standard error}} = \frac{b_1}{s/\sqrt{S_{xx}}}.$$

The failure to reject H_0: $\beta_1 = 0$ suggests that there is no linear relationship between Y and x. Figure 11.8 is an illustration of the implication of this result. It may mean that changing x has little impact on changes in Y, as seen in (a). However, it may also indicate that the true relationship is nonlinear, as indicated by (b).

When H_0: $\beta_1 = 0$ is rejected, there is an implication that the linear term in x residing in the model explains a significant portion of variability in Y. The two

11.5 Inferences Concerning the Regression Coefficients

```
Regression Analysis: COD versus Per_Red
The regression equation is COD = 3.83 + 0.904 Per_Red

Predictor      Coef    SE Coef       T       P
Constant      3.830     1.768     2.17   0.038
Per_Red     0.90364   0.05012    18.03   0.000

S = 3.22954   R-Sq = 91.3%   R-Sq(adj) = 91.0%
Analysis of Variance
Source           DF      SS       MS       F       P
Regression        1   3390.6   3390.6   325.08   0.000
Residual Error   31    323.3     10.4
Total            32   3713.9
```

Figure 11.7: *MINITAB* printout for *t*-test for data of Example 11.1.

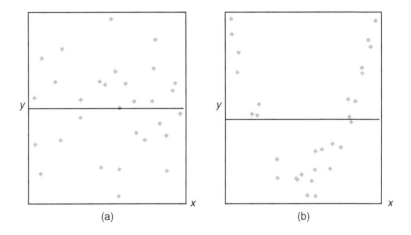

(a) (b)

Figure 11.8: The hypothesis H_0: $\beta_1 = 0$ is not rejected.

plots in Figure 11.9 illustrate possible scenarios. As depicted in (a) of the figure, rejection of H_0 may suggest that the relationship is, indeed, linear. As indicated in (b), it may suggest that while the model does contain a linear effect, a better representation may be found by including a polynomial (perhaps quadratic) term (i.e., terms that supplement the linear term).

Statistical Inference on the Intercept

Confidence intervals and hypothesis testing on the coefficient β_0 may be established from the fact that B_0 is also normally distributed. It is not difficult to show that

$$T = \frac{B_0 - \beta_0}{S\sqrt{\sum_{i=1}^{n} x_i^2/(nS_{xx})}}$$

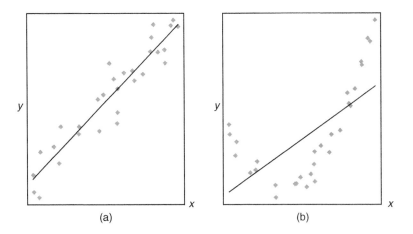

Figure 11.9: The hypothesis $H_0: \beta_1 = 0$ is rejected.

has a t-distribution with $n-2$ degrees of freedom from which we may construct a $100(1-\alpha)\%$ confidence interval for α.

Confidence Interval for β_0 A $100(1-\alpha)\%$ confidence interval for the parameter β_0 in the regression line $\mu_{Y|x} = \beta_0 + \beta_1 x$ is

$$b_0 - t_{\alpha/2}\frac{s}{\sqrt{nS_{xx}}}\sqrt{\sum_{i=1}^{n} x_i^2} < \beta_0 < b_0 + t_{\alpha/2}\frac{s}{\sqrt{nS_{xx}}}\sqrt{\sum_{i=1}^{n} x_i^2},$$

where $t_{\alpha/2}$ is a value of the t-distribution with $n-2$ degrees of freedom.

Example 11.4: Find a 95% confidence interval for β_0 in the regression line $\mu_{Y|x} = \beta_0 + \beta_1 x$, based on the data of Table 11.1.

Solution: In Examples 11.1 and 11.2, we found that

$$S_{xx} = 4152.18 \quad \text{and} \quad s = 3.2295.$$

From Example 11.1 we had

$$\sum_{i=1}^{n} x_i^2 = 41{,}086 \quad \text{and} \quad b_0 = 3.829633.$$

Using Table A.4, we find $t_{0.025} \approx 2.045$ for 31 degrees of freedom. Therefore, a 95% confidence interval for β_0 is

$$3.829633 - \frac{(2.045)(3.2295)\sqrt{41{,}086}}{\sqrt{(33)(4152.18)}} < \beta_0 < 3.829633 + \frac{(2.045)(3.2295)\sqrt{41{,}086}}{\sqrt{(33)(4152.18)}},$$

which simplifies to $0.2132 < \beta_0 < 7.4461$.

11.5 Inferences Concerning the Regression Coefficients

To test the null hypothesis H_0 that $\beta_0 = \beta_{00}$ against a suitable alternative, we can use the t-distribution with $n-2$ degrees of freedom to establish a critical region and then base our decision on the value of

$$t = \frac{b_0 - \beta_{00}}{s\sqrt{\sum_{i=1}^{n} x_i^2/(nS_{xx})}}.$$

Example 11.5: Using the estimated value $b_0 = 3.829633$ of Example 11.1, test the hypothesis that $\beta_0 = 0$ at the 0.05 level of significance against the alternative that $\beta_0 \neq 0$.

Solution: The hypotheses are H_0: $\beta_0 = 0$ and H_1: $\beta_0 \neq 0$. So

$$t = \frac{3.829633 - 0}{3.2295\sqrt{41{,}086/[(33)(4152.18)]}} = 2.17,$$

with 31 degrees of freedom. Thus, $P = P$-value ≈ 0.038 and we conclude that $\beta_0 \neq 0$. Note that this is merely Coef/StDev, as we see in the MINITAB printout in Figure 11.7. The SE Coef is the standard error of the estimated intercept.

A Measure of Quality of Fit: Coefficient of Determination

Note in Figure 11.7 that an item denoted by R-Sq is given with a value of 91.3%. This quantity, R^2, is called the **coefficient of determination**. This quantity is a measure of the **proportion of variability explained by the fitted model**. In Section 11.8, we shall introduce the notion of an analysis-of-variance approach to hypothesis testing in regression. The analysis-of-variance approach makes use of the error sum of squares $SSE = \sum_{i=1}^{n}(y_i - \hat{y}_i)^2$ and the **total corrected sum of squares** $SST = \sum_{i=1}^{n}(y_i - \bar{y}_i)^2$. The latter represents the variation in the response values that *ideally* would be explained by the model. The SSE value is the variation due to error, or **variation unexplained**. Clearly, if $SSE = 0$, all variation is explained. The quantity that represents variation explained is $SST - SSE$. The R^2 is

$$\text{Coeff. of determination:} \quad R^2 = 1 - \frac{SSE}{SST}.$$

Note that if the fit is perfect, *all residuals are zero*, and thus $R^2 = 1.0$. But if SSE is only slightly smaller than SST, $R^2 \approx 0$. Note from the printout in Figure 11.7 that the coefficient of determination suggests that the model fit to the data explains 91.3% of the variability observed in the response, the reduction in chemical oxygen demand.

Figure 11.10 provides an illustration of a good fit ($R^2 \approx 1.0$) in plot (a) and a poor fit ($R^2 \approx 0$) in plot (b).

Pitfalls in the Use of R^2

Analysts quote values of R^2 quite often, perhaps due to its simplicity. However, there are pitfalls in its interpretation. The reliability of R^2 is a function of the

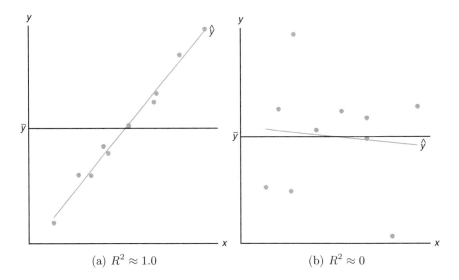

Figure 11.10: Plots depicting a very good fit and a poor fit.

size of the regression data set and the type of application. Clearly, $0 \leq R^2 \leq 1$ and the upper bound is achieved when the fit to the data is perfect (i.e., all of the residuals are zero). What is an acceptable value for R^2? This is a difficult question to answer. A chemist, charged with doing a linear calibration of a high-precision piece of equipment, certainly expects to experience a very high R^2-value (perhaps exceeding 0.99), while a behavioral scientist, dealing in data impacted by variability in human behavior, may feel fortunate to experience an R^2 as large as 0.70. An experienced model fitter senses when a value is large enough, given the situation confronted. Clearly, some scientific phenomena lend themselves to modeling with more precision than others.

The R^2 criterion is dangerous to use for comparing *competing models* for the same data set. Adding additional terms to the model (e.g., an additional regressor) decreases SSE and thus increases R^2 (or at least does not decrease it). This implies that R^2 can be made artificially high by an unwise practice of **overfitting** (i.e., the inclusion of too many model terms). Thus, the inevitable increase in R^2 enjoyed by adding an additional term does not imply the additional term was needed. In fact, the simple model may be superior for predicting response values. The role of overfitting and its influence on prediction capability will be discussed at length in Chapter 12 as we visit the notion of models involving **more than a single regressor**. Suffice it to say at this point that one *should not subscribe to a model selection process that solely involves the consideration of R^2*.

11.6 Prediction

There are several reasons for building a linear regression. One, of course, is to predict response values at one or more values of the independent variable. In this

11.6 Prediction

section, the focus is on errors associated with prediction.

The equation $\hat{y} = b_0 + b_1 x$ may be used to predict or estimate the **mean response** $\mu_{Y|x_0}$ at $x = x_0$, where x_0 is not necessarily one of the prechosen values, or it may be used to predict a single value y_0 of the variable Y_0, when $x = x_0$. We would expect the error of prediction to be higher in the case of a single predicted value than in the case where a mean is predicted. This, then, will affect the width of our intervals for the values being predicted.

Suppose that the experimenter wishes to construct a confidence interval for $\mu_{Y|x_0}$. We shall use the point estimator $\hat{Y}_0 = B_0 + B_1 x_0$ to estimate $\mu_{Y|x_0} = \beta_0 + \beta_1 x$. It can be shown that the sampling distribution of $\hat{Y}_0$ is normal with mean

$$\mu_{Y|x_0} = E(\hat{Y}_0) = E(B_0 + B_1 x_0) = \beta_0 + \beta_1 x_0 = \mu_{Y|x_0}$$

and variance

$$\sigma^2_{\hat{Y}_0} = \sigma^2_{B_0 + B_1 x_0} = \sigma^2_{\bar{Y} + B_1(x_0 - \bar{x})} = \sigma^2 \left[\frac{1}{n} + \frac{(x_0 - \bar{x})^2}{S_{xx}} \right],$$

the latter following from the fact that $\text{Cov}(\bar{Y}, B_1) = 0$ (see Review Exercise 11.61 on page 438). Thus, a $100(1-\alpha)\%$ confidence interval on the mean response $\mu_{Y|x_0}$ can now be constructed from the statistic

$$T = \frac{\hat{Y}_0 - \mu_{Y|x_0}}{S\sqrt{1/n + (x_0 - \bar{x})^2/S_{xx}}},$$

which has a t-distribution with $n - 2$ degrees of freedom.

Confidence Interval for $\mu_{Y|x_0}$ A $100(1-\alpha)\%$ confidence interval for the mean response $\mu_{Y|x_0}$ is

$$\hat{y}_0 - t_{\alpha/2} s \sqrt{\frac{1}{n} + \frac{(x_0 - \bar{x})^2}{S_{xx}}} < \mu_{Y|x_0} < \hat{y}_0 + t_{\alpha/2} s \sqrt{\frac{1}{n} + \frac{(x_0 - \bar{x})^2}{S_{xx}}},$$

where $t_{\alpha/2}$ is a value of the t-distribution with $n - 2$ degrees of freedom.

Example 11.6: Using the data of Table 11.1, construct 95% confidence limits for the mean response $\mu_{Y|x_0}$.

Solution: From the regression equation we find for $x_0 = 20\%$ solids reduction, say,

$$\hat{y}_0 = 3.829633 + (0.903643)(20) = 21.9025.$$

In addition, $\bar{x} = 33.4545$, $S_{xx} = 4152.18$, $s = 3.2295$, and $t_{0.025} \approx 2.045$ for 31 degrees of freedom. Therefore, a 95% confidence interval for $\mu_{Y|20}$ is

$$21.9025 - (2.045)(3.2295)\sqrt{\frac{1}{33} + \frac{(20 - 33.4545)^2}{4152.18}} < \mu_{Y|20}$$

$$< 21.9025 + (2.045)(3.2295)\sqrt{\frac{1}{33} + \frac{(20 - 33.4545)^2}{4152.18}},$$

or simply $20.1071 < \mu_{Y|20} < 23.6979$.

Repeating the previous calculations for each of several different values of x_0, one can obtain the corresponding confidence limits on each $\mu_{Y|x_0}$. Figure 11.11 displays the data points, the estimated regression line, and the upper and lower confidence limits on the mean of $Y|x$.

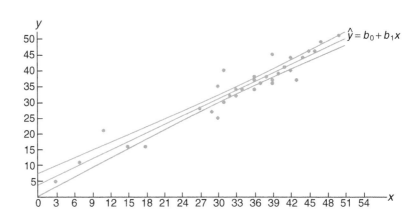

Figure 11.11: Confidence limits for the mean value of $Y|x$.

In Example 11.6, we are 95% confident that the population mean reduction in chemical oxygen demand is between 20.1071% and 23.6979% when solid reduction is 20%.

Prediction Interval

Another type of interval that is often misinterpreted and confused with that given for $\mu_{Y|x}$ is the prediction interval for a future observed response. Actually in many instances, the prediction interval is more relevant to the scientist or engineer than the confidence interval on the mean. In the tar content and inlet temperature example cited in Section 11.1, there would certainly be interest not only in estimating the mean tar content at a specific temperature but also in constructing an interval that reflects the error in predicting a future observed amount of tar content at the given temperature.

To obtain a **prediction interval** for any single value y_0 of the variable Y_0, it is necessary to estimate the variance of the differences between the ordinates $\hat{y}_0$, obtained from the computed regression lines in repeated sampling when $x = x_0$, and the corresponding true ordinate y_0. We can think of the difference $\hat{y}_0 - y_0$ as a value of the random variable $\hat{Y}_0 - Y_0$, whose sampling distribution can be shown to be normal with mean

$$\mu_{\hat{Y}_0 - Y_0} = E(\hat{Y}_0 - Y_0) = E[B_0 + B_1 x_0 - (\beta_0 + \beta_1 x_0 + \epsilon_0)] = 0$$

and variance

$$\sigma^2_{\hat{Y}_0 - Y_0} = \sigma^2_{B_0 + B_1 x_0 - \epsilon_0} = \sigma^2_{\bar{Y} + B_1(x_0 - \bar{x}) - \epsilon_0} = \sigma^2 \left[1 + \frac{1}{n} + \frac{(x_0 - \bar{x})^2}{S_{xx}} \right].$$

Thus, a $100(1-\alpha)\%$ prediction interval for a single predicted value y_0 can be constructed from the statistic

$$T = \frac{\hat{Y}_0 - Y_0}{S\sqrt{1 + 1/n + (x_0 - \bar{x})^2/S_{xx}}},$$

which has a t-distribution with $n-2$ degrees of freedom.

Prediction Interval for y_0 A $100(1-\alpha)\%$ prediction interval for a single response y_0 is given by

$$\hat{y}_0 - t_{\alpha/2} s\sqrt{1 + \frac{1}{n} + \frac{(x_0 - \bar{x})^2}{S_{xx}}} < y_0 < \hat{y}_0 + t_{\alpha/2} s\sqrt{1 + \frac{1}{n} + \frac{(x_0 - \bar{x})^2}{S_{xx}}},$$

where $t_{\alpha/2}$ is a value of the t-distribution with $n-2$ degrees of freedom.

Clearly, there is a distinction between the concept of a confidence interval and the prediction interval described previously. The interpretation of the confidence interval is identical to that described for all confidence intervals on population parameters discussed throughout the book. Indeed, $\mu_{Y|x_0}$ is a population parameter. The computed prediction interval, however, represents an interval that has a probability equal to $1-\alpha$ of containing not a parameter but a future value y_0 of the random variable Y_0.

Example 11.7: Using the data of Table 11.1, construct a 95% prediction interval for y_0 when $x_0 = 20\%$.

Solution: We have $n = 33$, $x_0 = 20$, $\bar{x} = 33.4545$, $\hat{y}_0 = 21.9025$, $S_{xx} = 4152.18$, $s = 3.2295$, and $t_{0.025} \approx 2.045$ for 31 degrees of freedom. Therefore, a 95% prediction interval for y_0 is

$$21.9025 - (2.045)(3.2295)\sqrt{1 + \frac{1}{33} + \frac{(20 - 33.4545)^2}{4152.18}} < y_0$$

$$< 21.9025 + (2.045)(3.2295)\sqrt{1 + \frac{1}{33} + \frac{(20 - 33.4545)^2}{4152.18}},$$

which simplifies to $15.0585 < y_0 < 28.7464$.

Figure 11.12 shows another plot of the chemical oxygen demand reduction data, with both the confidence interval on the mean response and the prediction interval on an individual response plotted. The plot reflects a much tighter interval around the regression line in the case of the mean response.

Exercises

11.15 With reference to Exercise 11.1 on page 398,
(a) evaluate s^2;
(b) test the hypothesis that $\beta_1 = 0$ against the alternative that $\beta_1 \neq 0$ at the 0.05 level of significance and interpret the resulting decision.

11.16 With reference to Exercise 11.2 on page 398,
(a) evaluate s^2;
(b) construct a 95% confidence interval for β_0;
(c) construct a 95% confidence interval for β_1.

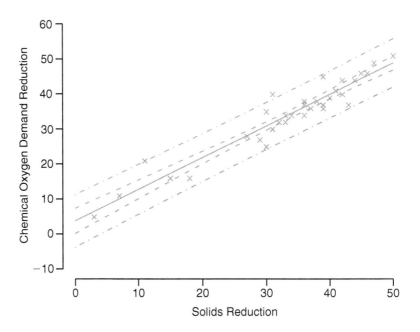

Figure 11.12: Confidence and prediction intervals for the chemical oxygen demand reduction data; inside bands indicate the confidence limits for the mean responses and outside bands indicate the prediction limits for the future responses.

11.17 With reference to Exercise 11.5 on page 398,
(a) evaluate s^2;
(b) construct a 95% confidence interval for β_0;
(c) construct a 95% confidence interval for β_1.

11.18 With reference to Exercise 11.6 on page 399,
(a) evaluate s^2;
(b) construct a 99% confidence interval for β_0;
(c) construct a 99% confidence interval for β_1.

11.19 With reference to Exercise 11.3 on page 398,
(a) evaluate s^2;
(b) construct a 99% confidence interval for β_0;
(c) construct a 99% confidence interval for β_1.

11.20 Test the hypothesis that $\beta_0 = 10$ in Exercise 11.8 on page 399 against the alternative that $\beta_0 < 10$. Use a 0.05 level of significance.

11.21 Test the hypothesis that $\beta_1 = 6$ in Exercise 11.9 on page 399 against the alternative that $\beta_1 < 6$. Use a 0.025 level of significance.

11.22 Using the value of s^2 found in Exercise 11.16(a), construct a 95% confidence interval for $\mu_{Y|85}$ in Exercise 11.2 on page 398.

11.23 With reference to Exercise 11.6 on page 399, use the value of s^2 found in Exercise 11.18(a) to compute
(a) a 95% confidence interval for the mean shear resistance when $x = 24.5$;
(b) a 95% prediction interval for a single predicted value of the shear resistance when $x = 24.5$.

11.24 Using the value of s^2 found in Exercise 11.17(a), graph the regression line and the 95% confidence bands for the mean response $\mu_{Y|x}$ for the data of Exercise 11.5 on page 398.

11.25 Using the value of s^2 found in Exercise 11.17(a), construct a 95% confidence interval for the amount of converted sugar corresponding to $x = 1.6$ in Exercise 11.5 on page 398.

11.26 With reference to Exercise 11.3 on page 398, use the value of s^2 found in Exercise 11.19(a) to compute
(a) a 99% confidence interval for the average amount

Exercises

of chemical that will dissolve in 100 grams of water at 50°C;

(b) a 99% prediction interval for the amount of chemical that will dissolve in 100 grams of water at 50°C.

11.27 Consider the regression of mileage for certain automobiles, measured in miles per gallon (mpg) on their weight in pounds (wt). The data are from *Consumer Reports* (April 1997). Part of the *SAS* output from the procedure is shown in Figure 11.13.

(a) Estimate the mileage for a vehicle weighing 4000 pounds.

(b) Suppose that Honda engineers claim that, on average, the Civic (or any other model weighing 2440 pounds) gets more than 30 mpg. Based on the results of the regression analysis, would you believe that claim? Why or why not?

(c) The design engineers for the Lexus ES300 targeted 18 mpg as being ideal for this model (or any other model weighing 3390 pounds), although it is expected that some variation will be experienced. Is it likely that this target value is realistic? Discuss.

11.28 There are important applications in which, due to known scientific constraints, the regression line **must go through the origin** (i.e., the intercept must be zero). In other words, the model should read

$$Y_i = \beta_1 x_i + \epsilon_i, \quad i = 1, 2, \ldots, n,$$

and only a simple parameter requires estimation. The model is often called the **regression through the origin model**.

(a) Show that the least squares estimator of the slope is

$$b_1 = \left(\sum_{i=1}^{n} x_i y_i\right) \bigg/ \left(\sum_{i=1}^{n} x_i^2\right).$$

(b) Show that $\sigma_{B_1}^2 = \sigma^2 \bigg/ \left(\sum_{i=1}^{n} x_i^2\right)$.

(c) Show that b_1 in part (a) is an unbiased estimator for β_1. That is, show $E(B_1) = \beta_1$.

11.29 Use the data set

y	x
7	2
50	15
100	30
40	10
70	20

(a) Plot the data.

(b) Fit a regression line through the origin.

(c) Plot the regression line on the graph with the data.

(d) Give a general formula (in terms of the y_i and the slope b_1) for the estimator of σ^2.

(e) Give a formula for Var($\hat{y}_i$), $i = 1, 2, \ldots, n$, for this case.

(f) Plot 95% confidence limits for the mean response on the graph around the regression line.

11.30 For the data in Exercise 11.29, find a 95% prediction interval at $x = 25$.

```
              Root MSE              1.48794    R-Square    0.9509
              Dependent Mean       21.50000    Adj R-Sq    0.9447
                           Parameter Estimates
                              Parameter      Standard
              Variable   DF   Estimate       Error       t Value    Pr > |t|
              Intercept   1    44.78018      1.92919      23.21     <.0001
              WT          1    -0.00686      0.00055133  -12.44     <.0001
```

MODEL	WT	MPG	Predict	LMean	UMean	Lpred	Upred	Residual
GMC	4520	15	13.7720	11.9752	15.5688	9.8988	17.6451	1.22804
Geo	2065	29	30.6138	28.6063	32.6213	26.6385	34.5891	-1.61381
Honda	2440	31	28.0412	26.4143	29.6681	24.2439	31.8386	2.95877
Hyundai	2290	28	29.0703	27.2967	30.8438	25.2078	32.9327	-1.07026
Infiniti	3195	23	22.8618	21.7478	23.9758	19.2543	26.4693	0.13825
Isuzu	3480	21	20.9066	19.8160	21.9972	17.3062	24.5069	0.09341
Jeep	4090	15	16.7219	15.3213	18.1224	13.0158	20.4279	-1.72185
Land	4535	13	13.6691	11.8570	15.4811	9.7888	17.5493	-0.66905
Lexus	3390	22	21.5240	20.4390	22.6091	17.9253	25.1227	0.47599
Lincoln	3930	18	17.8195	16.5379	19.1011	14.1568	21.4822	0.18051

Figure 11.13: *SAS* printout for Exercise 11.27.

11.7 Choice of a Regression Model

Much of what has been presented thus far on regression involving a single independent variable depends on the assumption that the model chosen is correct, the presumption that $\mu_{Y|x}$ is related to x linearly in the parameters. Certainly, one cannot expect the prediction of the response to be good if there are several independent variables, not considered in the model, that are affecting the response and are varying in the system. In addition, the prediction will certainly be inadequate if the true structure relating $\mu_{Y|x}$ to x is extremely nonlinear in the range of the variables considered.

Often the simple linear regression model is used even though it is known that the model is something other than linear or that the true structure is unknown. This approach is often sound, particularly when the range of x is narrow. Thus, the model used becomes an approximating function that one hopes is an adequate representation of the true picture in the region of interest. One should note, however, the effect of an inadequate model on the results presented thus far. For example, if the true model, unknown to the experimenter, is linear in more than one x, say

$$\mu_{Y|x_1,x_2} = \beta_0 + \beta_1 x_1 + \beta_2 x_2,$$

then the ordinary least squares estimate $b_1 = S_{xy}/S_{xx}$, calculated by only considering x_1 in the experiment, is, under general circumstances, a biased estimate of the coefficient β_1, the bias being a function of the additional coefficient β_2 (see Review Exercise 11.65 on page 438). Also, the estimate s^2 for σ^2 is biased due to the additional variable.

11.8 Analysis-of-Variance Approach

Often the problem of analyzing the quality of the estimated regression line is handled by an **analysis-of-variance** (ANOVA) approach: a procedure whereby the total variation in the dependent variable is subdivided into meaningful components that are then observed and treated in a systematic fashion. The analysis of variance, discussed in Chapter 13, is a powerful resource that is used for many applications.

Suppose that we have n experimental data points in the usual form (x_i, y_i) and that the regression line is estimated. In our estimation of σ^2 in Section 11.4, we established the identity

$$S_{yy} = b_1 S_{xy} + SSE.$$

An alternative and perhaps more informative formulation is

$$\sum_{i=1}^{n}(y_i - \bar{y})^2 = \sum_{i=1}^{n}(\hat{y}_i - \bar{y})^2 + \sum_{i=1}^{n}(y_i - \hat{y}_i)^2.$$

We have achieved a partitioning of the **total corrected sum of squares of y** into two components that should reflect particular meaning to the experimenter. We shall indicate this partitioning symbolically as

$$SST = SSR + SSE.$$

11.8 Analysis-of-Variance Approach

The first component on the right, SSR, is called the **regression sum of squares**, and it reflects the amount of variation in the y-values **explained by the model**, in this case the postulated straight line. The second component is the familiar error sum of squares, which reflects variation about the regression line.

Suppose that we are interested in testing the hypothesis

$$H_0: \beta_1 = 0 \text{ versus } H_1: \beta_1 \neq 0,$$

where the null hypothesis says essentially that the model is $\mu_{Y|x} = \beta_0$. That is, the variation in Y results from chance or random fluctuations which are independent of the values of x. This condition is reflected in Figure 11.10(b). Under the conditions of this null hypothesis, it can be shown that SSR/σ^2 and SSE/σ^2 are values of independent chi-squared variables with 1 and $n-2$ degrees of freedom, respectively, and then by Theorem 7.12 it follows that SST/σ^2 is also a value of a chi-squared variable with $n-1$ degrees of freedom. To test the hypothesis above, we compute

$$f = \frac{SSR/1}{SSE/(n-2)} = \frac{SSR}{s^2}$$

and reject H_0 at the α-level of significance when $f > f_\alpha(1, n-2)$.

The computations are usually summarized by means of an **analysis-of-variance table**, as in Table 11.2. It is customary to refer to the various sums of squares divided by their respective degrees of freedom as the **mean squares**.

Table 11.2: Analysis of Variance for Testing $\beta_1 = 0$

Source of Variation	Sum of Squares	Degrees of Freedom	Mean Square	Computed f
Regression	SSR	1	SSR	$\frac{SSR}{s^2}$
Error	SSE	$n-2$	$s^2 = \frac{SSE}{n-2}$	
Total	SST	$n-1$		

When the null hypothesis is rejected, that is, when the computed F-statistic exceeds the critical value $f_\alpha(1, n-2)$, we conclude that **there is a significant amount of variation in the response accounted for by the postulated model, the straight-line function**. If the F-statistic is in the fail to reject region, we conclude that the data did not reflect sufficient evidence to support the model postulated.

In Section 11.5, a procedure was given whereby the statistic

$$T = \frac{B_1 - \beta_{10}}{S/\sqrt{S_{xx}}}$$

is used to test the hypothesis

$$H_0: \beta_1 = \beta_{10} \text{ versus } H_1: \beta_1 \neq \beta_{10},$$

where T follows the t-distribution with $n-2$ degrees of freedom. The hypothesis is rejected if $|t| > t_{\alpha/2}$ for an α-level of significance. It is interesting to note that in the special case in which we are testing

$$H_0: \beta_1 = 0 \text{ versus } H_1: \beta_1 \neq 0,$$

the value of our T-statistic becomes

$$t = \frac{b_1}{s/\sqrt{S_{xx}}},$$

and the hypothesis under consideration is identical to that being tested in Table 11.2. Namely, the null hypothesis states that the variation in the response is due merely to chance. The analysis of variance uses the F-distribution rather than the t-distribution. For the two-sided alternative, the two approaches are identical. This we can see by writing

$$t^2 = \frac{b_1^2 S_{xx}}{s^2} = \frac{b_1 S_{xy}}{s^2} = \frac{SSR}{s^2},$$

which is identical to the f-value used in the analysis of variance. The basic relationship between the t-distribution with v degrees of freedom and the F-distribution with 1 and v degrees of freedom is

$$t^2 = f(1, v).$$

Of course, the t-test allows for testing against a one-sided alternative while the F-test is restricted to testing against a two-sided alternative.

Annotated Computer Printout for Simple Linear Regression

Consider again the chemical oxygen demand reduction data of Table 11.1. Figures 11.14 and 11.15 show more complete annotated computer printouts. Again we illustrate it with *MINITAB* software. The t-ratio column indicates tests for null hypotheses of zero values on the parameter. The term "Fit" denotes $\hat{y}$-values, often called **fitted values**. The term "SE Fit" is used in computing confidence intervals on mean response. The item R^2 is computed as $(SSR/SST) \times 100$ and signifies the proportion of variation in y explained by the straight-line regression. Also shown are confidence intervals on the mean response and prediction intervals on a new observation.

11.9 Test for Linearity of Regression: Data with Repeated Observations

In certain kinds of experimental situations, the researcher has the capability of obtaining repeated observations on the response for each value of x. Although it is not necessary to have these repetitions in order to estimate β_0 and β_1, nevertheless repetitions enable the experimenter to obtain quantitative information concerning the appropriateness of the model. In fact, if repeated observations are generated, the experimenter can make a significance test to aid in determining whether or not the model is adequate.

```
The regression equation is COD = 3.83 + 0.904 Per_Red
Predictor     Coef    SE Coef      T       P
Constant     3.830     1.768     2.17   0.038
  Per_Red  0.90364   0.05012    18.03   0.000
S = 3.22954    R-Sq = 91.3%    R-Sq(adj) = 91.0%
              Analysis of Variance
Source           DF        SS        MS        F       P
Regression        1    3390.6    3390.6   325.08   0.000
Residual Error   31     323.3      10.4
Total            32    3713.9

Obs   Per_Red     COD      Fit   SE Fit   Residual   St Resid
 1       3.0    5.000    6.541    1.627     -1.541      -0.55
 2      36.0   34.000   36.361    0.576     -2.361      -0.74
 3       7.0   11.000   10.155    1.440      0.845       0.29
 4      37.0   36.000   37.264    0.590     -1.264      -0.40
 5      11.0   21.000   13.770    1.258      7.230       2.43
 6      38.0   38.000   38.168    0.607     -0.168      -0.05
 7      15.0   16.000   17.384    1.082     -1.384      -0.45
 8      39.0   37.000   39.072    0.627     -2.072      -0.65
 9      18.0   16.000   20.095    0.957     -4.095      -1.33
10      39.0   36.000   39.072    0.627     -3.072      -0.97
11      27.0   28.000   28.228    0.649     -0.228      -0.07
12      39.0   45.000   39.072    0.627      5.928       1.87
13      29.0   27.000   30.035    0.605     -3.035      -0.96
14      40.0   39.000   39.975    0.651     -0.975      -0.31
15      30.0   25.000   30.939    0.588     -5.939      -1.87
16      41.0   41.000   40.879    0.678      0.121       0.04
17      30.0   35.000   30.939    0.588      4.061       1.28
18      42.0   40.000   41.783    0.707     -1.783      -0.57
19      31.0   30.000   31.843    0.575     -1.843      -0.58
20      42.0   44.000   41.783    0.707      2.217       0.70
21      31.0   40.000   31.843    0.575      8.157       2.57
22      43.0   37.000   42.686    0.738     -5.686      -1.81
23      32.0   32.000   32.746    0.567     -0.746      -0.23
24      44.0   44.000   43.590    0.772      0.410       0.13
25      33.0   34.000   33.650    0.563      0.350       0.11
26      45.0   46.000   44.494    0.807      1.506       0.48
27      33.0   32.000   33.650    0.563     -1.650      -0.52
28      46.0   46.000   45.397    0.843      0.603       0.19
29      34.0   34.000   34.554    0.563     -0.554      -0.17
30      47.0   49.000   46.301    0.881      2.699       0.87
31      36.0   37.000   36.361    0.576      0.639       0.20
32      50.0   51.000   49.012    1.002      1.988       0.65
33      36.0   38.000   36.361    0.576      1.639       0.52
```

Figure 11.14: *MINITAB* printout of simple linear regression for chemical oxygen demand reduction data; part I.

Let us select a random sample of n observations using k distinct values of x, say $x_1, x_2, \ldots, x_n$, such that the sample contains n_1 observed values of the random variable Y_1 corresponding to x_1, n_2 observed values of Y_2 corresponding to $x_2, \ldots,$ n_k observed values of Y_k corresponding to x_k. Of necessity, $n = \sum_{i=1}^{k} n_i$.

```
Obs      Fit    SE Fit        95% CI                95% PI
  1    6.541    1.627    ( 3.223,  9.858)      (-0.834, 13.916)
  2   36.361    0.576    (35.185, 37.537)      (29.670, 43.052)
  3   10.155    1.440    ( 7.218, 13.092)      ( 2.943, 17.367)
  4   37.264    0.590    (36.062, 38.467)      (30.569, 43.960)
  5   13.770    1.258    (11.204, 16.335)      ( 6.701, 20.838)
  6   38.168    0.607    (36.931, 39.405)      (31.466, 44.870)
  7   17.384    1.082    (15.177, 19.592)      (10.438, 24.331)
  8   39.072    0.627    (37.793, 40.351)      (32.362, 45.781)
  9   20.095    0.957    (18.143, 22.047)      (13.225, 26.965)
 10   39.072    0.627    (37.793, 40.351)      (32.362, 45.781)
 11   28.228    0.649    (26.905, 29.551)      (21.510, 34.946)
 12   39.072    0.627    (37.793, 40.351)      (32.362, 45.781)
 13   30.035    0.605    (28.802, 31.269)      (23.334, 36.737)
 14   39.975    0.651    (38.648, 41.303)      (33.256, 46.694)
 15   30.939    0.588    (29.739, 32.139)      (24.244, 37.634)
 16   40.879    0.678    (39.497, 42.261)      (34.149, 47.609)
 17   30.939    0.588    (29.739, 32.139)      (24.244, 37.634)
 18   41.783    0.707    (40.341, 43.224)      (35.040, 48.525)
 19   31.843    0.575    (30.669, 33.016)      (25.152, 38.533)
 20   41.783    0.707    (40.341, 43.224)      (35.040, 48.525)
 21   31.843    0.575    (30.669, 33.016)      (25.152, 38.533)
 22   42.686    0.738    (41.181, 44.192)      (35.930, 49.443)
 23   32.746    0.567    (31.590, 33.902)      (26.059, 39.434)
 24   43.590    0.772    (42.016, 45.164)      (36.818, 50.362)
 25   33.650    0.563    (32.502, 34.797)      (26.964, 40.336)
 26   44.494    0.807    (42.848, 46.139)      (37.704, 51.283)
 27   33.650    0.563    (32.502, 34.797)      (26.964, 40.336)
 28   45.397    0.843    (43.677, 47.117)      (38.590, 52.205)
 29   34.554    0.563    (33.406, 35.701)      (27.868, 41.239)
 30   46.301    0.881    (44.503, 48.099)      (39.473, 53.128)
 31   36.361    0.576    (35.185, 37.537)      (29.670, 43.052)
 32   49.012    1.002    (46.969, 51.055)      (42.115, 55.908)
 33   36.361    0.576    (35.185, 37.537)      (29.670, 43.052)
```

Figure 11.15: *MINITAB* printout of simple linear regression for chemical oxygen demand reduction data; part II.

We define

$$y_{ij} = \text{the } j\text{th value of the random variable } Y_i,$$

$$y_{i.} = T_{i.} = \sum_{j=1}^{n_i} y_{ij},$$

$$\bar{y}_{i.} = \frac{T_{i.}}{n_i}.$$

Hence, if $n_4 = 3$ measurements of Y were made corresponding to $x = x_4$, we would indicate these observations by $y_{41}, y_{42},$ and y_{43}. Then

$$T_{i.} = y_{41} + y_{42} + y_{43}.$$

Concept of Lack of Fit

The error sum of squares consists of two parts: the amount due to the variation between the values of Y within given values of x and a component that is normally called the **lack-of-fit** contribution. The first component reflects mere random variation, or **pure experimental error**, while the second component is a measure of the systematic variation brought about by higher-order terms. In our case, these are terms in x other than the linear, or first-order, contribution. Note that in choosing a linear model we are essentially assuming that this second component does not exist and hence our error sum of squares is completely due to random errors. If this should be the case, then $s^2 = SSE/(n-2)$ is an unbiased estimate of σ^2. However, if the model does not adequately fit the data, then the error sum of squares is inflated and produces a biased estimate of σ^2. Whether or not the model fits the data, an unbiased estimate of σ^2 can always be obtained when we have repeated observations simply by computing

$$s_i^2 = \frac{\sum_{j=1}^{n_i}(y_{ij} - \bar{y}_{i.})^2}{n_i - 1}, \quad i = 1, 2, \ldots, k,$$

for each of the k distinct values of x and then pooling these variances to get

$$s^2 = \frac{\sum_{i=1}^{k}(n_i - 1)s_i^2}{n-k} = \frac{\sum_{i=1}^{k}\sum_{j=1}^{n_i}(y_{ij} - \bar{y}_{i.})^2}{n-k}.$$

The numerator of s^2 is a **measure of the pure experimental error**. A computational procedure for separating the error sum of squares into the two components representing pure error and lack of fit is as follows:

Computation of Lack-of-Fit Sum of Squares

1. Compute the pure error sum of squares

$$\sum_{i=1}^{k}\sum_{j=1}^{n_i}(y_{ij} - \bar{y}_{i.})^2.$$

This sum of squares has $n - k$ degrees of freedom associated with it, and the resulting mean square is our unbiased estimate s^2 of σ^2.

2. Subtract the pure error sum of squares from the error sum of squares SSE, thereby obtaining the sum of squares due to lack of fit. The degrees of freedom for lack of fit are obtained by simply subtracting $(n-2) - (n-k) = k-2$.

The computations required for testing hypotheses in a regression problem with repeated measurements on the response may be summarized as shown in Table 11.3.

Figures 11.16 and 11.17 display the sample points for the "correct model" and "incorrect model" situations. In Figure 11.16, where the $\mu_{Y|x}$ fall on a straight line, there is no lack of fit when a linear model is assumed, so the sample variation around the regression line is a pure error resulting from the variation that occurs among repeated observations. In Figure 11.17, where the $\mu_{Y|x}$ clearly do not fall on a straight line, the lack of fit from erroneously choosing a linear model accounts for a large portion of the variation around the regression line, supplementing the pure error.

Table 11.3: Analysis of Variance for Testing Linearity of Regression

Source of Variation	Sum of Squares	Degrees of Freedom	Mean Square	Computed f
Regression	SSR	1	SSR	$\frac{SSR}{s^2}$
Error	SSE	$n-2$		
Lack of fit	$\begin{cases} SSE - SSE\text{ (pure)} \\ SSE\text{ (pure)} \end{cases}$	$\begin{cases} k-2 \\ n-k \end{cases}$	$\frac{SSE-SSE(\text{pure})}{k-2}$	$\frac{SSE-SSE(\text{pure})}{s^2(k-2)}$
Pure error			$s^2 = \frac{SSE(\text{pure})}{n-k}$	
Total	SST	$n-1$		

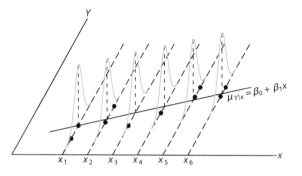
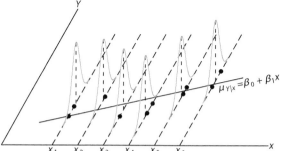

Figure 11.16: Correct linear model with no lack-of-fit component.

Figure 11.17: Incorrect linear model with lack-of-fit component.

What Is the Importance in Detecting Lack of Fit?

The concept of lack of fit is extremely important in applications of regression analysis. In fact, the need to construct or design an experiment that will account for lack of fit becomes more critical as the problem and the underlying mechanism involved become more complicated. Surely, one cannot always be certain that his or her postulated structure, in this case the linear regression model, is correct or even an adequate representation. The following example shows how the error sum of squares is partitioned into the two components representing pure error and lack of fit. The adequacy of the model is tested at the α-level of significance by comparing the lack-of-fit mean square divided by s^2 with $f_\alpha(k-2, n-k)$.

Example 11.8: Observations of the yield of a chemical reaction taken at various temperatures were recorded in Table 11.4. Estimate the linear model $\mu_{Y|x} = \beta_0 + \beta_1 x$ and test for lack of fit.

Solution: Results of the computations are shown in Table 11.5.

Conclusion: The partitioning of the total variation in this manner reveals a significant variation accounted for by the linear model and an insignificant amount of variation due to lack of fit. Thus, the experimental data do not seem to suggest the need to consider terms higher than first order in the model, and the null hypothesis is not rejected.

Table 11.4: Data for Example 11.8

y (%)	x (°C)	y (%)	x (°C)
77.4	150	88.9	250
76.7	150	89.2	250
78.2	150	89.7	250
84.1	200	94.8	300
84.5	200	94.7	300
83.7	200	95.9	300

Table 11.5: Analysis of Variance on Yield-Temperature Data

Source of Variation	Sum of Squares	Degrees of Freedom	Mean Square	Computed f	P-Values
Regression	509.2507	1	509.2507	1531.58	<0.0001
Error	3.8660	10			
Lack of fit	1.2060	2	0.6030	1.81	0.2241
Pure error	2.6600	8	0.3325		
Total	513.1167	11			

Annotated Computer Printout for Test for Lack of Fit

Figure 11.18 is an annotated computer printout showing analysis of the data of Example 11.8 with *SAS*. Note the "LOF" with 2 degrees of freedom, representing the quadratic and cubic contribution to the model, and the P-value of 0.22, suggesting that the linear (first-order) model is adequate.

```
Dependent Variable: yield
                              Sum of
Source                  DF    Squares         Mean Square    F Value    Pr > F
Model                    3    510.4566667     170.1522222    511.74     <.0001
Error                    8      2.6600000       0.3325000
Corrected Total         11    513.1166667

            R-Square    Coeff Var       Root MSE    yield Mean
            0.994816     0.666751       0.576628      86.48333

Source                  DF    Type I SS       Mean Square    F Value    Pr > F
temperature              1    509.2506667     509.2506667    1531.58    <.0001
LOF                      2      1.2060000       0.6030000       1.81    0.2241
```

Figure 11.18: *SAS* printout, showing analysis of data of Example 11.8.

Exercises

11.31 Test for linearity of regression in Exercise 11.3 on page 398. Use a 0.05 level of significance. Comment.

11.32 Test for linearity of regression in Exercise 11.8 on page 399. Comment.

11.33 Suppose we have a linear equation through the origin (Exercise 11.28) $\mu_{Y|x} = \beta x$.

(a) Estimate the regression line passing through the origin for the following data:

x	0.5	1.5	3.2	4.2	5.1	6.5
y	1.3	3.4	6.7	8.0	10.0	13.2

(b) Suppose it is not known whether the true regression should pass through the origin. Estimate the linear model $\mu_{Y|x} = \beta_0 + \beta_1 x$ and test the hypothesis that $\beta_0 = 0$, at the 0.10 level of significance, against the alternative that $\beta_0 \neq 0$.

11.34 Use an analysis-of-variance approach to test the hypothesis that $\beta_1 = 0$ against the alternative hypothesis $\beta_1 \neq 0$ in Exercise 11.5 on page 398 at the 0.05 level of significance.

11.35 The following data are a result of an investigation as to the effect of reaction temperature x on percent conversion of a chemical process y. (See Myers, Montgomery and Anderson-Cook, 2009.) Fit a simple linear regression, and use a lack-of-fit test to determine if the model is adequate. Discuss.

Observation	Temperature (°C), x	Conversion (%), y
1	200	43
2	250	78
3	200	69
4	250	73
5	189.65	48
6	260.35	78
7	225	65
8	225	74
9	225	76
10	225	79
11	225	83
12	225	81

11.36 Transistor gain between emitter and collector in an integrated circuit device (hFE) is related to two variables (Myers, Montgomery and Anderson-Cook, 2009) that can be controlled at the deposition process, emitter drive-in time (x_1, in minutes) and emitter dose (x_2, in ions $\times 10^{14}$). Fourteen samples were observed following deposition, and the resulting data are shown in the table below. We will consider linear regression models using gain as the response and emitter drive-in time or emitter dose as the regressor variable.

Obs.	x_1 (drive-in time, min)	x_2 (dose, ions $\times 10^{14}$)	y (gain, or hFE)
1	195	4.00	1004
2	255	4.00	1636
3	195	4.60	852
4	255	4.60	1506
5	255	4.20	1272
6	255	4.10	1270
7	255	4.60	1269
8	195	4.30	903
9	255	4.30	1555
10	255	4.00	1260
11	255	4.70	1146
12	255	4.30	1276
13	255	4.72	1225
14	340	4.30	1321

(a) Determine if emitter drive-in time influences gain in a linear relationship. That is, test H_0: $\beta_1 = 0$, where β_1 is the slope of the regressor variable.

(b) Do a lack-of-fit test to determine if the linear relationship is adequate. Draw conclusions.

(c) Determine if emitter dose influences gain in a linear relationship. Which regressor variable is the better predictor of gain?

11.37 Organophosphate (OP) compounds are used as pesticides. However, it is important to study their effect on species that are exposed to them. In the laboratory study *Some Effects of Organophosphate Pesticides on Wildlife Species*, by the Department of Fisheries and Wildlife at Virginia Tech, an experiment was conducted in which different dosages of a particular OP pesticide were administered to 5 groups of 5 mice (*peromysius leucopus*). The 25 mice were females of similar age and condition. One group received no chemical. The basic response y was a measure of activity in the brain. It was postulated that brain activity would decrease with an increase in OP dosage. The data are as follows:

Animal	Dose, x (mg/kg body weight)	Activity, y (moles/liter/min)
1	0.0	10.9
2	0.0	10.6
3	0.0	10.8
4	0.0	9.8
5	0.0	9.0
6	2.3	11.0
7	2.3	11.3
8	2.3	9.9
9	2.3	9.2
10	2.3	10.1
11	4.6	10.6
12	4.6	10.4
13	4.6	8.8
14	4.6	11.1
15	4.6	8.4
16	9.2	9.7
17	9.2	7.8
18	9.2	9.0
19	9.2	8.2
20	9.2	2.3
21	18.4	2.9
22	18.4	2.2
23	18.4	3.4
24	18.4	5.4
25	18.4	8.2

(a) Using the model

$$Y_i = \beta_0 + \beta_1 x_i + \epsilon_i, \quad i = 1, 2, \ldots, 25,$$

find the least squares estimates of β_0 and β_1.

(b) Construct an analysis-of-variance table in which the lack of fit and pure error have been separated.

Determine if the lack of fit is significant at the 0.05 level. Interpret the results.

11.38 Heat treating is often used to carburize metal parts such as gears. The thickness of the carburized layer is considered an important feature of the gear, and it contributes to the overall reliability of the part. Because of the critical nature of this feature, a lab test is performed on each furnace load. The test is a destructive one, where an actual part is cross sectioned and soaked in a chemical for a period of time. This test involves running a carbon analysis on the surface of both the gear pitch (top of the gear tooth) and the gear root (between the gear teeth). The data below are the results of the pitch carbon-analysis test for 19 parts.

Soak Time	Pitch	Soak Time	Pitch
0.58	0.013	1.17	0.021
0.66	0.016	1.17	0.019
0.66	0.015	1.17	0.021
0.66	0.016	1.20	0.025
0.66	0.015	2.00	0.025
0.66	0.016	2.00	0.026
1.00	0.014	2.20	0.024
1.17	0.021	2.20	0.025
1.17	0.018	2.20	0.024
1.17	0.019		

(a) Fit a simple linear regression relating the pitch carbon analysis y against soak time. Test H_0: $\beta_1 = 0$.
(b) If the hypothesis in part (a) is rejected, determine if the linear model is adequate.

11.39 A regression model is desired relating temperature and the proportion of impurities passing through solid helium. Temperature is listed in degrees centigrade. The data are as follows:

Temperature (°C)	Proportion of Impurities
−260.5	0.425
−255.7	0.224
−264.6	0.453
−265.0	0.475
−270.0	0.705
−272.0	0.860
−272.5	0.935
−272.6	0.961
−272.8	0.979
−272.9	0.990

(a) Fit a linear regression model.
(b) Does it appear that the proportion of impurities passing through helium increases as the temperature approaches −273 degrees centigrade?
(c) Find R^2.
(d) Based on the information above, does the linear model seem appropriate? What additional information would you need to better answer that question?

11.40 It is of interest to study the effect of population size in various cities in the United States on ozone concentrations. The data consist of the 1999 population in millions and the amount of ozone present per hour in ppb (parts per billion). The data are as follows.

Ozone (ppb/hour), y	Population, x
126	0.6
135	4.9
124	0.2
128	0.5
130	1.1
128	0.1
126	1.1
128	2.3
128	0.6
129	2.3

(a) Fit the linear regression model relating ozone concentration to population. Test H_0: $\beta_1 = 0$ using the ANOVA approach.
(b) Do a test for lack of fit. Is the linear model appropriate based on the results of your test?
(c) Test the hypothesis of part (a) using the pure mean square error in the F-test. Do the results change? Comment on the advantage of each test.

11.41 Evaluating nitrogen deposition from the atmosphere is a major role of the National Atmospheric Deposition Program (NADP), a partnership of many agencies. NADP is studying atmospheric deposition and its effect on agricultural crops, forest surface waters, and other resources. Nitrogen oxides may affect the ozone in the atmosphere and the amount of pure nitrogen in the air we breathe. The data are as follows:

Year	Nitrogen Oxide
1978	0.73
1979	2.55
1980	2.90
1981	3.83
1982	2.53
1983	2.77
1984	3.93
1985	2.03
1986	4.39
1987	3.04
1988	3.41
1989	5.07
1990	3.95
1991	3.14
1992	3.44
1993	3.63
1994	4.50
1995	3.95
1996	5.24
1997	3.30
1998	4.36
1999	3.33

(a) Plot the data.
(b) Fit a linear regression model and find R^2.
(c) What can you say about the trend in nitrogen oxide across time?

11.42 For a particular variety of plant, researchers wanted to develop a formula for predicting the quantity of seeds (in grams) as a function of the density of plants. They conducted a study with four levels of the factor x, the number of plants per plot. Four replications were used for each level of x. The data are shown as follows:

Plants per Plot, x	Quantity of Seeds, y (grams)			
10	12.6	11.0	12.1	10.9
20	15.3	16.1	14.9	15.6
30	17.9	18.3	18.6	17.8
40	19.2	19.6	18.9	20.0

Is a simple linear regression model adequate for analyzing this data set?

11.10 Data Plots and Transformations

In this chapter, we deal with building regression models where there is one independent, or regressor, variable. In addition, we are assuming, through model formulation, that both x and y enter the model in a *linear fashion*. Often it is advisable to work with an alternative model in which either x or y (or both) enters in a nonlinear way. A **transformation** of the data may be indicated because of theoretical considerations inherent in the scientific study, or a simple plotting of the data may suggest the need to *reexpress* the variables in the model. The need to perform a transformation is rather simple to diagnose in the case of simple linear regression because two-dimensional plots give a true pictorial display of how each variable enters the model.

A model in which x or y is transformed should not be viewed as a *nonlinear regression model*. We normally refer to a regression model as linear when it is **linear in the parameters**. In other words, suppose the complexion of the data or other scientific information suggests that we should **regress y^* against x^***, where each is a transformation on the natural variables x and y. Then the model of the form

$$y_i^* = \beta_0 + \beta_1 x_i^* + \epsilon_i$$

is a linear model since it is linear in the parameters β_0 and β_1. The material given in Sections 11.2 through 11.9 remains intact, with y_i^* and x_i^* replacing y_i and x_i. A simple and useful example is the log-log model

$$\log y_i = \beta_0 + \beta_1 \log x_i + \epsilon_i.$$

Although this model is not linear in x and y, it is linear in the parameters and is thus treated as a linear model. On the other hand, an example of a truly nonlinear model is

$$y_i = \beta_0 + \beta_1 x^{\beta_2} + \epsilon_i,$$

where the parameter β_2 (as well as β_0 and β_1) is to be estimated. The model is not linear in β_2.

Transformations that may enhance the fit and predictability of a model are many in number. For a thorough discussion of transformations, the reader is referred to Myers (1990, see the Bibliography). We choose here to indicate a few of them and show the appearance of the graphs that serve as a diagnostic tool. Consider Table 11.6. Several functions are given describing relationships between y and x that can produce a *linear regression* through the transformation indicated.

11.10 Data Plots and Transformations

In addition, for the sake of completeness the reader is given the dependent and independent variables to use in the resulting *simple linear regression*. Figure 11.19 depicts functions listed in Table 11.6. These serve as a guide for the analyst in choosing a transformation from the observation of the plot of y against x.

Table 11.6: Some Useful Transformations to Linearize

Functional Form Relating y to x	Proper Transformation	Form of Simple Linear Regression
Exponential: $y = \beta_0 e^{\beta_1 x}$	$y^* = \ln y$	Regress y^* against x
Power: $y = \beta_0 x^{\beta_1}$	$y^* = \log y;\quad x^* = \log x$	Regress y^* against x^*
Reciprocal: $y = \beta_0 + \beta_1 \left(\frac{1}{x}\right)$	$x^* = \frac{1}{x}$	Regress y against x^*
Hyperbolic: $y = \frac{x}{\beta_0 + \beta_1 x}$	$y^* = \frac{1}{y};\quad x^* = \frac{1}{x}$	Regress y^* against x^*

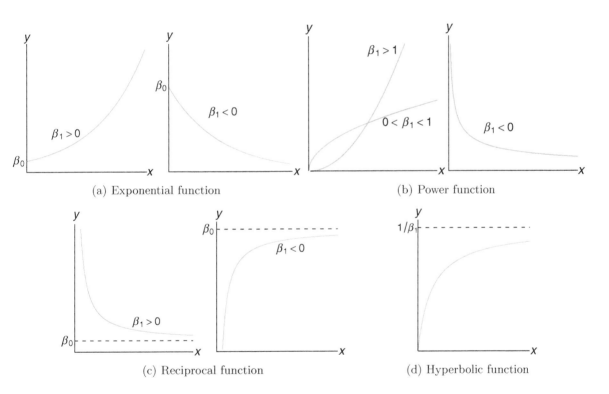

Figure 11.19: Diagrams depicting functions listed in Table 11.6.

What Are the Implications of a Transformed Model?

The foregoing is intended as an aid for the analyst when it is apparent that a transformation will provide an improvement. However, before we provide an example, two important points should be made. The first one revolves around the formal writing of the model when the data are transformed. Quite often the analyst does not think about this. He or she merely performs the transformation without any concern about the model form *before* and *after* the transformation. The exponential model serves as a good illustration. The model in the natural (untransformed) variables that produces an *additive error model* in the transformed variables is given by

$$y_i = \beta_0 e^{\beta_1 x_i} \cdot \epsilon_i,$$

which is a *multiplicative error model.* Clearly, taking logs produces

$$\ln y_i = \ln \beta_0 + \beta_1 x_i + \ln \epsilon_i.$$

As a result, it is on $\ln \epsilon_i$ that the basic assumptions are made. The purpose of this presentation is merely to remind the reader that one should not view a transformation as merely an algebraic manipulation with an error added. Often a model in the transformed variables that has a proper *additive error structure* is a result of a model in the natural variables with a different type of error structure.

The second important point deals with the notion of measures of improvement. Obvious measures of comparison are, of course, R^2 and the residual mean square, s^2. (Other measures of performance used to compare competing models are given in Chapter 12.) Now, if the response y is not transformed, then clearly s^2 and R^2 can be used in measuring the utility of the transformation. The residuals will be in the same units for both the transformed and the untransformed models. But when y is transformed, performance criteria for the transformed model should be based on values of the residuals in the metric of the untransformed response so that comparisons that are made are proper. The example that follows provides an illustration.

Example 11.9: The pressure P of a gas corresponding to various volumes V is recorded, and the data are given in Table 11.7.

Table 11.7: Data for Example 11.9

V (cm^3)	50	60	70	90	100
P (kg/cm^2)	64.7	51.3	40.5	25.9	7.8

The ideal gas law is given by the functional form $PV^\gamma = C$, where γ and C are constants. Estimate the constants C and γ.

Solution: Let us take natural logs of both sides of the model

$$P_i V^\gamma = C \cdot \epsilon_i, \quad i = 1, 2, 3, 4, 5.$$

As a result, a linear model can be written

$$\ln P_i = \ln C - \gamma \ln V_i + \epsilon_i^*, \quad i = 1, 2, 3, 4, 5,$$

where $\epsilon_i^* = \ln \epsilon_i$. The following represents results of the simple linear regression:

Intercept: $\widehat{\ln C} = 14.7589$, $\widehat{C} = 2{,}568{,}862.88$, Slope: $\hat{\gamma} = 2.65347221$.

The following represents information taken from the regression analysis.

P_i	V_i	$\ln P_i$	$\ln V_i$	$\widehat{\ln P_i}$	$\widehat{P_i}$	$e_i = P_i - \widehat{P_i}$
64.7	50	4.16976	3.91202	4.37853	79.7	−15.0
51.3	60	3.93769	4.09434	3.89474	49.1	2.2
40.5	70	3.70130	4.24850	3.48571	32.6	7.9
25.9	90	3.25424	4.49981	2.81885	16.8	9.1
7.8	100	2.05412	4.60517	2.53921	12.7	−4.9

It is instructive to plot the data and the regression equation. Figure 11.20 shows a plot of the data in the untransformed pressure and volume and the curve representing the regression equation.

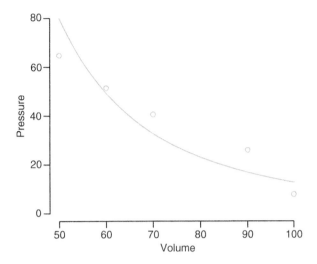

Figure 11.20: Pressure and volume data and fitted regression.

Diagnostic Plots of Residuals: Graphical Detection of Violation of Assumptions

Plots of the raw data can be extremely helpful in determining the nature of the model that should be fit to the data when there is a single independent variable. We have attempted to illustrate this in the foregoing. Detection of proper model form is, however, not the only benefit gained from diagnostic plotting. As in much of the material associated with significance testing in Chapter 10, plotting methods can illustrate and detect violation of assumptions. The reader should recall that much of what is illustrated in this chapter requires assumptions made on the model errors, the ϵ_i. In fact, we assume that the ϵ_i are independent $N(0, \sigma)$ random variables. Now, of course, the ϵ_i are not observed. However, the $e_i = y_i - \hat{y}_i$, the *residuals*, are the error in the fit of the regression line and thus serve to mimic the ϵ_i. Thus, the general complexion of these residuals can often highlight difficulties. Ideally, of course, the plot of the residuals is as depicted in Figure 11.21. That is, they should truly show random fluctuations around a value of zero.

Nonhomogeneous Variance

Homogeneous variance is an important assumption made in regression analysis. Violations can often be detected through the appearance of the residual plot. Increasing error variance with an increase in the regressor variable is a common condition in scientific data. Large error variance produces large residuals, and hence a residual plot like the one in Figure 11.22 is a signal of nonhomogeneous variance. More discussion regarding these residual plots and information regard-

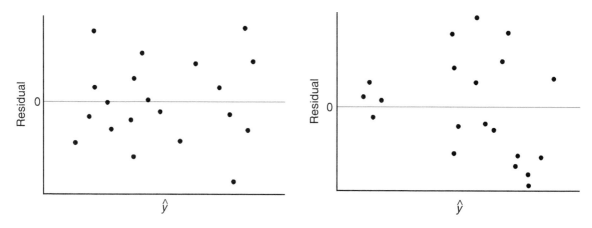

Figure 11.21: Ideal residual plot.

Figure 11.22: Residual plot depicting heterogeneous error variance.

ing different types of residuals appears in Chapter 12, where we deal with multiple linear regression.

Normal Probability Plotting

The assumption that the model errors are normal is made when the data analyst deals in either hypothesis testing or confidence interval estimation. Again, the numerical counterpart to the ϵ_i, namely the residuals, are subjects of diagnostic plotting to detect any extreme violations. In Chapter 8, we introduced normal quantile-quantile plots and briefly discussed normal probability plots. These plots on residuals are illustrated in the case study introduced in the next section.

11.11 Simple Linear Regression Case Study

In the manufacture of commercial wood products, it is important to estimate the relationship between the density of a wood product and its stiffness. A relatively new type of particleboard is being considered that can be formed with considerably more ease than the accepted commercial product. It is necessary to know at what density the stiffness is comparable to that of the well-known, well-documented commercial product. A study was done by Terrance E. Conners, *Investigation of Certain Mechanical Properties of a Wood-Foam Composite* (M.S. Thesis, Department of Forestry and Wildlife Management, University of Massachusetts). Thirty particleboards were produced at densities ranging from roughly 8 to 26 pounds per cubic foot, and the stiffness was measured in pounds per square inch. Table 11.8 shows the data.

It is necessary for the data analyst to focus on an appropriate fit to the data and use inferential methods discussed in this chapter. Hypothesis testing on the slope of the regression, as well as confidence or prediction interval estimation, may well be appropriate. We begin by demonstrating a simple scatter plot of the raw data with a simple linear regression superimposed. Figure 11.23 shows this plot.

The simple linear regression fit to the data produced the fitted model

$$\hat{y} = -25{,}433.739 + 3884.976x \quad (R^2 = 0.7975),$$

11.11 Simple Linear Regression Case Study

Table 11.8: Density and Stiffness for 30 Particleboards

Density, x	Stiffness, y	Density, x	Stiffness, y
9.50	14,814.00	8.40	17,502.00
9.80	14,007.00	11.00	19,443.00
8.30	7573.00	9.90	14,191.00
8.60	9714.00	6.40	8076.00
7.00	5304.00	8.20	10,728.00
17.40	43,243.00	15.00	25,319.00
15.20	28,028.00	16.40	41,792.00
16.70	49,499.00	15.40	25,312.00
15.00	26,222.00	14.50	22,148.00
14.80	26,751.00	13.60	18,036.00
25.60	96,305.00	23.40	104,170.00
24.40	72,594.00	23.30	49,512.00
19.50	32,207.00	21.20	48,218.00
22.80	70,453.00	21.70	47,661.00
19.80	38,138.00	21.30	53,045.00

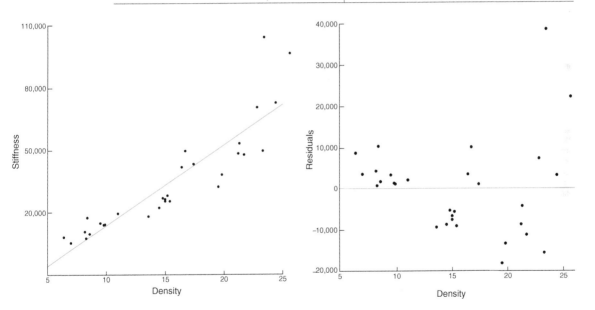

Figure 11.23: Scatter plot of the wood density data.   Figure 11.24: Residual plot for the wood density data.

and the residuals were computed. Figure 11.24 shows the residuals plotted against the measurements of density. This is hardly an ideal or healthy set of residuals. They do not show a random scatter around a value of zero. In fact, clusters of positive and negative values suggest that a curvilinear trend in the data should be investigated.

To gain some type of idea regarding the normal error assumption, a normal probability plot of the residuals was generated. This is the type of plot discussed in

Section 8.8 in which the horizontal axis represents the empirical normal distribution function on a scale that produces a straight-line plot when plotted against the residuals. Figure 11.25 shows the normal probability plot of the residuals. The normal probability plot does not reflect the straight-line appearance that one would like to see. This is another symptom of a faulty, perhaps overly simplistic choice of a regression model.

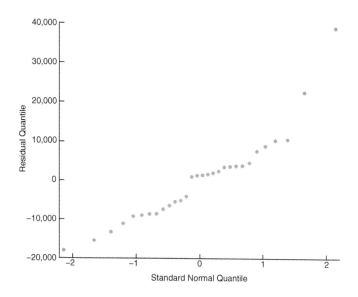

Figure 11.25: Normal probability plot of residuals for wood density data.

Both types of residual plots and, indeed, the scatter plot itself suggest here that a somewhat complicated model would be appropriate. One possible approach is to use a natural log transformation. In other words, one might choose to regress $\ln y$ against x. This produces the regression

$$\widehat{\ln y} = 8.257 + 0.125x \quad (R^2 = 0.9016).$$

To gain some insight into whether the transformed model is more appropriate, consider Figures 11.26 and 11.27, which reveal plots of the residuals in stiffness [i.e., y_i-antilog $(\widehat{\ln y})$] against density. Figure 11.26 appears to be closer to a random pattern around zero, while Figure 11.27 is certainly closer to a straight line. This in addition to the higher R^2-value would suggest that the transformed model is more appropriate.

11.12 Correlation

Up to this point we have assumed that the independent regressor variable x is a physical or scientific variable but not a random variable. In fact, in this context, x is often called a **mathematical variable**, which, in the sampling process, is measured with negligible error. In many applications of regression techniques, it is more realistic to assume that both X and Y are random variables and the measurements $\{(x_i, y_i); i = 1, 2, \ldots, n\}$ are observations from a population having

11.12 Correlation

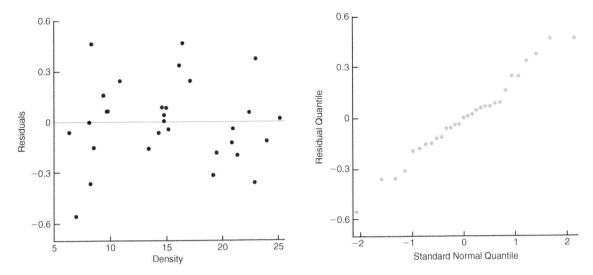

Figure 11.26: Residual plot using the log transformation for the wood density data.

Figure 11.27: Normal probability plot of residuals using the log transformation for the wood density data.

the joint density function $f(x,y)$. We shall consider the problem of measuring the relationship between the two variables X and Y. For example, if X and Y represent the length and circumference of a particular kind of bone in the adult body, we might conduct an anthropological study to determine whether large values of X are associated with large values of Y, and vice versa.

On the other hand, if X represents the age of a used automobile and Y represents the retail book value of the automobile, we would expect large values of X to correspond to small values of Y and small values of X to correspond to large values of Y. **Correlation analysis** attempts to measure the strength of such relationships between two variables by means of a single number called a **correlation coefficient**.

In theory, it is often assumed that the conditional distribution $f(y|x)$ of Y, for fixed values of X, is normal with mean $\mu_{Y|x} = \beta_0 + \beta_1 x$ and variance $\sigma^2_{Y|x} = \sigma^2$ and that X is likewise normally distributed with mean μ and variance σ_x^2. The joint density of X and Y is then

$$f(x,y) = n(y|x; \beta_0 + \beta_1 x, \sigma) n(x; \mu_X, \sigma_X)$$
$$= \frac{1}{2\pi\sigma_x\sigma} \exp\left\{-\frac{1}{2}\left[\left(\frac{y - \beta_0 - \beta_1 x}{\sigma}\right)^2 + \left(\frac{x - \mu_X}{\sigma_X}\right)^2\right]\right\},$$

for $-\infty < x < \infty$ and $-\infty < y < \infty$.

Let us write the random variable Y in the form

$$Y = \beta_0 + \beta_1 X + \epsilon,$$

where X is now a random variable independent of the random error ϵ. Since the mean of the random error ϵ is zero, it follows that

$$\mu_Y = \beta_0 + \beta_1 \mu_X \quad \text{and} \quad \sigma_Y^2 = \sigma^2 + \beta_1^2 \sigma_X^2.$$

Substituting for α and σ^2 into the preceding expression for $f(x,y)$, we obtain the **bivariate normal distribution**

$$f(x,y) = \frac{1}{2\pi\sigma_X\sigma_Y\sqrt{1-\rho^2}}$$
$$\times \exp\left\{-\frac{1}{2(1-\rho^2)}\left[\left(\frac{x-\mu_X}{\sigma_X}\right)^2 - 2\rho\left(\frac{x-\mu_X}{\sigma_X}\right)\left(\frac{y-\mu_Y}{\sigma_Y}\right) + \left(\frac{y-\mu_Y}{\sigma_Y}\right)^2\right]\right\},$$

for $-\infty < x < \infty$ and $-\infty < y < \infty$, where

$$\rho^2 = 1 - \frac{\sigma^2}{\sigma_Y^2} = \beta_1^2\frac{\sigma_X^2}{\sigma_Y^2}.$$

The constant ρ (rho) is called the **population correlation coefficient** and plays a major role in many bivariate data analysis problems. It is important for the reader to understand the physical interpretation of this correlation coefficient and the distinction between correlation and regression. The term *regression* still has meaning here. In fact, the straight line given by $\mu_{Y|x} = \beta_0 + \beta_1 x$ is still called the regression line as before, and the estimates of β_0 and β_1 are identical to those given in Section 11.3. The value of ρ is 0 when $\beta_1 = 0$, which results when there essentially is no linear regression; that is, the regression line is horizontal and any knowledge of X is useless in predicting Y. Since $\sigma_Y^2 \geq \sigma^2$, we must have $\rho^2 \leq 1$ and hence $-1 \leq \rho \leq 1$. Values of $\rho = \pm 1$ only occur when $\sigma^2 = 0$, in which case we have a perfect linear relationship between the two variables. Thus, a value of ρ equal to +1 implies a perfect linear relationship with a positive slope, while a value of ρ equal to -1 results from a perfect linear relationship with a negative slope. It might be said, then, that sample estimates of ρ close to unity in magnitude imply good correlation, or **linear association**, between X and Y, whereas values near zero indicate little or no correlation.

To obtain a sample estimate of ρ, recall from Section 11.4 that the error sum of squares is

$$SSE = S_{yy} - b_1 S_{xy}.$$

Dividing both sides of this equation by S_{yy} and replacing S_{xy} by $b_1 S_{xx}$, we obtain the relation

$$b_1^2 \frac{S_{xx}}{S_{yy}} = 1 - \frac{SSE}{S_{yy}}.$$

The value of $b_1^2 S_{xx}/S_{yy}$ is zero when $b_1 = 0$, which will occur when the sample points show no linear relationship. Since $S_{yy} \geq SSE$, we conclude that $b_1^2 S_{xx}/S_{xy}$ must be between 0 and 1. Consequently, $b_1\sqrt{S_{xx}/S_{yy}}$ must range from -1 to $+1$, negative values corresponding to lines with negative slopes and positive values to lines with positive slopes. A value of -1 or $+1$ will occur when $SSE = 0$, but this is the case where all sample points lie in a straight line. Hence, a perfect linear relationship appears in the sample data when $b_1\sqrt{S_{xx}/S_{yy}} = \pm 1$. Clearly, the quantity $b_1\sqrt{S_{xx}/S_{yy}}$, which we shall henceforth designate as r, can be used as an estimate of the population correlation coefficient ρ. It is customary to refer to the estimate r as the **Pearson product-moment correlation coefficient** or simply the **sample correlation coefficient**.

Correlation Coefficient The measure ρ of linear association between two variables X and Y is estimated by the **sample correlation coefficient** r, where

$$r = b_1\sqrt{\frac{S_{xx}}{S_{yy}}} = \frac{S_{xy}}{\sqrt{S_{xx}S_{yy}}}.$$

11.12 Correlation

For values of r between -1 and $+1$ we must be careful in our interpretation. For example, values of r equal to 0.3 and 0.6 only mean that we have two positive correlations, one somewhat stronger than the other. It is wrong to conclude that $r = 0.6$ indicates a linear relationship twice as good as that indicated by the value $r = 0.3$. On the other hand, if we write

$$r^2 = \frac{S_{xy}^2}{S_{xx}S_{yy}} = \frac{SSR}{S_{yy}},$$

then r^2, which is usually referred to as the **sample coefficient of determination**, represents the proportion of the variation of S_{yy} explained by the regression of Y on x, namely SSR. That is, r^2 expresses the proportion of the total variation in the values of the variable Y that can be accounted for or explained by a linear relationship with the values of the random variable X. Thus, a correlation of 0.6 means that 0.36, or 36%, of the total variation of the values of Y in our sample is accounted for by a linear relationship with values of X.

Example 11.10: It is important that scientific researchers in the area of forest products be able to study correlation among the anatomy and mechanical properties of trees. For the study *Quantitative Anatomical Characteristics of Plantation Grown Loblolly Pine (Pinus Taeda L.) and Cottonwood (Populus deltoides Bart. Ex Marsh.) and Their Relationships to Mechanical Properties*, conducted by the Department of Forestry and Forest Products at Virginia Tech, 29 loblolly pines were randomly selected for investigation. Table 11.9 shows the resulting data on the specific gravity in grams/cm³ and the modulus of rupture in kilopascals (kPa). Compute and interpret the sample correlation coefficient.

Table 11.9: Data on 29 Loblolly Pines for Example 11.10

Specific Gravity, x (g/cm³)	Modulus of Rupture, y (kPa)	Specific Gravity, x (g/cm³)	Modulus of Rupture, y (kPa)
0.414	29,186	0.581	85,156
0.383	29,266	0.557	69,571
0.399	26,215	0.550	84,160
0.402	30,162	0.531	73,466
0.442	38,867	0.550	78,610
0.422	37,831	0.556	67,657
0.466	44,576	0.523	74,017
0.500	46,097	0.602	87,291
0.514	59,698	0.569	86,836
0.530	67,705	0.544	82,540
0.569	66,088	0.557	81,699
0.558	78,486	0.530	82,096
0.577	89,869	0.547	75,657
0.572	77,369	0.585	80,490
0.548	67,095		

Solution: From the data we find that
$$S_{xx} = 0.11273, \quad S_{yy} = 11{,}807{,}324{,}805, \quad S_{xy} = 34{,}422.27572.$$

Therefore,
$$r = \frac{34{,}422.27572}{\sqrt{(0.11273)(11{,}807{,}324{,}805)}} = 0.9435.$$

A correlation coefficient of 0.9435 indicates a good linear relationship between X and Y. Since $r^2 = 0.8902$, we can say that approximately 89% of the variation in the values of Y is accounted for by a linear relationship with X.

A test of the special hypothesis $\rho = 0$ versus an appropriate alternative is equivalent to testing $\beta_1 = 0$ for the simple linear regression model, and therefore the procedures of Section 11.8 using either the t-distribution with $n-2$ degrees of freedom or the F-distribution with 1 and $n-2$ degrees of freedom are applicable. However, if one wishes to avoid the analysis-of-variance procedure and compute only the sample correlation coefficient, it can be verified (see Review Exercise 11.66 on page 438) that the t-value

$$t = \frac{b_1}{s/\sqrt{S_{xx}}}$$

can also be written as

$$t = \frac{r\sqrt{n-2}}{\sqrt{1-r^2}},$$

which, as before, is a value of the statistic T having a t-distribution with $n-2$ degrees of freedom.

Example 11.11: For the data of Example 11.10, test the hypothesis that there is no linear association among the variables.

Solution:
1. H_0: $\rho = 0$.
2. H_1: $\rho \neq 0$.
3. $\alpha = 0.05$.
4. Critical region: $t < -2.052$ or $t > 2.052$.
5. Computations: $t = \frac{0.9435\sqrt{27}}{\sqrt{1-0.9435^2}} = 14.79$, $P < 0.0001$.
6. Decision: Reject the hypothesis of no linear association.

A test of the more general hypothesis $\rho = \rho_0$ against a suitable alternative is easily conducted from the sample information. If X and Y follow the bivariate normal distribution, the quantity

$$\frac{1}{2}\ln\left(\frac{1+r}{1-r}\right)$$

is the value of a random variable that follows approximately the normal distribution with mean $\frac{1}{2}\ln\frac{1+\rho}{1-\rho}$ and variance $1/(n-3)$. Thus, the test procedure is to compute

$$z = \frac{\sqrt{n-3}}{2}\left[\ln\left(\frac{1+r}{1-r}\right) - \ln\left(\frac{1+\rho_0}{1-\rho_0}\right)\right] = \frac{\sqrt{n-3}}{2}\ln\left[\frac{(1+r)(1-\rho_0)}{(1-r)(1+\rho_0)}\right]$$

and compare it with the critical points of the standard normal distribution.

Example 11.12: For the data of Example 11.10, test the null hypothesis that $\rho = 0.9$ against the alternative that $\rho > 0.9$. Use a 0.05 level of significance.

Solution:
1. H_0: $\rho = 0.9$.
2. H_1: $\rho > 0.9$.
3. $\alpha = 0.05$.
4. Critical region: $z > 1.645$.

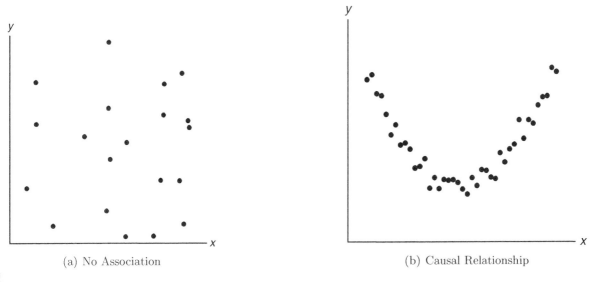

Figure 11.28: Scatter diagram showing zero correlation.

5. Computations:
$$z = \frac{\sqrt{26}}{2} \ln\left[\frac{(1+0.9435)(0.1)}{(1-0.9435)(1.9)}\right] = 1.51, \quad P = 0.0655.$$

6. Decision: There is certainly some evidence that the correlation coefficient does not exceed 0.9.

It should be pointed out that in correlation studies, as in linear regression problems, the results obtained are only as good as the model that is assumed. In the correlation techniques studied here, a bivariate normal density is assumed for the variables X and Y, with the mean value of Y at each x-value being linearly related to x. To observe the suitability of the linearity assumption, a preliminary plotting of the experimental data is often helpful. A value of the sample correlation coefficient close to zero will result from data that display a strictly random effect as in Figure 11.28(a), thus implying little or no causal relationship. It is important to remember that the correlation coefficient between two variables is a measure of their linear relationship and that a value of $r = 0$ implies *a lack of linearity and not a lack of association*. Hence, if a strong quadratic relationship exists between X and Y, as indicated in Figure 11.28(b), we can still obtain a zero correlation indicating a nonlinear relationship.

Exercises

11.43 Compute and interpret the correlation coefficient for the following grades of 6 students selected at random:

Mathematics grade	70	92	80	74	65	83
English grade	74	84	63	87	78	90

11.44 With reference to Exercise 11.1 on page 398, assume that x and y are random variables with a bivariate normal distribution.

(a) Calculate r.

(b) Test the hypothesis that $\rho = 0$ against the alternative that $\rho \neq 0$ at the 0.05 level of significance.

11.45 With reference to Exercise 11.13 on page 400, assume a bivariate normal distribution for x and y.

(a) Calculate r.

(b) Test the null hypothesis that $\rho = -0.5$ against the alternative that $\rho < -0.5$ at the 0.025 level of significance.

(c) Determine the percentage of the variation in the amount of particulate removed that is due to changes in the daily amount of rainfall.

11.46 Test the hypothesis that $\rho = 0$ in Exercise 11.43 against the alternative that $\rho \neq 0$. Use a 0.05 level of significance.

11.47 The following data were obtained in a study of the relationship between the weight and chest size of infants at birth.

Weight (kg)	Chest Size (cm)
2.75	29.5
2.15	26.3
4.41	32.2
5.52	36.5
3.21	27.2
4.32	27.7
2.31	28.3
4.30	30.3
3.71	28.7

(a) Calculate r.

(b) Test the null hypothesis that $\rho = 0$ against the alternative that $\rho > 0$ at the 0.01 level of significance.

(c) What percentage of the variation in infant chest sizes is explained by difference in weight?

Review Exercises

11.48 With reference to Exercise 11.8 on page 399, construct

(a) a 95% confidence interval for the average course grade of students who make a 35 on the placement test;

(b) a 95% prediction interval for the course grade of a student who made a 35 on the placement test.

11.49 The Laboratory for Interdisciplinary Statistical Analysis at Virginia Tech analyzed data on normal woodchucks for the Department of Veterinary Medicine. The variables of interest were body weight in grams and heart weight in grams. It was desired to develop a linear regression equation in order to determine if there is a significant linear relationship between heart weight and total body weight.

Body Weight (grams)	Heart Weight (grams)
4050	11.2
2465	12.4
3120	10.5
5700	13.2
2595	9.8
3640	11.0
2050	10.8
4235	10.4
2935	12.2
4975	11.2
3690	10.8
2800	14.2
2775	12.2
2170	10.0
2370	12.3
2055	12.5
2025	11.8
2645	16.0
2675	13.8

Use heart weight as the independent variable and body weight as the dependent variable and fit a simple linear regression using the following data. In addition, test the hypothesis H_0: $\beta_1 = 0$ versus H_1: $\beta_1 \neq 0$. Draw conclusions.

11.50 The amounts of solids removed from a particular material when exposed to drying periods of different lengths are as shown.

x (hours)	y (grams)	
4.4	13.1	14.2
4.5	9.0	11.5
4.8	10.4	11.5
5.5	13.8	14.8
5.7	12.7	15.1
5.9	9.9	12.7
6.3	13.8	16.5
6.9	16.4	15.7
7.5	17.6	16.9
7.8	18.3	17.2

(a) Estimate the linear regression line.

(b) Test at the 0.05 level of significance whether the linear model is adequate.

11.51 With reference to Exercise 11.9 on page 399, construct

(a) a 95% confidence interval for the average weekly sales when $45 is spent on advertising;

(b) a 95% prediction interval for the weekly sales when $45 is spent on advertising.

11.52 An experiment was designed for the Department of Materials Science and Engineering at Virginia Tech to study hydrogen embrittlement properties based on electrolytic hydrogen pressure measurements.

Review Exercises

The solution used was 0.1 N NaOH, and the material was a certain type of stainless steel. The cathodic charging current density was controlled and varied at four levels. The effective hydrogen pressure was observed as the response. The data follow.

Run	Charging Current Density, x (mA/cm^2)	Effective Hydrogen Pressure, y (atm)
1	0.5	86.1
2	0.5	92.1
3	0.5	64.7
4	0.5	74.7
5	1.5	223.6
6	1.5	202.1
7	1.5	132.9
8	2.5	413.5
9	2.5	231.5
10	2.5	466.7
11	2.5	365.3
12	3.5	493.7
13	3.5	382.3
14	3.5	447.2
15	3.5	563.8

(a) Run a simple linear regression of y against x.

(b) Compute the pure error sum of squares and make a test for lack of fit.

(c) Does the information in part (b) indicate a need for a model in x beyond a first-order regression? Explain.

11.53 The following data represent the chemistry grades for a random sample of 12 freshmen at a certain college along with their scores on an intelligence test administered while they were still seniors in high school.

Student	Test Score, x	Chemistry Grade, y
1	65	85
2	50	74
3	55	76
4	65	90
5	55	85
6	70	87
7	65	94
8	70	98
9	55	81
10	70	91
11	50	76
12	55	74

(a) Compute and interpret the sample correlation coefficient.

(b) State necessary assumptions on random variables.

(c) Test the hypothesis that $\rho = 0.5$ against the alternative that $\rho > 0.5$. Use a P-value in the conclusion.

11.54 The business section of the *Washington Times* in March of 1997 listed 21 different used computers and printers and their sale prices. Also listed was the average hover bid. Partial results from regression analysis using *SAS* software are shown in Figure 11.29 on page 439.

(a) Explain the difference between the confidence interval on the mean and the prediction interval.

(b) Explain why the standard errors of prediction vary from observation to observation.

(c) Which observation has the lowest standard error of prediction? Why?

11.55 Consider the vehicle data from *Consumer Reports* in Figure 11.30 on page 440. Weight is in tons, mileage in miles per gallon, and drive ratio is also indicated. A regression model was fitted relating weight x to mileage y. A partial *SAS* printout in Figure 11.30 on page 440 shows some of the results of that regression analysis, and Figure 11.31 on page 441 gives a plot of the residuals and weight for each vehicle.

(a) From the analysis and the residual plot, does it appear that an improved model might be found by using a transformation? Explain.

(b) Fit the model by replacing weight with log weight. Comment on the results.

(c) Fit a model by replacing mpg with gallons per 100 miles traveled, as mileage is often reported in other countries. Which of the three models is preferable? Explain.

11.56 Observations on the yield of a chemical reaction taken at various temperatures were recorded as follows:

x (°C)	y (%)	x (°C)	y (%)
150	75.4	150	77.7
150	81.2	200	84.4
200	85.5	200	85.7
250	89.0	250	89.4
250	90.5	300	94.8
300	96.7	300	95.3

(a) Plot the data.

(b) Does it appear from the plot as if the relationship is linear?

(c) Fit a simple linear regression and test for lack of fit.

(d) Draw conclusions based on your result in (c).

11.57 Physical fitness testing is an important aspect of athletic training. A common measure of the magnitude of cardiovascular fitness is the maximum volume of oxygen uptake during strenuous exercise. A study was conducted on 24 middle-aged men to determine the influence on oxygen uptake of the time required to complete a two-mile run. Oxygen uptake was measured with standard laboratory methods as the subjects performed on a treadmill. The work was published in "Maximal Oxygen Intake Prediction in Young and Middle Aged Males," *Journal of Sports Medicine* **9**, 1969, 17–22. The data are as follows:

Subject	y, Maximum Volume of O_2	x, Time in Seconds
1	42.33	918
2	53.10	805
3	42.08	892
4	50.06	962
5	42.45	968
6	42.46	907
7	47.82	770
8	49.92	743
9	36.23	1045
10	49.66	810
11	41.49	927
12	46.17	813
13	46.18	858
14	43.21	860
15	51.81	760
16	53.28	747
17	53.29	743
18	47.18	803
19	56.91	683
20	47.80	844
21	48.65	755
22	53.67	700
23	60.62	748
24	56.73	775

(a) Estimate the parameters in a simple linear regression model.

(b) Does the time it takes to run two miles have a significant influence on maximum oxygen uptake? Use H_0: $\beta_1 = 0$ versus H_1: $\beta_1 \neq 0$.

(c) Plot the residuals on a graph against x and comment on the appropriateness of the simple linear model.

11.58 Suppose a scientist postulates a model
$$Y_i = \beta_0 + \beta_1 x_i + \epsilon_i, \quad i = 1, 2, \ldots, n,$$
and β_0 is a **known value**, not necessarily zero.

(a) What is the appropriate least squares estimator of β_1? Justify your answer.

(b) What is the variance of the slope estimator?

11.59 For the simple linear regression model, prove that $E(s^2) = \sigma^2$.

11.60 Assuming that the ϵ_i are independent and normally distributed with zero means and common variance σ^2, show that B_0, the least squares estimator of β_0 in $\mu_{Y|x} = \beta_0 + \beta_1 x$, is normally distributed with mean β_0 and variance

$$\sigma_{B_0}^2 = \frac{\sum_{i=1}^{n} x_i^2}{n \sum_{i=1}^{n} (x_i - \bar{x})^2} \sigma^2.$$

11.61 For a simple linear regression model
$$Y_i = \beta_0 + \beta_1 x_i + \epsilon_i, \quad i = 1, 2, \ldots, n,$$
where the ϵ_i are independent and normally distributed with zero means and equal variances σ^2, show that $\bar{Y}$ and

$$B_1 = \frac{\sum_{i=1}^{n}(x_i - \bar{x})Y_i}{\sum_{i=1}^{n}(x_i - \bar{x})^2}$$

have zero covariance.

11.62 Show, in the case of a least squares fit to the simple linear regression model
$$Y_i = \beta_0 + \beta_1 x_i + \epsilon_i, \quad i = 1, 2, \ldots, n,$$
that $\sum_{i=1}^{n}(y_i - \hat{y}_i) = \sum_{i=1}^{n} e_i = 0$.

11.63 Consider the situation of Review Exercise 11.62 but suppose $n = 2$ (i.e., only two data points are available). Give an argument that the least squares regression line will result in $(y_1 - \hat{y}_1) = (y_2 - \hat{y}_2) = 0$. Also show that for this case $R^2 = 1.0$.

11.64 In Review Exercise 11.62, the student was required to show that $\sum_{i=1}^{n}(y_i - \hat{y}_i) = 0$ for a standard simple linear regression model. Does the same hold for a model with zero intercept? Show why or why not.

11.65 Suppose that an experimenter postulates a model of the type
$$Y_i = \beta_0 + \beta_1 x_{1i} + \epsilon_i, \quad i = 1, 2, \ldots, n,$$
when in fact an additional variable, say x_2, also contributes linearly to the response. The true model is then given by
$$Y_i = \beta_0 + \beta_1 x_{1i} + \beta_2 x_{2i} + \epsilon_i, \quad i = 1, 2, \ldots, n.$$
Compute the expected value of the estimator
$$B_1 = \frac{\sum_{i=1}^{n}(x_{1i} - \bar{x}_1)Y_i}{\sum_{i=1}^{n}(x_{1i} - \bar{x}_1)^2}.$$

11.66 Show the necessary steps in converting the equation $r = \frac{b_1}{s/\sqrt{S_{xx}}}$ to the equivalent form $t = \frac{r\sqrt{n-2}}{\sqrt{1-r^2}}$.

11.67 Consider the fictitious set of data shown below, where the line through the data is the fitted simple linear regression line. Sketch a residual plot.

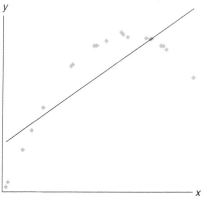

11.68 Project: This project can be done in groups or as individuals. Each group or person must find a set of data, preferably but not restricted to their field of study. The data need to fit the regression framework with a regression variable x and a response variable y. Carefully make the assignment as to which variable is x and which y. It may be necessary to consult a journal or periodical from your field if you do not have other research data available.

(a) Plot y versus x. Comment on the relationship as seen from the plot.

(b) Fit an appropriate regression model from the data. Use simple linear regression or fit a polynomial model to the data. Comment on measures of quality.

(c) Plot residuals as illustrated in the text. Check possible violation of assumptions. Show graphically a plot of confidence intervals on a mean response plotted against x. Comment.

```
    R-Square      Coeff Var        Root MSE    Price Mean
    0.967472      7.923338         70.83841    894.0476
                                   Standard
Parameter         Estimate         Error       t Value   Pr > |t|
Intercept         59.93749137      38.34195754    1.56   0.1345
Buyer              1.04731316       0.04405635   23.77   <.0001
                                       Predict Std Err Lower 95% Upper 95% Lower 95% Upper 95%
product                    Buyer Price  Value  Predict   Mean      Mean    Predict   Predict
IBM PS/1 486/66 420MB        325   375  400.31 25.8906  346.12    454.50   242.46    558.17
IBM ThinkPad 500             450   625  531.23 21.7232  485.76    576.70   376.15    686.31
IBM Think-Dad 755CX         1700  1850 1840.37 42.7041 1750.99   1929.75  1667.25   2013.49
AST Pentium 90 540MB         800   875  897.79 15.4590  865.43    930.14   746.03   1049.54
Dell Pentium 75 1GB          650   700  740.69 16.7503  705.63    775.75   588.34    893.05
Gateway 486/75 320MB         700   750  793.06 16.0314  759.50    826.61   641.04    945.07
Clone 586/133 1GB            500   600  583.59 20.2363  541.24    625.95   429.40    737.79
Compaq Contura 4/25 120MB    450   600  531.23 21.7232  485.76    576.70   376.15    686.31
Compaq Deskpro P90 1.2GB     800   850  897.79 15.4590  865.43    930.14   746.03   1049.54
Micron P75 810MB             800   675  897.79 15.4590  865.43    930.14   746.03   1049.54
Micron P100 1.2GB            900   975 1002.52 16.1176  968.78   1036.25   850.46   1154.58
Mac Quadra 840AV 500MB       450   575  531.23 21.7232  485.76    576.70   376.15    686.31
Mac Performer 6116 700MB     700   775  793.06 16.0314  759.50    826.61   641.04    945.07
PowerBook 540c 320MB        1400  1500 1526.18 30.7579 1461.80   1590.55  1364.54   1687.82
PowerBook 5300 500MB        1350  1575 1473.81 28.8747 1413.37   1534.25  1313.70   1633.92
Power Mac 7500/100 1GB      1150  1325 1264.35 21.9454 1218.42   1310.28  1109.13   1419.57
NEC Versa 486 340MB          800   900  897.79 15.4590  865.43    930.14   746.03   1049.54
Toshiba 1960CS 320MB         700   825  793.06 16.0314  759.50    826.61   641.04    945.07
Toshiba 4800VCT 500MB       1000  1150 1107.25 17.8715 1069.85   1144.66   954.34   1260.16
HP Laser jet III             350   475  426.50 25.0157  374.14    478.86   269.26    583.74
Apple Laser Writer Pro 63    750   800  845.42 15.5930  812.79    878.06   693.61    997.24
```

Figure 11.29: *SAS* printout, showing partial analysis of data of Review Exercise 11.54.

Obs	Model	WT	MPG	DR_RATIO
1	Buick Estate Wagon	4.360	16.9	2.73
2	Ford Country Squire Wagon	4.054	15.5	2.26
3	Chevy Ma libu Wagon	3.605	19.2	2.56
4	Chrysler LeBaron Wagon	3.940	18.5	2.45
5	Chevette	2.155	30.0	3.70
6	Toyota Corona	2.560	27.5	3.05
7	Datsun 510	2.300	27.2	3.54
8	Dodge Omni	2.230	30.9	3.37
9	Audi 5000	2.830	20.3	3.90
10	Volvo 240 CL	3.140	17.0	3.50
11	Saab 99 GLE	2.795	21.6	3.77
12	Peugeot 694 SL	3.410	16.2	3.58
13	Buick Century Special	3.380	20.6	2.73
14	Mercury Zephyr	3.070	20.8	3.08
15	Dodge Aspen	3.620	18.6	2.71
16	AMC Concord D/L	3.410	18.1	2.73
17	Chevy Caprice Classic	3.840	17.0	2.41
18	Ford LTP	3.725	17.6	2.26
19	Mercury Grand Marquis	3.955	16.5	2.26
20	Dodge St Regis	3.830	18.2	2.45
21	Ford Mustang 4	2.585	26.5	3.08
22	Ford Mustang Ghia	2.910	21.9	3.08
23	Mazda GLC	1.975	34.1	3.73
24	Dodge Colt	1.915	35.1	2.97
25	AMC Spirit	2.670	27.4	3.08
26	VW Scirocco	1.990	31.5	3.78
27	Honda Accord LX	2.135	29.5	3.05
28	Buick Skylark	2.570	28.4	2.53
29	Chevy Citation	2.595	28.8	2.69
30	Olds Omega	2.700	26.8	2.84
31	Pontiac Phoenix	2.556	33.5	2.69
32	Plymouth Horizon	2.200	34.2	3.37
33	Datsun 210	2.020	31.8	3.70
34	Fiat Strada	2.130	37.3	3.10
35	VW Dasher	2.190	30.5	3.70
36	Datsun 810	2.815	22.0	3.70
37	BMW 320i	2.600	21.5	3.64
38	VW Rabbit	1.925	31.9	3.78

R-Square	Coeff Var	Root MSE	MPG Mean
0.817244	11.46010	2.837580	24.76053

Parameter	Estimate	Standard Error	t Value	Pr > |t|
Intercept	48.67928080	1.94053995	25.09	<.0001
WT	-8.36243141	0.65908398	-12.69	<.0001

Figure 11.30: *SAS* printout, showing partial analysis of data of Review Exercise 11.55.

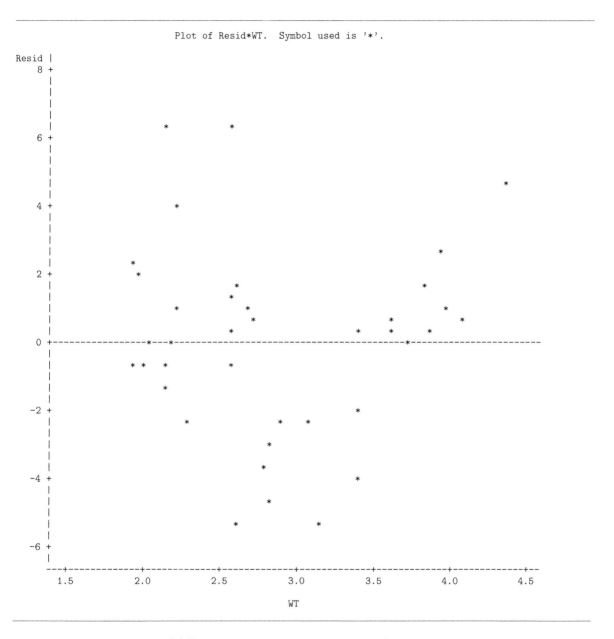

Figure 11.31: *SAS* printout, showing residual plot of Review Exercise 11.55.

11.13 Potential Misconceptions and Hazards; Relationship to Material in Other Chapters

Anytime one is considering the use of simple linear regression, a plot of the data is not only recommended but essential. A plot of the ordinary residuals and a normal probability plot of these residuals are always edifying. In addition, we introduce and illustrate an additional type of residual in Chapter 12 that is in a standardized form. All of these plots are designed to detect violation of assumptions.

The use of t-statistics for tests on regression coefficients is reasonably robust to the normality assumption. The homogeneous variance assumption is crucial, and residual plots are designed to detect a violation.

The material in this chapter is used heavily in Chapters 12 and 15. All of the information involving the method of least squares in the development of regression models carries over into Chapter 12. The difference is that Chapter 12 deals with the scientific conditions in which there is more than a single x variable, i.e., more than one regression variable. However, material in the current chapter that deals with regression diagnostics, types of residual plots, measures of model quality, and so on, applies and will carry over. The student will realize that more complications occur in Chapter 12 because the problems in multiple regression models often involve the backdrop of questions regarding how the various regression variables enter the model and even issues of which variables should remain in the model. Certainly Chapter 15 heavily involves the use of regression modeling, but we will preview the connection in the summary at the end of Chapter 12.

Chapter 12

Multiple Linear Regression and Certain Nonlinear Regression Models

12.1 Introduction

In most research problems where regression analysis is applied, more than one independent variable is needed in the regression model. The complexity of most scientific mechanisms is such that in order to be able to predict an important response, a **multiple regression model** is needed. When this model is linear in the coefficients, it is called a **multiple linear regression model**. For the case of k independent variables $x_1, x_2, \ldots, x_k$, the mean of $Y|x_1, x_2, \ldots, x_k$ is given by the multiple linear regression model

$$\mu_{Y|x_1, x_2, \ldots, x_k} = \beta_0 + \beta_1 x_1 + \cdots + \beta_k x_k,$$

and the estimated response is obtained from the sample regression equation

$$\hat{y} = b_0 + b_1 x_1 + \cdots + b_k x_k,$$

where each regression coefficient β_i is estimated by b_i from the sample data using the method of least squares. As in the case of a single independent variable, the multiple linear regression model can often be an adequate representation of a more complicated structure within certain ranges of the independent variables.

Similar least squares techniques can also be applied for estimating the coefficients when the linear model involves, say, powers and products of the independent variables. For example, when $k = 1$, the experimenter may believe that the means $\mu_{Y|x}$ do not fall on a straight line but are more appropriately described by the **polynomial regression model**

$$\mu_{Y|x} = \beta_0 + \beta_1 x + \beta_2 x^2 + \cdots + \beta_r x^r,$$

and the estimated response is obtained from the polynomial regression equation

$$\hat{y} = b_0 + b_1 x + b_2 x^2 + \cdots + b_r x^r.$$

Confusion arises occasionally when we speak of a polynomial model as a linear model. However, statisticians normally refer to a linear model as one in which the parameters occur linearly, regardless of how the independent variables enter the model. An example of a nonlinear model is the **exponential relationship**

$$\mu_{Y|x} = \alpha \beta^x,$$

whose response is estimated by the regression equation

$$\hat{y} = ab^x.$$

There are many phenomena in science and engineering that are inherently nonlinear in nature, and when the true structure is known, an attempt should certainly be made to fit the actual model. The literature on estimation by least squares of nonlinear models is voluminous. The nonlinear models discussed in this chapter deal with nonideal conditions in which the analyst is certain that the response and hence the response model error are not normally distributed but, rather, have a binomial or Poisson distribution. These situations do occur extensively in practice.

A student who wants a more general account of nonlinear regression should consult *Classical and Modern Regression with Applications* by Myers (1990; see the Bibliography).

12.2 Estimating the Coefficients

In this section, we obtain the least squares estimators of the parameters $\beta_0, \beta_1, \ldots, \beta_k$ by fitting the multiple linear regression model

$$\mu_{Y|x_1,x_2,\ldots,x_k} = \beta_0 + \beta_1 x_1 + \cdots + \beta_k x_k$$

to the data points

$$\{(x_{1i}, x_{2i}, \ldots, x_{ki}, y_i); \quad i = 1, 2, \ldots, n \text{ and } n > k\},$$

where y_i is the observed response to the values $x_{1i}, x_{2i}, \ldots, x_{ki}$ of the k independent variables $x_1, x_2, \ldots, x_k$. Each observation $(x_{1i}, x_{2i}, \ldots, x_{ki}, y_i)$ is assumed to satisfy the following equation.

Multiple Linear Regression Model

$$y_i = \beta_0 + \beta_1 x_{1i} + \beta_2 x_{2i} + \cdots + \beta_k x_{ki} + \epsilon_i$$

or

$$y_i = \hat{y}_i + e_i = b_0 + b_1 x_{1i} + b_2 x_{2i} + \cdots + b_k x_{ki} + e_i,$$

where ϵ_i and e_i are the random error and residual, respectively, associated with the response y_i and fitted value $\hat{y}_i$.

As in the case of simple linear regression, it is assumed that the ϵ_i are independent and identically distributed with mean 0 and common variance σ^2.

In using the concept of least squares to arrive at estimates $b_0, b_1, \ldots, b_k$, we minimize the expression

$$SSE = \sum_{i=1}^{n} e_i^2 = \sum_{i=1}^{n} (y_i - b_0 - b_1 x_{1i} - b_2 x_{2i} - \cdots - b_k x_{ki})^2.$$

Differentiating SSE in turn with respect to $b_0, b_1, \ldots, b_k$ and equating to zero, we generate the set of $k+1$ **normal equations for multiple linear regression**.

12.2 Estimating the Coefficients

Normal Estimation Equations for Multiple Linear Regression

$$nb_0 + b_1\sum_{i=1}^{n} x_{1i} + b_2\sum_{i=1}^{n} x_{2i} + \cdots + b_k\sum_{i=1}^{n} x_{ki} = \sum_{i=1}^{n} y_i$$

$$b_0\sum_{i=1}^{n} x_{1i} + b_1\sum_{i=1}^{n} x_{1i}^2 + b_2\sum_{i=1}^{n} x_{1i}x_{2i} + \cdots + b_k\sum_{i=1}^{n} x_{1i}x_{ki} = \sum_{i=1}^{n} x_{1i}y_i$$

$$\vdots \qquad \vdots \qquad \vdots \qquad \vdots \qquad \vdots$$

$$b_0\sum_{i=1}^{n} x_{ki} + b_1\sum_{i=1}^{n} x_{ki}x_{1i} + b_2\sum_{i=1}^{n} x_{ki}x_{2i} + \cdots + b_k\sum_{i=1}^{n} x_{ki}^2 = \sum_{i=1}^{n} x_{ki}y_i$$

These equations can be solved for $b_0, b_1, b_2, \ldots, b_k$ by any appropriate method for solving systems of linear equations. Most statistical software can be used to obtain numerical solutions of the above equations.

Example 12.1: A study was done on a diesel-powered light-duty pickup truck to see if humidity, air temperature, and barometric pressure influence emission of nitrous oxide (in ppm). Emission measurements were taken at different times, with varying experimental conditions. The data are given in Table 12.2. The model is

$$\mu_{Y|x_1,x_2,x_3} = \beta_0 + \beta_1 x_1 + \beta_2 x_2 + \beta_3 x_3,$$

or, equivalently,

$$y_i = \beta_0 + \beta_1 x_{1i} + \beta_2 x_{2i} + \beta_3 x_{3i} + \epsilon_i, \quad i = 1, 2, \ldots, 20.$$

Fit this multiple linear regression model to the given data and then estimate the amount of nitrous oxide emitted for the conditions where humidity is 50%, temperature is 76°F, and barometric pressure is 29.30.

Table 12.1: Data for Example 12.1

Nitrous Oxide, y	Humidity, x_1	Temp., x_2	Pressure, x_3	Nitrous Oxide, y	Humidity, x_1	Temp., x_2	Pressure, x_3
0.90	72.4	76.3	29.18	1.07	23.2	76.8	29.38
0.91	41.6	70.3	29.35	0.94	47.4	86.6	29.35
0.96	34.3	77.1	29.24	1.10	31.5	76.9	29.63
0.89	35.1	68.0	29.27	1.10	10.6	86.3	29.56
1.00	10.7	79.0	29.78	1.10	11.2	86.0	29.48
1.10	12.9	67.4	29.39	0.91	73.3	76.3	29.40
1.15	8.3	66.8	29.69	0.87	75.4	77.9	29.28
1.03	20.1	76.9	29.48	0.78	96.6	78.7	29.29
0.77	72.2	77.7	29.09	0.82	107.4	86.8	29.03
1.07	24.0	67.7	29.60	0.95	54.9	70.9	29.37

Source: Charles T. Hare, "Light-Duty Diesel Emission Correction Factors for Ambient Conditions," EPA-600/2-77-116. U.S. Environmental Protection Agency.

Solution: The solution of the set of estimating equations yields the unique estimates

$$b_0 = -3.507778, \quad b_1 = -0.002625, \quad b_2 = 0.000799, \quad b_3 = 0.154155.$$

Therefore, the regression equation is

$$\hat{y} = -3.507778 - 0.002625\,x_1 + 0.000799\,x_2 + 0.154155\,x_3.$$

For 50% humidity, a temperature of 76°F, and a barometric pressure of 29.30, the estimated amount of nitrous oxide emitted is

$$\hat{y} = -3.507778 - 0.002625(50.0) + 0.000799(76.0) + 0.1541553(29.30)$$
$$= 0.9384 \text{ ppm.}$$

Polynomial Regression

Now suppose that we wish to fit the polynomial equation

$$\mu_{Y|x} = \beta_0 + \beta_1 x + \beta_2 x^2 + \cdots + \beta_r x^r$$

to the n pairs of observations $\{(x_i, y_i);\ i = 1, 2, \ldots, n\}$. Each observation, y_i, satisfies the equation

$$y_i = \beta_0 + \beta_1 x_i + \beta_2 x_i^2 + \cdots + \beta_r x_i^r + \epsilon_i$$

or

$$y_i = \hat{y}_i + e_i = b_0 + b_1 x_i + b_2 x_i^2 + \cdots + b_r x_i^r + e_i,$$

where r is the degree of the polynomial and ϵ_i and e_i are again the random error and residual associated with the response y_i and fitted value $\hat{y}_i$, respectively. Here, the number of pairs, n, must be at least as large as $r+1$, the number of parameters to be estimated.

Notice that the polynomial model can be considered a special case of the more general multiple linear regression model, where we set $x_1 = x, x_2 = x^2, \ldots, x_r = x^r$. The normal equations assume the same form as those given on page 445. They are then solved for $b_0, b_1, b_2, \ldots, b_r$.

Example 12.2: Given the data

x	0	1	2	3	4	5	6	7	8	9
y	9.1	7.3	3.2	4.6	4.8	2.9	5.7	7.1	8.8	10.2

fit a regression curve of the form $\mu_{Y|x} = \beta_0 + \beta_1 x + \beta_2 x^2$ and then estimate $\mu_{Y|2}$.

Solution: From the data given, we find that

$$10 b_0 + 45\,b_1 + 285\,b_2 = 63.7,$$
$$45 b_0 + 285 b_1 + 2025\,b_2 = 307.3,$$
$$285 b_0 + 2025\,b_1 + 15{,}333 b_2 = 2153.3.$$

Solving these normal equations, we obtain

$$b_0 = 8.698, \quad b_1 = -2.341, \quad b_2 = 0.288.$$

Therefore,

$$\hat{y} = 8.698 - 2.341 x + 0.288 x^2.$$

When $x = 2$, our estimate of $\mu_{Y|2}$ is

$$\hat{y} = 8.698 - (2.341)(2) + (0.288)(2^2) = 5.168.$$

Example 12.3: The data in Table 12.2 represent the percent of impurities that resulted for various temperatures and sterilizing times during a reaction associated with the manufacturing of a certain beverage. Estimate the regression coefficients in the polynomial model

$$y_i = \beta_0 + \beta_1 x_{1i} + \beta_2 x_{2i} + \beta_{11} x_{1i}^2 + \beta_{22} x_{2i}^2 + \beta_{12} x_{1i} x_{2i} + \epsilon_i,$$

for $i = 1, 2, \ldots, 18$.

Table 12.2: Data for Example 12.3

Sterilizing	Temperature, x_1 (°C)		
Time, x_2 (min)	75	100	125
15	14.05	10.55	7.55
	14.93	9.48	6.59
20	16.56	13.63	9.23
	15.85	11.75	8.78
25	22.41	18.55	15.93
	21.66	17.98	16.44

Solution: Using the normal equations, we obtain

$$b_0 = 56.4411, \quad b_1 = -0.36190, \quad b_2 = -2.75299,$$
$$b_{11} = 0.00081, \quad b_{22} = 0.08173, \quad b_{12} = 0.00314,$$

and our estimated regression equation is

$$\hat{y} = 56.4411 - 0.36190 x_1 - 2.75299 x_2 + 0.00081 x_1^2 + 0.08173 x_2^2 + 0.00314 x_1 x_2.$$

Many of the principles and procedures associated with the estimation of polynomial regression functions fall into the category of **response surface methodology**, a collection of techniques that have been used quite successfully by scientists and engineers in many fields. The x_i^2 are called **pure quadratic terms**, and the $x_i x_j$ ($i \neq j$) are called **interaction terms**. Such problems as selecting a proper experimental design, particularly in cases where a large number of variables are in the model, and choosing optimum operating conditions for $x_1, x_2, \ldots, x_k$ are often approached through the use of these methods. For an extensive exposure, the reader is referred to *Response Surface Methodology: Process and Product Optimization Using Designed Experiments* by Myers, Montgomery, and Anderson-Cook (2009; see the Bibliography).

12.3 Linear Regression Model Using Matrices

In fitting a multiple linear regression model, particularly when the number of variables exceeds two, a knowledge of matrix theory can facilitate the mathematical manipulations considerably. Suppose that the experimenter has k independent variables $x_1, x_2, \ldots, x_k$ and n observations $y_1, y_2, \ldots, y_n$, each of which can be expressed by the equation

$$y_i = \beta_0 + \beta_1 x_{1i} + \beta_2 x_{2i} + \cdots + \beta_k x_{ki} + \epsilon_i.$$

This model essentially represents n equations describing how the response values are generated in the scientific process. Using matrix notation, we can write the following equation:

General Linear Model

$$\mathbf{y} = \mathbf{X}\boldsymbol{\beta} + \boldsymbol{\epsilon},$$

where

$$\mathbf{y} = \begin{bmatrix} y_1 \\ y_2 \\ \vdots \\ y_n \end{bmatrix}, \quad \mathbf{X} = \begin{bmatrix} 1 & x_{11} & x_{21} & \cdots & x_{k1} \\ 1 & x_{12} & x_{22} & \cdots & x_{k2} \\ \vdots & \vdots & \vdots & & \vdots \\ 1 & x_{1n} & x_{2n} & \cdots & x_{kn} \end{bmatrix}, \quad \boldsymbol{\beta} = \begin{bmatrix} \beta_0 \\ \beta_1 \\ \vdots \\ \beta_k \end{bmatrix}, \quad \boldsymbol{\epsilon} = \begin{bmatrix} \epsilon_1 \\ \epsilon_2 \\ \vdots \\ \epsilon_n \end{bmatrix}.$$

Then the least squares method for estimation of β, illustrated in Section 12.2, involves finding $\mathbf{b}$ for which

$$SSE = (\mathbf{y} - \mathbf{Xb})'(\mathbf{y} - \mathbf{Xb})$$

is minimized. This minimization process involves solving for $\mathbf{b}$ in the equation

$$\frac{\partial}{\partial \mathbf{b}}(SSE) = \mathbf{0}.$$

We will not present the details regarding solution of the equations above. The result reduces to the solution of $\mathbf{b}$ in

$$(\mathbf{X}'\mathbf{X})\mathbf{b} = \mathbf{X}'\mathbf{y}.$$

Notice the nature of the $\mathbf{X}$ matrix. Apart from the initial element, the ith row represents the x-values that give rise to the response y_i. Writing

$$\mathbf{A} = \mathbf{X}'\mathbf{X} = \begin{bmatrix} n & \sum_{i=1}^{n} x_{1i} & \sum_{i=1}^{n} x_{2i} & \cdots & \sum_{i=1}^{n} x_{ki} \\ \sum_{i=1}^{n} x_{1i} & \sum_{i=1}^{n} x_{1i}^2 & \sum_{i=1}^{n} x_{1i}x_{2i} & \cdots & \sum_{i=1}^{n} x_{1i}x_{ki} \\ \vdots & \vdots & \vdots & & \vdots \\ \sum_{i=1}^{n} x_{ki} & \sum_{i=1}^{n} x_{ki}x_{1i} & \sum_{i=1}^{n} x_{ki}x_{2i} & \cdots & \sum_{i=1}^{n} x_{ki}^2 \end{bmatrix}$$

and

$$\mathbf{g} = \mathbf{X}'\mathbf{y} = \begin{bmatrix} g_0 = \sum_{i=1}^{n} y_i \\ g_1 = \sum_{i=1}^{n} x_{1i} y_i \\ \vdots \\ g_k = \sum_{i=1}^{n} x_{ki} y_i \end{bmatrix}$$

allows the normal equations to be put in the matrix form

$$\mathbf{Ab} = \mathbf{g}.$$

12.3 Linear Regression Model Using Matrices

If the matrix $\mathbf{A}$ is nonsingular, we can write the solution for the regression coefficients as

$$\mathbf{b} = \mathbf{A}^{-1}\mathbf{g} = (\mathbf{X}'\mathbf{X})^{-1}\mathbf{X}'\mathbf{y}.$$

Thus, we can obtain the prediction equation or regression equation by solving a set of $k+1$ equations in a like number of unknowns. This involves the inversion of the $k+1$ by $k+1$ matrix $\mathbf{X}'\mathbf{X}$. Techniques for inverting this matrix are explained in most textbooks on elementary determinants and matrices. Of course, there are many high-speed computer packages available for multiple regression problems, packages that not only print out estimates of the regression coefficients but also provide other information relevant to making inferences concerning the regression equation.

Example 12.4: The percent survival rate of sperm in a certain type of animal semen, after storage, was measured at various combinations of concentrations of three materials used to increase chance of survival. The data are given in Table 12.3. Estimate the multiple linear regression model for the given data.

Table 12.3: Data for Example 12.4

y (% survival)	x_1 (weight %)	x_2 (weight %)	x_3 (weight %)
25.5	1.74	5.30	10.80
31.2	6.32	5.42	9.40
25.9	6.22	8.41	7.20
38.4	10.52	4.63	8.50
18.4	1.19	11.60	9.40
26.7	1.22	5.85	9.90
26.4	4.10	6.62	8.00
25.9	6.32	8.72	9.10
32.0	4.08	4.42	8.70
25.2	4.15	7.60	9.20
39.7	10.15	4.83	9.40
35.7	1.72	3.12	7.60
26.5	1.70	5.30	8.20

Solution: The least squares estimating equations, $(\mathbf{X}'\mathbf{X})\mathbf{b} = \mathbf{X}'\mathbf{y}$, are

$$\begin{bmatrix} 13.0 & 59.43 & 81.82 & 115.40 \\ 59.43 & 394.7255 & 360.6621 & 522.0780 \\ 81.82 & 360.6621 & 576.7264 & 728.3100 \\ 115.40 & 522.0780 & 728.3100 & 1035.9600 \end{bmatrix} \begin{bmatrix} b_0 \\ b_1 \\ b_2 \\ b_3 \end{bmatrix} = \begin{bmatrix} 377.5 \\ 1877.567 \\ 2246.661 \\ 3337.780 \end{bmatrix}.$$

From a computer readout we obtain the elements of the inverse matrix

$$(\mathbf{X}'\mathbf{X})^{-1} = \begin{bmatrix} 8.0648 & -0.0826 & -0.0942 & -0.7905 \\ -0.0826 & 0.0085 & 0.0017 & 0.0037 \\ -0.0942 & 0.0017 & 0.0166 & -0.0021 \\ -0.7905 & 0.0037 & -0.0021 & 0.0886 \end{bmatrix},$$

and then, using the relation $\mathbf{b} = (\mathbf{X}'\mathbf{X})^{-1}\mathbf{X}'\mathbf{y}$, the estimated regression coefficients are obtained as

$$b_0 = 39.1574, \ b_1 = 1.0161, \ b_2 = -1.8616, \ b_3 = -0.3433.$$

Hence, our estimated regression equation is

$$\hat{y} = 39.1574 + 1.0161x_1 - 1.8616x_2 - 0.3433x_3.$$

Exercises

12.1 A set of experimental runs was made to determine a way of predicting cooking time y at various values of oven width x_1 and flue temperature x_2. The coded data were recorded as follows:

y	x_1	x_2
6.40	1.32	1.15
15.05	2.69	3.40
18.75	3.56	4.10
30.25	4.41	8.75
44.85	5.35	14.82
48.94	6.20	15.15
51.55	7.12	15.32
61.50	8.87	18.18
100.44	9.80	35.19
111.42	10.65	40.40

Estimate the multiple linear regression equation

$$\mu_{Y|x_1,x_2} = \beta_0 + \beta_1 x_1 + \beta_2 x_2.$$

12.2 In *Applied Spectroscopy*, the infrared reflectance spectra properties of a viscous liquid used in the electronics industry as a lubricant were studied. The designed experiment consisted of the effect of band frequency x_1 and film thickness x_2 on optical density y using a Perkin-Elmer Model 621 infrared spectrometer. (*Source*: Pacansky, J., England, C. D., and Wattman, R., 1986.)

y	x_1	x_2
0.231	740	1.10
0.107	740	0.62
0.053	740	0.31
0.129	805	1.10
0.069	805	0.62
0.030	805	0.31
1.005	980	1.10
0.559	980	0.62
0.321	980	0.31
2.948	1235	1.10
1.633	1235	0.62
0.934	1235	0.31

Estimate the multiple linear regression equation

$$\hat{y} = b_0 + b_1 x_1 + b_2 x_2.$$

12.3 Suppose in Review Exercise 11.53 on page 437 that we were also given the number of class periods missed by the 12 students taking the chemistry course. The complete data are shown.

Student	Chemistry Grade, y	Test Score, x_1	Classes Missed, x_2
1	85	65	1
2	74	50	7
3	76	55	5
4	90	65	2
5	85	55	6
6	87	70	3
7	94	65	2
8	98	70	5
9	81	55	4
10	91	70	3
11	76	50	1
12	74	55	4

(a) Fit a multiple linear regression equation of the form $\hat{y} = b_0 + b_1 x_1 + b_2 x_2$.

(b) Estimate the chemistry grade for a student who has an intelligence test score of 60 and missed 4 classes.

12.4 An experiment was conducted to determine if the weight of an animal can be predicted after a given period of time on the basis of the initial weight of the animal and the amount of feed that was eaten. The following data, measured in kilograms, were recorded:

Final Weight, y	Initial Weight, x_1	Feed Weight, x_2
95	42	272
77	33	226
80	33	259
100	45	292
97	39	311
70	36	183
50	32	173
80	41	236
92	40	230
84	38	235

(a) Fit a multiple regression equation of the form

$$\mu_{Y|x_1,x_2} = \beta_0 + \beta_1 x_1 + \beta_2 x_2.$$

(b) Predict the final weight of an animal having an initial weight of 35 kilograms that is given 250 kilograms of feed.

12.5 The electric power consumed each month by a chemical plant is thought to be related to the average ambient temperature x_1, the number of days in the month x_2, the average product purity x_3, and the tons of product produced x_4. The past year's historical data are available and are presented in the following table.

Exercises

y	x_1	x_2	x_3	x_4
240	25	24	91	100
236	31	21	90	95
290	45	24	88	110
274	60	25	87	88
301	65	25	91	94
316	72	26	94	99
300	80	25	87	97
296	84	25	86	96
267	75	24	88	110
276	60	25	91	105
288	50	25	90	100
261	38	23	89	98

(a) Fit a multiple linear regression model using the above data set.

(b) Predict power consumption for a month in which $x_1 = 75°F$, $x_2 = 24$ days, $x_3 = 90\%$, and $x_4 = 98$ tons.

12.6 An experiment was conducted on a new model of a particular make of automobile to determine the stopping distance at various speeds. The following data were recorded.

Speed, v (km/hr)	35	50	65	80	95	110
Stopping Distance, d (m)	16	26	41	62	88	119

(a) Fit a multiple regression curve of the form $\mu_{D|v} = \beta_0 + \beta_1 v + \beta_2 v^2$.

(b) Estimate the stopping distance when the car is traveling at 70 kilometers per hour.

12.7 An experiment was conducted in order to determine if cerebral blood flow in human beings can be predicted from arterial oxygen tension (millimeters of mercury). Fifteen patients participated in the study, and the following data were collected:

Blood Flow, y	Arterial Oxygen Tension, x
84.33	603.40
87.80	582.50
82.20	556.20
78.21	594.60
78.44	558.90
80.01	575.20
83.53	580.10
79.46	451.20
75.22	404.00
76.58	484.00
77.90	452.40
78.80	448.40
80.67	334.80
86.60	320.30
78.20	350.30

Estimate the quadratic regression equation

$$\mu_{Y|x} = \beta_0 + \beta_1 x + \beta_2 x^2.$$

12.8 The following is a set of coded experimental data on the compressive strength of a particular alloy at various values of the concentration of some additive:

Concentration, x	Compressive Strength, y		
10.0	25.2	27.3	28.7
15.0	29.8	31.1	27.8
20.0	31.2	32.6	29.7
25.0	31.7	30.1	32.3
30.0	29.4	30.8	32.8

(a) Estimate the quadratic regression equation $\mu_{Y|x} = \beta_0 + \beta_1 x + \beta_2 x^2$.

(b) Test for lack of fit of the model.

12.9 (a) Fit a multiple regression equation of the form $\mu_{Y|x} = \beta_0 + \beta_1 x_1 + \beta_2 x^2$ to the data of Example 11.8 on page 420.

(b) Estimate the yield of the chemical reaction for a temperature of 225°C.

12.10 The following data are given:

x	0	1	2	3	4	5	6
y	1	4	5	3	2	3	4

(a) Fit the cubic model $\mu_{Y|x} = \beta_0 + \beta_1 x + \beta_2 x^2 + \beta_3 x^3$.

(b) Predict Y when $x = 2$.

12.11 An experiment was conducted to study the size of squid eaten by sharks and tuna. The regressor variables are characteristics of the beaks of the squid. The data are given as follows:

x_1	x_2	x_3	x_4	x_5	y
1.31	1.07	0.44	0.75	0.35	1.95
1.55	1.49	0.53	0.90	0.47	2.90
0.99	0.84	0.34	0.57	0.32	0.72
0.99	0.83	0.34	0.54	0.27	0.81
1.01	0.90	0.36	0.64	0.30	1.09
1.09	0.93	0.42	0.61	0.31	1.22
1.08	0.90	0.40	0.51	0.31	1.02
1.27	1.08	0.44	0.77	0.34	1.93
0.99	0.85	0.36	0.56	0.29	0.64
1.34	1.13	0.45	0.77	0.37	2.08
1.30	1.10	0.45	0.76	0.38	1.98
1.33	1.10	0.48	0.77	0.38	1.90
1.86	1.47	0.60	1.01	0.65	8.56
1.58	1.34	0.52	0.95	0.50	4.49
1.97	1.59	0.67	1.20	0.59	8.49
1.80	1.56	0.66	1.02	0.59	6.17
1.75	1.58	0.63	1.09	0.59	7.54
1.72	1.43	0.64	1.02	0.63	6.36
1.68	1.57	0.72	0.96	0.68	7.63
1.75	1.59	0.68	1.08	0.62	7.78
2.19	1.86	0.75	1.24	0.72	10.15
1.73	1.67	0.64	1.14	0.55	6.88

In the study, the regressor variables and response considered are

x_1 = rostral length, in inches,
x_2 = wing length, in inches,
x_3 = rostral to notch length, in inches,
x_4 = notch to wing length, in inches,
x_5 = width, in inches,
y = weight, in pounds.

Estimate the multiple linear regression equation

$$\mu_{Y|x_1,x_2,x_3,x_4,x_5} = \beta_0 + \beta_1 x_1 + \beta_2 x_2 + \beta_3 x_3 + \beta_4 x_4 + \beta_5 x_5.$$

12.12 The following data reflect information from 17 U.S. Naval hospitals at various sites around the world. The regressors are workload variables, that is, items that result in the need for personnel in a hospital. A brief description of the variables is as follows:

y = monthly labor-hours,
x_1 = average daily patient load,
x_2 = monthly X-ray exposures,
x_3 = monthly occupied bed-days,
x_4 = eligible population in the area/1000,
x_5 = average length of patient's stay, in days.

Site	x_1	x_2	x_3	x_4	x_5	y
1	15.57	2463	472.92	18.0	4.45	566.52
2	44.02	2048	1339.75	9.5	6.92	696.82
3	20.42	3940	620.25	12.8	4.28	1033.15
4	18.74	6505	568.33	36.7	3.90	1003.62
5	49.20	5723	1497.60	35.7	5.50	1611.37
6	44.92	11,520	1365.83	24.0	4.60	1613.27
7	55.48	5779	1687.00	43.3	5.62	1854.17
8	59.28	5969	1639.92	46.7	5.15	2160.55
9	94.39	8461	2872.33	78.7	6.18	2305.58
10	128.02	20,106	3655.08	180.5	6.15	3503.93
11	96.00	13,313	2912.00	60.9	5.88	3571.59
12	131.42	10,771	3921.00	103.7	4.88	3741.40
13	127.21	15,543	3865.67	126.8	5.50	4026.52
14	252.90	36,194	7684.10	157.7	7.00	10,343.81
15	409.20	34,703	12,446.33	169.4	10.75	11,732.17
16	463.70	39,204	14,098.40	331.4	7.05	15,414.94
17	510.22	86,533	15,524.00	371.6	6.35	18,854.45

The goal here is to produce an empirical equation that will estimate (or predict) personnel needs for Naval hospitals. Estimate the multiple linear regression equation

$$\mu_{Y|x_1,x_2,x_3,x_4,x_5} = \beta_0 + \beta_1 x_1 + \beta_2 x_2 + \beta_3 x_3 + \beta_4 x_4 + \beta_5 x_5.$$

12.13 A study was performed on a type of bearing to find the relationship of amount of wear y to x_1 = oil viscosity and x_2 = load. The following data were obtained. (From *Response Surface Methodology*, Myers, Montgomery, and Anderson-Cook, 2009.)

y	x_1	x_2	y	x_1	x_2
193	1.6	851	230	15.5	816
172	22.0	1058	91	43.0	1201
113	33.0	1357	125	40.0	1115

(a) Estimate the unknown parameters of the multiple linear regression equation

$$\mu_{Y|x_1,x_2} = \beta_0 + \beta_1 x_1 + \beta_2 x_2.$$

(b) Predict wear when oil viscosity is 20 and load is 1200.

12.14 Eleven student teachers took part in an evaluation program designed to measure teacher effectiveness and determine what factors are important. The response measure was a quantitative evaluation of the teacher. The regressor variables were scores on four standardized tests given to each teacher. The data are as follows:

y	x_1	x_2	x_3	x_4
410	69	125	59.00	55.66
569	57	131	31.75	63.97
425	77	141	80.50	45.32
344	81	122	75.00	46.67
324	0	141	49.00	41.21
505	53	152	49.35	43.83
235	77	141	60.75	41.61
501	76	132	41.25	64.57
400	65	157	50.75	42.41
584	97	166	32.25	57.95
434	76	141	54.50	57.90

Estimate the multiple linear regression equation

$$\mu_{Y|x_1,x_2,x_3,x_4} = \beta_0 + \beta_1 x_1 + \beta_2 x_2 + \beta_3 x_3 + \beta_4 x_4.$$

12.15 The personnel department of a certain industrial firm used 12 subjects in a study to determine the relationship between job performance rating (y) and scores on four tests. The data are as follows:

y	x_1	x_2	x_3	x_4
11.2	56.5	71.0	38.5	43.0
14.5	59.5	72.5	38.2	44.8
17.2	69.2	76.0	42.5	49.0
17.8	74.5	79.5	43.4	56.3
19.3	81.2	84.0	47.5	60.2
24.5	88.0	86.2	47.4	62.0
21.2	78.2	80.5	44.5	58.1
16.9	69.0	72.0	41.8	48.1
14.8	58.1	68.0	42.1	46.0
20.0	80.5	85.0	48.1	60.3
13.2	58.3	71.0	37.5	47.1
22.5	84.0	87.2	51.0	65.2

12.4 Properties of the Least Squares Estimators

Estimate the regression coefficients in the model

$$\hat{y} = b_0 + b_1 x_1 + b_2 x_2 + b_3 x_3 + b_4 x_4.$$

12.16 An engineer at a semiconductor company wants to model the relationship between the gain or hFE of a device (y) and three parameters: emitter-RS (x_1), base-RS (x_2), and emitter-to-base-RS (x_3). The data are shown below:

x_1, Emitter-RS	x_2, Base-RS	x_3, E-B-RS	y, hFE
14.62	226.0	7.000	128.40
15.63	220.0	3.375	52.62
14.62	217.4	6.375	113.90
15.00	220.0	6.000	98.01
14.50	226.5	7.625	139.90
15.25	224.1	6.000	102.60
			(cont.)
16.12	220.5	3.375	48.14
15.13	223.5	6.125	109.60
15.50	217.6	5.000	82.68
15.13	228.5	6.625	112.60
15.50	230.2	5.750	97.52
16.12	226.5	3.750	59.06
15.13	226.6	6.125	111.80
15.63	225.6	5.375	89.09
15.38	234.0	8.875	171.90
15.50	230.0	4.000	66.80
14.25	224.3	8.000	157.10
14.50	240.5	10.870	208.40
14.62	223.7	7.375	133.40

(Data from Myers, Montgomery, and Anderson-Cook, 2009.)

(a) Fit a multiple linear regression to the data.
(b) Predict hFE when $x_1 = 14$, $x_2 = 220$, and $x_3 = 5$.

12.4 Properties of the Least Squares Estimators

The means and variances of the estimators $b_0, b_1, \ldots, b_k$ are readily obtained under certain assumptions on the random errors $\epsilon_1, \epsilon_2, \ldots, \epsilon_k$ that are identical to those made in the case of simple linear regression. When we assume these errors to be independent, each with mean 0 and variance σ^2, it can be shown that $b_0, b_1, \ldots, b_k$ are, respectively, unbiased estimators of the regression coefficients $\beta_0, \beta_1, \ldots, \beta_k$. In addition, the variances of the b's are obtained through the elements of the inverse of the $\mathbf{A}$ matrix. Note that the off-diagonal elements of $\mathbf{A} = \mathbf{X}'\mathbf{X}$ represent sums of products of elements in the columns of $\mathbf{X}$, while the diagonal elements of $\mathbf{A}$ represent sums of squares of elements in the columns of $\mathbf{X}$. The inverse matrix, $\mathbf{A}^{-1}$, apart from the multiplier σ^2, represents the **variance-covariance matrix** of the estimated regression coefficients. That is, the elements of the matrix $\mathbf{A}^{-1}\sigma^2$ display the variances of $b_0, b_1, \ldots, b_k$ on the main diagonal and covariances on the off-diagonal. For example, in a $k = 2$ multiple linear regression problem, we might write

$$(\mathbf{X}'\mathbf{X})^{-1} = \begin{bmatrix} c_{00} & c_{01} & c_{02} \\ c_{10} & c_{11} & c_{12} \\ c_{20} & c_{21} & c_{22} \end{bmatrix}$$

with the elements below the main diagonal determined through the symmetry of the matrix. Then we can write

$$\sigma_{b_i}^2 = c_{ii}\sigma^2, \qquad i = 0, 1, 2,$$
$$\sigma_{b_i b_j} = \text{Cov}(b_i, b_j) = c_{ij}\sigma^2, \ i \neq j.$$

Of course, the estimates of the variances and hence the standard errors of these estimators are obtained by replacing σ^2 with the appropriate estimate obtained through experimental data. An unbiased estimate of σ^2 is once again defined in terms of the error sum of squares, which is computed using the formula established in Theorem 12.1. In the theorem, we are making the assumptions on the ϵ_i described above.

Theorem 12.1: For the linear regression equation

$$\mathbf{y} = \mathbf{X}\boldsymbol{\beta} + \boldsymbol{\epsilon},$$

an unbiased estimate of σ^2 is given by the error or residual mean square

$$s^2 = \frac{SSE}{n-k-1}, \quad \text{where} \quad SSE = \sum_{i=1}^{n} e_i^2 = \sum_{i=1}^{n}(y_i - \hat{y}_i)^2.$$

We can see that Theorem 12.1 represents a generalization of Theorem 11.1 for the simple linear regression case. The proof is left for the reader. As in the simpler linear regression case, the estimate s^2 is a measure of the variation in the prediction errors or residuals. Other important inferences regarding the fitted regression equation, based on the values of the individual residuals $e_i = y_i - \hat{y}_i$, $i = 1, 2, \ldots, n$, are discussed in Sections 12.10 and 12.11.

The error and regression sums of squares take on the same form and play the same role as in the simple linear regression case. In fact, the sum-of-squares identity

$$\sum_{i=1}^{n}(y_i - \bar{y})^2 = \sum_{i=1}^{n}(\hat{y}_i - \bar{y})^2 + \sum_{i=1}^{n}(y_i - \hat{y}_i)^2$$

continues to hold, and we retain our previous notation, namely

$$SST = SSR + SSE,$$

with

$$SST = \sum_{i=1}^{n}(y_i - \bar{y})^2 = \text{total sum of squares}$$

and

$$SSR = \sum_{i=1}^{n}(\hat{y}_i - \bar{y})^2 = \text{regression sum of squares}.$$

There are k degrees of freedom associated with SSR, and, as always, SST has $n-1$ degrees of freedom. Therefore, after subtraction, SSE has $n-k-1$ degrees of freedom. Thus, our estimate of σ^2 is again given by the error sum of squares divided by its degrees of freedom. All three of these sums of squares will appear on the printouts of most multiple regression computer packages. Note that the condition $n > k$ in Section 12.2 guarantees that the degrees of freedom of SSE cannot be negative.

Analysis of Variance in Multiple Regression

The partition of the total sum of squares into its components, the regression and error sums of squares, plays an important role. An **analysis of variance** can be conducted to shed light on the quality of the regression equation. A useful hypothesis that determines if a significant amount of variation is explained by the model is

$$H_0: \beta_1 = \beta_2 = \beta_3 = \cdots = \beta_k = 0.$$

The analysis of variance involves an F-test via a table given as follows:

Source	Sum of Squares	Degrees of Freedom	Mean Squares	F
Regression	SSR	k	$MSR = \frac{SSR}{k}$	$f = \frac{MSR}{MSE}$
Error	SSE	$n-(k+1)$	$MSE = \frac{SSE}{n-(k+1)}$	
Total	SST	$n-1$		

This test is an **upper-tailed test**. Rejection of H_0 implies that the **regression equation differs from a constant**. That is, at least one regressor variable is important. More discussion of the use of analysis of variance appears in subsequent sections.

Further utility of the mean square error (or residual mean square) lies in its use in hypothesis testing and confidence interval estimation, which is discussed in Section 12.5. In addition, the mean square error plays an important role in situations where the scientist is searching for the best from a set of competing models. Many model-building criteria involve the statistic s^2. Criteria for comparing competing models are discussed in Section 12.11.

12.5 Inferences in Multiple Linear Regression

A knowledge of the distributions of the individual coefficient estimators enables the experimenter to construct confidence intervals for the coefficients and to test hypotheses about them. Recall from Section 12.4 that the b_j $(j = 0, 1, 2, \ldots, k)$ are normally distributed with mean β_j and variance $c_{jj}\sigma^2$. Thus, we can use the statistic

$$t = \frac{b_j - \beta_{j0}}{s\sqrt{c_{jj}}}$$

with $n - k - 1$ degrees of freedom to test hypotheses and construct confidence intervals on β_j. For example, if we wish to test

$$H_0: \beta_j = \beta_{j0},$$
$$H_1: \beta_j \neq \beta_{j0},$$

we compute the above t-statistic and do not reject H_0 if $-t_{\alpha/2} < t < t_{\alpha/2}$, where $t_{\alpha/2}$ has $n - k - 1$ degrees of freedom.

Example 12.5: For the model of Example 12.4, test the hypothesis that $\beta_2 = -2.5$ at the 0.05 level of significance against the alternative that $\beta_2 > -2.5$.

Solution:
$$H_0: \beta_2 = -2.5,$$
$$H_1: \beta_2 > -2.5.$$

Computations:
$$t = \frac{b_2 - \beta_{20}}{s\sqrt{c_{22}}} = \frac{-1.8616 + 2.5}{2.073\sqrt{0.0166}} = 2.390,$$
$$P = P(T > 2.390) = 0.04.$$

Decision: Reject H_0 and conclude that $\beta_2 > -2.5$.

Individual *t*-Tests for Variable Screening

The *t*-test most often used in multiple regression is the one that tests the importance of individual coefficients (i.e., $H_0: \beta_j = 0$ against the alternative $H_1: \beta_j \neq 0$). These tests often contribute to what is termed **variable screening**, where the analyst attempts to arrive at the most useful model (i.e., the choice of which regressors to use). It should be emphasized here that if a coefficient is found insignificant (i.e., the hypothesis $H_0: \beta_j = 0$ **is not rejected**), the conclusion drawn is that the **variable** is insignificant (i.e., explains an insignificant amount of variation in y), **in the presence of the other regressors in the model**. This point will be reaffirmed in a future discussion.

Inferences on Mean Response and Prediction

One of the most useful inferences that can be made regarding the quality of the predicted response y_0 corresponding to the values $x_{10}, x_{20}, \ldots, x_{k0}$ is the confidence interval on the mean response $\mu_{Y|x_{10},x_{20},\ldots,x_{k0}}$. We are interested in constructing a confidence interval on the mean response for the set of conditions given by

$$\mathbf{x}_0' = [1, x_{10}, x_{20}, \ldots, x_{k0}].$$

We augment the conditions on the x's by the number 1 in order to facilitate the matrix notation. Normality in the ϵ_i produces normality in the b_j and the mean and variance are still the same as indicated in Section 12.4. So is the covariance between b_i and b_j, for $i \neq j$. Hence,

$$\hat{y} = b_0 + \sum_{j=1}^{k} b_j x_{j0}$$

is likewise normally distributed and is, in fact, an unbiased estimator for the **mean response** on which we are attempting to attach a confidence interval. The variance of $\hat{y}_0$, written in matrix notation simply as a function of σ^2, $(\mathbf{X}'\mathbf{X})^{-1}$, and the condition vector $\mathbf{x}_0'$, is

$$\sigma_{\hat{y}_0}^2 = \sigma^2 \mathbf{x}_0'(\mathbf{X}'\mathbf{X})^{-1}\mathbf{x}_0.$$

12.5 Inferences in Multiple Linear Regression

If this expression is expanded for a given case, say $k = 2$, it is readily seen that it appropriately accounts for the variance of the b_j and the covariance of b_i and b_j, for $i \neq j$. After σ^2 is replaced by s^2 as given by Theorem 12.1, the $100(1 - \alpha)\%$ confidence interval on $\mu_{Y|x_{10},x_{20},\ldots,x_{k0}}$ can be constructed from the statistic

$$T = \frac{\hat{y}_0 - \mu_{Y|x_{10},x_{20},\ldots,x_{k0}}}{s\sqrt{\mathbf{x}_0'(\mathbf{X}'\mathbf{X})^{-1}\mathbf{x}_0}},$$

which has a t-distribution with $n - k - 1$ degrees of freedom.

Confidence Interval for $\mu_{Y|x_{10},x_{20},\ldots,x_{k0}}$ A $100(1 - \alpha)\%$ confidence interval for the **mean response** $\mu_{Y|x_{10},x_{20},\ldots,x_{k0}}$ is

$$\hat{y}_0 - t_{\alpha/2} s \sqrt{\mathbf{x}_0'(\mathbf{X}'\mathbf{X})^{-1}\mathbf{x}_0} < \mu_{Y|x_{10},x_{20},\ldots,x_{k0}} < \hat{y}_0 + t_{\alpha/2} s \sqrt{\mathbf{x}_0'(\mathbf{X}'\mathbf{X})^{-1}\mathbf{x}_0},$$

where $t_{\alpha/2}$ is a value of the t-distribution with $n - k - 1$ degrees of freedom.

The quantity $s\sqrt{\mathbf{x}_0'(\mathbf{X}'\mathbf{X})^{-1}\mathbf{x}_0}$ is often called the **standard error of prediction** and appears on the printout of many regression computer packages.

Example 12.6: Using the data of Example 12.4, construct a 95% confidence interval for the mean response when $x_1 = 3\%$, $x_2 = 8\%$, and $x_3 = 9\%$.

Solution: From the regression equation of Example 12.4, the estimated percent survival when $x_1 = 3\%$, $x_2 = 8\%$, and $x_3 = 9\%$ is

$$\hat{y} = 39.1574 + (1.0161)(3) - (1.8616)(8) - (0.3433)(9) = 24.2232.$$

Next, we find that

$$\mathbf{x}_0'(\mathbf{X}'\mathbf{X})^{-1}\mathbf{x}_0 = [1, 3, 8, 9] \begin{bmatrix} 8.0648 & -0.0826 & -0.0942 & -0.7905 \\ -0.0826 & 0.0085 & 0.0017 & 0.0037 \\ -0.0942 & 0.0017 & 0.0166 & -0.0021 \\ -0.7905 & 0.0037 & -0.0021 & 0.0886 \end{bmatrix} \begin{bmatrix} 1 \\ 3 \\ 8 \\ 9 \end{bmatrix}$$

$$= 0.1267.$$

Using the mean square error, $s^2 = 4.298$ or $s = 2.073$, and Table A.4, we see that $t_{0.025} = 2.262$ for 9 degrees of freedom. Therefore, a 95% confidence interval for the mean percent survival for $x_1 = 3\%$, $x_2 = 8\%$, and $x_3 = 9\%$ is given by

$$24.2232 - (2.262)(2.073)\sqrt{0.1267} < \mu_{Y|3,8,9}$$
$$< 24.2232 + (2.262)(2.073)\sqrt{0.1267},$$

or simply $22.5541 < \mu_{Y|3,8,9} < 25.8923$.

As in the case of simple linear regression, we need to make a clear distinction between the confidence interval on a mean response and the prediction interval on an *observed response*. The latter provides a bound within which we can say with a preselected degree of certainty that a new observed response will fall.

A prediction interval for a single predicted response y_0 is once again established by considering the difference $\hat{y}_0 - y_0$. The sampling distribution can be shown to be normal with mean

$$\mu_{\hat{y}_0 - y_0} = 0$$

and variance

$$\sigma^2_{\hat{y}_0 - y_0} = \sigma^2[1 + \mathbf{x}'_0(\mathbf{X}'\mathbf{X})^{-1}\mathbf{x}_0].$$

Thus, a $100(1-\alpha)\%$ prediction interval for a single prediction value y_0 can be constructed from the statistic

$$T = \frac{\hat{y}_0 - y_0}{s\sqrt{1 + \mathbf{x}'_0(\mathbf{X}'\mathbf{X})^{-1}\mathbf{x}_0}},$$

which has a t-distribution with $n - k - 1$ degrees of freedom.

Prediction Interval for y_0 A $100(1-\alpha)\%$ prediction interval for a **single response** y_0 is given by

$$\hat{y}_0 - t_{\alpha/2}s\sqrt{1 + \mathbf{x}'_0(\mathbf{X}'\mathbf{X})^{-1}\mathbf{x}_0} < y_0 < \hat{y}_0 + t_{\alpha/2}s\sqrt{1 + \mathbf{x}'_0(\mathbf{X}'\mathbf{X})^{-1}\mathbf{x}_0},$$

where $t_{\alpha/2}$ is a value of the t-distribution with $n - k - 1$ degrees of freedom.

Example 12.7: Using the data of Example 12.4, construct a 95% prediction interval for an individual percent survival response when $x_1 = 3\%$, $x_2 = 8\%$, and $x_3 = 9\%$.

Solution: Referring to the results of Example 12.6, we find that the 95% prediction interval for the response y_0, when $x_1 = 3\%$, $x_2 = 8\%$, and $x_3 = 9\%$, is

$$24.2232 - (2.262)(2.073)\sqrt{1.1267} < y_0 < 24.2232 + (2.262)(2.073)\sqrt{1.1267},$$

which reduces to $19.2459 < y_0 < 29.2005$. Notice, as expected, that the prediction interval is considerably wider than the confidence interval for mean percent survival found in Example 12.6.

Annotated Printout for Data of Example 12.4

Figure 12.1 shows an annotated computer printout for a multiple linear regression fit to the data of Example 12.4. The package used is *SAS*.

Note the model parameter estimates, the standard errors, and the t-statistics shown in the output. The standard errors are computed from square roots of diagonal elements of $(\mathbf{X}'\mathbf{X})^{-1}s^2$. In this illustration, the variable x_3 is insignificant in the presence of x_1 and x_2 based on the t-test and the corresponding P-value of 0.5916. The terms CLM and CLI are confidence intervals on mean response and prediction limits on an individual observation, respectively. The f-test in the analysis of variance indicates that a significant amount of variability is explained. As an example of the interpretation of CLM and CLI, consider observation 10. With an observation of 25.2000 and a predicted value of 26.0676, we are 95% confident that the mean response is between 24.5024 and 27.6329, and a new observation will fall between 21.1238 and 31.0114 with probability 0.95. The R^2 value of 0.9117 implies that the model explains 91.17% of the variability in the response. More discussion about R^2 appears in Section 12.6.

```
                         Sum of      Mean
Source              DF   Squares     Square    F Value   Pr > F
Model                3   399.45437   133.15146   30.98   <.0001
Error                9    38.67640     4.29738
Corrected Total     12   438.13077

Root MSE             2.07301   R-Square    0.9117
Dependent Mean      29.03846   Adj R-Sq    0.8823
Coeff Var            7.13885

                      Parameter    Standard
Variable      DF      Estimate     Error      t Value   Pr > |t|
Intercept      1      39.15735     5.88706      6.65    <.0001
x1             1       1.01610     0.19090      5.32     0.0005
x2             1      -1.86165     0.26733     -6.96    <.0001
x3             1      -0.34326     0.61705     -0.56     0.5916

     Dependent  Predicted   Std Error
Obs  Variable   Value       Mean Predict    95% CL Mean      95% CL Predict    Residual
 1   25.5000    27.3514     1.4152  24.1500  30.5528   21.6734  33.0294   -1.8514
 2   31.2000    32.2623     0.7846  30.4875  34.0371   27.2482  37.2764   -1.0623
 3   25.9000    27.3495     1.3588  24.2757  30.4234   21.7425  32.9566   -1.4495
 4   38.4000    38.3096     1.2818  35.4099  41.2093   32.7960  43.8232    0.0904
 5   18.4000    15.5447     1.5789  11.9730  19.1165    9.6499  21.4395    2.8553
 6   26.7000    26.1081     1.0358  23.7649  28.4512   20.8658  31.3503    0.5919
 7   26.4000    28.2532     0.8094  26.4222  30.0841   23.2189  33.2874   -1.8532
 8   25.9000    26.2219     0.9732  24.0204  28.4233   21.0414  31.4023   -0.3219
 9   32.0000    32.0882     0.7828  30.3175  33.8589   27.0755  37.1008   -0.0882
10   25.2000    26.0676     0.6919  24.5024  27.6329   21.1238  31.0114   -0.8676
11   39.7000    37.2524     1.3070  34.2957  40.2090   31.7086  42.7961    2.4476
12   35.7000    32.4879     1.4648  29.1743  35.8015   26.7459  38.2300    3.2121
13   26.5000    28.2032     0.9841  25.9771  30.4294   23.0122  33.3943   -1.7032
```

Figure 12.1: *SAS* printout for data in Example 12.4.

More on Analysis of Variance in Multiple Regression (Optional)

In Section 12.4, we discussed briefly the partition of the total sum of squares $\sum_{i=1}^{n}(y_i - \bar{y})^2$ into its two components, the regression model and error sums of squares (illustrated in Figure 12.1). The analysis of variance leads to a test of

$$H_0\colon \beta_1 = \beta_2 = \beta_3 = \cdots = \beta_k = 0.$$

Rejection of the null hypothesis has an important interpretation for the scientist or engineer. (For those who are interested in more extensive treatment of the subject using matrices, it is useful to discuss the development of these sums of squares used in ANOVA.)

First, recall in Section 12.3, **b**, the vector of least squares estimators, is given by

$$\mathbf{b} = (\mathbf{X}'\mathbf{X})^{-1}\mathbf{X}'\mathbf{y}.$$

A partition of the **uncorrected sum of squares**

$$\mathbf{y}'\mathbf{y} = \sum_{i=1}^{n} y_i^2$$

into two components is given by

$$\mathbf{y}'\mathbf{y} = \mathbf{b}'\mathbf{X}'\mathbf{y} + (\mathbf{y}'\mathbf{y} - \mathbf{b}'\mathbf{X}'\mathbf{y})$$
$$= \mathbf{y}'\mathbf{X}(\mathbf{X}'\mathbf{X})^{-1}\mathbf{X}'\mathbf{y} + [\mathbf{y}'\mathbf{y} - \mathbf{y}'\mathbf{X}(\mathbf{X}'\mathbf{X})^{-1}\mathbf{X}'\mathbf{y}].$$

The second term (in brackets) on the right-hand side is simply the error sum of squares $\sum_{i=1}^{n}(y_i - \hat{y}_i)^2$. The reader should see that an alternative expression for the error sum of squares is

$$SSE = \mathbf{y}'[\mathbf{I}_n - \mathbf{X}(\mathbf{X}'\mathbf{X})^{-1}\mathbf{X}']\mathbf{y}.$$

The term $\mathbf{y}'\mathbf{X}(\mathbf{X}'\mathbf{X})^{-1}\mathbf{X}'\mathbf{y}$ is called the **regression sum of squares**. However, it is not the expression $\sum_{i=1}^{n}(\hat{y}_i - \bar{y})^2$ used for testing the "importance" of the terms $b_1, b_2, \ldots, b_k$ but, rather,

$$\mathbf{y}'\mathbf{X}(\mathbf{X}'\mathbf{X})^{-1}\mathbf{X}'\mathbf{y} = \sum_{i=1}^{n} \hat{y}_i^2,$$

which is a regression sum of squares uncorrected for the mean. As such, it would only be used in testing if the regression equation differs significantly from zero, that is,

$$H_0\colon \beta_0 = \beta_1 = \beta_2 = \cdots = \beta_k = 0.$$

In general, this is not as important as testing

$$H_0\colon \beta_1 = \beta_2 = \cdots = \beta_k = 0,$$

since the latter states that the mean response is a constant, not necessarily zero.

Degrees of Freedom

Thus, the partition of sums of squares and degrees of freedom reduces to

Source	Sum of Squares	d.f.
Regression	$\sum_{i=1}^{n} \hat{y}_i^2 = \mathbf{y}'\mathbf{X}(\mathbf{X}'\mathbf{X})^{-1}\mathbf{X}'\mathbf{y}$	$k+1$
Error	$\sum_{i=1}^{n}(y_i - \hat{y}_i)^2 = \mathbf{y}'[\mathbf{I}_n - \mathbf{X}(\mathbf{X}'\mathbf{X})^{-1}\mathbf{X}']\mathbf{y}$	$n-(k+1)$
Total	$\sum_{i=1}^{n} y_i^2 = \mathbf{y}'\mathbf{y}$	n

Hypothesis of Interest

Now, of course, the hypotheses of interest for an ANOVA must eliminate the role of the intercept described previously. Strictly speaking, if $H_0\colon \beta_1 = \beta_2 = \cdots = \beta_k = 0$, then the estimated regression line is merely $\hat{y}_i = \bar{y}$. As a result, we are actually seeking evidence that the regression equation "varies from a constant." Thus, the total and regression sums of squares must be corrected for the mean. As a result, we have

$$\sum_{i=1}^{n}(y_i - \bar{y})^2 = \sum_{i=1}^{n}(\hat{y}_i - \bar{y})^2 + \sum_{i=1}^{n}(y_i - \hat{y}_i)^2.$$

In matrix notation this is simply

$$\mathbf{y}'[\mathbf{I}_n - \mathbf{1}(\mathbf{1}'\mathbf{1})^{-1}\mathbf{1}']\mathbf{y} = \mathbf{y}'[\mathbf{X}(\mathbf{X}'\mathbf{X})^{-1}\mathbf{X}' - \mathbf{1}(\mathbf{1}'\mathbf{1})^{-1}\mathbf{1}']\mathbf{y} + \mathbf{y}'[\mathbf{I}_n - \mathbf{X}(\mathbf{X}'\mathbf{X})^{-1}\mathbf{X}']\mathbf{y}.$$

In this expression, $\mathbf{1}$ is a vector of n ones. As a result, we are merely subtracting

$$\mathbf{y}'\mathbf{1}(\mathbf{1}'\mathbf{1})^{-1}\mathbf{1}'\mathbf{y} = \frac{1}{n}\left(\sum_{i=1}^{n} y_i\right)^2$$

from $\mathbf{y}'\mathbf{y}$ and from $\mathbf{y}'\mathbf{X}(\mathbf{X}'\mathbf{X})^{-1}\mathbf{X}'\mathbf{y}$ (i.e., correcting the total and regression sums of squares for the mean).

Finally, the appropriate partitioning of sums of squares with degrees of freedom is as follows:

Source	Sum of Squares	d.f.
Regression	$\sum_{i=1}^{n}(\hat{y}_i - \bar{y})^2 = \mathbf{y}'[\mathbf{X}(\mathbf{X}'\mathbf{X})^{-1}\mathbf{X}' - \mathbf{1}(\mathbf{1}'\mathbf{1})^{-1}\mathbf{1}']\mathbf{y}$	k
Error	$\sum_{i=1}^{n}(y_i - \hat{y}_i)^2 = \mathbf{y}'[\mathbf{I}_n - \mathbf{X}(\mathbf{X}'\mathbf{X})^{-1}\mathbf{X}']\mathbf{y}$	$n - (k+1)$
Total	$\sum_{i=1}^{n}(y_i - \bar{y})^2 = \mathbf{y}'[\mathbf{I}_n - \mathbf{1}(\mathbf{1}'\mathbf{1})^{-1}\mathbf{1}']\mathbf{y}$	$n - 1$

This is the ANOVA table that appears in the computer printout of Figure 12.1. The expression $\mathbf{y}'[\mathbf{1}(\mathbf{1}'\mathbf{1})^{-1}\mathbf{1}']\mathbf{y}$ is often called the **regression sum of squares associated with the mean**, and 1 degree of freedom is allocated to it.

Exercises

12.17 For the data of Exercise 12.2 on page 450, estimate σ^2.

12.18 For the data of Exercise 12.1 on page 450, estimate σ^2.

12.19 For the data of Exercise 12.5 on page 450, estimate σ^2.

12.20 Obtain estimates of the variances and the co- variance of the estimators b_1 and b_2 of Exercise 12.2 on page 450.

12.21 Referring to Exercise 12.5 on page 450, find the estimate of

(a) $\sigma_{b_2}^2$;

(b) $\text{Cov}(b_1, b_4)$.

12.22 For the model of Exercise 12.7 on page 451,

test the hypothesis that $\beta_2 = 0$ at the 0.05 level of significance against the alternative that $\beta_2 \neq 0$.

12.23 For the model of Exercise 12.2 on page 450, test the hypothesis that $\beta_1 = 0$ at the 0.05 level of significance against the alternative that $\beta_1 \neq 0$.

12.24 For the model of Exercise 12.1 on page 450, test the hypotheses that $\beta_1 = 2$ against the alternative that $\beta_1 \neq 2$. Use a P-value in your conclusion.

12.25 Using the data of Exercise 12.2 on page 450 and the estimate of σ^2 from Exercise 12.17, compute 95% confidence intervals for the predicted response and the mean response when $x_1 = 900$ and $x_2 = 1.00$.

12.26 For Exercise 12.8 on page 451, construct a 90% confidence interval for the mean compressive strength when the concentration is $x = 19.5$ and a quadratic model is used.

12.27 Using the data of Exercise 12.5 on page 450 and the estimate of σ^2 from Exercise 12.19, compute 95% confidence intervals for the predicted response and the mean response when $x_1 = 75$, $x_2 = 24$, $x_3 = 90$, and $x_4 = 98$.

12.28 Consider the following data from Exercise 12.13 on page 452.

y (wear)	x_1 (oil viscosity)	x_2 (load)
193	1.6	851
230	15.5	816
172	22.0	1058
91	43.0	1201
113	33.0	1357
125	40.0	1115

(a) Estimate σ^2 using multiple regression of y on x_1 and x_2.

(b) Compute predicted values, a 95% confidence interval for mean wear, and a 95% prediction interval for observed wear if $x_1 = 20$ and $x_2 = 1000$.

12.29 Using the data from Exercise 12.28, test the following at the 0.05 level.

(a) H_0: $\beta_1 = 0$ versus H_1: $\beta_1 \neq 0$;
(b) H_0: $\beta_2 = 0$ versus H_1: $\beta_2 \neq 0$.
(c) Do you have any reason to believe that the model in Exercise 12.28 should be changed? Why or why not?

12.30 Use the data from Exercise 12.16 on page 453.

(a) Estimate σ^2 using the multiple regression of y on x_1, x_2, and x_3.

(b) Compute a 95% prediction interval for the observed gain with the three regressors at $x_1 = 15.0$, $x_2 = 220.0$, and $x_3 = 6.0$.

12.6 Choice of a Fitted Model through Hypothesis Testing

In many regression situations, individual coefficients are of importance to the experimenter. For example, in an economics application, $\beta_1, \beta_2, \ldots$ might have some particular significance, and thus confidence intervals and tests of hypotheses on these parameters would be of interest to the economist. However, consider an industrial chemical situation in which the postulated model assumes that reaction yield is linearly dependent on reaction temperature and concentration of a certain catalyst. It is probably known that this is not the true model but an adequate approximation, so interest is likely to be not in the individual parameters but rather in the ability of the entire function to predict the true response in the range of the variables considered. Therefore, in this situation, one would put more emphasis on $\sigma^2_{\hat{Y}}$, confidence intervals on the mean response, and so forth, and likely deemphasize inferences on individual parameters.

The experimenter using regression analysis is also interested in deletion of variables when the situation dictates that, in addition to arriving at a workable prediction equation, he or she must find the "best regression" involving only variables that are useful predictors. There are a number of computer programs that sequentially arrive at the so-called best regression equation depending on certain criteria. We discuss this further in Section 12.9.

One criterion that is commonly used to illustrate the adequacy of a fitted regression model is the **coefficient of determination**, or R^2.

Coefficient of Determination, or R^2

$$R^2 = \frac{SSR}{SST} = \frac{\sum_{i=1}^{n}(\hat{y}_i - \bar{y})^2}{\sum_{i=1}^{n}(y_i - \bar{y})^2} = 1 - \frac{SSE}{SST}.$$

Note that this parallels the description of R^2 in Chapter 11. At this point the explanation might be clearer since we now focus on SSR as the **variability explained**. The quantity R^2 merely indicates what proportion of the total variation in the response Y is explained by the fitted model. Often an experimenter will report $R^2 \times 100\%$ and interpret the result as percent variation explained by the postulated model. The square root of R^2 is called the **multiple correlation coefficient** between Y and the set $x_1, x_2, \ldots, x_k$. The value of R^2 for the case in Example 12.4, indicating the proportion of variation explained by the three independent variables x_1, x_2, and x_3, is

$$R^2 = \frac{SSR}{SST} = \frac{399.45}{438.13} = 0.9117,$$

which means that 91.17% of the variation in percent survival has been explained by the linear regression model.

The regression sum of squares can be used to give some indication concerning whether or not the model is an adequate explanation of the true situation. We can test the hypothesis H_0 that the **regression is not significant** by merely forming the ratio

$$f = \frac{SSR/k}{SSE/(n-k-1)} = \frac{SSR/k}{s^2}$$

and rejecting H_0 at the α-level of significance when $f > f_\alpha(k, n-k-1)$. For the data of Example 12.4, we obtain

$$f = \frac{399.45/3}{4.298} = 30.98.$$

From the printout in Figure 12.1, the P-value is less than 0.0001. This should not be misinterpreted. Although it does indicate that the regression explained by the model is significant, this does not rule out the following possibilities:

1. The linear regression model for this set of x's is not the only model that can be used to explain the data; indeed, there may be other models with transformations on the x's that give a larger value of the F-statistic.

2. The model might have been more effective with the inclusion of other variables in addition to x_1, x_2, and x_3 or perhaps with the deletion of one or more of the variables in the model, say x_3, which has a $P = 0.5916$.

The reader should recall the discussion in Section 11.5 regarding the pitfalls in the use of R^2 as a criterion for comparing competing models. These pitfalls are certainly relevant in multiple linear regression. In fact, in its employment in multiple regression, the dangers are even more pronounced since the temptation to overfit is so great. One should always keep in mind that $R^2 \approx 1.0$ can always be achieved at the expense of error degrees of freedom when an excess of model terms is employed. However, $R^2 = 1$, describing a model with a near perfect fit, does not always result in a model that predicts well.

The Adjusted Coefficient of Determination (R^2_{adj})

In Chapter 11, several figures displaying computer printout from both *SAS* and *MINITAB* featured a statistic called *adjusted* R^2 or adjusted coefficient of determination. Adjusted R^2 is a variation on R^2 that provides an **adjustment for degrees of freedom**. The coefficient of determination as defined on page 407 cannot decrease as terms are added to the model. In other words, R^2 does not decrease as the error degrees of freedom $n - k - 1$ are reduced, the latter result being produced by an increase in k, the number of model terms. Adjusted R^2 is computed by dividing SSE and SST by their respective degrees of freedom as follows.

Adjusted R^2

$$R^2_{\text{adj}} = 1 - \frac{SSE/(n-k-1)}{SST/(n-1)}.$$

To illustrate the use of R^2_{adj}, Example 12.4 will be revisited.

How Are R^2 and R^2_{adj} Affected by Removal of x_3?

The *t*-test (or corresponding *F*-test) for x_3 suggests that a simpler model involving only x_1 and x_2 may well be an improvement. In other words, the complete model with all the regressors may be an overfitted model. It is certainly of interest to investigate R^2 and R^2_{adj} for both the full (x_1, x_2, x_3) and the reduced (x_1, x_2) models. We already know that $R^2_{\text{full}} = 0.9117$ from Figure 12.1. The SSE for the reduced model is 40.01, and thus $R^2_{\text{reduced}} = 1 - \frac{40.01}{438.13} = 0.9087$. Thus, more variability is explained with x_3 in the model. However, as we have indicated, this will occur even if the model is an overfitted model. Now, of course, R^2_{adj} is designed to provide a statistic that punishes an overfitted model, so we might expect it to favor the reduced model. Indeed, for the full model

$$R^2_{\text{adj}} = 1 - \frac{38.6764/9}{438.1308/12} = 1 - \frac{4.2974}{36.5109} = 0.8823,$$

whereas for the reduced model (deletion of x_3)

$$R^2_{\text{adj}} = 1 - \frac{40.01/10}{438.1308/12} = 1 - \frac{4.001}{36.5109} = 0.8904.$$

Thus, R^2_{adj} does indeed favor the reduced model and confirms the evidence produced by the *t*- and *F*-tests, suggesting that the reduced model is preferable to the model containing all three regressors. The reader may expect that other statistics would suggest rejection of the overfitted model. See Exercise 12.40 on page 471.

Test on an Individual Coefficient

The addition of any single variable to a regression system *will increase the regression sum of squares* and thus *reduce the error sum of squares*. Consequently, we must decide whether the increase in regression is sufficient to warrant using the variable in the model. As we might expect, the use of unimportant variables can reduce the effectiveness of the prediction equation by increasing the variance of the estimated response. We shall pursue this point further by considering the importance of x_3 in Example 12.4. Initially, we can test

$$H_0: \beta_3 = 0,$$
$$H_1: \beta_3 \neq 0$$

by using the t-distribution with 9 degrees of freedom. We have

$$t = \frac{b_3 - 0}{s\sqrt{c_{33}}} = \frac{-0.3433}{2.073\sqrt{0.0886}} = -0.556,$$

which indicates that β_3 does not differ significantly from zero, and hence we may very well feel justified in removing x_3 from the model. Suppose that we consider the regression of Y on the set (x_1, x_2), the least squares normal equations now reducing to

$$\begin{bmatrix} 13.0 & 59.43 & 81.82 \\ 59.43 & 394.7255 & 360.6621 \\ 81.82 & 360.6621 & 576.7264 \end{bmatrix} \begin{bmatrix} b_0 \\ b_1 \\ b_2 \end{bmatrix} = \begin{bmatrix} 377.50 \\ 1877.5670 \\ 2246.6610 \end{bmatrix}.$$

The estimated regression coefficients for this reduced model are

$$b_0 = 36.094, \quad b_1 = 1.031, \quad b_2 = -1.870,$$

and the resulting regression sum of squares with 2 degrees of freedom is

$$R(\beta_1, \beta_2) = 398.12.$$

Here we use the notation $R(\beta_1, \beta_2)$ to indicate the regression sum of squares of the restricted model; it should not be confused with SSR, the regression sum of squares of the original model with 3 degrees of freedom. The new error sum of squares is then

$$SST - R(\beta_1, \beta_2) = 438.13 - 398.12 = 40.01,$$

and the resulting mean square error with 10 degrees of freedom becomes

$$s^2 = \frac{40.01}{10} = 4.001.$$

Does a Single Variable t-Test Have an F Counterpart?

From Example 12.4, the amount of variation in the percent survival that is attributed to x_3, in the presence of the variables x_1 and x_2, is

$$R(\beta_3 \mid \beta_1, \beta_2) = SSR - R(\beta_1, \beta_2) = 399.45 - 398.12 = 1.33,$$

which represents a small proportion of the entire regression variation. This amount of added regression is statistically insignificant, as indicated by our previous test on β_3. An equivalent test involves the formation of the ratio

$$f = \frac{R(\beta_3 \mid \beta_1, \beta_2)}{s^2} = \frac{1.33}{4.298} = 0.309,$$

which is a value of the F-distribution with 1 and 9 degrees of freedom. Recall that the basic relationship between the t-distribution with v degrees of freedom and the F-distribution with 1 and v degrees of freedom is

$$t^2 = f(1, v),$$

and note that the f-value of 0.309 is indeed the square of the t-value of -0.56.

To generalize the concepts above, we can assess the work of an independent variable x_i in the general multiple linear regression model

$$\mu_{Y|x_1, x_2, \ldots, x_k} = \beta_0 + \beta_1 x_1 + \cdots + \beta_k x_k$$

by observing the amount of regression attributed to x_i **over and above that attributed to the other variables**, that is, the regression on x_i *adjusted for the other variables*. For example, we say that x_1 is assessed by calculating

$$R(\beta_1 \mid \beta_2, \beta_3, \ldots, \beta_k) = SSR - R(\beta_2, \beta_3, \ldots, \beta_k),$$

where $R(\beta_2, \beta_3, \ldots, \beta_k)$ is the regression sum of squares with $\beta_1 x_1$ removed from the model. To test the hypothesis

$$H_0\colon \beta_1 = 0,$$
$$H_1\colon \beta_1 \neq 0,$$

we compute

$$f = \frac{R(\beta_1 \mid \beta_2, \beta_3, \ldots, \beta_k)}{s^2},$$

and compare it with $f_\alpha(1, n - k - 1)$.

Partial F-Tests on Subsets of Coefficients

In a similar manner, we can test for the significance of a *set* of the variables. For example, to investigate simultaneously the importance of including x_1 and x_2 in the model, we test the hypothesis

$$H_0\colon \beta_1 = \beta_2 = 0,$$
$$H_1\colon \beta_1 \text{ and } \beta_2 \text{ are not both zero,}$$

by computing

$$f = \frac{[R(\beta_1, \beta_2 \mid \beta_3, \beta_4, \ldots, \beta_k)]/2}{s^2} = \frac{[SSR - R(\beta_3, \beta_4, \ldots, \beta_k)]/2}{s^2}$$

and comparing it with $f_\alpha(2, n-k-1)$. The number of degrees of freedom associated with the numerator, in this case 2, equals the number of variables in the set being investigated.

Suppose we wish to test the hypothesis

$$H_0: \beta_2 = \beta_3 = 0,$$
$$H_1: \beta_2 \text{ and } \beta_3 \text{ are not both zero}$$

for Example 12.4. If we develop the regression model

$$y = \beta_0 + \beta_1 x_1 + \epsilon,$$

we can obtain $R(\beta_1) = SSR_{\text{reduced}} = 187.31179$. From Figure 12.1 on page 459, we have $s^2 = 4.29738$ for the full model. Hence, the f-value for testing the hypothesis is

$$f = \frac{R(\beta_2, \beta_3 \mid \beta_1)/2}{s^2} = \frac{[R(\beta_1, \beta_2, \beta_3) - R(\beta_1)]/2}{s^2} = \frac{[SSR_{\text{full}} - SSR_{\text{reduced}}]/2}{s^2}$$
$$= \frac{(399.45437 - 187.31179)/2}{4.29738} = 24.68278.$$

This implies that β_2 and β_3 are not simultaneously zero. Using statistical software such as *SAS* one can directly obtain the above result with a P-value of 0.0002. Readers should note that in statistical software package output there are P-values associated with each individual model coefficient. The null hypothesis for each is that the coefficient is zero. However, it should be noted that the insignificance of any coefficient does not necessarily imply that it does not belong in the final model. It merely suggests that it is insignificant in the presence of all other variables in the problem. The case study at the end of this chapter illustrates this further.

12.7 Special Case of Orthogonality (Optional)

Prior to our original development of the general linear regression problem, the assumption was made that the independent variables are measured without error and are often controlled by the experimenter. Quite often they occur as a result of an *elaborately designed experiment*. In fact, we can increase the effectiveness of the resulting prediction equation with the use of a suitable experimental plan.

Suppose that we once again consider the **X** matrix as defined in Section 12.3. We can rewrite it as

$$\mathbf{X} = [\mathbf{1}, \mathbf{x}_1, \mathbf{x}_2, \ldots, \mathbf{x}_k],$$

where **1** represents a column of ones and $\mathbf{x}_j$ is a column vector representing the levels of x_j. If

$$\mathbf{x}_p' \mathbf{x}_q = \mathbf{0}, \quad \text{for } p \neq q,$$

the variables x_p and x_q are said to be *orthogonal* to each other. There are certain obvious advantages to having a completely orthogonal situation where $\mathbf{x}_p' \mathbf{x}_q = \mathbf{0}$

for all possible p and q, $p \neq q$, and, in addition,

$$\sum_{i=1}^{n} x_{ji} = 0, \quad j = 1, 2, \ldots, k.$$

The resulting $\mathbf{X'X}$ is a diagonal matrix, and the normal equations in Section 12.3 reduce to

$$nb_0 = \sum_{i=1}^{n} y_i, \quad b_1 \sum_{i=1}^{n} x_{1i}^2 = \sum_{i=1}^{n} x_{1i} y_i, \cdots, b_k \sum_{i=1}^{n} x_{ki}^2 = \sum_{i=1}^{n} x_{ki} y_i.$$

An important advantage is that one is easily able to partition SSR into **single-degree-of-freedom components**, each of which corresponds to the amount of variation in Y accounted for by a given controlled variable. In the orthogonal situation, we can write

$$SSR = \sum_{i=1}^{n} (\hat{y}_i - \bar{y})^2 = \sum_{i=1}^{n} (b_0 + b_1 x_{1i} + \cdots + b_k x_{ki} - b_0)^2$$
$$= b_1^2 \sum_{i=1}^{n} x_{1i}^2 + b_2^2 \sum_{i=1}^{n} x_{2i}^2 + \cdots + b_k^2 \sum_{i=1}^{n} x_{ki}^2$$
$$= R(\beta_1) + R(\beta_2) + \cdots + R(\beta_k).$$

The quantity $R(\beta_i)$ is the amount of the regression sum of squares associated with a model involving a single independent variable x_i.

To test simultaneously for the significance of a set of m variables in an orthogonal situation, the regression sum of squares becomes

$$R(\beta_1, \beta_2, \ldots, \beta_m \mid \beta_{m+1}, \beta_{m+2}, \ldots, \beta_k) = R(\beta_1) + R(\beta_2) + \cdots + R(\beta_m),$$

and thus we have the further simplification

$$R(\beta_1 \mid \beta_2, \beta_3, \ldots, \beta_k) = R(\beta_1)$$

when evaluating a single independent variable. Therefore, the contribution of a given variable or set of variables is essentially found by *ignoring* the other variables in the model. Independent evaluations of the worth of the individual variables are accomplished using analysis-of-variance techniques, as given in Table 12.4. The total variation in the response is partitioned into single-degree-of-freedom components plus the error term with $n-k-1$ degrees of freedom. Each computed f-value is used to test one of the hypotheses

$$\left. \begin{array}{l} H_0\colon \beta_i = 0 \\ H_1\colon \beta_i \neq 0 \end{array} \right\} \quad i = 1, 2, \ldots, k,$$

by comparing with the critical point $f_\alpha(1, n-k-1)$ or merely interpreting the P-value computed from the f-distribution.

12.7 Special Case of Orthogonality (Optional)

Table 12.4: Analysis of Variance for Orthogonal Variables

Source of Variation	Sum of Squares	Degrees of Freedom	Mean Square	Computed f
β_1	$R(\beta_1) = b_1^2 \sum_{i=1}^{n} x_{1i}^2$	1	$R(\beta_1)$	$\frac{R(\beta_1)}{s^2}$
β_2	$R(\beta_2) = b_2^2 \sum_{i=1}^{n} x_{2i}^2$	1	$R(\beta_2)$	$\frac{R(\beta_2)}{s^2}$
$\vdots$	$\vdots$	$\vdots$	$\vdots$	$\vdots$
β_k	$R(\beta_k) = b_k^2 \sum_{i=1}^{n} x_{ki}^2$	1	$R(\beta_k)$	$\frac{R(\beta_k)}{s^2}$
Error	SSE	$n-k-1$	$s^2 = \frac{SSE}{n-k-1}$	
Total	$SST = S_{yy}$	$n-1$		

Example 12.8: Suppose that a scientist takes experimental data on the radius of a propellant grain Y as a function of powder temperature x_1, extrusion rate x_2, and die temperature x_3. Fit a linear regression model for predicting grain radius, and determine the effectiveness of each variable in the model. The data are given in Table 12.5.

Table 12.5: Data for Example 12.8

Grain Radius	Powder Temperature	Extrusion Rate	Die Temperature
82	150 (−1)	12 (−1)	220 (−1)
93	190 (+1)	12 (−1)	220 (−1)
114	150 (−1)	24 (+1)	220 (−1)
124	150 (−1)	12 (−1)	250 (+1)
111	190 (+1)	24 (+1)	220 (−1)
129	190 (+1)	12 (−1)	250 (+1)
157	150 (−1)	24 (+1)	250 (+1)
164	190 (+1)	24 (+1)	250 (+1)

Solution: Note that each variable is controlled at two levels, and the experiment is composed of the eight possible combinations. The data on the independent variables are coded for convenience by means of the following formulas:

$$x_1 = \frac{\text{powder temperature} - 170}{20},$$

$$x_2 = \frac{\text{extrusion rate} - 18}{6},$$

$$x_3 = \frac{\text{die temperature} - 235}{15}.$$

The resulting levels of x_1, x_2, and x_3 take on the values -1 and $+1$ as indicated in the table of data. This particular experimental design affords the orthogonality that we want to illustrate here. (A more thorough treatment of this type of experimental layout appears in Chapter 15.) The **X** matrix is

$$\mathbf{X} = \begin{bmatrix} 1 & -1 & -1 & -1 \\ 1 & 1 & -1 & -1 \\ 1 & -1 & 1 & -1 \\ 1 & -1 & -1 & 1 \\ 1 & 1 & 1 & -1 \\ 1 & 1 & -1 & 1 \\ 1 & -1 & 1 & 1 \\ 1 & 1 & 1 & 1 \end{bmatrix},$$

and the orthogonality conditions are readily verified.

We can now compute coefficients

$$b_0 = \frac{1}{8}\sum_{i=1}^{8} y_i = 121.75, \qquad b_1 = \frac{1}{8}\sum_{i=1}^{8} x_{1i}y_i = \frac{20}{8} = 2.5,$$

$$b_2 = \frac{\sum_{i=1}^{8} x_{2i}y_i}{8} = \frac{118}{8} = 14.75, \qquad b_3 = \frac{\sum_{i=1}^{8} x_{3i}y_i}{8} = \frac{174}{8} = 21.75,$$

so in terms of the coded variables, the prediction equation is

$$\hat{y} = 121.75 + 2.5\,x_1 + 14.75\,x_2 + 21.75\,x_3.$$

The analysis of variance in Table 12.6 shows independent contributions to SSR for each variable. The results, when compared to the $f_{0.05}(1,4)$ critical point of 7.71, indicate that x_1 does not contribute significantly at the 0.05 level, whereas variables x_2 and x_3 are significant. In this example, the estimate for σ^2 is 23.1250. As for the single independent variable case, it should be pointed out that this estimate does not solely contain experimental error variation unless the postulated model is correct. Otherwise, the estimate is "contaminated" by lack of fit in addition to pure error, and the lack of fit can be separated out only if we obtain multiple experimental observations for the various (x_1, x_2, x_3) combinations.

Table 12.6: Analysis of Variance for Grain Radius Data

Source of Variation	Sum of Squares	Degrees of Freedom	Mean Squares	Computed f	P-Value
β_1	$(2.5)^2(8) = 50.00$	1	50.00	2.16	0.2156
β_2	$(14.75)^2(8) = 1740.50$	1	1740.50	75.26	0.0010
β_3	$(21.75)^2(8) = 3784.50$	1	3784.50	163.65	0.0002
Error	92.50	4	23.13		
Total	5667.50	7			

Since x_1 is not significant, it can simply be eliminated from the model without altering the effects of the other variables. Note that x_2 and x_3 both impact the grain radius in a positive fashion, with x_3 being the more important factor based on the smallness of its P-value.

Exercises

12.31 Compute and interpret the coefficient of multiple determination for the variables of Exercise 12.1 on page 450.

12.32 Test whether the regression explained by the model in Exercise 12.1 on page 450 is significant at the 0.01 level of significance.

12.33 Test whether the regression explained by the model in Exercise 12.5 on page 450 is significant at the 0.01 level of significance.

12.34 For the model of Exercise 12.5 on page 450, test the hypothesis

$H_0: \beta_1 = \beta_2 = 0$,
$H_1: \beta_1$ and β_2 are not both zero.

12.35 Repeat Exercise 12.17 on page 461 using an F-statistic.

12.36 A small experiment was conducted to fit a multiple regression equation relating the yield y to temperature x_1, reaction time x_2, and concentration of one of the reactants x_3. Two levels of each variable were chosen, and measurements corresponding to the coded independent variables were recorded as follows:

y	x_1	x_2	x_3
7.6	-1	-1	-1
8.4	1	-1	-1
9.2	-1	1	-1
10.3	-1	-1	1
9.8	1	1	-1
11.1	1	-1	1
10.2	-1	1	1
12.6	1	1	1

(a) Using the coded variables, estimate the multiple linear regression equation

$$\mu_{Y|x_1,x_2,x_3} = \beta_0 + \beta_1 x_1 + \beta_2 x_2 + \beta_3 x_3.$$

(b) Partition SSR, the regression sum of squares, into three single-degree-of-freedom components attributable to x_1, x_2, and x_3, respectively. Show an analysis-of-variance table, indicating significance tests on each variable.

12.37 Consider the electric power data of Exercise 12.5 on page 450. Test $H_0: \beta_1 = \beta_2 = 0$, making use of $R(\beta_1, \beta_2 \mid \beta_3, \beta_4)$. Give a P-value, and draw conclusions.

12.38 Consider the data for Exercise 12.36. Compute the following:

$R(\beta_1 \mid \beta_0)$, $\quad R(\beta_1 \mid \beta_0, \beta_2, \beta_3)$,
$R(\beta_2 \mid \beta_0, \beta_1)$, $\quad R(\beta_2 \mid \beta_0, \beta_1, \beta_3)$,
$R(\beta_3 \mid \beta_0, \beta_1, \beta_2)$, $R(\beta_1, \beta_2 \mid \beta_3)$.

Comment.

12.39 Consider the data of Exercise 11.55 on page 437. Fit a regression model using weight and drive ratio as explanatory variables. Compare this model with the SLR (simple linear regression) model using weight alone. Use R^2, R_{adj}^2, and any t-statistics (or F-statistics) you may need to compare the SLR with the multiple regression model.

12.40 Consider Example 12.4. Figure 12.1 on page 459 displays a SAS printout of an analysis of the model containing variables x_1, x_2, and x_3. Focus on the confidence interval of the mean response μ_Y at the (x_1, x_2, x_3) locations representing the 13 data points. Consider an item in the printout indicated by C.V. This is the **coefficient of variation**, which is defined by

$$\text{C.V.} = \frac{s}{\bar{y}} \cdot 100,$$

where $s = \sqrt{s^2}$ is the **root mean squared error**. The coefficient of variation is often used as yet another criterion for comparing competing models. It is a scale-free quantity which expresses the estimate of σ, namely s, as a percent of the average response $\bar{y}$. In competition for the "best" among a group of competing models, one strives for the model with a small value of C.V. Do a regression analysis of the data set shown in Example 12.4 but eliminate x_3. Compare the full (x_1, x_2, x_3) model with the restricted (x_1, x_2) model and focus on two criteria: (i) C.V.; (ii) the widths of the confidence intervals on μ_Y. For the second criterion you may want to use the average width. Comment.

12.41 Consider Example 12.3 on page 447. Compare the two competing models.

First order: $y_i = \beta_0 + \beta_1 x_{1i} + \beta_2 x_{2i} + \epsilon_i$,
Second order: $y_i = \beta_0 + \beta_1 x_{1i} + \beta_2 x_{2i}$
$\qquad + \beta_{11} x_{1i}^2 + \beta_{22} x_{2i}^2 + \beta_{12} x_{1i} x_{2i} + \epsilon_i.$

Use R_{adj}^2 in your comparison. Test $H_0: \beta_{11} = \beta_{22} = \beta_{12} = 0$. In addition, use the C.V. discussed in Exercise 12.40.

12.42 In Example 12.8, a case is made for eliminating x_1, powder temperature, from the model since the P-value based on the F-test is 0.2156 while P-values for x_2 and x_3 are near zero.

(a) Reduce the model by eliminating x_1, thereby producing a full and a restricted (or reduced) model, and compare them on the basis of R^2_{adj}.

(b) Compare the full and restricted models using the width of the 95% prediction intervals on a new observation. The better of the two models would be that with the tightened prediction intervals. Use the average of the width of the prediction intervals.

12.43 Consider the data of Exercise 12.13 on page 452. Can the response, wear, be explained adequately by a single variable (either viscosity or load) in an SLR rather than with the full two-variable regression? Justify your answer thoroughly through tests of hypotheses as well as comparison of the three competing models.

12.44 For the data set given in Exericse 12.16 on page 453, can the response be explained adequately by any two regressor variables? Discuss.

12.8 Categorical or Indicator Variables

An extremely important special-case application of multiple linear regression occurs when one or more of the regressor variables are **categorical**, **indicator**, or **dummy variables**. In a chemical process, the engineer may wish to model the process yield against regressors such as process temperature and reaction time. However, there is interest in using two different catalysts and somehow including "the catalyst" in the model. The catalyst effect cannot be measured on a continuum and is hence a categorical variable. An analyst may wish to model the price of homes against regressors that include square feet of living space x_1, the land acreage x_2, and age of the house x_3. These regressors are clearly continuous in nature. However, it is clear that cost of homes may vary substantially from one area of the country to another. If data are collected on homes in the east, midwest, south, and west, we have an indicator variable with **four categories**. In the chemical process example, if two catalysts are used, we have an indicator variable with two categories. In a biomedical example in which a drug is to be compared to a placebo, all subjects are evaluated on several continuous measurements such as age, blood pressure, and so on, as well as gender, which of course is categorical with two categories. So, included along with the continuous variables are two indicator variables: treatment with two categories (active drug and placebo) and gender with two categories (male and female).

Model with Categorical Variables

Let us use the chemical processing example to illustrate how indicator variables are involved in the model. Suppose $y =$ yield and $x_1 =$ temperature and $x_2 =$ reaction time. Now let us denote the indicator variable by z. Let $z = 0$ for catalyst 1 and $z = 1$ for catalyst 2. The assignment of the $(0, 1)$ indicator to the catalyst is arbitrary. As a result, the model becomes

$$y_i = \beta_0 + \beta_1 x_{1i} + \beta_2 x_{2i} + \beta_3 z_i + \epsilon_i, \quad i = 1, 2, \ldots, n.$$

Three Categories

The estimation of coefficients by the method of least squares continues to apply. In the case of three levels or categories of a single indicator variable, the model will

12.8 Categorical or Indicator Variables

include **two** regressors, say z_1 and z_2, where the $(0, 1)$ assignment is as follows:

$$\begin{matrix} z_1 & z_2 \\ \begin{bmatrix} 1 \\ 0 \\ 0 \end{bmatrix} & \begin{bmatrix} 0 \\ 1 \\ 0 \end{bmatrix} \end{matrix},$$

where **0** and **1** are vectors of 0's and 1's, respectively. In other words, if there are ℓ categories, the model includes $\ell - 1$ actual model terms.

It may be instructive to look at a graphical representation of the model with three categories. For the sake of simplicity, let us assume a single continuous variable x. As a result, the model is given by

$$y_i = \beta_0 + \beta_1 x_i + \beta_2 z_{1i} + \beta_3 z_{2i} + \epsilon_i.$$

Thus, Figure 12.2 reflects the nature of the model. The following are model expressions for the three categories.

$$\begin{aligned} E(Y) &= (\beta_0 + \beta_2) + \beta_1 x, & \text{category 1,} \\ E(Y) &= (\beta_0 + \beta_3) + \beta_1 x, & \text{category 2,} \\ E(Y) &= \beta_0 + \beta_1 x, & \text{category 3.} \end{aligned}$$

As a result, the model involving categorical variables essentially involves a **change in the intercept** as we change from one category to another. Here of course we are assuming that the **coefficients of continuous variables remain the same across the categories**.

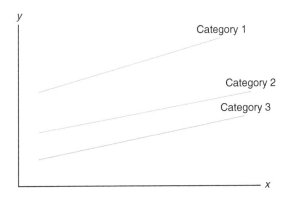

Figure 12.2: Case of three categories.

Example 12.9: Consider the data in Table 12.7. The response y is the amount of suspended solids in a coal cleansing system. The variable x is the pH of the system. Three different polymers are used in the system. Thus, "polymer" is categorical with three categories and hence produces two model terms. The model is given by

$$y_i = \beta_0 + \beta_1 x_i + \beta_2 z_{1i} + \beta_3 z_{2i} + \epsilon_i, \quad i = 1, 2, \ldots, 18.$$

Here we have

$$z_1 = \begin{cases} 1, & \text{for polymer 1,} \\ 0, & \text{otherwise,} \end{cases} \quad \text{and} \quad z_2 = \begin{cases} 1, & \text{for polymer 2,} \\ 0, & \text{otherwise.} \end{cases}$$

From the analysis in Figure 12.3, the following conclusions are drawn. The coefficient b_1 for pH is the estimate of the **common slope** that is assumed in the regression analysis. All model terms are statistically significant. Thus, pH and the nature of the polymer have an impact on the amount of cleansing. The signs and magnitudes of the coefficients of z_1 and z_2 indicate that polymer 1 is most effective (producing higher suspended solids) for cleansing, followed by polymer 2. Polymer 3 is least effective.

Table 12.7: Data for Example 12.9

x, (pH)	y, (amount of suspended solids)	Polymer
6.5	292	1
6.9	329	1
7.8	352	1
8.4	378	1
8.8	392	1
9.2	410	1
6.7	198	2
6.9	227	2
7.5	277	2
7.9	297	2
8.7	364	2
9.2	375	2
6.5	167	3
7.0	225	3
7.2	247	3
7.6	268	3
8.7	288	3
9.2	342	3

Slope May Vary with Indicator Categories

In the discussion given here, we have assumed that the indicator variable model terms enter the model in an additive fashion. This suggests that the slopes, as in Figure 12.2, are constant across categories. Obviously, this is not always going to be the case. We can account for the possibility of varying slopes and indeed test for this condition of **parallelism** by including product or **interaction** terms between indicator terms and continuous variables. For example, suppose a model with one continuous regressor and an indicator variable with two levels is chosen. The model is given by

$$y = \beta_0 + \beta_1 x + \beta_2 z + \beta_3 xz + \epsilon.$$

12.8 Categorical or Indicator Variables

Source	DF	Sum of Squares	Mean Square	F Value	Pr > F
Model	3	80181.73127	26727.24376	73.68	<.0001
Error	14	5078.71318	362.76523		
Corrected Total	17	85260.44444			

R-Square	Coeff Var	Root MSE	y Mean
0.940433	6.316049	19.04640	301.5556

Parameter	Estimate	Standard Error	t Value	Pr > |t|
Intercept	-161.8973333	37.43315576	-4.32	0.0007
x	54.2940260	4.75541126	11.42	<.0001
z1	89.9980606	11.05228237	8.14	<.0001
z2	27.1656970	11.01042883	2.47	0.0271

Figure 12.3: *SAS* printout for Example 12.9.

This model suggests that for category 1 ($z = 1$),

$$E(y) = (\beta_0 + \beta_2) + (\beta_1 + \beta_3)x,$$

while for category 2 ($z = 0$),

$$E(y) = \beta_0 + \beta_1 x.$$

Thus, we allow for varying intercepts and slopes for the two categories. Figure 12.4 displays the regression lines with varying slopes for the two categories.

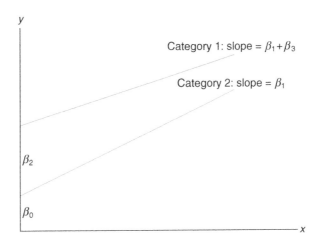

Figure 12.4: Nonparallelism in categorical variables.

In this case, β_0, β_1, and β_2 are positive while β_3 is negative with $|\beta_3| < \beta_1$. Obviously, if the interaction coefficient β_3 is insignificant, we are back to the common slope model.

Exercises

12.45 A study was done to assess the cost effectiveness of driving a four-door sedan instead of a van or an SUV (sports utility vehicle). The continuous variables are odometer reading and octane of the gasoline used. The response variable is miles per gallon. The data are presented here.

MPG	Car Type	Odometer	Octane
34.5	sedan	75,000	87.5
33.3	sedan	60,000	87.5
30.4	sedan	88,000	78.0
32.8	sedan	15,000	78.0
35.0	sedan	25,000	90.0
29.0	sedan	35,000	78.0
32.5	sedan	102,000	90.0
29.6	sedan	98,000	87.5
16.8	van	56,000	87.5
19.2	van	72,000	90.0
22.6	van	14,500	87.5
24.4	van	22,000	90.0
20.7	van	66,500	78.0
25.1	van	35,000	90.0
18.8	van	97,500	87.5
15.8	van	65,500	78.0
17.4	van	42,000	78.0
15.6	SUV	65,000	78.0
17.3	SUV	55,500	87.5
20.8	SUV	26,500	87.5
22.2	SUV	11,500	90.0
16.5	SUV	38,000	78.0
21.3	SUV	77,500	90.0
20.7	SUV	19,500	78.0
24.1	SUV	87,000	90.0

(a) Fit a linear regression model including two indicator variables. Use $(0,0)$ to denote the four-door sedan.

(b) Which type of vehicle appears to get the best gas mileage?

(c) Discuss the difference between a van and an SUV in terms of gas mileage.

12.46 A study was done to determine whether the gender of the credit card holder was an important factor in generating profit for a certain credit card company. The variables considered were income, the number of family members, and the gender of the card holder. The data are as follows:

Profit	Income	Gender	Family Members
157	45,000	M	1
−181	55,000	M	2
−253	45,800	M	4
158	38,000	M	3
75	75,000	M	4
202	99,750	M	4
−451	28,000	M	1
146	39,000	M	2
89	54,350	M	1
−357	32,500	M	1
522	36,750	F	1
78	42,500	F	3
5	34,250	F	2
−177	36,750	F	3
123	24,500	F	2
251	27,500	F	1
−56	18,000	F	1
453	24,500	F	1
288	88,750	F	1
−104	19,750	F	2

(a) Fit a linear regression model using the variables available. Based on the fitted model, would the company prefer male or female customers?

(b) Would you say that income was an important factor in explaining the variability in profit?

12.9 Sequential Methods for Model Selection

At times, the significance tests outlined in Section 12.6 are quite adequate for determining which variables should be used in the final regression model. These tests are certainly effective if the experiment can be planned and the variables are orthogonal to each other. Even if the variables are not orthogonal, the individual t-tests can be of some use in many problems where the number of variables under investigation is small. However, there are many problems where it is necessary to use more elaborate techniques for screening variables, particularly when the experiment exhibits a substantial deviation from orthogonality. Useful measures of **multicollinearity** (linear dependency) among the independent variables are provided by the sample correlation coefficients $r_{x_i x_j}$. Since we are concerned only with linear dependency among independent variables, no confusion will result if we drop the x's from our notation and simply write $r_{x_i x_j} = r_{ij}$, where

$$r_{ij} = \frac{S_{ij}}{\sqrt{S_{ii}S_{jj}}}.$$

Note that the r_{ij} do not give true estimates of population correlation coefficients in the strict sense, since the x's are actually not random variables in the context discussed here. Thus, the term *correlation*, although standard, is perhaps a misnomer.

When one or more of these sample correlation coefficients deviate substantially from zero, it can be quite difficult to find the most effective subset of variables for inclusion in our prediction equation. In fact, for some problems the multicollinearity will be so extreme that a suitable predictor cannot be found unless all possible subsets of the variables are investigated. Informative discussions of model selection in regression by Hocking (1976) are cited in the Bibliography. Procedures for detection of multicollinearity are discussed in the textbook by Myers (1990), also cited.

The user of multiple linear regression attempts to accomplish one of three objectives:

1. Obtain estimates of individual coefficients in a complete model.
2. Screen variables to determine which have a significant effect on the response.
3. Arrive at the most effective prediction equation.

In (1) it is known a priori that all variables are to be included in the model. In (2) prediction is secondary, while in (3) individual regression coefficients are not as important as the quality of the estimated response $\hat{y}$. For each of the situations above, multicollinearity in the experiment can have a profound effect on the success of the regression.

In this section, some standard sequential procedures for selecting variables are discussed. They are based on the notion that a single variable or a collection of variables should not appear in the estimating equation unless the variables result in a significant increase in the regression sum of squares or, equivalently, a significant increase in R^2, the coefficient of multiple determination.

Illustration of Variable Screening in the Presence of Collinearity

Example 12.10: Consider the data of Table 12.8, where measurements were taken for nine infants. The purpose of the experiment was to arrive at a suitable estimating equation relating the length of an infant to all or a subset of the independent variables. The sample correlation coefficients, indicating the linear dependency among the independent variables, are displayed in the symmetric matrix

$$\begin{array}{c c} & \begin{array}{cccc} x_1 & x_2 & x_3 & x_4 \end{array} \\ & \begin{bmatrix} 1.0000 & 0.9523 & 0.5340 & 0.3900 \\ 0.9523 & 1.0000 & 0.2626 & 0.1549 \\ 0.5340 & 0.2626 & 1.0000 & 0.7847 \\ 0.3900 & 0.1549 & 0.7847 & 1.0000 \end{bmatrix} \end{array}$$

Table 12.8: Data Relating to Infant Length*

Infant Length, y (cm)	Age, x_1 (days)	Length at Birth, x_2 (cm)	Weight at Birth, x_3 (kg)	Chest Size at Birth, x_4 (cm)
57.5	78	48.2	2.75	29.5
52.8	69	45.5	2.15	26.3
61.3	77	46.3	4.41	32.2
67.0	88	49.0	5.52	36.5
53.5	67	43.0	3.21	27.2
62.7	80	48.0	4.32	27.7
56.2	74	48.0	2.31	28.3
68.5	94	53.0	4.30	30.3
69.2	102	58.0	3.71	28.7

*Data analyzed by the Laboratory for Interdisciplinary Statistical Analysis, Virginia Tech, Blacksburg, Virginia.

Note that there appears to be an appreciable amount of multicollinearity. Using the least squares technique outlined in Section 12.2, the estimated regression equation was fitted using the complete model and is

$$\hat{y} = 7.1475 + 0.1000x_1 + 0.7264x_2 + 3.0758x_3 - 0.0300x_4.$$

The value of s^2 with 4 degrees of freedom is 0.7414, and the value for the coefficient of determination for this model is found to be 0.9908. Regression sums of squares, measuring the variation attributed to each individual variable in the presence of the others, and the corresponding t-values are given in Table 12.9.

Table 12.9: t-Values for the Regression Data of Table 12.8

Variable x_1	Variable x_2	Variable x_3	Variable x_4
$R(\beta_1 \mid \beta_2, \beta_3, \beta_4)$ = 0.0644	$R(\beta_2 \mid \beta_1, \beta_3, \beta_4)$ = 0.6334	$R(\beta_3 \mid \beta_1, \beta_2, \beta_4)$ = 6.2523	$R(\beta_4 \mid \beta_1, \beta_2, \beta_3)$ = 0.0241
$t = 0.2947$	$t = 0.9243$	$t = 2.9040$	$t = -0.1805$

A two-tailed critical region with 4 degrees of freedom at the 0.05 level of significance is given by $|t| > 2.776$. Of the four computed t-values, **only variable x_3 appears to be significant**. However, recall that although the t-statistic described in Section 12.6 measures the worth of a variable adjusted for all other variables, it does not detect the potential importance of a variable in combination with a subset of the variables. For example, consider the model with only the variables x_2 and x_3 in the equation. The data analysis gives the regression function

$$\hat{y} = 2.1833 + 0.9576x_2 + 3.3253x_3,$$

with $R^2 = 0.9905$, certainly not a substantial reduction from $R^2 = 0.9907$ for the complete model. However, unless the performance characteristics of this particular combination had been observed, one would not be aware of its predictive potential. This, of course, lends support for a methodology that observes *all possible regressions* or a systematic sequential procedure designed to test subsets.

Stepwise Regression

One standard procedure for searching for the "optimum subset" of variables in the absence of orthogonality is a technique called **stepwise regression**. It is based on the procedure of sequentially introducing the variables into the model one at a time. Given a predetermined size α, the description of the stepwise routine will be better understood if the methods of **forward selection** and **backward elimination** are described first.

Forward selection is based on the notion that variables should be inserted one at a time until a satisfactory regression equation is found. The procedure is as follows:

STEP 1. Choose the variable that gives the largest regression sum of squares when performing a simple linear regression with y or, equivalently, that which gives the largest value of R^2. We shall call this initial variable x_1. If x_1 is insignificant, the procedure is terminated.

STEP 2. Choose the variable that, when inserted in the model, gives the largest increase in R^2, in the presence of x_1, over the R^2 found in step 1. This, of course, is the variable x_j for which

$$R(\beta_j|\beta_1) = R(\beta_1, \beta_j) - R(\beta_1)$$

is largest. Let us call this variable x_2. The regression model with x_1 and x_2 is then fitted and R^2 observed. If x_2 is insignificant, the procedure is terminated.

STEP 3. Choose the variable x_j that gives the largest value of

$$R(\beta_j \mid \beta_1, \beta_2) = R(\beta_1, \beta_2, \beta_j) - R(\beta_1, \beta_2),$$

again resulting in the largest increase of R^2 over that given in step 2. Calling this variable x_3, we now have a regression model involving x_1, x_2, and x_3. If x_3 is insignificant, the procedure is terminated.

This process is continued until the most recent variable inserted fails to induce a significant increase in the explained regression. Such an increase can be determined at each step by using the appropriate partial F-test or t-test. For example, in step 2 the value

$$f = \frac{R(\beta_2|\beta_1)}{s^2}$$

can be determined to test the appropriateness of x_2 in the model. Here the value of s^2 is the mean square error for the model containing the variables x_1 and x_2. Similarly, in step 3 the ratio

$$f = \frac{R(\beta_3 \mid \beta_1, \beta_2)}{s^2}$$

tests the appropriateness of x_3 in the model. Now, however, the value for s^2 is the mean square error for the model that contains the three variables x_1, x_2, and x_3. If $f < f_\alpha(1, n-3)$ at step 2, for a prechosen significance level, x_2 is not included and the process is terminated, resulting in a simple linear equation relating y and x_1. However, if $f > f_\alpha(1, n-3)$, we proceed to step 3. Again, if $f < f_\alpha(1, n-4)$ at step 3, x_3 is not included and the process is terminated with the appropriate regression equation containing the variables x_1 and x_2.

Backward elimination involves the same concepts as forward selection except that one begins with all the variables in the model. Suppose, for example, that there are five variables under consideration. The steps are as follows:

STEP 1. Fit a regression equation with all five variables included in the model. Choose the variable that gives the smallest value of the regression sum of squares **adjusted for the others**. Suppose that this variable is x_2. Remove x_2 from the model if

$$f = \frac{R(\beta_2 \mid \beta_1, \beta_3, \beta_4, \beta_5)}{s^2}$$

is insignificant.

STEP 2. Fit a regression equation using the remaining variables x_1, x_3, x_4, and x_5, and repeat step 1. Suppose that variable x_5 is chosen this time. Once again, if

$$f = \frac{R(\beta_5 \mid \beta_1, \beta_3, \beta_4)}{s^2}$$

is insignificant, the variable x_5 is removed from the model. At each step, the s^2 used in the F-test is the mean square error for the regression model at that stage.

This process is repeated until at some step the variable with the smallest adjusted regression sum of squares results in a significant f-value for some predetermined significance level.

Stepwise regression is accomplished with a slight but important modification of the forward selection procedure. The modification involves further testing at each stage to ensure the continued effectiveness of variables that had been inserted into the model at an earlier stage. This represents an improvement over forward selection, since it is quite possible that a variable entering the regression equation at an early stage might have been rendered unimportant or redundant because of relationships that exist between it and other variables entering at later stages. Therefore, at a stage in which a new variable has been entered into the regression equation through a significant increase in R^2 as determined by the F-test, all the variables already in the model are subjected to F-tests (or, equivalently, to t-tests) in light of this new variable and are deleted if they do not display a significant f-value. The procedure is continued until a stage is reached where no additional variables can be inserted or deleted. We illustrate the stepwise procedure in the following example.

Example 12.11: Using the techniques of stepwise regression, find an appropriate linear regression model for predicting the length of infants for the data of Table 12.8.

Solution: STEP 1. Considering each variable separately, four individual simple linear regression equations are fitted. The following pertinent regression sums of

12.9 Sequential Methods for Model Selection

squares are computed:

$$R(\beta_1) = 288.1468, \quad R(\beta_2) = 215.3013,$$
$$R(\beta_3) = 186.1065, \quad R(\beta_4) = 100.8594.$$

Variable x_1 clearly gives the largest regression sum of squares. The mean square error for the equation involving only x_1 is $s^2 = 4.7276$, and since

$$f = \frac{R(\beta_1)}{s^2} = \frac{288.1468}{4.7276} = 60.9500,$$

which exceeds $f_{0.05}(1,7) = 5.59$, the variable x_1 is significant and is entered into the model.

STEP 2. Three regression equations are fitted at this stage, all containing x_1. The important results for the combinations (x_1, x_2), (x_1, x_3), and (x_1, x_4) are

$$R(\beta_2|\beta_1) = 23.8703, \quad R(\beta_3|\beta_1) = 29.3086, \quad R(\beta_4|\beta_1) = 13.8178.$$

Variable x_3 displays the largest regression sum of squares in the presence of x_1. The regression involving x_1 and x_3 gives a new value of $s^2 = 0.6307$, and since

$$f = \frac{R(\beta_3|\beta_1)}{s^2} = \frac{29.3086}{0.6307} = 46.47,$$

which exceeds $f_{0.05}(1,6) = 5.99$, the variable x_3 is significant and is included along with x_1 in the model. Now we must subject x_1 in the presence of x_3 to a significance test. We find that $R(\beta_1 \mid \beta_3) = 131.349$, and hence

$$f = \frac{R(\beta_1|\beta_3)}{s^2} = \frac{131.349}{0.6307} = 208.26,$$

which is highly significant. Therefore, x_1 is retained along with x_3.

STEP 3. With x_1 and x_3 already in the model, we now require $R(\beta_2 \mid \beta_1, \beta_3)$ and $R(\beta_4 \mid \beta_1, \beta_3)$ in order to determine which, if any, of the remaining two variables is entered at this stage. From the regression analysis using x_2 along with x_1 and x_3, we find $R(\beta_2 \mid \beta_1, \beta_3) = 0.7948$, and when x_4 is used along with x_1 and x_3, we obtain $R(\beta_4 \mid \beta_1, \beta_3) = 0.1855$. The value of s^2 is 0.5979 for the (x_1, x_2, x_3) combination and 0.7198 for the (x_1, x_2, x_4) combination. Since neither f-value is significant at the $\alpha = 0.05$ level, the final regression model includes only the variables x_1 and x_3. The estimating equation is found to be

$$\hat{y} = 20.1084 + 0.4136x_1 + 2.0253x_3,$$

and the coefficient of determination for this model is $R^2 = 0.9882$.

Although (x_1, x_3) is the combination chosen by stepwise regression, it is not necessarily the combination of two variables that gives the largest value of R^2. In fact, we have already observed that the combination (x_2, x_3) gives $R^2 = 0.9905$. Of course, the stepwise procedure never observed this combination. A rational argument could be made that there is actually a negligible difference in performance between these two estimating equations, at least in terms of percent variation explained. It is interesting to observe, however, that the backward elimination procedure gives the combination (x_2, x_3) in the final equation (see Exercise 12.49 on page 494).

Summary

The main function of each of the procedures explained in this section is to expose the variables to a systematic methodology designed to ensure the eventual inclusion of the best combinations of the variables. Obviously, there is no assurance that this will happen in all problems, and, of course, it is possible that the multicollinearity is so extensive that one has no alternative but to resort to estimation procedures other than least squares. These estimation procedures are discussed in Myers (1990), listed in the Bibliography.

The sequential procedures discussed here represent three of many such methods that have been put forth in the literature and appear in various regression computer packages that are available. These methods are designed to be computationally efficient but, of course, do not give results for all possible subsets of the variables. As a result, the procedures are most effective for data sets that involve a **large number of variables**. For regression problems involving a relatively small number of variables, modern regression computer packages allow for the computation and summarization of quantitative information on all models for every possible subset of the variables. Illustrations are provided in Section 12.11.

Choice of *P*-Values

As one might expect, the choice of the final model with these procedures may depend dramatically on what *P*-value is chosen. In addition, a procedure is most successful when it is forced to test a large number of candidate variables. For this reason, any forward procedure will be most useful when a relatively large *P*-value is used. Thus, some software packages use a default *P*-value of 0.50.

12.10 Study of Residuals and Violation of Assumptions (Model Checking)

It was suggested earlier in this chapter that the residuals, or errors in the regression fit, often carry information that can be very informative to the data analyst. The $e_i = y_i - \hat{y}_i$, $i = 1, 2, \ldots, n$, which are the numerical counterpart to the ϵ_i, the model errors, often shed light on the possible violation of assumptions or the presence of "suspect" data points. Suppose that we let the vector $\mathbf{x}_i$ denote the values of the regressor variables corresponding to the ith data point, supplemented by a 1 in the initial position. That is,

$$\mathbf{x}'_i = [1, x_{1i}, x_{2i}, \ldots, x_{ki}].$$

Consider the quantity

$$h_{ii} = \mathbf{x}'_i (\mathbf{X}'\mathbf{X})^{-1} \mathbf{x}_i, \quad i = 1, 2, \ldots, n.$$

12.10 Study of Residuals and Violation of Assumptions (Model Checking)

The reader should recognize that h_{ii} was used in the computation of the confidence intervals on the mean response in Section 12.5. Apart from σ^2, h_{ii} represents the variance of the fitted value $\hat{y}_i$. The h_{ii} values are the diagonal elements of the **HAT matrix**

$$\mathbf{H} = \mathbf{X}(\mathbf{X}'\mathbf{X})^{-1}\mathbf{X}',$$

which plays an important role in any study of residuals and in other modern aspects of regression analysis (see Myers, 1990, listed in the Bibliography). The term *HAT matrix* is derived from the fact that $\mathbf{H}$ generates the "y-hats," or the fitted values when multiplied by the vector $\mathbf{y}$ of observed responses. That is, $\hat{\mathbf{y}} = \mathbf{Xb}$, and thus

$$\hat{\mathbf{y}} = \mathbf{X}(\mathbf{X}'\mathbf{X})^{-1}\mathbf{X}'\mathbf{y} = \mathbf{H}\mathbf{y},$$

where $\hat{\mathbf{y}}$ is the vector whose ith element is $\hat{y}_i$.

If we make the usual assumptions that the ϵ_i are independent and normally distributed with mean 0 and variance σ^2, the statistical properties of the residuals are readily characterized. Then

$$E(e_i) = E(y_i - \hat{y}_i) = 0 \quad \text{and} \quad \sigma^2_{\epsilon_i} = (1 - h_{ii})\sigma^2,$$

for $i = 1, 2, \ldots, n$. (See Myers, 1990, for details.) It can be shown that the HAT diagonal values are bounded according to the inequality

$$\frac{1}{n} \leq h_{ii} \leq 1.$$

In addition, $\sum_{i=1}^{n} h_{ii} = k + 1$, the number of regression parameters. As a result, any data point whose HAT diagonal element is large, that is, well above the average value of $(k+1)/n$, is in a position in the data set where the variance of $\hat{y}_i$ is relatively large and the variance of a residual is relatively small. As a result, the data analyst can gain some insight into how large a residual may become before its deviation from zero can be attributed to something other than mere chance. Many of the commercial regression computer packages produce the set of **studentized residuals**.

Studentized Residual
$$r_i = \frac{e_i}{s\sqrt{1 - h_{ii}}}, \quad i = 1, 2, \ldots, n$$

Here each residual has been **divided by an estimate of its standard deviation**, creating a *t-like* statistic that is designed to give the analyst a scale-free quantity providing information regarding the *size* of the residual. In addition, standard computer packages often provide values of another set of studentized-type residuals called the **R-Student values**.

R-Student Residual
$$t_i = \frac{e_i}{s_{-i}\sqrt{1 - h_{ii}}}, \quad i = 1, 2, \ldots, n,$$

where s_{-i} is an estimate of the error standard deviation, calculated with the i**th data point deleted**.

There are three types of violations of assumptions that are readily detected through use of residuals or *residual plots*. While plots of the raw residuals, the e_i, can be helpful, it is often more informative to plot the studentized residuals. The three violations are as follows:

1. Presence of outliers
2. Heterogeneous error variance
3. Model misspecification

In case 1, we choose to define an **outlier** as a data point where there is a deviation from the usual assumption $E(\epsilon_i) = 0$ for a specific value of i. If there is a reason to believe that a specific data point is an outlier exerting a large influence on the fitted model, r_i or t_i may be informative. The R-Student values can be expected to be more sensitive to outliers than the r_i values.

In fact, under the condition that $E(\epsilon_i) = 0$, t_i is a value of a random variable following a t-distribution with $n - 1 - (k+1) = n - k - 2$ degrees of freedom. Thus, a two-sided t-test can be used to provide information for detecting whether or not the ith point is an outlier.

Although the R-Student statistic t_i produces an exact t-test for detection of an outlier at a specific data location, the t-distribution would not apply for simultaneously testing for outliers at all locations. As a result, the studentized residuals or R-Student values should be used strictly as diagnostic tools *without* formal hypothesis testing as the mechanism. The implication is that these statistics highlight data points where the error of fit is larger than what is expected by chance. R-Student values large in magnitude suggest a need for "checking" the data with whatever resources are possible. The practice of eliminating observations from regression data sets should not be done indiscriminately. (For further information regarding the use of outlier diagnostics, see Myers, 1990, in the Bibliography.)

Illustration of Outlier Detection

Case Study 12.1: **Method for Capturing Grasshoppers**: In a biological experiment conducted at Virginia Tech by the Department of Entomology, n experimental runs were made with two different methods for capturing grasshoppers. The methods were drop net catch and sweep net catch. The average number of grasshoppers caught within a set of field quadrants on a given date was recorded for each of the two methods. An additional regressor variable, the average plant height in the quadrants, was also recorded. The experimental data are given in Table 12.10.

The goal is to be able to estimate grasshopper catch by using only the sweep net method, which is less costly. There was some concern about the validity of the fourth data point. The observed catch that was reported using the net drop method seemed unusually high given the other conditions and, indeed, it was felt that the figure might be erroneous. Fit a model of the type

$$y_i = \beta_0 + \beta_1 x_1 + \beta_2 x_2$$

to the 17 data points and study the residuals to determine if data point 4 is an outlier.

12.10 Study of Residuals and Violation of Assumptions (Model Checking)

Table 12.10: Data Set for Case Study 12.1

Observation	Drop Net Catch, y	Sweep Net Catch, x_1	Plant Height, x_2 (cm)
1	18.0000	4.15476	52.705
2	8.8750	2.02381	42.069
3	2.0000	0.15909	34.766
4	20.0000	2.32812	27.622
5	2.3750	0.25521	45.879
6	2.7500	0.57292	97.472
7	3.3333	0.70139	102.062
8	1.0000	0.13542	97.790
9	1.3333	0.12121	88.265
10	1.7500	0.10937	58.737
11	4.1250	0.56250	42.386
12	12.8750	2.45312	31.274
13	5.3750	0.45312	31.750
14	28.0000	6.68750	35.401
15	4.7500	0.86979	64.516
16	1.7500	0.14583	25.241
17	0.1333	0.01562	36.354

Solution: A computer package generated the fitted regression model

$$\hat{y} = 3.6870 + 4.1050x_1 - 0.0367x_2$$

along with the statistics $R^2 = 0.9244$ and $s^2 = 5.580$. The residuals and other diagnostic information were also generated and recorded in Table 12.11.

As expected, the residual at the fourth location appears to be unusually high, namely 7.769. The vital issue here is whether or not this residual is larger than one would expect by chance. The residual standard error for point 4 is 2.209. The R-Student value t_4 is found to be 9.9315. Viewing this as a value of a random variable having a t-distribution with 13 degrees of freedom, one would certainly conclude that the residual of the fourth observation is estimating something greater than 0 and that the suspected measurement error is supported by the study of residuals. Notice that no other residual results in an R-Student value that produces any cause for alarm.

Plotting Residuals for Case Study 12.1

In Chapter 11, we discussed, in some detail, the usefulness of plotting residuals in regression analysis. Violation of model assumptions can often be detected through these plots. In multiple regression, normal probability plotting of residuals or plotting of residuals against $\hat{y}$ may be useful. However, it is often preferable to plot studentized residuals.

Keep in mind that the preference for the studentized residuals over ordinary residuals for plotting purposes stems from the fact that since the variance of the

Table 12.11: Residual Information for the Data Set of Case Study 12.1

Obs.	y_i	$\hat{y}_i$	$y_i - \hat{y}_i$	h_{ii}	$s\sqrt{1-h_{ii}}$	r_i	t_i
1	18.000	18.809	−0.809	0.2291	2.074	−0.390	−0.3780
2	8.875	10.452	−1.577	0.0766	2.270	−0.695	−0.6812
3	2.000	3.065	−1.065	0.1364	2.195	−0.485	−0.4715
4	20.000	12.231	7.769	0.1256	2.209	3.517	9.9315
5	2.375	3.052	−0.677	0.0931	2.250	−0.301	−0.2909
6	2.750	2.464	0.286	0.2276	2.076	0.138	0.1329
7	3.333	2.823	0.510	0.2669	2.023	0.252	0.2437
8	1.000	0.656	0.344	0.2318	2.071	0.166	0.1601
9	1.333	0.947	0.386	0.1691	2.153	0.179	0.1729
10	1.750	1.982	−0.232	0.0852	2.260	−0.103	−0.0989
11	4.125	4.442	−0.317	0.0884	2.255	−0.140	−0.1353
12	12.875	12.610	0.265	0.1152	2.222	0.119	0.1149
13	5.375	4.383	0.992	0.1339	2.199	0.451	0.4382
14	28.000	29.841	−1.841	0.6233	1.450	−1.270	−1.3005
15	4.750	4.891	−0.141	0.0699	2.278	−0.062	−0.0598
16	1.750	3.360	−1.610	0.1891	2.127	−0.757	−0.7447
17	0.133	2.418	−2.285	0.1386	2.193	−1.042	−1.0454

ith residual depends on the ith HAT diagonal, variances of residuals will differ if there is a dispersion in the HAT diagonals. Thus, the appearance of a plot of residuals may seem to suggest heterogeneity because the residuals themselves do not behave, in general, in an ideal way. The purpose of using studentized residuals is to provide a type of *standardization*. Clearly, if σ were known, then under ideal conditions (i.e., a correct model and homogeneous variance), we would have

$$E\left(\frac{e_i}{\sigma\sqrt{1-h_{ii}}}\right) = 0 \quad \text{and} \quad \text{Var}\left(\frac{e_i}{\sigma\sqrt{1-h_{ii}}}\right) = 1.$$

So the studentized residuals produce a set of statistics that behave in a standard way under ideal conditions. Figure 12.5 shows a plot of the **R-Student** values for the grasshopper data of Case Study 12.1. Note how the value for observation 4 stands out from the rest. The R-Student plot was generated by *SAS* software. The plot shows the residuals against the $\hat{y}$-values.

Normality Checking

The reader should recall the importance of normality checking through the use of normal probability plotting, as discussed in Chapter 11. The same recommendation holds for the case of multiple linear regression. Normal probability plots can be generated using standard regression software. Again, however, they can be more effective when one does not use ordinary residuals but, rather, studentized residuals or R-Student values.

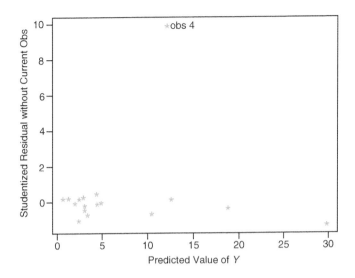

Figure 12.5: *R*-Student values plotted against predicted values for grasshopper data of Case Study 12.1.

12.11 Cross Validation, C_p, and Other Criteria for Model Selection

For many regression problems, the experimenter must choose among various alternative models or model forms that are developed from the same data set. Quite often, the model that best predicts or estimates mean response is required. The experimenter should take into account the relative sizes of the s^2-values for the candidate models and certainly the general nature of the confidence intervals on the mean response. One must also consider how well the model predicts response values that were **not used in building the candidate models**. The models should be subjected to **cross validation**. What are required, then, are cross-validation errors rather than fitting errors. Such errors in prediction are the **PRESS residuals**

$$\delta_i = y_i - \hat{y}_{i,-i}, \quad i = 1, 2, \ldots, n,$$

where $\hat{y}_{i,-i}$ is the prediction of the ith data point by a model that did not make use of the ith point in the calculation of the coefficients. These PRESS residuals are calculated from the formula

$$\delta_i = \frac{e_i}{1 - h_{ii}}, \quad i = 1, 2, \ldots, n.$$

(The derivation can be found in Myers, 1990.)

Use of the PRESS Statistic

The motivation for PRESS and the utility of PRESS residuals are very simple to understand. The purpose of extracting or *setting aside* data points one at a time is to allow the use of separate methodologies for fitting and assessment of a specific model. For assessment of a model, the "$-i$" indicates that the PRESS residual gives a prediction error where the observation being predicted is *independent of the model fit*.

Criteria that make use of the PRESS residuals are given by

$$\sum_{i=1}^{n} |\delta_i| \quad \text{and} \quad \text{PRESS} = \sum_{i=1}^{n} \delta_i^2.$$

(The term **PRESS** is an acronym for **prediction sum of squares**.) We suggest that both of these criteria be used. It is possible for PRESS to be dominated by one or only a few large PRESS residuals. Clearly, the criterion on $\sum_{i=1}^{n} |\delta_i|$ is less sensitive to a small number of large values.

In addition to the PRESS statistic itself, the analyst can simply compute an R^2-like statistic reflecting prediction performance. The statistic is often called R^2_{pred} and is given as follows:

R^2 of Prediction Given a fitted model with a specific value for PRESS, R^2_{pred} is given by

$$R^2_{\text{pred}} = 1 - \frac{\text{PRESS}}{\sum_{i=1}^{n}(y_i - \bar{y})^2}.$$

Note that R^2_{pred} is merely the ordinary R^2 statistic with SSE replaced by the PRESS statistic.

In the following case study, an illustration is provided in which many candidate models are fit to a set of data and the best model is chosen. The sequential procedures described in Section 12.9 are not used. Rather, the role of the PRESS residuals and other statistical values in selecting the best regression equation is illustrated.

Case Study 12.2: **Football Punting**: Leg strength is a necessary characteristic of a successful punter in American football. One measure of the quality of a good punt is the "hang time." This is the time that the ball hangs in the air before being caught by the punt returner. To determine what leg strength factors influence hang time and to develop an empirical model for predicting this response, a study on *The Relationship Between Selected Physical Performance Variables and Football Punting Ability* was conducted by the Department of Health, Physical Education, and Recreation at Virginia Tech. Thirteen punters were chosen for the experiment, and each punted a football 10 times. The average hang times, along with the strength measures used in the analysis, were recorded in Table 12.12.

Each regressor variable is defined as follows:

1. **RLS**, right leg strength (pounds)
2. **LLS**, left leg strength (pounds)
3. **RHF**, right hamstring muscle flexibility (degrees)
4. **LHF**, left hamstring muscle flexibility (degrees)

12.11 Cross Validation, C_p, and Other Criteria for Model Selection

5. **Power**, overall leg strength (foot-pounds)

Determine the most appropriate model for predicting hang time.

Table 12.12: Data for Case Study 12.2

Punter	Hang Time, y (sec)	RLS, x_1	LLS, x_2	RHF, x_3	LHF, x_4	Power, x_5
1	4.75	170	170	106	106	240.57
2	4.07	140	130	92	93	195.49
3	4.04	180	170	93	78	152.99
4	4.18	160	160	103	93	197.09
5	4.35	170	150	104	93	266.56
6	4.16	150	150	101	87	260.56
7	4.43	170	180	108	106	219.25
8	3.20	110	110	86	92	132.68
9	3.02	120	110	90	86	130.24
10	3.64	130	120	85	80	205.88
11	3.68	120	140	89	83	153.92
12	3.60	140	130	92	94	154.64
13	3.85	160	150	95	95	240.57

Solution: In the search for the best of the candidate models for predicting hang time, the information in Table 12.13 was obtained from a regression computer package. The models are ranked in ascending order of the values of the PRESS statistic. This display provides enough information on all possible models to enable the user to eliminate from consideration all but a few models. The model containing x_2 and x_5 (LLS and Power), denoted by x_2x_5, appears to be superior for predicting punter hang time. Also note that all models with low PRESS, low s^2, low $\sum_{i=1}^{n} |\delta_i|$, and high R^2-values contain these two variables.

In order to gain some insight from the residuals of the fitted regression

$$\hat{y}_i = b_0 + b_2 x_{2i} + b_5 x_{5i},$$

the residuals and PRESS residuals were generated. The actual prediction model (see Exercise 12.47 on page 494) is given by

$$\hat{y} = 1.10765 + 0.01370 x_2 + 0.00429 x_5.$$

Residuals, HAT diagonal values, and PRESS values are listed in Table 12.14.

Note the relatively good fit of the two-variable regression model to the data. The PRESS residuals reflect the capability of the regression equation to predict hang time if independent predictions were to be made. For example, for punter number 4, the hang time of 4.180 would encounter a prediction error of 0.039 if the model constructed by using the remaining 12 punters were used. For this model, the average prediction error or cross-validation error is

$$\frac{1}{13} \sum_{i=1}^{n} |\delta_i| = 0.1489 \text{ second},$$

Table 12.13: Comparing Different Regression Models

Model	s^2	$\sum |\delta_i|$	PRESS	R^2
x_2x_5	0.036907	1.93583	0.54683	0.871300
$x_1x_2x_5$	0.041001	2.06489	0.58998	0.871321
$x_2x_4x_5$	0.037708	2.18797	0.59915	0.881658
$x_2x_3x_5$	0.039636	2.09553	0.66182	0.875606
$x_1x_2x_4x_5$	0.042265	2.42194	0.67840	0.882093
$x_1x_2x_3x_5$	0.044578	2.26283	0.70958	0.875642
$x_2x_3x_4x_5$	0.042421	2.55789	0.86236	0.881658
$x_1x_3x_5$	0.053664	2.65276	0.87325	0.831580
$x_1x_4x_5$	0.056279	2.75390	0.89551	0.823375
x_1x_5	0.059621	2.99434	0.97483	0.792094
x_2x_3	0.056153	2.95310	0.98815	0.804187
x_1x_3	0.059400	3.01436	0.99697	0.792864
$x_1x_2x_3x_4x_5$	0.048302	2.87302	1.00920	0.882096
x_2	0.066894	3.22319	1.04564	0.743404
x_3x_5	0.065678	3.09474	1.05708	0.770971
x_1x_2	0.068402	3.09047	1.09726	0.761474
x_3	0.074518	3.06754	1.13555	0.714161
$x_1x_3x_4$	0.065414	3.36304	1.15043	0.794705
$x_2x_3x_4$	0.062082	3.32392	1.17491	0.805163
x_2x_4	0.063744	3.59101	1.18531	0.777716
$x_1x_2x_3$	0.059670	3.41287	1.26558	0.812730
x_3x_4	0.080605	3.28004	1.28314	0.718921
x_1x_4	0.069965	3.64415	1.30194	0.756023
x_1	0.080208	3.31562	1.30275	0.692334
$x_1x_3x_4x_5$	0.059169	3.37362	1.36867	0.834936
$x_1x_2x_4$	0.064143	3.89402	1.39834	0.798692
$x_3x_4x_5$	0.072505	3.49695	1.42036	0.772450
$x_1x_2x_3x_4$	0.066088	3.95854	1.52344	0.815633
x_5	0.111779	4.17839	1.72511	0.571234
x_4x_5	0.105648	4.12729	1.87734	0.631593
x_4	0.186708	4.88870	2.82207	0.283819

which is small compared to the average hang time for the 13 punters.

We indicated in Section 12.9 that the use of all possible subset regressions is often advisable when searching for the best model. Most commercial statistics software packages contain an *all possible regressions* routine. These algorithms compute various criteria for all subsets of model terms. Obviously, criteria such as R^2, s^2, and PRESS are reasonable for choosing among candidate subsets. Another very popular and useful statistic, particularly for areas in the physical sciences and engineering, is the C_p statistic, described below.

12.11 Cross Validation, C_p, and Other Criteria for Model Selection

Table 12.14: PRESS Residuals

Punter	y_i	$\hat{y}_i$	$e_i = y_i - \hat{y}_i$	h_{ii}	δ_i
1	4.750	4.470	0.280	0.198	0.349
2	4.070	3.728	0.342	0.118	0.388
3	4.040	4.094	−0.054	0.444	−0.097
4	4.180	4.146	0.034	0.132	0.039
5	4.350	4.307	0.043	0.286	0.060
6	4.160	4.281	−0.121	0.250	−0.161
7	4.430	4.515	−0.085	0.298	−0.121
8	3.200	3.184	0.016	0.294	0.023
9	3.020	3.174	−0.154	0.301	−0.220
10	3.640	3.636	0.004	0.231	0.005
11	3.680	3.687	−0.007	0.152	−0.008
12	3.600	3.553	0.047	0.142	0.055
13	3.850	4.196	−0.346	0.154	−0.409

The C_p Statistic

Quite often, the choice of the most appropriate model involves many considerations. Obviously, the number of model terms is important; the matter of parsimony is a consideration that cannot be ignored. On the other hand, the analyst cannot be pleased with a model that is too simple, to the point where there is serious underspecification. A single statistic that represents a nice compromise in this regard is the C_p statistic. (See Mallows, 1973, in the Bibliography.)

The C_p statistic appeals nicely to common sense and is developed from considerations of the proper compromise between excessive bias incurred when one underfits (chooses too few model terms) and excessive prediction variance produced when one overfits (has redundancies in the model). The C_p statistic is a simple function of the total number of parameters in the candidate model and the mean square error s^2.

We will not present the entire development of the C_p statistic. (For details, the reader is referred to Myers, 1990, in the Bibliography.) The C_p for a particular subset model is *an estimate* of the following:

$$\Gamma_{(p)} = \frac{1}{\sigma^2} \sum_{i=1}^{n} \text{Var}(\hat{y}_i) + \frac{1}{\sigma^2} \sum_{i=1}^{n} (\text{Bias } \hat{y}_i)^2.$$

It turns out that under the standard least squares assumptions indicated earlier in this chapter, and assuming that the "true" model is the model containing all candidate variables,

$$\frac{1}{\sigma^2} \sum_{i=1}^{n} \text{Var}(\hat{y}_i) = p \quad \text{(number of parameters in the candidate model)}$$

(see Review Exercise 12.63) and an unbiased estimate of

$$\frac{1}{\sigma^2}\sum_{i=1}^{n}(\text{Bias }\hat{y}_i)^2 \text{ is given by } \frac{1}{\sigma^2}\sum_{i=1}^{n}(\widehat{\text{Bias }}\hat{y}_i)^2 = \frac{(s^2-\sigma^2)(n-p)}{\sigma^2}.$$

In the above, s^2 is the mean square error for the candidate model and σ^2 is the population error variance. Thus, if we assume that some estimate $\hat{\sigma}^2$ is available for σ^2, C_p is given by the following equation:

C_p Statistic

$$C_p = p + \frac{(s^2-\hat{\sigma}^2)(n-p)}{\hat{\sigma}^2},$$

where p is the number of model parameters, s^2 is the mean square error for the candidate model, and $\hat{\sigma}^2$ is an estimate of σ^2.

Obviously, the scientist should adopt models with small values of C_p. The reader should note that, unlike the PRESS statistic, C_p is scale-free. In addition, one can gain some insight concerning the adequacy of a candidate model by observing its value of C_p. For example, $C_p > p$ indicates a model that is biased due to being an underfitted model, whereas $C_p \approx p$ indicates a reasonable model.

There is often confusion concerning where $\hat{\sigma}^2$ comes from in the formula for C_p. Obviously, the scientist or engineer does not have access to the population quantity σ^2. In applications where replicated runs are available, say in an experimental design situation, a model-independent estimate of σ^2 is available (see Chapters 11 and 15). However, most software packages use $\hat{\sigma}^2$ as the *mean square error from the most complete model*. Obviously, if this is not a good estimate, the bias portion of the C_p statistic can be negative. Thus, C_p can be less than p.

Example 12.12: Consider the data set in Table 12.15, in which a maker of asphalt shingles is interested in the relationship between sales for a particular year and factors that influence sales. (The data were taken from Kutner et al., 2004, in the Bibliography.)

Of the possible subset models, three are of particular interest. These three are x_2x_3, $x_1x_2x_3$, and $x_1x_2x_3x_4$. The following represents pertinent information for comparing the three models. We include the PRESS statistics for the three models to supplement the decision making.

Model	R^2	R^2_{pred}	s^2	PRESS	C_p
x_2x_3	0.9940	0.9913	44.5552	782.1896	11.4013
$x_1x_2x_3$	0.9970	0.9928	24.7956	643.3578	3.4075
$x_1x_2x_3x_4$	0.9971	0.9917	26.2073	741.7557	5.0

It seems clear from the information in the table that the model x_1, x_2, x_3 is preferable to the other two. Notice that, for the full model, $C_p = 5.0$. This occurs since the *bias portion* is zero, and $\hat{\sigma}^2 = 26.2073$ is the mean square error from the full model.

Figure 12.6 is a *SAS PROC REG* printout showing information for all possible regressions. Here we are able to show comparisons of other models with (x_1, x_2, x_3). Note that (x_1, x_2, x_3) appears to be quite good when compared to all models.

As a final check on the model (x_1, x_2, x_3), Figure 12.7 shows a normal probability plot of the residuals for this model.

12.11 Cross Validation, C_p, and Other Criteria for Model Selection

Table 12.15: Data for Example 12.12

District	Promotional Accounts, x_1	Active Accounts, x_2	Competing Brands, x_3	Potential, x_4	Sales, y (thousands)
1	5.5	31	10	8	$ 79.3
2	2.5	55	8	6	200.1
3	8.0	67	12	9	163.2
4	3.0	50	7	16	200.1
5	3.0	38	8	15	146.0
6	2.9	71	12	17	177.7
7	8.0	30	12	8	30.9
8	9.0	56	5	10	291.9
9	4.0	42	8	4	160.0
10	6.5	73	5	16	339.4
11	5.5	60	11	7	159.6
12	5.0	44	12	12	86.3
13	6.0	50	6	6	237.5
14	5.0	39	10	4	107.2
15	3.5	55	10	4	155.0

```
                    Dependent Variable: sales
Number in                      Adjusted
 Model      C(p)   R-Square    R-Square       MSE     Variables in Model

    3       3.4075   0.9970     0.9961      24.79560   x1 x2 x3
    4       5.0000   0.9971     0.9959      26.20728   x1 x2 x3 x4
    2      11.4013   0.9940     0.9930      44.55518   x2 x3
    3      13.3770   0.9940     0.9924      48.54787   x2 x3 x4
    3    1053.643    0.6896     0.6049    2526.96144   x1 x3 x4
    2    1082.670    0.6805     0.6273    2384.14286   x3 x4
    2    1215.316    0.6417     0.5820    2673.83349   x1 x3
    1    1228.460    0.6373     0.6094    2498.68333   x3
    3    1653.770    0.5140     0.3814    3956.75275   x1 x2 x4
    2    1668.699    0.5090     0.4272    3663.99357   x1 x2
    2    1685.024    0.5042     0.4216    3699.64814   x2 x4
    1    1693.971    0.5010     0.4626    3437.12846   x2
    2    3014.641    0.1151    -.0324     6603.45109   x1 x4
    1    3088.650    0.0928     0.0231    6248.72283   x4
    1    3364.884    0.0120    -.0640     6805.59568   x1
```

Figure 12.6: *SAS* printout of all possible subsets on sales data for Example 12.12.

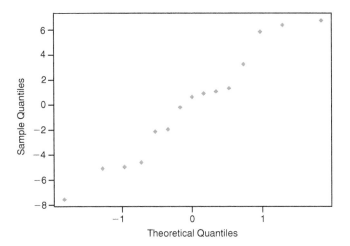

Figure 12.7: Normal probability plot of residuals using the model $x_1x_2x_3$ for Example 12.12.

Exercises

12.47 Consider the "hang time" punting data given in Case Study 12.2, using only the variables x_2 and x_3.

(a) Verify the regression equation shown on page 489.

(b) Predict punter hang time for a punter with LLS = 180 pounds and Power = 260 foot-pounds.

(c) Construct a 95% confidence interval for the mean hang time of a punter with LLS = 180 pounds and Power = 260 foot-pounds.

12.48 For the data of Exercise 12.15 on page 452, use the techniques of

(a) *forward selection* with a 0.05 level of significance to choose a linear regression model;

(b) *backward elimination* with a 0.05 level of significance to choose a linear regression model;

(c) *stepwise regression* with a 0.05 level of significance to choose a linear regression model.

12.49 Use the techniques of *backward elimination* with $\alpha = 0.05$ to choose a prediction equation for the data of Table 12.8.

12.50 For the punter data in Case Study 12.2, an additional response, "punting distance," was also recorded. The average distance values for each of the 13 punters are given.

(a) Using the distance data rather than the hang times, estimate a multiple linear regression model of the type

$$\mu_{Y|x_1,x_2,x_3,x_4,x_5} = \beta_0 + \beta_1 x_1 + \beta_2 x_2 + \beta_3 x_3 + \beta_4 x_4 + \beta_5 x_5$$

for predicting punting distance.

(b) Use stepwise regression with a significance level of 0.10 to select a combination of variables.

(c) Generate values for s^2, R^2, PRESS, and $\sum_{i=1}^{13} |\delta_i|$ for the entire set of 31 models. Use this information to determine the best combination of variables for predicting punting distance.

(d) For the final model you choose, plot the standardized residuals against Y and do a normal probability plot of the ordinary residuals. Comment.

Punter	Distance, y (ft)
1	162.50
2	144.00
3	147.50
4	163.50
5	192.00
6	171.75
7	162.00
8	104.93
9	105.67
10	117.59
11	140.25
12	150.17
13	165.16

12.51 The following is a set of data for y, the amount of money (in thousands of dollars) contributed to the alumni association at Virginia Tech by the Class of 1960, and x, the number of years following graduation:

y	x	y	x
812.52	1	2755.00	11
822.50	2	4390.50	12
1211.50	3	5581.50	13
1348.00	4	5548.00	14
1301.00	8	6086.00	15
2567.50	9	5764.00	16
2526.50	10	8903.00	17

(a) Fit a regression model of the type

$$\mu_{Y|x} = \beta_0 + \beta_1 x.$$

(b) Fit a quadratic model of the type

$$\mu_{Y|x} = \beta_0 + \beta_1 x + \beta_{11} x^2.$$

(c) Determine which of the models in (a) or (b) is preferable. Use s^2, R^2, and the PRESS residuals to support your decision.

12.52 For the model of Exercise 12.50(a), test the hypothesis

$$H_0: \beta_4 = 0,$$
$$H_1: \beta_4 \neq 0.$$

Use a P-value in your conclusion.

12.53 For the quadratic model of Exercise 12.51(b), give estimates of the variances and covariances of the estimates of β_1 and β_{11}.

12.54 A client from the Department of Mechanical Engineering approached the Laboratory for Interdisciplinary Statistical Analysis at Virginia Tech for help in analyzing an experiment dealing with gas turbine engines. The voltage output of engines was measured at various combinations of blade speed and sensor extension.

y (volts)	Speed, x_1 (in./sec)	Extension, x_2 (in.)
1.95	6336	0.000
2.50	7099	0.000
2.93	8026	0.000
1.69	6230	0.000
1.23	5369	0.000
3.13	8343	0.000
1.55	6522	0.006
1.94	7310	0.006
2.18	7974	0.006
2.70	8501	0.006
1.32	6646	0.012
1.60	7384	0.012
1.89	8000	0.012
2.15	8545	0.012
1.09	6755	0.018
1.26	7362	0.018
1.57	7934	0.018
1.92	8554	0.018

(a) Fit a multiple linear regression to the data.
(b) Compute t-tests on coefficients. Give P-values.
(c) Comment on the quality of the fitted model.

12.55 Rayon whiteness is an important factor for scientists dealing in fabric quality. Whiteness is affected by pulp quality and other processing variables. Some of the variables include acid bath temperature, °C (x_1); cascade acid concentration, % (x_2); water temperature, °C (x_3); sulfide concentration, % (x_4); amount of chlorine bleach, lb/min (x_5); and blanket finish temperature, °C (x_6). A set of data from rayon specimens is given here. The response, y, is the measure of whiteness.

y	x_1	x_2	x_3	x_4	x_5	x_6
88.7	43	0.211	85	0.243	0.606	48
89.3	42	0.604	89	0.237	0.600	55
75.5	47	0.450	87	0.198	0.527	61
92.1	46	0.641	90	0.194	0.500	65
83.4	52	0.370	93	0.198	0.485	54
44.8	50	0.526	85	0.221	0.533	60
50.9	43	0.486	83	0.203	0.510	57
78.0	49	0.504	93	0.279	0.489	49
86.8	51	0.609	90	0.220	0.462	64
47.3	51	0.702	86	0.198	0.478	63
53.7	48	0.397	92	0.231	0.411	61
92.0	46	0.488	88	0.211	0.387	88
87.9	43	0.525	85	0.199	0.437	63
90.3	45	0.486	84	0.189	0.499	58
94.2	53	0.527	87	0.245	0.530	65
89.5	47	0.601	95	0.208	0.500	67

(a) Use the criteria MSE, C_p, and PRESS to find the "best" model from among all subset models.
(b) Plot standardized residuals against Y and do a normal probability plot of residuals for the "best" model. Comment.

12.56 In an effort to model executive compensation for the year 1979, 33 firms were selected, and data were gathered on compensation, sales, profits, and employment. The following data were gathered for the year 1979.

Firm	Compensation, y (thousands)	Sales, x_1 (millions)	Profits, x_2 (millions)	Employment, x_3
1	$450	$4600.6	$128.1	48,000
2	387	9255.4	783.9	55,900
3	368	1526.2	136.0	13,783
4	277	1683.2	179.0	27,765
5	676	2752.8	231.5	34,000
6	454	2205.8	329.5	26,500
7	507	2384.6	381.8	30,800
8	496	2746.0	237.9	41,000
9	487	1434.0	222.3	25,900
				(cont.)

Firm	Compensation, y (thousands)	Sales, x_1 (millions)	Profits, x_2 (millions)	Employment, x_3
10	$383	$470.6	$63.7	8600
11	311	1508.0	149.5	21,075
12	271	464.4	30.0	6874
13	524	9329.3	577.3	39,000
14	498	2377.5	250.7	34,300
15	343	1174.3	82.6	19,405
16	354	409.3	61.5	3586
17	324	724.7	90.8	3905
18	225	578.9	63.3	4139
19	254	966.8	42.8	6255
20	208	591.0	48.5	10,605
21	518	4933.1	310.6	65,392
22	406	7613.2	491.6	89,400
23	332	3457.4	228.0	55,200
24	340	545.3	54.6	7800
25	698	22,862.8	3011.3	337,119
26	306	2361.0	203.0	52,000
27	613	2614.1	201.0	50,500
28	302	1013.2	121.3	18,625
29	540	4560.3	194.6	97,937
30	293	855.7	63.4	12,300
31	528	4211.6	352.1	71,800
32	456	5440.4	655.2	87,700
33	417	1229.9	97.5	14,600

Consider the model

$$y_i = \beta_0 + \beta_1 \ln x_{1i} + \beta_2 \ln x_{2i} + \beta_3 \ln x_{3i} + \epsilon_i, \quad i = 1, 2, \ldots, 33.$$

(a) Fit the regression with the model above.
(b) Is a model with a subset of the variables preferable to the full model?

12.57 The pull strength of a wire bond is an important characteristic. The following data give information on pull strength y, die height x_1, post height x_2, loop height x_3, wire length x_4, bond width on the die x_5, and bond width on the post x_6. (From Myers, Montgomery, and Anderson-Cook, 2009.)

(a) Fit a regression model using all independent variables.
(b) Use stepwise regression with input significance level 0.25 and removal significance level 0.05. Give your final model.
(c) Use all possible regression models and compute R^2, C_p, s^2, and adjusted R^2 for all models.

(d) Give the final model.
(e) For your model in part (d), plot studentized residuals (or R-Student) and comment.

y	x_1	x_2	x_3	x_4	x_5	x_6
8.0	5.2	19.6	29.6	94.9	2.1	2.3
8.3	5.2	19.8	32.4	89.7	2.1	1.8
8.5	5.8	19.6	31.0	96.2	2.0	2.0
8.8	6.4	19.4	32.4	95.6	2.2	2.1
9.0	5.8	18.6	28.6	86.5	2.0	1.8
9.3	5.2	18.8	30.6	84.5	2.1	2.1
9.3	5.6	20.4	32.4	88.8	2.2	1.9
9.5	6.0	19.0	32.6	85.7	2.1	1.9
9.8	5.2	20.8	32.2	93.6	2.3	2.1
10.0	5.8	19.9	31.8	86.0	2.1	1.8
10.3	6.4	18.0	32.6	87.1	2.0	1.6
10.5	6.0	20.6	33.4	93.1	2.1	2.1
10.8	6.2	20.2	31.8	83.4	2.2	2.1
11.0	6.2	20.2	32.4	94.5	2.1	1.9
11.3	6.2	19.2	31.4	83.4	1.9	1.8
11.5	5.6	17.0	33.2	85.2	2.1	2.1
11.8	6.0	19.8	35.4	84.1	2.0	1.8
12.3	5.8	18.8	34.0	86.9	2.1	1.8
12.5	5.6	18.6	34.2	83.0	1.9	2.0

12.58 For Exercise 12.57, test $H_0: \beta_1 = \beta_6 = 0$. Give P-values and comment.

12.59 In Exercise 12.28, page 462, we have the following data concerning wear of a bearing:

y (wear)	x_1 (oil viscosity)	x_2 (load)
193	1.6	851
230	15.5	816
172	22.0	1058
91	43.0	1201
113	33.0	1357
125	40.0	1115

(a) The following model may be considered to describe the data:

$$y_i = \beta_0 + \beta_1 x_{1i} + \beta_2 x_{2i} + \beta_{12} x_{1i} x_{2i} + \epsilon_i,$$

for $i = 1, 2, \ldots, 6$. The $x_1 x_2$ is an "interaction" term. Fit this model and estimate the parameters.
(b) Use the models (x_1), (x_1, x_2), (x_2), $(x_1, x_2, x_1 x_2)$ and compute PRESS, C_p, and s^2 to determine the "best" model.

12.12 Special Nonlinear Models for Nonideal Conditions

In much of the preceding material in this chapter and in Chapter 11, we have benefited substantially from the assumption that the model errors, the ϵ_i, are normal with mean 0 and constant variance σ^2. However, there are many real-life situations in which the response is clearly nonnormal. For example, a wealth of applications exist where the **response is binary** (0 or 1) and hence Bernoulli in nature. In the social sciences, the problem may be to develop a model to predict whether or not an individual is a good credit risk (0 or 1) as a function of certain socioeconomic regressors such as income, age, gender, and level of education. In a biomedical drug trial, the response is often whether or not the patient responds positively to a drug while regressors may include drug dosage as well as biological factors such as age, weight, and blood pressure. Again the response is binary in nature. Applications are also abundant in manufacturing areas where certain controllable factors influence whether a manufactured item is **defective or not**.

A second type of nonnormal application on which we will touch briefly has to do with **count data**. Here the assumption of a Poisson response is often convenient. In biomedical applications, the number of cancer cell colonies may be the response which is modeled against drug dosages. In the textile industry, the number of imperfections per yard of cloth may be a reasonable response which is modeled against certain process variables.

Nonhomogeneous Variance

The reader should note the comparison of the ideal (i.e., the normal response) situation with that of the Bernoulli (or binomial) or the Poisson response. We have become accustomed to the fact that the normal case is very special in that the variance is **independent of the mean**. Clearly this is not the case for either Bernoulli or Poisson responses. For example, if the response is 0 or 1, suggesting a Bernoulli response, then the model is of the form

$$p = f(\mathbf{x}, \beta),$$

where p is the **probability of a success** (say response $= 1$). The parameter p plays the role of $\mu_{Y|x}$ in the normal case. However, the Bernoulli variance is $p(1-p)$, which, of course, is also a function of the regressor $\mathbf{x}$. As a result, the variance is not constant. This rules out the use of standard least squares, which we have utilized in our linear regression work up to this point. The same is true for the Poisson case since the model is of the form

$$\lambda = f(\mathbf{x}, \beta),$$

with $\text{Var}(y) = \mu_y = \lambda$, which varies with $\mathbf{x}$.

Binary Response (Logistic Regression)

The most popular approach to modeling binary responses is a technique entitled **logistic regression**. It is used extensively in the biological sciences, biomedical research, and engineering. Indeed, even in the social sciences binary responses are found to be plentiful. The basic distribution for the response is either Bernoulli or binomial. The former is found in observational studies where there are no repeated runs at each regressor level, while the latter will be the case when an experiment is designed. For example, in a clinical trial in which a new drug is being evaluated, the goal might be to determine the dose of the drug that provides efficacy. So certain doses will be employed in the experiment, and more than one subject will be used for each dose. This case is called the **grouped case**.

What Is the Model for Logistic Regression?

In the case of binary responses, the mean response is a probability. In the preceding clinical trial illustration, we might say that we wish to estimate the probability that the patient responds properly to the drug, $P(\text{success})$. Thus, the model is written in terms of a probability. Given regressors $\mathbf{x}$, the logistic function is given by

$$p = \frac{1}{1 + e^{-\mathbf{x}'\beta}}.$$

The portion $\mathbf{x}'\beta$ is called the **linear predictor**, and in the case of a single regressor x it might be written $\mathbf{x}'\beta = \beta_0 + \beta_1 x$. Of course, we do not rule out involving multiple regressors and polynomial terms in the so-called linear predictor. In the grouped case, the model involves modeling the mean of a binomial rather than a Bernoulli, and thus we have the mean given by

$$np = \frac{n}{1 + e^{-\mathbf{x}'\beta}}.$$

Characteristics of Logistic Function

A plot of the logistic function reveals a great deal about its characteristics and why it is utilized for this type of problem. First, the function is nonlinear. In addition, the plot in Figure 12.8 reveals the S-shape with the function approaching $p = 1.0$ as an asymptote. In this case, $\beta_1 > 0$. Thus, we would never experience an estimated probability exceeding 1.0.

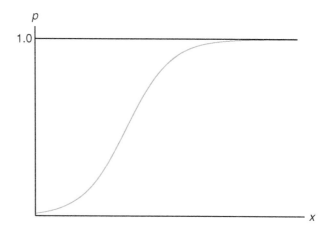

Figure 12.8: The logistic function.

The regression coefficients in the linear predictor can be estimated by the method of maximum likelihood, as described in Chapter 9. The solution to the

12.12 Special Nonlinear Models for Nonideal Conditions

likelihood equations involves an iterative methodology that will not be described here. However, we will present an example and discuss the computer printout and conclusions.

Example 12.13: The data set in Table 12.16 will be used to illustrate the use of logistic regression to analyze a single-agent quantal bioassay of a toxicity experiment. The results show the effect of different doses of nicotine on the common fruit fly.

Table 12.16: Data Set for Example 12.13

x Concentration (grams/100 cc)	n_i Number of Insects	y Number Killed	Percent Killed
0.10	47	8	17.0
0.15	53	14	26.4
0.20	55	24	43.6
0.30	52	32	61.5
0.50	46	38	82.6
0.70	54	50	92.6
0.95	52	50	96.2

The purpose of the experiment was to arrive at an appropriate model relating probability of "kill" to concentration. In addition, the analyst sought the so-called **effective dose** (ED), that is, the concentration of nicotine that results in a certain probability. Of particular interest was the ED_{50}, the concentration that produces a 0.5 probability of "insect kill."

This example is grouped, and thus the model is given by

$$E(Y_i) = n_i p_i = \frac{n_i}{1 + e^{-(\beta_0 + \beta_1 x_i)}}.$$

Estimates of β_0 and β_1 and their standard errors are found by the method of maximum likelihood. Tests on individual coefficients are found using χ^2-statistics rather than t-statistics since there is no common variance σ^2. The χ^2-statistic is derived from $\left(\frac{\text{coeff}}{\text{standard error}}\right)^2$.

Thus, we have the following from a *SAS* PROC LOGIST printout.

		Analysis of Parameter Estimates			
	df	Estimate	Standard Error	Chi-Squared	P-Value
β_0	1	-1.7361	0.2420	51.4482	< 0.0001
β_1	1	6.2954	0.7422	71.9399	< 0.0001

Both coefficients are significantly different from zero. Thus, the fitted model used to predict the probability of "kill" is given by

$$\hat{p} = \frac{1}{1 + e^{-(-1.7361 + 6.2954x)}}.$$

Estimate of Effective Dose

The estimate of ED_{50} for Example 12.13 is found very simply from the estimates b_0 for β_0 and b_1 for β_1. From the logistic function, we see that

$$\log\left(\frac{p}{1-p}\right) = \beta_0 + \beta_1 x.$$

As a result, for $p = 0.5$, an estimate of x is found from

$$b_0 + b_1 x = 0.$$

Thus, ED_{50} is given by

$$x = -\left(\frac{b_0}{b_1}\right) = 0.276 \text{ gram}/100 \text{ cc}.$$

Concept of Odds Ratio

Another form of inference that is conveniently accomplished using logistic regression is derived from the use of the odds ratio. The odds ratio is designed to determine how the **odds of success**, $\frac{p}{1-p}$, increases as certain changes in regressor values occur. For example, in the case of Example 12.13 we may wish to know how the odds would increase if one were to increase dosage by, say, 0.2 gram/100 cc.

Definition 12.1: In logistic regression, an **odds ratio** is the ratio of odds of success at condition 2 to that of condition 1 in the regressors, that is,

$$\frac{[p/(1-p)]_2}{[p/(1-p)]_1}.$$

This allows the analyst to ascertain a sense of the utility of changing the regressor by a certain number of units. Now, since $\left(\frac{p}{1-p}\right) = e^{\beta_0 + \beta_1 x}$, for Example 12.13, the ratio reflecting the increase in odds of success when the dosage of nicotine is increased by 0.2 gram/100 cc is given by

$$e^{0.2 b_1} = e^{(0.2)(6.2954)} = 3.522.$$

The implication of an odds ratio of 3.522 is that the odds of success is enhanced by a factor of 3.522 when the nicotine dose is increased by 0.2 gram/100 cc.

Exercises

12.60 From a set of streptonignic dose-response data, an experimenter desires to develop a relationship between the proportion of lymphoblasts sampled that contain aberrations and the dosage of streptonignic. Five dosage levels were applied to the rabbits used for the experiment. The data are as follows (see Myers, 1990, in the Bibliography):

Dose (mg/kg)	Number of Lymphoblasts	Number with Aberrations
0	600	15
30	500	96
60	600	187
75	300	100
90	300	145

(a) Fit a logistic regression to the data set and thus estimate β_0 and β_1 in the model

$$p = \frac{1}{1 + e^{-(\beta_0 + \beta_1 x)}},$$

where n is the number of lymphoblasts, x is the dose, and p is the probability of an aberration.

(b) Show results of χ^2-tests revealing the significance of the regression coefficients β_0 and β_1.

(c) Estimate ED_{50} and give an interpretation.

12.61 In an experiment to ascertain the effect of load, x, in lb/inches^2, on the probability of failure of specimens of a certain fabric type, an experiment was conducted in which numbers of specimens were exposed to loads ranging from 5 lb/in.^{2} to 90 lb/in.^{2}. The numbers of "failures" were observed. The data are as follows:

Load	Number of Specimens	Number of Failures
5	600	13
35	500	95
70	600	189
80	300	95
90	300	130

(a) Use logistic regression to fit the model

$$p = \frac{1}{1 + e^{-(\beta_0 + \beta_1 x)}},$$

where p is the probability of failure and x is load.

(b) Use the odds ratio concept to determine the increase in odds of failure that results by increasing the load from 20 lb/in.^{2}.

Review Exercises

12.62 In the Department of Fisheries and Wildlife Conservation at Virginia Tech, an experiment was conducted to study the effect of stream characteristics on fish biomass. The regressor variables are as follows: average depth (of 50 cells), x_1; area of in-stream cover (i.e., undercut banks, logs, boulders, etc.), x_2; percent canopy cover (average of 12), x_3; and area ≥ 25 centimeters in depth, x_4. The response is y, the fish biomass. The data are as follows:

Obs.	y	x_1	x_2	x_3	x_4
1	100	14.3	15.0	12.2	48.0
2	388	19.1	29.4	26.0	152.2
3	755	54.6	58.0	24.2	469.7
4	1288	28.8	42.6	26.1	485.9
5	230	16.1	15.9	31.6	87.6
6	0	10.0	56.4	23.3	6.9
7	551	28.5	95.1	13.0	192.9
8	345	13.8	60.6	7.5	105.8
9	0	10.7	35.2	40.3	0.0
10	348	25.9	52.0	40.3	116.6

(a) Fit a multiple linear regression including all four regression variables.

(b) Use C_p, R^2, and s^2 to determine the best subset of variables. Compute these statistics for all possible subsets.

(c) Compare the appropriateness of the models in parts (a) and (b) for predicting fish biomass.

12.63 Show that, in a multiple linear regression data set,

$$\sum_{i=1}^{n} h_{ii} = p.$$

12.64 A small experiment was conducted to fit a multiple regression equation relating the yield y to temperature x_1, reaction time x_2, and concentration of one of the reactants x_3. Two levels of each variable were chosen, and measurements corresponding to the coded independent variables were recorded as follows:

y	x_1	x_2	x_3
7.6	-1	-1	-1
5.5	1	-1	-1
9.2	-1	1	-1
10.3	-1	-1	1
11.6	1	1	-1
11.1	1	-1	1
10.2	-1	1	1
14.0	1	1	1

(a) Using the coded variables, estimate the multiple linear regression equation

$$\mu_{Y|x_1, x_2, x_3} = \beta_0 + \beta_1 x_1 + \beta_2 x_2 + \beta_3 x_3.$$

(b) Partition SSR, the regression sum of squares, into three single-degree-of-freedom components attributable to x_1, x_2, and x_3, respectively. Show an analysis-of-variance table, indicating significance tests on each variable. Comment on the results.

12.65 In a chemical engineering experiment dealing with heat transfer in a shallow fluidized bed, data are collected on the following four regressor variables: fluidizing gas flow rate, lb/hr (x_1); supernatant gas flow rate, lb/hr (x_2); supernatant gas inlet nozzle opening, millimeters (x_3); and supernatant gas inlet temperature, °F (x_4). The responses measured are heat transfer efficiency (y_1) and thermal efficiency (y_2). The data are as follows:

Obs.	y_1	y_2	x_1	x_2	x_3	x_4
1	41.852	38.75	69.69	170.83	45	219.74
2	155.329	51.87	113.46	230.06	25	181.22
3	99.628	53.79	113.54	228.19	65	179.06
4	49.409	53.84	118.75	117.73	65	281.30
5	72.958	49.17	119.72	117.69	25	282.20
6	107.702	47.61	168.38	173.46	45	216.14
7	97.239	64.19	169.85	169.85	45	223.88
8	105.856	52.73	169.85	170.86	45	222.80
9	99.348	51.00	170.89	173.92	80	218.84
10	111.907	47.37	171.31	173.34	25	218.12
11	100.008	43.18	171.43	171.43	45	219.20
12	175.380	71.23	171.59	263.49	45	168.62
13	117.800	49.30	171.63	171.63	45	217.58
14	217.409	50.87	171.93	170.91	10	219.92
15	41.725	54.44	173.92	71.73	45	296.60
16	151.139	47.93	221.44	217.39	65	189.14
17	220.630	42.91	222.74	221.73	25	186.08
18	131.666	66.60	228.90	114.40	25	285.80
19	80.537	64.94	231.19	113.52	65	286.34
20	152.966	43.18	236.84	167.77	45	221.72

Consider the model for predicting the heat transfer coefficient response

$$y_{1i} = \beta_0 + \sum_{j=1}^{4} \beta_j x_{ji} + \sum_{i=1}^{4} \beta_{jj} x_{ji}^2$$
$$+ \sum\sum_{j \neq l} \beta_{jl} x_{ji} x_{li} + \epsilon_i, \quad i = 1, 2, \ldots, 20.$$

(a) Compute PRESS and $\sum_{i=1}^{n} |y_i - \hat{y}_{i,-i}|$ for the least squares regression fit to the model above.

(b) Fit a second-order model with x_4 completely eliminated (i.e., deleting all terms involving x_4). Compute the prediction criteria for the reduced model. Comment on the appropriateness of x_4 for prediction of the heat transfer coefficient.

(c) Repeat parts (a) and (b) for thermal efficiency.

12.66 In exercise physiology, an objective measure of aerobic fitness is the oxygen consumption in volume per unit body weight per unit time. Thirty-one individuals were used in an experiment in order to be able to model oxygen consumption against age in years (x_1), weight in kilograms (x_2), time to run $1\frac{1}{2}$ miles (x_3), resting pulse rate (x_4), pulse rate at the end of run (x_5), and maximum pulse rate during run (x_6).

(a) Do a stepwise regression with input significance level 0.25. Quote the final model.

(b) Do all possible subsets using s^2, C_p, R^2, and R^2_{adj}. Make a decision and quote the final model.

ID	y	x_1	x_2	x_3	x_4	x_5	x_6
1	44.609	44	89.47	11.37	62	178	182
2	45.313	40	75.07	10.07	62	185	185
3	54.297	44	85.84	8.65	45	156	168
4	59.571	42	68.15	8.17	40	166	172
5	49.874	38	89.02	9.22	55	178	180
6	44.811	47	77.45	11.63	58	176	176
7	45.681	40	75.98	11.95	70	176	180
8	49.091	43	81.19	10.85	64	162	170
9	39.442	44	81.42	13.08	63	174	176
10	60.055	38	81.87	8.63	48	170	186
11	50.541	44	73.03	10.13	45	168	168
12	37.388	45	87.66	14.03	56	186	192
13	44.754	45	66.45	11.12	51	176	176
14	47.273	47	79.15	10.60	47	162	164
15	51.855	54	83.12	10.33	50	166	170
16	49.156	49	81.42	8.95	44	180	185
17	40.836	51	69.63	10.95	57	168	172
18	46.672	51	77.91	10.00	48	162	168
19	46.774	48	91.63	10.25	48	162	164
20	50.388	49	73.37	10.08	76	168	168
21	39.407	57	73.37	12.63	58	174	176
22	46.080	54	79.38	11.17	62	156	165
23	45.441	52	76.32	9.63	48	164	166
24	54.625	50	70.87	8.92	48	146	155
25	45.118	51	67.25	11.08	48	172	172
26	39.203	54	91.63	12.88	44	168	172
27	45.790	51	73.71	10.47	59	186	188
28	50.545	57	59.08	9.93	49	148	155
29	48.673	49	76.32	9.40	56	186	188
30	47.920	48	61.24	11.50	52	170	176
31	47.467	52	82.78	10.50	53	170	172

12.67 Consider the data of Review Exercise 12.64. Suppose it is of interest to add some "interaction" terms. Namely, consider the model

$$y_i = \beta_0 + \beta_1 x_{1i} + \beta_2 x_{2i} + \beta_3 x_{3i} + \beta_{12} x_{1i} x_{2i}$$
$$+ \beta_{13} x_{1i} x_{3i} + \beta_{23} x_{2i} x_{3i} + \beta_{123} x_{1i} x_{2i} x_{3i} + \epsilon_i.$$

(a) Do we still have orthogonality? Comment.

(b) With the fitted model in part (a), can you find prediction intervals and confidence intervals on the mean response? Why or why not?

(c) Consider a model with $\beta_{123} x_1 x_2 x_3$ removed. To determine if interactions (as a whole) are needed, test

$$H_0: \beta_{12} = \beta_{13} = \beta_{23} = 0.$$

Give the P-value and conclusions.

12.68 A carbon dioxide (CO_2) flooding technique is used to extract crude oil. The CO_2 floods oil pockets and displaces the crude oil. In an experiment, flow tubes are dipped into sample oil pockets containing a known amount of oil. Using three different values of flow pressure and three different values of dipping angles, the oil pockets are flooded with CO_2, and the percentage of oil displaced recorded. Consider the model

$$y_i = \beta_0 + \beta_1 x_{1i} + \beta_2 x_{2i} + \beta_{11} x_{1i}^2 \\ + \beta_{22} x_{2i}^2 + \beta_{12} x_{1i} x_{2i} + \epsilon_i.$$

Fit the model above to the data, and suggest any model editing that may be needed.

Pressure (lb/in^2), x_1	Dipping Angle, x_2	Oil Recovery (%), y
1000	0	60.58
1000	15	72.72
1000	30	79.99
1500	0	66.83
1500	15	80.78
1500	30	89.78
2000	0	69.18
2000	15	80.31
2000	30	91.99

Source: Wang, G. C. "Microscopic Investigations of CO_2 Flooding Process," *Journal of Petroleum Technology*, Vol. 34, No. 8, Aug. 1982.

12.69 An article in the *Journal of Pharmaceutical Sciences* (Vol. 80, 1991) presents data on the mole fraction solubility of a solute at a constant temperature. Also measured are the dispersion x_1 and dipolar and hydrogen bonding solubility parameters x_2 and x_3. A portion of the data is shown in the table below. In the model, y is the negative logarithm of the mole fraction. Fit the model

$$y_i = \beta_0 + \beta_1 x_{1i} + \beta_2 x_{2i} + \beta_3 x_{3i} + \epsilon_i,$$

for $i = 1, 2, \ldots, 20$.

Obs.	y	x_1	x_2	x_3
1	0.2220	7.3	0.0	0.0
2	0.3950	8.7	0.0	0.3
3	0.4220	8.8	0.7	1.0
4	0.4370	8.1	4.0	0.2
5	0.4280	9.0	0.5	1.0
6	0.4670	8.7	1.5	2.8
7	0.4440	9.3	2.1	1.0
8	0.3780	7.6	5.1	3.4
9	0.4940	10.0	0.0	0.3
10	0.4560	8.4	3.7	4.1
11	0.4520	9.3	3.6	2.0
12	0.1120	7.7	2.8	7.1
13	0.4320	9.8	4.2	2.0
14	0.1010	7.3	2.5	6.8
15	0.2320	8.5	2.0	6.6
16	0.3060	9.5	2.5	5.0
17	0.0923	7.4	2.8	7.8
18	0.1160	7.8	2.8	7.7
19	0.0764	7.7	3.0	8.0
20	0.4390	10.3	1.7	4.2

(a) Test H_0: $\beta_1 = \beta_2 = \beta_3 = 0$.
(b) Plot studentized residuals against x_1, x_2, and x_3 (three plots). Comment.
(c) Consider two additional models that are competitors to the models above:

Model 2: Add x_1^2, x_2^2, x_3^2.

Model 3: Add $x_1^2, x_2^2, x_3^2, x_1 x_2, x_1 x_3, x_2 x_3$.

Use PRESS and C_p with these three models to arrive at the best among the three.

12.70 A study was conducted to determine whether lifestyle changes could replace medication in reducing blood pressure among hypertensives. The factors considered were a healthy diet with an exercise program, the typical dosage of medication for hypertension, and no intervention. The pretreatment body mass index (BMI) was also calculated because it is known to affect blood pressure. The response considered in this study was change in blood pressure. The variable "group" had the following levels.

1 = Healthy diet and an exercise program
2 = Medication
3 = No intervention

(a) Fit an appropriate model using the data below. Does it appear that exercise and diet could be effectively used to lower blood pressure? Explain your answer from the results.
(b) Would exercise and diet be an effective alternative to medication?

(*Hint*: You may wish to form the model in more than one way to answer both of these questions.)

Change in Blood Pressure	Group	BMI
−32	1	27.3
−21	1	22.1
−26	1	26.1
−16	1	27.8
−11	2	19.2
−19	2	26.1
−23	2	28.6
−5	2	23.0
−6	3	28.1
5	3	25.3
−11	3	26.7
14	3	22.3

12.71 Show that in choosing the so-called best subset model from a series of candidate models, choosing the model with the smallest s^2 is equivalent to choosing the model with the smallest R_{adj}^2.

12.72 Case Study: Consider the data set for Exercise 12.12, page 452 (hospital data), repeated here.

Site	x_1	x_2	x_3	x_4	x_5	y
1	15.57	2463	472.92	18.0	4.45	566.52
2	44.02	2048	1339.75	9.5	6.92	696.82
3	20.42	3940	620.25	12.8	4.28	1033.15
4	18.74	6505	568.33	36.7	3.90	1003.62
5	49.20	5723	1497.60	35.7	5.50	1611.37
6	44.92	11,520	1365.83	24.0	4.60	1613.27
7	55.48	5779	1687.00	43.3	5.62	1854.17
8	59.28	5969	1639.92	46.7	5.15	2160.55
9	94.39	8461	2872.33	78.7	6.18	2305.58
10	128.02	20,106	3655.08	180.5	6.15	3503.93
11	96.00	13,313	2912.00	60.9	5.88	3571.59
12	131.42	10,771	3921.00	103.7	4.88	3741.40
13	127.21	15,543	3865.67	126.8	5.50	4026.52
14	252.90	36,194	7684.10	157.7	7.00	10,343.81
15	409.20	34,703	12,446.33	169.4	10.75	11,732.17
16	463.70	39,204	14,098.40	331.4	7.05	15,414.94
17	510.22	86,533	15,524.00	371.6	6.35	18,854.45

(a) The *SAS* PROC REG outputs provided in Figures 12.9 and 12.10 supply a considerable amount of information. Goals are to do outlier detection and eventually determine which model terms are to be used in the final model.

(b) Often the role of a single regressor variable is not apparent when it is studied in the presence of several other variables. This is due to multicollinearity. With this in mind, comment on the importance of x_2 and x_3 in the full model as opposed to their importance in a model in which they are the only variables.

(c) Comment on what other analyses should be run.

(d) Run appropriate analyses and write your conclusions concerning the final model.

```
Dependent Variable: y
                      Analysis of Variance
                              Sum of       Mean
Source              DF       Squares      Square    F Value    Pr > F
Model                5     490177488    98035498     237.79    <.0001
Error               11       4535052      412277
Corrected Total     16     494712540

              Root MSE              642.08838    R-Square    0.9908
              Dependent Mean       4978.48000    Adj R-Sq    0.9867
              Coeff Var              12.89728
                           Parameter Estimates
                                          Parameter    Standard
Variable  Label                      DF    Estimate       Error  t Value  Pr > |t|
Intercept Intercept                   1  1962.94816  1071.36170     1.83    0.0941
x1        Average Daily Patient Load  1   -15.85167    97.65299    -0.16    0.8740
x2        Monthly X-Ray Exposure      1     0.05593     0.02126     2.63    0.0234
x3        Monthly Occupied Bed Days   1     1.58962     3.09208     0.51    0.6174
x4        Eligible Population in the  1    -4.21867     7.17656    -0.59    0.5685
          Area/100
x5        Average Length of Patients  1  -394.31412   209.63954    -1.88    0.0867
          Stay in Days
```

Figure 12.9: *SAS* output for Review Exercise 12.72; part I.

Obs	Dependent Variable	Predicted Value	Std Error Mean Predict	95% CL Mean		95% CL Predict	
1	566.5200	775.0251	241.2323	244.0765	1306	-734.6494	2285
2	696.8200	740.6702	331.1402	11.8355	1470	-849.4275	2331
3	1033	1104	278.5116	490.9234	1717	-436.5244	2644
4	1604	1240	268.1298	650.3459	1831	-291.0028	2772
5	1611	1564	211.2372	1099	2029	76.6816	3052
6	1613	2151	279.9293	1535	2767	609.5796	3693
7	1854	1690	218.9976	1208	2172	196.5345	3183
8	2161	1736	468.9903	703.9948	2768	-13.8306	3486
9	2306	2737	290.4749	2098	3376	1186	4288
10	3504	3682	585.2517	2394	4970	1770	5594
11	3572	3239	189.0989	2823	3655	1766	4713
12	3741	4353	328.8507	3630	5077	2766	5941
13	4027	4257	314.0481	3566	4948	2684	5830
14	10344	8768	252.2617	8213	9323	7249	10286
15	11732	12237	573.9168	10974	13500	10342	14133
16	15415	15038	585.7046	13749	16328	13126	16951
17	18854	19321	599.9780	18000	20641	17387	21255

Obs	Residual	Std Error Residual	Student Residual	-2-1 0 1 2
1	-208.5051	595.0	-0.350	| | |
2	-43.8502	550.1	-0.0797	| | |
3	-70.7734	578.5	-0.122	| | |
4	363.1244	583.4	0.622	| |* |
5	46.9483	606.3	0.0774	| | |
6	-538.0017	577.9	-0.931	| *| |
7	164.4696	603.6	0.272	| | |
8	424.3145	438.5	0.968	| |* |
9	-431.4090	572.6	-0.753	| *| |
10	-177.9234	264.1	-0.674	| *| |
11	332.6011	613.6	0.542	| |* |
12	-611.9330	551.5	-1.110	| **| |
13	-230.5684	560.0	-0.412	| | |
14	1576	590.5	2.669	| |***** |
15	-504.8574	287.9	-1.753	| ***| |
16	376.5491	263.1	1.431	| |** |
17	-466.2470	228.7	-2.039	| ****| |

Figure 12.10: *SAS* output for Review Exercise 12.72; part II.

12.13 Potential Misconceptions and Hazards; Relationship to Material in Other Chapters

There are several procedures discussed in this chapter for use in the "attempt" to find the best model. However, one of the most important misconceptions under which naïve scientists or engineers labor is that there is a **true linear model** and that it can be found. In most scientific phenomena, relationships between scientific variables are nonlinear in nature and the true model is unknown. Linear statistical models are **empirical approximations**.

At times, the choice of the model to be adopted may depend on what information needs to be derived from the model. Is it to be used for prediction? Is it to be used for the purpose of explaining the role of each regressor? This "choice" can be made difficult in the presence of collinearity. It is true that for many regression problems there are multiple models that are very similar in performance. See the Myers reference (1990) for details.

One of the most damaging misuses of the material in this chapter is to assign too much importance to R^2 in the choice of the so-called best model. It is important to remember that for any data set, one can obtain an R^2 as large as one desires, within the constraint $0 \leq R^2 \leq 1$. **Too much attention to R^2 often leads to overfitting**.

Much attention was given in this chapter to outlier detection. A classical serious misuse of statistics centers around the decision made concerning the detection of outliers. We hope it is clear that the analyst should absolutely not carry out the exercise of detecting outliers, eliminating them from the data set, fitting a new model, reporting outlier detection, and so on. This is a tempting and disastrous procedure for arriving at a model that fits the data well, with the result being an example of **how to lie with statistics**. If an outlier is detected, the history of the data should be checked for possible clerical or procedural error before it is eliminated from the data set. One must remember that an outlier by definition is a data point that the model did not fit well. The problem may not be in the data but rather in the model selection. A changed model may result in the point not being detected as an outlier.

There are many types of responses that occur naturally in practice but can't be used in an analysis of standard least squares because classic least squares assumptions do not hold. The assumptions that often fail are those of normal errors and homogeneous variance. For example, if the response is a proportion, say proportion defective, the response distribution is related to the binomial distribution. A second response that occurs often in practice is that of Poisson counts. Clearly the distribution is not normal, and the response variance, which is equal to the Poisson mean, will vary from observation to observation. For more details on these nonideal conditions, see Myers et al. (2008) in the Bibliography.

Chapter 13

One-Factor Experiments: General

13.1 Analysis-of-Variance Technique

In the estimation and hypothesis testing material covered in Chapters 9 and 10, we were restricted in each case to considering no more than two population parameters. Such was the case, for example, in testing for the equality of two population means using independent samples from normal populations with common but unknown variance, where it was necessary to obtain a pooled estimate of σ^2.

This material dealing in two-sample inference represents a special case of what we call the *one-factor problem*. For example, in Exercise 10.35 on page 357, the survival time was measured for two samples of mice, where one sample received a new serum for leukemia treatment and the other sample received no treatment. In this case, we say that there is *one factor*, namely *treatment*, and the factor is at *two levels*. If several competing treatments were being used in the sampling process, more samples of mice would be necessary. In this case, the problem would involve one factor with more than two levels and thus more than two samples.

In the $k > 2$ sample problem, it will be assumed that there are k samples from k populations. One very common procedure used to deal with testing population means is called the **analysis of variance**, or **ANOVA**.

The analysis of variance is certainly not a new technique to the reader who has followed the material on regression theory. We used the analysis-of-variance approach to partition the total sum of squares into a portion due to regression and a portion due to error.

Suppose in an industrial experiment that an engineer is interested in how the mean absorption of moisture in concrete varies among 5 different concrete aggregates. The samples are exposed to moisture for 48 hours. It is decided that 6 samples are to be tested for each aggregate, requiring a total of 30 samples to be tested. The data are recorded in Table 13.1.

The model for this situation may be set up as follows. There are 6 observations taken from each of 5 populations with means $\mu_1, \mu_2, \ldots, \mu_5$, respectively. We may wish to test

H_0: $\mu_1 = \mu_2 = \cdots = \mu_5$,

H_1: At least two of the means are not equal.

Table 13.1: Absorption of Moisture in Concrete Aggregates

Aggregate:	1	2	3	4	5	
	551	595	639	417	563	
	457	580	615	449	631	
	450	508	511	517	522	
	731	583	573	438	613	
	499	633	648	415	656	
	632	517	677	555	679	
Total	3320	3416	3663	2791	3664	16,854
Mean	553.33	569.33	610.50	465.17	610.67	561.80

In addition, we may be interested in making individual comparisons among these 5 population means.

Two Sources of Variability in the Data

In the analysis-of-variance procedure, it is assumed that whatever variation exists among the aggregate averages is attributed to (1) variation in absorption among observations *within* aggregate types and (2) variation *among* aggregate types, that is, due to differences in the chemical composition of the aggregates. The **within-aggregate variation** is, of course, brought about by various causes. Perhaps humidity and temperature conditions were not kept entirely constant throughout the experiment. It is possible that there was a certain amount of heterogeneity in the batches of raw materials that were used. At any rate, we shall consider the within-sample variation to be **chance or random variation**. Part of the goal of the analysis of variance is to determine if the differences among the 5 sample means are what we would expect due to random variation alone or, rather, due to variation beyond merely random effects, i.e., differences in the chemical composition of the aggregates.

Many pointed questions appear at this stage concerning the preceding problem. For example, how many samples must be tested for each aggregate? This is a question that continually haunts the practitioner. In addition, what if the within-sample variation is so large that it is difficult for a statistical procedure to detect the systematic differences? Can we systematically control extraneous sources of variation and thus remove them from the portion we call random variation? We shall attempt to answer these and other questions in the following sections.

13.2 The Strategy of Experimental Design

In Chapters 9 and 10, the notions of estimation and testing for the two-sample case were covered under the important backdrop of the way the experiment is conducted. This falls into the broad category of design of experiments. For example, for the **pooled *t*-test** discussed in Chapter 10, it is assumed that the factor levels (treatments in the mice example) are assigned randomly to the experimental units (mice). The notion of experimental units was discussed in Chapters 9 and 10 and illustrated through examples. Simply put, experimental units are the units (mice, patients, concrete specimens, time) that **provide the heterogeneity that leads to experimental error** in a scientific investigation. The random assignment eliminates bias that could result with systematic assignment. The goal is to distribute uniformly among the factor levels the risks brought about by the heterogeneity of the experimental units. Random assignment best simulates the conditions that are assumed by the model. In Section 13.7, we discuss **blocking** in experiments. The notion of blocking was presented in Chapters 9 and 10, when comparisons between means were accomplished with **pairing**, that is, the division of the experimental units into homogeneous pairs called **blocks**. The factor levels or treatments are then assigned randomly within blocks. The purpose of blocking is to reduce the effective experimental error. In this chapter, we naturally extend the pairing to larger block sizes, with analysis of variance being the primary analytical tool.

13.3 One-Way Analysis of Variance: Completely Randomized Design (One-Way ANOVA)

Random samples of size n are selected from each of k populations. The k different populations are classified on the basis of a single criterion such as different treatments or groups. Today the term **treatment** is used generally to refer to the various classifications, whether they be different aggregates, different analysts, different fertilizers, or different regions of the country.

Assumptions and Hypotheses in One-Way ANOVA

It is assumed that the k populations are independent and normally distributed with means $\mu_1, \mu_2, \ldots, \mu_k$ and common variance σ^2. As indicated in Section 13.2, these assumptions are made more palatable by randomization. We wish to derive appropriate methods for testing the hypothesis

$$H_0:\ \mu_1 = \mu_2 = \cdots = \mu_k,$$
$$H_1:\ \text{At least two of the means are not equal.}$$

Let y_{ij} denote the jth observation from the ith treatment and arrange the data as in Table 13.2. Here, $Y_{i.}$ is the total of all observations in the sample from the ith treatment, $\bar{y}_{i.}$ is the mean of all observations in the sample from the ith treatment, $Y_{..}$ is the total of all nk observations, and $\bar{y}_{..}$ is the mean of all nk observations.

Model for One-Way ANOVA

Each observation may be written in the form

$$Y_{ij} = \mu_i + \epsilon_{ij},$$

where ϵ_{ij} measures the deviation of the jth observation of the ith sample from the corresponding treatment mean. The ϵ_{ij}-term represents random error and plays the same role as the error terms in the regression models. An alternative and

Table 13.2: k Random Samples

Treatment:	1	2	$\cdots$	i	$\cdots$	k	
	y_{11}	y_{21}	$\cdots$	y_{i1}	$\cdots$	y_{k1}	
	y_{12}	y_{22}	$\cdots$	y_{i2}	$\cdots$	y_{k2}	
	$\vdots$	$\vdots$		$\vdots$		$\vdots$	
	y_{1n}	y_{2n}	$\cdots$	y_{in}	$\cdots$	y_{kn}	
Total	$Y_{1.}$	$Y_{2.}$	$\cdots$	$Y_{i.}$	$\cdots$	$Y_{k.}$	$Y_{..}$
Mean	$\bar{y}_{1.}$	$\bar{y}_{2.}$	$\cdots$	$\bar{y}_{i.}$	$\cdots$	$\bar{y}_{k.}$	$\bar{y}_{..}$

preferred form of this equation is obtained by substituting $\mu_i = \mu + \alpha_i$, subject to the constraint $\sum_{i=1}^{k} \alpha_i = 0$. Hence, we may write

$$Y_{ij} = \mu + \alpha_i + \epsilon_{ij},$$

where μ is just the **grand mean** of all the μ_i, that is,

$$\mu = \frac{1}{k} \sum_{i=1}^{k} \mu_i,$$

and α_i is called the **effect** of the ith treatment.

The null hypothesis that the k population means are equal against the alternative that at least two of the means are unequal may now be replaced by the equivalent hypothesis

H_0: $\alpha_1 = \alpha_2 = \cdots = \alpha_k = 0$,

H_1: At least one of the α_i is not equal to zero.

Resolution of Total Variability into Components

Our test will be based on a comparison of two independent estimates of the common population variance σ^2. These estimates will be obtained by partitioning the total variability of our data, designated by the double summation

$$\sum_{i=1}^{k} \sum_{j=1}^{n} (y_{ij} - \bar{y}_{..})^2,$$

into two components.

Theorem 13.1: **Sum-of-Squares Identity**

$$\sum_{i=1}^{k} \sum_{j=1}^{n} (y_{ij} - \bar{y}_{..})^2 = n \sum_{i=1}^{k} (\bar{y}_{i.} - \bar{y}_{..})^2 + \sum_{i=1}^{k} \sum_{j=1}^{n} (y_{ij} - \bar{y}_{i.})^2$$

It will be convenient in what follows to identify the terms of the sum-of-squares identity by the following notation:

13.3 One-Way Analysis of Variance: Completely Randomized Design

Three Important Measures of Variability

$$SST = \sum_{i=1}^{k}\sum_{j=1}^{n}(y_{ij} - \bar{y}_{..})^2 = \text{total sum of squares,}$$

$$SSA = n\sum_{i=1}^{k}(\bar{y}_{i.} - \bar{y}_{..})^2 = \text{treatment sum of squares,}$$

$$SSE = \sum_{i=1}^{k}\sum_{j=1}^{n}(y_{ij} - \bar{y}_{i.})^2 = \text{error sum of squares.}$$

The sum-of-squares identity can then be represented symbolically by the equation

$$SST = SSA + SSE.$$

The identity above expresses how between-treatment and within-treatment variation add to the total sum of squares. However, much insight can be gained by investigating the **expected value of both SSA and SSE**. Eventually, we shall develop variance estimates that formulate the ratio to be used to test the equality of population means.

Theorem 13.2:

$$E(SSA) = (k-1)\sigma^2 + n\sum_{i=1}^{k}\alpha_i^2$$

The proof of the theorem is left as an exercise (see Review Exercise 13.53 on page 556).

If H_0 is true, an estimate of σ^2, based on $k-1$ degrees of freedom, is provided by this expression:

Treatment Mean Square

$$s_1^2 = \frac{SSA}{k-1}$$

If H_0 is true and thus each α_i in Theorem 13.2 is equal to zero, we see that

$$E\left(\frac{SSA}{k-1}\right) = \sigma^2,$$

and s_1^2 is an unbiased estimate of σ^2. However, if H_1 is true, we have

$$E\left(\frac{SSA}{k-1}\right) = \sigma^2 + \frac{n}{k-1}\sum_{i=1}^{k}\alpha_i^2,$$

and s_1^2 estimates σ^2 plus an additional term, which measures variation due to the systematic effects.

A second and independent estimate of σ^2, based on $k(n-1)$ degrees of freedom, is this familiar formula:

Error Mean Square

$$s^2 = \frac{SSE}{k(n-1)}$$

It is instructive to point out the importance of the expected values of the mean squares indicated above. In the next section, we discuss the use of an **F-ratio** with the treatment mean square residing in the numerator. It turns out that when H_1 is true, the presence of the condition $E(s_1^2) > E(s^2)$ suggests that the F-ratio be used in the context of a **one-sided upper-tailed test**. That is, when H_1 is true, we would expect the numerator s_1^2 to exceed the denominator.

Use of *F*-Test in ANOVA

The estimate s^2 is unbiased regardless of the truth or falsity of the null hypothesis (see Review Exercise 13.52 on page 556). It is important to note that the sum-of-squares identity has partitioned not only the total variability of the data, but also the total number of degrees of freedom. That is,

$$nk - 1 = k - 1 + k(n - 1).$$

F-Ratio for Testing Equality of Means

When H_0 is true, the ratio $f = s_1^2/s^2$ is a value of the random variable F having the F-distribution with $k-1$ and $k(n-1)$ degrees of freedom (see Theorem 8.8). Since s_1^2 overestimates σ^2 when H_0 is false, we have a one-tailed test with the critical region entirely in the right tail of the distribution.

The null hypothesis H_0 is rejected at the α-level of significance when

$$f > f_\alpha[k - 1, k(n - 1)].$$

Another approach, the *P*-value approach, suggests that the evidence in favor of or against H_0 is

$$P = P\{f[k - 1, k(n - 1)] > f\}.$$

The computations for an analysis-of-variance problem are usually summarized in tabular form, as shown in Table 13.3.

Table 13.3: Analysis of Variance for the One-Way ANOVA

Source of Variation	Sum of Squares	Degrees of Freedom	Mean Square	Computed f
Treatments	SSA	$k - 1$	$s_1^2 = \dfrac{SSA}{k-1}$	$\dfrac{s_1^2}{s^2}$
Error	SSE	$k(n - 1)$	$s^2 = \dfrac{SSE}{k(n-1)}$	
Total	SST	$kn - 1$		

Example 13.1: Test the hypothesis $\mu_1 = \mu_2 = \cdots = \mu_5$ at the 0.05 level of significance for the data of Table 13.1 on absorption of moisture by various types of cement aggregates.

13.3 One-Way Analysis of Variance: Completely Randomized Design

Solution: The hypotheses are

$$H_0: \mu_1 = \mu_2 = \cdots = \mu_5,$$
$$H_1: \text{At least two of the means are not equal.}$$
$$\alpha = 0.05.$$

Critical region: $f > 2.76$ with $v_1 = 4$ and $v_2 = 25$ degrees of freedom. The sum-of-squares computations give

$$SST = 209{,}377, \quad SSA = 85{,}356,$$
$$SSE = 209{,}377 - 85{,}356 = 124{,}021.$$

These results and the remaining computations are exhibited in Figure 13.1 in the *SAS* ANOVA procedure.

```
                        The GLM Procedure
Dependent Variable: moisture

                                    Sum of
Source              DF             Squares      Mean Square    F Value    Pr > F
Model                4          85356.4667       21339.1167       4.30    0.0088
Error               25         124020.3333        4960.8133
Corrected Total     29         209376.8000

    R-Square     Coeff Var      Root MSE    moisture Mean
    0.407669     12.53703       70.43304          561.8000

Source              DF           Type I SS      Mean Square    F Value    Pr > F
aggregate            4         85356.46667       21339.11667      4.30    0.0088
```

Figure 13.1: *SAS* output for the analysis-of-variance procedure.

Decision: Reject H_0 and conclude that the aggregates do not have the same mean absorption. The P-value for $f = 4.30$ is 0.0088, which is smaller than 0.05.

In addition to the ANOVA, a box plot was constructed for each aggregate. The plots are shown in Figure 13.2. From these plots it is evident that the absorption is not the same for all aggregates. In fact, it appears as if aggregate 4 stands out from the rest. A more formal analysis showing this result will appear in Exercise 13.21 on page 531.

During experimental work, one often loses some of the desired observations. Experimental animals may die, experimental material may be damaged, or human subjects may drop out of a study. The previous analysis for equal sample size will still be valid if we slightly modify the sum of squares formulas. We now assume the k random samples to be of sizes $n_1, n_2, \ldots, n_k$, respectively.

Sum of Squares, Unequal Sample Sizes

$$SST = \sum_{i=1}^{k}\sum_{j=1}^{n_i}(y_{ij} - \bar{y}_{..})^2, \quad SSA = \sum_{i=1}^{k} n_i(\bar{y}_{i.} - \bar{y}_{..})^2, \quad SSE = SST - SSA$$

514 Chapter 13 One-Factor Experiments: General

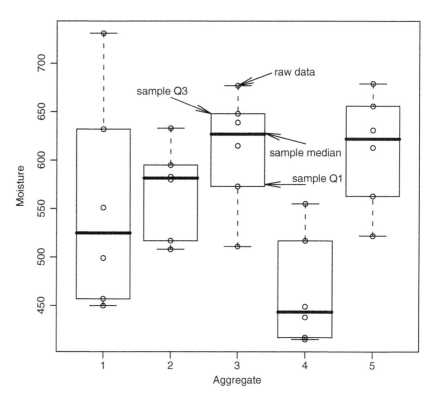

Figure 13.2: Box plots for the absorption of moisture in concrete aggregates.

The degrees of freedom are then partitioned as before: $N-1$ for SST, $k-1$ for SSA, and $N-1-(k-1) = N-k$ for SSE, where $N = \sum_{i=1}^{k} n_i$.

Example 13.2: Part of a study conducted at Virginia Tech was designed to measure serum alkaline phosphatase activity levels (in Bessey-Lowry units) in children with seizure disorders who were receiving anticonvulsant therapy under the care of a private physician. Forty-five subjects were found for the study and categorized into four drug groups:

G-1: Control (not receiving anticonvulsants and having no history of seizure disorders)

G-2: Phenobarbital

G-3: Carbamazepine

G-4: Other anticonvulsants

From blood samples collected from each subject, the serum alkaline phosphatase activity level was determined and recorded as shown in Table 13.4. Test the hypothesis at the 0.05 level of significance that the average serum alkaline phosphatase activity level is the same for the four drug groups.

13.3 One-Way Analysis of Variance: Completely Randomized Design

Table 13.4: Serum Alkaline Phosphatase Activity Level

G-1		G-2	G-3	G-4
49.20	97.50	97.07	62.10	110.60
44.54	105.00	73.40	94.95	57.10
45.80	58.05	68.50	142.50	117.60
95.84	86.60	91.85	53.00	77.71
30.10	58.35	106.60	175.00	150.00
36.50	72.80	0.57	79.50	82.90
82.30	116.70	0.79	29.50	111.50
87.85	45.15	0.77	78.40	
105.00	70.35	0.81	127.50	
95.22	77.40			

Solution: With the level of significance at 0.05, the hypotheses are

$$H_0: \mu_1 = \mu_2 = \mu_3 = \mu_4,$$
$$H_1: \text{At least two of the means are not equal.}$$

Critical region: $f > 2.836$, from interpolating in Table A.6.
Computations: $Y_{1.} = 1460.25$, $Y_{2.} = 440.36$, $Y_{3.} = 842.45$, $Y_{4.} = 707.41$, and $Y_{..} = 3450.47$. The analysis of variance is shown in the $MINITAB$ output of Figure 13.3.

```
One-way ANOVA: G-1, G-2, G-3, G-4

Source  DF     SS    MS     F      P
Factor   3  13939  4646  3.57  0.022
Error   41  53376  1302
Total   44  67315

S = 36.08   R-Sq = 20.71%   R-Sq(adj) = 14.90%

                        Individual 95% CIs For Mean Based on
                        Pooled StDev
Level   N    Mean  StDev  --+---------+---------+---------+-------
G-1    20   73.01  25.75                (----*-----)
G-2     9   48.93  47.11  (-------*-------)
G-3     9   93.61  46.57                      (-------*-------)
G-4     7  101.06  30.76                         (--------*--------)
                          --+---------+---------+---------+-------
                           30        60        90       120

Pooled StDev = 36.08
```

Figure 13.3: $MINITAB$ analysis of data in Table 13.4.

Decision: Reject H_0 and conclude that the average serum alkaline phosphatase activity levels for the four drug groups are not all the same. The calculated P-value is 0.022.

In concluding our discussion on the analysis of variance for the one-way classification, we state the advantages of choosing equal sample sizes over the choice of unequal sample sizes. The first advantage is that the f-ratio is insensitive to slight departures from the assumption of equal variances for the k populations when the samples are of equal size. Second, the choice of equal sample sizes minimizes the probability of committing a type II error.

13.4 Tests for the Equality of Several Variances

Although the f-ratio obtained from the analysis-of-variance procedure is insensitive to departures from the assumption of equal variances for the k normal populations when the samples are of equal size, we may still prefer to exercise caution and run a preliminary test for homogeneity of variances. Such a test would certainly be advisable in the case of unequal sample sizes if there was a reasonable doubt concerning the homogeneity of the population variances. Suppose, therefore, that we wish to test the null hypothesis

$$H_0: \sigma_1^2 = \sigma_2^2 = \cdots = \sigma_k^2$$

against the alternative

$$H_1: \text{The variances are not all equal.}$$

The test that we shall use, called **Bartlett's test**, is based on a statistic whose sampling distribution provides exact critical values when the sample sizes are equal. These critical values for equal sample sizes can also be used to yield highly accurate approximations to the critical values for unequal sample sizes.

First, we compute the k sample variances $s_1^2, s_2^2, \ldots, s_k^2$ from samples of size $n_1, n_2, \ldots, n_k$, with $\sum_{i=1}^{k} n_i = N$. Second, we combine the sample variances to give the pooled estimate

$$s_p^2 = \frac{1}{N-k} \sum_{i=1}^{k} (n_i - 1) s_i^2.$$

Now

$$b = \frac{[(s_1^2)^{n_1-1}(s_2^2)^{n_2-1} \cdots (s_k^2)^{n_k-1}]^{1/(N-k)}}{s_p^2}$$

is a value of a random variable B having the **Bartlett distribution**. For the special case where $n_1 = n_2 = \cdots = n_k = n$, we reject H_0 at the α-level of significance if

$$b < b_k(\alpha; n),$$

13.4 Tests for the Equality of Several Variances

where $b_k(\alpha; n)$ is the critical value leaving an area of size α in the left tail of the Bartlett distribution. Table A.10 gives the critical values, $b_k(\alpha; n)$, for $\alpha = 0.01$ and 0.05; $k = 2, 3, \ldots, 10$; and selected values of n from 3 to 100.

When the sample sizes are unequal, the null hypothesis is rejected at the α-level of significance if

$$b < b_k(\alpha; n_1, n_2, \ldots, n_k),$$

where

$$b_k(\alpha; n_1, n_2, \ldots, n_k) \approx \frac{n_1 b_k(\alpha; n_1) + n_2 b_k(\alpha; n_2) + \cdots + n_k b_k(\alpha; n_k)}{N}.$$

As before, all the $b_k(\alpha; n_i)$ for sample sizes $n_1, n_2, \ldots, n_k$ are obtained from Table A.10.

Example 13.3: Use Bartlett's test to test the hypothesis at the 0.01 level of significance that the population variances of the four drug groups of Example 13.2 are equal.

Solution: We have the hypotheses

$$H_0: \sigma_1^2 = \sigma_2^2 = \sigma_3^2 = \sigma_4^2,$$
$$H_1: \text{The variances are not equal},$$

with $\alpha = 0.01$.

Critical region: Referring to Example 13.2, we have $n_1 = 20$, $n_2 = 9$, $n_3 = 9$, $n_4 = 7$, $N = 45$, and $k = 4$. Therefore, we reject when

$$b < b_4(0.01; 20, 9, 9, 7)$$
$$\approx \frac{(20)(0.8586) + (9)(0.6892) + (9)(0.6892) + (7)(0.6045)}{45}$$
$$= 0.7513.$$

Computations: First compute

$$s_1^2 = 662.862, \quad s_2^2 = 2219.781, \quad s_3^2 = 2168.434, \quad s_4^2 = 946.032,$$

and then

$$s_p^2 = \frac{(19)(662.862) + (8)(2219.781) + (8)(2168.434) + (6)(946.032)}{41}$$
$$= 1301.861.$$

Now

$$b = \frac{[(662.862)^{19}(2219.781)^8(2168.434)^8(946.032)^6]^{1/41}}{1301.861} = 0.8557.$$

Decision: Do not reject the hypothesis, and conclude that the population variances of the four drug groups are not significantly different.

Although Bartlett's test is most often used for testing of homogeneity of variances, other methods are available. A method due to Cochran provides a computationally simple procedure, but it is restricted to situations in which the sample sizes are equal. **Cochran's test** is particularly useful for detecting if one variance is much larger than the others. The statistic that is used is

$$G = \frac{\text{largest } S_i^2}{\sum_{i=1}^{k} S_i^2},$$

and the hypothesis of equality of variances is rejected if $g > g_\alpha$, where the value of g_α is obtained from Table A.11.

To illustrate Cochran's test, let us refer again to the data of Table 13.1 on moisture absorption in concrete aggregates. Were we justified in assuming equal variances when we performed the analysis of variance in Example 13.1? We find that

$$s_1^2 = 12{,}134, \quad s_2^2 = 2303, \quad s_3^2 = 3594, \quad s_4^2 = 3319, \quad s_5^2 = 3455.$$

Therefore,

$$g = \frac{12{,}134}{24{,}805} = 0.4892,$$

which does not exceed the table value $g_{0.05} = 0.5065$. Hence, we conclude that the assumption of equal variances is reasonable.

Exercises

13.1 Six different machines are being considered for use in manufacturing rubber seals. The machines are being compared with respect to tensile strength of the product. A random sample of four seals from each machine is used to determine whether the mean tensile strength varies from machine to machine. The following are the tensile-strength measurements in kilograms per square centimeter $\times 10^{-1}$:

		Machine			
1	2	3	4	5	6
17.5	16.4	20.3	14.6	17.5	18.3
16.9	19.2	15.7	16.7	19.2	16.2
15.8	17.7	17.8	20.8	16.5	17.5
18.6	15.4	18.9	18.9	20.5	20.1

Perform the analysis of variance at the 0.05 level of significance and indicate whether or not the mean tensile strengths differ significantly for the six machines.

13.2 The data in the following table represent the number of hours of relief provided by five different brands of headache tablets administered to 25 subjects experiencing fevers of $38°C$ or more. Perform the analysis of variance and test the hypothesis at the 0.05 level of significance that the mean number of hours of relief provided by the tablets is the same for all five brands. Discuss the results.

		Tablet		
A	B	C	D	E
5.2	9.1	3.2	2.4	7.1
4.7	7.1	5.8	3.4	6.6
8.1	8.2	2.2	4.1	9.3
6.2	6.0	3.1	1.0	4.2
3.0	9.1	7.2	4.0	7.6

13.3 In an article "Shelf-Space Strategy in Retailing," published in *Proceedings: Southern Marketing Association*, the effect of shelf height on the supermarket sales of canned dog food is investigated. An experiment was conducted at a small supermarket for a period of 8 days on the sales of a single brand of dog food, referred to as Arf dog food, involving three levels of shelf height: knee level, waist level, and eye level. During each day, the shelf height of the canned dog food was randomly changed on three different occasions. The remaining sections of the gondola that housed the given brand were filled with a mixture of dog food brands that were both familiar and unfamiliar to customers in this particular geographic area. Sales, in hundreds of dollars, of Arf dog food per day for the three shelf heights are given. Based on the data, is there a significant difference in the average daily sales of this dog food based on shelf height? Use a 0.01 level of significance.

Shelf Height		
Knee Level	Waist Level	Eye Level
77	88	85
82	94	85
86	93	87
78	90	81
81	91	80
86	94	79
77	90	87
81	87	93

13.4 Immobilization of free-ranging white-tailed deer by drugs allows researchers the opportunity to closely examine the deer and gather valuable physiological information. In the study *Influence of Physical Restraint and Restraint Facilitating Drugs on Blood Measurements of White-Tailed Deer and Other Selected Mammals*, conducted at Virginia Tech, wildlife biologists tested the "knockdown" time (time from injection to immobilization) of three different immobilizing drugs. Immobilization, in this case, is defined as the point where the animal no longer has enough muscle control to remain standing. Thirty male white-tailed deer were randomly assigned to each of three treatments. Group A received 5 milligrams of liquid succinylcholine chloride (SCC); group B received 8 milligrams of powdered SCC; and group C received 200 milligrams of phencyclidine hydrochloride. Knockdown times, in minutes, were recorded. Perform an analysis of variance at the 0.01 level of significance and determine whether or not the average knockdown time for the three drugs is the same.

Group		
A	B	C
11	10	4
5	7	4
14	16	6
7	7	3
10	7	5
7	5	6
23	10	8
4	10	3
11	6	7
11	12	3

13.5 The mitochondrial enzyme NADPH:NAD transhydrogenase of the common rat tapeworm (*Hymenolepiasis diminuta*) catalyzes hydrogen in the transfer from NADPH to NAD, producing NADH. This enzyme is known to serve a vital role in the tapeworm's anaerobic metabolism, and it has recently been hypothesized that it may serve as a proton exchange pump, transferring protons across the mitochondrial membrane. A study on *Effect of Various Substrate Concentrations on the Conformational Variation of the NADPH:NAD Transhydrogenase of Hymenolepiasis diminuta*, conducted at Bowling Green State University, was designed to assess the ability of this enzyme to undergo conformation or shape changes. Changes in the specific activity of the enzyme caused by variations in the concentration of NADP could be interpreted as supporting the theory of conformational change. The enzyme in question is located in the inner membrane of the tapeworm's mitochondria. Tapeworms were homogenized, and through a series of centrifugations, the enzyme was isolated. Various concentrations of NADP were then added to the isolated enzyme solution, and the mixture was then incubated in a water bath at 56°C for 3 minutes. The enzyme was then analyzed on a dual-beam spectrophotometer, and the results shown were calculated, with the specific activity of the enzyme given in nanomoles per minute per milligram of protein. Test the hypothesis at the 0.01 level that the average specific activity is the same for the four concentrations.

NADP Concentration (nm)				
0	80	160	360	
11.01	11.38	11.02	6.04	10.31
12.09	10.67	10.67	8.65	8.30
10.55	12.33	11.50	7.76	9.48
11.26	10.08	10.31	10.13	8.89
			9.36	

13.6 A study measured the sorption (either absorption or adsorption) rates of three different types of organic chemical solvents. These solvents are used to clean industrial fabricated-metal parts and are potential hazardous waste. Independent samples from each type of solvent were tested, and their sorption rates were recorded as a mole percentage. (See McClave, Dietrich, and Sincich, 1997.)

Aromatics		Chloroalkanes		Esters		
1.06	0.95	1.58	1.12	0.29	0.43	0.06
0.79	0.65	1.45	0.91	0.06	0.51	0.09
0.82	1.15	0.57	0.83	0.44	0.10	0.17
0.89	1.12	1.16	0.43	0.55	0.53	0.17
1.05				0.61	0.34	0.60

Is there a significant difference in the mean sorption rates for the three solvents? Use a *P*-value for your conclusions. Which solvent would you use?

13.7 It has been shown that the fertilizer magnesium ammonium phosphate, $MgNH_4PO_4$, is an effective supplier of the nutrients necessary for plant growth. The compounds supplied by this fertilizer are highly soluble in water, allowing the fertilizer to be applied directly on the soil surface or mixed with the growth substrate during the potting process. A study on the *Effect of Magnesium Ammonium Phosphate on Height of Chrysanthemums* was conducted at George Mason University to determine a possible optimum level of fertilization, based on the enhanced vertical growth response of the chrysanthemums. Forty chrysanthemum seedlings were divided into four groups, each containing 10 plants. Each was planted in a similar pot containing a uniform growth medium. To each group of plants an increasing concentration of $MgNH_4PO_4$, measured in grams per bushel, was added. The four groups of plants were grown under uniform conditions in a greenhouse for a period of four weeks. The treatments and the respective changes in heights, measured in centimeters, are shown next.

Treatment							
50 g/bu		100 g/bu		200 g/bu		400 g/bu	
13.2	12.4	16.0	12.6	7.8	14.4	21.0	14.8
12.8	17.2	14.8	13.0	20.0	15.8	19.1	15.8
13.0	14.0	14.0	23.6	17.0	27.0	18.0	26.0
14.2	21.6	14.0	17.0	19.6	18.0	21.1	22.0
15.0	20.0	22.2	24.4	20.2	23.2	25.0	18.2

Can we conclude at the 0.05 level of significance that different concentrations of $MgNH_4PO_4$ affect the average attained height of chrysanthemums? How much $MgNH_4PO_4$ appears to be best?

13.8 For the data set in Exercise 13.7, use Bartlett's test to check whether the variances are equal. Use $\alpha = 0.05$.

13.9 Use Bartlett's test at the 0.01 level of significance to test for homogeneity of variances in Exercise 13.5 on page 519.

13.10 Use Cochran's test at the 0.01 level of significance to test for homogeneity of variances in Exercise 13.4 on page 519.

13.11 Use Bartlett's test at the 0.05 level of significance to test for homogeneity of variances in Exercise 13.6 on page 519.

13.5 Single-Degree-of-Freedom Comparisons

The analysis of variance in a one-way classification, or a one-factor experiment, as it is often called, merely indicates whether or not the hypothesis of equal treatment means can be rejected. Usually, an experimenter would prefer his or her analysis to probe deeper. For instance, in Example 13.1, by rejecting the null hypothesis we concluded that the means are not all equal, but we still do not know where the differences exist among the aggregates. The engineer might have the feeling *a priori* that aggregates 1 and 2 should have similar absorption properties and that the same is true for aggregates 3 and 5. However, it is of interest to study the difference between the two groups. It would seem, then, appropriate to test the hypothesis

$$H_0: \mu_1 + \mu_2 - \mu_3 - \mu_5 = 0,$$
$$H_1: \mu_1 + \mu_2 - \mu_3 - \mu_5 \neq 0.$$

We notice that the hypothesis is a linear function of the population means where the coefficients sum to zero.

Definition 13.1: Any linear function of the form

$$\omega = \sum_{i=1}^{k} c_i \mu_i,$$

where $\sum_{i=1}^{k} c_i = 0$, is called a **comparison** or **contrast** in the treatment means.

The experimenter can often make multiple comparisons by testing the significance of contrasts in the treatment means, that is, by testing a hypothesis of the following type:

13.5 Single-Degree-of-Freedom Comparisons

Hypothesis for a Contrast

$$H_0: \sum_{i=1}^{k} c_i \mu_i = 0,$$

$$H_1: \sum_{i=1}^{k} c_i \mu_i \neq 0,$$

where $\sum_{i=1}^{k} c_i = 0$.

The test is conducted by first computing a similar contrast in the sample means,

$$w = \sum_{i=1}^{k} c_i \bar{y}_{i.}.$$

Since $\bar{Y}_{1.}, \bar{Y}_{2.}, \ldots, \bar{Y}_{k.}$ are independent random variables having normal distributions with means $\mu_1, \mu_2, \ldots, \mu_k$ and variances $\sigma_1^2/n_1, \sigma_2^2/n_2, \ldots, \sigma_k^2/n_k$, respectively, Theorem 7.11 assures us that w is a value of the normal random variable W with

$$\text{mean } \mu_W = \sum_{i=1}^{k} c_i \mu_i \text{ and variance } \sigma_W^2 = \sigma^2 \sum_{i=1}^{k} \frac{c_i^2}{n_i}.$$

Therefore, when H_0 is true, $\mu_W = 0$ and, by Example 7.5, the statistic

$$\frac{W^2}{\sigma_W^2} = \frac{\left(\sum_{i=1}^{k} c_i \bar{Y}_{i.}\right)^2}{\sigma^2 \sum_{i=1}^{k} (c_i^2/n_i)}$$

is distributed as a chi-squared random variable with 1 degree of freedom.

Test Statistic for Testing a Contrast

Our hypothesis is tested at the α-level of significance by computing

$$f = \frac{\left(\sum_{i=1}^{k} c_i \bar{y}_{i.}\right)^2}{s^2 \sum_{i=1}^{k} (c_i^2/n_i)} = \frac{\left[\sum_{i=1}^{k} (c_i Y_{i.}/n_i)\right]^2}{s^2 \sum_{i=1}^{k} (c_i^2/n_i)} = \frac{SSw}{s^2}.$$

Here f is a value of the random variable F having the F-distribution with 1 and $N - k$ degrees of freedom.

When the sample sizes are all equal to n,

$$SSw = \frac{\left(\sum_{i=1}^{k} c_i Y_{i.}\right)^2}{n \sum_{i=1}^{k} c_i^2}.$$

The quantity SSw, called the **contrast sum of squares**, indicates the portion of SSA that is explained by the contrast in question.

This sum of squares will be used to test the hypothesis that

$$\sum_{i=1}^{k} c_i \mu_i = 0.$$

It is often of interest to test multiple contrasts, particularly contrasts that are linearly independent or orthogonal. As a result, we need the following definition:

Definition 13.2: The two contrasts

$$\omega_1 = \sum_{i=1}^{k} b_i \mu_i \quad \text{and} \quad \omega_2 = \sum_{i=1}^{k} c_i \mu_i$$

are said to be **orthogonal** if $\sum_{i=1}^{k} b_i c_i / n_i = 0$ or, when the n_i are all equal to n, if

$$\sum_{i=1}^{k} b_i c_i = 0.$$

If ω_1 and ω_2 are orthogonal, then the quantities SSw_1 and SSw_2 are components of SSA, each with a single degree of freedom. The treatment sum of squares with $k-1$ degrees of freedom can be partitioned into at most $k-1$ independent single-degree-of-freedom contrast sums of squares satisfying the identity

$$SSA = SSw_1 + SSw_2 + \cdots + SSw_{k-1},$$

if the contrasts are orthogonal to each other.

Example 13.4: Referring to Example 13.1, find the contrast sum of squares corresponding to the orthogonal contrasts

$$\omega_1 = \mu_1 + \mu_2 - \mu_3 - \mu_5, \quad \omega_2 = \mu_1 + \mu_2 + \mu_3 - 4\mu_4 + \mu_5,$$

and carry out appropriate tests of significance. In this case, it is of interest a priori to compare the two groups $(1,2)$ and $(3,5)$. An important and independent contrast is the comparison between the set of aggregates $(1,2,3,5)$ and aggregate 4.

Solution: It is obvious that the two contrasts are orthogonal, since

$$(1)(1) + (1)(1) + (-1)(1) + (0)(-4) + (-1)(1) = 0.$$

The second contrast indicates a comparison between aggregates $(1, 2, 3,$ and $5)$ and aggregate 4. We can write two additional contrasts orthogonal to the first two, namely

$$\omega_3 = \mu_1 - \mu_2 \quad \text{(aggregate 1 versus aggregate 2)},$$
$$\omega_4 = \mu_3 - \mu_5 \quad \text{(aggregate 3 versus aggregate 5)}.$$

From the data of Table 13.1, we have

$$SSw_1 = \frac{(3320 + 3416 - 3663 - 3664)^2}{6[(1)^2 + (1)^2 + (-1)^2 + (-1)^2]} = 14{,}553,$$

$$SSw_2 = \frac{[3320 + 3416 + 3663 + 3664 - 4(2791)]^2}{6[(1)^2 + (1)^2 + (1)^2 + (1)^2 + (-4)^2]} = 70{,}035.$$

A more extensive analysis-of-variance table is shown in Table 13.5. We note that the two contrast sums of squares account for nearly all the aggregate sum of squares. There is a significant difference between aggregates in their absorption properties, and the contrast ω_1 is marginally significant. However, the f-value of 14.12 for ω_2 is highly significant, and the hypothesis

$$H_0: \mu_1 + \mu_2 + \mu_3 + \mu_5 = 4\mu_4$$

is rejected.

Table 13.5: Analysis of Variance Using Orthogonal Contrasts

Source of Variation	Sum of Squares	Degrees of Freedom	Mean Square	Computed f
Aggregates	85,356	4	21,339	4.30
(1, 2) vs. (3, 5)	14,553	1	14,553	2.93
(1, 2, 3, 5) vs. 4	70,035	1	70,035	14.12
Error	124,021	25	4961	
Total	209,377	29		

Orthogonal contrasts allow the practitioner to partition the treatment variation into independent components. Normally, the experimenter would have certain contrasts that were of interest to him or her. Such was the case in our example, where *a priori* considerations suggested that aggregates (1, 2) and (3, 5) constituted distinct groups with different absorption properties, a postulation that was not strongly supported by the significance test. However, the second comparison supported the conclusion that aggregate 4 seemed to "stand out" from the rest. In this case, the complete partitioning of SSA was not necessary, since two of the four possible independent comparisons accounted for a majority of the variation in treatments.

Figure 13.4 shows a *SAS* GLM procedure that displays a complete set of orthogonal contrasts. Note that the sums of squares for the four contrasts add to the aggregate sum of squares. Also, note that the latter two contrasts (1 versus 2, 3 versus 5) reveal insignificant comparisons.

13.6 Multiple Comparisons

The analysis of variance is a powerful procedure for testing the homogeneity of a set of means. However, if we reject the null hypothesis and accept the stated alternative—that the means are not all equal—we still do not know which of the population means are equal and which are different.

```
                           The GLM Procedure
Dependent Variable: moisture
                            Sum of
Source                 DF   Squares        Mean Square   F Value   Pr > F
Model                   4   85356.4667     21339.1167      4.30    0.0088
Error                  25  124020.3333      4960.8133
Corrected Total        29  209376.8000

        R-Square        Coeff Var       Root MSE      moisture Mean
        0.407669        12.53703         70.43304         561.8000

Source              DF   Type I SS       Mean Square   F Value   Pr > F
aggregate            4   85356.46667     21339.11667     4.30    0.0088

Source              DF   Type III SS     Mean Square   F Value   Pr > F
aggregate            4   85356.46667     21339.11667     4.30    0.0088

Contrast            DF   Contrast SS     Mean Square   F Value   Pr > F
(1,2,3,5) vs. 4      1   70035.00833     70035.00833    14.12    0.0009
(1,2) vs. (3,5)      1   14553.37500     14553.37500     2.93    0.0991
1 vs. 2              1     768.00000       768.00000     0.15    0.6973
3 vs. 5              1       0.08333         0.08333     0.00    0.9968
```

Figure 13.4: A set of orthogonal contrasts

Often it is of interest to make several (perhaps all possible) **paired comparisons** among the treatments. Actually, a paired comparison may be viewed as a simple contrast, namely, a test of

$$H_0: \mu_i - \mu_j = 0,$$
$$H_1: \mu_i - \mu_j \neq 0,$$

for all $i \neq j$. Making all possible paired comparisons among the means can be very beneficial when particular complex contrasts are not known *a priori*. For example, in the aggregate data of Table 13.1, suppose that we wish to test

$$H_0: \mu_1 - \mu_5 = 0,$$
$$H_1: \mu_1 - \mu_5 \neq 0.$$

The test is developed through use of an F, t, or confidence interval approach. Using t, we have

$$t = \frac{\bar{y}_{1.} - \bar{y}_{5.}}{s\sqrt{2/n}},$$

where s is the square root of the mean square error and $n = 6$ is the sample size per treatment. In this case,

$$t = \frac{553.33 - 610.67}{\sqrt{4961}\sqrt{1/3}} = -1.41.$$

Relationship between T and F

In the foregoing, we displayed the use of a pooled t-test along the lines of that discussed in Chapter 10. The pooled estimate was taken from the mean squared error in order to enjoy the degrees of freedom that are pooled across all five samples. In addition, we have tested a contrast. The reader should note that if the t-value is squared, the result is exactly of the same form as the value of f for a test on a contrast, discussed in the preceding section. In fact,

$$f = \frac{(\bar{y}_{1.} - \bar{y}_{5.})^2}{s^2(1/6 + 1/6)} = \frac{(553.33 - 610.67)^2}{4961(1/3)} = 1.988,$$

which, of course, is t^2.

Confidence Interval Approach to a Paired Comparison

It is straightforward to solve the same problem of a paired comparison (or a contrast) using a confidence interval approach. Clearly, if we compute a $100(1-\alpha)\%$ confidence interval on $\mu_1 - \mu_5$, we have

$$\bar{y}_{1.} - \bar{y}_{5.} \pm t_{\alpha/2} s \sqrt{\frac{2}{6}},$$

where $t_{\alpha/2}$ is the upper $100(1-\alpha/2)\%$ point of a t-distribution with 25 degrees of freedom (degrees of freedom coming from s^2). This straightforward connection between hypothesis testing and confidence intervals should be obvious from discussions in Chapters 9 and 10. The test of the simple contrast $\mu_1 - \mu_5$ involves no more than observing whether or not the confidence interval above covers zero. Substituting the numbers, we have as the 95% confidence interval

$$(553.33 - 610.67) \pm 2.060\sqrt{4961}\sqrt{\frac{1}{3}} = -57.34 \pm 83.77.$$

Thus, since the interval covers zero, the contrast is not significant. In other words, we do not find a significant difference between the means of aggregates 1 and 5.

Experiment-wise Error Rate

Serious difficulties occur when the analyst attempts to make many or all possible paired comparisons. For the case of k means, there will be, of course, $r = k(k-1)/2$ possible paired comparisons. Assuming independent comparisons, the **experiment-wise error rate** or **family error rate** (i.e., the probability of false rejection of at least one of the hypotheses) is given by $1 - (1-\alpha)^r$, where α is the selected probability of a type I error for a specific comparison. Clearly, this measure of experiment-wise type I error can be quite large. For example, even if there are only 6 comparisons, say, in the case of 4 means, and $\alpha = 0.05$, the experiment-wise rate is

$$1 - (0.95)^6 \approx 0.26.$$

When many paired comparisons are being tested, there is usually a need to make the effective contrast on a single comparison more conservative. That is, with the confidence interval approach, the confidence intervals would be much wider than the $\pm t_{\alpha/2} s \sqrt{2/n}$ used for the case where only a single comparison is being made.

Tukey's Test

There are several standard methods for making paired comparisons that sustain the credibility of the type I error rate. We shall discuss and illustrate two of them here. The first one, called **Tukey's procedure**, allows formation of simultaneous $100(1-\alpha)\%$ confidence intervals for all paired comparisons. The method is based on the *studentized* range distribution. The appropriate percentile point is a function of α, k, and $v =$ degrees of freedom for s^2. A list of upper percentage points for $\alpha = 0.05$ is shown in Table A.12. The method of paired comparisons by Tukey involves finding a significant difference between means i and j $(i \neq j)$ if $|\bar{y}_{i.} - \bar{y}_{j.}|$ exceeds $q(\alpha, k, v)\sqrt{\frac{s^2}{n}}$.

Tukey's procedure is easily illustrated. Consider a hypothetical example where we have 6 treatments in a one-factor completely randomized design, with 5 observations taken per treatment. Suppose that the mean square error taken from the analysis-of-variance table is $s^2 = 2.45$ (24 degrees of freedom). The sample means are in ascending order:

$\bar{y}_{2.}$	$\bar{y}_{5.}$	$\bar{y}_{1.}$	$\bar{y}_{3.}$	$\bar{y}_{6.}$	$\bar{y}_{4.}$
14.50	16.75	19.84	21.12	22.90	23.20

With $\alpha = 0.05$, the value of $q(0.05, 6, 24)$ is 4.37. Thus, all absolute differences are to be compared to

$$4.37\sqrt{\frac{2.45}{5}} = 3.059.$$

As a result, the following represent means found to be significantly different using Tukey's procedure:

4 and 1, 4 and 5, 4 and 2, 6 and 1, 6 and 5,
6 and 2, 3 and 5, 3 and 2, 1 and 5, 1 and 2.

Where Does the α-Level Come From in Tukey's Test?

We briefly alluded to the concept of **simultaneous confidence intervals** being employed for Tukey's procedure. The reader will gain a useful insight into the notion of multiple comparisons if he or she gains an understanding of what is meant by simultaneous confidence intervals.

In Chapter 9, we saw that if we compute a 95% confidence interval on, say, a mean μ, then the probability that the interval covers the true mean μ is 0.95.

However, as we have discussed, for the case of multiple comparisons, the effective probability of interest is tied to the experiment-wise error rate, and it should be emphasized that the confidence intervals of the type $\bar{y}_{i.} - \bar{y}_{j.} \pm q(\alpha, k, v) s \sqrt{1/n}$ are not independent since they all involve s and many involve the use of the same averages, the $\bar{y}_{i.}$. Despite the difficulties, if we use $q(0.05, k, v)$, the simultaneous confidence level is controlled at 95%. The same holds for $q(0.01, k, v)$; namely, the confidence level is controlled at 99%. In the case of $\alpha = 0.05$, there is a probability of 0.05 that at least one pair of measures will be falsely found to be different (false rejection of at least one null hypothesis). In the $\alpha = 0.01$ case, the corresponding probability will be 0.01.

Duncan's Test

The second procedure we shall discuss is called **Duncan's procedure** or **Duncan's multiple-range test**. This procedure is also based on the general notion of studentized range. The range of any subset of p sample means must exceed a certain value before any of the p means are found to be different. This value is called the **least significant range** for the p means and is denoted by R_p, where

$$R_p = r_p \sqrt{\frac{s^2}{n}}.$$

The values of the quantity r_p, called the **least significant studentized range**, depend on the desired level of significance and the number of degrees of freedom of the mean square error. These values may be obtained from Table A.13 for $p = 2, 3, \ldots, 10$ means.

To illustrate the multiple-range test procedure, let us consider the hypothetical example where 6 treatments are compared, with 5 observations per treatment. This is the same example used to illustrate Tukey's test. We obtain R_p by multiplying each r_p by 0.70. The results of these computations are summarized as follows:

p	2	3	4	5	6
r_p	2.919	3.066	3.160	3.226	3.276
R_p	2.043	2.146	2.212	2.258	2.293

Comparing these least significant ranges with the differences in ordered means, we arrive at the following conclusions:

1. Since $\bar{y}_{4.} - \bar{y}_{2.} = 8.70 > R_6 = 2.293$, we conclude that μ_4 and μ_2 are significantly different.

2. Comparing $\bar{y}_{4.} - \bar{y}_{5.}$ and $\bar{y}_{6.} - \bar{y}_{2.}$ with R_5, we conclude that μ_4 is significantly greater than μ_5 and μ_6 is significantly greater than μ_2.

3. Comparing $\bar{y}_{4.} - \bar{y}_{1.}$, $\bar{y}_{6.} - \bar{y}_{5.}$, and $\bar{y}_{3.} - \bar{y}_{2.}$ with R_4, we conclude that each difference is significant.

4. Comparing $\bar{y}_{4.} - \bar{y}_{3.}$, $\bar{y}_{6.} - \bar{y}_{1.}$, $\bar{y}_{3.} - \bar{y}_{5.}$, and $\bar{y}_{1.} - \bar{y}_{2.}$ with R_3, we find all differences significant except for $\mu_4 - \mu_3$. Therefore, μ_3, μ_4, and μ_6 constitute a subset of homogeneous means.

5. Comparing $\bar{y}_{3.} - \bar{y}_{1.}$, $\bar{y}_{1.} - \bar{y}_{5.}$, and $\bar{y}_{5.} - \bar{y}_{2.}$ with R_2, we conclude that only μ_3 and μ_1 are not significantly different.

It is customary to summarize the conclusions above by drawing a line under any subsets of adjacent means that are not significantly different. Thus, we have

$$\begin{array}{cccccc} \bar{y}_{2.} & \bar{y}_{5.} & \bar{y}_{1.} & \bar{y}_{3.} & \bar{y}_{6.} & \bar{y}_{4.} \\ 14.50 & 16.75 & 19.84 & 21.12 & 22.90 & 23.20 \end{array}$$

It is clear that in this case the results from Tukey's and Duncan's procedures are very similar. Tukey's procedure did not detect a difference between 2 and 5, whereas Duncan's did.

Dunnett's Test: Comparing Treatment with a Control

In many scientific and engineering problems, one is not interested in drawing inferences regarding all possible comparisons among the treatment means of the type $\mu_i - \mu_j$. Rather, the experiment often dictates the need to simultaneously compare each *treatment* with a *control*. A test procedure developed by C. W. Dunnett determines significant differences between each treatment mean and the control, at a single joint significance level α. To illustrate Dunnett's procedure, let us consider the experimental data of Table 13.6 for a one-way classification where the effect of three catalysts on the yield of a reaction is being studied. A fourth treatment, no catalyst, is used as a control.

Table 13.6: Yield of Reaction

Control	Catalyst 1	Catalyst 2	Catalyst 3
50.7	54.1	52.7	51.2
51.5	53.8	53.9	50.8
49.2	53.1	57.0	49.7
53.1	52.5	54.1	48.0
52.7	54.0	52.5	47.2
$\bar{y}_{0.} = 51.44$	$\bar{y}_{1.} = 53.50$	$\bar{y}_{2.} = 54.04$	$\bar{y}_{3.} = 49.38$

In general, we wish to test the k hypotheses

$$\left. \begin{array}{l} H_0\colon \mu_0 = \mu_i \\ H_1\colon \mu_0 \neq \mu_i \end{array} \right\} \quad i = 1, 2, \ldots, k,$$

where μ_0 represents the mean yield for the population of measurements in which the control is used. The usual analysis-of-variance assumptions, as outlined in Section 13.3, are expected to remain valid. To test the null hypotheses specified by H_0 against two-sided alternatives for an experimental situation in which there are k treatments, excluding the control, and n observations per treatment, we first calculate the values

$$d_i = \frac{\bar{y}_{i.} - \bar{y}_{0.}}{\sqrt{2s^2/n}}, \quad i = 1, 2, \ldots, k.$$

The sample variance s^2 is obtained, as before, from the mean square error in the analysis of variance. Now, the critical region for rejecting H_0, at the α-level of significance, is established by the inequality

$$|d_i| > d_{\alpha/2}(k, v),$$

where v is the number of degrees of freedom for the mean square error. The values of the quantity $d_{\alpha/2}(k, v)$ for a two-tailed test are given in Table A.14 for $\alpha = 0.05$ and $\alpha = 0.01$ for various values of k and v.

Example 13.5: For the data of Table 13.6, test hypotheses comparing each catalyst with the control, using two-sided alternatives. Choose $\alpha = 0.05$ as the joint significance level.

Solution: The mean square error with 16 degrees of freedom is obtained from the analysis-of-variance table, using all $k + 1$ treatments. The mean square error is given by

$$s^2 = \frac{36.812}{16} = 2.30075 \text{ and } \sqrt{\frac{2s^2}{n}} = \sqrt{\frac{(2)(2.30075)}{5}} = 0.9593.$$

Hence,

$$d_1 = \frac{53.50 - 51.44}{0.9593} = 2.147, \quad d_2 = \frac{54.04 - 51.44}{0.9593} = 2.710,$$

$$d_3 = \frac{49.38 - 51.44}{0.9593} = -2.147.$$

From Table A.14 the critical value for $\alpha = 0.05$ is found to be $d_{0.025}(3, 16) = 2.59$. Since $|d_1| < 2.59$ and $|d_3| < 2.59$, we conclude that only the mean yield for catalyst 2 is significantly different from the mean yield of the reaction using the control.

Many practical applications dictate the need for a one-tailed test for comparing treatments with a control. Certainly, when a pharmacologist is concerned with the effect of various dosages of a drug on cholesterol level and his control is zero dosage, it is of interest to determine if each dosage produces a significantly larger reduction than the control. Table A.15 shows the critical values of $d_\alpha(k, v)$ for one-sided alternatives.

Exercises

13.12 Consider the data of Review Exercise 13.45 on page 555. Make significance tests on the following contrasts:

(a) B versus A, C, and D;

(b) C versus A and D;

(c) A versus D.

13.13 The purpose of the study *The Incorporation of a Chelating Agent into a Flame Retardant Finish of a Cotton Flannelette and the Evaluation of Selected Fabric Properties* conducted at Virginia Tech was to evaluate the use of a chelating agent as part of the flame-retardant finish of cotton flannelette by determining its effects upon flammability after the fabric is laundered under specific conditions. Two baths were prepared, one with carboxymethyl cellulose and one without. Twelve pieces of fabric were laundered 5 times in bath I, and 12 other pieces of fabric were laundered 10 times in bath I. This was repeated using 24 additional pieces of cloth in bath II. After the washings the lengths of fabric that burned and the burn times were measured. For convenience, let us define the following treatments:

Treatment 1: 5 launderings in bath I,

Treatment 2: 5 launderings in bath II,

Treatment 3: 10 launderings in bath I,

Treatment 4: 10 launderings in bath II.

Burn times, in seconds, were recorded as follows:

	Treatment		
1	2	3	4
13.7	6.2	27.2	18.2
23.0	5.4	16.8	8.8
15.7	5.0	12.9	14.5
25.5	4.4	14.9	14.7
15.8	5.0	17.1	17.1
14.8	3.3	13.0	13.9
14.0	16.0	10.8	10.6
29.4	2.5	13.5	5.8
9.7	1.6	25.5	7.3
14.0	3.9	14.2	17.7
12.3	2.5	27.4	18.3
12.3	7.1	11.5	9.9

(a) Perform an analysis of variance, using a 0.01 level of significance, and determine whether there are any significant differences among the treatment means.

(b) Use single-degree-of-freedom contrasts with $\alpha = 0.01$ to compare the mean burn time of treatment 1 versus treatment 2 and also treatment 3 versus treatment 4.

13.14 The study *Loss of Nitrogen Through Sweat by Preadolescent Boys Consuming Three Levels of Dietary Protein* was conducted by the Department of Human Nutrition and Foods at Virginia Tech to determine perspiration nitrogen loss at various dietary protein levels. Twelve preadolescent boys ranging in age from 7 years, 8 months to 9 years, 8 months, all judged to be clinically healthy, were used in the experiment. Each boy was subjected to one of three controlled diets in which 29, 54, or 84 grams of protein were consumed per day. The following data represent the body perspiration nitrogen loss, in milligrams, during the last two days of the experimental period:

	Protein Level	
29 Grams	54 Grams	84 Grams
190	318	390
266	295	321
270	271	396
	438	399
	402	

(a) Perform an analysis of variance at the 0.05 level of significance to show that the mean perspiration nitrogen losses at the three protein levels are different.

(b) Use Tukey's test to determine which protein levels are significantly different from each other in mean nitrogen loss.

13.15 Use Tukey's test, with a 0.05 level of significance, to analyze the means of the five different brands of headache tablets in Exercise 13.2 on page 518.

13.16 An investigation was conducted to determine the source of reduction in yield of a certain chemical product. It was known that the loss in yield occurred in the mother liquor, that is, the material removed at the filtration stage. It was thought that different blends of the original material might result in different yield reductions at the mother liquor stage. The following are the percent reductions for 3 batches at each of 4 preselected blends:

	Blend		
1	2	3	4
25.6	25.2	20.8	31.6
24.3	28.6	26.7	29.8
27.9	24.7	22.2	34.3

(a) Perform the analysis of variance at the $\alpha = 0.05$ level of significance.

(b) Use Duncan's multiple-range test to determine which blends differ.

(c) Do part (b) using Tukey's test.

13.17 In the study *An Evaluation of the Removal Method for Estimating Benthic Populations and Diversity* conducted by Virginia Tech on the Jackson River, 5 different sampling procedures were used to determine the species counts. Twenty samples were selected at random, and each of the 5 sampling procedures was repeated 4 times. The species counts were recorded as follows:

		Sampling Procedure		
			Substrate	
Deple-tion	Modified Hess	Surber	Removal Kicknet	Kick-net
85	75	31	43	17
55	45	20	21	10
40	35	9	15	8
77	67	37	27	15

(a) Is there a significant difference in the average species counts for the different sampling procedures? Use a P-value in your conclusion.

(b) Use Tukey's test with $\alpha = 0.05$ to find which sampling procedures differ.

13.18 The following data are values of pressure (psi) in a torsion spring for several settings of the angle between the legs of the spring in a free position:

		Angle (°)		
67	71	75	79	83
83	84	86 87	89	90
85	85	87 87	90	92
	85	88 88	90	
	86	88 88	91	
	86	88 89		
	87	90		

Compute a one-way analysis of variance for this experiment and state your conclusion concerning the effect of angle on the pressure in the spring. (From C. R. Hicks, *Fundamental Concepts in the Design of Experiments*, Holt, Rinehart and Winston, New York, 1973.)

13.19 It is suspected that the environmental temperature at which batteries are activated affects their life. Thirty homogeneous batteries were tested, six at each of five temperatures, and the data are shown below (activated life in seconds). Analyze and interpret the data. (From C. R. Hicks, *Fundamental Concepts in Design of Experiments*, Holt, Rinehart and Winston, New York, 1973.)

\multicolumn{5}{c}{Temperature (°C)}
0	25	50	75	100
55	60	70	72	65
55	61	72	72	66
57	60	72	72	60
54	60	68	70	64
54	60	77	68	65
56	60	77	69	65

13.20 The following table (from A. Hald, *Statistical Theory with Engineering Applications*, John Wiley & Sons, New York, 1952) gives tensile strengths (in deviations from 340) for wires taken from nine cables to be used for a high-voltage network. Each cable is made from 12 wires. We want to know whether the mean strengths of the wires in the nine cables are the same. If the cables are different, which ones differ? Use a P-value in your analysis of variance.

Cable	Tensile Strength
1	5 −13 −5 −2 −10 −6 −5 0 −3 2 −7 −5
2	−11 −13 −8 8 −3 −12 −12 −10 5 −6 −12 −10
3	0 −10 −15 −12 −2 −8 −5 0 −4 −1 −5 −11
4	−12 4 2 10 −5 −8 −12 0 −5 −3 −3 0
5	7 1 5 0 10 6 5 2 0 −1 −10 −2
6	1 0 −5 −4 −1 0 2 5 1 −2 6 7
7	−1 0 2 1 −4 2 7 5 1 0 −4 2
8	−1 0 7 5 10 8 1 2 −3 6 0 5
9	2 6 7 8 15 11 −7 7 10 7 8 1

13.21 The printout in Figure 13.5 on page 532 gives information on Duncan's test, using PROC GLM in SAS, for the aggregate data in Example 13.1. Give conclusions regarding paired comparisons using Duncan's test results.

13.22 Do Duncan's test for paired comparisons for the data of Exercise 13.6 on page 519. Discuss the results.

13.23 In a biological experiment, four concentrations of a certain chemical are used to enhance the growth of a certain type of plant over time. Five plants are used at each concentration, and the growth in each plant is measured in centimeters. The following growth data are taken. A control (no chemical) is also applied.

	\multicolumn{4}{c}{Concentration}			
Control	1	2	3	4
6.8	8.2	7.7	6.9	5.9
7.3	8.7	8.4	5.8	6.1
6.3	9.4	8.6	7.2	6.9
6.9	9.2	8.1	6.8	5.7
7.1	8.6	8.0	7.4	6.1

Use Dunnett's two-sided test at the 0.05 level of significance to simultaneously compare the concentrations with the control.

13.24 The financial structure of a firm refers to the way the firm's assets are divided into equity and debt, and the financial leverage refers to the percentage of assets financed by debt. In the paper *The Effect of Financial Leverage on Return*, Tai Ma of Virginia Tech claims that financial leverage can be used to increase the rate of return on equity. To say it another way, stockholders can receive higher returns on equity with the same amount of investment through the use of financial leverage. The following data show the rates of return on equity using 3 different levels of financial leverage and a control level (zero debt) for 24 randomly selected firms:

	\multicolumn{3}{c}{Financial Leverage}		
Control	Low	Medium	High
2.1	6.2	9.6	10.3
5.6	4.0	8.0	6.9
3.0	8.4	5.5	7.8
7.8	2.8	12.6	5.8
5.2	4.2	7.0	7.2
2.6	5.0	7.8	12.0

Source: Standard & Poor's *Machinery Industry Survey*, 1975.

(a) Perform the analysis of variance at the 0.05 level of significance.

(b) Use Dunnett's test at the 0.01 level of significance to determine whether the mean rates of return on equity are higher at the low, medium, and high levels of financial leverage than at the control level.

```
                    The GLM Procedure
            Duncan's Multiple Range Test for moisture
NOTE: This test controls the Type I comparisonwise error rate,
      not the experimentwise error rate.
            Alpha                              0.05
            Error Degrees of Freedom             25
            Error Mean Square              4960.813
  Number of Means          2           3          4          5
  Critical Range        83.75       87.97      90.69      92.61

Means with the same letter are not significantly different.
      Duncan Grouping          Mean        N     aggregate
                    A         610.67       6     5
                    A
                    A         610.50       6     3
                    A
                    A         569.33       6     2
                    A
                    A         553.33       6     1

                    B         465.17       6     4
```

Figure 13.5: *SAS* printout for Exercise 13.21.

13.7 Comparing a Set of Treatments in Blocks

In Section 13.2, we discussed the idea of blocking, that is, isolating sets of experimental units that are reasonably homogeneous and randomly assigning treatments to these units. This is an extension of the "pairing" concept discussed in Chapters 9 and 10, and it is done to reduce experimental error, since the units in a block have more common characteristics than units in different blocks.

The reader should not view blocks as a second factor, although this is a tempting way of visualizing the design. In fact, the main factor (treatments) still carries the major thrust of the experiment. Experimental units are still the source of error, just as in the completely randomized design. We merely treat sets of these units more systematically when blocking is accomplished. In this way, we say there are restrictions in randomization. Before we turn to a discussion of blocking, let us look at two examples of a **completely randomized design**. The first example is a chemical experiment designed to determine if there is a difference in mean reaction yield among four catalysts. Samples of materials to be tested are drawn from the same batches of raw materials, while other conditions, such as temperature and concentration of reactants, are held constant. In this case, the time of day for the experimental runs might represent the experimental units, and if the experimenter believed that there could possibly be a slight time effect, he or she would randomize the assignment of the catalysts to the runs to counteract the possible trend. As a second example of such a design, consider an experiment to compare four methods of measuring a particular physical property of a fluid substance. Suppose the sampling process is destructive; that is, once a sample of the substance has been measured by one method, it cannot be measured again by any of the other methods. If it is decided that five measurements are to be taken for each method, then 20 samples of the material are selected from a large batch at random and are used in the experiment to compare the four measuring methods. The experimental units are the randomly selected samples. Any variation from sample to sample will appear in the error variation, as measured by s^2 in the analysis.

What Is the Purpose of Blocking?

If the variation due to heterogeneity in experimental units is so large that the sensitivity with which treatment differences are detected is reduced due to an inflated value of s^2, a better plan might be to "block off" variation due to these units and thus reduce the extraneous variation to that accounted for by smaller or more homogeneous blocks. For example, suppose that in the previous catalyst illustration it is known *a priori* that there definitely is a significant day-to-day effect on the yield and that we can measure the yield for four catalysts on a given day. Rather than assign the four catalysts to the 20 test runs completely at random, we choose, say, five days and run each of the four catalysts on each day, randomly assigning the catalysts to the runs within days. In this way, the day-to-day variation is removed from the analysis, and consequently the experimental error, which still includes any time trend *within days*, more accurately represents chance variation. Each day is referred to as a **block**.

The most straightforward of the randomized block designs is one in which we randomly assign each treatment once to every block. Such an experimental layout is called a **randomized complete block (RCB) design**, each block constituting a single replication of the treatments.

13.8 Randomized Complete Block Designs

A typical layout for the randomized complete block design using 3 measurements in 4 blocks is as follows:

Block 1	Block 2	Block 3	Block 4
t_2	t_1	t_3	t_2
t_1	t_3	t_2	t_1
t_3	t_2	t_1	t_3

The t's denote the assignment to blocks of each of the 3 treatments. Of course, the true allocation of treatments to units within blocks is done at random. Once the experiment has been completed, the data can be recorded in the following 3×4 array:

Treatment	Block:	1	2	3	4
1		y_{11}	y_{12}	y_{13}	y_{14}
2		y_{21}	y_{22}	y_{23}	y_{24}
3		y_{31}	y_{32}	y_{33}	y_{34}

where y_{11} represents the response obtained by using treatment 1 in block l, y_{12} represents the response obtained by using treatment 1 in block 2, ..., and y_{34} represents the response obtained by using treatment 3 in block 4.

Let us now generalize and consider the case of k treatments assigned to b blocks. The data may be summarized as shown in the $k \times b$ rectangular array of Table 13.7. It will be assumed that the y_{ij}, $i = 1, 2, \ldots, k$ and $j = 1, 2, \ldots, b$, are values of independent random variables having normal distributions with mean μ_{ij} and common variance σ^2.

Table 13.7: $k \times b$ Array for the RCB Design

Treatment	Block					Total	Mean
	1	2	$\cdots$	j	$\cdots$ b		
1	y_{11}	y_{12}	$\cdots$	y_{1j}	$\cdots$ y_{1b}	$T_{1.}$	$\bar{y}_{1.}$
2	y_{21}	y_{22}	$\cdots$	y_{2j}	$\cdots$ y_{2b}	$T_{2.}$	$\bar{y}_{2.}$
$\vdots$	$\vdots$	$\vdots$		$\vdots$	$\vdots$	$\vdots$	$\vdots$
i	y_{i1}	y_{i2}	$\cdots$	y_{ij}	$\cdots$ y_{ib}	$T_{i.}$	$\bar{y}_{i.}$
$\vdots$	$\vdots$	$\vdots$		$\vdots$	$\vdots$	$\vdots$	$\vdots$
k	y_{k1}	y_{k2}	$\cdots$	y_{kj}	$\cdots$ y_{kb}	$T_{k.}$	$\bar{y}_{k.}$
Total	$T_{.1}$	$T_{.2}$	$\cdots$	$T_{.j}$	$\cdots$ $T_{.b}$	$T_{..}$	
Mean	$\bar{y}_{.1}$	$\bar{y}_{.2}$	$\cdots$	$\bar{y}_{.j}$	$\cdots$ $\bar{y}_{.b}$		$\bar{y}_{..}$

Let $\mu_{i.}$ represent the average (rather than the total) of the b population means for the ith treatment. That is,

$$\mu_{i.} = \frac{1}{b} \sum_{j=1}^{b} \mu_{ij}, \text{ for } i = 1, \ldots, k.$$

Similarly, the average of the population means for the jth block, $\mu_{.j}$, is defined by

$$\mu_{.j} = \frac{1}{k} \sum_{i=1}^{k} \mu_{ij}, \text{ for } j = 1, \ldots, b$$

and the average of the bk population means, μ, is defined by

$$\mu = \frac{1}{bk} \sum_{i=1}^{k} \sum_{j=1}^{b} \mu_{ij}.$$

To determine if part of the variation in our observations is due to differences among the treatments, we consider the following test:

Hypothesis of Equal Treatment Means	H_0: $\mu_{1.} = \mu_{2.} = \cdots \mu_{k.} = \mu,$ H_1: The $\mu_{i.}$ are not all equal.

Model for the RCB Design

Each observation may be written in the form

$$y_{ij} = \mu_{ij} + \epsilon_{ij},$$

where ϵ_{ij} measures the deviation of the observed value y_{ij} from the population mean μ_{ij}. The preferred form of this equation is obtained by substituting

$$\mu_{ij} = \mu + \alpha_i + \beta_j,$$

where α_i is, as before, the effect of the ith treatment and β_j is the effect of the jth block. It is assumed that the treatment and block effects are additive. Hence, we may write

$$y_{ij} = \mu + \alpha_i + \beta_j + \epsilon_{ij}.$$

Notice that the model resembles that of the one-way classification, the essential difference being the introduction of the block effect β_j. The basic concept is much like that of the one-way classification except that we must account in the analysis for the additional effect due to blocks, since we are now systematically controlling variation *in two directions*. If we now impose the restrictions that

$$\sum_{i=1}^{k} \alpha_i = 0 \quad \text{and} \quad \sum_{j=1}^{b} \beta_j = 0,$$

then

$$\mu_{i.} = \frac{1}{b} \sum_{j=1}^{b} (\mu + \alpha_i + \beta_j) = \mu + \alpha_i, \text{ for } i = 1, \ldots, k,$$

and

$$\mu_{.j} = \frac{1}{k} \sum_{i=1}^{k} (\mu + \alpha_i + \beta_j) = \mu + \beta_j, \text{ for } j = 1, \ldots, b.$$

The null hypothesis that the k treatment means $\mu_{i.}$ are equal, and therefore equal to μ, is now **equivalent to testing the hypothesis**

H_0: $\alpha_1 = \alpha_2 = \cdots = \alpha_k = 0,$
H_1: At least one of the α_i is not equal to zero.

Each of the tests on treatments will be based on a comparison of independent estimates of the common population variance σ^2. These estimates will be obtained by splitting the total sum of squares of our data into three components by means of the following identity.

Theorem 13.3: Sum-of-Squares Identity

$$\sum_{i=1}^{k}\sum_{j=1}^{b}(y_{ij}-\bar{y}_{..})^2 = b\sum_{i=1}^{k}(\bar{y}_{i.}-\bar{y}_{..})^2 + k\sum_{j=1}^{b}(\bar{y}_{.j}-\bar{y}_{..})^2$$
$$+ \sum_{i=1}^{k}\sum_{j=1}^{b}(y_{ij}-\bar{y}_{i.}-\bar{y}_{.j}+\bar{y}_{..})^2$$

The proof is left to the reader.

The sum-of-squares identity may be presented symbolically by the equation

$$SST = SSA + SSB + SSE,$$

where

$$SST = \sum_{i=1}^{k}\sum_{j=1}^{b}(y_{ij}-\bar{y}_{..})^2 \qquad = \text{total sum of squares,}$$

$$SSA = b\sum_{i=1}^{k}(\bar{y}_{i.}-\bar{y}_{..})^2 \qquad = \text{treatment sum of squares,}$$

$$SSB = k\sum_{j=1}^{b}(\bar{y}_{.j}-\bar{y}_{..})^2 \qquad = \text{block sum of squares,}$$

$$SSE = \sum_{i=1}^{k}\sum_{j=1}^{b}(y_{ij}-\bar{y}_{i.}-\bar{y}_{.j}+\bar{y}_{..})^2 = \text{error sum of squares.}$$

Following the procedure outlined in Theorem 13.2, where we interpreted the sums of squares as functions of the independent random variables $Y_{11}, Y_{12}, \ldots, Y_{kb}$, we can show that the expected values of the treatment, block, and error sums of squares are given by

$$E(SSA) = (k-1)\sigma^2 + b\sum_{i=1}^{k}\alpha_i^2, \quad E(SSB) = (b-1)\sigma^2 + k\sum_{j=1}^{b}\beta_j^2,$$
$$E(SSE) = (b-1)(k-1)\sigma^2.$$

As in the case of the one-factor problem, we have the treatment mean square

$$s_1^2 = \frac{SSA}{k-1}.$$

If the treatment effects $\alpha_1 = \alpha_2 = \cdots = \alpha_k = 0$, s_1^2 is an unbiased estimate of σ^2. However, if the treatment effects are not all zero, we have the following:

Expected Treatment Mean Square

$$E\left(\frac{SSA}{k-1}\right) = \sigma^2 + \frac{b}{k-1}\sum_{i=1}^{k}\alpha_i^2$$

In this case, s_1^2 overestimates σ^2. A second estimate of σ^2, based on $b-1$ degrees of freedom, is

$$s_2^2 = \frac{SSB}{b-1}.$$

The estimate s_2^2 is an unbiased estimate of σ^2 if the block effects $\beta_1 = \beta_2 = \cdots = \beta_b = 0$. If the block effects are not all zero, then

$$E\left(\frac{SSB}{b-1}\right) = \sigma^2 + \frac{k}{b-1}\sum_{j=1}^{b}\beta_j^2,$$

and s_2^2 will overestimate σ^2. A third estimate of σ^2, based on $(k-1)(b-1)$ degrees of freedom and independent of s_1^2 and s_2^2, is

$$s^2 = \frac{SSE}{(k-1)(b-1)},$$

which is unbiased regardless of the truth or falsity of either null hypothesis.

To test the null hypothesis that the treatment effects are all equal to zero, we compute the ratio $f_1 = s_1^2/s^2$, which is a value of the random variable F_1 having an F-distribution with $k-1$ and $(k-1)(b-1)$ degrees of freedom when the null hypothesis is true. The null hypothesis is rejected at the α-level of significance when

$$f_1 > f_\alpha[k-1, (k-1)(b-1)].$$

In practice, we first compute SST, SSA, and SSB and then, using the sum-of-squares identity, obtain SSE by subtraction. The degrees of freedom associated with SSE are also usually obtained by subtraction; that is,

$$(k-1)(b-1) = kb - 1 - (k-1) - (b-1).$$

The computations in an analysis-of-variance problem for a randomized complete block design may be summarized as shown in Table 13.8.

Example 13.6: Four different machines, M_1, M_2, M_3, and M_4, are being considered for the assembling of a particular product. It was decided that six different operators would be used in a randomized block experiment to compare the machines. The machines were assigned in a random order to each operator. The operation of the machines requires physical dexterity, and it was anticipated that there would be a difference among the operators in the speed with which they operated the machines. The amounts of time (in seconds) required to assemble the product are shown in Table 13.9.

Test the hypothesis H_0, at the 0.05 level of significance, that the machines perform at the same mean rate of speed.

Table 13.8: Analysis of Variance for the Randomized Complete Block Design

Source of Variation	Sum of Squares	Degrees of Freedom	Mean Square	Computed f
Treatments	SSA	$k-1$	$s_1^2 = \dfrac{SSA}{k-1}$	$f_1 = \dfrac{s_1^2}{s^2}$
Blocks	SSB	$b-1$	$s_2^2 = \dfrac{SSB}{b-1}$	
Error	SSE	$(k-1)(b-1)$	$s^2 = \dfrac{SSE}{(k-1)(b-1)}$	
Total	SST	$kb-1$		

Table 13.9: Time, in Seconds, to Assemble Product

Machine	Operator 1	2	3	4	5	6	Total
1	42.5	39.3	39.6	39.9	42.9	43.6	247.8
2	39.8	40.1	40.5	42.3	42.5	43.1	248.3
3	40.2	40.5	41.3	43.4	44.9	45.1	255.4
4	41.3	42.2	43.5	44.2	45.9	42.3	259.4
Total	163.8	162.1	164.9	169.8	176.2	174.1	1010.9

Solution: The hypotheses are

$H_0: \alpha_1 = \alpha_2 = \alpha_3 = \alpha_4 = 0$ (machine effects are zero),

$H_1:$ At least one of the α_i is not equal to zero.

The sum-of-squares formulas shown on page 536 and the degrees of freedom are used to produce the analysis of variance in Table 13.10. The value $f = 3.34$ is significant at $P = 0.048$. If we use $\alpha = 0.05$ as at least an approximate yardstick, we conclude that the machines do not perform at the same mean rate of speed.

Table 13.10: Analysis of Variance for the Data of Table 13.9

Source of Variation	Sum of Squares	Degrees of Freedom	Mean Square	Computed f
Machines	15.93	3	5.31	3.34
Operators	42.09	5	8.42	
Error	23.84	15	1.59	
Total	81.86	23		

Further Comments Concerning Blocking

In Chapter 10, we presented a procedure for comparing means when the observations were *paired*. The procedure involved "subtracting out" the effect due to the homogeneous pair and thus working with differences. This is a special case of a randomized complete block design with $k = 2$ treatments. The n homogeneous units to which the treatments were assigned take on the role of blocks.

If there is heterogeneity in the experimental units, the experimenter should not be misled into believing that it is always advantageous to reduce the experimental error through the use of small homogeneous blocks. Indeed, there may be instances where it would not be desirable to block. The purpose in reducing the error variance is to increase the *sensitivity* of the test for detecting differences in the treatment means. This is reflected in the power of the test procedure. (The power of the analysis-of-variance test procedure is discussed more extensively in Section 13.11.) The power to detect certain differences among the treatment means increases with a decrease in the error variance. However, the power is also affected by the degrees of freedom with which this variance is estimated, and blocking reduces the degrees of freedom that are available from $k(b-1)$ for the one-way classification to $(k-1)(b-1)$. So one could lose power by blocking if there is not a significant reduction in the error variance.

Interaction between Blocks and Treatments

Another important assumption that is implicit in writing the model for a randomized complete block design is that the treatment and block effects are additive. This is equivalent to stating that

$$\mu_{ij} - \mu_{ij'} = \mu_{i'j} - \mu_{i'j'} \quad \text{or} \quad \mu_{ij} - \mu_{i'j} = \mu_{ij'} - \mu_{i'j'},$$

for every value of i, i', j, and j'. That is, the difference between the population means for blocks j and j' is the same for every treatment and the difference between the population means for treatments i and i' is the same for every block. The parallel lines of Figure 13.6(a) illustrate a set of mean responses for which the treatment and block effects are additive, whereas the intersecting lines of Figure 13.6(b) show a situation in which treatment and block effects are said to **interact**. Referring to Example 13.6, if operator 3 is 0.5 second faster on the average than operator 2 when machine 1 is used, then operator 3 will still be 0.5 second faster on the average than operator 2 when machine 2, 3, or 4 is used. In many experiments, the assumption of additivity does not hold and the analysis described in this section leads to erroneous conclusions. Suppose, for instance, that operator 3 is 0.5 second faster on the average than operator 2 when machine 1 is used but is 0.2 second slower on the average than operator 2 when machine 2 is used. The operators and machines are now interacting.

An inspection of Table 13.9 suggests the possible presence of interaction. This apparent interaction may be real or it may be due to experimental error. The analysis of Example 13.6 was based on the assumption that the apparent interaction was due entirely to experimental error. If the total variability of our data was in part due to an interaction effect, this source of variation remained a part of the error sum of squares, **causing the mean square error to overestimate σ^2** and thereby increasing the probability of committing a type II error. We have, in fact, assumed an incorrect model. If we let $(\alpha\beta)_{ij}$ denote the interaction effect of the ith treatment and the jth block, we can write a more appropriate model in the

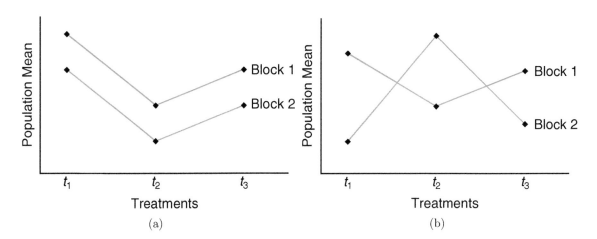

Figure 13.6: Population means for (a) additive results and (b) interacting effects.

form

$$y_{ij} = \mu + \alpha_i + \beta_j + (\alpha\beta)_{ij} + \epsilon_{ij},$$

on which we impose the additional restrictions

$$\sum_{i=1}^{k}(\alpha\beta)_{ij} = \sum_{j=1}^{b}(\alpha\beta)_{ij} = 0, \text{ for } i = 1, \ldots, k \text{ and } j = 1, \ldots, b.$$

We can now readily verify that

$$E\left[\frac{SSE}{(b-1)(k-1)}\right] = \sigma^2 + \frac{1}{(b-1)(k-1)}\sum_{i=1}^{k}\sum_{j=1}^{b}(\alpha\beta)_{ij}^2.$$

Thus, the mean square error is seen to be a **biased estimate of σ^2 when existing interaction has been ignored**. It would seem necessary at this point to arrive at a procedure for the detection of interaction for cases where there is suspicion that it exists. Such a procedure requires the availability of an unbiased and independent estimate of σ^2. Unfortunately, the randomized block design does not lend itself to such a test unless the experimental setup is altered. This subject is discussed extensively in Chapter 14.

13.9 Graphical Methods and Model Checking

In several chapters, we make reference to graphical procedures displaying data and analytical results. In early chapters, we used stem-and-leaf and box-and-whisker plots as visuals to aid in summarizing samples. We used similar diagnostics to better understand the data in two sample problems in Chapter 10. In Chapter 11 we introduced the notion of residual plots to detect violations of standard assumptions. In recent years, much attention in data analysis has centered on **graphical**

13.9 Graphical Methods and Model Checking

methods. Like regression, analysis of variance lends itself to graphics that aid in summarizing data as well as detecting violations. For example, a simple plotting of the raw observations around each treatment mean can give the analyst a feel for variability between sample means and within samples. Figure 13.7 depicts such a plot for the aggregate data of Table 13.1. From the appearance of the plot one may even gain a graphical insight into which aggregates (if any) stand out from the others. It is clear that aggregate 4 stands out from the others. Aggregates 3 and 5 certainly form a homogeneous group, as do aggregates 1 and 2.

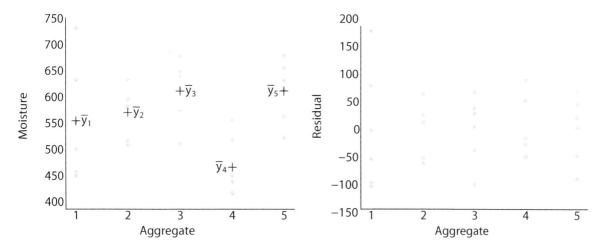

Figure 13.7: Plot of data around the mean for the aggregate data of Table 13.1.

Figure 13.8: Plot of residuals for five aggregates, using data in Table 13.1.

As in the case of regression, residuals can be helpful in analysis of variance in providing a diagnostic that may detect violations of assumptions. To form the residuals, we merely need to consider the model of the one-factor problem, namely

$$y_{ij} = \mu_i + \epsilon_{ij}.$$

It is straightforward to determine that the estimate of μ_i is $\bar{y}_{i.}$. Hence, the ijth residual is $y_{ij} - \bar{y}_{i.}$. This is easily extendable to the randomized complete block model. It may be instructive to have the residuals plotted for each aggregate in order to gain some insight regarding the homogeneous variance assumption. This plot is shown in Figure 13.8.

Trends in plots such as these may reveal difficulties in some situations, particularly when the violation of a particular assumption is graphic. In the case of Figure 13.8, the residuals seem to indicate that the *within-treatment* variances are reasonably homogeneous apart from aggregate 1. There is some graphical evidence that the variance for aggregate 1 is larger than the rest.

What Is a Residual for an RCB Design?

The randomized complete block design is another experimental situation in which graphical displays can make the analyst feel comfortable with an "ideal picture" or perhaps highlight difficulties. Recall that the model for the randomized complete block design is

$$y_{ij} = \mu + \alpha_i + \beta_j + \epsilon_{ij}, \quad i = 1, \ldots, k, \quad j = 1, \ldots, b,$$

with the imposed constraints

$$\sum_{i=1}^{k} \alpha_i = 0, \quad \sum_{j=1}^{b} \beta_j = 0.$$

To determine what indeed constitutes a residual, consider that

$$\alpha_i = \mu_{i.} - \mu, \quad \beta_j = \mu_{.j} - \mu$$

and that μ is estimated by $\bar{y}_{..}$, $\mu_{i.}$ is estimated by $\bar{y}_{i.}$, and $\mu_{.j}$ is estimated by $\bar{y}_{.j}$. As a result, the predicted or *fitted value* $\hat{y}_{ij}$ is given by

$$\hat{y}_{ij} = \hat{\mu} + \hat{\alpha}_i + \hat{\beta}_j = \bar{y}_{i.} + \bar{y}_{.j} - \bar{y}_{..},$$

and thus the residual at the (i, j) observation is given by

$$y_{ij} - \hat{y}_{ij} = y_{ij} - \bar{y}_{i.} - \bar{y}_{.j} + \bar{y}_{..}.$$

Note that $\hat{y}_{ij}$, the fitted value, is an estimate of the mean μ_{ij}. This is consistent with the partitioning of variability given in Theorem 13.3, where the error sum of squares is

$$SSE = \sum_{i}^{k} \sum_{j}^{b} (y_{ij} - \bar{y}_{i.} - \bar{y}_{.j} + \bar{y}_{..})^2.$$

The visual displays in the randomized complete block design involve plotting the residuals separately for each treatment and for each block. The analyst should expect roughly equal variability if the homogeneous variance assumption holds. The reader should recall that in Chapter 12 we discussed plotting residuals for the purpose of detecting model misspecification. In the case of the randomized complete block design, the serious model misspecification may be related to our assumption of additivity (i.e., no interaction). If no interaction is present, a random pattern should appear.

Consider the data of Example 13.6, in which treatments are four machines and blocks are six operators. Figures 13.9 and 13.10 give the residual plots for separate treatments and separate blocks. Figure 13.11 shows a plot of the residuals against the fitted values. Figure 13.9 reveals that the error variance may not be the same for all machines. The same may be true for error variance for each of the six operators. However, two unusually large residuals appear to produce the apparent difficulty. Figure 13.11 is a plot of residuals that shows reasonable evidence of random behavior. However, the two large residuals displayed earlier still stand out.

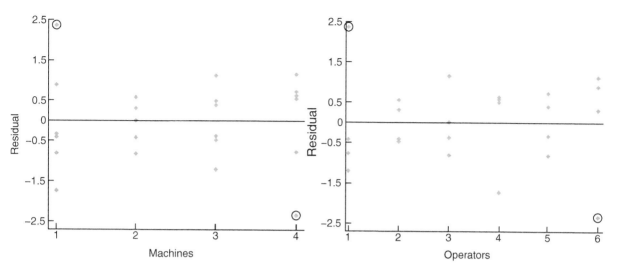

Figure 13.9: Residual plot for the four machines for the data of Example 13.6.

Figure 13.10: Residual plot for the six operators for the data of Example 13.6.

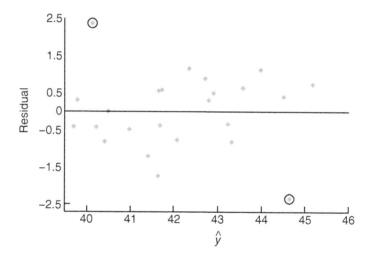

Figure 13.11: Residuals plotted against fitted values for the data of Example 13.6.

13.10 Data Transformations in Analysis of Variance

In Chapter 11, considerable attention was given to transformation of the response y in situations where a linear regression model was being fit to a set of data. Obviously, the same concept applies to multiple linear regression, though it was not discussed in Chapter 12. In the regression modeling discussion, emphasis was placed on the transformations of y that would produce a model that fit the data better than the model in which y enters linearly. For example, if the "time" structure is exponential in nature, then a log transformation on y linearizes the structure and thus more success is anticipated when one uses the transformed response.

While the primary purpose for data transformation discussed thus far has been to improve the fit of the model, there are certainly other reasons to transform or reexpress the response y, and many of them are related to assumptions that are being made (i.e., assumptions on which the validity of the analysis depends). One very important assumption in analysis of variance is the homogeneous variance assumption discussed early in Section 13.4. We assume a **common variance σ^2**. If the variance differs a great deal from treatment to treatment and we perform the standard ANOVA discussed in this chapter (and future chapters), the results can be substantially flawed. In other words, the analysis of variance is not **robust** to the assumption of homogeneous variance. As we have discussed thus far, this is the centerpiece of motivation for the residual plots discussed in the previous section and illustrated in Figures 13.9, 13.10, and 13.11. These plots allow us to detect nonhomogeneous variance problems. However, what do we do about them? How can we accommodate them?

Where Does Nonhomogeneous Variance Come From?

Often, but not always, nonhomogeneous variance in ANOVA is present because of the distribution of the responses. Now, of course we assume normality in the response. But there certainly are situations in which tests on means are needed even though the distribution of the response is one of the nonnormal distributions discussed in Chapters 5 and 6, such as Poisson, lognormal, exponential, or gamma. ANOVA-type problems certainly exist with count data, time to failure data, and so on.

We demonstrated in Chapters 5 and 6 that, apart from the normal case, the variance of a distribution will often be a function of the mean, say $\sigma_i^2 = g(\mu_i)$. For example, in the Poisson case $\text{Var}(Y_i) = \mu_i = \sigma_i^2$ (i.e., the *variance is equal to the mean*). In the case of the exponential distribution, $\text{Var}(Y_i) = \sigma_i^2 = \mu_i^2$ (i.e., the *variance is equal to the square of the mean*). For the case of the lognormal, a log transformation produces a normal distribution with constant variance σ^2.

The same concepts that we used in Chapter 4 to determine the variance of a nonlinear function can be used as an aid to determine the nature of the *variance stabilizing transformation $g(y_i)$*. Recall that the first order Taylor series expansion of $g(y_i)$ around $y_i = \mu_i$ where $g'(\mu_i) = \left[\dfrac{\partial g(y_i)}{\partial y_i}\right]_{y_i = \mu_i}$. The transformation function $g(y)$ must be independent of μ in order to suffice as the variance stabilizing transformation. From the above,

$$\text{Var}[g(y_i)] \approx [g'(\mu_i)]^2 \sigma_i^2.$$

As a result, $g(y_i)$ must be such that $g'(\mu_i) \propto \frac{1}{\sigma}$. Thus, if we suspect that the response is Poisson distributed, $\sigma_i = \mu_i^{1/2}$, so $g'(\mu_i) \propto \frac{1}{\mu_i^{1/2}}$. Thus, the variance stabilizing transformation is $g(y_i) = y_i^{1/2}$. From this illustration and similar manipulation for the exponential and gamma distributions, we have the following.

Distribution	Variance Stabilizing Transformations
Poisson	$g(y) = y^{1/2}$
Exponential	$g(y) = \ln y$
Gamma	$g(y) = \ln y$

Exercises

13.25 Four kinds of fertilizer f_1, f_2, f_3, and f_4 are used to study the yield of beans. The soil is divided into 3 blocks, each containing 4 homogeneous plots. The yields in kilograms per plot and the corresponding treatments are as follows:

Block 1	Block 2	Block 3
$f_1 = 42.7$	$f_3 = 50.9$	$f_4 = 51.1$
$f_3 = 48.5$	$f_1 = 50.0$	$f_2 = 46.3$
$f_4 = 32.8$	$f_2 = 38.0$	$f_1 = 51.9$
$f_2 = 39.3$	$f_4 = 40.2$	$f_3 = 53.5$

Conduct an analysis of variance at the 0.05 level of significance using the randomized complete block model.

13.26 Three varieties of potatoes are being compared for yield. The experiment is conducted by assigning each variety at random to one of 3 equal-size plots at each of 4 different locations. The following yields for varieties A, B, and C, in 100 kilograms per plot, were recorded:

Location 1	Location 2	Location 3	Location 4
B: 13	C: 21	C: 9	A: 11
A: 18	A: 20	B: 12	C: 10
C: 12	B: 23	A: 14	B: 17

Perform a randomized complete block analysis of variance to test the hypothesis that there is no difference in the yielding capabilities of the 3 varieties of potatoes. Use a 0.05 level of significance. Draw conclusions.

13.27 The following data are the percents of foreign additives measured by 5 analysts for 3 similar brands of strawberry jam, A, B, and C:

Analyst 1	Analyst 2	Analyst 3	Analyst 4	Analyst 5
B: 2.7	C: 7.5	B: 2.8	A: 1.7	C: 8.1
C: 3.6	A: 1.6	A: 2.7	B: 1.9	A: 2.0
A: 3.8	B: 5.2	C: 6.4	C: 2.6	B: 4.8

Perform a randomized complete block analysis of variance to test the hypothesis, at the 0.05 level of significance, that the percent of foreign additives is the same for all 3 brands of jam. Which brand of jam appears to have fewer additives?

13.28 The following data represent the final grades obtained by 5 students in mathematics, English, French, and biology:

Student	Subject			
	Math	English	French	Biology
1	68	57	73	61
2	83	94	91	86
3	72	81	63	59
4	55	73	77	66
5	92	68	75	87

Test the hypothesis that the courses are of equal difficulty. Use a P-value in your conclusions and discuss your findings.

13.29 In a study on *The Periphyton of the South River, Virginia: Mercury Concentration, Productivity, and Autotropic Index Studies*, conducted by the Department of Environmental Sciences and Engineering at Virginia Tech, the total mercury concentration in periphyton total solids was measured at 6 different stations on 6 different days. Determine whether the mean mercury content is significantly different between the stations by using the following recorded data. Use a P-value and discuss your findings.

Date	Station					
	CA	CB	E1	E2	E3	E4
April 8	0.45	3.24	1.33	2.04	3.93	5.93
June 23	0.10	0.10	0.99	4.31	9.92	6.49
July 1	0.25	0.25	1.65	3.13	7.39	4.43
July 8	0.09	0.06	0.92	3.66	7.88	6.24
July 15	0.15	0.16	2.17	3.50	8.82	5.39
July 23	0.17	0.39	4.30	2.91	5.50	4.29

13.30 A nuclear power facility produces a vast amount of heat, which is usually discharged into aquatic systems. This heat raises the temperature of the aquatic system, resulting in a greater concentration of chlorophyll a, which in turn extends the growing season. To study this effect, water samples were collected monthly at 3 stations for a period of 12 months. Station A is located closest to a potential heated water discharge, station C is located farthest away from the discharge, and station B is located halfway between stations A and C. The following concentrations of chlorophyll a were recorded.

	Station		
Month	A	B	C
January	9.867	3.723	4.410
February	14.035	8.416	11.100
March	10.700	20.723	4.470
April	13.853	9.168	8.010
May	7.067	4.778	34.080
June	11.670	9.145	8.990
July	7.357	8.463	3.350
August	3.358	4.086	4.500
September	4.210	4.233	6.830
October	3.630	2.320	5.800
November	2.953	3.843	3.480
December	2.640	3.610	3.020

Perform an analysis of variance and test the hypothesis, at the 0.05 level of significance, that there is no difference in the mean concentrations of chlorophyll a at the 3 stations.

13.31 In a study conducted by the Department of Human Nutrition, Foods and Exercise at Virginia Tech, 3 diets were assigned for a period of 3 days to each of 6 subjects in a randomized complete block design. The subjects, playing the role of blocks, were assigned the following 3 diets in a random order:

- Diet 1:   mixed fat and carbohydrates,
- Diet 2:   high fat,
- Diet 3:   high carbohydrates.

At the end of the 3-day period, each subject was put on a treadmill and the time to exhaustion, in seconds, was measured. Perform the analysis of variance, separating out the diet, subject, and error sum of squares. Use a P-value to determine if there are significant differences among the diets, using the following recorded data.

	Subject					
Diet	1	2	3	4	5	6
1	84	35	91	57	56	45
2	91	48	71	45	61	61
3	122	53	110	71	91	122

13.32 Organic arsenicals are used by forestry personnel as silvicides. The amount of arsenic that the body takes in when exposed to these silvicides is a major health problem. It is important that the amount of exposure be determined quickly so that a field worker with a high level of arsenic can be removed from the job. In an experiment reported in the paper "A Rapid Method for the Determination of Arsenic Concentrations in Urine at Field Locations," published in the *American Industrial Hygiene Association Journal* (Vol. 37, 1976), urine specimens from 4 forest service personnel were divided equally into 3 samples each so that each individual's urine could be analyzed for arsenic by a university laboratory, by a chemist using a portable system, and by a forest-service employee after a brief orientation. The following arsenic levels, in parts per million, were recorded:

	Analyst		
Individual	Employee	Chemist	Laboratory
1	0.05	0.05	0.04
2	0.05	0.05	0.04
3	0.04	0.04	0.03
4	0.15	0.17	0.10

Perform an analysis of variance and test the hypothesis, at the 0.05 level of significance, that there is no difference in the arsenic levels for the 3 methods of analysis.

13.33 Scientists in the Department of Plant Pathology, Physiology, and Weed Science at Virginia Tech devised an experiment in which 5 different treatments were applied to 6 different locations in an apple orchard to determine if there were significant differences in growth among the treatments. Treatments 1 through 4 represent different herbicides and treatment 5 represents a control. The growth period was from May to November in 1982, and the amounts of new growth, measured in centimeters, for samples selected from the 6 locations in the orchard were recorded as follows:

	Locations					
Treatment	1	2	3	4	5	6
1	455	72	61	215	695	501
2	622	82	444	170	437	134
3	695	56	50	443	701	373
4	607	650	493	257	490	262
5	388	263	185	103	518	622

Perform an analysis of variance, separating out the treatment, location, and error sum of squares. Determine if there are significant differences among the treatment means. Quote a P-value.

13.34 In the paper "Self-Control and Therapist Control in the Behavioral Treatment of Overweight Women," published in *Behavioral Research and Therapy* (Vol. 10, 1972), two reduction treatments and a control treatment were studied for their effects on the weight change of obese women. The two reduction treatments were a self-induced weight reduction program and a therapist-controlled reduction program. Each of 10 subjects was assigned to one of the 3 treatment programs in a random order and measured for weight loss. The following weight changes were recorded:

	Treatment		
Subject	Control	Self-induced	Therapist
1	1.00	−2.25	−10.50
2	3.75	−6.00	−13.50
3	0.00	−2.00	0.75
4	−0.25	−1.50	−4.50
5	−2.25	−3.25	−6.00
6	−1.00	−1.50	4.00
7	−1.00	−10.75	−12.25
8	3.75	−0.75	−2.75
9	1.50	0.00	−6.75
10	0.50	−3.75	−7.00

Perform an analysis of variance and test the hypothesis, at the 0.01 level of significance, that there is no difference in the mean weight losses for the 3 treatments. Which treatment was best?

13.35 In the book *Design of Experiments for the Quality Improvement*, published by the Japanese Standards Association (1989), a study on the amount of dye needed to get the best color for a certain type of fabric was reported. The three amounts of dye, $\frac{1}{3}$% wof ($\frac{1}{3}$% of the weight of a fabric), 1% wof, and 3% wof, were each administered at two different plants. The color density of the fabric was then observed four times for each level of dye at each plant.

	Amount of Dye					
	1/3%		1%		3%	
Plant 1	5.2	6.0	12.3	10.5	22.4	17.8
	5.9	5.9	12.4	10.9	22.5	18.4
Plant 2	6.5	5.5	14.5	11.8	29.0	23.2
	6.4	5.9	16.0	13.6	29.7	24.0

Perform an analysis of variance to test the hypothesis, at the 0.05 level of significance, that there is no difference in the color density of the fabric for the three levels of dye. Consider plants to be blocks.

13.36 An experiment was conducted to compare three types of coating materials for copper wire. The purpose of the coating is to eliminate "flaws" in the wire. Ten different specimens of length 5 millimeters were randomly assigned to receive each coating, and the thirty specimens were subjected to an abrasive wear type process. The number of flaws was measured for each, and the results are as follows:

	Material								
1				2				3	
6	8	4	5	3	3	5	4	12	8	7	14
7	7	9	6	2	4	4	5	18	6	7	18
7	8			4	3			8	5

Suppose it is assumed that the Poisson process applies and thus the model is $Y_{ij} = \mu_i + \epsilon_{ij}$, where μ_i is the mean of a Poisson distribution and $\sigma^2_{Y_{ij}} = \mu_i$.

(a) Do an appropriate transformation on the data and perform an analysis of variance.
(b) Determine whether or not there is sufficient evidence to choose one coating material over the other. Show whatever findings suggest a conclusion.
(c) Do a plot of the residuals and comment.
(d) Give the purpose of your data transformation.
(e) What additional assumption is made here that may not have been completely satisfied by your transformation?
(f) Comment on (e) after doing a normal probability plot on the residuals.

13.11 Random Effects Models

Throughout this chapter, we deal with analysis-of-variance procedures in which the primary goal is to study the effect on some response of certain fixed or predetermined treatments. Experiments in which the treatments or treatment levels are preselected by the experimenter as opposed to being chosen randomly are called **fixed effects experiments**. For the fixed effects model, inferences are made only on those particular treatments used in the experiment.

It is often important that the experimenter be able to draw inferences about a population of treatments by means of an experiment in which the treatments used are chosen randomly from the population. For example, a biologist may be interested in whether or not there is significant variance in some physiological characteristic due to animal type. The animal types actually used in the experiment are then chosen randomly and represent the treatment effects. A chemist may be interested in studying the effect of analytical laboratories on the chemical analysis of a substance. She is not concerned with particular laboratories but rather with a large population of laboratories. She might then select a group of laboratories at random and allocate samples to each for analysis. The statistical inference would then involve (1) testing whether or not the laboratories contribute a nonzero variance to the analytical results and (2) estimating the variance due to laboratories and the variance within laboratories.

Model and Assumptions for Random Effects Model

The one-way **random effects model** is written like the fixed effects model but with the terms taking on different meanings. The response $y_{ij} = \mu + \alpha_i + \epsilon_{ij}$ is now a value of the random variable

$$Y_{ij} = \mu + A_i + \epsilon_{ij}, \text{ with } i = 1, 2, \ldots, k \text{ and } j = 1, 2, \ldots, n,$$

where the A_i are independently and normally distributed with mean 0 and variance σ_α^2 and are independent of the ϵ_{ij}. As for the fixed effects model, the ϵ_{ij} are also independently and normally distributed with mean 0 and variance σ^2. Note that for a random effects experiment, the constraint that $\sum_{i=1}^{k} \alpha_i = 0$ no longer applies.

Theorem 13.4: For the one-way random effects analysis-of-variance model,

$$E(SSA) = (k-1)\sigma^2 + n(k-1)\sigma_\alpha^2 \quad \text{and} \quad E(SSE) = k(n-1)\sigma^2.$$

Table 13.11 shows the expected mean squares for both a fixed effects and a random effects experiment. The computations for a random effects experiment are carried out in exactly the same way as for a fixed effects experiment. That is, the sum-of-squares, degrees-of-freedom, and mean-square columns in an analysis-of-variance table are the same for both models.

Table 13.11: Expected Mean Squares for the One-Factor Experiment

Source of Variation	Degrees of Freedom	Mean Squares	Expected Mean Squares Fixed Effects	Expected Mean Squares Random Effects
Treatments	$k-1$	s_1^2	$\sigma^2 + \dfrac{n}{k-1}\sum_i \alpha_i^2$	$\sigma^2 + n\sigma_\alpha^2$
Error	$k(n-1)$	s^2	σ^2	σ^2
Total	$nk-1$			

For the random effects model, the hypothesis that the treatment effects are all zero is written as follows:

Hypothesis for a Random Effects Experiment

$$H_0: \sigma_\alpha^2 = 0,$$
$$H_1: \sigma_\alpha^2 \neq 0.$$

This hypothesis says that the different treatments contribute nothing to the variability of the response. It is obvious from Table 13.11 that s_1^2 and s^2 are both

13.11 Random Effects Models

estimates of σ^2 when H_0 is true and that the ratio

$$f = \frac{s_1^2}{s^2}$$

is a value of the random variable F having the F-distribution with $k-1$ and $k(n-1)$ degrees of freedom. The null hypothesis is rejected at the α-level of significance when

$$f > f_\alpha[k-1, k(n-1)].$$

In many scientific and engineering studies, interest is not centered on the F-test. The scientist knows that the random effect does, indeed, have a significant effect. What is more important is estimation of the various variance components. This produces a *ranking* in terms of what factors produce the most variability and by how much. In the present context, it may be of interest to quantify how much larger the *single-factor variance component* is than that produced by chance (random variation).

Estimation of Variance Components

Table 13.11 can also be used to estimate the **variance components** σ^2 and σ_α^2. Since s_1^2 estimates $\sigma^2 + n\sigma_\alpha^2$ and s^2 estimates σ^2,

$$\hat{\sigma}^2 = s^2, \qquad \hat{\sigma}_\alpha^2 = \frac{s_1^2 - s^2}{n}.$$

Example 13.7: The data in Table 13.12 are coded observations on the yield of a chemical process, using five batches of raw material selected randomly. Show that the batch variance component is significantly greater than zero and obtain its estimate.

Table 13.12: Data for Example 13.7

Batch:	1	2	3	4	5	
	9.7	10.4	15.9	8.6	9.7	
	5.6	9.6	14.4	11.1	12.8	
	8.4	7.3	8.3	10.7	8.7	
	7.9	6.8	12.8	7.6	13.4	
	8.2	8.8	7.9	6.4	8.3	
	7.7	9.2	11.6	5.9	11.7	
	8.1	7.6	9.8	8.1	10.7	
Total	55.6	59.7	80.7	58.4	75.3	329.7

Solution: The total, batch, and error sums of squares are, respectively,

$$SST = 194.64, \ SSA = 72.60, \text{ and } SSE = 194.64 - 72.60 = 122.04.$$

These results, with the remaining computations, are shown in Table 13.13.

Table 13.13: Analysis of Variance for Example 13.7

Source of Variation	Sum of Squares	Degrees of Freedom	Mean Square	Computed f
Batches	72.60	4	18.15	4.46
Error	122.04	30	4.07	
Total	194.64	34		

The f-ratio is significant at the $\alpha = 0.05$ level, indicating that the hypothesis of a zero batch component is rejected. An estimate of the batch variance component is

$$\hat{\sigma}_\alpha^2 = \frac{18.15 - 4.07}{7} = 2.01.$$

Note that while the **batch variance component** is significantly different from zero, when gauged against the estimate of σ^2, namely $\hat{\sigma}^2 = MSE = 4.07$, it appears as if the batch variance component is not appreciably large.

If the result using the formula for σ_α^2 appears negative, (i.e., when s_1^2 is smaller than s^2), $\hat{\sigma}_\alpha^2$ is then set to zero. This is a biased estimator. In order to have a better estimator of σ_α^2, a method called **restricted** (or **residual**) **maximum likelihood (REML)** is commonly used (see Harville, 1977, in the Bibliography). Such an estimator can be found in many statistical software packages. The details for this estimation procedure are beyond the scope of this text.

Randomized Block Design with Random Blocks

In a randomized complete block experiment where the blocks represent days, it is conceivable that the experimenter would like the results to apply not only to the actual days used in the analysis but to every day in the year. He or she would then select at random the days on which to run the experiment as well as the treatments and use the random effects model

$$Y_{ij} = \mu + A_i + B_j + \epsilon_{ij}, \text{ for } i = 1, 2, \ldots, k \text{ and } j = 1, 2, \ldots, b,$$

with the A_i, B_j, and ϵ_{ij} being independent random variables with means 0 and variances σ_α^2, σ_β^2, and σ^2, respectively. The expected mean squares for a random effects randomized complete block design are obtained, using the same procedure as for the one-factor problem, and are presented along with those for a fixed effects experiment in Table 13.14.

Again the computations for the individual sums of squares and degrees of freedom are identical to those of the fixed effects model. The hypothesis

$$H_0: \sigma_\alpha^2 = 0,$$
$$H_1: \sigma_\alpha^2 \neq 0$$

is carried out by computing

$$f = \frac{s_1^2}{s^2}$$

Table 13.14: Expected Mean Squares for the Randomized Complete Block Design

Source of Variation	Degrees of Freedom	Mean Squares	Expected Mean Squares Fixed Effects	Expected Mean Squares Random Effects
Treatments	$k-1$	s_1^2	$\sigma^2 + \dfrac{b}{k-1}\sum_i \alpha_i^2$	$\sigma^2 + b\sigma_\alpha^2$
Blocks	$b-1$	s_2^2	$\sigma^2 + \dfrac{k}{b-1}\sum_j \beta_j^2$	$\sigma^2 + k\sigma_\beta^2$
Error	$(k-1)(b-1)$	s^2	σ^2	σ^2
Total	$kb-1$			

and rejecting H_0 when $f > f_\alpha[k-1,(b-1)(k-1)]$.

The unbiased estimates of the variance components are

$$\hat{\sigma}^2 = s^2, \qquad \hat{\sigma}_\alpha^2 = \frac{s_1^2 - s^2}{b}, \qquad \hat{\sigma}_\beta^2 = \frac{s_2^2 - s^2}{k}.$$

Tests of hypotheses concerning the various variance components are made by computing the ratios of appropriate mean squares, as indicated in Table 13.14, and comparing them with corresponding f-values from Table A.6.

13.12 Case Study

Case Study 13.1: Chemical Analysis: Personnel in the Chemistry Department of Virginia Tech were called upon to analyze a data set that was produced to compare 4 different methods of analysis of aluminum in a certain solid igniter mixture. To get a broad range of analytical laboratories involved, 5 laboratories were used in the experiment. These laboratories were selected because they are generally adept in doing these types of analyses. Twenty samples of igniter material containing 2.70% aluminum were assigned randomly, 4 to each laboratory, and directions were given on how to carry out the chemical analysis using all 4 methods. The data retrieved are as follows:

Method	Laboratory 1	2	3	4	5	Mean
A	2.67	2.69	2.62	2.66	2.70	2.668
B	2.71	2.74	2.69	2.70	2.77	2.722
C	2.76	2.76	2.70	2.76	2.81	2.758
D	2.65	2.69	2.60	2.64	2.73	2.662

The laboratories are not considered as random effects since they were not selected randomly from a larger population of laboratories. The data were analyzed as a randomized complete block design. Plots of the data were sought to determine if an additive model of the type

$$y_{ij} = \mu + m_i + l_j + \epsilon_{ij}$$

is appropriate: in other words, a model with additive effects. The randomized block is not appropriate when interaction between laboratories and methods exists. Consider the plot shown in Figure 13.12. Although this plot is a bit difficult to interpret because each point is a single observation, there appears to be no appreciable interaction between methods and laboratories.

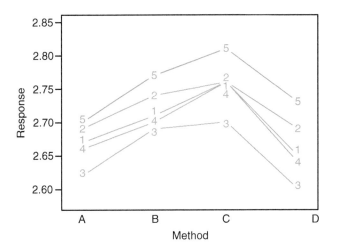

Figure 13.12: Interaction plot for data of Case Study 13.1.

Residual Plots

Residual plots were used as diagnostic indicators regarding the homogeneous variance assumption. Figure 13.13 shows a plot of residuals against analytical methods. The variability depicted in the residuals seems to be remarkably homogeneous. For completeness, a normal probability plot of the residuals is shown in Figure 13.14.

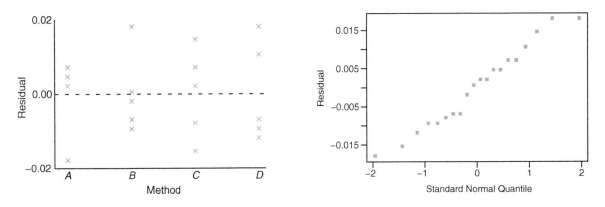

Figure 13.13: Plot of residuals against method for the data of Case Study 13.1.

Figure 13.14: Normal probability plot of residuals for the data of Case Study 13.1.

The residual plots show no difficulty with either the assumption of normal errors or the assumption of homogeneous variance. *SAS* PROC GLM was used to conduct the analysis of variance. Figure 13.15 shows the annotated computer printout.

The computed f- and P-values do indicate a significant difference between analytical methods. This analysis can be followed by a multiple comparison analysis to determine where the differences are among the methods.

Exercises

13.37 Testing patient blood samples for HIV antibodies, a spectrophotometer determines the optical density of each sample. Optical density is measured as the absorbance of light at a particular wavelength. The blood sample is positive if it exceeds a certain cutoff value that is determined by the control samples for that run. Researchers are interested in comparing the laboratory variability for the positive control values. The data represent positive control values for 10 different runs at 4 randomly selected laboratories.

Run	Laboratory 1	2	3	4
1	0.888	1.065	1.325	1.232
2	0.983	1.226	1.069	1.127
3	1.047	1.332	1.219	1.051
4	1.087	0.958	0.958	0.897
5	1.125	0.816	0.819	1.222
6	0.997	1.015	1.140	1.125
7	1.025	1.071	1.222	0.990
8	0.969	0.905	0.995	0.875
9	0.898	1.140	0.928	0.930
10	1.018	1.051	1.322	0.775

(a) Write an appropriate model for this experiment.
(b) Estimate the laboratory variance component and the variance within laboratories.

13.38 An experiment is conducted in which 4 treatments are to be compared in 5 blocks. The data are given below.

Treatment	Block 1	2	3	4	5
1	12.8	10.6	11.7	10.7	11.0
2	11.7	14.2	11.8	9.9	13.8
3	11.5	14.7	13.6	10.7	15.9
4	12.6	16.5	15.4	9.6	17.1

(a) Assuming a random effects model, test the hypothesis, at the 0.05 level of significance, that there is no difference between treatment means.
(b) Compute estimates of the treatment and block variance components.

13.39 The following data show the effect of 4 operators, chosen randomly, on the output of a particular machine.

Operator 1	2	3	4
175.4	168.5	170.1	175.2
171.7	162.7	173.4	175.7
173.0	165.0	175.7	180.1
170.5	164.1	170.7	183.7

(a) Perform a random effects analysis of variance at the 0.05 level of significance.
(b) Compute an estimate of the operator variance component and the experimental error variance component.

13.40 Five "pours" of metals have had 5 core samples each analyzed for the amount of a trace element. The data for the 5 randomly selected pours are as follows:

Core	Pour 1	2	3	4	5
1	0.98	0.85	1.12	1.21	1.00
2	1.02	0.92	1.68	1.19	1.21
3	1.57	1.16	0.99	1.32	0.93
4	1.25	1.43	1.26	1.08	0.86
5	1.16	0.99	1.05	0.94	1.41

(a) The intent is that the pours be identical. Thus, test that the "pour" variance component is zero. Draw conclusions.
(b) Show a complete ANOVA along with an estimate of the within-pour variance.

13.41 A textile company weaves a certain fabric on a large number of looms. The managers would like the looms to be homogeneous so that their fabric is of uniform strength. It is suspected that there may be significant variation in strength among looms. Consider the following data for 4 randomly selected looms. Each observation is a determination of strength of the fabric in pounds per square inch.

Loom 1	2	3	4
99	97	94	93
97	96	95	94
97	92	90	90
96	98	92	92

(a) Write a model for the experiment.
(b) Does the loom variance component differ significantly from zero?
(c) Comment on the managers' suspicion.

```
                    The GLM Procedure
                  Class Level Information
            Class         Levels    Values
            Method           4      A B C D
            Lab              5      1 2 3 4 5
         Number of Observations Read        20
         Number of Observations Used        20
Dependent Variable: Response
                        Sum of
Source              DF    Squares     Mean Square   F Value   Pr > F
Model                7    0.05340500  0.00762929    42.19     <.0001
Error               12    0.00217000  0.00018083
Corrected Total     19    0.05557500

R-Square     Coeff Var      Root MSE      Response Mean
0.960954     0.497592       0.013447         2.702500

Source              DF    Type III SS   Mean Square   F Value   Pr > F
Method               3    0.03145500    0.01048500    57.98     <.0001
Lab                  4    0.02195000    0.00548750    30.35     <.0001

Observation        Observed         Predicted           Residual
      1           2.67000000       2.66300000          0.00700000
      2           2.71000000       2.71700000         -0.00700000
      3           2.76000000       2.75300000          0.00700000
      4           2.65000000       2.65700000         -0.00700000
      5           2.69000000       2.68550000          0.00450000
      6           2.74000000       2.73950000          0.00050000
      7           2.76000000       2.77550000         -0.01550000
      8           2.69000000       2.67950000          0.01050000
      9           2.62000000       2.61800000          0.00200000
     10           2.69000000       2.67200000          0.01800000
     11           2.70000000       2.70800000         -0.00800000
     12           2.60000000       2.61200000         -0.01200000
     13           2.66000000       2.65550000          0.00450000
     14           2.70000000       2.70950000         -0.00950000
     15           2.76000000       2.74550000          0.01450000
     16           2.64000000       2.64950000         -0.00950000
     17           2.70000000       2.71800000         -0.01800000
     18           2.77000000       2.77200000         -0.00200000
     19           2.81000000       2.80800000          0.00200000
     20           2.73000000       2.71200000          0.01800000
```

Figure 13.15: *SAS* printout for data of Case Study 13.1.

Review Exercises

13.42 An analysis was conducted by the Laboratory for Interdisciplinary Statistical Analysis at Virginia Tech in conjunction with the Department of Forestry. A certain treatment was applied to a set of tree stumps in which the chemical Garlon was used with the purpose of regenerating the roots of the stumps. A spray was used with four levels of Garlon concentration. After a period of time, the height of the shoots was observed. Perform a one-factor analysis of variance on the following data. Test to see if the concentration of Garlon has a significant impact on the height of the shoots. Use $\alpha = 0.05$.

Garlon Level							
1		2		3		4	
2.87	2.31	3.27	2.66	2.39	1.91	3.05	0.91
3.91	2.04	3.15	2.00	2.89	1.89	2.43	0.01

13.43 Consider the aggregate data of Example 13.1. Perform Bartlett's test, at level $\alpha = 0.1$, to determine if there is heterogeneity of variance among the aggregates.

13.44 Three catalysts are used in a chemical process; a control (no catalyst) is also included. The following are yield data from the process:

	Catalyst		
Control	1	2	3
74.5	77.5	81.5	78.1
76.1	82.0	82.3	80.2
75.9	80.6	81.4	81.5
78.1	84.9	79.5	83.0
76.2	81.0	83.0	82.1

Use Dunnett's test at the $\alpha = 0.01$ level of significance to determine if a significantly higher yield is obtained with the catalysts than with no catalyst.

13.45 Four laboratories are being used to perform chemical analysis. Samples of the same material are sent to the laboratories for analysis as part of a study to determine whether or not they give, on the average, the same results. The analytical results for the four laboratories are as follows:

Laboratory			
A	B	C	D
58.7	62.7	55.9	60.7
61.4	64.5	56.1	60.3
60.9	63.1	57.3	60.9
59.1	59.2	55.2	61.4
58.2	60.3	58.1	62.3

(a) Use Bartlett's test to show that the within-laboratory variances are not significantly different at the $\alpha = 0.05$ level of significance.

(b) Perform the analysis of variance and give conclusions concerning the laboratories.

(c) Do a normal probability plot of residuals.

13.46 An experiment was designed for personnel in the Department of Animal and Poultry Science at Virginia Tech to study urea and aqueous ammonia treatment of wheat straw. The purpose was to improve nutritional value for male sheep. The diet treatments were control, urea at feeding, ammonia-treated straw, and urea-treated straw. Twenty-four sheep were used in the experiment, and they were separated according to relative weight. There were four sheep in each homogeneous group (by weight) and each of them was given one of the four diets in random order. For each of the 24 sheep, the percent dry matter digested was measured. The data follow.

Diet	Group by Weight (block)					
	1	2	3	4	5	6
Control	32.68	36.22	36.36	40.95	34.99	33.89
Urea at feeding	35.90	38.73	37.55	34.64	37.36	34.35
Ammonia treated	49.43	53.50	52.86	45.00	47.20	49.76
Urea treated	46.58	42.82	45.41	45.08	43.81	47.40

(a) Use a randomized complete block type of analysis to test for differences between the diets. Use $\alpha = 0.05$.

(b) Use Dunnett's test to compare the three diets with the control. Use $\alpha = 0.05$.

(c) Do a normal probability plot of residuals.

13.47 In a study that was analyzed for personnel in the Department of Biochemistry at Virginia Tech, three diets were given to groups of rats in order to study the effect of each on dietary residual zinc in the bloodstream. Five pregnant rats were randomly assigned to each diet group, and each was given the diet on day 22 of pregnancy. The amount of zinc in parts per million was measured. The data are as follows:

	1	0.50	0.42	0.65	0.47	0.44
Diet	2	0.42	0.40	0.73	0.47	0.69
	3	1.06	0.82	0.72	0.72	0.82

Determine if there is a significant difference in residual dietary zinc among the three diets. Use $\alpha = 0.05$. Perform a one-way ANOVA.

13.48 An experiment was conducted to compare three types of paint for evidence of differences in their wearing qualities. They were exposed to abrasive action and the time in hours until abrasion was noticed was observed. Six specimens were used for each type of paint. The data are as follows.

Paint Type		
1	2	3
158 97 282	515 264 544	317 662 213
315 220 115	525 330 525	536 175 614

(a) Do an analysis of variance to determine if the evidence suggests that wearing quality differs for the three paints. Use a P-value in your conclusion.

(b) If significant differences are found, characterize what they are. Is there one paint that stands out? Discuss your findings.

(c) Do whatever graphical analysis you need to determine if assumptions used in (a) are valid. Discuss your findings.

(d) Suppose it is determined that the data for each treatment follow an exponential distribution. Does this suggest an alternative analysis? If so, do the alternative analysis and give findings.

13.49 A company that stamps gaskets out of sheets of rubber, plastic, and cork wants to compare the mean number of gaskets produced per hour for the three types of material. Two randomly selected stamping machines are chosen as blocks. The data represent the number of gaskets (in thousands) produced per hour. The data is given below. In addition, the printout analysis is given in Figure 13.16 on page 557.

Machine	Cork	Material Rubber	Plastic
A	4.31 4.27 4.40	3.36 3.42 3.48	4.01 3.94 3.89
B	3.94 3.81 3.99	3.91 3.80 3.85	3.48 3.53 3.42

(a) Why would the stamping machines be chosen as blocks?

(b) Plot the six means for machine and material combinations.

(c) Is there a single material that is best?

(d) Is there an interaction between treatments and blocks? If so, is the interaction causing any serious difficulty in arriving at a proper conclusion? Explain.

13.50 A study is conducted to compare gas mileage for 3 competing brands of gasoline. Four different automobile models of varying size are randomly selected. The data, in miles per gallon, follow. The order of testing is random for each model.

Model	Gasoline Brand		
	A	B	C
A	32.4	35.6	38.7
B	28.8	28.6	29.9
C	36.5	37.6	39.1
D	34.4	36.2	37.9

(a) Discuss the need for the use of more than a single model of car.

(b) Consider the ANOVA from the SAS printout in Figure 13.17 on page 558. Does brand of gasoline matter?

(c) Which brand of gasoline would you select? Consult the result of Duncan's test.

13.51 Four different locations in the northeast were used for collecting ozone measurements in parts per million. Amounts of ozone were collected in 5 samples at each location.

Location			
1	2	3	4
0.09	0.15	0.10	0.10
0.10	0.12	0.13	0.07
0.08	0.17	0.08	0.05
0.08	0.18	0.08	0.08
0.11	0.14	0.09	0.09

(a) Is there sufficient information here to suggest that there are differences in the mean ozone levels across locations? Be guided by a P-value.

(b) If significant differences are found in (a), characterize the nature of the differences. Use whatever methods you have learned.

13.52 Show that the mean square error

$$s^2 = \frac{SSE}{k(n-1)}$$

for the analysis of variance in a one-way classification is an unbiased estimate of σ^2.

13.53 Prove Theorem 13.2.

13.54 Show that the computing formula for SSB, in the analysis of variance of the randomized complete block design, is equivalent to the corresponding term in the identity of Theorem 13.3.

13.55 For the randomized block design with k treatments and b blocks, show that

$$E(SSB) = (b-1)\sigma^2 + k\sum_{j=1}^{b} \beta_j^2.$$

```
                           The GLM Procedure

Dependent Variable: gasket
                                Sum of
Source                  DF     Squares      Mean Square    F Value    Pr > F
Model                    5    1.68122778    0.33624556      76.52     <.0001
Error                   12    0.05273333    0.00439444
Corrected Total         17    1.73396111
R-Square        Coeff Var         Root MSE       gasket Mean
0.969588        1.734095          0.066291        3.822778


Source                  DF    Type III SS   Mean Square    F Value    Pr > F
material                 2    0.81194444    0.40597222      92.38     <.0001
machine                  1    0.10125000    0.10125000      23.04     0.0004
material*machine         2    0.76803333    0.38401667      87.39     <.0001
 Level of      Level of                    ------------gasket----------
 material      machine         N              Mean              Std Dev
 cork          A               3           4.32666667         0.06658328
 cork          B               3           3.91333333         0.09291573
 plastic       A               3           3.94666667         0.06027714
 plastic       B               3           3.47666667         0.05507571
 rubber        A               3           3.42000000         0.06000000
 rubber        B               3           3.85333333         0.05507571

 Level of                 ------------gasket-----------
 material      N                Mean              Std Dev
 cork          6           4.12000000         0.23765521
 plastic       6           3.71166667         0.26255793
 rubber        6           3.63666667         0.24287171

 Level of                 ------------gasket-----------
 machine       N                Mean              Std Dev
 A             9           3.89777778         0.39798800
 B             9           3.74777778         0.21376259
```

Figure 13.16: *SAS* printout for Review Exercise 13.49.

```
                         The GLM Procedure
Dependent Variable: MPG
                           Sum of
Source              DF    Squares      Mean Square   F Value   Pr > F
Model                5    153.2508333  30.6501667    24.66     0.0006
Error                6    7.4583333    1.2430556
Corrected Total     11    160.7091667

R-Square      Coeff Var      Root MSE       MPG Mean
0.953591      3.218448       1.114924       34.64167

Source              DF    Type III SS   Mean Square   F Value   Pr > F
Model                3    130.3491667   43.4497222    34.95     0.0003
Brand                2    22.9016667    11.4508333    9.21      0.0148
                    Duncan's Multiple Range Test for MPG
NOTE: This test controls the Type I comparisonwise error rate, not
the experimentwise error rate.

                  Alpha                          0.05
                  Error Degrees of Freedom          6
                  Error Mean Square           1.243056

                  Number of Means         2           3
                  Critical Range       1.929       1.999
        Means with the same letter are not significantly different.
                  Duncan Grouping        Mean       N    Brand
                              A         36.4000     4    C
                              A
                          B   A         34.5000     4    B
                          B
                          B             33.0250     4    A
```

Figure 13.17: SAS printout for Review Exercise 13.50.

13.56 Group Project: It is of interest to determine which type of sports ball can be thrown the longest distance. The competition involves a tennis ball, a baseball, and a softball. Divide the class into teams of five individuals. Each team should design and conduct a separate experiment. Each team should also analyze the data from its own experiment. For a given team, each of the five individuals will throw each ball (after sufficient arm warmup). The experimental response will be the distance (in feet) that the ball is thrown. The data for each team will involve 15 observations. Important points:

(a) This is not a competition among teams. The competition is among the three types of sports balls. One would expect that the conclusion drawn by each team would be similar.

(b) Each team should be gender mixed.

(c) The experimental design for each team should be a randomized complete block design. The five individuals throwing are the blocks.

(d) Be sure to incorporate the appropriate randomization in conducting the experiment.

(e) The results should contain a description of the experiment with an ANOVA table complete with a P-value and appropriate conclusions. Use graphical techniques where appropriate. Use multiple comparisons where appropriate. Draw practical conclusions concerning differences between the ball types. Be thorough.

13.13 Potential Misconceptions and Hazards; Relationship to Material in Other Chapters

As in other procedures covered in previous chapters, the analysis of variance is reasonably robust to the normality assumption but less robust to the homogeneous variance assumption. Also we note here that Bartlett's test for equal variance is extremely nonrobust to normality.

This chapter is an extremely pivotal chapter in that it is essentially an "entry level" point for important topics such as design of experiments and analysis of variance. Chapter 14 will concern itself with the same topics, but the expansion will be to more than one factor, with the total analysis further complicated by the interpretation of interaction among factors. There are times when the role of interaction in a scientific experiment is more important than the role of the main factors (main effects). The presence of interaction results in even more emphasis placed on graphical displays. In Chapters 14 and 15, it will be necessary to give more details regarding the randomization process since the number of factor combinations can be large.

Chapter 14

Factorial Experiments (Two or More Factors)

14.1 Introduction

Consider a situation where it is of interest to study the effects of **two factors**, A and B, on some response. For example, in a chemical experiment, we would like to vary simultaneously the reaction pressure and reaction time and study the effect of each on the yield. In a biological experiment, it is of interest to study the effects of drying time and temperature on the amount of solids (percent by weight) left in samples of yeast. As in Chapter 13, the term **factor** is used in a general sense to denote any feature of the experiment such as temperature, time, or pressure that may be varied from trial to trial. We define the **levels** of a factor to be the actual values used in the experiment.

For each of these cases, it is important to determine not only if each of the two factors has an influence on the response, but also if there is a significant interaction between the two factors. As far as terminology is concerned, the experiment described here is a two-factor experiment and the experimental design may be either a completely randomized design, in which the various treatment combinations are assigned randomly to all the experimental units, or a randomized complete block design, in which factor combinations are assigned randomly within blocks. In the case of the yeast example, the various treatment combinations of temperature and drying time would be assigned randomly to the samples of yeast if we were using a completely randomized design.

Many of the concepts studied in Chapter 13 are extended in this chapter to two and three factors. The main thrust of this material is the use of the completely randomized design with a *factorial experiment*. A factorial experiment in two factors involves experimental trials (or a single trial) with all factor combinations. For example, in the temperature-drying-time example with, say, 3 levels of each and $n = 2$ runs at each of the 9 combinations, we have a *two-factor factorial experiment in a completely randomized design*. Neither factor is a blocking factor; we are interested in how each influences percent solids in the samples and whether or not they interact. The biologist would have available 18 physical samples of material which are experimental units. These would then be assigned randomly to the 18 combinations (9 treatment combinations, each duplicated).

Before we launch into analytical details, sums of squares, and so on, it may be of interest for the reader to observe the obvious connection between what we have described and the situation with the one-factor problem. Consider the yeast experiment. Explanation of degrees of freedom aids the reader or the analyst in visualizing the extension. We should initially view the 9 treatment combinations as if they represented one factor with 9 levels (8 degrees of freedom). Thus, an initial look at degrees of freedom gives

Treatment combinations	8
Error	9
Total	17

Main Effects and Interaction

The experiment could be analyzed as described in the above table. However, the F-test for combinations would probably not give the analyst the information he or she desires, namely, that which considers the role of temperature and drying time. Three drying times have 2 associated degrees of freedom; three temperatures have 2 degrees of freedom. The main factors, temperature and drying time, are called **main effects**. The main effects represent 4 of the 8 degrees of freedom for *factor combinations*. The additional 4 degrees of freedom are associated with *interaction* between the two factors. As a result, the analysis involves

Combinations	8
Temperature	2
Drying time	2
Interaction	4
Error	9
Total	17

Recall from Chapter 13 that factors in an analysis of variance may be viewed as fixed or random, depending on the type of inference desired and how the levels were chosen. Here we must consider fixed effects, random effects, and even cases where effects are mixed. Most attention will be directed toward expected mean squares when we advance to these topics. In the following section, we focus on the concept of interaction.

14.2 Interaction in the Two-Factor Experiment

In the randomized block model discussed previously, it was assumed that one observation on each treatment is taken in each block. If the model assumption is correct, that is, if blocks and treatments are the only real effects and interaction does not exist, the expected value of the mean square error is the experimental error variance σ^2. Suppose, however, that there is interaction occurring between treatments and blocks as indicated by the model

$$y_{ij} = \mu + \alpha_i + \beta_j + (\alpha\beta)_{ij} + \epsilon_{ij}$$

of Section 13.8. The expected value of the mean square error is then given as

$$E\left[\frac{SSE}{(b-1)(k-1)}\right] = \sigma^2 + \frac{1}{(b-1)(k-1)}\sum_{i=1}^{k}\sum_{j=1}^{b}(\alpha\beta)_{ij}^2.$$

The treatment and block effects do not appear in the expected mean square error, but the interaction effects do. Thus, if there is interaction in the model, the mean square error reflects variation due to experimental error plus an interaction contribution, and for this experimental plan, there is no way of separating them.

Interaction and the Interpretation of Main Effects

From an experimenter's point of view it should seem necessary to arrive at a significance test on the existence of interaction by separating true error variation from that due to interaction. The main effects, A and B, take on a different meaning in the presence of interaction. In the previous biological example, the effect that drying time has on the amount of solids left in the yeast might very well depend on the temperature to which the samples are exposed. In general, there could be experimental situations in which factor A has a positive effect on the response at one level of factor B, while at a different level of factor B the effect of A is negative. We use the term **positive effect** here to indicate that the yield or response increases as the levels of a given factor increase according to some defined order. In the same sense, a **negative effect** corresponds to a decrease in response for increasing levels of the factor.

Consider, for example, the following data on temperature (factor A at levels t_1, t_2, and t_3 in increasing order) and drying time d_1, d_2, and d_3 (also in increasing order). The response is percent solids. These data are completely hypothetical and given to illustrate a point.

		B		
A	d_1	d_2	d_3	Total
t_1	4.4	8.8	5.2	18.4
t_2	7.5	8.5	2.4	18.4
t_3	9.7	7.9	0.8	18.4
Total	21.6	25.2	8.4	55.2

Clearly the effect of temperature on percent solids is positive at the low drying time d_1 but negative for high drying time d_3. This **clear interaction** between temperature and drying time is obviously of interest to the biologist, but, based on the totals of the responses for temperatures t_1, t_2, and t_3, the temperature sum of squares, SSA, will yield a value of zero. We say then that the presence of interaction is **masking** the effect of temperature. Thus, if we consider the average effect of temperature, averaged over drying time, **there is no effect**. This then defines the main effect. But, of course, this is likely not what is pertinent to the biologist.

Before drawing any final conclusions resulting from tests of significance on the main effects and interaction effects, the **experimenter should first observe whether or not the test for interaction is significant**. If interaction is not significant, then the results of the tests on the main effects are meaningful. However, if interaction should be significant, then only those tests on the main effects that turn out to be significant are meaningful. Nonsignificant main effects in the presence of interaction might well be a result of masking and dictate the need to observe the influence of each factor at fixed levels of the other.

A Graphical Look at Interaction

The presence of interaction as well as its scientific impact can be interpreted nicely through the use of **interaction plots**. The plots clearly give a pictorial view of the tendency in the data to show the effect of changing one factor as one moves from one level to another of a second factor. Figure 14.1 illustrates the strong temperature by drying time interaction. The interaction is revealed in nonparallel lines.

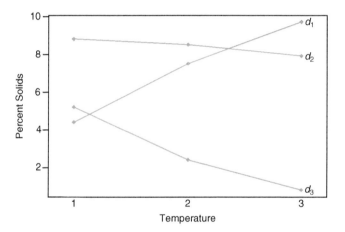

Figure 14.1: Interaction plot for temperature–drying time data.

The relatively strong *temperature effect* on percent solids at the lower drying time is reflected in the steep slope at d_1. At the middle drying time d_2 the temperature has very little effect, while at the high drying time d_3 the negative slope illustrates a negative effect of temperature. Interaction plots such as this set give the scientist a quick and meaningful interpretation of the interaction that is present. It should be apparent that **parallelism** in the plots signals an **absence of interaction**.

Need for Multiple Observations

Interaction and experimental error are separated in the two-factor experiment only if multiple observations are taken at the various treatment combinations. For maximum efficiency, there should be the same number n of observations at each combination. These should be true replications, not just repeated measurements. For example, in the yeast illustration, if we take $n = 2$ observations at each combination of temperature and drying time, there should be two separate samples and not merely repeated measurements on the same sample. This allows variability due to experimental units to appear in "error," so the variation is not merely measurement error.

14.3 Two-Factor Analysis of Variance

To present general formulas for the analysis of variance of a two-factor experiment using repeated observations in a completely randomized design, we shall consider the case of n replications of the treatment combinations determined by a levels of factor A and b levels of factor B. The observations may be classified by means of a rectangular array where the rows represent the levels of factor A and the columns represent the levels of factor B. Each treatment combination defines a cell in our array. Thus, we have ab cells, each cell containing n observations. Denoting the kth observation taken at the ith level of factor A and the jth level of factor B by y_{ijk}, Table 14.1 shows the abn observations.

Table 14.1: Two-Factor Experiment with n Replications

A	B 1	2	$\cdots$	b	Total	Mean
1	y_{111}	y_{121}	$\cdots$	y_{1b1}	$Y_{1..}$	$\bar{y}_{1..}$
	y_{112}	y_{122}		y_{1b2}		
	$\vdots$	$\vdots$		$\vdots$		
	y_{11n}	y_{12n}	$\cdots$	y_{1bn}		
2	y_{211}	y_{221}	$\cdots$	y_{2b1}	$Y_{2..}$	$\bar{y}_{2..}$
	y_{212}	y_{222}	$\cdots$	y_{2b2}		
	$\vdots$	$\vdots$		$\vdots$		
	y_{21n}	y_{22n}	$\cdots$	y_{2bn}		
$\vdots$	$\vdots$	$\vdots$		$\vdots$	$\vdots$	$\vdots$
a	y_{a11}	y_{a21}	$\cdots$	y_{ab1}	$Y_{a..}$	$\bar{y}_{a..}$
	y_{a12}	y_{a22}	$\cdots$	y_{ab2}		
	$\vdots$	$\vdots$		$\vdots$		
	y_{a1n}	y_{a2n}	$\cdots$	y_{abn}		
Total	$Y_{.1.}$	$Y_{.2.}$	$\cdots$	$Y_{.b.}$	$Y_{...}$	
Mean	$\bar{y}_{.1.}$	$\bar{y}_{.2.}$	$\cdots$	$\bar{y}_{.b.}$		$\bar{y}_{...}$

The observations in the (ij)th cell constitute a random sample of size n from a population that is assumed to be normally distributed with mean μ_{ij} and variance σ^2. All ab populations are assumed to have the same variance σ^2. Let us define the following useful symbols, some of which are used in Table 14.1:

$Y_{ij.}$ = sum of the observations in the (ij)th cell,
$Y_{i..}$ = sum of the observations for the ith level of factor A,
$Y_{.j.}$ = sum of the observations for the jth level of factor B,
$Y_{...}$ = sum of all abn observations,
$\bar{y}_{ij.}$ = mean of the observations in the (ij)th cell,
$\bar{y}_{i..}$ = mean of the observations for the ith level of factor A,
$\bar{y}_{.j.}$ = mean of the observations for the jth level of factor B,
$\bar{y}_{...}$ = mean of all abn observations.

Unlike in the one-factor situation covered at length in Chapter 13, here we are assuming that the **populations**, where n independent identically distributed observations are taken, are **combinations** of factors. Also we will assume throughout that an equal number (n) of observations are taken at each factor combination. In cases in which the sample sizes per combination are unequal, the computations are more complicated but the concepts are transferable.

Model and Hypotheses for the Two-Factor Problem

Each observation in Table 14.1 may be written in the form

$$y_{ijk} = \mu_{ij} + \epsilon_{ijk},$$

where ϵ_{ijk} measures the deviations of the observed y_{ijk} values in the (ij)th cell from the population mean μ_{ij}. If we let $(\alpha\beta)_{ij}$ denote the interaction effect of the ith level of factor A and the jth level of factor B, α_i the effect of the ith level of factor A, β_j the effect of the jth level of factor B, and μ the overall mean, we can write

$$\mu_{ij} = \mu + \alpha_i + \beta_j + (\alpha\beta)_{ij},$$

and then

$$y_{ijk} = \mu + \alpha_i + \beta_j + (\alpha\beta)_{ij} + \epsilon_{ijk},$$

on which we impose the restrictions

$$\sum_{i=1}^{a} \alpha_i = 0, \quad \sum_{j=1}^{b} \beta_j = 0, \quad \sum_{i=1}^{a} (\alpha\beta)_{ij} = 0, \quad \sum_{j=1}^{b} (\alpha\beta)_{ij} = 0.$$

The three hypotheses to be tested are as follows:

1. H_0': $\alpha_1 = \alpha_2 = \cdots = \alpha_a = 0$,
   H_1': At least one of the α_i is not equal to zero.
2. H_0'': $\beta_1 = \beta_2 = \cdots = \beta_b = 0$,
   H_1'': At least one of the β_j is not equal to zero.

3. H_0''': $(\alpha\beta)_{11} = (\alpha\beta)_{12} = \cdots = (\alpha\beta)_{ab} = 0$,
H_1''': At least one of the $(\alpha\beta)_{ij}$ is not equal to zero.

We warned the reader about the problem of masking of main effects when interaction is a heavy contributor in the model. It is recommended that the interaction test result be considered first. The interpretation of the main effect test follows, and the nature of the scientific conclusion depends on whether interaction is found. If interaction is ruled out, then hypotheses 1 and 2 above can be tested and the interpretation is quite simple. However, if interaction is found to be present the interpretation can be more complicated, as we have seen from the discussion of the drying time and temperature in the previous section. In what follows, the structure of the tests of hypotheses 1, 2, and 3 will be discussed. Interpretation of results will be incorporated in the discussion of the analysis in Example 14.1.

The tests of the hypotheses above will be based on a comparison of independent estimates of σ^2 provided by splitting the total sum of squares of our data into four components by means of the following identity.

Partitioning of Variability in the Two-Factor Case

Theorem 14.1: **Sum-of-Squares Identity**

$$\sum_{i=1}^{a}\sum_{j=1}^{b}\sum_{k=1}^{n}(y_{ijk}-\bar{y}_{...})^2 = bn\sum_{i=1}^{a}(\bar{y}_{i..}-\bar{y}_{...})^2 + an\sum_{j=1}^{b}(\bar{y}_{.j.}-\bar{y}_{...})^2$$
$$+ n\sum_{i=1}^{a}\sum_{j=1}^{b}(\bar{y}_{ij.}-\bar{y}_{i..}-\bar{y}_{.j.}+\bar{y}_{...})^2 + \sum_{i=1}^{a}\sum_{j=1}^{b}\sum_{k=1}^{n}(y_{ijk}-\bar{y}_{ij.})^2$$

Symbolically, we write the sum-of-squares identity as

$$SST = SSA + SSB + SS(AB) + SSE,$$

where SSA and SSB are called the sums of squares for the main effects A and B, respectively, $SS(AB)$ is called the interaction sum of squares for A and B, and SSE is the error sum of squares. The degrees of freedom are partitioned according to the identity

$$abn - 1 = (a-1) + (b-1) + (a-1)(b-1) + ab(n-1).$$

Formation of Mean Squares

If we divide each of the sums of squares on the right side of the sum-of-squares identity by its corresponding number of degrees of freedom, we obtain the four statistics

$$S_1^2 = \frac{SSA}{a-1}, \quad S_2^2 = \frac{SSB}{b-1}, \quad S_3^2 = \frac{SS(AB)}{(a-1)(b-1)}, \quad S^2 = \frac{SSE}{ab(n-1)}.$$

All of these variance estimates are independent estimates of σ^2 under the condition that there are no effects α_i, β_j, and, of course, $(\alpha\beta)_{ij}$. If we interpret the sums of squares as functions of the independent random variables $y_{111}, y_{112}, \ldots, y_{abn}$, it is not difficult to verify that

$$E(S_1^2) = E\left[\frac{SSA}{a-1}\right] = \sigma^2 + \frac{nb}{a-1}\sum_{i=1}^{a}\alpha_i^2,$$

$$E(S_2^2) = E\left[\frac{SSB}{b-1}\right] = \sigma^2 + \frac{na}{b-1}\sum_{j=1}^{b}\beta_j^2,$$

$$E(S_3^2) = E\left[\frac{SS(AB)}{(a-1)(b-1)}\right] = \sigma^2 + \frac{n}{(a-1)(b-1)}\sum_{i=1}^{a}\sum_{j=1}^{b}(\alpha\beta)_{ij}^2,$$

$$E(S^2) = E\left[\frac{SSE}{ab(n-1)}\right] = \sigma^2,$$

from which we immediately observe that all four estimates of σ^2 are unbiased when H_0', H_0'', and H_0''' are true.

To test the hypothesis H_0', that the effects of factors A are all equal to zero, we compute the following ratio:

F-Test for Factor A

$$f_1 = \frac{s_1^2}{s^2},$$

which is a value of the random variable F_1 having the F-distribution with $a-1$ and $ab(n-1)$ degrees of freedom when H_0' is true. The null hypothesis is rejected at the α-level of significance when $f_1 > f_\alpha[a-1, ab(n-1)]$.

Similarly, to test the hypothesis H_0'' that the effects of factor B are all equal to zero, we compute the following ratio:

F-Test for Factor B

$$f_2 = \frac{s_2^2}{s^2},$$

which is a value of the random variable F_2 having the F-distribution with $b-1$ and $ab(n-1)$ degrees of freedom when H_0'' is true. This hypothesis is rejected at the α-level of significance when $f_2 > f_\alpha[b-1, ab(n-1)]$.

Finally, to test the hypothesis H_0''', that the interaction effects are all equal to zero, we compute the following ratio:

F-Test for Interaction

$$f_3 = \frac{s_3^2}{s^2},$$

which is a value of the random variable F_3 having the F-distribution with $(a-1)(b-1)$ and $ab(n-1)$ degrees of freedom when H_0''' is true. We conclude that, at the α-level of significance, interaction is present when $f_3 > f_\alpha[(a-1)(b-1), ab(n-1)]$.

As indicated in Section 14.2, it is advisable to interpret the test for interaction before attempting to draw inferences on the main effects. If interaction is not significant, there is certainly evidence that the tests on main effects are interpretable. Rejection of hypothesis 1 on page 566 implies that the response means at the levels

14.3 Two-Factor Analysis of Variance

of factor A are significantly different, while rejection of hypothesis 2 implies a similar condition for the means at levels of factor B. However, a significant interaction could very well imply that the data should be analyzed in a somewhat different manner—**perhaps observing the effect of factor A at fixed levels of factor B**, and so forth.

The computations in an analysis-of-variance problem, for a two-factor experiment with n replications, are usually summarized as in Table 14.2.

Table 14.2: Analysis of Variance for the Two-Factor Experiment with n Replications

Source of Variation	Sum of Squares	Degrees of Freedom	Mean Square	Computed f
Main effect:				
A	SSA	$a-1$	$s_1^2 = \frac{SSA}{a-1}$	$f_1 = \frac{s_1^2}{s^2}$
B	SSB	$b-1$	$s_2^2 = \frac{SSB}{b-1}$	$f_2 = \frac{s_2^2}{s^2}$
Two-factor interactions:				
AB	SS(AB)	$(a-1)(b-1)$	$s_3^2 = \frac{SS(AB)}{(a-1)(b-1)}$	$f_3 = \frac{s_3^2}{s^2}$
Error	SSE	$ab(n-1)$	$s^2 = \frac{SSE}{ab(n-1)}$	
Total	SST	$abn-1$		

Example 14.1: In an experiment conducted to determine which of 3 different missile systems is preferable, the propellant burning rate for 24 static firings was measured. Four different propellant types were used. The experiment yielded duplicate observations of burning rates at each combination of the treatments.

The data, after coding, are given in Table 14.3. Test the following hypotheses: (a) H_0': there is no difference in the mean propellant burning rates when different missile systems are used, (b) H_0'': there is no difference in the mean propellant burning rates of the 4 propellant types, (c) H_0''': there is no interaction between the different missile systems and the different propellant types.

Table 14.3: Propellant Burning Rates

Missile System	Propellant Type			
	b_1	b_2	b_3	b_4
a_1	34.0	30.1	29.8	29.0
	32.7	32.8	26.7	28.9
a_2	32.0	30.2	28.7	27.6
	33.2	29.8	28.1	27.8
a_3	28.4	27.3	29.7	28.8
	29.3	28.9	27.3	29.1

Solution: 1. (a) H_0': $\alpha_1 = \alpha_2 = \alpha_3 = 0$.
(b) H_0'': $\beta_1 = \beta_2 = \beta_3 = \beta_4 = 0$.

(c) H_0''': $(\alpha\beta)_{11} = (\alpha\beta)_{12} = \cdots = (\alpha\beta)_{34} = 0$.

2. (a) H_1': At least one of the α_i is not equal to zero.
   (b) H_1'': At least one of the β_j is not equal to zero.
   (c) H_1''': At least one of the $(\alpha\beta)_{ij}$ is not equal to zero.

The sum-of-squares formula is used as described in Theorem 14.1. The analysis of variance is shown in Table 14.4.

Table 14.4: Analysis of Variance for the Data of Table 14.3

Source of Variation	Sum of Squares	Degrees of Freedom	Mean Square	Computed f
Missile system	14.52	2	7.26	5.84
Propellant type	40.08	3	13.36	10.75
Interaction	22.16	6	3.69	2.97
Error	14.91	12	1.24	
Total	91.68	23		

The reader is directed to a *SAS* GLM Procedure (General Linear Models) for analysis of the burning rate data in Figure 14.2. Note how the "model" (11 degrees of freedom) is initially tested and the system, type, and system by type interaction are tested separately. The F-test on the model ($P = 0.0030$) is testing the accumulation of the two main effects and the interaction.

(a) Reject H_0' and conclude that different missile systems result in different mean propellant burning rates. The P-value is approximately 0.0169.

(b) Reject H_0'' and conclude that the mean propellant burning rates are not the same for the four propellant types. The P-value is approximately 0.0010.

(c) Interaction is barely insignificant at the 0.05 level, but the P-value of approximately 0.0513 would indicate that interaction must be taken seriously.

At this point we should draw some type of interpretation of the interaction. It should be emphasized that statistical significance of a main effect merely implies that *marginal means are significantly different*. However, consider the two-way table of averages in Table 14.5.

Table 14.5: Interpretation of Interaction

	b_1	b_2	b_3	b_4	Average
a_1	33.35	31.45	28.25	28.95	30.50
a_2	32.60	30.00	28.40	27.70	29.68
a_3	28.85	28.10	28.50	28.95	28.60
Average	31.60	29.85	28.38	28.53	

It is apparent that more important information exists in the body of the table—trends that are inconsistent with the trend depicted by marginal averages. Table 14.5 certainly suggests that the effect of propellant type depends on the system

14.3 Two-Factor Analysis of Variance

```
                     The GLM Procedure
Dependent Variable: rate
                         Sum of
Source              DF   Squares       Mean Square    F Value    Pr > F
Model               11   76.76833333   6.97893939     5.62       0.0030
Error               12   14.91000000   1.24250000
Corrected Total     23   91.67833333

R-Square     Coeff Var      Root MSE      rate Mean
0.837366     3.766854       1.114675      29.59167

Source          DF    Type III SS    Mean Square    F Value    Pr > F
system          2     14.52333333    7.26166667     5.84       0.0169
type            3     40.08166667    13.36055556    10.75      0.0010
system*type     6     22.16333333    3.69388889     2.97       0.0512
```

Figure 14.2: *SAS* printout of the analysis of the propellant rate data of Table 14.3.

being used. For example, for system 3 the propellant-type effect does not appear to be important, although it does have a large effect if either system 1 or system 2 is used. This explains the "significant" interaction between these two factors. More will be revealed subsequently concerning this interaction.

Example 14.2: Referring to Example 14.1, choose two orthogonal contrasts to partition the sum of squares for the missile systems into single-degree-of-freedom components to be used in comparing systems 1 and 2 versus 3, and system 1 versus system 2.

Solution: The contrast for comparing systems 1 and 2 with 3 is

$$w_1 = \mu_{1.} + \mu_{2.} - 2\mu_{3.}.$$

A second contrast, orthogonal to w_1, for comparing system 1 with system 2, is given by $w_2 = \mu_{1.} - \mu_{2.}$. The single-degree-of-freedom sums of squares are

$$SSw_1 = \frac{[244.0 + 237.4 - (2)(228.8)]^2}{(8)[(1)^2 + (1)^2 + (-2)^2]} = 11.80$$

and

$$SSw_2 = \frac{(244.0 - 237.4)^2}{(8)[(1)^2 + (-1)^2]} = 2.72.$$

Notice that $SSw_1 + SSw_2 = SSA$, as expected. The computed *f*-values corresponding to w_1 and w_2 are, respectively,

$$f_1 = \frac{11.80}{1.24} = 9.5 \quad \text{and} \quad f_2 = \frac{2.72}{1.24} = 2.2.$$

Compared to the critical value $f_{0.05}(1, 12) = 4.75$, we find f_1 to be significant. In fact, the *P*-value is less than 0.01. Thus, the first contrast indicates that the hypothesis

$$H_0: \frac{1}{2}(\mu_{1.} + \mu_{2.}) = \mu_{3.}$$

is rejected. Since $f_2 < 4.75$, the mean burning rates of the first and second systems are not significantly different.

Impact of Significant Interaction in Example 14.1

If the hypothesis of no interaction in Example 14.1 is true, we could make the *general* comparisons of Example 14.2 regarding our missile systems rather than separate comparisons for each propellant. Similarly, we might make general comparisons among the propellants rather than separate comparisons for each missile system. For example, we could compare propellants 1 and 2 with 3 and 4 and also propellant 1 versus propellant 2. The resulting *f*-ratios, each with 1 and 12 degrees of freedom, turn out to be 24.81 and 7.39, respectively, and both are quite significant at the 0.05 level.

From propellant averages there appears to be evidence that propellant 1 gives the highest mean burning rate. A prudent experimenter might be somewhat cautious in drawing overall conclusions in a problem such as this one, where the *f*-ratio for interaction is barely below the 0.05 critical value. For example, the overall evidence, 31.60 versus 29.85 on the average for the two propellants, certainly indicates that propellant 1 is superior, in terms of a higher burning rate, to propellant 2. However, if we restrict ourselves to system 3, where we have an average of 28.85 for propellant 1 as opposed to 28.10 for propellant 2, there appears to be little or no difference between these two propellants. In fact, there appears to be a stabilization of burning rates for the different propellants if we operate with system 3. There is certainly overall evidence which indicates that system 1 gives a higher burning rate than system 3, but if we restrict ourselves to propellant 4, this conclusion does not appear to hold.

The analyst can conduct a simple *t*-test using average burning rates for system 3 in order to display conclusive evidence that interaction is *producing considerable difficulty in allowing broad conclusions on main effects*. Consider a comparison of propellant 1 against propellant 2 only using system 3. Borrowing an estimate of σ^2 from the overall analysis, that is, using $s^2 = 1.24$ with 12 degrees of freedom, we have

$$|t| = \frac{0.75}{\sqrt{2s^2/n}} = \frac{0.75}{\sqrt{1.24}} = 0.67,$$

which is not even close to being significant. This illustration suggests that one must be cautious about strict interpretation of main effects in the presence of interaction.

Graphical Analysis for the Two-Factor Problem of Example 14.1

Many of the same types of graphical displays that were suggested in the one-factor problems certainly apply in the two-factor case. Two-dimensional plots of cell means or treatment combination means can provide insight into the presence of

14.3 Two-Factor Analysis of Variance

interactions between the two factors. In addition, a plot of residuals against fitted values may well provide an indication of whether or not the homogeneous variance assumption holds. Often, of course, a violation of the homogeneous variance assumption involves an increase in the error variance as *the response numbers get larger*. As a result, this plot may point out the violation.

Figure 14.3 shows the plot of cell means in the case of the missile system propellant illustration in Example 14.1. Notice how graphically (in this case) the lack of parallelism shows through. Note the flatness of the part of the figure showing the propellant effect for system 3. This illustrates interaction among the factors. Figure 14.4 shows the plot of residuals against fitted values for the same data. There is no apparent sign of difficulty with the homogeneous variance assumption.

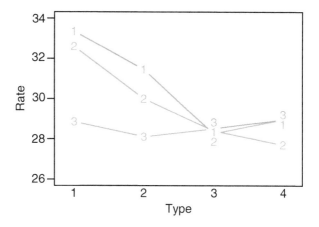

Figure 14.3: Plot of cell means for data of Example 14.1. Numbers represent missile systems.

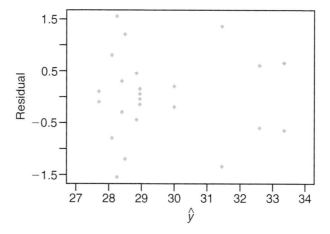

Figure 14.4: Residual plot of data of Example 14.1.

Example 14.3: An electrical engineer is investigating a plasma etching process used in semiconductor manufacturing. It is of interest to study the effects of two factors, the C_2F_6 gas flow rate (A) and the power applied to the cathode (B). The response is the etch rate. Each factor is run at 3 levels, and 2 experimental runs on etch rate are made for each of the 9 combinations. The setup is that of a completely randomized design. The data are given in Table 14.6. The etch rate is in $A°/\min$.

Table 14.6: Data for Example 14.3

C_2F_6 Flow Rate	Power Supplied		
	1	2	3
1	288	488	670
	360	465	720
2	385	482	692
	411	521	724
3	488	595	761
	462	612	801

The levels of the factors are in ascending order, with level 1 being low level and level 3 being the highest.

(a) Show an analysis of variance table and draw conclusions, beginning with the test on interaction.

(b) Do tests on main effects and draw conclusions.

Solution: A *SAS* output is given in Figure 14.5. From the output we learn the following.

```
                          The GLM Procedure
Dependent Variable: etchrate
                             Sum of
Source                 DF    Squares       Mean Square   F Value   Pr > F
Model                   8    379508.7778   47438.5972    61.00     <.0001
Error                   9    6999.5000     777.7222
Corrected Total        17    386508.2778

R-Square     Coeff Var     Root MSE      etchrate Mean
0.981890     5.057714      27.88767      551.3889

Source                 DF    Type III SS   Mean Square   F Value   Pr > F
c2f6                    2    46343.1111    23171.5556    29.79     0.0001
power                   2    330003.4444   165001.7222   212.16    <.0001
c2f6*power              4    3162.2222     790.5556      1.02      0.4485
```

Figure 14.5: *SAS* printout for Example 14.3.

(a) The *P*-value for the test of interaction is 0.4485. We can conclude that there is no significant interaction.

(b) There is a significant difference in mean etch rate for the 3 levels of C_2F_6 flow rate. Duncan's test shows that the mean etch rate for level 3 is significantly higher than that for level 2 and the rate for level 2 is significantly higher than that for level 1. See Figure 14.6(a).

There is a significant difference in mean etch rate based on the level of power to the cathode. Duncan's test revealed that the etch rate for level 3 is significantly higher than that for level 2 and the rate for level 2 is significantly higher than that for level 1. See Figure 14.6(b).

Duncan Grouping	Mean	N	c2f6
A	619.83	6	3
B	535.83	6	2
C	498.50	6	1

(a)

Duncan Grouping	Mean	N	power
A	728.00	6	3
B	527.17	6	2
C	399.00	6	1

(b)

Figure 14.6: *SAS* output, for Example 14.3. (a) Duncan's test on gas flow rate; (b) Duncan's test on power.

Exercises

14.1 An experiment was conducted to study the effects of temperature and type of oven on the life of a particular component. Four types of ovens and 3 temperature levels were used in the experiment. Twenty-four pieces were assigned randomly, two to each combination of treatments, and the following results recorded.

Temperature (°F)	O_1	O_2	O_3	O_4
500	227	214	225	260
	221	259	236	229
550	187	181	232	246
	208	179	198	273
600	174	198	178	206
	202	194	213	219

Using a 0.05 level of significance, test the hypothesis that

(a) different temperatures have no effect on the life of the component;

(b) different ovens have no effect on the life of the component;

(c) the type of oven and temperature do not interact.

14.2 To ascertain the stability of vitamin C in reconstituted frozen orange juice concentrate stored in a refrigerator for a period of up to one week, the study *Vitamin C Retention in Reconstituted Frozen Orange Juice* was conducted by the Department of Human Nutrition, Foods, and Exercise at Virginia Tech. Three types of frozen orange juice concentrate were tested using 3 different time periods. The time periods refer to the number of days from when the orange juice was blended until it was tested. The results, in milligrams of ascorbic acid per liter, were recorded. Use a 0.05 level of significance to test the hypothesis that

(a) there is no difference in ascorbic acid contents among the different brands of orange juice concentrate;

(b) there is no difference in ascorbic acid contents for the different time periods;

(c) the brands of orange juice concentrate and the number of days from the time the juice was blended until it was tested do not interact.

Brand	Time (days)		
	0	3	7
Richfood	52.6 54.2	49.4 49.2	42.7 48.8
	49.8 46.5	42.8 53.2	40.4 47.6
Sealed-Sweet	56.0 48.0	48.8 44.0	49.2 44.0
	49.6 48.4	44.0 42.4	42.0 43.2
Minute Maid	52.5 52.0	48.0 47.0	48.5 43.3
	51.8 53.6	48.2 49.6	45.2 47.6

14.3 Three strains of rats were studied under 2 environmental conditions for their performance in a maze test. The error scores for the 48 rats were recorded.

Environment	Strain		
	Bright	Mixed	Dull
Free	28 12	33 83	101 94
	22 23	36 14	33 56
	25 10	41 76	122 83
	36 86	22 58	35 23
Restricted	72 32	60 89	136 120
	48 93	35 126	38 153
	25 31	83 110	64 128
	91 19	99 118	87 140

Use a 0.01 level of significance to test the hypothesis that
(a) there is no difference in error scores for different environments;
(b) there is no difference in error scores for different strains;
(c) the environments and strains of rats do not interact.

14.4 Corrosion fatigue in metals has been defined as the simultaneous action of cyclic stress and chemical attack on a metal structure. A widely used technique for minimizing corrosion fatigue damage in aluminum involves the application of a protective coating. A study conducted by the Department of Mechanical Engineering at Virginia Tech used 3 different levels of humidity

Low: 20–25% relative humidity

Medium: 55–60% relative humidity

High: 86–91% relative humidity

and 3 types of surface coatings

Uncoated: no coating

Anodized: sulfuric acid anodic oxide coating

Conversion: chromate chemical conversion coating

The corrosion fatigue data, expressed in thousands of cycles to failure, were recorded as follows:

Coating	Relative Humidity					
	Low		Medium		High	
Uncoated	361	469	314	522	1344	1216
	466	937	244	739	1027	1097
	1069	1357	261	134	1011	1011
Anodized	114	1032	322	471	78	466
	1236	92	306	130	387	107
	533	211	68	398	130	327
Conversion	130	1482	252	874	586	524
	841	529	105	755	402	751
	1595	754	847	573	846	529

(a) Perform an analysis of variance with $\alpha = 0.05$ to test for significant main and interaction effects.
(b) Use Duncan's multiple-range test at the 0.05 level of significance to determine which humidity levels result in different corrosion fatigue damage.

14.5 To determine which muscles need to be subjected to a conditioning program in order to improve one's performance on the flat serve used in tennis, a study was conducted by the Human Nutrition, Foods, and Exercise Department at Virginia Tech. Five different muscles

1: anterior deltoid   4: middle deltoid
2: pectoral major   5: triceps
3: posterior deltoid

were tested on each of 3 subjects, and the experiment was carried out 3 times for each treatment combination. The electromyographic data, recorded during the serve, are presented here.

Subject	Muscle				
	1	2	3	4	5
1	32	5	58	10	19
	59	1.5	61	10	20
	38	2	66	14	23
2	63	10	64	45	43
	60	9	78	61	61
	50	7	78	71	42
3	43	41	26	63	61
	54	43	29	46	85
	47	42	23	55	95

Use a 0.01 level of significance to test the hypothesis that
(a) different subjects have equal electromyographic measurements;
(b) different muscles have no effect on electromyographic measurements;
(c) subjects and types of muscle do not interact.

14.6 An experiment was conducted to determine whether additives increase the adhesiveness of rubber products. Sixteen products were made with the new additive and another 16 without the new additive. The observed adhesiveness was as recorded below.

	Temperature (°C)			
	50	60	70	80
Without Additive	2.3	3.4	3.8	3.9
	2.9	3.7	3.9	3.2
	3.1	3.6	4.1	3.0
	3.2	3.2	3.8	2.7
With Additive	4.3	3.8	3.9	3.5
	3.9	3.8	4.0	3.6
	3.9	3.9	3.7	3.8
	4.2	3.5	3.6	3.9

Perform an analysis of variance to test for significant main and interaction effects.

14.7 The extraction rate of a certain polymer is known to depend on the reaction temperature and the amount of catalyst used. An experiment was conducted at four levels of temperature and five levels of the catalyst, and the extraction rate was recorded in the following table.

	Amount of Catalyst				
	0.5%	0.6%	0.7%	0.8%	0.9%
50°C	38	45	57	59	57
	41	47	59	61	58
60°C	44	56	70	73	61
	43	57	69	72	58
70°C	44	56	70	73	61
	47	60	67	61	59
80°C	49	62	70	62	53
	47	65	55	69	58

Perform an analysis of variance. Test for significant main and interaction effects.

14.8 In Myers, Montgomery, and Anderson-Cook (2009), a scenario is discussed involving an auto bumper plating process. The response is the thickness of the material. Factors that may impact the thickness include amount of nickel (A) and pH (B). A two-factor experiment is designed. The plan is a completely randomized design in which the individual bumpers are assigned randomly to the factor combinations. Three levels of pH and two levels of nickel content are involved in the experiment. The thickness data, in cm $\times 10^{-3}$, are as follows:

Nickel Content		pH	
(grams)	5	5.5	6
18	250	211	221
	195	172	150
	188	165	170
10	115	88	69
	165	112	101
	142	108	72

(a) Display the analysis-of-variance table with tests for both main effects and interaction. Show P-values.

(b) Give engineering conclusions. What have you learned from the analysis of the data?

(c) Show a plot that depicts either a presence or an absence of interaction.

14.9 An engineer is interested in the effects of cutting speed and tool geometry on the life in hours of a machine tool. Two cutting speeds and two different geometries are used. Three experimental tests are accomplished at each of the four combinations. The data are as follows.

Tool	Cutting Speed	
Geometry	Low	High
1	22 28 20	34 37 29
2	18 15 16	11 10 10

(a) Show an analysis-of-variance table with tests on interaction and main effects.

(b) Comment on the effect that interaction has on the test on cutting speed.

(c) Do secondary tests that will allow the engineer to learn the true impact of cutting speed.

(d) Show a plot that graphically displays the interaction effect.

14.10 Two factors in a manufacturing process for an integrated circuit are studied in a two-factor experiment. The purpose of the experiment is to learn their effect on the resistivity of the wafer. The factors are implant dose (2 levels) and furnace position (3 levels). Experimentation is costly so only one experimental run is made at each combination. The data are as follows.

Dose	Position		
1	15.5	14.8	21.3
2	27.2	24.9	26.1

It is to be assumed that no interaction exists between these two factors.

(a) Write the model and explain terms.

(b) Show the analysis-of-variance table.

(c) Explain the 2 "error" degrees of freedom.

(d) Use Tukey's test to do multiple-comparison tests on furnace position. Explain what the results show.

14.11 A study was done to determine the impact of two factors, method of analysis and the laboratory doing the analysis, on the level of sulfur content in coal. Twenty-eight coal specimens were randomly assigned to 14 factor combinations, the structure of the experimental units represented by combinations of seven laboratories and two methods of analysis with two specimens per factor combination. The data, expressed in percent of sulfur, are as follows:

	Method			
Laboratory	1		2	
1	0.109	0.105	0.105	0.108
2	0.129	0.122	0.127	0.124
3	0.115	0.112	0.109	0.111
4	0.108	0.108	0.117	0.118
5	0.097	0.096	0.110	0.097
6	0.114	0.119	0.116	0.122
7	0.155	0.145	0.164	0.160

(The data are taken from G. Taguchi, "Signal to Noise Ratio and Its Applications to Testing Material," *Reports of Statistical Application Research*, Union of Japanese Scientists and Engineers, Vol. 18, No. 4, 1971.)

(a) Do an analysis of variance and show results in an analysis-of-variance table.

(b) Is interaction significant? If so, discuss what it means to the scientist. Use a P-value in your conclusion.

(c) Are the individual main effects, laboratory, and method of analysis statistically significant? Discuss what is learned and let your answer be couched in the context of any significant interaction.

(d) Do an interaction plot that illustrates the effect of interaction.

(e) Do a test comparing methods 1 and 2 at laboratory 1 and do the same test at laboratory 7. Comment on what these results illustrate.

14.12 In an experiment conducted in the Civil Engineering Department at Virginia Tech, growth of a certain type of algae in water was observed as a function of time and the dosage of copper added to the water. The data are as follows. Response is in units of algae.

	Time in Days		
Copper	5	12	18
1	0.30	0.37	0.25
	0.34	0.36	0.23
	0.32	0.35	0.24
2	0.24	0.30	0.27
	0.23	0.32	0.25
	0.22	0.31	0.25
3	0.20	0.30	0.27
	0.28	0.31	0.29
	0.24	0.30	0.25

(a) Do an analysis of variance and show the analysis-of-variance table.

(b) Comment concerning whether the data are sufficient to show a time effect on algae concentration.

(c) Do the same for copper content. Does the level of copper impact algae concentration?

(d) Comment on the results of the test for interaction. How is the effect of copper content influenced by time?

14.13 In Myers, *Classical and Modern Regression with Applications* (Duxbury Classic Series, 2nd edition, 1990), an experiment is described in which the Environmental Protection Agency seeks to determine the effect of two water treatment methods on magnesium uptake. Magnesium levels in grams per cubic centimeter (cc) are measured, and two different time levels are incorporated into the experiment. The data are as follows:

	Treatment	
Time (hr)	1	2
1	2.19 2.15 2.16	2.03 2.01 2.04
2	2.01 2.03 2.04	1.88 1.86 1.91

(a) Do an interaction plot. What is your impression?

(b) Do an analysis of variance and show tests for the main effects and interaction.

(c) Give scientific findings regarding how time and treatment influence magnesium uptake.

(d) Fit the appropriate regression model with treatment as a categorical variable. Include interaction in the model.

(e) Is interaction significant in the regression model?

14.14 Consider the data set in Exercise 14.12 and answer the following questions.

(a) Both factors, copper and time, are quantitative in nature. As a result, a regression model may be of interest. Describe what might be an appropriate model using x_1 = copper content and x_2 = time. Fit the model to the data, showing regression coefficients and a t-test on each.

(b) Fit the model

$$Y = \beta_0 + \beta_1 x_1 + \beta_2 x_2 + \beta_{12} x_1 x_2 + \beta_{11} x_1^2 + \beta_{22} x_2^2 + \epsilon,$$

and compare it to the one you chose in (a). Which is more appropriate? Use R^2_{adj} as a criterion.

14.15 The purpose of the study *The Incorporation of a Chelating Agent into a Flame Retardant Finish of a Cotton Flannelette and the Evaluation of Selected Fabric Properties*, conducted at Virginia Tech, was to evaluate the use of a chelating agent as part of the flame retardant finish of cotton flannelette by determining its effect upon flammability after the fabric is laundered under specific conditions. There were two treatments at two levels. Two baths were prepared, one with carboxymethyl cellulose (bath I) and one without (bath II). Half of the fabric was laundered 5 times and half was laundered 10 times. There were 12 pieces of fabric in each bath/number of launderings combination. After the washings, the lengths of fabric that burned and the burn times were measured. Burn times (in seconds) were recorded as follows:

Launderings	Bath I			Bath II		
5	13.7	23.0	15.7	6.2	5.4	5.0
	25.5	15.8	14.8	4.4	5.0	3.3
	14.0	29.4	9.7	16.0	2.5	1.6
	14.0	12.3	12.3	3.9	2.5	7.1
10	27.2	16.8	12.9	18.2	8.8	14.5
	14.9	17.1	13.0	14.7	17.1	13.9
	10.8	13.5	25.5	10.6	5.8	7.3
	14.2	27.4	11.5	17.7	18.3	9.9

(a) Perform an analysis of variance. Is there a significant interaction term?

(b) Are there main effect differences? Discuss.

14.4 Three-Factor Experiments

In this section, we consider an experiment with three factors, A, B, and C, at a, b, and c levels, respectively, in a completely randomized experimental design. Assume again that we have n observations for each of the abc treatment combinations. We shall proceed to outline significance tests for the three main effects and interactions involved. It is hoped that the reader can then use the description given here to generalize the analysis to $k > 3$ factors.

Model for the Three-Factor Experiment

The model for the three-factor experiment is

$$y_{ijkl} = \mu + \alpha_i + \beta_j + \gamma_k + (\alpha\beta)_{ij} + (\alpha\gamma)_{ik} + (\beta\gamma)_{jk} + (\alpha\beta\gamma)_{ijk} + \epsilon_{ijkl},$$

$i = 1, 2, \ldots, a$; $j = 1, 2, \ldots, b$; $k = 1, 2, \ldots, c$; and $l = 1, 2, \ldots, n$, where α_i, β_j, and γ_k are the main effects and $(\alpha\beta)_{ij}$, $(\alpha\gamma)_{ik}$, and $(\beta\gamma)_{jk}$ are the two-factor interaction effects that have the same interpretation as in the two-factor experiment.

The term $(\alpha\beta\gamma)_{ijk}$ is called the **three-factor interaction effect**, a term that represents a nonadditivity of the $(\alpha\beta)_{ij}$ over the different levels of the factor C. As before, the sum of all main effects is zero and the sum over any subscript of the two- and three-factor interaction effects is zero. In many experimental situations, these higher-order interactions are insignificant and their mean squares reflect only random variation, but we shall outline the analysis in its most general form.

Again, in order that valid significance tests can be made, we must assume that the errors are values of independent and normally distributed random variables, each with mean 0 and common variance σ^2.

The general philosophy concerning the analysis is the same as that discussed for the one- and two-factor experiments. The sum of squares is partitioned into eight terms, each representing a source of variation from which we obtain independent estimates of σ^2 when all the main effects and interaction effects are zero. If the effects of any given factor or interaction are not all zero, then the mean square will estimate the error variance plus a component due to the systematic effect in question.

Sum of Squares for a Three-Factor Experiment

$$SSA = bcn \sum_{i=1}^{a} (\bar{y}_{i...} - \bar{y}_{....})^2 \qquad SS(AB) = cn \sum_i \sum_j (\bar{y}_{ij..} - \bar{y}_{i...} - \bar{y}_{.j..} + \bar{y}_{....})^2$$

$$SSB = acn \sum_{j=1}^{b} (\bar{y}_{.j..} - \bar{y}_{....})^2 \qquad SS(AC) = bn \sum_i \sum_k (\bar{y}_{i.k.} - \bar{y}_{i...} - \bar{y}_{..k.} + \bar{y}_{....})^2$$

$$SSC = abn \sum_{k=1}^{c} (\bar{y}_{..k.} - \bar{y}_{....})^2 \qquad SS(BC) = an \sum_j \sum_k (\bar{y}_{.jk.} - \bar{y}_{.j..} - \bar{y}_{..k.} + \bar{y}_{....})^2$$

$$SS(ABC) = n \sum_i \sum_j \sum_k (\bar{y}_{ijk.} - \bar{y}_{ij..} - \bar{y}_{i.k.} - \bar{y}_{.jk.} + \bar{y}_{i...} + \bar{y}_{.j..} + \bar{y}_{..k.} - \bar{y}_{....})^2$$

$$SST = \sum_i \sum_j \sum_k \sum_l (y_{ijkl} - \bar{y}_{....})^2 \qquad SSE = \sum_i \sum_j \sum_k \sum_l (y_{ijkl} - \bar{y}_{ijk.})^2$$

Although we emphasize interpretation of annotated computer printout in this section rather than being concerned with laborious computation of sums of squares, we do offer the following as the sums of squares for the three main effects and interactions. Notice the obvious extension from the two- to three-factor problem.

The averages in the formulas are defined as follows:

$\bar{y}_{....}$ = average of all $abcn$ observations,

$\bar{y}_{i...}$ = average of the observations for the ith level of factor A,

$\bar{y}_{.j..}$ = average of the observations for the jth level of factor B,

$\bar{y}_{..k.}$ = average of the observations for the kth level of factor C,

$\bar{y}_{ij..}$ = average of the observations for the ith level of A and the jth level of B,

$\bar{y}_{i.k.}$ = average of the observations for the ith level of A and the kth level of C,

$\bar{y}_{.jk.}$ = average of the observations for the jth level of B and the kth level of C,

$\bar{y}_{ijk.}$ = average of the observations for the (ijk)th treatment combination.

The computations in an analysis-of-variance table for a three-factor problem with n replicated runs at each factor combination are summarized in Table 14.7.

Table 14.7: ANOVA for the Three-Factor Experiment with n Replications

Source of Variation	Sum of Squares	Degrees of Freedom	Mean Square	Computed f
Main effect:				
A	SSA	$a-1$	s_1^2	$f_1 = \frac{s_1^2}{s^2}$
B	SSB	$b-1$	s_2^2	$f_2 = \frac{s_2^2}{s^2}$
C	SSC	$c-1$	s_3^2	$f_3 = \frac{s_3^2}{s^2}$
Two-factor interaction:				
AB	$SS(AB)$	$(a-1)(b-1)$	s_4^2	$f_4 = \frac{s_4^2}{s^2}$
AC	$SS(AC)$	$(a-1)(c-1)$	s_5^2	$f_5 = \frac{s_5^2}{s^2}$
BC	$SS(BC)$	$(b-1)(c-1)$	s_6^2	$f_6 = \frac{s_6^2}{s^2}$
Three-factor interaction:				
ABC	$SS(ABC)$	$(a-1)(b-1)(c-1)$	s_7^2	$f_7 = \frac{s_7^2}{s^2}$
Error	SSE	$abc(n-1)$	s^2	
Total	SST	$abcn-1$		

For the three-factor experiment with a single experimental run per combination, we may use the analysis of Table 14.7 by setting $n = 1$ and using the ABC interaction sum of squares for SSE. In this case, we are assuming that the $(\alpha\beta\gamma)_{ijk}$ interaction effects are all equal to zero so that

$$E\left[\frac{SS(ABC)}{(a-1)(b-1)(c-1)}\right] = \sigma^2 + \frac{n}{(a-1)(b-1)(c-1)}\sum_{i=1}^{a}\sum_{j=1}^{b}\sum_{k=1}^{c}(\alpha\beta\gamma)_{ijk}^2 = \sigma^2.$$

14.4 Three-Factor Experiments

That is, $SS(ABC)$ represents variation due only to experimental error. Its mean square thereby provides an unbiased estimate of the error variance. With $n = 1$ and $SSE = SS(ABC)$, the error sum of squares is found by subtracting the sums of squares of the main effects and two-factor interactions from the total sum of squares.

Example 14.4: In the production of a particular material, three variables are of interest: A, the operator effect (three operators); B, the catalyst used in the experiment (three catalysts); and C, the washing time of the product following the cooling process (15 minutes and 20 minutes). Three runs were made at each combination of factors. It was felt that all interactions among the factors should be studied. The coded yields are in Table 14.8. Perform an analysis of variance to test for significant effects.

Table 14.8: Data for Example 14.4

				Washing Time, C			
		15 Minutes			20 Minutes		
		Catalyst, B			Catalyst, B		
Operator, A		1	2	3	1	2	3
1		10.7	10.3	11.2	10.9	10.5	12.2
		10.8	10.2	11.6	12.1	11.1	11.7
		11.3	10.5	12.0	11.5	10.3	11.0
2		11.4	10.2	10.7	9.8	12.6	10.8
		11.8	10.9	10.5	11.3	7.5	10.2
		11.5	10.5	10.2	10.9	9.9	11.5
3		13.6	12.0	11.1	10.7	10.2	11.9
		14.1	11.6	11.0	11.7	11.5	11.6
		14.5	11.5	11.5	12.7	10.9	12.2

Solution: Table 14.9 shows an analysis of variance of the data given above. None of the interactions show a significant effect at the $\alpha = 0.05$ level. However, the P-value for BC is 0.0610; thus, it should not be ignored. The operator and catalyst effects are significant, while the effect of washing time is not significant.

Impact of Interaction BC

More should be discussed regarding Example 14.4, particularly about dealing with the effect that the interaction between catalyst and washing time is having on the test on the washing time main effect (factor C). Recall our discussion in Section 14.2. Illustrations were given of how the presence of interaction could change the interpretation that we make regarding main effects. In Example 14.4, the BC interaction is significant at approximately the 0.06 level. Suppose, however, that we observe a two-way table of means as in Table 14.10.

It is clear why washing time was found not to be significant. A non-thorough analyst may get the impression that washing time can be eliminated from any future study in which yield is being measured. However, it is obvious how the

Table 14.9: ANOVA for a Three-Factor Experiment in a Completely Randomized Design

Source	df	Sum of Squares	Mean Square	F-Value	P-Value
A	2	13.98	6.99	11.64	0.0001
B	2	10.18	5.09	8.48	0.0010
AB	4	4.77	1.19	1.99	0.1172
C	1	1.19	1.19	1.97	0.1686
AC	2	2.91	1.46	2.43	0.1027
BC	2	3.63	1.82	3.03	0.0610
ABC	4	4.91	1.23	2.04	0.1089
Error	36	21.61	0.60		
Total	53	63.19			

Table 14.10: Two-Way Table of Means for Example 14.4

	Washing Time, C	
Catalyst, B	15 min	20 min
1	12.19	11.29
2	10.86	10.50
3	11.09	11.46
Means	11.38	11.08

effect of washing time changes from a negative effect for the first catalyst to what appears to be a positive effect for the third catalyst. If we merely focus on the data for catalyst 1, a simple comparison between the means at the two washing times will produce a simple t-statistic:

$$t = \frac{12.19 - 11.29}{\sqrt{0.6(2/9)}} = 2.5,$$

which is significant at a level less than 0.02. Thus, an important negative effect of washing time for catalyst 1 might very well be ignored if the analyst makes the incorrect broad interpretation of the insignificant F-ratio for washing time.

Pooling in Multifactor Models

We have described the three-factor model and its analysis in the most general form by including all possible interactions in the model. Of course, there are many situations where it is known *a priori* that the model should not contain certain interactions. We can then take advantage of this knowledge by combining or pooling the sums of squares corresponding to negligible interactions with the error sum of squares to form a new estimator for σ^2 with a larger number of degrees of freedom. For example, in a metallurgy experiment designed to study the effect on film thickness of three important processing variables, suppose it is known that factor A, acid concentration, does not interact with factors B and C. The

14.4 Three-Factor Experiments

Table 14.11: ANOVA with Factor A Noninteracting

Source of Variation	Sum of Squares	Degrees of Freedom	Mean Square	Computed f
Main effect:				
A	SSA	$a-1$	s_1^2	$f_1 = \frac{s_1^2}{s^2}$
B	SSB	$b-1$	s_2^2	$f_2 = \frac{s_2^2}{s^2}$
C	SSC	$c-1$	s_3^2	$f_3 = \frac{s_3^2}{s^2}$
Two-factor interaction:				
BC	$SS(BC)$	$(b-1)(c-1)$	s_4^2	$f_4 = \frac{s_4^2}{s^2}$
Error	SSE	Subtraction	s^2	
Total	SST	$abcn-1$		

sums of squares SSA, SSB, SSC, and $SS(BC)$ are computed using the methods described earlier in this section. The mean squares for the remaining effects will now all independently estimate the error variance σ^2. Therefore, we form our new **mean square error by pooling** $SS(AB)$, $SS(AC)$, $SS(ABC)$, and SSE, along with the corresponding degrees of freedom. The resulting denominator for the significance tests is then the mean square error given by

$$s^2 = \frac{SS(AB) + SS(AC) + SS(ABC) + SSE}{(a-1)(b-1) + (a-1)(c-1) + (a-1)(b-1)(c-1) + abc(n-1)}.$$

Computationally, of course, one obtains the pooled sum of squares and the pooled degrees of freedom by subtraction once SST and the sums of squares for the existing effects are computed. The analysis-of-variance table would then take the form of Table 14.11.

Factorial Experiments in Blocks

In this chapter, we have assumed that the experimental design used is a completely randomized design. By interpreting the levels of factor A in Table 14.11 **as different blocks**, we then have the analysis-of-variance procedure for a two-factor experiment in a randomized block design. For example, if we interpret the operators in Example 14.4 as blocks and assume no interaction between blocks and the other two factors, the analysis of variance takes the form of Table 14.12 rather than that of Table 14.9. The reader can verify that the mean square error is also

$$s^2 = \frac{4.77 + 2.91 + 4.91 + 21.61}{4 + 2 + 4 + 36} = 0.74,$$

which demonstrates the pooling of the sums of squares for the nonexisting interaction effects. Note that factor B, catalyst, has a significant effect on yield.

Table 14.12: ANOVA for a Two-Factor Experiment in a Randomized Block Design

Source of Variation	Sum of Squares	Degrees of Freedom	Mean Square	Computed f	P-Value
Blocks	13.98	2	6.99		
Main effect:					
B	10.18	2	5.09	6.88	0.0024
C	1.18	1	1.18	1.59	0.2130
Two-factor interaction:					
BC	3.64	2	1.82	2.46	0.0966
Error	34.21	46	0.74		
Total	63.19	53			

Example 14.5: An experiment was conducted to determine the effects of temperature, pressure, and stirring rate on product filtration rate. This was done in a pilot plant. The experiment was run at two levels of each factor. In addition, it was decided that two batches of raw materials should be used, where batches were treated as blocks. Eight experimental runs were made in random order for each batch of raw materials. It is thought that all two-factor interactions may be of interest. No interactions with batches are assumed to exist. The data appear in Table 14.13. "L" and "H" imply low and high levels, respectively. The filtration rate is in gallons per hour.

(a) Show the complete ANOVA table. Pool all "interactions" with blocks into error.

(b) What interactions appear to be significant?

(c) Create plots to reveal and interpret the significant interactions. Explain what the plot means to the engineer.

Table 14.13: Data for Example 14.5

	Batch 1					
	Low Stirring Rate			High Stirring Rate		
Temp.	Pressure L	Pressure H	Temp.	Pressure L	Pressure H	
L	43	49	L	44	47	
H	64	68	H	97	102	
	Batch 2					
	Low Stirring Rate			High Stirring Rate		
Temp.	Pressure L	Pressure H	Temp.	Pressure L	Pressure H	
L	49	57	L	51	55	
H	70	76	H	103	106	

14.4 Three-Factor Experiments

Solution: (a) The *SAS* printout is given in Figure 14.7.

(b) As seen in Figure 14.7, the temperature by stirring rate (strate) interaction appears to be highly significant. The pressure by stirring rate interaction also appears to be significant. Incidentally, if one were to do further pooling by combining the insignificant interactions with error, the conclusions would remain the same and the *P*-value for the pressure by stirring rate interaction would become stronger, namely 0.0517.

(c) The main effects for both stirring rate and temperature are highly significant, as shown in Figure 14.7. A look at the interaction plot of Figure 14.8(a) shows that the effect of stirring rate is dependent upon the level of temperature. At the low level of temperature the stirring rate effect is negligible, whereas at the high level of temperature stirring rate has a strong positive effect on mean filtration rate. In Figure 14.8(b), the interaction between pressure and stirring rate, though not as pronounced as that of Figure 14.8(a), still shows a slight inconsistency of the stirring rate effect across pressure.

Source	DF	Type III SS	Mean Square	F Value	Pr > F
batch	1	175.562500	175.562500	177.14	<.0001
pressure	1	95.062500	95.062500	95.92	<.0001
temp	1	5292.562500	5292.562500	5340.24	<.0001
pressure*temp	1	0.562500	0.562500	0.57	0.4758
strate	1	1040.062500	1040.062500	1049.43	<.0001
pressure*strate	1	5.062500	5.062500	5.11	0.0583
temp*strate	1	1072.562500	1072.562500	1082.23	<.0001
pressure*temp*strate	1	1.562500	1.562500	1.58	0.2495
Error	7	6.937500	0.991071		
Corrected Total	15	7689.937500			

Figure 14.7: ANOVA for Example 14.5, batch interaction pooled with error.

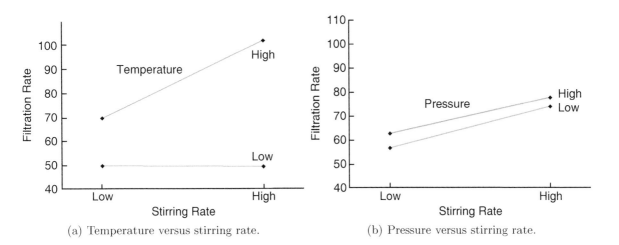

(a) Temperature versus stirring rate.   (b) Pressure versus stirring rate.

Figure 14.8: Interaction plots for Example 14.5.

Exercises

14.16 Consider an experimental situation involving factors A, B, and C, where we assume a three-way fixed effects model of the form $y_{ijkl} = \mu + \alpha_i + \beta_j + \gamma_k + (\beta\gamma)_{jk} + \epsilon_{ijkl}$. All other interactions are considered to be nonexistent or negligible. The data are presented here.

	B_1			B_2		
	C_1	C_2	C_3	C_1	C_2	C_3
A_1	4.0	3.4	3.9	4.4	3.1	3.1
	4.9	4.1	4.3	3.4	3.5	3.7
A_2	3.6	2.8	3.1	2.7	2.9	3.7
	3.9	3.2	3.5	3.0	3.2	4.2
A_3	4.8	3.3	3.6	3.6	2.9	2.9
	3.7	3.8	4.2	3.8	3.3	3.5
A_4	3.6	3.2	3.2	2.2	2.9	3.6
	3.9	2.8	3.4	3.5	3.2	4.3

(a) Perform a test of significance on the BC interaction at the $\alpha = 0.05$ level.

(b) Perform tests of significance on the main effects A, B, and C using a pooled mean square error at the $\alpha = 0.05$ level.

14.17 The following data are measurements from an experiment conducted using three factors A, B, and C, all fixed effects:

		C_1			C_2			C_3	
	B_1	B_2	B_3	B_1	B_2	B_3	B_1	B_2	B_3
A_1	15.0	14.8	15.9	16.8	14.2	13.2	15.8	15.5	19.2
	18.5	13.6	14.8	15.4	12.9	11.6	14.3	13.7	13.5
	22.1	12.2	13.6	14.3	13.0	10.1	13.0	12.6	11.1
A_2	11.3	17.2	16.1	18.9	15.4	12.4	12.7	17.3	7.8
	14.6	15.5	14.7	17.3	17.0	13.6	14.2	15.8	11.5
	18.2	14.2	13.4	16.1	18.6	15.2	15.9	14.6	12.2

(a) Perform tests of significance on all interactions at the $\alpha = 0.05$ level.

(b) Perform tests of significance on the main effects at the $\alpha = 0.05$ level.

(c) Give an explanation of how a significant interaction has masked the effect of factor C.

14.18 The method of X-ray fluorescence is an important analytical tool for determining the concentration of material in solid missile propellants. In the paper *An X-ray Fluorescence Method for Analyzing Polybutadiene Acrylic Acid (PBAA) Propellants* (Quarterly Report, RK-TR-62-1, Army Ordinance Missile Command, 1962), it is postulated that the propellant mixing process and analysis time have an influence on the homogeneity of the material and hence on the accuracy of X-ray intensity measurements. An experiment was conducted using 3 factors: A, the mixing conditions (4 levels); B, the analysis time (2 levels); and C, the method of loading propellant into sample holders (hot and room temperature). The following data, which represent the weight percent of ammonium perchlorate in a particular propellant, were recorded.

		Method of Loading, C			
		Hot		Room Temp.	
A		B_1	B_2	B_1	B_2
1		38.62	38.45	39.82	39.82
		37.20	38.64	39.15	40.26
		38.02	38.75	39.78	39.72
2		37.67	37.81	39.53	39.56
		37.57	37.75	39.76	39.25
		37.85	37.91	39.90	39.04
3		37.51	37.21	39.34	39.74
		37.74	37.42	39.60	39.49
		37.58	37.79	39.62	39.45
4		37.52	37.60	40.09	39.36
		37.15	37.55	39.63	39.38
		37.51	37.91	39.67	39.00

(a) Perform an analysis of variance with $\alpha = 0.01$ to test for significant main and interaction effects.

(b) Discuss the influence of the three factors on the weight percent of ammonium perchlorate. Let your discussion involve the role of any significant interaction.

14.19 Corrosion fatigue in metals has been defined as the simultaneous action of cyclic stress and chemical attack on a metal structure. In the study *Effect of Humidity and Several Surface Coatings on the Fatigue Life of 2024-T351 Aluminum Alloy*, conducted by the Department of Mechanical Engineering at Virginia Tech, a technique involving the application of a protective chromate coating was used to minimize corrosion fatigue damage in aluminum. Three factors were used in the investigation, with 5 replicates for each treatment combination: coating, at 2 levels, and humidity and shear stress, both with 3 levels. The fatigue data, recorded in thousands of cycles to failure, are presented here.

(a) Perform an analysis of variance with $\alpha = 0.01$ to test for significant main and interaction effects.

(b) Make a recommendation for combinations of the three factors that would result in low fatigue damage.

Coating	Humidity	Shear Stress (psi)		
		13,000	17,000	20,000
Uncoated	Low	4580	5252	361
	(20–25% RH)	10,126	897	466
		1341	1465	1069
		6414	2694	469
		3549	1017	937
	Medium	2858	799	314
	(50–60% RH)	8829	3471	244
		10,914	685	261
		4067	810	522
		2595	3409	739
	High	6489	1862	1344
	(86–91% RH)	5248	2710	1027
		6816	2632	663
		5860	2131	1216
		5901	2470	1097
Chromated	Low	5395	4035	130
	(20–25% RH)	2768	2022	841
		1821	914	1595
		3604	2036	1482
		4106	3524	529
	Medium	4833	1847	252
	(50–60% RH)	7414	1684	105
		10,022	3042	847
		7463	4482	874
		21,906	996	755
	High	3287	1319	586
	(86–91% RH)	5200	929	402
		5493	1263	846
		4145	2236	524
		3336	1392	751

14.20 For a study of the hardness of gold dental fillings, five randomly chosen dentists were assigned combinations of three methods of condensation and two types of gold. The hardness was measured. (See Hoaglin, Mosteller, and Tukey, 1991.) Let the dentists play the role of blocks. The data are presented here.

(a) State the appropriate model with the assumptions.
(b) Is there a significant interaction between method of condensation and type of gold filling material?
(c) Is there one method of condensation that seems to be best? Explain.

Dentist	Method	Type	
		Gold Foil	Goldent
1	1	792	824
	2	772	772
	3	782	803
2	1	803	803
	2	752	772
	3	715	707

(cont.)

Dentist	Method	Type	
		Gold Foil	Goldent
3	1	715	724
	2	792	715
	3	762	606
4	1	673	946
	2	657	743
	3	690	245
5	1	634	715
	2	649	724
	3	724	627

14.21 Electronic copiers make copies by gluing black ink on paper, using static electricity. Heating and gluing the ink on the paper comprise the final stage of the copying process. The gluing power during this final process determines the quality of the copy. It is postulated that temperature, surface state of the gluing roller, and hardness of the press roller influence the gluing power of the copier. An experiment is run with treatments consisting of a combination of these three factors at each of three levels. The following data show the gluing power for each treatment combination. Perform an analysis of variance with $\alpha = 0.05$ to test for significant main and interaction effects.

	Surface State of Gluing Roller	Hardness of the Press Roller		
		20	40	60
Low Temp.	Soft	0.52 0.44	0.54 0.52	0.60 0.55
		0.57 0.53	0.65 0.56	0.78 0.68
	Medium	0.64 0.59	0.79 0.73	0.49 0.48
		0.58 0.64	0.79 0.78	0.74 0.50
	Hard	0.67 0.77	0.58 0.68	0.55 0.65
		0.74 0.65	0.57 0.59	0.57 0.58
Medium Temp.	Soft	0.46 0.40	0.31 0.49	0.56 0.42
		0.58 0.37	0.48 0.66	0.49 0.49
	Medium	0.60 0.43	0.66 0.57	0.64 0.54
		0.62 0.61	0.72 0.56	0.74 0.56
	Hard	0.53 0.65	0.53 0.45	0.56 0.66
		0.66 0.56	0.59 0.47	0.71 0.67
High Temp.	Soft	0.52 0.44	0.54 0.52	0.65 0.49
		0.57 0.53	0.65 0.56	0.65 0.52
	Medium	0.53 0.65	0.53 0.45	0.49 0.48
		0.66 0.56	0.59 0.47	0.74 0.50
	Hard	0.43 0.43	0.48 0.31	0.55 0.65
		0.47 0.44	0.43 0.27	0.57 0.58

14.22 Consider the data set in Exercise 14.21.
(a) Construct an interaction plot for any two-factor interaction that is significant.
(b) Do a normal probability plot of residuals and comment.

14.23 Consider combinations of three factors in the

removal of dirt from standard loads of laundry. The first factor is the brand of the detergent, X, Y, or Z. The second factor is the type of detergent, liquid or powder. The third factor is the temperature of the water, hot or warm. The experiment was replicated three times. Response is percent dirt removal. The data are as follows:

Brand	Type	Temperature			
X	Powder	Hot	85	88	80
		Warm	82	83	85
	Liquid	Hot	78	75	72
		Warm	75	75	73
Y	Powder	Hot	90	92	92
		Warm	88	86	88
	Liquid	Hot	78	76	70
		Warm	76	77	76
Z	Powder	Hot	85	87	88
		Warm	76	74	78
	Liquid	Hot	60	70	68
		Warm	55	57	54

(a) Are there significant interaction effects at the $\alpha = 0.05$ level?

(b) Are there significant differences between the three brands of detergent?

(c) Which combination of factors would you prefer to use?

14.24 A scientist collects experimental data on the radius of a propellant grain, y, as a function of powder temperature, extrusion rate, and die temperature. Results of the three-factor experiment are as follows:

	Powder Temp			
	150		190	
	Die Temp		Die Temp	
Rate	220	250	220	250
12	82	124	88	129
24	114	157	121	164

Resources are not available to make repeated experimental trials at the eight combinations of factors. It is believed that extrusion rate does not interact with die temperature and that the three-factor interaction should be negligible. Thus, these two interactions may be pooled to produce a 2 d.f. "error" term.

(a) Do an analysis of variance that includes the three main effects and two two-factor interactions. Determine what effects influence the radius of the propellant grain.

(b) Construct interaction plots for the powder temperature by die temperature and powder temperature by extrusion rate interactions.

(c) Comment on the consistency in the appearance of the interaction plots and the tests on the two interactions in the ANOVA.

14.25 In the book *Design of Experiments for Quality Improvement*, published by the Japanese Standards Association (1989), a study is reported on the extraction of polyethylene by using a solvent and how the amount of gel (proportion) is influenced by three factors: the type of solvent, extraction temperature, and extraction time. A factorial experiment was designed, and the following data were collected on proportion of gel.

Solvent	Temp.	Time					
		4		8		16	
Ethanol	120	94.0	94.0	93.8	94.2	91.1	90.5
	80	95.3	95.1	94.9	95.3	92.5	92.4
Toluene	120	94.6	94.5	93.6	94.1	91.1	91.0
	80	95.4	95.4	95.6	96.0	92.1	92.1

(a) Do an analysis of variance and determine what factors and interactions influence the proportion of gel.

(b) Construct an interaction plot for any two-factor interaction that is significant. In addition, explain what conclusion can be drawn from the presence of the interaction.

(c) Do a normal probability plot of residuals and comment.

14.5 Factorial Experiments for Random Effects and Mixed Models

In a two-factor experiment with random effects, we have the model

$$Y_{ijk} = \mu + A_i + B_j + (AB)_{ij} + \epsilon_{ijk},$$

for $i = 1, 2, \ldots, a$; $j = 1, 2, \ldots, b$; and $k = 1, 2, \ldots, n$, where the A_i, B_j, $(AB)_{ij}$, and ϵ_{ijk} are independent random variables with means 0 and variances σ_α^2, σ_β^2, $\sigma_{\alpha\beta}^2$, and σ^2, respectively. The sums of squares for random effects experiments are computed in exactly the same way as for fixed effects experiments. We are now

14.5 Factorial Experiments for Random Effects and Mixed Models

interested in testing hypotheses of the form

$$H_0': \sigma_\alpha^2 = 0, \quad H_0'': \sigma_\beta^2 = 0, \quad H_0''': \sigma_{\alpha\beta}^2 = 0,$$
$$H_1': \sigma_\alpha^2 \neq 0, \quad H_1'': \sigma_\beta^2 \neq 0, \quad H_1''': \sigma_{\alpha\beta}^2 \neq 0,$$

where the denominator in the *f*-ratio is not necessarily the mean square error. The appropriate denominator can be determined by examining the expected values of the various mean squares. These are shown in Table 14.14.

Table 14.14: Expected Mean Squares for a Two-Factor Random Effects Experiment

Source of Variation	Degrees of Freedom	Mean Square	Expected Mean Square
A	$a-1$	s_1^2	$\sigma^2 + n\sigma_{\alpha\beta}^2 + bn\sigma_\alpha^2$
B	$b-1$	s_2^2	$\sigma^2 + n\sigma_{\alpha\beta}^2 + an\sigma_\beta^2$
AB	$(a-1)(b-1)$	s_3^2	$\sigma^2 + n\sigma_{\alpha\beta}^2$
Error	$ab(n-1)$	s^2	σ^2
Total	$abn-1$		

From Table 14.14 we see that H_0' and H_0'' are tested by using s_3^2 in the denominator of the *f*-ratio, whereas H_0''' is tested using s^2 in the denominator. The unbiased estimates of the variance components are

$$\hat{\sigma}^2 = s^2, \quad \hat{\sigma}_{\alpha\beta}^2 = \frac{s_3^2 - s^2}{n}, \quad \hat{\sigma}_\alpha^2 = \frac{s_1^2 - s_3^2}{bn}, \quad \hat{\sigma}_\beta^2 = \frac{s_2^2 - s_3^2}{an}.$$

Table 14.15: Expected Mean Squares for a Three-Factor Random Effects Experiment

Source of Variation	Degrees of Freedom	Mean Square	Expected Mean Square
A	$a-1$	s_1^2	$\sigma^2 + n\sigma_{\alpha\beta\gamma}^2 + cn\sigma_{\alpha\beta}^2 + bn\sigma_{\alpha\gamma}^2 + bcn\sigma_\alpha^2$
B	$b-1$	s_2^2	$\sigma^2 + n\sigma_{\alpha\beta\gamma}^2 + cn\sigma_{\alpha\beta}^2 + an\sigma_{\beta\gamma}^2 + acn\sigma_\beta^2$
C	$c-1$	s_3^2	$\sigma^2 + n\sigma_{\alpha\beta\gamma}^2 + bn\sigma_{\alpha\gamma}^2 + an\sigma_{\beta\gamma}^2 + abn\sigma_\gamma^2$
AB	$(a-1)(b-1)$	s_4^2	$\sigma^2 + n\sigma_{\alpha\beta\gamma}^2 + cn\sigma_{\alpha\beta}^2$
AC	$(a-1)(c-1)$	s_5^2	$\sigma^2 + n\sigma_{\alpha\beta\gamma}^2 + bn\sigma_{\alpha\gamma}^2$
BC	$(b-1)(c-1)$	s_6^2	$\sigma^2 + n\sigma_{\alpha\beta\gamma}^2 + an\sigma_{\beta\gamma}^2$
ABC	$(a-1)(b-1)(c-1)$	s_7^2	$\sigma^2 + n\sigma_{\alpha\beta\gamma}^2$
Error	$abc(n-1)$	s^2	σ^2
Total	$abcn-1$		

The expected mean squares for the three-factor experiment with random effects in a completely randomized design are shown in Table 14.15. It is evident from the expected mean squares of Table 14.15 that one can form appropriate *f*-ratios for testing all two-factor and three-factor interaction variance components. However, to test a hypothesis of the form

$$H_0: \sigma_\alpha^2 = 0,$$
$$H_1: \sigma_\alpha^2 \neq 0,$$

there appears to be no appropriate f-ratio unless we have found one or more of the two-factor interaction variance components not significant. Suppose, for example, that we have compared s_5^2 (mean square AC) with s_7^2 (mean square ABC) and found $\sigma_{\alpha\gamma}^2$ to be negligible. We could then argue that the term $\sigma_{\alpha\gamma}^2$ should be dropped from all the expected mean squares of Table 14.15; then the ratio s_1^2/s_4^2 provides a test for the significance of the variance component σ_α^2. Therefore, if we are to test hypotheses concerning the variance components of the main effects, it is necessary first to investigate the significance of the two-factor interaction components. An approximate test derived by Satterthwaite (1946; see the Bibliography) may be used when certain two-factor interaction variance components are found to be significant and hence must remain a part of the expected mean square.

Example 14.6: In a study to determine which are the important sources of variation in an industrial process, 3 measurements are taken on yield for 3 operators chosen randomly and 4 batches of raw materials chosen randomly. It is decided that a statistical test should be made at the 0.05 level of significance to determine if the variance components due to batches, operators, and interaction are significant. In addition, estimates of variance components are to be computed. The data are given in Table 14.16, with the response being percent by weight.

Table 14.16: Data for Example 14.6

| | Batch | | | |
Operator	1	2	3	4
1	66.9	68.3	69.0	69.3
	68.1	67.4	69.8	70.9
	67.2	67.7	67.5	71.4
2	66.3	68.1	69.7	69.4
	65.4	66.9	68.8	69.6
	65.8	67.6	69.2	70.0
3	65.6	66.0	67.1	67.9
	66.3	66.9	66.2	68.4
	65.2	67.3	67.4	68.7

Solution: The sums of squares are found in the usual way, with the following results:

SST (total) $= 84.5564,$ SSE (error) $= 10.6733,$
SSA (operators) $= 18.2106,$ SSB (batches) $= 50.1564,$
$SS(AB)$ (interaction) $= 5.5161.$

All other computations are carried out and exhibited in Table 14.17. Since

$f_{0.05}(2,6) = 5.14,$ $f_{0.05}(3,6) = 4.76,$ and $f_{0.05}(6,24) = 2.51,$

we find the operator and batch variance components to be significant. Although the interaction variance is not significant at the $\alpha = 0.05$ level, the P-value is 0.095. Estimates of the main effect variance components are

$$\hat{\sigma}_\alpha^2 = \frac{9.1053 - 0.9194}{12} = 0.68, \qquad \hat{\sigma}_\beta^2 = \frac{16.7188 - 0.9194}{9} = 1.76.$$

Table 14.17: Analysis of Variance for Example 14.6

Source of Variation	Sum of Squares	Degrees of Freedom	Mean Square	Computed f
Operators	18.2106	2	9.1053	9.90
Batches	50.1564	3	16.7188	18.18
Interaction	5.5161	6	0.9194	2.07
Error	10.6733	24	0.4447	
Total	84.5564	35		

Mixed Model Experiment

There are situations where the experiment dictates the assumption of a **mixed model** (i.e., a mixture of random and fixed effects). For example, for the case of two factors, we may have

$$Y_{ijk} = \mu + A_i + B_j + (AB)_{ij} + \epsilon_{ijk},$$

for $i = 1, 2, \ldots, a$; $j = 1, 2, \ldots, b$; $k = 1, 2, \ldots, n$. The A_i may be independent random variables, independent of ϵ_{ijk}, and the B_j may be fixed effects. The mixed nature of the model requires that the interaction terms be random variables. As a result, the relevant hypotheses are of the form

$$H_0': \sigma_\alpha^2 = 0, \quad H_0'': B_1 = B_2 = \cdots = B_b = 0, \quad H_0''': \sigma_{\alpha\beta}^2 = 0,$$
$$H_1': \sigma_\alpha^2 \neq 0, \quad H_1'': \text{At least one the } B_j \text{ is not zero}, \quad H_1''': \sigma_{\alpha\beta}^2 \neq 0.$$

Again, the computations of sums of squares are identical to those of fixed and random effects situations, and the F-test is dictated by the expected mean squares. Table 14.18 provides the expected mean squares for the two-factor mixed model problem.

Table 14.18: Expected Mean Squares for Two-Factor Mixed Model Experiment

Factor	Expected Mean Square
A (random)	$\sigma^2 + bn\sigma_\alpha^2$
B (fixed)	$\sigma^2 + n\sigma_{\alpha\beta}^2 + \frac{an}{b-1}\sum_j B_j^2$
AB (random)	$\sigma^2 + n\sigma_{\alpha\beta}^2$
Error	σ^2

From the nature of the expected mean squares it becomes clear that the **test on the random effect employs the mean square error** s^2 as the denominator, whereas the **test on the fixed effect** uses the interaction mean square. Suppose we now consider three factors. Here, of course, we must take into account the situation where one factor is fixed and the situation in which two factors are fixed. Table 14.19 covers both situations.

Table 14.19: Expected Mean Squares for Mixed Model Factorial Experiments in Three Factors

	A Random	A Random, B Random
A	$\sigma^2 + bcn\sigma_\alpha^2$	$\sigma^2 + cn\sigma_{\alpha\beta}^2 + bcn\sigma_\alpha^2$
B	$\sigma^2 + cn\sigma_{\alpha\beta}^2 + acn\sum_{j=1}^{b}\frac{B_j^2}{b-1}$	$\sigma^2 + cn\sigma_{\alpha\beta}^2 + acn\sigma_\beta^2$
C	$\sigma^2 + bn\sigma_{\alpha\gamma}^2 + abn\sum_{k=1}^{c}\frac{C_k^2}{c-1}$	$\sigma^2 + n\sigma_{\alpha\beta\gamma}^2 + an\sigma_{\beta\gamma}^2 + bn\sigma_{\alpha\gamma}^2 + abn\sum_{k=1}^{c}\frac{C_k^2}{c-1}$
AB	$\sigma^2 + cn\sigma_{\alpha\beta}^2$	$\sigma^2 + cn\sigma_{\alpha\beta}^2$
AC	$\sigma^2 + bn\sigma_{\alpha\gamma}^2$	$\sigma^2 + n\sigma_{\alpha\beta\gamma}^2 + bn\sigma_{\alpha\gamma}^2$
BC	$\sigma^2 + n\sigma_{\alpha\beta\gamma}^2 + an\sum_j\sum_k\frac{(BC)_{jk}^2}{(b-1)(c-1)}$	$\sigma^2 + n\sigma_{\alpha\beta\gamma}^2 + an\sigma_{\beta\gamma}^2$
ABC	$\sigma^2 + n\sigma_{\alpha\beta\gamma}^2$	$\sigma^2 + n\sigma_{\alpha\beta\gamma}^2$
Error	σ^2	σ^2

Note that in the case of A random, all effects have proper f-tests. But in the case of A and B random, the main effect C must be tested using a Satterthwaite-type procedure similar to that used in the random effects experiment.

Exercises

14.26 Assuming a random effects experiment for Exercise 14.2 on page 575, estimate the variance components for brand of orange juice concentrate, for number of days from when orange juice was blended until it was tested, and for experimental error.

14.27 To estimate the various components of variability in a filtration process, the percent of material lost in the mother liquor is measured for 12 experimental conditions, with 3 runs on each condition. Three filters and 4 operators are selected at random for use in the experiment.

(a) Test the hypothesis of no interaction variance component between filters and operators at the $\alpha = 0.05$ level of significance.
(b) Test the hypotheses that the operators and the filters have no effect on the variability of the filtration process at the $\alpha = 0.05$ level of significance.
(c) Estimate the components of variance due to filters, operators, and experimental error.

	\multicolumn{4}{c}{Operator}
Filter	1	2	3	4
1	16.2	15.9	15.6	14.9
	16.8	15.1	15.9	15.2
	17.1	14.5	16.1	14.9
2	16.6	16.0	16.1	15.4
	16.9	16.3	16.0	14.6
	16.8	16.5	17.2	15.9
3	16.7	16.5	16.4	16.1
	16.9	16.9	17.4	15.4
	17.1	16.8	16.9	15.6

14.28 A defense contractor is interested in studying an inspection process to detect failure or fatigue of transformer parts. Three levels of inspections are used by three randomly chosen inspectors. Five lots are used for each combination in the study. The factor levels are given in the data. The response is in failures per 1000 pieces.

(a) Write an appropriate model, with assumptions.
(b) Use analysis of variance to test the appropriate hypothesis for inspector, inspection level, and interaction.

	Inspection Level		
Inspector	Full Military Inspection	Reduced Military Inspection	Commercial
A	7.50 7.42 5.85 5.89 5.35	7.08 6.17 5.65 5.30 5.02	6.15 5.52 5.48 5.48 5.98
B	7.58 6.52 6.54 5.64 5.12	7.68 5.86 5.28 5.38 4.87	6.17 6.20 5.44 5.75 5.68
C	7.70 6.82 6.42 5.39 5.35	7.19 6.19 5.85 5.35 5.01	6.21 5.66 5.36 5.90 6.12

14.29 Consider the following analysis of variance for a random effects experiment:

Source of Variation	Degrees of Freedom	Mean Square
A	3	140
B	1	480
C	2	325
AB	3	15
AC	6	24
BC	2	18
ABC	6	2
Error	24	5
Total	47	

Test for significant variance components among all main effects and interaction effects at the 0.01 level of significance

(a) by using a pooled estimate of error when appropriate;
(b) by not pooling sums of squares of insignificant effects.

14.30 A plant manager would like to show that the yield of a woven fabric in the plant does not depend on machine operator or time of day and is consistently high. Four randomly selected operators and 3 randomly selected hours of the day are chosen for the study. The yield is measured in yards produced per minute. Samples are taken on 3 randomly chosen days.

(a) Write the appropriate model.
(b) Evaluate the variance components for operator and time.
(c) Draw conclusions.

		Operator		
Time	1	2	3	4
1	9.5 9.8 10.0	9.8 10.1 9.6	9.8 10.3 9.7	10.0 9.7 10.2
2	10.2 9.9 9.5	10.1 9.8 9.7	10.2 9.8 9.7	10.3 10.1 9.9
3	10.5 10.2 9.3	10.4 10.2 9.8	9.9 10.3 10.2	10.0 10.1 9.7

14.31 A manufacturer of latex house paint (brand A) would like to show that its paint is more robust to the material being painted than that of its two closest competitors. The response is the time, in years, until chipping occurs. The study involves the three brands of paint and three randomly chosen materials. Two pieces of material are used for each combination.

Material	Brand of Paint		
	A	B	C
A	5.50 5.15	4.75 4.60	5.10 5.20
B	5.60 5.55	5.50 5.60	5.40 5.50
C	5.40 5.48	5.05 4.95	4.50 4.55

(a) What is this type of model called?
(b) Analyze the data, using the appropriate model.
(c) Did the manufacturer of brand A support its claim with the data?

14.32 A process engineer wants to determine if the power setting on the machines used to fill certain types of cereal boxes results in a significant effect on the actual weight of the product. The study consists of 3 randomly chosen types of cereal manufactured by the company and 3 fixed power settings. Weight is measured for 4 different randomly selected boxes of cereal at each combination. The desired weight is 400 grams. The data are presented here.

Power Setting	Cereal Type		
	1	2	3
Low	395 390 401 400	392 392 394 401	402 405 399 399
Current	396 399 400 402	390 392 395 502	404 403 400 399
High	410 408 408 407	404 406 401 400	415 412 413 415

(a) Give the appropriate model, and list the assumption being made.
(b) Is there a significant effect due to the power setting?
(c) Is there a significant variance component due to cereal type?

Review Exercises

14.33 The Laboratory for Interdisciplinary Statistical Analysis at Virginia Tech at Virginia Tech was involved in analyzing a set of data taken by personnel in the Human Nutrition, Foods, and Exercise Department in which it was of interest to study the effects of flour type and percent sweetener on certain physical attributes of a type of cake. All-purpose flour and cake flour were used, and the percent sweetener was varied at four levels. The following data show information on specific gravity of cake samples. Three cakes were prepared at each of the eight factor combinations.

Sweetener	Flour	
Concentration	All-Purpose	Cake
0	0.90 0.87 0.90	0.91 0.90 0.80
50	0.86 0.89 0.91	0.88 0.82 0.83
75	0.93 0.88 0.87	0.86 0.85 0.80
100	0.79 0.82 0.80	0.86 0.85 0.85

(a) Treat the analysis as a two-factor analysis of variance. Test for differences between flour type. Test for differences between sweetener concentration.

(b) Discuss the effect of interaction, if any. Give P-values on all tests.

14.34 An experiment was conducted in the Department of Food Science and Technology at Virginia Tech. It was of interest to characterize the texture of certain types of fish in the herring family. The effect of sauce types used in preparing the fish was also studied. The response in the experiment was "texture value," measured with a machine that sliced the fish product. The following are data on texture values:

	Fish Type		
Sauce Type	Unbleached Menhaden	Bleached Menhaden	Herring
Sour Cream	27.6 57.4	64.0 66.9	107.0 83.9
	47.8 71.1	66.5 66.8	110.4 93.4
	53.8	53.8	83.1
Wine Sauce	49.8 31.0	48.3 62.2	88.0 95.2
	11.8 35.1	54.6 43.6	108.2 86.7
	16.1	41.8	105.2

(a) Do an analysis of variance. Determine whether or not there is an interaction between sauce type and fish type.

(b) Based on your results from part (a) and on F-tests on main effects, determine if there is a significant difference in texture due to sauce types, and determine whether there is a significant difference due to fish types.

14.35 A study was made to determine if humidity conditions have an effect on the force required to pull apart pieces of glued plastic. Three types of plastic were tested using 4 different levels of humidity. The results, in kilograms, are as follows:

	Humidity			
Plastic Type	30%	50%	70%	90%
A	39.0	33.1	33.8	33.0
	42.8	37.8	30.7	32.9
B	36.9	27.2	29.7	28.5
	41.0	26.8	29.1	27.9
C	27.4	29.2	26.7	30.9
	30.3	29.9	32.0	31.5

(a) Assuming a fixed effects experiment, perform an analysis of variance and test the hypothesis of no interaction between humidity and plastic type at the 0.05 level of significance.

(b) Using only plastics A and B and the value of s^2 from part (a), once again test for the presence of interaction at the 0.05 level of significance.

14.36 Personnel in the Materials Science and Engineering Department at Virginia Tech conducted an experiment to study the effects of environmental factors on the stability of a certain type of copper-nickel alloy. The basic response was the fatigue life of the material. The factors are level of stress and environment. The data are as follows:

	Stress Level		
Environment	Low	Medium	High
Dry	11.08	13.12	14.18
Hydrogen	10.98	13.04	14.90
	11.24	13.37	15.10
High	10.75	12.73	14.15
Humidity	10.52	12.87	14.42
(95%)	10.43	12.95	14.25

(a) Do an analysis of variance to test for interaction between the factors. Use $\alpha = 0.05$.

(b) Based on part (a), do an analysis on the two main effects and draw conclusions. Use a P-value approach in drawing conclusions.

14.37 In the experiment of Review Exercise 14.33, cake volume was also used as a response. The units are cubic inches. Test for interaction between factors and discuss main effects. Assume that both factors are fixed effects.

Sweetener	Flour	
Concentration	All-Purpose	Cake
0	4.48 3.98 4.42	4.12 4.92 5.10
50	3.68 5.04 3.72	5.00 4.26 4.34
75	3.92 3.82 4.06	4.82 4.34 4.40
100	3.26 3.80 3.40	4.32 4.18 4.30

14.38 A control valve needs to be very sensitive to the input voltage, thus generating a good output voltage. An engineer turns the control bolts to change the input voltage. The book *SN-Ratio for the Quality Evaluation*, published by the Japanese Standards Association (1988), described a study on how these three factors (relative position of control bolts, control range of bolts, and input voltage) affect the sensitivity of a control valve. The factors and their levels are shown below. The data show the sensitivity of a control valve.
Factor A, relative position of control bolts:
 center -0.5, center, and center $+0.5$
Factor B, control range of bolts:
 2, 4.5, and 7 (mm)
Factor C, input voltage:
 100, 120, and 150 (V)

		C_1	C_2	C_3
A_1	B_1	151 135	151 135	151 138
A_1	B_2	178 171	180 173	181 174
A_1	B_3	204 190	205 190	206 192
A_2	B_1	156 148	158 149	158 150
A_2	B_2	183 168	183 170	183 172
A_2	B_3	210 204	211 203	213 204
A_3	B_1	161 145	162 148	163 148
A_3	B_2	189 182	191 184	192 183
A_3	B_3	215 202	216 203	217 205

Perform an analysis of variance with $\alpha = 0.05$ to test for significant main and interaction effects. Draw conclusions.

14.39 Exercise 14.25 on page 588 describes an experiment involving the extraction of polyethylene through use of a solvent.

Solvent	Temp.	Time 4	8	16
Ethanol	120	94.0 94.0	93.8 94.2	91.1 90.5
	80	95.3 95.1	94.9 95.3	92.5 92.4
Toluene	120	94.6 94.5	93.6 94.1	91.1 91.0
	80	95.4 95.4	95.6 96.0	92.1 92.1

(a) Do a different sort of analysis on the data. Fit an appropriate regression model with a solvent categorical variable, a temperature term, a time term, a temperature by time interaction, a solvent by temperature interaction, and a solvent by time interaction. Do t-tests on all coefficients and report your findings.

(b) Do your findings suggest that different models are appropriate for ethanol and toluene, or are they equivalent apart from the intercepts? Explain.

(c) Do you find any conclusions here that contradict conclusions drawn in your solution of Exercise 14.25? Explain.

14.40 In the book *SN-Ratio for the Quality Evaluation*, published by the Japanese Standards Association (1988), a study on how tire air pressure affects the maneuverability of an automobile was described. Three different tire air pressures were compared on three different driving surfaces. The three air pressures were both left- and right-side tires inflated to 6 kgf/cm^2, left-side tires inflated to 6 kgf/cm^2 and right-side tires inflated to 3 kgf/cm^2, and both left- and right-side tires inflated to 3 kgf/cm^2. The three driving surfaces were asphalt, dry asphalt, and dry cement. The turning radius of a test vehicle was observed twice for each level of tire pressure on each of the three different driving surfaces.

Driving Surface	Tire Air Pressure 1	2	3
Asphalt	44.0 25.5	34.2 37.2	27.4 42.8
Dry Asphalt	31.9 33.7	31.8 27.6	43.7 38.2
Dry Cement	27.3 39.5	46.6 28.1	35.5 34.6

Perform an analysis of variance of the above data. Comment on the interpretation of the main and interaction effects.

14.41 The manufacturer of a certain brand of freeze-dried coffee hopes to shorten the process time without jeopardizing the integrity of the product. The process engineer wants to use 3 temperatures for the drying chamber and 4 drying times. The current drying time is 3 hours at a temperature of $-15°C$. The flavor response is an average of scores of 4 professional judges. The score is on a scale from 1 to 10, with 10 being the best. The data are as shown in the following table.

	Temperature		
Time	$-20°C$	$-15°C$	$-10°C$
1 hr	9.60 9.63	9.55 9.50	9.40 9.43
1.5 hr	9.75 9.73	9.60 9.61	9.55 9.48
2 hr	9.82 9.93	9.81 9.78	9.50 9.52
3 hr	9.78 9.81	9.80 9.75	9.55 9.58

(a) What type of model should be used? State assumptions.

(b) Analyze the data appropriately.

(c) Write a brief report to the vice-president in charge and make a recommendation for future manufacturing of this product.

14.42 To ascertain the number of tellers needed during peak hours of operation, data were collected by an urban bank. Four tellers were studied during three "busy" times: (1) weekdays between 10:00 and 11:00 A.M., (2) weekday afternoons between 2:00 and 3:00 P.M., and (3) Saturday mornings between 11:00 A.M. and 12:00 noon. An analyst chose four randomly selected times within each of the three time periods for each of the four teller positions over a period of months, and the numbers of customers serviced were observed. The data are as follows:

	Time Period		
Teller	1	2	3
1	18 24 17 22	25 29 23 32	29 30 21 34
2	16 11 19 14	23 32 25 17	27 29 18 16
3	12 19 11 22	27 33 27 24	25 20 29 15
4	11 9 13 8	10 7 19 8	11 9 17 9

It is assumed that the number of customers served is a Poisson random variable.

(a) Discuss the danger in doing a standard analysis of variance on the data above. What assumptions, if any, would be violated?

(b) Construct a standard ANOVA table that includes F-tests on main effects and interactions. If interactions and main effects are found to be significant, give scientific conclusions. What have we learned? Be sure to interpret any significant interaction. Use your own judgment regarding P-values.

(c) Do the entire analysis again using an appropriate transformation on the response. Do you see any differences in your findings? Comment.

14.6 Potential Misconceptions and Hazards; Relationship to Material in Other Chapters

One of the most confusing issues in the analysis of factorial experiments resides in the interpretation of main effects in the presence of interaction. The presence of a relatively large P-value for a main effect when interactions are clearly present may tempt the analyst to conclude "no significant main effect." However, one must understand that if a main effect is involved in a significant interaction, then the main effect is **influencing the response**. The nature of the effect is inconsistent across levels of other effects. The nature of the role of the main effect can be deduced from **interaction plots**.

In light of what is communicated in the preceding paragraph, there is danger of a substantial misuse of statistics when one employs a multiple comparison test on main effects in the clear presence of interaction among the factors.

One must be cautious in the analysis of a factorial experiment when the assumption of a complete randomized design is made when in fact complete randomization is not carried out. For example, it is common to encounter factors that are very **difficult to change**. As a result, factor levels may need to be held without change for long periods of time throughout the experiment. For instance, a temperature factor is a common example. Moving temperature up and down in a randomization scheme is a costly plan, and most experimenters will refuse to do it. Experimental designs with *restrictions in randomization* are quite common and are called **split plot designs**. They are beyond the scope of the book, but presentations are found in Montgomery (2008a).

Many of the concepts discussed in this chapter carry over into Chapter 15 (e.g., the importance of randomization and the role of interaction in the interpretation of results). However, there are two areas covered in Chapter 15 that represent an expansion of principles dealt with both in Chapter 13 and in this chapter. In Chapter 15, problem solving through the use of factorial experiments is done with regression analysis since most of the factors are assumed to be quantitative and measured on a continuum (e.g., temperature and time). Prediction equations are developed from the data of the designed experiment, and they are used for process improvement or even process optimization. In addition, development is given on the topic of fractional factorials, in which only a portion or fraction of the entire factorial experiment is implemented due to the prohibitive cost of doing the entire experiment.

Chapter 15

2^k Factorial Experiments and Fractions

15.1 Introduction

We have already been exposed to certain experimental design concepts. The sampling plan for the simple t-test on the mean of a normal population and the analysis of variance involve randomly allocating pre-chosen treatments to experimental units. The randomized block design, where treatments are assigned to units within relatively homogeneous blocks, involves restricted randomization.

In this chapter, we give special attention to experimental designs in which the experimental plan calls for the study of the effect on a response of k factors, each at two levels. These are commonly known as **2^k factorial experiments**. We often denote the levels as "high" and "low" even though this notation may be arbitrary in the case of qualitative variables. The complete factorial design requires that each level of every factor occur with each level of every other factor, giving a total of **2^k treatment combinations**.

Factor Screening and Sequential Experimentation

Often, when experimentation is conducted either on a research or on a development level, a well-planned experimental design is a **stage** of what is truly a **sequential plan** of experimentation. More often than not, the scientists and engineers at the outset of a study may not be aware of which factors are important or what are appropriate ranges for the potential factors on which experimentation should be conducted. For example, in the text *Response Surface Methodology* by Myers, Montgomery, and Anderson-Cook (2009), one example is given of an investigation of a pilot plant experiment in which four factors—temperature, pressure, concentration of formaldehyde, and steering rate—are varied in order to establish their influence on the response, filtration rate of a certain chemical product. Even at the pilot plant level, the scientists are not certain if all four factors should be involved in the model. In addition, the eventual goal is to determine the proper settings of contributing factors that maximize the filtration rate. Thus, there is a need to determine the **proper region of experimentation**. These questions can be answered only if the total experimental plan is done sequentially. Many experimental endeavors are plans that feature *iterative learning*, the type of learning that is consistent with the scientific method, with the word *iterative* implying stage-wise experimentation.

Generally, the initial stage of the ideal sequential plan is variable or **factor screening**, a procedure that involves an inexpensive experimental design using the **candidate factors**. This is particularly important when the plan involves a complex system like a manufacturing process. The information received from the results of a *screening design* is used to design one or more subsequent experiments in which adjustments in the important factors are made, the adjustments that provide improvements in the system or process.

The 2^k factorial experiments and fractions of the 2^k are powerful tools that are ideal screening designs. They are simple, practical, and intuitively appealing. Many of the general concepts discussed in Chapter 14 continue to apply. However, there are graphical methods that provide useful intuition in the analysis of the two-level designs.

Screening Designs for Large Numbers of Factors

When k is small, say $k = 2$ or even $k = 3$, the utility of the 2^k factorial for factor screening is clear. Analysis of variance and/or regression analysis as discussed and illustrated in Chapters 12, 13, and 14 remain useful as tools. In addition, graphical approaches are helpful.

If k is large, say as large as 6, 7, or 8, the number of factor combinations and thus experimental runs required for the 2^k factorial often becomes prohibitive. For example, suppose one is interested in carrying out a screening design involving $k = 8$ factors. There may be interest in gaining information on all $k = 8$ main effects as well as the $\frac{k(k-1)}{2} = 28$ two-factor interactions. However, including $2^8 = 256$ runs would appear to make the study much too large and be wasteful for studying $28 + 8 = 36$ effects. But, as we will illustrate in future sections, when k is large we can gain considerable information in an efficient manner by using only a fraction of the complete 2^k factorial experiment. This class of designs is the class of *fractional factorial designs*. The goal is to retain high-quality information on main effects and interesting interactions even though the size of the design is reduced considerably.

15.2 The 2^k Factorial: Calculation of Effects and Analysis of Variance

Consider initially a 2^2 factorial with factors A and B and n *experimental observations per factor combination*. It is useful to use the symbols (1), a, b, and ab to signify the design points, where the presence of a lowercase letter implies that the factor (A or B) is at the *high level*. Thus, absence of the lower case implies that the factor is at the *low level*. So ab is the design point $(+, +)$, a is $(+, -)$, b is $(-, +)$ and (1) is $(-, -)$. There are situations in which the notation also stands

15.2 The 2^k Factorial: Calculation of Effects and Analysis of Variance

for the response data at the design point in question. As an introduction to the calculation of important **effects** that aid in the determination of the influence of the factors and **sums of squares** that are incorporated into analysis of variance computations, we have Table 15.1.

Table 15.1: A 2^2 Factorial Experiment

		A		Mean
B		b	ab	$\frac{b+ab}{2n}$
		(1)	a	$\frac{(1)+a}{2n}$
Mean		$\frac{(1)+b}{2n}$	$\frac{a+ab}{2n}$	

In this table, (1), a, b, and ab signify totals of the n response values at the individual design points. The simplicity of the 2^2 factorial lies in the fact that apart from experimental error, important information comes to the analyst in single-degree-of-freedom components, one each for the two main effects A and B and one degree of freedom for interaction AB. The information retrieved on all these takes the form of three **contrasts**. Let us define the following contrasts among the treatment totals:

$$A \text{ contrast} = ab + a - b - (1),$$
$$B \text{ contrast} = ab - a + b - (1),$$
$$AB \text{ contrast} = ab - a - b + (1).$$

The three **effects** from the experiment involve these contrasts and appeal to common sense and intuition. The two computed main effects are of the form

$$\text{effect} = \bar{y}_H - \bar{y}_L,$$

where $\bar{y}_H$ and $\bar{y}_L$ are average response at the high, or "+" level and average response at the low, or "−" level, respectively. As a result,

Calculation of Main Effects

$$A = \frac{ab + a - b - (1)}{2n} = \frac{A \text{ contrast}}{2n}$$

and

$$B = \frac{ab - a + b - (1)}{2n} = \frac{B \text{ contrast}}{2n}.$$

The quantity A is seen to be the *difference between the mean responses at the low and high levels of factor A*. In fact, we call A the **main effect** of factor A. Similarly, B is the main effect of factor B. Apparent interaction in the data is observed by inspecting the difference between $ab - b$ and $a - (1)$ or between $ab - a$ and $b - (1)$ in Table 15.1. If, for example,

$$ab - a \approx b - (1) \quad \text{or} \quad ab - a - b + (1) \approx 0,$$

a line connecting the responses for each level of factor A at the high level of factor B will be approximately parallel to a line connecting the responses for each level of factor A at the low level of factor B. The nonparallel lines of Figure 15.1 suggest the presence of interaction. To test whether this apparent interaction is significant, a third contrast in the treatment totals orthogonal to the main effect contrasts, called the **interaction effect**, is constructed by evaluating

Interaction Effect
$$AB = \frac{ab - a - b + (1)}{2n} = \frac{AB \text{ contrast}}{2n}.$$

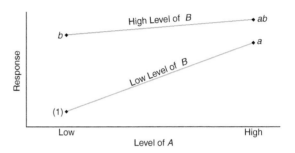

Figure 15.1: Response suggesting apparent interaction.

Example 15.1: Consider the data in Tables 15.2 and 15.3 with $n = 1$ for a 2^2 factorial experiment.

Table 15.2: 2^2 Factorial with No Interaction

A	B	
	$-$	$+$
$+$	50	70
$-$	80	100

Table 15.3: 2^2 Factorial with Interaction

A	B	
	$-$	$+$
$+$	50	70
$-$	80	40

The numbers in the cells in Tables 15.2 and 15.3 clearly illustrate how contrasts and the resulting calculation of the two main effects and resulting conclusions can be highly influenced by the presence of interaction. In Table 15.2, the effect of A is -30 at both the low and high levels of factor B and the effect of B is 20 at both the low and high levels of factor A. This "consistency of effect" (no interaction) can be very important information to the analyst. The main effects are

$$A = \frac{70 + 50}{2} - \frac{100 + 80}{2} = 60 - 90 = -30,$$
$$B = \frac{100 + 70}{2} - \frac{80 + 50}{2} = 85 - 65 = 20,$$

while the interaction effect is

$$AB = \frac{100 + 50}{2} - \frac{80 + 70}{2} = 75 - 75 = 0.$$

On the other hand, in Table 15.3 the effect of A is once again -30 at the low level of B but $+30$ at the high level of B. This "inconsistency of effect" (interaction) also is present for B across levels of A. In these cases, the main effects can be meaningless and, in fact, highly misleading. For example, the effect of A is

$$A = \frac{50 + 70}{2} - \frac{80 + 40}{2} = 0,$$

since there is a complete "masking" of the effect as one averages over levels of B. The strong interaction is illustrated by the calculated effect

$$AB = \frac{70 + 80}{2} - \frac{50 + 40}{2} = 30.$$

Here it is convenient to illustrate the scenarios of Tables 15.2 and 15.3 with interaction plots. Note the parallelism in the plot of Figure 15.2 and the interaction that is apparent in Figure 15.3.

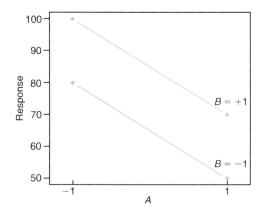

Figure 15.2: Interaction plot for data of Table 15.2.

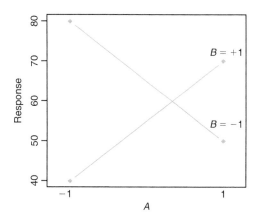

Figure 15.3: Interaction plot for data of Table 15.3.

Computation of Sums of Squares

We take advantage of the fact that in the 2^2 factorial, or for that matter in the general 2^k factorial experiment, each main effect and interaction effect has an associated **single degree of freedom**. Therefore, we can write $2^k - 1$ orthogonal single-degree-of-freedom contrasts in the treatment combinations, each accounting for variation due to some main or interaction effect. Thus, under the usual independence and normality assumptions in the experimental model, we can make tests to determine if the contrast reflects systematic variation or merely chance or random variation. The sums of squares for each contrast are found by following the procedures given in Section 13.5. Writing

$$Y_{1..} = b + (1), \qquad Y_{2..} = ab + a, \qquad c_1 = -1, \quad \text{and} \quad c_2 = 1,$$

where $Y_{1..}$ and $Y_{2..}$ are the total of $2n$ observations, we have

$$SSA = SSw_A = \frac{\left(\sum_{i=1}^{2} c_i Y_{i..}\right)^2}{2n \sum_{i=1}^{2} c_i^2} = \frac{[ab + a - b - (1)]^2}{2^2 n} = \frac{(A \text{ contrast})^2}{2^2 n},$$

with 1 degree of freedom. Similarly, we find that

$$SSB = \frac{[ab + b - a - (1)]^2}{2^2 n} = \frac{(B \text{ contrast})^2}{2^2 n}$$

and

$$SS(AB) = \frac{[ab + (1) - a - b]^2}{2^2 n} = \frac{(AB \text{ contrast})^2}{2^2 n}.$$

Each contrast has 1 degree of freedom, whereas the error sum of squares, with $2^2(n-1)$ degrees of freedom, is obtained by subtraction from the formula

$$SSE = SST - SSA - SSB - SS(AB).$$

In computing the sums of squares for the main effects A and B and the interaction effect AB, it is convenient to present the total responses of the treatment combinations along with the appropriate algebraic signs for each contrast, as in Table 15.4. The main effects are obtained as simple comparisons between the low and high levels. Therefore, we assign a positive sign to the treatment combination that is at the high level of a given factor and a negative sign to the treatment combination at the low level. The positive and negative signs for the interaction effect are obtained by multiplying the corresponding signs of the contrasts of the interacting factors.

Table 15.4: Signs for Contrasts in a 2^2 Factorial Experiment

Treatment Combination	Factorial Effect		
	A	B	AB
(1)	−	−	+
a	+	−	−
b	−	+	−
ab	+	+	+

The 2^3 Factorial

Let us now consider an experiment using three factors, A, B, and C, each with levels -1 and $+1$. This is a 2^3 factorial experiment giving the eight treatment combinations (1), a, b, c, ab, ac, bc, and abc. The treatment combinations and the appropriate algebraic signs for each contrast used in computing the sums of squares for the main effects and interaction effects are presented in Table 15.5.

15.2 The 2^k Factorial: Calculation of Effects and Analysis of Variance

Table 15.5: Signs for Contrasts in a 2^3 Factorial Experiment

Treatment Combination	Factorial Effect (symbolic)						
	A	B	C	AB	AC	BC	ABC
(1)	−	−	−	+	+	+	−
a	+	−	−	−	−	+	+
b	−	+	−	−	+	−	+
c	−	−	+	+	−	−	+
ab	+	+	−	+	−	−	−
ac	+	−	+	−	+	−	−
bc	−	+	+	−	−	+	−
abc	+	+	+	+	+	+	+

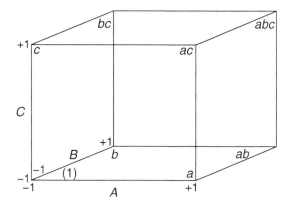

Figure 15.4: Geometric view of 2^3.

It is helpful to discuss and illustrate the geometry of the 2^3 factorial much as we illustrated that of the 2^2 factorial in Figure 15.1. For the 2^3, the **eight design points** represent the vertices of a cube, as shown in Figure 15.4.

The columns of Table 15.5 represent the signs that are used for the contrasts and thus computation of seven effects and corresponding sums of squares. These columns are analogous to those given in Table 15.4 for the case of the 2^2. Seven effects are available since there are eight design points. For example,

$$A = \frac{a + ab + ac + abc - (1) - b - c - bc}{4n},$$

$$AB = \frac{(1) + c + ab + abc - a - b - ac - bc}{4n},$$

and so on. The sums of squares are merely given by

$$SS(\text{effect}) = \frac{(\text{contrast})^2}{2^3 n}.$$

An inspection of Table 15.5 reveals that for the 2^3 experiment all contrasts among the seven are mutually orthogonal, and therefore the seven effects are assessed independently.

Effects and Sums of Squares for the 2^k

For a 2^k factorial experiment the single-degree-of-freedom sums of squares for the main effects and interaction effects are obtained by squaring the appropriate contrasts in the treatment totals and dividing by $2^k n$, where n is the number of replications of the treatment combinations.

As before, an effect is always calculated by subtracting the average response at the "low" level from the average response at the "high" level. The high and low for main effects are quite clear. The symbolic high and low for interactions are evident from information as in Table 15.5.

The orthogonality property has the same importance here as it does for the material on comparisons discussed in Chapter 13. Orthogonality of contrasts implies that the estimated effects and thus the sums of squares are independent. This independence is readily illustrated in the 2^3 factorial experiment if the responses, with factor A at its high level, are increased by an amount x in Table 15.5. Only the A contrast leads to a larger sum of squares, since the x effect cancels out in the formation of the six remaining contrasts as a result of the two positive and two negative signs associated with treatment combinations in which A is at the high level.

There are additional advantages produced by orthogonality. These are pointed out when we discuss the 2^k factorial experiment in regression situations.

15.3 Nonreplicated 2^k Factorial Experiment

The full 2^k factorial may often involve considerable experimentation, particularly when k is large. As a result, replication of each factor combination is often not feasible. If all effects, including all interactions, are included in the model of the experiment, no degrees of freedom are allowed for error. Often, when k is large, the data analyst will *pool* sums of squares and corresponding degrees of freedom for high-order interactions that are known or assumed to be negligible. This will produce F-tests for main effects and lower-order interactions.

Diagnostic Plotting with Nonreplicated 2^k Factorial Experiments

Normal probability plotting can be a very useful methodology for determining the relative importance of effects in a reasonably large two-level factored experiment when there is no replication. This type of diagnostic plot can be particularly useful when the data analyst is hesitant to pool high-order interactions for fear that some of the effects pooled in the "error" may truly be real effects and not merely random. The reader should bear in mind that all effects that are not real (i.e., they are independent *estimates of zero*) follow a normal distribution with mean near zero and constant variance. For example, in a 2^4 factorial experiment, we are reminded that all effects (keep in mind that $n=1$) are of the form

$$AB = \frac{\text{contrast}}{8} = \bar{y}_H - \bar{y}_L,$$

15.3 Nonreplicated 2^k Factorial Experiment

where $\bar{y}_H$ is the average of eight independent experimental runs at the high, or "+," level and $\bar{y}_L$ is the average of eight independent runs at the low, or "−," level. Thus, the variance of each contrast is $\text{Var}(\bar{y}_H - \bar{y}_L) = \sigma^2/4$. For any real effects, $E(\bar{y}_H - \bar{y}_L) \neq 0$. Thus, normal probability plotting should reveal "significant" effects as those that fall off the straight line that depicts realizations of independent, identically distributed normal random variables.

The probability plotting can take one of many forms. The reader is referred to Chapter 8, where these plots were first presented. The empirical normal quantile-quantile plot may be used. The plotting procedure that makes use of normal probability paper may also be used. In addition, there are several other types of diagnostic normal probability plots. In summary, the procedure for diagnostic effect plots is as follows.

Probability Effect Plots for Nonreplicated 2^k Factorial Experiments

1. Calculate effects as

$$\text{effect} = \frac{\text{contrast}}{2^{k-1}}.$$

2. Construct a normal probability plot of all effects.

3. Effects that fall off the straight line should be considered real effects.

Further comments regarding normal probability plotting of effects are in order. First, the data analyst may feel frustrated if he or she uses these plots with a small experiment. On the other hand, the plotting is likely to give satisfying results when there is *effect sparsity*—many effects that are truly not real. This sparsity will be evident in large experiments where high-order interactions are not likely to be real.

Case Study 15.1: Injection Molding: Many manufacturing companies in the United States and abroad use molded parts as components. Shrinkage is often a major problem. Often, a molded die for a part is built larger than nominal to allow for part shrinkage. In the following experimental situation, a new die is being produced, and ultimately it is important to find the proper process settings to minimize shrinkage. In the following experiment, the response values are deviations from nominal (i.e., shrinkage). The factors and levels are as follows:

	Coded Levels	
	−1	+1
A. Injection velocity (ft/sec)	1.0	2.0
B. Mold temperature (°C)	100	150
C. Mold pressure (psi)	500	1000
D. Back pressure (psi)	75	120

The purpose of the experiment was to determine what effects (main effects and interaction effects) influence shrinkage. The experiment was considered a preliminary screening experiment from which the factors for a more complete analysis might be determined. Also, it was hoped that some insight might be gained into how the important factors impact shrinkage. The data from a nonreplicated 2^4 factorial experiment are given in Table 15.6.

Table 15.6: Data for Case Study 15.1

Factor Combination	Response (cm × 10^4)	Factor Combination	Response (cm × 10^4)
(1)	72.68	d	73.52
a	71.74	ad	75.97
b	76.09	bd	74.28
ab	93.19	abd	92.87
c	71.25	cd	79.34
ac	70.59	acd	75.12
bc	70.92	bcd	79.67
abc	104.96	abcd	97.80

Initially, effects were calculated and placed on a normal probability plot. The calculated effects are as follows:

$$A = 10.5613, \quad BD = -2.2787, \quad B = 12.4463,$$
$$C = 2.4138, \quad D = 2.1438, \quad AB = 11.4038,$$
$$AC = 1.2613, \quad AD = -1.8238, \quad BC = 1.8163,$$
$$CD = 1.4088, \quad ABC = 2.8588, \quad ABD = -1.7813,$$
$$ACD = -3.0438, \quad BCD = -0.4788, \quad ABCD = -1.3063.$$

The normal quantile-quantile plot is shown in Figure 15.5. The plot seems to imply that effects A, B, and AB stand out as being important. The signs of the important effects indicate the preliminary conclusions.

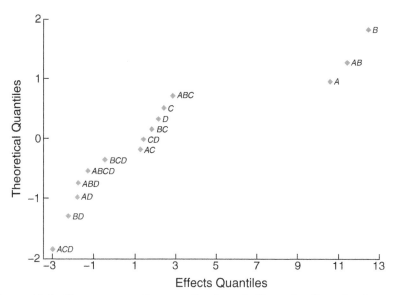

Figure 15.5: Normal quantile-quantile plot of effects for Case Study 15.1.

1. An increase in injection velocity from 1.0 to 2.0 increases shrinkage.
2. An increase in mold temperature from 100°C to 150°C increases shrinkage.
3. There is an interaction between injection velocity and mold temperature; although both main effects are important, it is crucial that we understand the impact of the two-factor interaction.

Interpretation of Two-Factor Interaction

As one would expect, a two-way table of means provides ease in interpretation of the AB interaction. Consider the two-factor situation in Table 15.7.

Table 15.7: Illustration of Two-Factor Interaction

	B (temperature)	
A (velocity)	100	150
2	73.355	97.205
1	74.1975	75.240

Notice that the large sample mean at high velocity and high temperature created the significant interaction. The **shrinkage increases in a nonadditive manner**. Mold temperature appears to have a positive effect despite the velocity level. But the effect is greatest at high velocity. The velocity effect is very slight at low temperature but clearly is positive at high mold temperature. To control shrinkage at a low level, *one should avoid using high injection velocity and high mold temperature simultaneously.* All of these results are illustrated graphically in Figure 15.6.

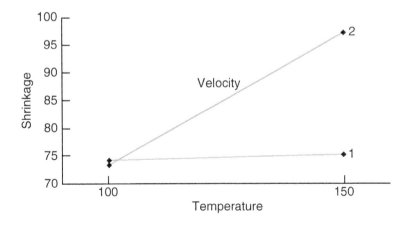

Figure 15.6: Interaction plot for Case Study 15.1.

Analysis with Pooled Mean Square Error: Annotated Computer Printout

It may be of interest to observe an analysis of variance of the injection molding data with high-order interactions pooled to form a mean square error. Interactions of order three and four are pooled. Figure 15.7 shows a *SAS* PROC GLM printout. The analysis of variance reveals essentially the same conclusion as that of the normal probability plot.

The tests and *P*-values shown in Figure 15.7 require interpretation. A significant *P*-value suggests that the effect differs significantly from zero. The tests on main effects (which in the presence of interactions may be regarded as the effects averaged over the levels of the other factors) indicate significance for effects *A* and *B*. The signs of the effects are also important. An increase in the level from low

```
                         The GLM Procedure
Dependent Variable: y
                              Sum of
Source                  DF   Squares    Mean Square   F Value   Pr > F
Model                   10   1689.237462  168.923746    9.37    0.0117
Error                    5     90.180831   18.036166
Corrected Total         15   1779.418294
R-Square      Coeff Var       Root MSE        y Mean
0.949320      5.308667        4.246901       79.99938
Source         DF    Type III SS    Mean Square    F Value     Pr > F
A               1    446.1600062    446.1600062     24.74      0.0042
B               1    619.6365563    619.6365563     34.36      0.0020
C               1     23.3047563     23.3047563      1.29      0.3072
D               1     18.3826563     18.3826563      1.02      0.3590
A*B             1    520.1820562    520.1820562     28.84      0.0030
A*C             1      6.3630063      6.3630063      0.35      0.5784
A*D             1     13.3042562     13.3042562      0.74      0.4297
B*C             1     13.1950562     13.1950562      0.73      0.4314
B*D             1     20.7708062     20.7708062      1.15      0.3322
C*D             1      7.9383063      7.9383063      0.44      0.5364
                              Standard
Parameter       Estimate        Error    t Value     Pr > |t|
Intercept      79.99937500    1.06172520   75.35      <.0001
A               5.28062500    1.06172520    4.97      0.0042
B               6.22312500    1.06172520    5.86      0.0020
C               1.20687500    1.06172520    1.14      0.3072
D               1.07187500    1.06172520    1.01      0.3590
A*B             5.70187500    1.06172520    5.37      0.0030
A*C             0.63062500    1.06172520    0.59      0.5784
A*D            -0.91187500    1.06172520   -0.86      0.4297
B*C             0.90812500    1.06172520    0.86      0.4314
B*D            -1.13937500    1.06172520   -1.07      0.3322
C*D             0.70437500    1.06172520    0.66      0.5364
```

Figure 15.7: *SAS* printout for data of Case Study 15.1.

Exercises

to high of A, injection velocity, results in increased shrinkage. The same is true for B. However, because of the significant interaction AB, main effect interpretations may be viewed as trends across the levels of the other factors. The impact of the significant AB interaction is better understood by using a two-way table of means.

Exercises

15.1 The following data are obtained from a 2^3 factorial experiment replicated three times. Evaluate the sums of squares for all factorial effects by the contrast method. Draw conclusions.

Treatment Combination	Rep 1	Rep 2	Rep 3
(1)	12	19	10
a	15	20	16
b	24	16	17
ab	23	17	27
c	17	25	21
ac	16	19	19
bc	24	23	29
abc	28	25	20

15.2 In an experiment conducted by the Mining and Minerals Engineering Department at Virginia Tech to study a particular filtering system for coal, a coagulant was added to a solution in a tank containing coal and sludge, which was then placed in a recirculation system in order that the coal could be washed. Three factors were varied in the experimental process:

Factor A: percent solids circulated initially in the overflow
Factor B: flow rate of the polymer
Factor C: pH of the tank

The amount of solids in the underflow of the cleansing system determines how clean the coal has become. Two levels of each factor were used and two experimental runs were made for each of the $2^3 = 8$ combinations. The response measurements in percent solids by weight in the underflow of the circulation system are as specified in the following table:

Treatment Combination	Response Replication 1	Replication 2
(1)	4.65	5.81
a	21.42	21.35
b	12.66	12.56
ab	18.27	16.62
c	7.93	7.88
ac	13.18	12.87
bc	6.51	6.26
abc	18.23	17.83

Assuming that all interactions are potentially important, do a complete analysis of the data. Use P-values in your conclusion.

15.3 In a metallurgy experiment, it is desired to test the effect of four factors and their interactions on the concentration (percent by weight) of a particular phosphorus compound in casting material. The variables are A, percent phosphorus in the refinement; B, percent remelted material; C, fluxing time; and D, holding time. The four factors are varied in a 2^4 factorial experiment with two castings taken at each factor combination. The 32 castings were made in random order. The following table shows the data and an ANOVA table is given in Figure 15.8 on page 610. Discuss the effects of the factors and their interactions on the concentration of the phosphorus compound.

Treatment Combination	Weight % of Phosphorus Compound		
	Rep 1	Rep 2	Total
(1)	30.3	28.6	58.9
a	28.5	31.4	59.9
b	24.5	25.6	50.1
ab	25.9	27.2	53.1
c	24.8	23.4	48.2
ac	26.9	23.8	50.7
bc	24.8	27.8	52.6
abc	22.2	24.9	47.1
d	31.7	33.5	65.2
ad	24.6	26.2	50.8
bd	27.6	30.6	58.2
abd	26.3	27.8	54.1
cd	29.9	27.7	57.6
acd	26.8	24.2	51.0
bcd	26.4	24.9	51.3
$abcd$	26.9	29.3	56.2
Total	428.1	436.9	865.0

15.4 A preliminary experiment is conducted to study the effects of four factors and their interactions on the output of a certain machining operation. Two runs are made at each of the treatment combinations in order to supply a measure of pure experimental error. Two levels of each factor are used, resulting in the data shown next page. Make tests on all main effects and interactions at the 0.05 level of significance. Draw conclusions.

Source of Variation	Effects	Sum of Squares	Degrees of Freedom	Mean Square	Computed f	P-Value
Main effect :						
A	−1.2000	11.52	1	11.52	4.68	0.0459
B	−1.2250	12.01	1	12.01	4.88	0.0421
C	−2.2250	39.61	1	39.61	16.10	0.0010
D	1.4875	17.70	1	17.70	7.20	0.0163
Two-factor interaction :						
AB	0.9875	7.80	1	7.80	3.17	0.0939
AC	0.6125	3.00	1	3.00	1.22	0.2857
AD	−1.3250	14.05	1	14.05	5.71	0.0295
BC	1.1875	11.28	1	11.28	4.59	0.0480
BD	0.6250	3.13	1	3.13	1.27	0.2763
CD	0.7000	3.92	1	3.92	1.59	0.2249
Three-factor interaction :						
ABC	−0.5500	2.42	1	2.42	0.98	0.3360
ABD	1.7375	24.15	1	24.15	9.82	0.0064
ACD	1.4875	17.70	1	17.70	7.20	0.0163
BCD	−0.8625	5.95	1	5.95	2.42	0.1394
Four-factor interaction :						
$ABCD$	0.7000	3.92	1	3.92	1.59	0.2249
Error		39.36	16	2.46		
Total		217.51	31			

Figure 15.8: ANOVA table for Exercise 15.3.

Treatment Combination	Replicate 1	Replicate 2
(1)	7.9	9.6
a	9.1	10.2
b	8.6	5.8
c	10.4	12.0
d	7.1	8.3
ab	11.1	12.3
ac	16.4	15.5
ad	7.1	8.7
bc	12.6	15.2
bd	4.7	5.8
cd	7.4	10.9
abc	21.9	21.9
abd	9.8	7.8
acd	13.8	11.2
bcd	10.2	11.1
$abcd$	12.8	14.3

15.5 In the study *An X-Ray Fluorescence Method for Analyzing Polybutadiene-Acrylic Acid (PBAA) Propellants* (Quarterly Reports, RK-TR-62-1, Army Ordnance Missile Command), an experiment was conducted to determine whether or not there was a significant difference in the amount of aluminum obtained in an analysis with certain levels of certain processing variables. The data are shown below.

Obs.	Phys. State	Mixing Time	Blade Speed	Nitrogen Condition	Aluminum
1	1	1	2	2	16.3
2	1	2	2	2	16.0
3	1	1	1	1	16.2
4	1	2	1	2	16.1
5	1	1	1	2	16.0
6	1	2	1	1	16.0
7	1	2	2	1	15.5
8	1	1	2	1	15.9
9	2	1	2	2	16.7
10	2	2	2	2	16.1
11	2	1	1	1	16.3
12	2	2	1	2	15.8
13	2	1	1	2	15.9
14	2	2	1	1	15.9
15	2	2	2	1	15.6
16	2	1	2	1	15.8

The variables in the data are given as below.
A: mixing time
  level 1: 2 hours
  level 2: 4 hours

B: blade speed
  level 1: 36 rpm
  level 2: 78 rpm
C: condition of nitrogen passed over propellant
  level 1: dry
  level 2: 72% relative humidity
D: physical state of propellant
  level 1: uncured
  level 2: cured

Assuming all three- and four-factor interactions to be negligible, analyze the data. Use a 0.05 level of significance. Write a brief report summarizing the findings.

15.6 It is important to study the effect of the concentration of the reactant and the feed rate on the viscosity of the product from a chemical process. Let the reactant concentration be factor A, at levels 15% and 25%. Let the feed rate be factor B, with levels 20 lb/hr and 30 lb/hr. The experiment involves two experimental runs at each of the four combinations (L = low and H = high). The viscosity readings are as follows.

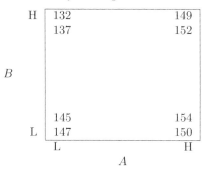

(a) Assuming a model containing two main effects and an interaction, calculate the three effects. Do you have any interpretation at this point?

(b) Do an analysis of variance and test for interaction. Give conclusions.

(c) Test for main effects and give final conclusions regarding the importance of all these effects.

15.7 Consider Exercise 15.3. It is of interest to the researcher to learn not only that AD, BC, and possibly AB are important, but also what they mean scientifically. Show two-dimensional interaction plots for all three and give an interpretation.

15.8 Consider Exercise 15.3 once again. Three-factor interactions are often not significant, and even if they are, they are difficult to interpret. The interaction ABD appears to be important. To gain some sense of interpretation, show two AD interaction plots, one for $B = -1$ and the other for $B = +1$. From the appearance of these, give an interpretation of the ABD interaction.

15.9 Consider Exercise 15.6. Use a $+1$ and -1 scaling for "high" and "low," respectively, and do a multiple linear regression with the model

$$Y_i = \beta_0 + \beta_1 x_{1i} + \beta_2 x_{2i} + \beta_{12} x_{1i} x_{2i} + \epsilon_i,$$

with x_{1i} = reactant concentration $(-1, +1)$ and x_{2i} = feed rate $(-1, +1)$.

(a) Compute regression coefficients.

(b) How do the coefficients b_1, b_2, and b_{12} relate to the effects you found in Exercise 15.6(a)?

(c) In your regression analysis, do t-tests on b_1, b_2, and b_{12}. How do these test results relate to those in Exercise 15.6(b) and (c)?

15.10 Consider Exercise 15.5. Compute all 15 effects and do normal probability plots of the effects.

(a) Does it appear as if your assumption of negligible three- and four-factor interactions has merit?

(b) Are the results of the effect plots consistent with what you communicated about the importance of main effects and two-factor interactions in your summary report?

15.11 In Myers, Montgomery, and Anderson-Cook (2009), a data set is discussed in which a 2^3 factorial is used by an engineer to study the effects of cutting speed (A), tool geometry (B), and cutting angle (C) on the life (in hours) of a machine tool. Two levels of each factor are chosen, and duplicates are run at each design point with the order of the runs being random. The data are presented here.

	A	B	C	Life
(1)	−	−	−	22, 31
a	+	−	−	32, 43
b	−	+	−	35, 34
ab	+	+	−	35, 47
c	−	−	+	44, 45
ac	+	−	+	40, 37
bc	−	+	+	60, 50
abc	+	+	+	39, 41

(a) Calculate all seven effects. Which appear, based on their magnitude, to be important?

(b) Do an analysis of variance and observe P-values.

(c) Do your results in (a) and (b) agree?

(d) The engineer felt confident that cutting speed and cutting angle should interact. If this interaction is significant, draw an interaction plot and discuss the engineering meaning of the interaction.

15.12 Consider Exercise 15.11. Suppose there was some experimental difficulty in making the runs. In fact, the total experiment had to be halted after only 4 runs. As a result, the abbreviated experiment is given by

	Life
a	43
b	35
c	44
abc	39

With only these runs, we have the signs for contrasts given by

	A	B	C	AB	AC	BC	ABC
a	+	−	−	−	−	+	+
b	−	+	−	−	+	−	+
c	−	−	+	+	−	−	+
abc	+	+	+	+	+	+	+

Comment. In your comments, determine whether or not the contrasts are orthogonal. Which are and which are not? Are main effects orthogonal to each other? In this abbreviated experiment (called a *fractional factorial*), can we study interactions independent of main effects? Is it a useful experiment if we are convinced that interactions are negligible? Explain.

15.4 Factorial Experiments in a Regression Setting

Thus far in this chapter, we have mostly confined our discussion of analysis of the data for a 2^k factorial to the method of analysis of variance. The only reference to an alternative analysis resides in Exercise 15.9. Indeed, this exercise introduces much of what motivates the present section. There are situations in which model fitting is important **and** the factors under study **can be controlled**. For example, a biologist may wish to study the growth of a certain type of algae in the water, and so a model that looks at units of algae as a function of the *amount of a pollutant* and, say, *time* would be very helpful. Thus, the study involves a factorial experiment in a laboratory setting in which concentration of the pollutant and time are the factors. As we shall discuss later in this section, a more precise model can be fitted if the factors are controlled in a factorial array, with the 2^k factorial often being a useful choice. In many biological and chemical processes, the levels of the regressor variables can and should be controlled.

Recall that the regression model employed in Chapter 12 can be written in matrix notation as

$$\mathbf{y} = \mathbf{X}\boldsymbol{\beta} + \boldsymbol{\epsilon}.$$

The $\mathbf{X}$ matrix is referred to as the **model matrix**. Suppose, for example, that a 2^3 factorial experiment is employed with the variables

Temperature:	150°C	200°C
Humidity:	15%	20%
Pressure (psi):	1000	1500

The familiar +1, −1 levels can be generated through the following centering and scaling to *design units*:

$$x_1 = \frac{\text{temperature} - 175}{25}, \quad x_2 = \frac{\text{humidity} - 17.5}{2.5}, \quad x_3 = \frac{\text{pressure} - 1250}{250}.$$

15.4 Factorial Experiments in a Regression Setting

As a result, the **X** matrix becomes

$$\mathbf{X} = \begin{bmatrix} 1 & -1 & -1 & -1 \\ 1 & 1 & -1 & -1 \\ 1 & -1 & 1 & -1 \\ 1 & -1 & -1 & 1 \\ 1 & 1 & 1 & -1 \\ 1 & 1 & -1 & 1 \\ 1 & -1 & 1 & 1 \\ 1 & 1 & 1 & 1 \end{bmatrix} \begin{matrix} x_1 & x_2 & x_3 & \text{Design Identification} \\ & & & (1) \\ & & & a \\ & & & b \\ & & & c \\ & & & ab \\ & & & ac \\ & & & bc \\ & & & abc \end{matrix}$$

It is now seen that the contrasts illustrated and discussed in Section 15.2 are directly related to regression coefficients. Notice that all the columns of the **X** matrix in our 2^3 example are *orthogonal*. As a result, the computation of regression coefficients as described in Section 12.3 becomes

$$b = \begin{bmatrix} b_0 \\ b_1 \\ b_2 \\ b_3 \end{bmatrix} = (\mathbf{X}'\mathbf{X})^{-1}\mathbf{X}'\mathbf{y} = \left(\frac{1}{8}\mathbf{I}\right)\mathbf{X}'\mathbf{y}$$

$$= \frac{1}{8}\begin{bmatrix} a + ab + ac + abc + (1) + b + c + bc \\ a + ab + ac + abc - (1) - b - c - bc \\ b + ab + bc + abc - (1) - a - c - ac \\ c + ac + bc + abc - (1) - a - b - ab \end{bmatrix},$$

where a, ab, and so on, are response measures.

One can now see that the notion of *calculated main effects*, which has been emphasized throughout this chapter with 2^k factorials, is related to coefficients in a fitted regression model when factors are quantitative. In fact, for a 2^k with, say, n experimental runs per design point, the relationships between effects and regression coefficients are as follows:

$$\text{Effect} = \frac{\text{contrast}}{2^{k-1}(n)}$$

$$\text{Regression coefficient} = \frac{\text{contrast}}{2^k(n)} = \frac{\text{effect}}{2}.$$

This relationship should make sense to the reader, since a regression coefficient b_j is an average rate of change in response *per unit change* in x_j. Of course, as one goes from -1 to $+1$ in x_j (low to high), the design variable changes by 2 units.

Example 15.2: Consider an experiment where an engineer desires to fit a linear regression of yield y against holding time x_1 and flexing time x_2 in a certain chemical system. All other factors are held fixed. The data in the natural units are given in Table 15.8. Estimate the multiple linear regression model.

Solution: The fitted regression model is

$$\hat{y} = b_0 + b_1 x_1 + b_2 x_2.$$

Table 15.8: Data for Example 15.2

Holding Time (hr)	Flexing Time (hr)	Yield (%)
0.5	0.10	28
0.8	0.10	39
0.5	0.20	32
0.8	0.20	46

The design units are

$$x_1 = \frac{\text{holding time} - 0.65}{0.15}, \qquad x_2 = \frac{\text{flexing time} - 0.15}{0.05}$$

and the **X** matrix is

$$\begin{array}{cc} & \begin{array}{cc} x_1 & x_2 \end{array} \\ & \begin{bmatrix} 1 & -1 & -1 \\ 1 & 1 & -1 \\ 1 & -1 & 1 \\ 1 & 1 & 1 \end{bmatrix} \end{array}$$

with the regression coefficients

$$\begin{bmatrix} b_0 \\ b_1 \\ b_2 \end{bmatrix} = (\mathbf{X}'\mathbf{X})^{-1}\mathbf{X}'\mathbf{y} = \begin{bmatrix} \frac{(1) + a + b + ab}{4} \\ \frac{a + ab - (1) - b}{4} \\ \frac{b + ab - (1) - a}{4} \end{bmatrix} = \begin{bmatrix} 36.25 \\ 6.25 \\ 2.75 \end{bmatrix}.$$

Thus, the least squares regression equation is

$$\hat{y} = 36.25 + 6.25 x_1 + 2.75 x_2.$$

This example provides an illustration of the use of the two-level factorial experiment in a regression setting. The four experimental runs in the 2^2 design were used to calculate a regression equation, with the obvious interpretation of the regression coefficients. The value $b_1 = 6.25$ represents the estimated increase in response (percent yield) per *design unit* change (0.15 hour) in holding time. The value $b_2 = 2.75$ represents a similar rate of change for flexing time.

Interaction in the Regression Model

The interaction contrasts discussed in Section 15.2 have definite interpretations in the regression context. In fact, interactions are accounted for in regression models by product terms. For example, in Example 15.2, the model with interaction is

$$y = b_0 + b_1 x_1 + b_2 x_2 + b_{12} x_1 x_2$$

with b_0, b_1, b_2 as before and

$$b_{12} = \frac{ab + (1) - a - b}{4} = \frac{46 + 28 - 39 - 32}{4} = 0.75.$$

15.4 Factorial Experiments in a Regression Setting

Thus, the regression equation expressing two *linear main effects* and interaction is

$$\hat{y} = 36.25 + 6.25x_1 + 2.75x_2 + 0.75x_1x_2.$$

The regression context provides a framework in which the reader should better understand the advantage of orthogonality that is enjoyed by the 2^k factorial. In Section 15.2, the merits of orthogonality were discussed from the point of view of *analysis of variance* of the data in a 2^k factorial experiment. It was pointed out that orthogonality among effects leads to independence among the sums of squares. Of course, the presence of regression variables certainly does not rule out the use of analysis of variance. In fact, f-tests are conducted just as they were described in Section 15.2. Of course, a distinction must be made. In the case of ANOVA, the hypotheses evolve from population means, while in the regression case, the hypotheses involve regression coefficients.

For instance, consider the experimental design in Exercise 15.2 on page 609. Each factor is continuous. Suppose that the levels are

$$
\begin{array}{lll}
A\ (x_1): & 20\% & 40\% \\
B\ (x_2): & 5\ \text{lb/sec} & 10\ \text{lb/sec} \\
C\ (x_3): & 5 & 5.5
\end{array}
$$

and we have, for design levels,

$$x_1 = \frac{\%\ \text{solids} - 30}{10}, \quad x_2 = \frac{\text{flow rate} - 7.5}{2.5}, \quad x_3 = \frac{\text{pH} - 5.25}{0.25}.$$

Suppose that it is of interest to fit a multiple regression model in which all linear coefficients and available interactions are to be considered. In addition, the engineer wants to obtain some insight into what levels of the factor will *maximize* cleansing (i.e., maximize the response). This problem will be the subject of Case Study 15.2.

Case Study 15.2: Coal Cleansing Experiment[1]: Figure 15.9 represents annotated computer printout for the regression analysis for the fitted model

$$\hat{y} = b_0 + b_1x_1 + b_2x_2 + b_3x_3 + b_{12}x_1x_2 + b_{13}x_1x_3 + b_{23}x_2x_3 + b_{123}x_1x_2x_3,$$

where x_1, x_2, and x_3 are percent solids, flow rate, and pH of the system, respectively. The computer system used is *SAS* PROC REG.

Note the parameter estimates, standard error, and P-values in the printout. The parameter estimates represent coefficients in the model. All model coefficients are significant except the x_2x_3 term (BC interaction). Note also that residuals, confidence intervals, and prediction intervals appear as discussed in the regression material in Chapters 11 and 12.

The reader can use the values of the model coefficients and predicted values from the printout to ascertain what combination of the factors results in **maximum cleansing efficiency**. Factor A (percent solids circulated) has a large positive coefficient, suggesting a high value for percent solids. In addition, a low value for factor C (pH of the tank) is suggested. Though the B main effect (flow rate of the polymer) coefficient is positive, the rather large positive coefficient of

---
[1] See Exercise 15.2.

```
Dependent Variable: Y
                    Analysis of Variance
                       Sum of        Mean
Source             DF  Squares      Square   F Value   Pr > F
Model               7  490.23499   70.03357   254.43   <.0001
Error               8    2.20205    0.27526
Corrected Total    15  492.43704
Root MSE              0.52465    R-Square    0.9955
Dependent Mean       12.75188    Adj R-Sq    0.9916
Coeff Var             4.11429
                    Parameter Estimates
                 Parameter    Standard
Variable    DF   Estimate      Error     t Value   Pr > |t|
Intercept    1   12.75188     0.13116     97.22    <.0001
A            1    4.71938     0.13116     35.98    <.0001
B            1    0.86563     0.13116      6.60     0.0002
C            1   -1.41563     0.13116    -10.79    <.0001
AB           1   -0.59938     0.13116     -4.57     0.0018
AC           1   -0.52813     0.13116     -4.03     0.0038
BC           1    0.00562     0.13116      0.04     0.9668
ABC          1    2.23063     0.13116     17.01    <.0001
     Dependent  Predicted       Std Error
Obs  Variable   Value  Mean Predict    95% CL Mean      95% CL Predict  Residual
 1    4.6500    5.2300     0.3710    4.3745   6.0855   3.7483   6.7117  -0.5800
 2   21.4200   21.3850     0.3710   20.5295  22.2405  19.9033  22.8667   0.0350
 3   12.6600   12.6100     0.3710   11.7545  13.4655  11.1283  14.0917   0.0500
 4   18.2700   17.4450     0.3710   16.5895  18.3005  15.9633  18.9267   0.8250
 5    7.9300    7.9050     0.3710    7.0495   8.7605   6.4233   9.3867   0.0250
 6   13.1800   13.0250     0.3710   12.1695  13.8805  11.5433  14.5067   0.1550
 7    6.5100    6.3850     0.3710    5.5295   7.2405   4.9033   7.8667   0.1250
 8   18.2300   18.0300     0.3710   17.1745  18.8855  16.5483  19.5117   0.2000
 9    5.8100    5.2300     0.3710    4.3745   6.0855   3.7483   6.7117   0.5800
10   21.3500   21.3850     0.3710   20.5295  22.2405  19.9033  22.8667  -0.0350
11   12.5600   12.6100     0.3710   11.7545  13.4655  11.1283  14.0917  -0.0500
12   16.6200   17.4450     0.3710   16.5895  18.3005  15.9633  18.9267  -0.8250
13    7.8800    7.9050     0.3710    7.0495   8.7605   6.4233   9.3867  -0.0250
14   12.8700   13.0250     0.3710   12.1695  13.8805  11.5433  14.5067  -0.1550
15    6.2600    6.3850     0.3710    5.5295   7.2405   4.9033   7.8667  -0.1250
16   17.8300   18.0300     0.3710   17.1745  18.8855  16.5483  19.5117  -0.2000
```

Figure 15.9: *SAS* printout for data of Case Study 15.2.

$x_1 x_2 x_3$ (ABC) suggests that flow rate should be at the low level to enhance efficiency. Indeed, the regression model generated in the *SAS* printout suggests that the combination of factors that may produce optimum results, or perhaps suggest direction for further experimentation, is given by

A: high level
B: low level
C: low level

15.5 The Orthogonal Design

In experimental situations where it is appropriate to fit models that are linear in the design variables and possibly should involve interactions or product terms, there are advantages gained from the two-level *orthogonal design*, or orthogonal array. By an orthogonal design we mean one in which there is orthogonality among the columns of the $\mathbf{X}$ matrix. For example, consider the $\mathbf{X}$ matrix for the 2^2 factorial of Example 15.2. Notice that all three columns are mutually orthogonal. The $\mathbf{X}$ matrix for the 2^3 factorial also contains orthogonal columns. The 2^3 factorial with interactions would yield an $\mathbf{X}$ matrix of the type

$$\mathbf{X} = \begin{bmatrix} & x_1 & x_2 & x_3 & x_1x_2 & x_1x_3 & x_2x_3 & x_1x_2x_3 \\ 1 & -1 & -1 & -1 & 1 & 1 & 1 & -1 \\ 1 & 1 & -1 & -1 & -1 & -1 & 1 & 1 \\ 1 & -1 & 1 & -1 & -1 & 1 & -1 & 1 \\ 1 & -1 & -1 & 1 & 1 & -1 & -1 & 1 \\ 1 & 1 & 1 & -1 & 1 & -1 & -1 & -1 \\ 1 & 1 & -1 & 1 & -1 & 1 & -1 & -1 \\ 1 & -1 & 1 & 1 & -1 & -1 & 1 & -1 \\ 1 & 1 & 1 & 1 & 1 & 1 & 1 & 1 \end{bmatrix}$$

The outline of degrees of freedom is

Source	d.f.	
Regression	3	
Lack of fit	4	(x_1x_2, x_1x_3, x_2x_3, $x_1x_2x_3$)
Error (pure)	8	
Total	15	

The 8 degrees of freedom for pure error are obtained from the *duplicate runs* at each design point. Lack-of-fit degrees of freedom may be viewed as the difference between the number of distinct design points and the number of total model terms; in this case, there are 8 points and 4 model terms.

Standard Error of Coefficients and *T*-Tests

In previous sections, we showed how the designer of an experiment may exploit the notion of orthogonality to design a regression experiment with coefficients that attain minimum variance on a per cost basis. We should be able to make use of our exposure to regression in Section 12.4 to compute estimates of variances of coefficients and hence their standard errors. It is also of interest to note the relationship between the *t*-statistic on a coefficient and the *F*-statistic described and illustrated in previous chapters.

Recall from Section 12.4 that the variances and covariances of coefficients appear in A^{-1}, or, in terms of present notation, the *variance-covariance matrix* of coefficients is

$$\sigma^2 A^{-1} = \sigma^2 (\mathbf{X}'\mathbf{X})^{-1}.$$

In the case of the 2^k factorial experiment, the columns of $\mathbf{X}$ are mutually orthogonal, imposing a very special structure. In general, for the 2^k we can write

$$\mathbf{X} = [\mathbf{1} \quad \underset{\pm 1}{x_1} \quad \underset{\pm 1}{x_2} \quad \cdots \quad \underset{\pm 1}{x_k} \quad \underset{\pm 1}{x_1 x_2} \quad \cdots],$$

where each column contains 2^k or $2^k n$ entries, where n is the number of replicate runs at each design point. Thus, formation of $\mathbf{X}'\mathbf{X}$ yields

$$\mathbf{X}'\mathbf{X} = 2^k n \mathbf{I}_p,$$

where $\mathbf{I}$ is the identity matrix of dimension p, the number of model parameters.

Example 15.3: Consider a 2^3 factorial design with duplicated runs fitted to the model

$$E(Y) = \beta_0 + \beta_1 x_1 + \beta_2 x_2 + \beta_3 x_3 + \beta_{12} x_1 x_2 + \beta_{13} x_1 x_3 + \beta_{23} x_2 x_3.$$

Give expressions for the standard errors of the least squares estimates of b_0, b_1, b_2, b_3, b_{12}, b_{13}, and b_{23}.

Solution:

$$\mathbf{X} = \begin{bmatrix} & x_1 & x_2 & x_3 & x_1 x_2 & x_1 x_3 & x_2 x_3 \\ 1 & -1 & -1 & -1 & 1 & 1 & 1 \\ 1 & 1 & -1 & -1 & -1 & -1 & 1 \\ 1 & -1 & 1 & -1 & -1 & 1 & -1 \\ 1 & -1 & -1 & 1 & 1 & -1 & -1 \\ 1 & 1 & 1 & -1 & 1 & -1 & -1 \\ 1 & 1 & -1 & 1 & -1 & 1 & -1 \\ 1 & -1 & 1 & 1 & -1 & -1 & 1 \\ 1 & 1 & 1 & 1 & 1 & 1 & 1 \end{bmatrix}$$

with each unit viewed as being *repeated* (i.e., each observation is duplicated). As a result,

$$\mathbf{X}'\mathbf{X} = 16 \mathbf{I}_7.$$

Thus,

$$(\mathbf{X}'\mathbf{X})^{-1} = \frac{1}{16} \mathbf{I}_7.$$

From the foregoing it should be clear that the variances of all coefficients for a 2^k factorial with n runs at each design point are

$$\text{Var}(b_j) = \frac{\sigma^2}{2^k n},$$

and, of course, all covariances are zero. As a result, standard errors of coefficients are calculated as

$$s_{b_j} = s \sqrt{\frac{1}{2^k n}},$$

where s is found from the square root of the mean square error (hopefully obtained from adequate replication). Thus, in our case with the 2^3,

$$s_{b_j} = s \left(\frac{1}{4} \right).$$

15.5 The Orthogonal Design

Example 15.4: Consider the metallurgy experiment in Exercise 15.3 on page 609. Suppose that the fitted model is

$$E(Y) = \beta_0 + \beta_1 x_1 + \beta_2 x_2 + \beta_3 x_3 + \beta_4 x_4 + \beta_{12} x_1 x_2 + \beta_{13} x_1 x_3 \\ + \beta_{14} x_1 x_4 + \beta_{23} x_2 x_3 + \beta_{24} x_2 x_4 + \beta_{34} x_3 x_4.$$

What are the standard errors of the least squares regression coefficients?

Solution: Standard errors of all coefficients for the 2^k factorial are equal and are

$$s_{b_j} = s\sqrt{\frac{1}{2^k n}},$$

which in this illustration is

$$s_{b_j} = s\sqrt{\frac{1}{(16)(2)}}.$$

In this case, the pure mean square error is given by $s^2 = 2.46$ (16 degrees of freedom). Thus,

$$s_{b_j} = 0.28.$$

The standard errors of coefficients can be used to construct t-statistics on all coefficients. These t-values are related to the F-statistics in the analysis of variance. We have already demonstrated that an F-statistic on a coefficient, using the 2^k factorial, is

$$f = \frac{(\text{contrast})^2}{(2^k n) s^2}.$$

This is the form of the F-statistics on page 610 for the metallurgy experiment (Exercise 15.3). It is easy to verify that if we write

$$t = \frac{b_j}{s_{b_j}}, \quad \text{where} \quad b_j = \frac{\text{contrast}}{2^k n},$$

then

$$t^2 = \frac{(\text{contrast})^2}{s^2 2^k n} = f.$$

As a result, the usual relationship holds between t-statistics on coefficients and the F-values. As we might expect, the only difference between the use of t and F in assessing significance lies in the fact that the t-statistic indicates the sign, or direction, of the effect of the coefficient.

It would appear that the 2^k factorial plan would handle many practical situations in which regression models are fitted. It can accommodate linear and interaction terms, providing optimal estimates of all coefficients (from a variance point of view). However, when k is large, the number of design points required is very large. Often, portions of the total design can be used and still allow orthogonality with all its advantages. These designs are discussed in Section 15.6.

A More Thorough Look at the Orthogonality Property in the 2^k Factorial

We have learned that for the case of the 2^k factorial all the information that is delivered to the analyst about the main effects and interactions is in the form of contrasts. These "$2^k - 1$ pieces of information" carry a single degree of freedom apiece and they are independent of each other. In an analysis of variance, they manifest themselves as *effects*, whereas if a regression model is being constructed, the effects turn out to be regression coefficients, apart from a factor of 2. With either form of analysis, significance tests can be carried out and the *t*-test for a given effect is numerically the same as that for the corresponding regression coefficient. In the case of ANOVA, variable screening and scientific interpretation of interactions are important, whereas in the case of a regression analysis, a model may be used to predict response and/or determine which factor/level combinations are optimum (e.g. maximize yield or maximize cleaning efficiency, as in the case of Case Study 15.2).

It turns out that the orthogonality property is important whether the analysis is to be ANOVA or regression. The orthogonality among the columns of **X**, the model matrix in, say, Example 15.3, provides special conditions that have an important impact on the **variance of effects** or **regression coefficients**. In fact, it has already become apparent that the orthogonal design results in equality of variance for all effects or coefficients. Thus, in this way, the precision, for purposes of estimation or testing, is the same for all coefficients, main effects, or interactions. In addition, if the regression model contains only linear terms and thus only main effects are of interest, the following conditions result in the minimization of variances of all effects (or, correspondingly, first-order regression coefficients).

Conditions for Minimum Variances of Coefficients: If the regression model contains terms no higher than first order, and if the ranges on the variables are given by $x_j \in [-1, +1]$ for $j = 1, 2, \ldots, k$, then $\mathrm{Var}(b_j)/\sigma^2$, for $j = 1, 2, \ldots, k$, is minimized if the design is orthogonal and all x_i levels in the design are at ± 1 for $i = 1, 2, \ldots, k$.

Thus, in terms of coefficients of model terms or main effects, orthogonality in the 2^k is a very desirable property.

Another approach to a better understanding of the "balance" provided by the 2^3 is to look at the situation graphically. All of the contrasts that are orthogonal and thus mutually independent are shown graphically in Figure 15.10. In the graphs, the planes of the squares whose vertices contain the responses labeled "+" are compared to those containing the responses labeled "−." Those given in (a) show contrasts for main effects and should be obvious to the reader. Those in (b) show the planes representing "+" vertices and "−" vertices for the three two-factor interaction contrasts. In (c), we see the geometric representation of the contrasts for the three-factor (ABC) interaction.

Center Runs with 2^k Designs

In the situation in which the 2^k design is implemented with **continuous** design variables and one is seeking to fit a linear regression model, the use of replicated runs in the **design center** can be extremely useful. In fact, quite apart from the advantages that will be discussed in what follows, a majority of scientists and

15.5 The Orthogonal Design

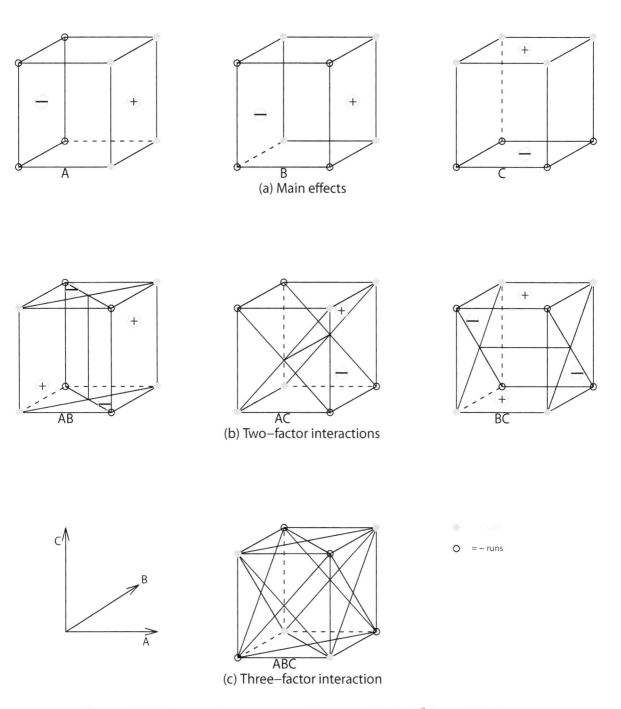

Figure 15.10: Geometric presentation of contrasts for the 2^3 factorial design.

engineers would consider center runs (i.e., the runs at $x_i = 0$ for $i = 1, 2, \ldots, k$) as not only a reasonable practice but something that was intuitively appealing. In many areas of application of the 2^k design, the scientist desires to determine if he or she might benefit from moving to a different region of interest in the factors. In many cases, the center (i.e., the point $(0, 0, \ldots, 0)$ in the coded factors) is often either the current operating conditions of the process or at least those conditions that are considered "currently optimum." So it is often the case that the scientist will require data on the response at the center.

Center Runs and Lack of Fit

In addition to the intuitive appeal of the augmentation of the 2^k with center runs, a second advantage is enjoyed that relates to the kind of model that is fitted to the data. Consider, for example, the case with $k = 2$, illustrated in Figure 15.11.

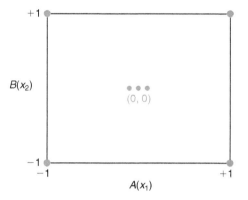

Figure 15.11: A 2^2 design with center runs.

It is clear that *without the center runs* the model terms are the intercept, x_1, x_2, $x_1 x_2$. These account for the four model degrees of freedom delivered by the four design points, apart from any replication. Since each factor has response information available *only at two locations* $\{-1, +1\}$, no "pure" second-order curvature terms can be accommodated in the model (i.e, x_1^2 or x_2^2). But the information at $(0, 0)$ produces an additional model degree of freedom. While this important degree of freedom does not allow both x_1^2 and x_2^2 to be used in the model, it does allow for testing the significance of a linear combination of x_1^2 and x_2^2. For n_c center runs, there are then $n_c - 1$ degrees of freedom available for replication or "pure" error. This allows an estimate of σ^2 for testing the model terms and significance of the 1 d.f. for **quadratic lack of fit**. The concept here is very much like that discussed in the lack-of-fit material in Chapter 11.

In order to gain a complete understanding of how the lack-of-fit test works, assume that for $k = 2$ the **true model** contains the full second-order complement of terms, including x_1^2 and x_2^2. In other words,

$$E(Y) = \beta_0 + \beta_1 x_1 + \beta_2 x_2 + \beta_{12} x_1 x_2 + \beta_{11} x_1^2 + \beta_{22} x_2^2.$$

15.5 The Orthogonal Design

Now, consider the contrast

$$\bar{y}_f - \bar{y}_0,$$

where $\bar{y}_f$ is the average response at the factorial locations and $\bar{y}_0$ is the average response at the center point. It can be shown easily (see Review Exercise 15.46) that

$$E(\bar{y}_f - \bar{y}_0) = \beta_{11} + \beta_{22},$$

and, in fact, for the general case with k factors,

$$E(\bar{y}_f - \bar{y}_0) = \sum_{i=1}^{k} \beta_{ii}.$$

As a result, the lack-of-fit test is a simple t-test (or $F = t^2$) with

$$t_{n_c-1} = \frac{\bar{y}_f - \bar{y}_0}{s_{\bar{y}_f - \bar{y}_0}} = \frac{\bar{y}_f - \bar{y}_0}{\sqrt{MSE(1/n_f + 1/n_c)}},$$

where n_f is the number of factorial points and MSE is simply the sample variance of the response values at $(0, 0, \ldots, 0)$.

Example 15.5: This example is taken from Myers, Montgomery, and Anderson-Cook (2009). A chemical engineer is attempting to model the percent conversion in a process. There are two variables of interest, reaction time and reaction temperature. In an attempt to arrive at the appropriate model, a preliminary experiment was conducted in a 2^2 factorial using the current region of interest in reaction time and temperature. Single runs were made at each of the four factorial points and five runs were made at the design center in order that a lack-of-fit test for curvature could be conducted. Figure 15.12 shows the design region and the experimental runs on yield.

The time and temperature readings at the center are, of course, 35 minutes and 145°C. The estimates of the main effects and single interaction coefficient are computed through contrasts, just as before. The center runs **play no role in the computation of** b_1, b_2, and b_{12}. This should be intuitively reasonable to the reader. The intercept is merely $\bar{y}$ for the entire experiment. This value is $\bar{y} = 40.4444$. The standard errors are found through the use of diagonal elements of $(\mathbf{X'X})^{-1}$, as discussed earlier. For this case,

$$\mathbf{X} = \begin{bmatrix} & x_1 & x_2 & x_1x_2 \\ 1 & -1 & -1 & 1 \\ 1 & -1 & 1 & -1 \\ 1 & 1 & -1 & -1 \\ 1 & 1 & 1 & 1 \\ 1 & 0 & 0 & 0 \\ 1 & 0 & 0 & 0 \\ 1 & 0 & 0 & 0 \\ 1 & 0 & 0 & 0 \\ 1 & 0 & 0 & 0 \end{bmatrix}$$

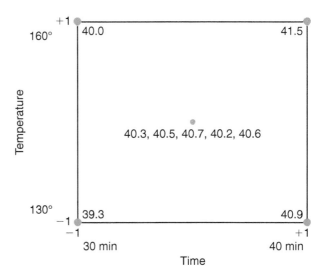

Figure 15.12: 2^2 factorial with 5 center runs.

After the computations, we have

$b_0 = 40.4444,$ $b_1 = 0.7750,$ $b_2 = 0.3250,$ $b_{12} = -0.0250,$
$s_{b_0} = 0.06231,$ $s_{b_1} = 0.09347,$ $s_{b_2} = 0.09347,$ $s_{b_{12}} = 0.09347,$
$t_{b_0} = 649.07$ $t_{b_1} = 8.29$ $t_{b_2} = 3.48$ $t_{b_{12}} = -0.27$ $(P = 0.800).$

The contrast $\bar{y}_f - \bar{y}_0 = 40.425 - 40.46 = -0.035$, and the t-statistic that *tests for curvature* is given by

$$t = \frac{40.425 - 40.46}{\sqrt{0.0430(1/4 + 1/5)}} = 0.251 \quad (P = 0.814).$$

As a result, it appears as if the appropriate model should contain only first-order terms (apart from the intercept).

An Intuitive Look at the Test on Curvature

If one considers the simple case of a single design variable with runs at -1 and $+1$, it should seem clear that the average response at -1 and $+1$ should be close to the response at 0, the center, if the model is first order in nature. Any deviation would certainly suggest curvature. This is simple to extend to two variables. Consider Figure 15.13.

The figure shows the plane on y that passes through the y values of the factorial points. This is the plane that would represent the perfect fit for the model containing x_1, x_2, and $x_1 x_2$. If the model contains no quadratic curvature (i.e., $\beta_{11} = \beta_{22} = 0$), we would expect the response at $(0, 0)$ to be at or near the plane. If the response is far away from the plane, as in the case of Figure 15.13, then it can be seen graphically that quadratic curvature is present.

Exercises

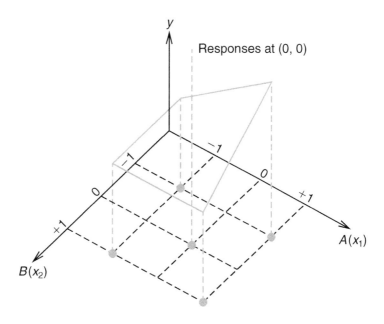

Figure 15.13: 2^2 factorial with runs at $(0,0)$.

Exercises

15.13 Consider a 2^5 experiment where the experimental runs are on 4 different machines. Use the machines as blocks, and assume that all main effects and two-factor interactions may be important.

(a) Which runs would be made on each of the 4 machines?

(b) Which effects are confounded with blocks?

15.14 An experiment is described in Myers, Montgomery, and Anderson-Cook (2009) in which optimum conditions are sought for storing bovine semen to obtain maximum survival. The variables are percent sodium citrate, percent glycerol, and equilibration time in hours. The response is percent survival of the motile spermatozoa. The natural levels are found in the above reference. The data with coded levels for the factorial portion of the design and the center runs are given.

(a) Fit a linear regression model to the data and determine which linear and interaction terms are significant. Assume that the $x_1x_2x_3$ interaction is negligible.

(b) Test for quadratic lack of fit and comment.

x_1, % Sodium Citrate	x_2, % Glycerol	x_3 Equilibration Time	% Survival
-1	-1	-1	57
1	-1	-1	40
-1	1	1	19
1	1	1	40
-1	-1	-1	54
1	-1	-1	41
-1	1	1	21
1	1	1	43
0	0	0	63
0	0	0	61

15.15 Oil producers are interested in nickel alloys that are strong and corrosion resistant. An experiment was conducted in which yield strengths were compared for nickel alloy tensile specimens charged in solutions of sulfuric acid saturated with carbon disulfide. Two alloys were compared: a 75% nickel composition and a 30% nickel composition. The alloys were tested under two different charging times, 25 and 50 days. A 2^3

factorial was conducted with the following factors:

% sulfuric acid: 4%, 6% (x_1)
charging time: 25 days, 50 days (x_2)
nickel composition: 30%, 75% (x_3)

A specimen was prepared for each of the 8 conditions. Since the engineers were not certain of the nature of the model (i.e., whether or not quadratic terms would be needed), a third level (middle level) was incorporated, and 4 center runs were employed using 4 specimens at 5% sulfuric acid, 37.5 days, and 52.5% nickel composition. The following are the yield strengths in kilograms per square inch.

	Charging Time			
	25 Days		50 Days	
Nickel	Sulfuric Acid		Sulfuric Acid	
Comp.	4%	6%	4%	6%
75%	52.5	56.5	47.9	47.2
30%	50.2	50.8	47.4	41.7

The center runs gave the following strengths:

51.6, 51.4, 52.4, 52.9

(a) Test to determine which main effects and interactions should be involved in the fitted model.
(b) Test for quadratic curvature.
(c) If quadratic curvature is significant, how many additional design points are needed to determine which quadratic terms should be included in the model?

15.16 Suppose a second replicate of the experiment in Exercise 15.13 could be performed.

(a) Would a second replication of the blocking scheme of Exercise 15.13 be the best choice?
(b) If the answer to part (a) is no, give the layout for a better choice for the second replicate.
(c) What concept did you use in your design selection?

15.17 Consider Figure 15.14, which represents a 2^2 factorial with 3 center runs. If quadratic curvature is significant, what additional design points would you select that might allow the estimation of the terms x_1^2, x_2^2? Explain.

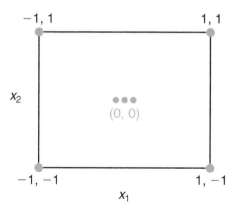

Figure 15.14: Graph for Exercise 15.17.

15.6 Fractional Factorial Experiments

The 2^k factorial experiment can become quite demanding, in terms of the number of experimental units required, when k is large. One of the real advantages of this experimental plan is that it allows a degree of freedom for each interaction. However, in many experimental situations, it is known that certain interactions are negligible, and thus it would be a waste of experimental effort to use the complete factorial experiment. In fact, the experimenter may have an economic constraint that disallows taking observations at all of the 2^k treatment combinations. When k is large, we can often make use of a **fractional factorial experiment** where per-

15.6 Fractional Factorial Experiments

haps one-half, one-fourth, or even one-eighth of the total factorial plan is actually carried out.

Construction of $\frac{1}{2}$ Fraction

The construction of the half-replicate design is identical to the allocation of the 2^k factorial experiment into two blocks. We begin by selecting a defining contrast that is to be completely sacrificed. We then construct the two blocks accordingly and choose either of them as the experimental plan.

A $\frac{1}{2}$ fraction of a 2^k factorial is often referred to as a 2^{k-1} design, the latter indicating the number of design points. Our first illustration of a 2^{k-1} will be a $\frac{1}{2}$ of 2^3, or a 2^{3-1}, design. In other words, the scientist or engineer cannot use the full complement (i.e., the full 2^3 with 8 design points) and hence must settle for a design with only four design points. The question is, of the design points (1), a, b, ab, ac, c, bc, and abc, which four design points would result in the most useful design? The answer, along with the important concepts involved, appears in the table of + and − signs displaying contrasts for the full 2^3. Consider Table 15.9.

Table 15.9: Contrasts for the Seven Available Effects for a 2^3 Factorial Experiment

	Treatment Combination	Effects							
		I	A	B	C	AB	AC	BC	ABC
2^{3-1}	a	+	+	−	−	−	−	+	+
	b	+	−	+	−	−	+	−	+
	c	+	−	−	+	+	−	−	+
	abc	+	+	+	+	+	+	+	+
2^{3-1}	ab	+	+	+	−	+	−	−	−
	ac	+	+	−	+	−	+	−	−
	bc	+	−	+	+	−	−	+	−
	(1)	+	−	−	−	+	+	+	−

Note that the two $\frac{1}{2}$ fractions are $\{a, b, c, abc\}$ and $\{ab, ac, bc, (1)\}$. Note also from Table 15.9 that in both designs ABC has no contrast but all other effects do have contrasts. In one of the fractions we have ABC containing all + signs, and in the other fraction the ABC effect contains all − signs. As a result, we say that the top design in the table is described by $\boldsymbol{ABC = I}$ and the bottom design by $\boldsymbol{ABC = -I}$. The interaction $\boldsymbol{ABC}$ is called the **design generator**, and $ABC = I$ (or $ABC = -I$ for the second design) is called the **defining relation**.

Aliases in the 2^{3-1}

If we focus on the $ABC = I$ design (the upper 2^{3-1}), it becomes apparent that six effects contain contrasts. This produces the initial appearance that all *effects* can be studied apart from ABC. However, the reader can certainly recall that with only four design points, even if points are replicated, the degrees of freedom available (apart from experimental error) are

Regression model terms	3
Intercept	1
	4

A closer look suggests that the seven effects are not orthogonal, and each contrast is represented in another effect. In fact, using ≡ to signify **identical contrasts**, we have

$$A \equiv BC; \quad B \equiv AC; \quad C \equiv AB.$$

As a result, within a pair an effect cannot be estimated independently of its alias "partner." The effects

$$A = \frac{a + abc - b - c}{2} \quad \text{and} \quad BC = \frac{a + abc - b - c}{2}$$

will produce the same numerical result and thus contain the same information. In fact, it is often said that they **share a degree of freedom**. In truth, the estimated effect actually estimates the sum, namely $A + BC$. We say that A and BC are aliases, B and AC are aliases, and C and AB are aliases.

For the $ABC = -I$ fraction we can observe that the aliases are the same as those for the $ABC = I$ fraction, apart from sign. Thus, we have

$$A \equiv -BC; \quad B \equiv -AC; \quad C \equiv -AB.$$

The two fractions appear on corners of the cubes in Figures 15.15(a) and 15.15(b).

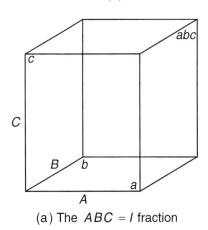
(a) The $ABC = I$ fraction

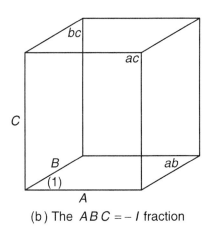
(b) The $ABC = -I$ fraction

Figure 15.15: The $\frac{1}{2}$ fractions of the 2^3 factorial.

How Aliases Are Determined in General

In general, for a 2^{k-1}, each effect, apart from that defined by the generator, will have a *single alias partner*. The effect defined by the generator will not be aliased

15.6 Fractional Factorial Experiments

by another effect but rather will be aliased with the mean since the least squares estimator will be the mean. To determine the alias for each effect, one merely begins with the defining relation, say $ABC = I$ for the 2^{3-1}. Then to find, say, the alias for effect A, multiply A by both sides of the equation $ABC = I$ and reduce any exponent by modulo 2. For example,

$$A \cdot ABC = A, \quad \text{thus,} \quad BC \equiv A.$$

In a similar fashion,

$$B \equiv B \cdot ABC \equiv AB^2 C \equiv AC,$$

and, of course,

$$C \equiv C \cdot ABC \equiv ABC^2 \equiv AB.$$

Now for the second fraction (i.e., defined by the relation $ABC = -I$),

$$A \equiv -BC; \quad B \equiv -AC; \quad C \equiv -AB.$$

As a result, the numerical value of effect A is actually estimating $A - BC$. Similarly, the value of B estimates $B - AC$, and the value of C estimates $C - AB$.

Formal Construction of the 2^{k-1}

A clear understanding of the concept of aliasing makes it very simple to understand the construction of the 2^{k-1}. We begin with investigation of the 2^{3-1}. There are three factors and four design points required. The procedure begins with a **full factorial** in $k - 1 = 2$ factors A and B. Then a third factor is added according to the desired alias structures. For example, with ABC as the generator, clearly $C = \pm AB$. Thus, $C = AB$ or $C = -AB$ is found to supplement the full factorial in A and B. Table 15.10 illustrates what is a very simple procedure.

Table 15.10: Construction of the Two 2^{3-1} Designs

Basic 2^2		2^{3-1}; $ABC = I$			2^{3-1}; $ABC = -I$		
A	B	A	B	C = AB	A	B	C = −AB
−	−	−	−	+	−	−	−
+	−	+	−	−	+	−	+
−	+	−	+	−	−	+	+
+	+	+	+	+	+	+	−

Note that we saw earlier that $ABC = I$ gives the design points a, b, c, and abc while $ABC = -I$ gives (1), ac, bc, and ab. Earlier we were able to construct the same designs using the table of contrasts in Table 15.9. However, as the design becomes more complicated with higher fractions, these contrast tables become more difficult to deal with.

Consider now a 2^{4-1} (i.e., a $\frac{1}{2}$ of a 2^4 factorial design) involving factors A, B, C, and D. As in the case of the 2^{3-1}, the highest-order interaction, in this case

$ABCD$, is used as the generator. We must keep in mind that $ABCD = I$; the defining relation suggests that the information on $ABCD$ is sacrificed. Here we begin with the full 2^3 in A, B, and C and form $D = \pm ABC$ to generate the two 2^{4-1} designs. Table 15.11 illustrates the construction of both designs.

Table 15.11: Construction of the Two 2^{4-1} Designs

Basic 2^3			2^{4-1}; $ABCD = I$				2^{4-1}; $ABCD = -I$			
A	B	C	A	B	C	$D = ABC$	A	B	C	$D = -ABC$
−	−	−	−	−	−	−	−	−	−	+
+	−	−	+	−	−	+	+	−	−	−
−	+	−	−	+	−	+	−	+	−	−
+	+	−	+	+	−	−	+	+	−	+
−	−	+	−	−	+	+	−	−	+	−
+	−	+	+	−	+	−	+	−	+	+
−	+	+	−	+	+	−	−	+	+	+
+	+	+	+	+	+	+	+	+	+	−

Here, using the notation a, b, c, and so on, we have the following designs:

$$ABCD = I, \; (1), \; ad, \; bd, \; ab, \; cd, \; ac, \; bc, \; abcd$$
$$ABCD = -I, \; d, \; a, \; b, \; abd, \; c, \; acd, \; bcd, \; abc.$$

The aliases in the case of the 2^{4-1} are found as illustrated earlier for the 2^{3-1}. Each effect has a single alias partner and is found by multiplication via the use of the defining relation. For example, the alias for A for the $ABCD = I$ design is given by

$$A = A \cdot ABCD = A^2 BCD = BCD.$$

The alias for AB is given by

$$AB = AB \cdot ABCD = A^2 B^2 CD = CD.$$

As we can observe easily, main effects are aliased with three-factor interactions and two-factor interactions are aliased with other two-factor interactions. A complete listing is given by

$$A = BCD \qquad AB = CD$$
$$B = ACD \qquad AC = BD$$
$$C = ABD \qquad AD = BC$$
$$D = ABC.$$

Construction of the $\frac{1}{4}$ Fraction

In the case of the $\frac{1}{4}$ fraction, two interactions are selected to be sacrificed rather than one, and the third results from finding the generalized interaction of the

15.6 Fractional Factorial Experiments

selected two. Note that this is very much like the construction of four blocks. The fraction used is simply one of the blocks. A simple example aids a great deal in seeing the connection to the construction of the $\frac{1}{2}$ fraction. Consider the construction of $\frac{1}{4}$ of a 2^5 factorial (i.e., a 2^{5-2}), with factors A, B, C, D, and E. One procedure that **avoids the confounding of two main effects** is the choice of ABD and ACE as the interactions that correspond to the two generators, giving $ABD = I$ and $ACE = I$ as the defining relations. The third interaction sacrificed would then be $(ABD)(ACE) = A^2BCDE = BCDE$. For the construction of the design, we begin with a $2^{5-2} = 2^3$ factorial in A, B, and C. We use the interactions ABD and ACE to supply the generators, so the 2^3 factorial in A, B, and C is supplemented by factor $D = \pm AB$ and $E = \pm AC$. Thus, one of the fractions is given by

$$
\begin{bmatrix}
A & B & C & D=AB & E=AC \\
- & - & - & + & + \\
+ & - & - & - & - \\
- & + & - & - & + \\
+ & + & - & + & - \\
- & - & + & + & - \\
+ & - & + & - & + \\
- & + & + & - & - \\
+ & + & + & + & +
\end{bmatrix}
\begin{matrix}
de \\ a \\ be \\ abd \\ cd \\ ace \\ bc \\ abcde
\end{matrix}
$$

The other three fractions are found by using the generators $\{D = -AB, E = AC\}$, $\{D = AB, E = -AC\}$, and $\{D = -AB, E = -AC\}$. Consider an analysis of the above 2^{5-2} design. It contains eight design points to study five factors. The aliases for main effects are given by

$$
\begin{array}{lll}
A(ABD) \equiv BD & A(ACE) \equiv CE & A(BCDE) \equiv ABCDE \\
B \equiv AD & \equiv ABCE & \equiv CDE \\
C \equiv ABCD & \equiv AE & \equiv BDE \\
D \equiv AB & \equiv ACDE & \equiv BCE \\
E \equiv ABDE & \equiv AC & \equiv BCD
\end{array}
$$

Aliases for other effects can be found in the same fashion. The breakdown of degrees of freedom is given by (apart from replication)

Main effects	5	
Lack of fit	2	($CD = BE$, $BC = DE$)
Total	7	

We list interactions only through degree 2 in the lack of fit.

Consider now the case of a 2^{6-2}, which allows 16 design points to study six factors. Once again two design generators are chosen. A pragmatic choice to supplement a $2^{6-2} = 2^4$ full factorial in A, B, C, and D is to use $E = \pm ABC$ and $F = \pm BCD$. The construction is given in Table 15.12.

Obviously, with eight more design points than in the 2^{5-2}, the aliases for main effects will not present as difficult a problem. In fact, note that with defining relations $ABCE = \pm I$, $BCDF = \pm I$, and $(ABCE)(BCDF) = ADEF = \pm I$,

Table 15.12: A 2^{6-2} Design

A	B	C	D	E = ABC	F = BCD	Treatment Combination
−	−	−	−	−	−	(1)
+	−	−	−	+	−	ae
−	+	−	−	+	+	bef
+	+	−	−	−	+	abf
−	−	+	−	+	+	cef
+	−	+	−	−	+	acf
−	+	+	−	−	−	bc
+	+	+	−	+	−	$abce$
−	−	−	+	−	+	df
+	−	−	+	+	+	$adef$
−	+	−	+	+	−	bde
+	+	−	+	−	−	abd
−	−	+	+	+	−	cde
+	−	+	+	−	−	acd
−	+	+	+	−	+	$bcdf$
+	+	+	+	+	+	$abcdef$

main effects will be aliased with interactions that are no less complex than those of third order. The alias structure for main effects is written

$$A \equiv BCE \equiv ABCDF \equiv DEF, \quad D \equiv ABCDE \equiv BCF \equiv AEF,$$
$$B \equiv ACE \equiv CDF \equiv ABDEF, \quad E \equiv ABC \equiv BCDEF \equiv ADF,$$
$$C \equiv ABE \equiv BDF \equiv ACDEF, \quad F \equiv ABCEF \equiv BCD \equiv ADE,$$

each with a single degree of freedom. For the two-factor interactions,

$$AB \equiv CE \equiv ACDF \equiv BDEF, \quad AF \equiv BCEF \equiv ABCD \equiv DE,$$
$$AC \equiv BE \equiv ABDF \equiv CDEF, \quad BD \equiv ACDE \equiv CF \equiv ABEF,$$
$$AD \equiv BCDE \equiv ABCF \equiv EF, \quad BF \equiv ACEF \equiv CD \equiv ABDE,$$
$$AE \equiv BC \equiv ABCDEF \equiv DF.$$

Here, of course, there is some aliasing among the two-factor interactions. The remaining 2 degrees of freedom are accounted for by the following groups:

$$ABD \equiv CDE \equiv ACF \equiv BEF, \quad ACD \equiv BDE \equiv ABF \equiv CEF.$$

It becomes evident that we should always be aware of what the alias structure is for a fractional experiment before we finally recommend the experimental plan. Proper choice in defining contrasts is important, since it dictates the alias structure.

15.7 Analysis of Fractional Factorial Experiments

The difficulty of making formal significance tests using data from fractional factorial experiments lies in the determination of the proper error term. Unless there are data available from prior experiments, the error must come from a pooling of contrasts representing effects that are presumed to be negligible.

Sums of squares for individual effects are found by using essentially the same procedures given for the complete factorial. We can form a contrast in the treatment combinations by constructing the table of positive and negative signs. For example, for a half-replicate of a 2^3 factorial experiment with ABC the defining contrast, one possible set of treatment combinations, along with the appropriate algebraic sign for each contrast used in computing effects and the sums of squares for the various effects, is presented in Table 15.13.

Table 15.13: Signs for Contrasts in a Half-Replicate of a 2^3 Factorial Experiment

Treatment Combination	A	B	C	AB	AC	BC	ABC
a	+	−	−	−	−	+	+
b	−	+	−	−	+	−	+
c	−	−	+	+	−	−	+
abc	+	+	+	+	+	+	+

Note that in Table 15.13 the A and BC contrasts are identical, illustrating the aliasing. Also, $B \equiv AC$ and $C \equiv AB$. In this situation, we have three orthogonal contrasts representing the 3 degrees of freedom available. If two observations were obtained for each of the four treatment combinations, we would then have an estimate of the error variance with 4 degrees of freedom. Assuming the interaction effects to be negligible, we could test all the main effects for significance.

An example effect and corresponding sum of squares is

$$A = \frac{a - b - c + abc}{2n}, \quad SSA = \frac{(a - b - c + abc)^2}{2^2 n}.$$

In general, the single-degree-of-freedom sum of squares for any effect in a 2^{-p} fraction of a 2^k factorial experiment ($p < k$) is obtained by squaring contrasts in the treatment totals selected and dividing by $2^{k-p}n$, where n is the number of replications of these treatment combinations.

Example 15.6: Suppose that we wish to use a half-replicate to study the effects of five factors, each at two levels, on some response, and it is known that whatever the effect of each factor, it will be constant for each level of the other factors. In other words, there are no interactions. Let the defining contrast be $ABCDE$, causing main effects to be aliased with four-factor interactions. The pooling of contrasts involving interactions provides $15 - 5 = 10$ degrees of freedom for error. Perform an analysis of variance on the data in Table 15.14, testing all main effects for significance at the 0.05 level.

Solution: The sums of squares and effects for the main effects are

$$SSA = \frac{(11.3 - 15.6 - \cdots - 14.7 + 13.2)^2}{2^{5-1}} = \frac{(-17.5)^2}{16} = 19.14,$$

Table 15.14: Data for Example 15.6

Treatment	Response	Treatment	Response
a	11.3	bcd	14.1
b	15.6	abe	14.2
c	12.7	ace	11.7
d	10.4	ade	9.4
e	9.2	bce	16.2
abc	11.0	bde	13.9
abd	8.9	cde	14.7
acd	9.6	$abcde$	13.2

$A = -\frac{17.5}{8} = -2.19,$

$$SSB = \frac{(-11.3 + 15.6 - \cdots - 14.7 + 13.2)^2}{2^{5-1}} = \frac{(18.1)^2}{16} = 20.48,$$

$B = \frac{18.1}{8} = 2.26,$

$$SSC = \frac{(-11.3 - 15.6 + \cdots + 14.7 + 13.2)^2}{2^{5-1}} = \frac{(10.3)^2}{16} = 6.63,$$

$C = \frac{10.3}{8} = 1.21,$

$$SSD = \frac{(-11.3 - 15.6 - \cdots + 14.7 + 13.2)^2}{2^{5-1}} = \frac{(-7.7)^2}{16} = 3.71,$$

$D = \frac{-7.7}{8} = -0.96,$

$$SSE = \frac{(-11.3 - 15.6 - \cdots + 14.7 + 13.2)^2}{2^{5-1}} = \frac{(8.9)^2}{16} = 4.95,$$

$E = \frac{8.9}{8} = 1.11.$

All other calculations and tests of significance are summarized in Table 15.15. The tests indicate that factor A has a significant negative effect on the response, whereas factor B has a significant positive effect. Factors C, D, and E are not significant at the 0.05 level.

Exercises

15.18 List the aliases for the various effects in a 2^5 factorial experiment when the defining contrast is $ACDE$.

15.19 (a) Obtain a $\frac{1}{2}$ fraction of a 2^4 factorial design using BCD as the defining contrast.
(b) Divide the $\frac{1}{2}$ fraction into 2 blocks of 4 units each by confounding ABC.
(c) Show the analysis-of-variance table (sources of variation and degrees of freedom) for testing all unconfounded main effects, assuming that all interaction effects are negligible.

15.20 Construct a $\frac{1}{4}$ fraction of a 2^6 factorial design using $ABCD$ and $BDEF$ as the defining contrasts. Show what effects are aliased with the six main effects.

Table 15.15: Analysis of Variance for the Data of a Half-Replicate of a 2^5 Factorial Experiment

Source of Variation	Sum of Squares	Degrees of Freedom	Mean Square	Computed f
Main effect:				
A	19.14	1	19.14	6.21
B	20.48	1	20.48	6.65
C	6.63	1	6.63	2.15
D	3.71	1	3.71	1.20
E	4.95	1	4.95	1.61
Error	30.83	10	3.08	
Total	85.74	15		

15.21 (a) Using the defining contrasts $ABCE$ and $ABDF$, obtain a $\frac{1}{4}$ fraction of a 2^6 design.

(b) Show the analysis-of-variance table (sources of variation and degrees of freedom) for all appropriate tests, assuming that E and F do not interact and all three-factor and higher interactions are negligible.

15.22 Seven factors are varied at two levels in an experiment involving only 16 trials. A $\frac{1}{8}$ fraction of a 2^7 factorial experiment is used, with the defining contrasts being ACD, BEF, and CEG. The data are as follows:

Treat. Comb.	Response	Treat. Comb.	Response
(1)	31.6	acg	31.1
ad	28.7	cdg	32.0
abce	33.1	beg	32.8
cdef	33.6	adefg	35.3
acef	33.7	efg	32.4
bcde	34.2	abdeg	35.3
abdf	32.5	bcdfg	35.6
bf	27.8	abcfg	35.1

Perform an analysis of variance on all seven main effects, assuming that interactions are negligible. Use a 0.05 level of significance.

15.23 An experiment is conducted so that an engineer can gain insight into the influence of sealing temperature A, cooling bar temperature B, percent polyethylene additive C, and pressure D on the seal strength (in grams per inch) of a bread-wrapper stock. A $\frac{1}{2}$ fraction of a 2^4 factorial experiment is used, with the defining contrast being $ABCD$. The data are presented here. Perform an analysis of variance on main effects only. Use $\alpha = 0.05$.

A	B	C	D	Response
−1	−1	−1	−1	6.6
1	−1	−1	1	6.9
−1	1	−1	1	7.9
1	1	−1	−1	6.1
−1	−1	1	1	9.2
1	−1	1	−1	6.8
−1	1	1	−1	10.4
1	1	1	1	7.3

15.24 In an experiment conducted at the Department of Mechanical Engineering and analyzed by the Laboratory for Interdisciplinary Statistical Analysis at Virginia Tech, a sensor detects an electrical charge each time a turbine blade makes one rotation. The sensor then measures the amplitude of the electrical current. Six factors are rpm A, temperature B, gap between blades C, gap between blade and casing D, location of input E, and location of detection F. A $\frac{1}{4}$ fraction of a 2^6 factorial experiment is used, with defining contrasts being $ABCE$ and $BCDF$. The data are as follows:

A	B	C	D	E	F	Response
−1	−1	−1	−1	−1	−1	3.89
1	−1	−1	−1	1	−1	10.46
−1	1	−1	−1	1	1	25.98
1	1	−1	−1	−1	1	39.88
−1	−1	1	−1	1	1	61.88
1	−1	1	−1	−1	1	3.22
−1	1	1	−1	−1	−1	8.94
1	1	1	−1	1	−1	20.29
−1	−1	−1	1	−1	1	32.07
1	−1	−1	1	1	1	50.76
−1	1	−1	1	1	−1	2.80
1	1	−1	1	−1	−1	8.15
−1	−1	1	1	1	−1	16.80
1	−1	1	1	−1	−1	25.47
−1	1	1	1	−1	1	44.44
1	1	1	1	1	1	2.45

Perform an analysis of variance on main effects and two-factor interactions, assuming that all three-factor and higher interactions are negligible. Use $\alpha = 0.05$.

15.25 In the study *Durability of Rubber to Steel Adhesively Bonded Joints*, conducted at the Department of Environmental Science and Mechanics and analyzed by the Laboratory for Interdisciplinary Statistical Analysis at Virginia Tech, an experimenter measured the number of breakdowns in an adhesive seal. It was postulated that concentration of seawater A, temperature B, pH C, voltage D, and stress E influence the breakdown of an adhesive seal. A $\frac{1}{2}$ fraction of a 2^5 factorial experiment was used, with the defining contrast being $ABCDE$. The data are as follows:

A	B	C	D	E	Response
−1	−1	−1	−1	1	462
1	−1	−1	−1	−1	746
−1	1	−1	−1	−1	714
1	1	−1	−1	1	1070
−1	−1	1	−1	−1	474
1	−1	1	−1	1	832
−1	1	1	−1	1	764
1	1	1	−1	−1	1087
−1	−1	−1	1	−1	522
1	−1	−1	1	1	854
−1	1	−1	1	1	773
1	1	−1	1	−1	1068
−1	−1	1	1	1	572
1	−1	1	1	−1	831
−1	1	1	1	−1	819
1	1	1	1	1	1104

Perform an analysis of variance on main effects and two factor interactions AD, AE, BD, BE, assuming that all three-factor and higher interactions are negligible. Use $\alpha = 0.05$.

15.26 Consider a 2^{5-1} design with factors A, B, C, D, and E. Construct the design by beginning with a 2^4 and use $E = ABCD$ as the generator. Show all aliases.

15.27 There are six factors and only eight design points can be used. Construct a 2^{6-3} by beginning with a 2^3 and use $D = AB$, $E = -AC$, and $F = BC$ as the generators.

15.28 Consider Exercise 15.27. Construct another 2^{6-3} that is different from the design chosen in Exercise 15.27.

15.29 For Exercise 15.27, give all aliases for the six main effects.

15.30 In Myers, Montgomery, and Anderson-Cook (2009), an application is discussed in which an engineer is concerned with the effects on the cracking of a titanium alloy. The three factors are A, temperature; B, titanium content; and C, amount of grain refiner. The following table gives a portion of the design and the response, crack length induced in the sample of the alloy.

A	B	C	Response
−1	−1	−1	0.5269
1	1	−1	2.3380
1	−1	1	4.0060
−1	1	1	3.3640

(a) What is the defining relation?
(b) Give aliases for all three main effects assuming that two-factor interactions may be real.
(c) Assuming that interactions are negligible, which main factor is most important?
(d) At what level would you suggest the factor named in (c) be for final production, high or low?
(e) At what levels would you suggest the other factors be for final production?
(f) What hazards lie in the recommendations you made in (d) and (e)? Be thorough in your answer.

15.8 Higher Fractions and Screening Designs

Some industrial situations require the analyst to determine which of a large number of controllable factors have an impact on some important response. The factors may be qualitative or class variables, regression variables, or a mixture of both. The analytical procedure may involve analysis of variance, regression, or both. Often the regression model used involves only linear main effects, although a few interactions may be estimated. The situation calls for variable screening and the resulting experimental designs are known as **screening designs**. Clearly, two-level orthogonal designs that are saturated or nearly saturated are viable candidates.

Design Resolution

Two-level orthogonal designs are often classified according to their **resolution**, the latter determined through the following definition.

Definition 15.1: The **resolution** of a two-level orthogonal design is the length of the smallest (least complex) interaction among the set of defining contrasts.

If the design is constructed as a full or fractional factorial (i.e., either a 2^k or a 2^{k-p} design, $p = 1, 2, \ldots, k-1$), the notion of design resolution is an aid in categorizing the impact of the aliasing. For example, a resolution II design would have little use, since there would be at least one instance of aliasing of one main effect with another. A resolution III design will have all main effects (linear effects) orthogonal to each other. However, there will be some aliasing among linear effects and two-factor interactions. Clearly, then, if the analyst is interested in studying main effects (linear effects in the case of regression) and there are no two-factor interactions, a design of resolution at least III is required.

15.9 Construction of Resolution III and IV Designs with 8, 16, and 32 Design Points

Useful designs of resolution III and IV can be constructed for 2 to 7 variables with 8 design points. We begin with a 2^3 factorial that has been symbolically saturated with interactions.

$$\begin{array}{ccccccc} x_1 & x_2 & x_3 & x_1x_2 & x_1x_3 & x_2x_3 & x_1x_2x_3 \\ -1 & -1 & -1 & 1 & 1 & 1 & -1 \\ 1 & -1 & -1 & -1 & -1 & 1 & 1 \\ -1 & 1 & -1 & -1 & 1 & -1 & 1 \\ -1 & -1 & 1 & 1 & -1 & -1 & 1 \\ 1 & 1 & -1 & 1 & -1 & -1 & -1 \\ 1 & -1 & 1 & -1 & 1 & -1 & -1 \\ -1 & 1 & 1 & -1 & -1 & 1 & -1 \\ 1 & 1 & 1 & 1 & 1 & 1 & 1 \end{array}$$

It is clear that a resolution III design can be constructed merely by replacing interaction columns by new main effects through 7 variables. For example, we may define

$$\begin{aligned} x_4 &= x_1x_2 \quad &\text{(defining contrast } ABD\text{)} \\ x_5 &= x_1x_3 \quad &\text{(defining contrast } ACE\text{)} \\ x_6 &= x_2x_3 \quad &\text{(defining contrast } BCF\text{)} \\ x_7 &= x_1x_2x_3 \quad &\text{(defining contrast } ABCG\text{)} \end{aligned}$$

and obtain a 2^{-4} fraction of a 2^7 factorial. The preceding expressions identify the chosen defining contrasts. Eleven additional defining contrasts result, and all defining contrasts contain at least three letters. Thus, the design is a resolution III design. Clearly, if we begin with a *subset* of the augmented columns and conclude

Table 15.16: Some Resolution III, IV, V, VI and VII 2^{k-p} Designs

Number of Factors	Design	Number of Points	Generators
3	2^{3-1}_{III}	4	$C = \pm AB$
4	2^{4-1}_{IV}	8	$D = \pm ABC$
5	2^{5-2}_{III}	8	$D = \pm AB;\ E = \pm AC$
6	2^{6-1}_{VI}	32	$F = \pm ABCDE$
	2^{6-2}_{IV}	16	$E = \pm ABC;\ F = \pm BCD$
	2^{6-3}_{III}	8	$D = \pm AB;\ F = \pm BC;\ E = \pm AC$
7	2^{7-1}_{VII}	64	$G = \pm ABCDEF$
	2^{7-2}_{IV}	32	$E = \pm ABC;\ G = \pm ABDE$
	2^{7-3}_{IV}	16	$E = \pm ABC;\ F = \pm BCD;\ G = \pm ACD$
	2^{7-4}_{III}	8	$D = \pm AB;\ E = \pm AC;\ F = \pm BC;\ G = \pm ABC$
8	2^{8-2}_{V}	64	$G = \pm ABCD;\ H = \pm ABEF$
	2^{8-3}_{IV}	32	$F = \pm ABC;\ G = \pm ABD;\ H = \pm BCDE$
	2^{8-4}_{IV}	16	$E = \pm BCD;\ F = \pm ACD;\ G = \pm ABC;\ H = \pm ABD$

with a design involving fewer than 7 design variables, the result is a resolution III design in fewer than 7 variables.

A similar set of possible designs can be constructed for 16 design points by beginning with a 2^4 saturated with interactions. Definitions of variables that correspond to these interactions produce resolution III designs through 15 variables. In a similar fashion, designs containing 32 runs can be constructed by beginning with a 2^5.

Table 15.16 provides guidelines for constructing 8, 16, 32, and 64 point designs that are resolution III, IV and even V. The table gives the number of factors, the number of runs, and the generators that are used to produce the 2^{k-p} designs. The generator given is used to **augment the full factorial** containing $k - p$ factors.

15.10 Other Two-Level Resolution III Designs; The Plackett-Burman Designs

A family of designs developed by Plackett and Burman (1946; see the Bibliography) fills sample size voids that exist with the fractional factorials. The latter are useful with sample sizes 2^r (i.e., they involve sample sizes $4, 8, 16, 32, 64, \dots$). The Plackett-Burman designs involve $4r$ design points, and thus designs of sizes 12, 20, 24, 28, and so on, are available. These two-level Plackett-Burman designs are resolution III designs and are very simple to construct. "Basic lines" are given for each sample size. These lines of $+$ and $-$ signs are $n - 1$ in number. To construct the columns of the design matrix, we begin with the basic line and do a cyclic permutation on the columns until k (the desired number of variables) columns are formed. Then we fill in the last row with negative signs. The result will be a resolution III design with k variables ($k = 1, 2, \ldots, N$). The basic lines are as follows:

$N = 12$ + + − + + + − − − + −
$N = 16$ + + + + − + − + + − − + − − −
$N = 20$ + + − − + + + + − + − + − − − − + + −
$N = 24$ + + + + + − + − + + − − + + − − + − + − − − −

Example 15.7: Construct a two-level screening design with 6 variables containing 12 design points.

Solution: Begin with the basic line in the initial column. The second column is formed by bringing the bottom entry of the first column to the top of the second column and repeating the first column. The third column is formed in the same fashion, using entries in the second column. When there is a sufficient number of columns, **simply fill in the last row with negative signs**. The resulting design is as follows:

x_1	x_2	x_3	x_4	x_5	x_6
+	−	+	−	−	−
+	+	−	+	−	−
−	+	+	−	+	−
+	−	+	+	−	+
+	+	−	+	+	−
+	+	+	−	+	+
−	+	+	+	−	+
−	−	+	+	+	−
−	−	−	+	+	+
+	−	−	−	+	+
−	+	−	−	−	+
−	−	−	−	−	−

The Plackett-Burman designs are popular in industry for screening situations. Because they are resolution III designs, all linear effects are orthogonal. For any sample size, the user has available a design for $k = 2, 3, \ldots, N - 1$ variables.

The alias structure for the Plackett-Burman design is very complicated, and thus the user cannot construct the design with complete control over the alias structure, as in the case of 2^k or 2^{k-p} designs. However, in the case of regression models, the Plackett-Burman design can accommodate interactions (although they will not be orthogonal) when sufficient degrees of freedom are available.

15.11 Introduction to Response Surface Methodology

In Case Study 15.2, a regression model was fitted to a set of data with the specific goal of finding conditions on those design variables that optimize (maximize) the cleansing efficiency of coal. The model contained three linear main effects, three two-factor interaction terms, and one three-factor interaction term. The model response was the cleansing efficiency, and the optimum conditions on x_1, x_2, and x_3

were found by using the signs and the magnitude of the model coefficients. In this example, a two-level design was employed for process improvement or process optimization. In many areas of science and engineering, the application is expanded to involve more complicated models and designs, and this collection of techniques is called **response surface methodology (RSM)**. It encompasses both graphical and analytical approaches. The term *response surface* is derived from the appearance of the multidimensional surface of constant estimated response from a second-order model, i.e., a model with first- and second-order terms. An example will follow.

The Second-Order Response Surface Model

In many industrial examples of process optimization, a *second-order response surface model* is used. For the case of, say, $k = 2$ process variables, or design variables, and a single response y, the model is given by

$$y = \beta_0 + \beta_1 x_1 + \beta_2 x_2 + \beta_{11} x_1^2 + \beta_{22} x_2^2 + \beta_{12} x_1 x_2 + \epsilon.$$

Here we have $k = 2$ first-order terms, two pure second-order, or quadratic, terms, and one interaction term given by $\beta_{12} x_1 x_2$. The terms x_1 and x_2 are taken to be in the familiar ± 1 coded form. The ϵ term designates the usual model error. In general, for k design variables the model will contain $1 + k + k + \binom{k}{2}$ model terms, and hence the experimental design must contain at least a like number of design points. In addition, the quadratic terms require that the design variables be fixed in the design with at least three levels. The resulting design is referred to as a *second-order design*. Illustrations will follow.

The following **central composite design (CCD)** and example is taken from Myers, Montgomery, and Anderson-Cook (2009). Perhaps the most popular class of second-order designs is the class of central composite designs. The example given in Table 15.17 involves a chemical process in which reaction temperature, ξ_1, and reactant concentration, ξ_2, are shown at their natural levels. They also appear in coded form. There are five levels of each of the two factors. In addition, we have the order in which the observations on x_1 and x_2 were run. The column on the right gives values of the response y, the percent conversion of the process. The first four design points represent the familiar factorial points at levels ± 1. The next four points are called axial points. They are followed by the center runs that were discussed and illustrated earlier in this chapter. Thus, the five levels of each of the two factors are $-1, +1, -1.414, +1.414,$ and 0. A clear picture of the geometry of the central composite design for this $k = 2$ example appears in Figure 15.16. This figure illustrates the source of the term **axial points**. These four points are on the factor axes at an axial distance of $\alpha = \sqrt{2} = 1.414$ from the design center. In fact, for this particular CCD, the perimeter points, axial and factorial, are all at the distance $\sqrt{2}$ from the design center, and as a result we have eight equally spaced points on a circle plus four replications at the design center.

Example 15.8: **Response Surface Analysis:** An analysis of the data in the two-variable example may involve the fitting of a second-order response function. The resulting response surface can be used analytically or graphically to determine the impact that x_1

15.11 Introduction to Response Surface Methodology

Table 15.17: Central Composite Design for Example 15.8

Observation	Run	Temperature (°C) ξ_1	Concentration (%) ξ_2	x_1	x_2	y
1	4	200	15	−1	−1	43
2	12	250	15	1	−1	78
3	11	200	25	−1	1	69
4	5	250	25	1	1	73
5	6	189.65	20	−1.414	0	48
6	7	260.35	20	1.414	0	78
7	1	225	12.93	0	−1.414	65
8	3	225	27.07	0	1.414	74
9	8	225	20	0	0	76
10	10	225	20	0	0	79
11	9	225	20	0	0	83
12	2	225	20	0	0	81

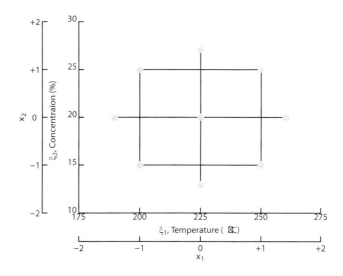

Figure 15.16: Central composite design for Example 15.8.

and x_2 have on percent conversion of the process. The coefficients in the response function are determined by the method of least squares developed in Chapter 12 and illustrated throughout this chapter. The resulting second-order response model is given in the coded variables as

$$\hat{y} = 79.75 + 10.18x_1 + 4.22x_2 - 8.50x_1^2 - 5.25x_2^2 - 7.75x_1x_2,$$

whereas in the natural variables it is given by

$$\hat{y} = -1080.22 + 7.7671\xi_1 + 23.1932\xi_2 - 0.0136\xi_1^2 - 0.2100\xi_2^2 - 0.0620\xi_1\xi_2.$$

Since the current example contains only two design variables, the most illuminating approach to determining the nature of the response surface in the design region is through two- or three-dimensional graphics. It is of interest to determine what levels of temperature x_1 and concentration x_2 produce a desirable estimated percent conversion, $\hat{y}$. The estimated response function above was plotted in three dimensions, and the resulting *response surface* is shown in Figure 15.17. The height of the surface is $\hat{y}$ in percent. It is readily seen from this figure why the term **response surface** is employed. In cases where only two design variables are used, two-dimensional contour plotting can be useful. Thus, make note of Figure 15.18. Contours of constant estimated conversion are seen as slices from the response surface. Note that the viewer of either figure can readily observe which coordinates of temperature and concentration produce the largest estimated percent conversion. In the plots, the coordinates are given in both coded units and natural units. Notice that the largest estimated conversion is at approximately 240°C and 20% concentration. The maximum estimated (or predicted) response at that location is 82.47%.

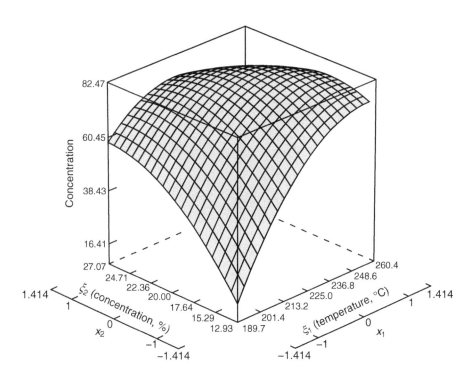

Figure 15.17: Plot for the response surface prediction conversion for Example 15.8.

Other Comments Concerning Response Surface Analysis

The book by Myers, Montgomery, and Anderson-Cook (2009) provides a great deal of information concerning both design and analysis of RSM. The graphical illustration we have used here can be augmented by analytical results that provide information about the nature of the response surface inside the design region.

15.12 Robust Parameter Design

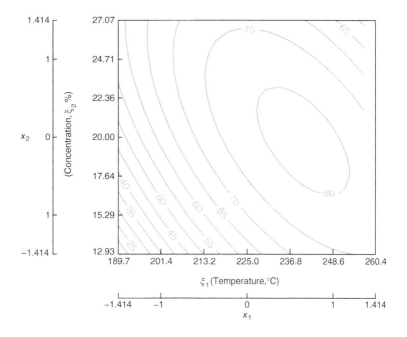

Figure 15.18: Contour plot of predicted conversion for Example 15.8.

Other computations can be used to determine whether the location of the optimum conditions is, in fact, inside or remote from the experimental design region. There are many important considerations when one is required to determine appropriate conditions for future operation of a process.

Other material in Myers, Montgomery, and Anderson-Cook (2009) deals with further experimental design issues. For example, the CCD, while the most generally useful design, is not the only class of design used in RSM. Many others are discussed in the aforementioned text. Also, the CCD discussed here is a special case in which $k = 2$. The more general $k > 2$ case is discussed in Myers, Montgomery, and Anderson-Cook (2009).

15.12 Robust Parameter Design

In this chapter, we have emphasized the notion of using design of experiments (DOE) to learn about engineering and scientific processes. In the case where the process involves a product, DOE can be used to provide product improvement or quality improvement. As we pointed out in Chapter 1, much importance has been attached to the use of statistical methods in product improvement. An important aspect of this quality improvement effort that surfaced in the 1980s and continued through the 1990s is to design quality into processes and products at the research stage or the process design stage. One often requires DOE in the development of processes that have the following properties:

1. Insensitive (robust) to environmental conditions

2. Insensitive (robust) to factors difficult to control

3. Provide minimum variation in performance

The methods used to attain the desirable characteristics in 1, 2, and 3 are a part of what is referred to as *robust parameter design*, or RPD (see Taguchi, 1991; Taguchi and Wu, 1985; and Kackar, 1985, in the Bibliography). The term *design* in this context refers to the design of the process or system; *parameter* refers to the parameters in the system. These are what we have been calling *factors* or *variables*.

It is very clear that goals 1, 2, and 3 above are quite noble. For example, a petroleum engineer may have a fine gasoline blend that performs quite well as long as conditions are ideal and stable. However, the performance may deteriorate because of changes in environmental conditions, such as type of driver, weather conditions, type of engine, and so forth. A scientist at a food company may have a cake mix that is quite good unless the user does not exactly follow directions on the box, directions that deal with oven temperature, baking time, and so forth. A product or process whose performance is consistent when exposed to these changing environmental conditions is called a **robust product** or **robust process**. (See Myers, Montgomery, and Anderson-Cook, 2009, in the Bibliography.)

Control and Noise Variables

Taguchi (1991) emphasized the notion of using two classes of design variables in a study involving RPD: *control factors* and *noise factors*.

Definition 15.2: **Control factors** are variables that can be controlled both in the experiment and in the process. **Noise factors** are variables that may or may not be controlled in the experiment but cannot be controlled in the process (or not controlled well in the process).

An important approach is to use control variables and noise variables in the same experiment as fixed effects. Orthogonal designs or orthogonal arrays are popular designs to use in this effort.

Goal of Robust Parameter Design: The goal of robust parameter design is to choose the levels of the control variables (i.e., the design of the process) that are most robust (insensitive) to changes in the noise variables.

It should be noted that *changes in the noise variables* actually imply changes during the process, changes in the field, changes in the environment, changes in handling or usage by the consumer, and so forth.

The Product Array

One approach to the design of experiments involving both control and noise variables is to use an experimental plan that calls for an orthogonal design for both the control and the noise variables separately. The complete experiment, then, is merely the product or crossing of these two orthogonal designs. The following is a simple example of a product array with two control and two noise variables.

15.12 Robust Parameter Design

Example 15.9: In the article "The Taguchi Approach to Parameter Design" in *Quality Progress*, December 1987, D. M. Byrne and S. Taguchi discuss an interesting example in which a method is sought for attaching an electrometric connector to a nylon tube so as to deliver the pull-off performance required for an automotive engine application. The objective is to find controllable conditions that maximize pull-off force. Among the controllable variables are A, connector wall thickness, and B, insertion depth. During routine operation there are several variables that cannot be controlled, although they will be controlled during the experiment. Among them are C, conditioning time, and D, conditioning temperature. Three levels are taken for each control variable and two for each noise variable. As a result, the crossed array is as follows. The control array is a 3×3 array, and the noise array is a familiar 2^2 factorial with (1), c, d, and cd representing the four factor combinations. The purpose of the noise factor is to create the *kind of variability in the response, pull-off force, that might be expected in day-to-day operation with the process*. The design is shown in Table 15.18.

Table 15.18: Design for Example 15.9

		\multicolumn{3}{c}{B (depth)}		
		Shallow	Medium	Deep
	Thin	(1)	(1)	(1)
		c	c	c
		d	d	d
		cd	cd	cd
	Medium	(1)	(1)	(1)
		c	c	c
A (wall thickness)		d	d	d
		cd	cd	cd
	Thick	(1)	(1)	(1)
		c	c	c
		d	d	d
		cd	cd	cd

Case Study 15.3: Solder Process Optimization: In an experiment described in *Understanding Industrial Designed Experiments* by Schmidt and Launsby (1991; see the Bibliography), solder process optimization is accomplished by a printed circuit-board assembly plant. Parts are inserted either manually or automatically into a bare board with a circuit printed on it. After the parts are inserted, the board is put through a wave solder machine, which is used to connect all the parts into the circuit. Boards are placed on a conveyor and taken through a series of steps. They are bathed in a flux mixture to remove oxide. To minimize warpage, they are preheated before the solder is applied. Soldering takes place as the boards move across the wave of solder. The object of the experiment is to minimize the number of solder defects per million joints. The control factors and levels are as given in Table 15.19.

Table 15.19: Control Factors for Case Study 15.3

Factor	(−1)	(+1)
A, solder pot temperature (°F)	480	510
B, conveyor speed (ft/min)	7.2	10
C, flux density	0.9°	1.0°
D, preheat temperature	150	200
E, wave height (in.)	0.5	0.6

These factors are easy to control at the experimental level but are more formidable at the plant or process level.

Noise Factors: Tolerances on Control Factors

Often in processes such as this one, the natural noise factors are tolerances on the control factors. For example, in the actual on-line process, solder pot temperature and conveyor-belt speed are difficult to control. It is known that the control of temperature is within ±5°F and the control of conveyor-belt speed is within ±0.2 ft/min. It is certainly conceivable that variability in the product response (soldering performance) is increased because of an inability to control these two factors at some nominal levels. The third noise factor is the type of assembly involved. In practice, one of two types of assemblies will be used. Thus, we have the noise factors given in Table 15.20.

Table 15.20: Noise Factors for Case Study 15.3

Factor	(−1)	(+1)
A^*, solder pot temperature tolerance (°F) (deviation from nominal)	−5	+5
B^*, conveyor speed tolerance (ft/min) (deviation from ideal)	−0.2	+0.2
C^*, assembly type	1	2

Both the control array (inner array) and the noise array (outer array) were chosen to be fractional factorials, the former a $\frac{1}{4}$ of a 2^5 and the latter a $\frac{1}{2}$ of a 2^3. The crossed array and the response values are shown in Table 15.21. The first three columns of the inner array represent a 2^3. The fourth and fifth columns are formed by $D = -AC$ and $E = -BC$. Thus, the defining interactions for the inner array are ACD, BCE, and $ABDE$. The outer array is a standard resolution III fraction of a 2^3. Notice that each inner array point contains runs from the outer array. Thus, four response values are observed at each combination of the control array. Figure 15.19 displays plots which reveal the effect of temperature and density on the mean response.

15.12 Robust Parameter Design

Table 15.21: Crossed Arrays and Response Values for Case Study 15.3

Inner Array					Outer Array					
A	B	C	D	E	(1)	a*b*	a*c*	b*c*	$\bar{y}$	s_y
1	1	1	−1	−1	194	197	193	275	214.75	40.20
1	1	−1	1	1	136	136	132	136	135.00	2.00
1	−1	1	−1	1	185	261	264	264	243.50	39.03
1	−1	−1	1	−1	47	125	127	42	85.25	47.11
−1	1	1	1	−1	295	216	204	293	252.00	48.75
−1	1	−1	−1	1	234	159	231	157	195.25	43.04
−1	−1	1	1	1	328	326	247	322	305.75	39.25
−1	−1	−1	−1	−1	186	187	105	104	145.50	47.35

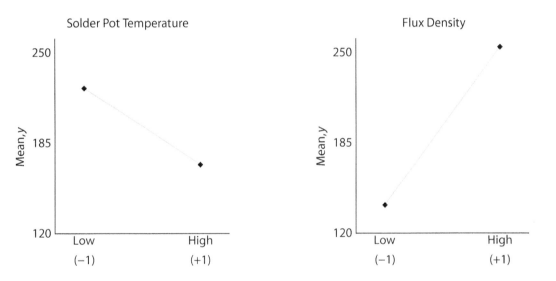

Figure 15.19: Plot showing the influence of factors on the mean response.

Simultaneous Analysis of Process Mean and Variance

In most examples using RPD, the analyst is interested in finding conditions on the control variables that give suitable values for the mean response $\bar{y}$. However, varying the noise variables produces information on the process variance σ_y^2 that might be anticipated in the process. Obviously a robust product is one for which the process is consistent and thus has a small process variance. RPD may involve the simultaneous analysis of $\bar{y}$ and s_y.

It turns out that temperature and flux density are the most important factors in Case Study 15.3. They seem to influence both s_y and $\bar{y}$. Fortunately, *high temperature* and *low flux density* are preferable for both. From Figure 15.19, the "optimum" conditions are

$$\text{solder temperature} = 510°F, \quad \text{flux density} = 0.9°.$$

Alternative Approaches to Robust Parameter Design

One approach suggested by many is to model the sample mean and sample variance separately. Separate modeling often helps the experimenter to obtain a better understanding of the process involved. In the following example, we illustrate this approach with the solder process experiment.

Case Study 15.4: Consider the data set of Case Study 15.3. An alternative approach is to fit separate models for the mean $\bar{y}$ and the sample standard deviation. Suppose that we use the usual $+1$ and -1 coding for the control factors. Based on the apparent importance of solder pot temperature x_1 and flux density x_2, linear regression on the response (number of errors per million joints) produces

$$\hat{y} = 197.125 - 27.5x_1 + 57.875x_2.$$

To find the most robust levels of temperature and flux density, it is important to procure a compromise between the mean response and variability, which requires a modeling of the variability. An important tool in this regard is the log transformation (see Bartlett and Kendall, 1946, or Carroll and Ruppert, 1988):

$$\ln s^2 = \gamma_0 + \gamma_1(x_1) + \gamma_2(x_2).$$

This modeling process produces the following result:

$$\widehat{\ln s^2} = 6.6975 - 0.7458x_1 + 0.6150x_2.$$

The *log linear* model finds extensive use for modeling sample variance, since the log transformation on the sample variance lends itself to use of the method of least squares. This results from the fact that normality and homogeneous variance assumptions are often quite good when one uses $\ln s^2$ rather than s^2 as the model response.

The analysis that is important to the scientist or engineer makes use of the two models simultaneously. A graphical approach can be very useful. Figure 15.20 shows simple plots of the mean and standard deviation models simultaneously. As one would expect, the location of temperature and flux density that minimizes the mean number of errors is the same as that which minimizes variability, namely high temperature and low flux density. The graphical *multiple response surface* approach allows the user to see tradeoffs between process mean and process variability. For this example, the engineer may be dissatisfied with the extreme conditions in solder temperature and flux density. The figure offers estimates of how much is lost as one moves away from the optimum mean and variability conditions to any intermediate conditions.

In Case Study 15.4, values for control variables were chosen that gave desirable conditions for both the mean and the variance of the process. The mean and variance were taken across the distribution of noise variables in the process and

15.12 Robust Parameter Design

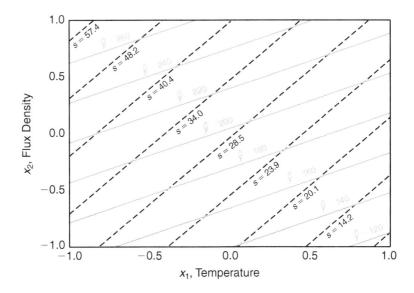

Figure 15.20: Mean and standard deviation for Case Study 15.4.

were modeled separately, and appropriate conditions were found through a dual response surface approach. Since Case Study 15.4 involved two models (mean and variance), this can be viewed as a dual response surface analysis. Fortunately, in this example the same conditions on the two relevant control variables, solder temperature and flux density, were optimal for both the process mean and the variance. Much of the time in practice some type of compromise between the mean and variance would need to be invoked.

The approach illustrated in Case Study 15.4 involves finding optimal process conditions when the data used are from a product array (or crossed array) type of experimental design. Often, using the product array, a cross between two designs, can be very costly. However, the development of dual response surface models, i.e., a model for the mean and a model for the variance, can be accomplished without a product array. A design that involves both control and noise variables is often called a *combined array*. This type of design and the resulting analysis can be used to determine what conditions on the control variables are most robust (insensitive) to variation in the noise variables. This can be viewed as tantamount to finding control levels that minimize the process variance produced by movement in the noise variables.

The Role of the Control-by-Noise Interaction

The structure of the process variance is greatly determined by the nature of the control-by-noise interaction. The nature of the nonhomogeneity of process variance is a function of which control variables interact with which noise variables. Specifically, as we will illustrate, those control variables that interact with one or more noise variables can be the object of the analysis. For example, let us consider an illustration used in Myers, Montgomery, and Anderson-Cook (2009) involving two control variables and a single noise variable with the data given in Table 15.22. A and B are control variables and C is a noise variable.

Table 15.22: Experimental Data in a Crossed Array

Inner Array		Outer Array		Response Mean
A	B	$C = -1$	$C = +1$	
−1	−1	11	15	13.0
−1	1	7	8	7.5
1	−1	10	26	18.0
1	1	10	14	12.0

One can illustrate the interactions AC and BC with plots, as given in Figure 15.21. One must understand that while A and B are held constant in the process C follows a probability distribution during the process. Given this information, it becomes clear that $A = -1$ and $B = +1$ are levels that produce smaller values for the process variance, while $A = +1$ and $B = -1$ give larger values. Thus, we say that $A = -1$ and $B = +1$ are robust values, i.e., insensitive to inevitable changes in the noise variable C during the process.

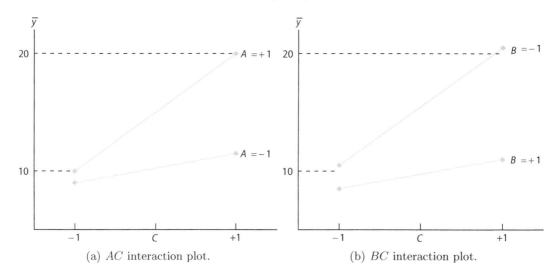

(a) AC interaction plot.　　(b) BC interaction plot.

Figure 15.21: Interaction plots for the data in Table 15.22.

In the above example, we say that both A and B are dispersion effects (i.e. both factors impact the process variance). In addition, both factors are location effects since the mean of y changes as both factors move from -1 to $+1$.

Analysis Involving the Model Containing Both Control and Noise Variables

While it has been emphasized that noise variables are not constant during the working of the process, analysis that results in desirable or even optimal conditions on the control variables is best accomplished through an experiment in which both control and noise variables are fixed effects. Thus, both main effects in the control and noise variables and all the important control-by-noise interactions can be evaluated. This model in x and z, often called a response model, can both directly and indirectly provide useful information regarding the process. The response model is actually a response surface model in vector **x** and vector **z**, where **x** contains control variables and **z** the noise variables. Certain operations allow models to be generated for the process mean and variance much as in Case Study 15.4. Details are supplied in Myers, Montgomery, and Anderson-Cook (2009); we will illustrate with a very simple example. Consider the data of Table 15.22 on page 650 with control variables A and B and noise variable C. There are eight experimental runs in a $2^2 \times 2$, or 2^3, factorial. Thus, the response model can be written

$$y(x, z) = \beta_0 + \beta_1 x_1 + \beta_2 x_2 + \beta_3 z + \beta_{12} x_1 x_2 + \beta_{1z} x_1 z + \beta_{2z} x_2 z + \epsilon.$$

We will not include the three-factor interaction in the regression model. A, B, and C in Table 15.22 are represented by x_1, x_2, and z, respectively, in the model. We assume that the error term ϵ has the usual independence and constant variance properties.

The Mean and Variance Response Surfaces

The process mean and variance response surfaces are best understood by considering the expectation and variance of z across the process. We assume that the noise variable C [denoted by z in $y(x, z)$] is continuous with mean 0 and variance σ_z^2. The process mean and variance models may be viewed as

$$E_z[y(x, z)] = \beta_0 + \beta_1 x_1 + \beta_2 x_2 + \beta_{12} x_1 x_2,$$
$$\text{Var}_z[y(x, z)] = \sigma^2 + \sigma_z^2 (\beta_3 + \beta_{1z} x_1 + \beta_{2z} x_2)^2 = \sigma^2 + \sigma_z^2 l_x^2,$$

where l_x is the slope $\frac{\partial y(x,z)}{\partial z}$ in the direction of z. As we indicated earlier, note how the interactions of factors A and B with the noise variable C are key components of the process variance.

Though we have already analyzed the current example through plots in Figure 15.21, which displayed the role of AB and AC interactions, it is instructive to look at the analysis in light of $E_z[y(x, z)]$ and $\text{Var}_z[y(x, z)]$ above. In this example, the reader can easily verify the estimate b_{1z} for β_{1z} is $15/8$ while the estimate b_{2z} for β_{2z} is $-15/8$. The coefficient $b_3 = 25/8$. Thus, the condition $x_1 = +1$ and $x_2 = -1$ results in a process variance estimate of

$$\widehat{\text{Var}}_z[y(x, z)] = \sigma^2 + \sigma_z^2 (b_3 + b_{1z} x_1 + b_{2z} x_2)^2$$
$$= \sigma^2 + \sigma_z^2 \left[\frac{25}{8} + \left(\frac{15}{8}\right)(1) + \left(\frac{-15}{8}\right)(-1) \right]^2 = \sigma^2 + \sigma_z^2 \left(\frac{55}{8}\right)^2,$$

whereas for $x_1 = -1$ and $x_2 = 1$, we have

$$\widehat{\text{Var}}_z[y(x, z)] = \sigma^2 + \sigma_z^2 (b_3 + b_{1z} x_1 + b_{2z} x_2)^2$$
$$= \sigma^2 + \sigma_z^2 \left[\frac{25}{8} + \left(\frac{15}{8}\right)(-1) + \left(\frac{15}{8}\right)(-1) \right]^2 = \sigma^2 + \sigma_z^2 \left(\frac{-5}{8}\right)^2.$$

Thus, for the most desirable (robust) condition of $x_1 = -1$ and $x_2 = 1$, the estimated process variance due to the noise variable C (or z) is $(25/64)\sigma_z^2$. The most undesirable condition, the condition of maximum process variance (i.e., $x_1 = +1$ and $x_2 = -1$), produces an estimated process variance of $(3025/64)\sigma_z^2$. As far as the mean response is concerned, Figure 15.21 indicates that if maximum response is desired $x_1 = +1$ and $x_2 = -1$ produce the best result.

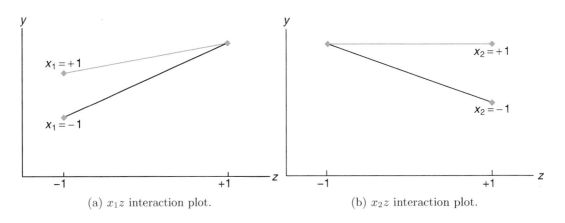

Figure 15.22: Interaction plots for the data in Exercise 15.31.

Exercises

15.31 Consider an example in which there are two control variables x_1 and x_2 and a single noise variable z. The goal is to determine the levels of x_1 and x_2 that are robust to changes in z, i.e., levels of x_1 and x_2 that minimize the variance produced in the response y as z moves between -1 and $+1$. The variables x_1 and x_2 are at two levels, -1 and $+1$, in the experiment. The data produce the plots in Figure 15.22 above. Note that x_1 and x_2 interact with the noise variable z. What settings on x_1 and x_2 (-1 or $+1$ for each) result in minimum variance in y? Explain.

15.32 Consider the following 2^3 factorial with control variables x_1 and x_2 and noise variable z. Can x_1 and x_2 be chosen at levels for which $\text{Var}(y)$ is minimized? Explain why or why not.

	$z = -1$		$z = +1$	
	$x_2 = -1$	$x_2 = +1$	$x_2 = -1$	$x_2 = +1$
$x_1 = -1$	4	6	8	10
$x_1 = +1$	1	3	3	5

15.33 Consider Case Study 15.1 involving the injection molding data. Suppose mold temperature is difficult to control and thus it can be assumed that in the process it follows a normal distribution with mean 0 and variance σ_z^2. Of concern is the variance of the shrinkage response in the process itself. In the analysis of Figure 15.7, it is clear that mold temperature, injection velocity, and the interaction between the two are the only important factors.

(a) Can the setting on velocity be used to create some type of control on the process variance in shrinkage which arises due to the inability to control temperature? Explain.

(b) Using parameter estimates from Figure 15.7, give an estimate of the following models:

(i) mean shrinkage across the distribution of temperature;

(ii) shrinkage variance as a function of σ_z^2.

(c) Use the estimated variance model to determine the level of velocity that minimizes the shrinkage variance.

(d) Use the mean shrinkage model to determine what value of velocity minimizes mean shrinkage.

(e) Are your results above consistent with your analysis from the interaction plot in Figure 15.6? Explain.

15.34 In Case Study 15.2 involving the coal cleansing data, the percent solids in the process system is known to vary uncontrollably during the process and is viewed as a noise factor with mean 0 and variance σ_z^2. The response, cleansing efficiency, has a mean and variance that change behavior during the process. Use only significant terms in the following parts.

(a) Use the estimates in Figure 15.9 to develop the process mean efficiency and variance models.

(b) What factor (or factors) might be controlled at certain levels to control or otherwise minimize the process variance?

(c) What conditions of factors B and C within the design region maximize the estimated mean?

(d) What level of C would you suggest for minimization of process variance when $B = 1$? When $B = -1$?

15.35 Use the coal cleansing data of Exercise 15.2 on page 609 to fit a model of the type

$$E(Y) = \beta_0 + \beta_1 x_1 + \beta_2 x_2 + \beta_3 x_3,$$

where the levels are

x_1, percent solids: 8, 12
x_2, flow rate: 150, 250 gal/min
x_3, pH: 5, 6

Center and scale the variables to design units. Also conduct a test for lack of fit, and comment concerning the adequacy of the linear regression model.

15.36 A 2^5 factorial plan is used to build a regression model containing first-order coefficients and model terms for all two-factor interactions. Duplicate runs are made for each factor. Outline the analysis-of-variance table, showing degrees of freedom for regression, lack of fit, and pure error.

15.37 Consider the $\frac{1}{16}$ of the 2^7 factorial discussed in Section 15.9. List the additional 11 defining contrasts.

15.38 Construct a Plackett-Burman design for 10 variables containing 24 experimental runs.

Review Exercises

15.39 A Plackett-Burman design was used to study the rheological properties of high-molecular-weight copolymers. Two levels of each of six variables were fixed in the experiment. The viscosity of the polymer is the response. The data were analyzed by the Laboratory for Interdisciplinary Statistical Analysis at Virginia Tech for personnel in the Chemical Engineering Department at the University. The variables are as follows: hard block chemistry x_1, nitrogen flow rate x_2, heat-up time x_3, percent compression x_4, scans (high and low) x_5, percent strain x_6. The data are presented here.

Obs.	x_1	x_2	x_3	x_4	x_5	x_6	y
1	1	−1	1	−1	−1	−1	194,700
2	1	1	−1	1	−1	−1	588,400
3	−1	1	1	−1	1	−1	7533
4	1	−1	1	1	−1	1	514,100
5	1	1	−1	1	1	−1	277,300
6	1	1	1	−1	1	1	493,500
7	−1	1	1	1	−1	1	8969
8	−1	−1	1	1	1	−1	18,340
9	−1	−1	−1	1	1	1	6793
10	1	−1	−1	−1	1	1	160,400
11	−1	1	−1	−1	−1	1	7008
12	−1	−1	−1	−1	−1	−1	3637

Build a regression equation relating viscosity to the levels of the six variables. Conduct t-tests for all main effects. Recommend factors that should be retained for future studies and those that should not. Use the residual mean square (5 degrees of freedom) as a measure of experimental error.

15.40 A large petroleum company in the Southwest regularly conducts experiments to test additives to drilling fluids. Plastic viscosity is a rheological measure reflecting the thickness of the fluid. Various polymers are added to the fluid to increase viscosity. The following is a data set in which two polymers are used at two levels each and the viscosity measured. The concentration of the polymers is indicated as "low" or "high." Conduct an analysis of the 2^2 factorial experiment. Test for effects for the two polymers and interaction.

Polymer 2	Polymer 1			
	Low		High	
	Low		High	
Low	3.0	3.5	11.3	12.0
High	11.7	12.0	21.7	22.4

15.41 A 2^2 factorial experiment is analyzed by the Laboratory for Interdisciplinary Statistical Analysis at Virginia Tech. The client is a member of the Department of Apparel, Housing, and Resource Management. The client is interested in comparing cold start to preheating ovens in terms of total energy delivered to the product. In addition, convection is being compared to regular mode. Four experimental runs are made at each of the four factor combinations. Following are the data from the experiment:

	Preheat		Cold	
Convection Mode	618	619.3	575	573.7
	629	611	574	572
Regular Mode	581	585.7	558	562
	581	595	562	566

Do an analysis of variance to study main effects and interaction. Draw conclusions.

15.42 In the study "The Use of Regression Analysis for Correcting Matrix Effects in the X-Ray Fluorescence Analysis of Pyrotechnic Compositions," published in the *Proceedings of the Tenth Conference on the Design of Experiments in Army Research Development and Testing*, ARO-D Report 65-3 (1965), an experiment was conducted in which the concentrations of four components of a propellant mixture and the weights of fine and coarse particles in the slurry were each allowed to vary. Factors A, B, C, and D, each at two levels, represent the concentrations of the four components, and factors E and F, also at two levels, represent the weights of the fine and coarse particles present in the slurry. The goal of the analysis was to determine if the X-ray intensity ratios associated with component 1 of the propellant were significantly influenced by varying the concentrations of the various components and the weights of the particles in the mixture. A $\frac{1}{8}$ fraction of a 2^6 factorial experiment was used, with the defining contrasts being ADE, BCE, and ACF. The data shown here represent the total of a pair of intensity readings.

The pooled mean square error with 8 degrees of freedom is given by 0.02005. Analyze the data using a 0.05 level of significance to determine if the concentrations of the components and the weights of the fine and coarse particles present in the slurry have a significant influence on the intensity ratios associated with component 1. Assume that no interaction exists among the six factors.

Batch	Treatment Combination	Intensity Ratio Total
1	$abef$	2.2480
2	$cdef$	1.8570
3	(1)	2.2428
4	ace	2.3270
5	bde	1.8830
6	$abcd$	1.8078
7	adf	2.1424
8	bcf	1.9122

15.43 Use Table 15.16 to construct a 16-run design with 8 factors that is resolution IV.

15.44 Verify that your design in Review Exercise 15.43 is indeed resolution IV.

15.45 Construct a design that contains 9 design points, is orthogonal, contains 12 total runs and 3 degrees of freedom for replication error, and allows for a lack-of-fit test for pure quadratic curvature.

15.46 Consider a design which is a 2^{3-1}_{III} with 2 center runs. Consider $\bar{y}_f$ as the average response at the design parameter and $\bar{y}_0$ as the average response at the design center. Suppose the true regression model is

$$E(Y) = \beta_0 + \beta_1 x_1 + \beta_2 x_2 + \beta_3 x_3 \\ + \beta_{11} x_1^2 + \beta_{22} x_2^2 + \beta_{33} x_3^2.$$

(a) Give (and verify) $E(\bar{y}_f - \bar{y}_0)$.

(b) Explain what you have learned from the result in (a).

15.13 Potential Misconceptions and Hazards; Relationship to Material in Other Chapters

In the use of fractional factorial experiments, one of the most important considerations that the analyst must be aware of is the *design resolution*. A design of low resolution is smaller (and hence cheaper) than one of higher resolution. However, a price is paid for the cheaper design. The design of lower resolution has heavier aliasing than one of higher resolution. For example, if the researcher has expectations that two-factor interactions may be important, then resolution III should not be used. A resolution III design is strictly a **main effects plan**.

Chapter 16

Nonparametric Statistics

16.1 Nonparametric Tests

Most of the hypothesis-testing procedures discussed in previous chapters are based on the assumption that the random samples are selected from normal populations. Fortunately, most of these tests are still reliable when we experience slight departures from normality, particularly when the sample size is large. Traditionally, these testing procedures have been referred to as **parametric methods**. In this chapter, we consider a number of alternative test procedures, called **nonparametric** or **distribution-free methods**, that often assume no knowledge whatsoever about the distributions of the underlying populations, except perhaps that they are continuous.

Nonparametric, or distribution-free procedures, are used with increasing frequency by data analysts. There are many applications in science and engineering where the data are reported as values not on a continuum but rather on an **ordinal scale** such that it is quite natural to assign ranks to the data. In fact, the reader may notice quite early in this chapter that the distribution-free methods described here involve an *analysis of ranks*. Most analysts find the computations involved in nonparametric methods to be very appealing and intuitive.

For an example where a nonparametric test is applicable, consider the situation in which two judges rank five brands of premium beer by assigning a rank of 1 to the brand believed to have the best overall quality, a rank of 2 to the second best, and so forth. A nonparametric test could then be used to determine whether there is any agreement between the two judges.

We should also point out that there are a number of disadvantages associated with nonparametric tests. Primarily, they do not utilize all the information provided by the sample, and thus a nonparametric test will be less efficient than the corresponding parametric procedure when both methods are applicable. Consequently, to achieve the same power, a nonparametric test will require a larger sample size than will the corresponding parametric test.

As we indicated earlier, slight departures from normality result in minor deviations from the ideal for the standard parametric tests. This is particularly true for the t-test and the F-test. In the case of the t-test and the F-test, the P-value quoted may be slightly in error if there is a moderate violation of the normality assumption.

In summary, if a parametric and a nonparametric test are both applicable to the same set of data, we should carry out the more efficient parametric technique. However, we should recognize that the assumptions of normality often cannot be justified and that we do not always have quantitative measurements. It is fortunate that statisticians have provided us with a number of useful nonparametric procedures. Armed with nonparametric techniques, the data analyst has more ammunition to accommodate a wider variety of experimental situations. It should be pointed out that even under the standard normal theory assumptions, the efficiencies of the nonparametric techniques are remarkably close to those of the corresponding parametric procedure. On the other hand, serious departures from normality will render the nonparametric method much more efficient than the parametric procedure.

Sign Test

The reader should recall that the procedures discussed in Section 10.4 for testing the null hypothesis that $\mu = \mu_0$ are valid only if the population is approximately normal or if the sample is large. If $n < 30$ and the population is decidedly nonnormal, we must resort to a nonparametric test.

The sign test is used to test hypotheses on a population *median*. In the case of many of the nonparametric procedures, the mean is replaced by the median as the pertinent **location parameter** under test. Recall that the sample median was defined in Section 1.3. The population counterpart, denoted by $\tilde{\mu}$, has an analogous definition. Given a random variable X, $\tilde{\mu}$ is defined such that $P(X > \tilde{\mu}) \leq 0.5$ and $P(X < \tilde{\mu}) \leq 0.5$. In the continuous case,

$$P(X > \tilde{\mu}) = P(X < \tilde{\mu}) = 0.5.$$

Of course, if the distribution is symmetric, the population mean and median are equal. In testing the null hypothesis H_0 that $\tilde{\mu} = \tilde{\mu}_0$ against an appropriate alternative, on the basis of a random sample of size n, we replace each sample value exceeding $\tilde{\mu}_0$ with a *plus* sign and each sample value less than $\tilde{\mu}_0$ with a *minus* sign. If the null hypothesis is true and the population is symmetric, the sum of the plus signs should be approximately equal to the sum of the minus signs. When one sign appears more frequently than it should based on chance alone, we reject the hypothesis that the population median $\tilde{\mu}$ is equal to $\tilde{\mu}_0$.

In theory, the sign test is applicable only in situations where $\tilde{\mu}_0$ cannot equal the value of any of the observations. Although there is a zero probability of obtaining a sample observation exactly equal to $\tilde{\mu}_0$ when the population is continuous, nevertheless, in practice a sample value equal to $\tilde{\mu}_0$ will often occur from a lack of precision in recording the data. When sample values equal to $\tilde{\mu}_0$ are observed, they are excluded from the analysis and the sample size is correspondingly reduced.

The appropriate test statistic for the sign test is the binomial random variable X, representing the number of plus signs in our random sample. If the null hypothesis that $\tilde{\mu} = \tilde{\mu}_0$ is true, the probability that a sample value results in either a plus or a minus sign is equal to $1/2$. Therefore, to test the null hypothesis that

$\tilde{\mu} = \tilde{\mu}_0$, we actually test the null hypothesis that the number of plus signs is a value of a random variable having the binomial distribution with the parameter $p = 1/2$. P-values for both one-sided and two-sided alternatives can then be calculated using this binomial distribution. For example, in testing

$$H_0: \tilde{\mu} = \tilde{\mu}_0,$$
$$H_1: \tilde{\mu} < \tilde{\mu}_0,$$

we shall reject H_0 in favor of H_1 only if the proportion of plus signs is sufficiently less than $1/2$, that is, when the value x of our random variable is small. Hence, if the computed P-value

$$P = P(X \leq x \text{ when } p = 1/2)$$

is less than or equal to some preselected significance level α, we reject H_0 in favor of H_1. For example, when $n = 15$ and $x = 3$, we find from Table A.1 that

$$P = P(X \leq 3 \text{ when } p = 1/2) = \sum_{x=0}^{3} b\left(x; 15, \frac{1}{2}\right) = 0.0176,$$

so the null hypothesis $\tilde{\mu} = \tilde{\mu}_0$ can certainly be rejected at the 0.05 level of significance but not at the 0.01 level.

To test the hypothesis

$$H_0: \tilde{\mu} = \tilde{\mu}_0,$$
$$H_1: \tilde{\mu} > \tilde{\mu}_0,$$

we reject H_0 in favor of H_1 only if the proportion of plus signs is sufficiently greater than $1/2$, that is, when x is large. Hence, if the computed P-value

$$P = P(X \geq x \text{ when } p = 1/2)$$

is less than α, we reject H_0 in favor of H_1. Finally, to test the hypothesis

$$H_0: \tilde{\mu} = \tilde{\mu}_0,$$
$$H_1: \tilde{\mu} \neq \tilde{\mu}_0,$$

we reject H_0 in favor of H_1 when the proportion of plus signs is significantly less than or greater than $1/2$. This, of course, is equivalent to x being sufficiently small or sufficiently large. Therefore, if $x < n/2$ and the computed P-value

$$P = 2P(X \leq x \text{ when } p = 1/2)$$

is less than or equal to α, or if $x > n/2$ and the computed P-value

$$P = 2P(X \geq x \text{ when } p = 1/2)$$

is less than or equal to α, we reject H_0 in favor of H_1.

Whenever $n > 10$, binomial probabilities with $p = 1/2$ can be approximated from the normal curve, since $np = nq > 5$. Suppose, for example, that we wish to test the hypothesis

$$H_0: \tilde{\mu} = \tilde{\mu}_0,$$
$$H_1: \tilde{\mu} < \tilde{\mu}_0,$$

at the $\alpha = 0.05$ level of significance, for a random sample of size $n = 20$ that yields $x = 6$ plus signs. Using the normal curve approximation with

$$\tilde{\mu} = np = (20)(0.5) = 10$$

and

$$\sigma = \sqrt{npq} = \sqrt{(20)(0.5)(0.5)} = 2.236,$$

we find that

$$z = \frac{6.5 - 10}{2.236} = -1.57.$$

Therefore,

$$P = P(X \leq 6) \approx P(Z < -1.57) = 0.0582,$$

which leads to the nonrejection of the null hypothesis.

Example 16.1: The following data represent the number of hours that a rechargeable hedge trimmer operates before a recharge is required:

$$1.5, 2.2, 0.9, 1.3, 2.0, 1.6, 1.8, 1.5, 2.0, 1.2, 1.7.$$

Use the sign test to test the hypothesis, at the 0.05 level of significance, that this particular trimmer operates a median of 1.8 hours before requiring a recharge.

Solution:
1. $H_0: \tilde{\mu} = 1.8$.
2. $H_1: \tilde{\mu} \neq 1.8$.
3. $\alpha = 0.05$.
4. Test statistic: Binomial variable X with $p = \frac{1}{2}$.
5. Computations: Replacing each value by the symbol "+" if it exceeds 1.8 and by the symbol "−" if it is less than 1.8 and discarding the one measurement that equals 1.8, we obtain the sequence

$$- + - - + - - + - -$$

for which $n = 10$, $x = 3$, and $n/2 = 5$. Therefore, from Table A.1 the computed P-value is

$$P = 2P\left(X \leq 3 \text{ when } p = \frac{1}{2}\right) = 2\sum_{x=0}^{3} b\left(x; 10, \frac{1}{2}\right) = 0.3438 > 0.05.$$

6. **Decision:** Do not reject the null hypothesis and conclude that the median operating time is not significantly different from 1.8 hours.

We can also use the sign test to test the null hypothesis $\tilde{\mu}_1 - \tilde{\mu}_2 = d_0$ for paired observations. Here we replace each difference, d_i, with a plus or minus sign depending on whether the adjusted difference, $d_i - d_0$, is positive or negative. Throughout this section, we have assumed that the populations are symmetric. However, even if populations are skewed, we can carry out the same test procedure, but the hypotheses refer to the population medians rather than the means.

Example 16.2: A taxi company is trying to decide whether the use of radial tires instead of regular belted tires improves fuel economy. Sixteen cars are equipped with radial tires and driven over a prescribed test course. Without changing drivers, the same cars are then equipped with the regular belted tires and driven once again over the test course. The gasoline consumption, in kilometers per liter, is given in Table 16.1. Can we conclude at the 0.05 level of significance that cars equipped with radial tires obtain better fuel economy than those equipped with regular belted tires?

Table 16.1: Data for Example 16.2

Car	1	2	3	4	5	6	7	8
Radial Tires	4.2	4.7	6.6	7.0	6.7	4.5	5.7	6.0
Belted Tires	4.1	4.9	6.2	6.9	6.8	4.4	5.7	5.8
Car	9	10	11	12	13	14	15	16
Radial Tires	7.4	4.9	6.1	5.2	5.7	6.9	6.8	4.9
Belted Tires	6.9	4.9	6.0	4.9	5.3	6.5	7.1	4.8

Solution: Let $\tilde{\mu}_1$ and $\tilde{\mu}_2$ represent the median kilometers per liter for cars equipped with radial and belted tires, respectively.

1. H_0: $\tilde{\mu}_1 - \tilde{\mu}_2 = 0$.
2. H_1: $\tilde{\mu}_1 - \tilde{\mu}_2 > 0$.
3. $\alpha = 0.05$.
4. Test statistic: Binomial variable X with $p = 1/2$.
5. Computations: After replacing each positive difference by a "+" symbol and each negative difference by a "−" symbol and then discarding the two zero differences, we obtain the sequence

$$+ \; - \; + \; + \; - \; + \; + \; + \; + \; + \; + \; + \; - \; +$$

for which $n = 14$ and $x = 11$. Using the normal curve approximation, we find

$$z = \frac{10.5 - 7}{\sqrt{(14)(0.5)(0.5)}} = 1.87,$$

and then

$$P = P(X \geq 11) \approx P(Z > 1.87) = 0.0307.$$

6. **Decision:** Reject H_0 and conclude that, on the average, radial tires do improve fuel economy.

Not only is the sign test one of the simplest nonparametric procedures to apply; it has the additional advantage of being applicable to dichotomous data that cannot be recorded on a numerical scale but can be represented by positive and negative responses. For example, the sign test is applicable in experiments where a qualitative response such as "hit" or "miss" is recorded, and in sensory-type experiments where a plus or minus sign is recorded depending on whether the taste tester correctly or incorrectly identifies the desired ingredient.

We shall attempt to make comparisons between many of the nonparametric procedures and the corresponding parametric tests. In the case of the sign test the competition is, of course, the t-test. If we are sampling from a normal distribution, the use of the t-test will result in a larger power for the test. If the distribution is merely symmetric, though not normal, the t-test is preferred in terms of power unless the distribution has extremely "heavy tails" compared to the normal distribution.

16.2 Signed-Rank Test

The reader should note that the sign test utilizes only the plus and minus signs of the differences between the observations and $\tilde{\mu}_0$ in the one-sample case, or the plus and minus signs of the differences between the pairs of observations in the paired-sample case; it does not take into consideration the magnitudes of these differences. A test utilizing both direction and magnitude, proposed in 1945 by Frank Wilcoxon, is now commonly referred to as the **Wilcoxon signed-rank test**.

The analyst can extract more information from the data in a nonparametric fashion if it is reasonable to invoke an additional restriction on the distribution from which the data were taken. The Wilcoxon signed-rank test applies in the case of a **symmetric continuous distribution**. Under this condition, we can test the null hypothesis $\tilde{\mu} = \tilde{\mu}_0$. We first subtract $\tilde{\mu}_0$ from each sample value, discarding all differences equal to zero. The remaining differences are then ranked without regard to sign. A rank of 1 is assigned to the smallest absolute difference (i.e., without sign), a rank of 2 to the next smallest, and so on. When the absolute value of two or more differences is the same, assign to each the average of the ranks that would have been assigned if the differences were distinguishable. For example, if the fifth and sixth smallest differences are equal in absolute value, each is assigned a rank of 5.5. If the hypothesis $\tilde{\mu} = \tilde{\mu}_0$ is true, the total of the ranks corresponding to the positive differences should nearly equal the total of the ranks corresponding to the negative differences. Let us represent these totals by w_+ and w_-, respectively. We designate the smaller of w_+ and w_- by w.

In selecting repeated samples, we would expect w_+ and w_-, and therefore w, to vary. Thus, we may think of w_+, w_-, and w as values of the corresponding random variables W_+, W_-, and W. The null hypothesis $\tilde{\mu} = \tilde{\mu}_0$ can be rejected in favor of the alternative $\tilde{\mu} < \tilde{\mu}_0$ only if w_+ is small and w_- is large. Likewise, the alternative $\tilde{\mu} > \tilde{\mu}_0$ can be accepted only if w_+ is large and w_- is small. For a two-sided alternative, we may reject H_0 in favor of H_1 if either w_+ or w_-, and hence w, is sufficiently small. Therefore, no matter what the alternative hypothesis

Two Samples with Paired Observations

To test the null hypothesis that we are sampling two continuous symmetric populations with $\tilde{\mu}_1 = \tilde{\mu}_2$ for the paired-sample case, we rank the differences of the paired observations without regard to sign and proceed as in the single-sample case. The various test procedures for both the single- and paired-sample cases are summarized in Table 16.2.

Table 16.2: Signed-Rank Test

H_0	H_1	Compute
$\tilde{\mu} = \tilde{\mu}_0$	$\tilde{\mu} < \tilde{\mu}_0$	w_+
	$\tilde{\mu} > \tilde{\mu}_0$	w_-
	$\tilde{\mu} \neq \tilde{\mu}_0$	w
$\tilde{\mu}_1 = \tilde{\mu}_2$	$\tilde{\mu}_1 < \tilde{\mu}_2$	w_+
	$\tilde{\mu}_1 > \tilde{\mu}_2$	w_-
	$\tilde{\mu}_1 \neq \tilde{\mu}_2$	w

It is not difficult to show that whenever $n < 5$ and the level of significance does not exceed 0.05 for a one-tailed test or 0.10 for a two-tailed test, all possible values of w_+, w_-, or w will lead to the acceptance of the null hypothesis. However, when $5 \leq n \leq 30$, Table A.16 shows approximate critical values of W_+ and W_- for levels of significance equal to 0.01, 0.025, and 0.05 for a one-tailed test and critical values of W for levels of significance equal to 0.02, 0.05, and 0.10 for a two-tailed test. The null hypothesis is rejected if the computed value w_+, w_-, or w is **less than or equal to** the appropriate tabled value. For example, when $n = 12$, Table A.16 shows that a value of $w_+ \leq 17$ is required for the one-sided alternative $\tilde{\mu} < \tilde{\mu}_0$ to be significant at the 0.05 level.

Example 16.3: Rework Example 16.1 by using the signed-rank test.

Solution:
1. H_0: $\tilde{\mu} = 1.8$.
2. H_1: $\tilde{\mu} \neq 1.8$.
3. $\alpha = 0.05$.
4. Critical region: Since $n = 10$ after discarding the one measurement that equals 1.8, Table A.16 shows the critical region to be $w \leq 8$.
5. Computations: Subtracting 1.8 from each measurement and then ranking the differences without regard to sign, we have

d_i	-0.3	0.4	-0.9	-0.5	0.2	-0.2	-0.3	0.2	-0.6	-0.1
Ranks	5.5	7	10	8	3	3	5.5	3	9	1

Now $w_+ = 13$ and $w_- = 42$, so $w = 13$, the smaller of w_+ and w_-.

6. Decision: As before, do not reject H_0 and conclude that the median operating time is not significantly different from 1.8 hours.

The signed-rank test can also be used to test the null hypothesis that $\tilde{\mu}_1 - \tilde{\mu}_2 = d_0$. In this case, the populations need not be symmetric. As with the sign test, we subtract d_0 from each difference, rank the adjusted differences without regard to sign, and apply the same procedure as above.

Example 16.4: It is claimed that a college senior can increase his or her score in the major field area of the graduate record examination by at least 50 points if he or she is provided with sample problems in advance. To test this claim, 20 college seniors are divided into 10 pairs such that the students in each matched pair have almost the same overall grade-point averages for their first 3 years in college. Sample problems and answers are provided at random to one member of each pair 1 week prior to the examination. The examination scores are given in Table 16.3.

Table 16.3: Data for Example 16.4

	\multicolumn{10}{c}{Pair}
	1	2	3	4	5	6	7	8	9	10
With Sample Problems	531	621	663	579	451	660	591	719	543	575
Without Sample Problems	509	540	688	502	424	683	568	748	530	524

Test the null hypothesis, at the 0.05 level of significance, that sample problems increase scores by 50 points against the alternative hypothesis that the increase is less than 50 points.

Solution: Let $\tilde{\mu}_1$ and $\tilde{\mu}_2$ represent the median scores of all students taking the test in question with and without sample problems, respectively.

1. H_0: $\tilde{\mu}_1 - \tilde{\mu}_2 = 50$.
2. H_1: $\tilde{\mu}_1 - \tilde{\mu}_2 < 50$.
3. $\alpha = 0.05$.
4. Critical region: Since $n = 10$, Table A.16 shows the critical region to be $w_+ \leq 11$.
5. Computations:

	Pair									
	1	2	3	4	5	6	7	8	9	10
d_i	22	81	−25	77	27	−23	23	−29	13	51
$d_i - d_0$	−28	31	−75	27	−23	−73	−27	−79	−37	1
Ranks	5	6	9	3.5	2	8	3.5	10	7	1

Now we find that $w_+ = 6 + 3.5 + 1 = 10.5$.

6. Decision: Reject H_0 and conclude that sample problems do not, on average, increase one's graduate record score by as much as 50 points.

Normal Approximation for Large Samples

When $n \geq 15$, the sampling distribution of W_+ (or W_-) approaches the normal distribution with mean and variance given by

$$\mu_{W_+} = \frac{n(n+1)}{4} \text{ and } \sigma^2_{W_+} = \frac{n(n+1)(2n+1)}{24}.$$

Therefore, when n exceeds the largest value in Table A.16, the statistic

$$Z = \frac{W_+ - \mu_{W_+}}{\sigma_{W_+}}$$

can be used to determine the critical region for the test.

Exercises

16.1 The following data represent the time, in minutes, that a patient has to wait during 12 visits to a doctor's office before being seen by the doctor:

17 15 20 20 32 28
12 26 25 25 35 24

Use the sign test at the 0.05 level of significance to test the doctor's claim that the median waiting time for her patients is not more than 20 minutes.

16.2 The following data represent the number of hours of flight training received by 18 student pilots from a certain instructor prior to their first solo flight:

9   12 18 14 12 14 12 10 16
11 9   11 13 11 13 15 13 14

Using binomial probabilities from Table A.1, perform a sign test at the 0.02 level of significance to test the instructor's claim that the median time required before his students' solo is 12 hours of flight training.

16.3 A food inspector examined 16 jars of a certain brand of jam to determine the percent of foreign impurities. The following data were recorded:

2.4 2.3 3.1 2.2 2.3 1.2 1.0 2.4
1.7 1.1 4.2 1.9 1.7 3.6 1.6 2.3

Using the normal approximation to the binomial distribution, perform a sign test at the 0.05 level of significance to test the null hypothesis that the median percent of impurities in this brand of jam is 2.5% against the alternative that the median percent of impurities is not 2.5%.

16.4 A paint supplier claims that a new additive will reduce the drying time of its acrylic paint. To test this claim, 12 panels of wood were painted, one-half of each panel with paint containing the regular additive and the other half with paint containing the new additive. The drying times, in hours, were recorded as follows:

	Drying Time (hours)	
Panel	New Additive	Regular Additive
1	6.4	6.6
2	5.8	5.8
3	7.4	7.8
4	5.5	5.7
5	6.3	6.0
6	7.8	8.4
7	8.6	8.8
8	8.2	8.4
9	7.0	7.3
10	4.9	5.8
11	5.9	5.8
12	6.5	6.5

Use the sign test at the 0.05 level to test the null hypothesis that the new additive is no better than the regular additive in reducing the drying time of this kind of paint.

16.5 It is claimed that a new diet will reduce a person's weight by 4.5 kilograms, on average, in a period of 2 weeks. The weights of 10 women were recorded before and after a 2-week period during which they followed this diet, yielding the following data:

Woman	Weight Before	Weight After
1	58.5	60.0
2	60.3	54.9
3	61.7	58.1
4	69.0	62.1
5	64.0	58.5
6	62.6	59.9
7	56.7	54.4
8	63.6	60.2
9	68.2	62.3
10	59.4	58.7

Use the sign test at the 0.05 level of significance to test the hypothesis that the diet reduces the median weight by 4.5 kilograms against the alternative hypothesis that the median weight loss is less than 4.5 kilograms.

16.6 Two types of instruments for measuring the amount of sulfur monoxide in the atmosphere are being compared in an air-pollution experiment. The following readings were recorded daily for a period of 2 weeks:

Day	Sulfur Monoxide	
	Instrument A	Instrument B
1	0.96	0.87
2	0.82	0.74
3	0.75	0.63
4	0.61	0.55
5	0.89	0.76
6	0.64	0.70
7	0.81	0.69
8	0.68	0.57
9	0.65	0.53
10	0.84	0.88
11	0.59	0.51
12	0.94	0.79
13	0.91	0.84
14	0.77	0.63

Using the normal approximation to the binomial distribution, perform a sign test to determine whether the different instruments lead to different results. Use a 0.05 level of significance.

16.7 The following figures give the systolic blood pressure of 16 joggers before and after an 8-kilometer run:

Jogger	Before	After
1	158	164
2	149	158
3	160	163
4	155	160
5	164	172
6	138	147
7	163	167
8	159	169
9	165	173
10	145	147
11	150	156
12	161	164
13	132	133
14	155	161
15	146	154
16	159	170

Use the sign test at the 0.05 level of significance to test the null hypothesis that jogging 8 kilometers increases the median systolic blood pressure by 8 points against the alternative that the increase in the median is less than 8 points.

16.8 Analyze the data of Exercise 16.1 by using the signed-rank test.

16.9 Analyze the data of Exercise 16.2 by using the signed-rank test.

16.10 The weights of 5 people before they stopped smoking and 5 weeks after they stopped smoking, in kilograms, are as follows:

	Individual				
	1	2	3	4	5
Before	66	80	69	52	75
After	71	82	68	56	73

Use the signed-rank test for paired observations to test the hypothesis, at the 0.05 level of significance, that giving up smoking has no effect on a person's weight against the alternative that one's weight increases if he or she quits smoking.

16.11 Rework Exercise 16.5 by using the signed-rank test.

16.12 The following are the numbers of prescriptions filled by two pharmacies over a 20-day period:

Day	Pharmacy A	Pharmacy B
1	19	17
2	21	15
3	15	12
4	17	12
5	24	16
6	12	15
7	19	11
8	14	13
9	20	14
10	18	21
11	23	19
12	21	15
13	17	11
14	12	10
15	16	20
16	15	12
17	20	13
18	18	17
19	14	16
20	22	18

Use the signed-rank test at the 0.01 level of significance to determine whether the two pharmacies, on average, fill the same number of prescriptions against the alternative that pharmacy A fills more prescriptions than pharmacy B.

16.13 Rework Exercise 16.7 by using the signed-rank test.

16.14 Rework Exercise 16.6 by using the signed-rank test.

16.3 Wilcoxon Rank-Sum Test

As we indicated earlier, the nonparametric procedure is generally an appropriate alternative to the normal theory test when the normality assumption does not hold. When we are interested in testing equality of means of two continuous distributions that are obviously nonnormal, and samples are independent (i.e., there is no pairing of observations), the **Wilcoxon rank-sum test** or **Wilcoxon two-sample test** is an appropriate alternative to the two-sample t-test described in Chapter 10.

We shall test the null hypothesis H_0 that $\tilde{\mu}_1 = \tilde{\mu}_2$ against some suitable alternative. First we select a random sample from each of the populations. Let n_1 be the number of observations in the smaller sample, and n_2 the number of observations in the larger sample. When the samples are of equal size, n_1 and n_2 may be randomly assigned. Arrange the $n_1 + n_2$ observations of the combined samples in ascending order and substitute a rank of $1, 2, \ldots, n_1 + n_2$ for each observation. In the case of ties (identical observations), we replace the observations by the mean of the ranks that the observations would have if they were distinguishable. For example, if the seventh and eighth observations were identical, we would assign a rank of 7.5 to each of the two observations.

The sum of the ranks corresponding to the n_1 observations in the smaller sample is denoted by w_1. Similarly, the value w_2 represents the sum of the n_2 ranks corresponding to the larger sample. The total $w_1 + w_2$ depends only on the number of observations in the two samples and is in no way affected by the results of the experiment. Hence, if $n_1 = 3$ and $n_2 = 4$, then $w_1 + w_2 = 1 + 2 + \cdots + 7 = 28$, regardless of the numerical values of the observations. In general,

$$w_1 + w_2 = \frac{(n_1 + n_2)(n_1 + n_2 + 1)}{2},$$

the arithmetic sum of the integers $1, 2, \ldots, n_1 + n_2$. Once we have determined w_1, it may be easier to find w_2 by the formula

$$w_2 = \frac{(n_1 + n_2)(n_1 + n_2 + 1)}{2} - w_1.$$

In choosing repeated samples of sizes n_1 and n_2, we would expect w_1, and therefore w_2, to vary. Thus, we may think of w_1 and w_2 as values of the random variables W_1 and W_2, respectively. The null hypothesis $\tilde{\mu}_1 = \tilde{\mu}_2$ will be rejected in favor of the alternative $\tilde{\mu}_1 < \tilde{\mu}_2$ only if w_1 is small and w_2 is large. Likewise, the alternative $\tilde{\mu}_1 > \tilde{\mu}_2$ can be accepted only if w_1 is large and w_2 is small. For a two-tailed test, we may reject H_0 in favor of H_1 if w_1 is small and w_2 is large or if w_1 is large and w_2 is small. In other words, the alternative $\tilde{\mu}_1 < \tilde{\mu}_2$ is accepted if w_1 is sufficiently small; the alternative $\tilde{\mu}_1 > \tilde{\mu}_2$ is accepted if w_2 is sufficiently small; and the alternative $\tilde{\mu}_1 \neq \tilde{\mu}_2$ is accepted if the minimum of w_1 and w_2 is sufficiently small. In actual practice, we usually base our decision on the value

$$u_1 = w_1 - \frac{n_1(n_1 + 1)}{2} \quad \text{or} \quad u_2 = w_2 - \frac{n_2(n_2 + 1)}{2}$$

of the related statistic U_1 or U_2 or on the value u of the statistic U, the minimum of U_1 and U_2. These statistics simplify the construction of tables of critical values, since both U_1 and U_2 have symmetric sampling distributions and assume values in the interval from 0 to $n_1 n_2$ such that $u_1 + u_2 = n_1 n_2$.

From the formulas for u_1 and u_2 we see that u_1 will be small when w_1 is small and u_2 will be small when w_2 is small. Consequently, the null hypothesis will be rejected whenever the appropriate statistic U_1, U_2, or U assumes a value less than or equal to the desired critical value given in Table A.17. The various test procedures are summarized in Table 16.4.

Table 16.4: Rank-Sum Test

H_0	H_1	Compute
$\tilde{\mu}_1 = \tilde{\mu}_2$	$\tilde{\mu}_1 < \tilde{\mu}_2$	u_1
	$\tilde{\mu}_1 > \tilde{\mu}_2$	u_2
	$\tilde{\mu}_1 \neq \tilde{\mu}_2$	u

Table A.17 gives critical values of U_1 and U_2 for levels of significance equal to 0.001, 0.01, 0.025, and 0.05 for a one-tailed test, and critical values of U for levels of significance equal to 0.002, 0.02, 0.05, and 0.10 for a two-tailed test. If the observed value of u_1, u_2, or u is **less than or equal** to the tabled critical value, the null hypothesis is rejected at the level of significance indicated by the table. Suppose, for example, that we wish to test the null hypothesis that $\tilde{\mu}_1 = \tilde{\mu}_2$ against the one-sided alternative that $\tilde{\mu}_1 < \tilde{\mu}_2$ at the 0.05 level of significance for random samples of sizes $n_1 = 3$ and $n_2 = 5$ that yield the value $w_1 = 8$. It follows that

$$u_1 = 8 - \frac{(3)(4)}{2} = 2.$$

Our one-tailed test is based on the statistic U_1. Using Table A.17, we reject the null hypothesis of equal means when $u_1 \leq 1$. Since $u_1 = 2$ does not fall in the rejection region, the null hypothesis cannot be rejected.

Example 16.5: The nicotine content of two brands of cigarettes, measured in milligrams, was found to be as follows:

Brand A	2.1	4.0	6.3	5.4	4.8	3.7	6.1	3.3		
Brand B	4.1	0.6	3.1	2.5	4.0	6.2	1.6	2.2	1.9	5.4

Test the hypothesis, at the 0.05 level of significance, that the median nicotine contents of the two brands are equal against the alternative that they are unequal.

Solution: 1. H_0: $\tilde{\mu}_1 = \tilde{\mu}_2$.

2. H_1: $\tilde{\mu}_1 \neq \tilde{\mu}_2$.

3. $\alpha = 0.05$.

4. Critical region: $u \leq 17$ (from Table A.17).

5. Computations: The observations are arranged in ascending order and ranks from 1 to 18 assigned.

16.3 Wilcoxon Rank-Sum Test

Original Data	Ranks	Original Data	Ranks
0.6	1	4.0	10.5*
1.6	2	4.0	10.5
1.9	3	4.1	12
2.1	4*	4.8	13*
2.2	5	5.4	14.5*
2.5	6	5.4	14.5
3.1	7	6.1	16*
3.3	8*	6.2	17
3.7	9*	6.3	18*

*The ranks marked with an asterisk belong to sample A.

Now

$$w_1 = 4 + 8 + 9 + 10.5 + 13 + 14.5 + 16 + 18 = 93$$

and

$$w_2 = \frac{(18)(19)}{2} - 93 = 78.$$

Therefore,

$$u_1 = 93 - \frac{(8)(9)}{2} = 57, \qquad u_2 = 78 - \frac{(10)(11)}{2} = 23.$$

6. *Decision:* Do not reject the null hypothesis H_0 and conclude that there is no significant difference in the median nicotine contents of the two brands of cigarettes.

Normal Theory Approximation for Two Samples

When both n_1 and n_2 exceed 8, the sampling distribution of U_1 (or U_2) approaches the normal distribution with mean and variance given by

$$\mu_{U_1} = \frac{n_1 n_2}{2} \quad \text{and} \quad \sigma_{U_1}^2 = \frac{n_1 n_2 (n_1 + n_2 + 1)}{12}.$$

Consequently, when n_2 is greater than 20, the maximum value in Table A.17, and n_1 is at least 9, we can use the statistic

$$Z = \frac{U_1 - \mu_{U_1}}{\sigma_{U_1}}$$

for our test, with the critical region falling in either or both tails of the standard normal distribution, depending on the form of H_1.

The use of the Wilcoxon rank-sum test is not restricted to nonnormal populations. It can be used in place of the two-sample t-test when the populations are normal, although the power will be smaller. The Wilcoxon rank-sum test is always superior to the t-test for decidedly nonnormal populations.

16.4 Kruskal-Wallis Test

In Chapters 13, 14, and 15, the technique of analysis of variance was prominent as an analytical technique for testing equality of $k \geq 2$ population means. Again, however, the reader should recall that normality must be assumed in order for the F-test to be theoretically correct. In this section, we investigate a nonparametric alternative to analysis of variance.

The **Kruskal-Wallis test**, also called the **Kruskal-Wallis H test**, is a generalization of the rank-sum test to the case of $k > 2$ samples. It is used to test the null hypothesis H_0 that k independent samples are from identical populations. Introduced in 1952 by W. H. Kruskal and W. A. Wallis, the test is a nonparametric procedure for testing the equality of means in the one-factor analysis of variance when the experimenter wishes to avoid the assumption that the samples were selected from normal populations.

Let n_i ($i = 1, 2, \ldots, k$) be the number of observations in the ith sample. First, we combine all k samples and arrange the $n = n_1 + n_2 + \cdots + n_k$ observations in ascending order, substituting the appropriate rank from $1, 2, \ldots, n$ for each observation. In the case of ties (identical observations), we follow the usual procedure of replacing the observations by the mean of the ranks that the observations would have if they were distinguishable. The sum of the ranks corresponding to the n_i observations in the ith sample is denoted by the random variable R_i. Now let us consider the statistic

$$H = \frac{12}{n(n+1)} \sum_{i=1}^{k} \frac{R_i^2}{n_i} - 3(n+1),$$

which is approximated very well by a chi-squared distribution with $k-1$ degrees of freedom when H_0 is true, provided each sample consists of at least 5 observations. The fact that h, the assumed value of H, is large when the independent samples come from populations that are not identical allows us to establish the following decision criterion for testing H_0:

Kruskal-Wallis Test To test the null hypothesis H_0 that k independent samples are from identical populations, compute

$$h = \frac{12}{n(n+1)} \sum_{i=1}^{k} \frac{r_i^2}{n_i} - 3(n+1),$$

where r_i is the assumed value of R_i, for $i = 1, 2, \ldots, k$. If h falls in the critical region $H > \chi_\alpha^2$ with $v = k - 1$ degrees of freedom, reject H_0 at the α-level of significance; otherwise, fail to reject H_0.

Example 16.6: In an experiment to determine which of three different missile systems is preferable, the propellant burning rate is measured. The data, after coding, are given in Table 16.5. Use the Kruskal-Wallis test and a significance level of $\alpha = 0.05$ to test the hypothesis that the propellant burning rates are the same for the three missile systems.

16.4 Kruskal-Wallis Test

Table 16.5: Propellant Burning Rates

Missile System								
1			2			3		
24.0	16.7	22.8	23.2	19.8	18.1	18.4	19.1	17.3
19.8	18.9		17.6	20.2	17.8	17.3	19.7	18.9
						18.8	19.3	

Solution:
1. H_0: $\mu_1 = \mu_2 = \mu_3$.
2. H_1: The three means are not all equal.
3. $\alpha = 0.05$.
4. Critical region: $h > \chi^2_{0.05} = 5.991$, for $v = 2$ degrees of freedom.
5. Computations: In Table 16.6, we convert the 19 observations to ranks and sum the ranks for each missile system.

Table 16.6: Ranks for Propellant Burning Rates

Missile System		
1	2	3
19	18	7
1	14.5	11
17	6	2.5
14.5	4	2.5
9.5	16	13
$r_1 = 61.0$	5	9.5
	$r_2 = 63.5$	8
		12
		$r_3 = 65.5$

Now, substituting $n_1 = 5$, $n_2 = 6$, $n_3 = 8$ and $r_1 = 61.0$, $r_2 = 63.5$, $r_3 = 65.5$, our test statistic H assumes the value

$$h = \frac{12}{(19)(20)} \left(\frac{61.0^2}{5} + \frac{63.5^2}{6} + \frac{65.5^2}{8} \right) - (3)(20) = 1.66.$$

6. Decision: Since $h = 1.66$ does not fall in the critical region $h > 5.991$, we have insufficient evidence to reject the hypothesis that the propellant burning rates are the same for the three missile systems.

Exercises

16.15 A cigarette manufacturer claims that the tar content of brand B cigarettes is lower than that of brand A cigarettes. To test this claim, the following determinations of tar content, in milligrams, were recorded:

Brand A	1	12	9	13	11	14
Brand B	8	10	7			

Use the rank-sum test with $\alpha = 0.05$ to test whether the claim is valid.

16.16 To find out whether a new serum will arrest leukemia, nine patients, who have all reached an advanced stage of the disease, are selected. Five patients receive the treatment and four do not. The survival times, in years, from the time the experiment commenced are

Treatment	2.1	5.3	1.4	4.6	0.9
No treatment	1.9	0.5	2.8	3.1	

Use the rank-sum test, at the 0.05 level of significance, to determine if the serum is effective.

16.17 The following data represent the number of hours that two different types of scientific pocket calculators operate before a recharge is required.

Calculator A	5.5 5.6 6.3 4.6 5.3 5.0 6.2 5.8 5.1
Calculator B	3.8 4.8 4.3 4.2 4.0 4.9 4.5 5.2 4.5

Use the rank-sum test with $\alpha = 0.01$ to determine if calculator A operates longer than calculator B on a full battery charge.

16.18 A fishing line is being manufactured by two processes. To determine if there is a difference in the mean breaking strength of the lines, 10 pieces manufactured by each process are selected and then tested for breaking strength. The results are as follows:

Process 1	10.4	9.8	11.5	10.0	9.9
	9.6	10.9	11.8	9.3	10.7
Process 2	8.7	11.2	9.8	10.1	10.8
	9.5	11.0	9.8	10.5	9.9

Use the rank-sum test with $\alpha = 0.1$ to determine if there is a difference between the mean breaking strengths of the lines manufactured by the two processes.

16.19 From a mathematics class of 12 equally capable students using programmed materials, 5 students are selected at random and given additional instruction by the teacher. The results on the final examination are as follows:

	Grade						
Additional Instruction	87	69	78	91	80		
No Additional Instruction	75	88	64	82	93	79	67

Use the rank-sum test with $\alpha = 0.05$ to determine if the additional instruction affects the average grade.

16.20 The following data represent the weights, in kilograms, of personal luggage carried on various flights by a member of a baseball team and a member of a basketball team.

Luggage Weight (kilograms)					
Baseball Player			Basketball Player		
16.3	20.0	18.6	15.4	16.3	
18.1	15.0	15.4	17.7	18.1	
15.9	18.6	15.6	18.6	16.8	
14.1	14.5	18.3	12.7	14.1	
17.7	19.1	17.4	15.0	13.6	
16.3	13.6	14.8	15.9	16.3	
13.2	17.2	16.5			

Use the rank-sum test with $\alpha = 0.05$ to test the null hypothesis that the two athletes carry the same amount of luggage on the average against the alternative hypothesis that the average weights of luggage for the two athletes are different.

16.21 The following data represent the operating times in hours for three types of scientific pocket calculators before a recharge is required:

Calculator								
A			B			C		
4.9	6.1	4.3	5.5	5.4	6.2	6.4	6.8	5.6
4.6	5.2		5.8	5.5	5.2	6.5	6.3	6.6
			4.8					

Use the Kruskal-Wallis test, at the 0.01 level of significance, to test the hypothesis that the operating times for all three calculators are equal.

16.22 In Exercise 13.6 on page 519, use the Kruskal-Wallis test at the 0.05 level of significance to determine if the organic chemical solvents differ significantly in sorption rate.

16.5 Runs Test

In applying the many statistical concepts discussed throughout this book, it was always assumed that the sample data had been collected by some randomization procedure. The **runs test**, based on the order in which the sample observations are obtained, is a useful technique for testing the null hypothesis H_0 that the observations have indeed been drawn at random.

To illustrate the runs test, let us suppose that 12 people are polled to find out if they use a certain product. We would seriously question the assumed randomness of the sample if all 12 people were of the same sex. We shall designate a male and a female by the symbols M and F, respectively, and record the outcomes according to their sex in the order in which they occur. A typical sequence for the experiment might be

$$\underbrace{M\ M}\ \underbrace{F\ F\ F}\ \underbrace{M}\ \underbrace{F\ F}\ \underbrace{M\ M\ M\ M},$$

where we have grouped subsequences of identical symbols. Such groupings are called **runs**.

Definition 16.1: A **run** is a subsequence of one or more identical symbols representing a common property of the data.

Regardless of whether the sample measurements represent qualitative or quantitative data, the runs test divides the data into two mutually exclusive categories: male or female; defective or nondefective; heads or tails; above or below the median; and so forth. Consequently, a sequence will always be limited to two distinct symbols. Let n_1 be the number of symbols associated with the category that occurs the least and n_2 be the number of symbols that belong to the other category. Then the sample size $n = n_1 + n_2$.

For the $n = 12$ symbols in our poll, we have five runs, with the first containing two M's, the second containing three F's, and so on. If the number of runs is larger or smaller than what we would expect by chance, the hypothesis that the sample was drawn at random should be rejected. Certainly, a sample resulting in only two runs,

$$M\ M\ M\ M\ M\ M\ M\ F\ F\ F\ F\ F$$

or the reverse, is most unlikely to occur from a random selection process. Such a result indicates that the first 7 people interviewed were all males, followed by 5 females. Likewise, if the sample resulted in the maximum number of 12 runs, as in the alternating sequence

$$M\ F\ M\ F\ M\ F\ M\ F\ M\ F\ M\ F,$$

we would again be suspicious of the order in which the individuals were selected for the poll.

The runs test for randomness is based on the random variable V, the total number of runs that occur in the complete sequence of the experiment. In Table A.18, values of $P(V \leq v^*$ when H_0 is true) are given for $v^* = 2, 3, \ldots, 20$ runs and values of n_1 and n_2 less than or equal to 10. The P-values for both one-tailed and two-tailed tests can be obtained using these tabled values.

For the poll taken previously, we exhibit a total of 5 F's and 7 M's. Hence, with $n_1 = 5$, $n_2 = 7$, and $v = 5$, we note from Table A.18 that the P-value for a two-tailed test is

$$P = 2P(V \leq 5 \text{ when } H_0 \text{ is true}) = 0.394 > 0.05.$$

That is, the value $v = 5$ is reasonable at the 0.05 level of significance when H_0 is true, and therefore we have insufficient evidence to reject the hypothesis of randomness in our sample.

When the number of runs is large (for example, if $v = 11$ while $n_1 = 5$ and $n_2 = 7$), the P-value for a two-tailed test is

$$P = 2P(V \geq 11 \text{ when } H_0 \text{ is true}) = 2[1 - P(V \leq 10 \text{ when } H_0 \text{ is true})]$$
$$= 2(1 - 0.992) = 0.016 < 0.05,$$

which leads us to reject the hypothesis that the sample values occurred at random.

The runs test can also be used to detect departures from randomness of a sequence of quantitative measurements over time, caused by trends or periodicities. Replacing each measurement, in the order in which it was collected, by a *plus* symbol if it falls above the median or by a *minus* symbol if it falls below the median and omitting all measurements that are exactly equal to the median, we generate a sequence of plus and minus symbols that is tested for randomness as illustrated in the following example.

Example 16.7: A machine dispenses acrylic paint thinner into containers. Would you say that the amount of paint thinner being dispensed by this machine varies randomly if the contents of the next 15 containers are measured and found to be 3.6, 3.9, 4.1, 3.6, 3.8, 3.7, 3.4, 4.0, 3.8, 4.1, 3.9, 4.0, 3.8, 4.2, and 4.1 liters? Use a 0.1 level of significance.

Solution: 1. H_0: Sequence is random.

2. H_1: Sequence is not random.

3. $\alpha = 0.1$.

4. Test statistic: V, the total number of runs.

5. Computations: For the given sample, we find $\tilde{x} = 3.9$. Replacing each measurement by the symbol "+" if it falls above 3.9 or by the symbol "−" if it falls below 3.9 and omitting the two measurements that equal 3.9, we obtain the sequence

$$- + - - - - + - + + - + +$$

for which $n_1 = 6$, $n_2 = 7$, and $v = 8$. Therefore, from Table A.18, the computed P-value is

$$P = 2P(V \geq 8 \text{ when } H_0 \text{ is true})$$
$$= 2[1 - P(V \leq 8 \text{ when } H_0 \text{ is true})] = 2(0.5) = 1.$$

6. Decision: Do not reject the hypothesis that the sequence of measurements varies randomly.

16.5 Runs Test

The runs test, although less powerful, can also be used as an alternative to the Wilcoxon two-sample test to test the claim that two random samples come from populations having the same distributions and therefore equal means. If the populations are symmetric, rejection of the claim of equal distributions is equivalent to accepting the alternative hypothesis that the means are not equal. In performing the test, we first combine the observations from both samples and arrange them in ascending order. Now assign the letter A to each observation taken from one of the populations and the letter B to each observation from the other population, thereby generating a sequence consisting of the symbols A and B. If observations from one population are tied with observations from the other population, the sequence of A and B symbols generated will not be unique and consequently the number of runs is unlikely to be unique. Procedures for breaking ties usually result in additional tedious computations, and for this reason we might prefer to apply the Wilcoxon rank-sum test whenever these situations occur.

To illustrate the use of runs in testing for equal means, consider the survival times of the leukemia patients of Exercise 16.16 on page 670, for which we have

$$\begin{array}{ccccccccc} 0.5 & 0.9 & 1.4 & 1.9 & 2.1 & 2.8 & 3.1 & 4.6 & 5.3 \\ B & A & A & B & A & B & B & A & A \end{array}$$

resulting in $v = 6$ runs. If the two symmetric populations have equal means, the observations from the two samples will be intermingled, resulting in many runs. However, if the population means are significantly different, we would expect most of the observations for one of the two samples to be smaller than those for the other sample. In the extreme case where the populations do not overlap, we would obtain a sequence of the form

$$A\ A\ A\ A\ B\ B\ B\ B \quad \text{or} \quad B\ B\ B\ B\ A\ A\ A\ A$$

and in either case there would be only two runs. Consequently, the hypothesis of equal population means will be rejected at the α-level of significance only when v is small enough so that

$$P = P(V \le v \text{ when } H_0 \text{ is true}) \le \alpha,$$

implying a one-tailed test.

Returning to the data of Exercise 16.16 on page 670, for which $n_1 = 4$, $n_2 = 5$, and $v = 6$, we find from Table A.18 that

$$P = P(V \le 6 \text{ when } H_0 \text{ is true}) = 0.786 > 0.05$$

and therefore fail to reject the null hypothesis of equal means. Hence, we conclude that the new serum does not prolong life by arresting leukemia.

When n_1 and n_2 increase in size, the sampling distribution of V approaches the normal distribution with mean and variance given by

$$\mu_V = \frac{2n_1 n_2}{n_1 + n_2} + 1 \quad \text{and} \quad \sigma_V^2 = \frac{2n_1 n_2 (2n_1 n_2 - n_1 - n_2)}{(n_1 + n_2)^2 (n_1 + n_2 - 1)}.$$

Consequently, when n_1 and n_2 are both greater than 10, we can use the statistic

$$Z = \frac{V - \mu_V}{\sigma_V}$$

to establish the critical region for the runs test.

16.6 Tolerance Limits

Tolerance limits for a normal distribution of measurements were discussed in Chapter 9. In this section, we consider a method for constructing tolerance intervals that is independent of the shape of the underlying distribution. As we might suspect, for a reasonable degree of confidence they will be substantially longer than those constructed assuming normality, and the sample size required is generally very large. Nonparametric tolerance limits are stated in terms of the smallest and largest observations in our sample.

Two-Sided Tolerance Limits For any distribution of measurements, two-sided tolerance limits are indicated by the smallest and largest observations in a sample of size n, where n is determined so that one can assert with $100(1-\gamma)\%$ confidence that **at least** the proportion $1-\alpha$ of the distribution is included between the sample extremes.

Table A.19 gives required sample sizes for selected values of γ and $1-\alpha$. For example, when $\gamma = 0.01$ and $1-\alpha = 0.95$, we must choose a random sample of size $n = 130$ in order to be 99% confident that at least 95% of the distribution of measurements is included between the sample extremes.

Instead of determining the sample size n such that a specified proportion of measurements is contained between the sample extremes, it is desirable in many industrial processes to determine the sample size such that a fixed proportion of the population falls below the largest (or above the smallest) observation in the sample. Such limits are called one-sided tolerance limits.

One-Sided Tolerance Limits For any distribution of measurements, a one-sided tolerance limit is determined by the smallest (largest) observation in a sample of size n, where n is determined so that one can assert with $100(1-\gamma)\%$ confidence that **at least** the proportion $1-\alpha$ of the distribution will exceed the smallest (be less than the largest) observation in the sample.

Table A.20 shows required sample sizes corresponding to selected values of γ and $1-\alpha$. Hence, when $\gamma = 0.05$ and $1-\alpha = 0.70$, we must choose a sample of size $n = 9$ in order to be 95% confident that 70% of our distribution of measurements will exceed the smallest observation in the sample.

16.7 Rank Correlation Coefficient

In Chapter 11, we used the sample correlation coefficient r to measure the population correlation coefficient ρ, the linear relationship between two continuous variables X and Y. If ranks $1, 2, \ldots, n$ are assigned to the x observations in order of magnitude and similarly to the y observations, and if these ranks are then substituted for the actual numerical values in the formula for the correlation coefficient in Chapter 11, we obtain the nonparametric counterpart of the conventional correlation coefficient. A correlation coefficient calculated in this manner is known as the **Spearman rank correlation coefficient** and is denoted by r_s. When there are no ties among either set of measurements, the formula for r_s reduces to a much simpler expression involving the differences d_i between the ranks assigned to the n pairs of x's and y's, which we now state.

16.7 Rank Correlation Coefficient

Rank Correlation Coefficient A nonparametric measure of association between two variables X and Y is given by the **rank correlation coefficient**

$$r_s = 1 - \frac{6}{n(n^2-1)} \sum_{i=1}^{n} d_i^2,$$

where d_i is the difference between the ranks assigned to x_i and y_i and n is the number of pairs of data.

In practice, the preceding formula is also used when there are ties among either the x or y observations. The ranks for tied observations are assigned as in the signed-rank test by averaging the ranks that would have been assigned if the observations were distinguishable.

The value of r_s will usually be close to the value obtained by finding r based on numerical measurements and is interpreted in much the same way. As before, the value of r_s will range from -1 to $+1$. A value of $+1$ or -1 indicates perfect association between X and Y, the plus sign occurring for identical rankings and the minus sign occurring for reverse rankings. When r_s is close to zero, we conclude that the variables are uncorrelated.

Example 16.8: The figures listed in Table 16.7, released by the Federal Trade Commission, show the milligrams of tar and nicotine found in 10 brands of cigarettes. Calculate the rank correlation coefficient to measure the degree of relationship between tar and nicotine content in cigarettes.

Table 16.7: Tar and Nicotine Contents

Cigarette Brand	Tar Content	Nicotine Content
Viceroy	14	0.9
Marlboro	17	1.1
Chesterfield	28	1.6
Kool	17	1.3
Kent	16	1.0
Raleigh	13	0.8
Old Gold	24	1.5
Philip Morris	25	1.4
Oasis	18	1.2
Players	31	2.0

Solution: Let X and Y represent the tar and nicotine contents, respectively. First we assign ranks to each set of measurements, with the rank of 1 assigned to the lowest number in each set, the rank of 2 to the second lowest number in each set, and so forth, until the rank of 10 is assigned to the largest number. Table 16.8 shows the individual rankings of the measurements and the differences in ranks for the 10 pairs of observations.

Table 16.8: Rankings for Tar and Nicotine Content

Cigarette Brand	x_i	y_i	d_i
Viceroy	2.0	2.0	0.0
Marlboro	4.5	4.0	0.5
Chesterfield	9.0	9.0	0.0
Kool	4.5	6.0	−1.5
Kent	3.0	3.0	0.0
Raleigh	1.0	1.0	0.0
Old Gold	7.0	8.0	−1.0
Philip Morris	8.0	7.0	1.0
Oasis	6.0	5.0	1.0
Players	10.0	10.0	0.0

Substituting into the formula for r_s, we find that

$$r_s = 1 - \frac{(6)(5.50)}{(10)(100-1)} = 0.967,$$

indicating a high positive correlation between the amounts of tar and nicotine found in cigarettes.

Some advantages to using r_s rather than r do exist. For instance, we no longer assume the underlying relationship between X and Y to be linear and therefore, when the data possess a distinct curvilinear relationship, the rank correlation coefficient will likely be more reliable than the conventional measure. A second advantage to using the rank correlation coefficient is the fact that no assumptions of normality are made concerning the distributions of X and Y. Perhaps the greatest advantage occurs when we are unable to make meaningful numerical measurements but nevertheless can establish rankings. Such is the case, for example, when different judges rank a group of individuals according to some attribute. The rank correlation coefficient can be used in this situation as a measure of the consistency of the two judges.

To test the hypothesis that $\rho = 0$ by using a rank correlation coefficient, one needs to consider the sampling distribution of the r_s-values under the assumption of no correlation. Critical values for $\alpha = 0.05, 0.025, 0.01$, and 0.005 have been calculated and appear in Table A.21. The setup of this table is similar to that of the table of critical values for the t-distribution except for the left column, which now gives the number of pairs of observations rather than the degrees of freedom. Since the distribution of the r_s-values is symmetric about zero when $\rho = 0$, the r_s-value that leaves an area of α to the left is equal to the negative of the r_s-value that leaves an area of α to the right. For a two-sided alternative hypothesis, the critical region of size α falls equally in the two tails of the distribution. For a test in which the alternative hypothesis is negative, the critical region is entirely in the left tail of the distribution, and when the alternative is positive, the critical region is placed entirely in the right tail.

Exercises

Example 16.9: Refer to Example 16.8 and test the hypothesis that the correlation between the amounts of tar and nicotine found in cigarettes is zero against the alternative that it is greater than zero. Use a 0.01 level of significance.

Solution:
1. H_0: $\rho = 0$.
2. H_1: $\rho > 0$.
3. $\alpha = 0.01$.
4. Critical region: $r_s > 0.745$ from Table A.21.
5. Computations: From Example 16.8, $r_s = 0.967$.
6. Decision: Reject H_0 and conclude that there is a significant correlation between the amounts of tar and nicotine found in cigarettes.

Under the assumption of no correlation, it can be shown that the distribution of the r_s-values approaches a normal distribution with a mean of 0 and a standard deviation of $1/\sqrt{n-1}$ as n increases. Consequently, when n exceeds the values given in Table A.21, one can test for a significant correlation by computing

$$z = \frac{r_s - 0}{1/\sqrt{n-1}} = r_s\sqrt{n-1}$$

and comparing with critical values of the standard normal distribution shown in Table A.3.

Exercises

16.23 A random sample of 15 adults living in a small town were selected to estimate the proportion of voters favoring a certain candidate for mayor. Each individual was also asked if he or she was a college graduate. By letting Y and N designate the responses of "yes" and "no" to the education question, the following sequence was obtained:

N N N N Y Y N Y Y N Y N N N N

Use the runs test at the 0.1 level of significance to determine if the sequence supports the contention that the sample was selected at random.

16.24 A silver-plating process is used to coat a certain type of serving tray. When the process is in control, the thickness of the silver on the trays will vary randomly following a normal distribution with a mean of 0.02 millimeter and a standard deviation of 0.005 millimeter. Suppose that the next 12 trays examined show the following thicknesses of silver: 0.019, 0.021, 0.020, 0.019, 0.020, 0.018, 0.023, 0.021, 0.024, 0.022, 0.023, 0.022. Use the runs test to determine if the fluctuations in thickness from one tray to another are random. Let $\alpha = 0.05$.

16.25 Use the runs test to test, at level 0.01, whether there is a difference in the average operating time for the two calculators of Exercise 16.17 on page 670.

16.26 In an industrial production line, items are inspected periodically for defectives. The following is a sequence of defective items, D, and nondefective items, N, produced by this production line:

D D N N N D N N D D N N N N
N D D D N N D N N N N D N D

Use the large-sample theory for the runs test, with a significance level of 0.05, to determine whether the defectives are occurring at random.

16.27 Assuming that the measurements of Exercise 1.14 on page 30 were recorded successively from left to right as they were collected, use the runs test, with $\alpha = 0.05$, to test the hypothesis that the data represent a random sequence.

16.28 How large a sample is required to be 95% confident that at least 85% of the distribution of measurements is included between the sample extremes?

16.29 What is the probability that the range of a random sample of size 24 includes at least 90% of the population?

16.30 How large a sample is required to be 99% confident that at least 80% of the population will be less than the largest observation in the sample?

16.31 What is the probability that at least 95% of a population will exceed the smallest value in a random sample of size $n = 135$?

16.32 The following table gives the recorded grades for 10 students on a midterm test and the final examination in a calculus course:

Student	Midterm Test	Final Examination
L.S.A.	84	73
W.P.B.	98	63
R.W.K.	91	87
J.R.L.	72	66
J.K.L.	86	78
D.L.P.	93	78
B.L.P.	80	91
D.W.M.	0	0
M.N.M.	92	88
R.H.S.	87	77

(a) Calculate the rank correlation coefficient.
(b) Test the null hypothesis that $\rho = 0$ against the alternative that $\rho > 0$. Use $\alpha = 0.025$.

16.33 With reference to the data of Exercise 11.1 on page 398,
(a) calculate the rank correlation coefficient;
(b) test the null hypothesis, at the 0.05 level of significance, that $\rho = 0$ against the alternative that $\rho \neq 0$. Compare your results with those obtained in Exercise 11.44 on page 435.

16.34 Calculate the rank correlation coefficient for the daily rainfall and amount of particulate removed in Exercise 11.13 on page 400.

16.35 With reference to the weights and chest sizes of infants in Exercise 11.47 on page 436,
(a) calculate the rank correlation coefficient;

(b) test the hypothesis, at the 0.025 level of significance, that $\rho = 0$ against the alternative that $\rho > 0$.

16.36 A consumer panel tests nine brands of television for overall quality. The ranks assigned by the panel and the suggested retail prices are as follows:

Manufacturer	Panel Rating	Suggested Price
A	6	$480
B	9	395
C	2	575
D	8	550
E	5	510
F	1	545
G	7	400
H	4	465
I	3	420

Is there a significant relationship between the quality and the price of a television? Use a 0.05 level of significance.

16.37 Two judges at a college homecoming parade rank eight floats in the following order:

	Float							
	1	2	3	4	5	6	7	8
Judge A	5	8	4	3	6	2	7	1
Judge B	7	5	4	2	8	1	6	3

(a) Calculate the rank correlation coefficient.
(b) Test the null hypothesis that $\rho = 0$ against the alternative that $\rho > 0$. Use $\alpha = 0.05$.

16.38 In the article called "Risky Assumptions" by Paul Slovic, Baruch Fischoff, and Sarah Lichtenstein, published in *Psychology Today* (June 1980), the risk of dying in the United States from 30 activities and technologies is ranked by members of the League of Women Voters and also by experts who are professionally involved in assessing risks. The rankings are as shown in Table 16.9.
(a) Calculate the rank correlation coefficient.
(b) Test the null hypothesis of zero correlation between the rankings of the League of Women Voters and the experts against the alternative that the correlation is not zero. Use a 0.05 level of significance.

Table 16.9: The Ranking Data for Exercise 16.38

Activity or Technology Risk	Voters	Experts	Activity or Technology Risk	Voters	Experts
Nuclear power	1	20	Motor vehicles	2	1
Handguns	3	4	Smoking	4	2
Motorcycles	5	6	Alcoholic beverages	6	3
Private aviation	7	12	Police work	8	17
Pesticides	9	8	Surgery	10	5
Fire fighting	11	18	Large construction	12	13
Hunting	13	23	Spray cans	14	26
Mountain climing	15	29	Bicycles	16	15
Commercial aviation	17	16	Electric power	18	9
Swimming	19	10	Contraceptives	20	11
Skiing	21	30	X-rays	22	7
Football	23	27	Railroads	24	19
Food preservatives	25	14	Food coloring	26	21
Power mowers	27	28	Antibiotics	28	24
Home appliances	29	22	Vaccinations	30	25

Review Exercises

16.39 A study by a chemical company compared the drainage properties of two different polymers. Ten different sludges were used, and both polymers were allowed to drain in each sludge. The free drainage was measured in mL/min.

Sludge Type	Polymer A	Polymer B
1	12.7	12.0
2	14.6	15.0
3	18.6	19.2
4	17.5	17.3
5	11.8	12.2
6	16.9	16.6
7	19.9	20.1
8	17.6	17.6
9	15.6	16.0
10	16.0	16.1

(a) Use the sign test at the 0.05 level to test the null hypothesis that polymer A has the same median drainage as polymer B.

(b) Use the signed-rank test to test the hypotheses of part (a).

16.40 In Review Exercise 13.45 on page 555, use the Kruskal-Wallis test, at the 0.05 level of significance, to determine if the chemical analyses performed by the four laboratories give, on average, the same results.

16.41 Use the data from Exercise 13.14 on page 530 to see if the median amount of nitrogen lost in perspiration is different for the three levels of dietary protein.

Chapter 17

Statistical Quality Control

17.1 Introduction

The notion of using sampling and statistical analysis techniques in a production setting had its beginning in the 1920s. The objective of this highly successful concept is the systematic reduction of variability and the accompanying isolation of sources of difficulties *during production*. In 1924, Walter A. Shewhart of the Bell Telephone Laboratories developed the concept of a control chart. However, it was not until World War II that the use of control charts became widespread. This was due to the importance of maintaining quality in production processes during that period. In the 1950s and 1960s, the development of quality control and the general area of quality assurance grew rapidly, particularly with the emergence of the space program in the United States. There has been widespread and successful use of quality control in Japan thanks to the efforts of W. Edwards Deming, who served as a consultant in Japan following World War II. Quality control has been, and is, an important ingredient in the development of Japan's industry and economy.

Quality control is receiving increasing attention as a management tool in which important characteristics of a product are observed, assessed, and compared with some type of standard. The various procedures in quality control involve considerable use of sampling procedures and statistical principles that have been presented in previous chapters. The primary users of quality control are, of course, industrial corporations. It has become clear that an effective quality control program enhances the quality of the product being produced and increases profits. This is particularly true today since products are produced in such high volume. Before the movement toward quality control methods, quality often suffered because of lack of efficiency, which, of course, increases cost.

The Control Chart

The purpose of a control chart is to determine if the performance of a process is maintaining an acceptable level of quality. It is expected, of course, that any process will experience natural variability, that is, variability due to essentially unimportant and uncontrollable sources of variation. On the other hand, a process may experience more serious types of variability in key performance measures.

These sources of variability may arise from one of several types of nonrandom "assignable causes," such as operator errors or improperly adjusted dials on a machine. A process operating in this state is called **out of control**. A process experiencing only chance variation is said to be in **statistical control**. Of course, a successful production process may operate in an in-control state for a long period. It is presumed that during this period, the process is producing an acceptable product. However, there may be either a gradual or a sudden "shift" that requires detection.

A control chart is intended as a device to detect the nonrandom or out-of-control state of a process. Typically, the control chart takes the form indicated in Figure 17.1. It is important that the shift be detected quickly so that the problem can be corrected. Obviously, if detection is slow, many defective or nonconforming items are produced, resulting in considerable waste and increased cost.

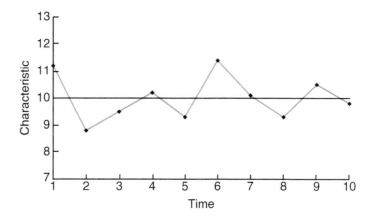

Figure 17.1: Typical control chart.

Some type of quality characteristic must be under consideration, and units of the process must be sampled over time. Say, for example, the characteristic is the circumference of an engine bearing. The centerline represents the average value of the characteristic when the process is in control. The points depicted in the figure represent results of, say, sample averages of this characteristic, with the samples taken over time. The upper control limit and the lower control limit are chosen in such a way that one would expect all sample points to be covered by these boundaries if the process is in control. As a result, the general complexion of the plotted points over time determines whether or not the process is concluded to be in control. The "in control" evidence is produced by a random pattern of points, with all plotted values being inside the control limits. When a point falls outside the control limits, this is taken to be evidence of a process that is out of control, and a search for the assignable cause is suggested. In addition, a nonrandom pattern of points may be considered suspicious and certainly an indication that an investigation for the appropriate corrective action is needed.

17.2 Nature of the Control Limits

The fundamental ideas on which control charts are based are similar in structure to those of hypothesis testing. Control limits are established to control the probability of making the error of concluding that the process is out of control when in fact it is not. This corresponds to the probability of making a type I error if we were testing the null hypothesis that the process is in control. On the other hand, we must be attentive to an error of the second kind, namely, not finding the process out of control when in fact it is (type II error). Thus, the choice of control limits is similar to the choice of a critical region.

As in the case of hypothesis testing, the sample size at each point is important. The choice of sample size depends to a large extent on the sensitivity or power of detection of the out-of-control state. In this application, the notion of *power* is very similar to that of the hypothesis-testing situation. Clearly, the larger the sample at each time period, the quicker the detection of an out-of-control process. In a sense, the control limits actually define what the user considers as being *in control*. In other words, the latitude given by the control limits must depend in some sense on the process variability. As a result, the computation of the control limits will naturally depend on data taken from the process results. Thus, any quality control application must have its beginning with computation from a preliminary sample or set of samples which will establish both the centerline and the quality control limits.

17.3 Purposes of the Control Chart

One obvious purpose of the control chart is mere surveillance of the process, that is, to determine if changes need to be made. In addition, the constant systematic gathering of data often allows management to assess process capability. Clearly, if a single performance characteristic is important, continual sampling and estimation of the mean and standard deviation of that performance characteristic provide an update on what the process can do in terms of mean performance and random variation. This is valuable even if the process stays in control for long periods. The systematic and formal structure of the control chart can often prevent overreaction to changes that represent only random fluctuations. Obviously, in many situations, changes brought about by overreaction can create serious problems that are difficult to solve.

Quality characteristics of control charts fall generally into *two* categories, **variables** and **attributes**. As a result, types of control charts often take the same classifications. In the case of the variables type of chart, the characteristic is usually a measurement on a continuum, such as diameter or weight. For the attribute chart, the characteristic reflects whether the individual product *conforms* (defective or not). Applications for these two distinct situations are obvious.

In the case of the variables chart, control must be exerted on both central tendency and variability. A quality control analyst must be concerned about whether there has been a shift in values of the performance characteristic *on average*. In addition, there will always be a concern about whether some change in process conditions results in a decrease in precision (i.e., an increase in variability). Separate control charts are essential for dealing with these two concepts. Central tendency is controlled by the $\bar{X}$-*chart*, where means of relatively small samples are plotted on a control chart. Variability around the mean is controlled by the *range* in the sample, or the sample *standard deviation*. In the case of attribute sampling, the *proportion defective* from a sample is often the quantity plotted on the chart. In the following section, we discuss the development of control charts for the variables type of performance characteristic.

17.4 Control Charts for Variables

Providing an example is a relatively easy way to explain the rudiments of the $\bar{X}$-chart for variables. Suppose that quality control charts are to be used on a process for manufacturing a certain engine part. Suppose the process mean is $\mu = 50$ mm and the standard deviation is $\sigma = 0.01$ mm. Suppose that groups of 5 are sampled every hour and the values of the *sample mean* $\bar{X}$ are recorded and plotted on a chart like the one in Figure 17.2. The limits for the $\bar{X}$-charts are based on the standard deviation of the random variable $\bar{X}$. We know from material in Chapter 8 that for the average of independent observations in a sample of size n,

$$\sigma_{\bar{X}} = \frac{\sigma}{\sqrt{n}},$$

where σ is the standard deviation of an individual observation. The control limits are designed to result in a small probability that a given value of $\bar{X}$ is outside the limits given that, indeed, the process is in control (i.e., $\mu = 50$). If we invoke the Central Limit Theorem, we have that under the condition that the process is in control,

$$\bar{X} \sim N\left(50, \frac{0.01}{\sqrt{5}}\right).$$

As a result, $100(1-\alpha)\%$ of the $\bar{X}$-values fall inside the limits when the process is in control if we use the limits

$$\text{LCL} = \mu - z_{\alpha/2}\frac{\sigma}{\sqrt{n}} = 50 - z_{\alpha/2}(0.0045), \qquad \text{UCL} = \mu + z_{\alpha/2}\frac{\sigma}{\sqrt{n}} = 50 + z_{\alpha/2}(0.0045).$$

Here LCL and UCL stand for lower control limit and upper control limit, respectively. Often the $\bar{X}$-charts are based on limits that are referred to as "three-sigma" limits, referring, of course, to $z_{\alpha/2} = 3$ and limits that become

$$\mu \pm 3\frac{\sigma}{\sqrt{n}}.$$

In our illustration, the upper and lower limits become

$$\text{LCL} = 50 - 3(0.0045) = 49.9865, \qquad \text{UCL} = 50 + 3(0.0045) = 50.0135.$$

Thus, if we view the structure of the 3σ limits from the point of view of hypothesis testing, for a given sample point, the probability is 0.0026 that the $\bar{X}$-value falls outside control limits, given that the process is in control. This is the probability

17.4 Control Charts for Variables

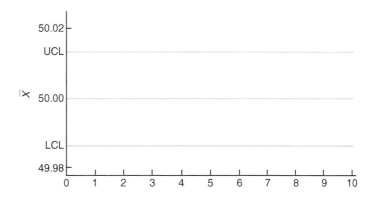

Figure 17.2: The 3σ control limits for the engine part example.

of the analyst *erroneously* determining that the process is out of control (see Table A.3).

The example above not only illustrates the $\bar{X}$-chart for variables, but also should provide the reader with insight into the nature of control charts in general. The centerline generally reflects the ideal value of an important parameter. Control limits are established from knowledge of the sampling properties of the statistic that estimates the parameter in question. They very often involve a multiple of the standard deviation of the statistic. It has become general practice to use 3σ limits. In the case of the $\bar{X}$-chart provided here, the Central Limit Theorem provides the user with a good approximation of the probability of falsely ruling that the process is out of control. In general, though, the user may not be able to rely on the normality of the statistic on the centerline. As a result, the exact probability of "type I error" may not be known. Despite this, it has become fairly standard to use the $k\sigma$ limits. While use of the 3σ limits is widespread, at times the user may wish to deviate from this approach. A smaller multiple of σ may be appropriate when it is important to quickly detect an out-of-control situation. Because of economic considerations, it may prove costly to allow a process to continue to run out of control for even short periods, while the cost of the search and correction of assignable causes may be relatively small. Clearly, in this case, control limits that are tighter than 3σ limits are appropriate.

Rational Subgroups

The sample values to be used in a quality control effort are divided into subgroups, with a *sample* representing a subgroup. As we indicated earlier, time order of production is certainly a natural basis for selection of the subgroups. We may view the quality control effort very simply as (1) sampling, (2) detection of an out-of-control state, and (3) a search for assignable causes that may be occurring over time. The selection of the basis for these sample groups would appear to be straightforward, but the choice of these subgroups of sampling information can have an important effect on the success of the quality control program. These subgroups are often called **rational subgroups**. Generally, if the analyst is interested in detecting a shift in location, the subgroups should be chosen so that within-subgroup variability is small and assignable causes, if they are present, have the greatest chance of being detected. Thus, we want to choose the subgroups in such a way as to maximize the between-subgroup variability. Choosing units in a subgroup that are produced close together in time, for example, is a reasonable approach. On the other hand, control charts are often used to control variability, in which case the performance statistic is *variability within the sample*. Thus, it is more important to choose the rational subgroups to maximize the within-sample variability. In this case, the observations in the subgroups should behave more like a random sample and the variability within samples needs to be a depiction of the variability of the process.

It is important to note that control charts on variability should be established before the development of charts on center of location (say, $\bar{X}$-charts). Any control chart on center of location will certainly depend on variability. For example, we have seen an illustration of the central tendency chart and it depends on σ. In the sections that follow, an estimate of σ from the data will be discussed.

$\bar{X}$-Chart with Estimated Parameters

In the foregoing, we have illustrated notions of the $\bar{X}$-chart that make use of the Central Limit Theorem and employ *known* values of the process mean and standard deviation. As we indicated earlier, the control limits

$$\text{LCL} = \mu - z_{\alpha/2}\frac{\sigma}{\sqrt{n}}, \qquad \text{UCL} = \mu + z_{\alpha/2}\frac{\sigma}{\sqrt{n}}$$

are used, and an $\bar{X}$-value falling outside these limits is viewed as evidence that the mean μ has changed and thus the process may be out of control.

In many practical situations, it is unreasonable to assume that we know μ and σ. As a result, estimates must be supplied from data taken when the process is in control. Typically, the estimates are determined during a period in which *background information* or *start-up information* is gathered. A basis for rational subgroups is chosen, and data are gathered with samples of size n in each subgroup. The sample sizes are usually small, say 4, 5, or 6, and k samples are taken, with k being at least 20. During this period in which it is assumed that the process is in control, the user establishes estimates of μ and σ on which the control chart is based. The important information gathered during this period includes the sample means in the subgroup, the overall mean, and the sample range in each subgroup. In the following paragraphs, we outline how this information is used to develop the control chart.

A portion of the sample information from these k samples takes the form $\bar{X}_1, \bar{X}_2, \ldots, \bar{X}_k$, where the random variable $\bar{X}_i$ is the average of the values in the ith sample. Obviously, the overall average is the random variable

$$\bar{\bar{X}} = \frac{1}{k}\sum_{i=1}^{k} \bar{X}_i.$$

This is the appropriate estimator of the process mean and, as a result, is the centerline in the $\bar{X}$ control chart. In quality control applications, it is often convenient to estimate σ from the information related to the *ranges* in the samples rather than sample standard deviations. Let us define

$$R_i = X_{\max,i} - X_{\min,i}$$

as the range for the data in the ith sample. Here $X_{\max,i}$ and $X_{\min,i}$ are the largest and smallest observations, respectively, in the sample. The appropriate estimate of σ is a function of the average range

$$\bar{R} = \frac{1}{k}\sum_{i=1}^{k} R_i.$$

An estimate of σ, say $\hat{\sigma}$, is obtained by

$$\hat{\sigma} = \frac{\bar{R}}{d_2},$$

where d_2 is a constant depending on the sample size. Values of d_2 are shown in Table A.22.

Use of the range in producing an estimate of σ has roots in quality-control-type applications, particularly since the range was so easy to compute, compared to other variability estimates, in the era when efficient computation was still an issue. The assumption of normality of the individual observations is implicit in the $\bar{X}$-chart. Of course, the existence of the Central Limit Theorem is certainly helpful in this regard. Under the assumption of normality, we make use of a random variable called the relative range, given by

$$W = \frac{R}{\sigma}.$$

It turns out that the moments of W are simple functions of the sample size n (see the reference to Montgomery, 2000b, in the Bibliography). The expected value of W is often referred to as d_2. Thus, by taking the expected value of W above, we have

$$\frac{E(R)}{\sigma} = d_2.$$

As a result, the rationale for the estimate $\hat{\sigma} = \bar{R}/d_2$ is readily understood. It is well known that the range method produces an efficient estimator of σ in relatively small samples. This makes the estimator particularly attractive in quality control applications, since the sample sizes in the subgroups are generally small. Using the range method for estimation of σ results in control charts with the following parameters:

$$\text{UCL} = \bar{\bar{X}} + \frac{3\bar{R}}{d_2\sqrt{n}}, \qquad \text{centerline} = \bar{\bar{X}}, \qquad \text{LCL} = \bar{\bar{X}} - \frac{3\bar{R}}{d_2\sqrt{n}}.$$

Defining the quantity

$$A_2 = \frac{3}{d_2\sqrt{n}},$$

we have that

$$\text{UCL} = \bar{\bar{X}} + A_2 \bar{R}, \qquad \text{LCL} = \bar{\bar{X}} - A_2 \bar{R}.$$

To simplify the structure, the user of $\bar{X}$-charts often finds values of A_2 tabulated. Values of A_2 are given for various sample sizes in Table A.22.

R-Charts to Control Variation

Up to this point, all illustrations and details have dealt with the quality control analysts' attempts at detection of out-of-control conditions produced by a *shift in the mean*. The control limits are based on the distribution of the random variable $\bar{X}$ and depend on the assumption of normality of the individual observations. It is important for control to be applied to variability as well as center of location. In fact, many experts believe that control of variability of the performance characteristic is more important and should be established before center of location is considered. Process variability can be controlled through the use of *plots of the sample range*. A plot over time of the sample ranges is called an **R-chart**. The same general structure can be used as in the case of the $\bar{X}$-chart, with $\bar{R}$ *being the centerline* and the control limits depending on an estimate of the standard deviation of the random variable R. Thus, as in the case of the $\bar{X}$-chart, 3σ limits are established where "3σ" implies $3\sigma_R$. The quantity σ_R must be estimated from the data just as $\sigma_{\bar{X}}$ is estimated.

The estimate of σ_R, the standard deviation, is also based on the distribution of the relative range

$$W = \frac{R}{\sigma}.$$

The standard deviation of W is a known function of the sample size and is generally denoted by d_3. As a result,

$$\sigma_R = \sigma d_3.$$

We can now replace σ by $\hat{\sigma} = \bar{R}/d_2$, and thus the estimator of σ_R is

$$\hat{\sigma}_R = \frac{\bar{R} d_3}{d_2}.$$

Thus, the quantities that define the R-chart are

$$\text{UCL} = \bar{R} D_4, \qquad \text{centerline} = \bar{R}, \qquad \text{LCL} = \bar{R} D_3,$$

where the constants D_4 and D_3 (depending only on n) are

$$D_4 = 1 + 3\frac{d_3}{d_2}, \qquad D_3 = 1 - 3\frac{d_3}{d_2}.$$

The constants D_4 and D_3 are tabulated in Table A.22.

$\bar{X}$- and R-Charts for Variables

A process manufacturing missile component parts is being controlled, with the performance characteristic being the tensile strength in pounds per square inch. Samples of size 5 each are taken every hour and 25 samples are reported. The data are shown in Table 17.1.

Table 17.1: Sample Information on Tensile Strength Data

Sample Number	Observations	$\bar{X}_i$	R_i
1	1515 1518 1512 1498 1511	1510.8	20
2	1504 1511 1507 1499 1502	1504.6	12
3	1517 1513 1504 1521 1520	1515.0	17
4	1497 1503 1510 1508 1502	1504.0	13
5	1507 1502 1497 1509 1512	1505.4	15
6	1519 1522 1523 1517 1511	1518.4	12
7	1498 1497 1507 1511 1508	1504.2	14
8	1511 1518 1507 1503 1509	1509.6	15
9	1506 1503 1498 1508 1506	1504.2	10
10	1503 1506 1511 1501 1500	1504.2	11
11	1499 1503 1507 1503 1501	1502.6	8
12	1507 1503 1502 1500 1501	1502.6	7
13	1500 1506 1501 1498 1507	1502.4	9
14	1501 1509 1503 1508 1503	1504.8	8
15	1507 1508 1502 1509 1501	1505.4	8
16	1511 1509 1503 1510 1507	1508.0	8
17	1508 1511 1513 1509 1506	1509.4	7
18	1508 1509 1512 1515 1519	1512.6	11
19	1520 1517 1519 1522 1516	1518.8	6
20	1506 1511 1517 1516 1508	1511.6	11
21	1500 1498 1503 1504 1508	1502.6	10
22	1511 1514 1509 1508 1506	1509.6	8
23	1505 1508 1500 1509 1503	1505.0	9
24	1501 1498 1505 1502 1505	1502.2	7
25	1509 1511 1507 1500 1499	1505.2	12

As we indicated earlier, it is important initially to establish "in control" conditions on variability. The calculated centerline for the R-chart is

$$\bar{R} = \frac{1}{25} \sum_{i=1}^{25} R_i = 10.72.$$

We find from Table A.22 that for $n = 5$, $D_3 = 0$ and $D_4 = 2.114$. As a result, the control limits for the R-chart are

$$\text{LCL} = \bar{R} D_3 = (10.72)(0) = 0,$$
$$\text{UCL} = \bar{R} D_4 = (10.72)(2.114) = 22.6621.$$

The R-chart is shown in Figure 17.3. None of the plotted ranges fall outside the control limits. As a result, there is no indication of an out-of-control situation.

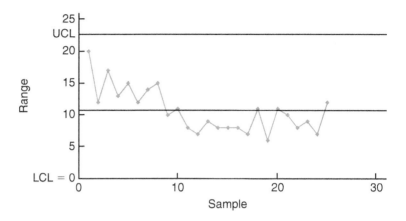

Figure 17.3: R-chart for the tensile strength example.

The $\bar{X}$-chart can now be constructed for the tensile strength readings. The centerline is

$$\bar{\bar{X}} = \frac{1}{25} \sum_{i=1}^{25} \bar{X}_i = 1507.328.$$

For samples of size 5, we find $A_2 = 0.577$ from Table A.22. Thus, the control limits are

$$\text{UCL} = \bar{\bar{X}} + A_2 \bar{R} = 1507.328 + (0.577)(10.72) = 1513.5134,$$
$$\text{LCL} = \bar{\bar{X}} - A_2 \bar{R} = 1507.328 - (0.577)(10.72) = 1501.1426.$$

The $\bar{X}$-chart is shown in Figure 17.4. As the reader can observe, three values fall outside the control limits. As a result, the control limits for $\bar{X}$ should not be used for line quality control.

Further Comments about Control Charts for Variables

A process may appear to be in control and, in fact, may stay in control for a long period. Does this necessarily mean that the process is operating successfully? A process that is operating *in control* is merely one in which the process mean and variability are stable. Apparently, no serious changes have occurred. "In control" implies that the process remains consistent with natural variability. Quality control charts may be viewed as a method in which the inherent natural variability governs the width of the control limits. There is no implication, however, to what extent an in-control process satisfies predetermined *specifications* required of the process. Specifications are limits that are established by the consumer. If the current natural

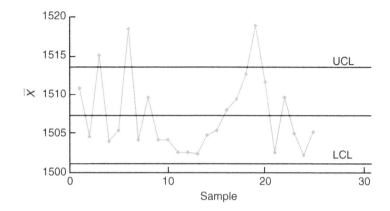

Figure 17.4: $\bar{X}$-chart for the tensile strength example.

variability of the process is larger than that dictated by the specifications, the process will not produce items that meet specifications with high frequency, even though the process is stable and in control.

We have alluded to the normality assumption on the individual observations in a variables control chart. For the $\bar{X}$-chart, if the individual observations are normal, the statistic $\bar{X}$ is normal. As a result, the quality control analyst has control over the probability of type I error in this case. If the individual X's are not normal, $\bar{X}$ is approximately normal and thus there is approximate control over the probability of type I error for the case in which σ is known. However, the use of the range method for estimating the standard deviation also depends on the normality assumption. Studies regarding the robustness of the $\bar{X}$-chart to departures from normality indicate that for samples of size $k \geq 4$ the $\bar{X}$ chart results in an α-risk close to that advertised (see the work by Montgomery, 2000b, and Schilling and Nelson, 1976, in the Bibliography). We indicated earlier that the $\pm k\sigma_R$ approach to the R-chart is a matter of convenience and tradition. Even if the distribution of individual observations is normal, the distribution of R is not normal. In fact, the distribution of R is not even symmetric. The symmetric control limits of $\pm k\sigma_R$ only give an approximation to the α-risk, and in some cases the approximation is not particularly good.

Choice of Sample Size (Operating Characteristic Function) in the Case of the $\bar{X}$-Chart

Scientists and engineers dealing in quality control often refer to factors that affect the *design of the control chart*. Components that determine the design of the chart include the sample size taken in each subgroup, the width of the control limits, and the frequency of sampling. All of these factors depend to a large extent on economic and practical considerations. Frequency of sampling obviously depends on the cost of sampling and the cost incurred if the process continues out of control for a long period. These same factors affect the width of the "in-control" region. The cost that is associated with investigation and search for assignable causes has an impact on the width of the region and on frequency of sampling. A considerable amount of attention has been devoted to optimal design of control charts, and extensive details will not be given here. The reader should refer to the work by Montgomery (2000b) cited in the Bibliography for an excellent historical account of much of this research.

Choice of sample size and frequency of sampling involves balancing available resources allocated to these two efforts. In many cases, the analyst may need to make changes in the strategy until the proper balance is achieved. The analyst should always be aware that if the cost of producing nonconforming items is great, a high sampling frequency with relatively small sample size is a proper strategy.

Many factors must be taken into consideration in the choice of a sample size. In the illustrations and discussion, we have emphasized the use of $n = 4$, 5, or 6. These values are considered relatively small for general problems in statistical inference but perhaps proper sample sizes for quality control. One justification, of course, is that quality control is a continuing process and the results produced by one sample or set of units will be followed by results from many more. Thus, the "effective" sample size of the entire quality control effort is many times larger than that used in a subgroup. It is generally considered to be more effective to *sample frequently* with a small sample size.

The analyst can make use of the notion of the power of a test to gain some insight into the effectiveness of the sample size chosen. This is particularly important since small sample sizes are usually used in each subgroup. Refer to Chapters 10 and 13 for a discussion of the power of formal tests on means and the analysis of variance. Although formal tests of hypotheses are not actually being conducted in quality control, one can treat the sampling information as if the strategy at each subgroup were to test a hypothesis, either on the population mean μ or on the standard deviation σ. Of interest is the *probability of detection* of an out-of-control condition for a given sample and, perhaps more important, the expected number of runs required for detection. The probability of detection of a specified out-of-control condition corresponds to the power of a test. It is not our intention to show development of the power for all of the types of control charts presented here, but rather to show the development for the $\bar{X}$-chart and present power results for the R-chart.

Consider the $\bar{X}$-chart for σ known. Suppose that the in-control state has $\mu = \mu_0$. A study of the role of the subgroup sample size is tantamount to investigating the β-risk, that is, the probability that an $\bar{X}$-value remains inside the control limits given that, indeed, a shift in the mean has occurred. Suppose that the form the shift takes is

$$\mu = \mu_0 + r\sigma.$$

Again, making use of the normality of $\bar{X}$, we have

$$\beta = P(\text{LCL} \leq \bar{X} \leq \text{UCL} \mid \mu = \mu_0 + r\sigma).$$

For the case of $k\sigma$ limits,

$$\text{LCL} = \mu_0 - \frac{k\sigma}{\sqrt{n}} \quad \text{and} \quad \text{UCL} = \mu_0 + \frac{k\sigma}{\sqrt{n}}.$$

17.4 Control Charts for Variables

As a result, if we denote by Z the standard normal random variable,

$$\beta = P\left[Z < \left(\frac{\mu_0 + k\sigma/\sqrt{n} - \mu}{\sigma/\sqrt{n}}\right)\right] - P\left[Z < \left(\frac{\mu_0 - k\sigma/\sqrt{n} - \mu}{\sigma/\sqrt{n}}\right)\right]$$

$$= P\left\{Z < \left[\frac{\mu_0 + k\sigma/\sqrt{n} - (\mu + r\sigma)}{\sigma/\sqrt{n}}\right]\right\} - P\left\{Z < \left[\frac{\mu_0 - k\sigma/\sqrt{n} - (\mu + r\sigma)}{\sigma/\sqrt{n}}\right]\right\}$$

$$= P(Z < k - r\sqrt{n}) - P(Z < -k - r\sqrt{n}).$$

Notice the role of n, r, and k in the expression for the β-risk. The probability of not detecting a specific shift clearly increases with an increase in k, as expected. β decreases with an increase in r, the magnitude of the shift, and decreases with an increase in the sample size n.

It should be emphasized that the expression above results in the β-risk (probability of type II error) for the case of a *single sample*. For example, suppose that in the case of a sample of size 4, a shift of σ occurs in the mean. The probability of detecting the shift (power) *in the first sample following the shift* is (assuming 3σ limits)

$$1 - \beta = 1 - [P(Z < 1) - P(Z < -5)] = 0.1587.$$

On the other hand, the probability of detecting a shift of 2σ is

$$1 - \beta = 1 - [P(Z < -1) - P(Z < -7)] = 0.8413.$$

The results above illustrate a fairly modest probability of detecting a shift of magnitude σ and a fairly high probability of detecting a shift of magnitude 2σ. The complete picture of how, say, 3σ control limits perform for the $\bar{X}$-chart described here is depicted in Figure 17.5. Rather than plotting the power functions, a plot is given of β against r, where the shift in the mean is of magnitude $r\sigma$. Of course, the sample sizes of $n = 4, 5, 6$ result in a small probability of detecting a shift of 1.0σ or even 1.5σ on the first sample after the shift.

But if sampling is done frequently, the probability may not be as important as the average or expected number of runs required before detection of the shift. Quick detection is important and is certainly possible even though the probability of detection on the first sample is not high. It turns out that $\bar{X}$-charts with these small samples will result in relatively rapid detection. If β is the probability of not detecting a shift on the first sample following the shift, then the probability of detecting the shift on the sth sample after the shift is (assuming independent samples)

$$P_s = (1 - \beta)\beta^{s-1}.$$

The reader should recognize this as an application of the geometric distribution. The average or expected value of the number of samples required for detection is

$$\sum_{s=1}^{\infty} s\beta^{s-1}(1 - \beta) = \frac{1}{1 - \beta}.$$

Thus, the expected number of samples required to detect the shift in the mean is the *reciprocal of the power* (i.e., the probability of detection on the first sample following the shift).

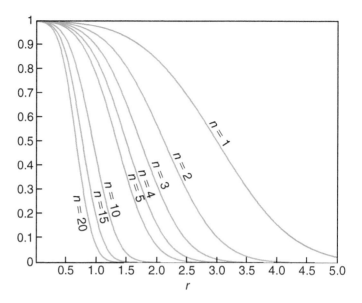

Figure 17.5: Operating characteristic curves for the $\bar{X}$-chart with 3σ limits. Here β is the type II probability error on the first sample after a shift in the mean of $r\sigma$.

Example 17.1: In a certain quality control effort, it is important for the quality control analyst to quickly detect shifts in the mean of $\pm\sigma$ while using a 3σ control chart with a sample size $n = 4$. The expected number of samples that are required following the shift for the detection of the out-of-control state can be an aid in the assessment of the quality control procedure.

From Figure 17.5, for $n = 4$ and $r = 1$, it can be seen that $\beta \approx 0.84$. If we allow s to denote the number of samples required to detect the shift, the mean of s is

$$E(s) = \frac{1}{1-\beta} = \frac{1}{0.16} = 6.25.$$

Thus, on the average, seven subgroups are required before detection of a shift of $\pm\sigma$.

Choice of Sample Size for the R-Chart

The OC curve for the R-chart is shown in Figure 17.6. Since the R-chart is used for control of the process standard deviation, the β-risk is plotted as a function of the in-control standard deviation, σ_0, and the standard deviation after the process goes out of control. The latter standard deviation will be denoted σ_1. Let

$$\lambda = \frac{\sigma_1}{\sigma_0}.$$

For various sample sizes, β is plotted against λ.

17.4 Control Charts for Variables

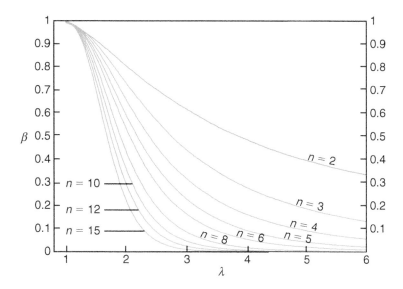

Figure 17.6: Operating characteristic curve for the R-charts with 3σ limits.

$\bar{X}$- and S-Charts for Variables

It is natural for the student of statistics to anticipate use of the sample variance in the $\bar{X}$-chart and in a chart to control variability. The range is efficient as an estimator for σ, but this efficiency decreases as the sample size gets larger. For n as large as 10, the familiar statistic

$$S = \sqrt{\frac{1}{n-1} \sum_{i=1}^{n} (X_i - \bar{X})^2}$$

should be used in the control chart for both the mean and the variability. The reader should recall from Chapter 9 that S^2 is an unbiased estimator for σ^2 but that S is not unbiased for σ. It has become customary to correct S for bias in control chart applications. We know, in general, that

$$E(S) \neq \sigma.$$

In the case in which the X_i are independent and normally distributed with mean μ and variance σ^2,

$$E(S) = c_4 \sigma, \quad \text{where} \quad c_4 = \left(\frac{2}{n-1}\right)^{1/2} \frac{\Gamma(n/2)}{\Gamma[(n-1)/2]}$$

and $\Gamma(\cdot)$ refers to the gamma function (see Chapter 6). For example, for $n = 5$, $c_4 = (3/8)\sqrt{2\pi}$. In addition, the variance of the estimator S is

$$\text{Var}(S) = \sigma^2 (1 - c_4^2).$$

We have established the properties of S that will allow us to write control limits for both $\bar{X}$ and S. To build a proper structure, we begin by assuming that σ is known. Later we discuss estimating σ from a preliminary set of samples.

If the statistic S is plotted, the obvious control chart parameters are

$$\text{UCL} = c_4\sigma + 3\sigma\sqrt{1 - c_4^2}, \quad \text{centerline} = c_4\sigma, \quad \text{LCL} = c_4\sigma - 3\sigma\sqrt{1 - c_4^2}.$$

As usual, the control limits are defined more succinctly through use of tabulated constants. Let

$$B_5 = c_4 - 3\sqrt{1 - c_4^2}, \qquad B_6 = c_4 + 3\sqrt{1 - c_4^2},$$

and thus we have

$$\text{UCL} = B_6\sigma, \quad \text{centerline} = c_4\sigma, \quad \text{LCL} = B_5\sigma.$$

The values of B_5 and B_6 for various sample sizes are tabulated in Table A.22.

Now, of course, the control limits above serve as a basis for the development of the quality control parameters for the situation that is most often seen in practice, namely, that in which σ is unknown. We must once again assume that a set of *base samples* or preliminary samples is taken to produce an estimate of σ during what is assumed to be an "in-control" period. Sample standard deviations $S_1, S_2, \ldots, S_m$ are obtained from samples that are each of size n. An unbiased estimator of the type

$$\frac{\bar{S}}{c_4} = \left(\frac{1}{m}\sum_{i=1}^{m} S_i\right) \Big/ c_4$$

is often used for σ. Here, of course, $\bar{S}$, the average value of the sample standard deviation in the preliminary sample, is the logical centerline in the control chart to control variability. The upper and lower control limits are unbiased estimators of the control limits that are appropriate for the case where σ is known. Since

$$E\left(\frac{\bar{S}}{c_4}\right) = \sigma,$$

the statistic $\bar{S}$ is an appropriate centerline (as an unbiased estimator of $c_4\sigma$) and the quantities

$$\bar{S} - 3\frac{\bar{S}}{c_4}\sqrt{1 - c_4^2} \quad \text{and} \quad \bar{S} + 3\frac{\bar{S}}{c_4}\sqrt{1 - c_4^2}$$

are the appropriate lower and upper 3σ control limits, respectively. As a result, the centerline and limits for the S-chart to control variability are

$$\text{LCL} = B_3\bar{S}, \quad \text{centerline} = \bar{S}, \quad \text{UCL} = B_4\bar{S},$$

where

$$B_3 = 1 - \frac{3}{c_4}\sqrt{1 - c_4^2}, \quad B_4 = 1 + \frac{3}{c_4}\sqrt{1 - c_4^2}.$$

17.5 Control Charts for Attributes

The constants B_3 and B_4 appear in Table A.22.

We can now write the parameters of the corresponding $\bar{X}$-chart involving the use of the sample standard deviation. Let us assume that S and $\bar{X}$ are available from the base preliminary sample. The centerline remains $\bar{\bar{X}}$ and the 3σ limits are merely of the form $\bar{\bar{X}} \pm 3\hat{\sigma}/\sqrt{n}$, where $\hat{\sigma}$ is an unbiased estimator. We simply supply $\bar{S}/c_4$ as an estimator for σ, and thus we have

$$\text{LCL} = \bar{\bar{X}} - A_3\bar{S}, \qquad \text{centerline} = \bar{\bar{X}}, \qquad \text{UCL} = \bar{\bar{X}} + A_3\bar{S},$$

where

$$A_3 = \frac{3}{c_4\sqrt{n}}.$$

The constant A_3 appears for various sample sizes in Table A.22.

Example 17.2: Containers are produced by a process where the volume of the containers is subject to quality control. Twenty-five samples of size 5 each were used to establish the quality control parameters. Information from these samples is documented in Table 17.2.

From Table A.22, $B_3 = 0$, $B_4 = 2.089$, and $A_3 = 1.427$. As a result, the control limits for $\bar{X}$ are given by

$$\text{UCL} = \bar{\bar{X}} + A_3\bar{S} = 62.3771, \qquad \text{LCL} = \bar{\bar{X}} - A_3\bar{S} = 62.2741,$$

and the control limits for the S-chart are

$$\text{LCL} = B_3\bar{S} = 0, \qquad \text{UCL} = B_4\bar{S} = 0.0754.$$

Figures 17.7 and 17.8 show the $\bar{X}$ and S control charts, respectively, for this example. Sample information for all 25 samples in the preliminary data set is plotted on the charts. Control seems to have been established after the first few samples.

17.5 Control Charts for Attributes

As we indicated earlier in this chapter, many industrial applications of quality control require that the quality characteristic indicate no more than that the item "conforms." In other words, there is no continuous measurement that is crucial to the performance of the item. An obvious illustration of this type of sampling, called **sampling for attributes**, is the performance of a light bulb, which either performs satisfactorily or does not. The item is either **defective or not defective**. Manufactured metal pieces may contain deformities. Containers from a production line may leak. In both of these cases, a defective item negates usage by the customer. The standard control chart for this situation is the p-chart, or *chart for fraction defective*. As we might expect, the probability distribution involved is the binomial distribution. The reader is referred to Chapter 5 for background on the binomial distribution.

Table 17.2: Volume of Containers for 25 Samples in a Preliminary Sample (in cubic centimeters)

Sample	Observations					$\bar{X}_i$	S_i
1	62.255	62.301	62.289	62.189	62.311	62.269	0.0495
2	62.187	62.225	62.337	62.297	62.307	62.271	0.0622
3	62.421	62.377	62.257	62.295	62.222	62.314	0.0829
4	62.301	62.315	62.293	62.317	62.409	62.327	0.0469
5	62.400	62.375	62.295	62.272	62.372	62.343	0.0558
6	62.372	62.275	62.315	62.372	62.302	62.327	0.0434
7	62.297	62.303	62.337	62.392	62.344	62.335	0.0381
8	62.325	62.362	62.351	62.371	62.397	62.361	0.0264
9	62.327	62.297	62.318	62.342	62.318	62.320	0.0163
10	62.297	62.325	62.303	62.307	62.333	62.313	0.0153
11	62.315	62.366	62.308	62.318	62.319	62.325	0.0232
12	62.297	62.322	62.344	62.342	62.313	62.324	0.0198
13	62.375	62.287	62.362	62.319	62.382	62.345	0.0406
14	62.317	62.321	62.297	62.372	62.319	62.325	0.0279
15	62.299	62.307	62.383	62.341	62.394	62.345	0.0431
16	62.308	62.319	62.344	62.319	62.378	62.334	0.0281
17	62.319	62.357	62.277	62.315	62.295	62.313	0.0300
18	62.333	62.362	62.292	62.327	62.314	62.326	0.0257
19	62.313	62.387	62.315	62.318	62.341	62.335	0.0313
20	62.375	62.321	62.354	62.342	62.375	62.353	0.0230
21	62.399	62.308	62.292	62.372	62.299	62.334	0.0483
22	62.309	62.403	62.318	62.295	62.317	62.328	0.0427
23	62.293	62.293	62.342	62.315	62.349	62.318	0.0264
24	62.388	62.308	62.315	62.392	62.303	62.341	0.0448
25	62.324	62.318	62.315	62.295	62.319	62.314	0.0111

$$\bar{\bar{X}} = 62.3256$$
$$\bar{S} = 0.0361$$

The p-Chart for Fraction Defective

Any manufactured item may have several characteristics that are important and should be examined by an inspector. However, the entire development here focuses on a single characteristic. Suppose that for all items the probability of a defective item is p, and that all items are being produced independently. Then, in a random sample of n items produced, allowing X to be the number of defective items, we have

$$P(X = x) = \binom{n}{x} p^x (1-p)^{n-x}, \qquad x = 0, 1, 2, \ldots, n.$$

As one might suspect, the mean and variance of the binomial random variable will play an important role in the development of the control chart. The reader should recall that

$$E(X) = np \qquad \text{and} \qquad \text{Var}(X) = np(1-p).$$

17.5 Control Charts for Attributes

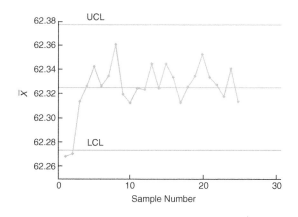

Figure 17.7: The $\bar{X}$-chart with control limits established by the data of Example 17.2.

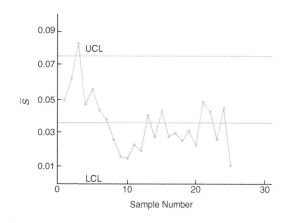

Figure 17.8: The S-chart with control limits established by the data of Example 17.2.

An unbiased estimator of p is the **fraction defective** or the **proportion defective**, $\hat{p}$, where

$$\hat{p} = \frac{\text{number of defectives in the sample of size } n}{n}.$$

As in the case of the variables control charts, the distributional properties of p are important in the development of the control chart. We know that

$$E(\hat{p}) = p, \qquad \text{Var}(\hat{p}) = \frac{p(1-p)}{n}.$$

Here we apply the same 3σ principles that we use for the variables charts. Let us assume initially that p is known. The structure, then, of the control charts involves the use of 3σ limits with

$$\hat{\sigma} = \sqrt{\frac{p(1-p)}{n}}.$$

Thus, the limits are

$$\text{LCL} = p - 3\sqrt{\frac{p(1-p)}{n}}, \qquad \text{UCL} = p + 3\sqrt{\frac{p(1-p)}{n}},$$

with the process considered in control when the $\hat{p}$-values from the sample lie inside the control limits.

Generally, of course, the value of p is not known and must be estimated from a base set of samples very much like the case of μ and σ in the variables charts. Assume that there are m preliminary samples of size n. For a given sample, each of the n observations is reported as either "defective" or "not defective." The obvious unbiased estimator for p to use in the control chart is

$$\bar{p} = \frac{1}{m}\sum_{i=1}^{m}\hat{p}_i,$$

where $\hat{p}_i$ is the proportion defective in the ith sample. As a result, the control limits are

$$\text{LCL} = \bar{p} - 3\sqrt{\frac{\bar{p}(1-\bar{p})}{n}}, \qquad \text{centerline} = \bar{p}, \qquad \text{UCL} = \bar{p} + 3\sqrt{\frac{\bar{p}(1-\bar{p})}{n}}.$$

Example 17.3: Consider the data shown in Table 17.3 on the number of defective electronic components in samples of size 50. Twenty samples were taken in order to establish preliminary control chart values. The control charts determined by this preliminary period will have centerline $\bar{p} = 0.088$ and control limits

$$\text{LCL} = \bar{p} - 3\sqrt{\frac{\bar{p}(1-\bar{p})}{50}} = -0.0322 \quad \text{and} \quad \text{UCL} = \bar{p} + 3\sqrt{\frac{\bar{p}(1-\bar{p})}{50}} = 0.2082.$$

Table 17.3: Data for Example 17.3 to Establish Control Limits for p-Charts, Samples of Size 50

Sample	Number of Defective Components	Fraction Defective $\hat{p}_i$
1	8	0.16
2	6	0.12
3	5	0.10
4	7	0.14
5	2	0.04
6	5	0.10
7	3	0.06
8	8	0.16
9	4	0.08
10	4	0.08
11	3	0.06
12	1	0.02
13	5	0.10
14	4	0.08
15	4	0.08
16	2	0.04
17	3	0.06
18	5	0.10
19	6	0.12
20	3	0.06
		$\bar{p} = 0.088$

Obviously, with a computed value that is negative, the LCL will be set to zero. It is apparent from the values of the control limits that the process is in control during this preliminary period.

Choice of Sample Size for the p-Chart

The choice of sample size for the p-chart for attributes involves the same general types of considerations as that of the chart for variables. A sample size is required

17.5 Control Charts for Attributes

that is sufficiently large to have a high probability of detection of an out-of-control condition when, in fact, a specified change in p has occurred. There is *no best method* for choice of sample size. However, one reasonable approach, suggested by Duncan (1986; see the Bibliography), is to choose n so that there is probability 0.5 that we detect a shift in p of a particular amount. The resulting solution for n is quite simple. Suppose that the normal approximation to the binomial distribution applies. We wish, under the condition that p has shifted to, say, $p_1 > p_0$, that

$$P(\hat{p} \geq \text{UCL}) = P\left[Z \geq \frac{\text{UCL} - p_1}{\sqrt{p_1(1-p_1)/n}}\right] = 0.5.$$

Since $P(Z > 0) = 0.5$, we set

$$\frac{\text{UCL} - p_1}{\sqrt{p_1(1-p_1)/n}} = 0.$$

Substituting

$$p + 3\sqrt{\frac{p(1-p)}{n}} = \text{UCL},$$

we have

$$(p - p_1) + 3\sqrt{\frac{p(1-p)}{n}} = 0.$$

We can now solve for n, the size of each sample:

$$n = \frac{9}{\Delta^2}p(1-p),$$

where, of course, Δ is the "shift" in the value of p, and p is the probability of a defective on which the control limits are based. However, if the control charts are based on $k\sigma$ limits, then

$$n = \frac{k^2}{\Delta^2}p(1-p).$$

Example 17.4: Suppose that an attribute quality control chart is being designed with a value of $p = 0.01$ for the in-control probability of a defective. What is the sample size per subgroup producing a probability of 0.5 that a process shift to $p = p_1 = 0.05$ will be detected? The resulting p-chart will involve 3σ limits.
Solution: Here we have $\Delta = 0.04$. The appropriate sample size is

$$n = \frac{9}{(0.04)^2}(0.01)(0.99) = 55.69 \approx 56.$$

Control Charts for Defects (Use of the Poisson Model)

In the preceding development, we assumed that the item under consideration is one that is either defective (i.e., nonfunctional) or not defective. In the latter case, it is functional and thus acceptable to the consumer. In many situations, this "defective or not" approach is too simplistic. Units may contain defects or nonconformities but still function quite well for the consumer. Indeed, in this case, it may be important to exert control on the *number of defects* or *number of nonconformities*. This type of quality control effort finds application when the units are either not simplistic or large. For example, the number of defects may be quite useful as the object of control when the single item or unit is, say, a personal computer. Another example is a unit defined by 50 feet of manufactured pipeline, where the number of defective welds is the object of quality control; the number of defects in 50 feet of manufactured carpeting; or the number of "bubbles" in a large manufactured sheet of glass.

It is clear from what we describe here that the binomial distribution is not appropriate. The total number of nonconformities in a unit or the average number per unit can be used as the measure for the control chart. Often it is assumed that the number of nonconformities in a sample of items follows the Poisson distribution. This type of chart is often called a **C-chart**.

Suppose that the number of defects X in one unit of product follows the Poisson distribution with parameter λ. (Here $t = 1$ for the Poisson model.) Recall that for the Poisson distribution,

$$P(X = x) = \frac{e^{-\lambda}\lambda^x}{x!}, \qquad x = 0, 1, 2, \ldots.$$

Here, the random variable X is the number of nonconformities. In Chapter 5, we learned that the mean and variance of the Poisson random variable are both λ. Thus, if the quality control chart were to be structured according to the usual 3σ limits, we could have, for λ known,

$$\text{UCL} = \lambda + 3\sqrt{\lambda}, \qquad \text{centerline} = \lambda, \qquad \text{LCL} = \lambda - 3\sqrt{\lambda}.$$

As usual, λ often must come from an estimator from the data. An unbiased estimate of λ is the *average* number of nonconformities per sample. Denote this estimate by $\hat{\lambda}$. Thus, the control chart has the limits

$$\text{UCL} = \hat{\lambda} + 3\sqrt{\hat{\lambda}}, \qquad \text{centerline} = \hat{\lambda}, \qquad \text{LCL} = \hat{\lambda} - 3\sqrt{\hat{\lambda}}.$$

Example 17.5: Table 17.4 represents the number of defects in 20 successive samples of sheet metal rolls each 100 feet long. A control chart is to be developed from these preliminary data for the purpose of controlling the number of defects in such samples. The estimate of the Poisson parameter λ is given by $\hat{\lambda} = 5.95$. As a result, the control limits suggested by these preliminary data are

$$\text{UCL} = \hat{\lambda} + 3\sqrt{\hat{\lambda}} = 13.2678 \quad \text{and} \quad \text{LCL} = \hat{\lambda} - 3\sqrt{\hat{\lambda}} = -1.3678,$$

with LCL being set to zero.

17.5 Control Charts for Attributes

Table 17.4: Data for Example 17.5; Control Involves Number of Defects in Sheet Metal Rolls

Sample Number	Number of Defects	Sample Number	Number of Defects
1	8	11	3
2	7	12	7
3	5	13	5
4	4	14	9
5	4	15	7
6	7	16	7
7	6	17	8
8	4	18	6
9	5	19	7
10	6	20	4
			Ave. 5.95

Figure 17.9 shows a plot of the preliminary data with the control limits revealed.

Table 17.5 shows additional data taken from the production process. For each sample, the unit on which the chart was based, namely 100 feet of the metal, was inspected. The information on 20 samples is included. Figure 17.10 shows a plot of the additional production data. It is clear that the process is in control, at least through the period for which the data were taken.

Table 17.5: Additional Data from the Production Process of Example 17.5

Sample Number	Number of Defects	Sample Number	Number of Defects
1	3	11	7
2	5	12	5
3	8	13	9
4	5	14	4
5	8	15	6
6	4	16	5
7	3	17	3
8	6	18	2
9	5	19	1
10	2	20	6

In Example 17.5, we have made very clear what the sampling or inspection unit is, namely, 100 feet of metal. In many cases where the item is a specific one (e.g., a personal computer or a specific type of electronic device), the inspection unit may be a *set of items*. For example, the analyst may decide to use 10 computers in each subgroup and observe a count of the total number of defects found. Thus, the preliminary sample for construction of the control chart would involve several samples, each containing 10 computers. The choice of the sample size may depend on many factors. Often, we may want a sample size that will ensure an LCL that is positive.

The analyst may wish to use the average number of defects per sampling unit as the basic measure in the control chart. For example, for the case of the personal

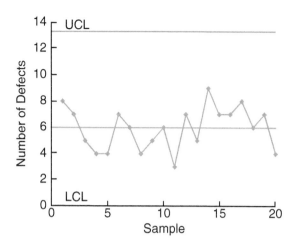
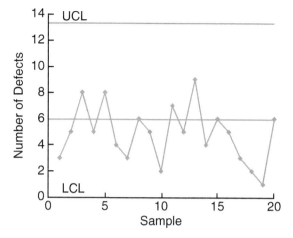

Figure 17.9: Preliminary data plotted on the control chart for Example 17.5.

Figure 17.10: Additional production data for Example 17.5.

computer, let the random variable total number of defects

$$U = \frac{\text{total number of defects}}{n}$$

be measured for each sample of, say, $n = 10$. We can use the method of moment-generating functions to show that U is a Poisson random variable (see Review Exercise 17.1) if we assume that the number of defects per sampling unit is Poisson with parameter λ. Thus, the control chart for this situation is characterized by the following:

$$\text{UCL} = \bar{U} + 3\sqrt{\frac{\bar{U}}{n}}, \qquad \text{centerline} = \bar{U}, \qquad \text{LCL} = \bar{U} - 3\sqrt{\frac{\bar{U}}{n}}.$$

Here, of course, $\bar{U}$ is the average of the U-values in the preliminary or base data set. The term $\bar{U}/n$ is derived from the result that

$$E(U) = \lambda, \qquad \text{Var}(U) = \frac{\lambda}{n},$$

and thus $\bar{U}$ is an unbiased estimate of $E(U) = \lambda$ and $\bar{U}/n$ is an unbiased estimate of $\text{Var}(U) = \lambda/n$. This type of control chart is often called a **U-chart**.

In this section, we based our entire development of control charts on the Poisson probability model. This model has been used in combination with the 3σ concept. As we implied earlier in this chapter, the notion of 3σ limits has its roots in the normal approximation, although many users feel that the concept works well as a pragmatic tool even if normality is not even approximately correct. The difficulty, of course, is that in the absence of normality, we cannot control the probability of incorrect specification of an out-of-control state. In the case of the Poisson model, when λ is small the distribution is quite asymmetric, a condition that may produce undesirable results if we hold to the 3σ approach.

17.6 Cusum Control Charts

The disadvantage of the Shewhart-type control charts, developed and illustrated in the preceding sections, lies in their inability to detect small changes in the mean. A quality control mechanism that has received considerable attention in the statistics literature and usage in industry is the **cumulative sum (cusum) chart**. The method for the cusum chart is simple and its appeal is intuitive. It should become obvious to the reader why it is more responsive to small changes in the mean. Consider a control chart for the mean with a reference level established at value W. Consider particular observations $X_1, X_2, \ldots, X_r$. The first r cusums are

$$S_1 = X_1 - W$$
$$S_2 = S_1 + (X_2 - W)$$
$$S_3 = S_2 + (X_3 - W)$$
$$\vdots$$
$$S_r = S_{r-1} + (X_r - W).$$

It becomes clear that the cusum is merely the accumulation of differences from the reference level. That is,

$$S_k = \sum_{i=1}^{k}(X_i - W), \quad k = 1, 2, \ldots.$$

The cusum chart is, then, a plot of S_k against time.

Suppose that we consider the reference level W to be an acceptable value of the mean μ. Clearly, if there is no shift in μ, the cusum chart should be approximately horizontal, with some minor fluctuations balanced around zero. Now, if there is only a moderate change in the mean, a relatively large change in the *slope* of the cusum chart should result, since each new observation has a chance of contributing a shift and the measure being plotted is accumulating these shifts. Of course, the signal that the mean has shifted lies in the nature of the slope of the cusum chart. The purpose of the chart is to detect changes that are moving away from the reference level. A nonzero slope (in either direction) represents a change away from the reference level. A positive slope indicates an increase in the mean above the reference level, while a negative slope signals a decrease.

Cusum charts are often devised with a defined *acceptable quality level* (AQL) and *rejectable quality level* (RQL) preestablished by the user. Both represent values of the mean. These may be viewed as playing roles somewhat similar to those of the null and alternative mean of hypothesis testing. Consider a situation where the analyst hopes to detect an increase in the value of the process mean. We shall use the notation μ_0 for AQL and μ_1 for RQL and let $\mu_1 > \mu_0$. The reference level is now set at

$$W = \frac{\mu_0 + \mu_1}{2}.$$

The values of S_r $(r = 1, 2, \ldots)$ will have a negative slope if the process mean is at μ_0 and a positive slope if the process mean is at μ_1.

Decision Rule for Cusum Charts

As indicated earlier, the slope of the cusum chart provides the signal for action by the quality control analyst. The decision rule calls for action if, at the rth sampling period,

$$d_r > h,$$

where h is a prespecified value called the **length of the decision interval** and

$$d_r = S_r - \min_{1 \leq i \leq r-1} S_i.$$

In other words, action is taken if the data reveal that the current cusum value exceeds by a specified amount the previous smallest cusum value.

A modification in the mechanics described above makes employing the method easier. We have described a procedure that plots the cusums and computes differences. A simple modification involves plotting the differences directly and allows for checking against the decision interval. The general expression for d_r is quite simple. For the cusum procedure where we are detecting increases in the mean,

$$d_r = \max[0, d_{r-1} + (X_r - W)].$$

The choice of the value of h is, of course, very important. We do not choose in this book to provide the many details in the literature dealing with this choice. The reader is referred to Ewan and Kemp, 1960, and Montgomery, 2000b (see the Bibliography) for a thorough discussion. One important consideration is the **expected run length**. Ideally, the expected run length is quite large under $\mu = \mu_0$ and quite small when $\mu = \mu_1$.

Review Exercises

17.1 Consider $X_1, X_2, \ldots, X_n$ independent Poisson random variables with parameters $\mu_1, \mu_2, \ldots, \mu_n$. Use the properties of moment-generating functions to show that the random variable $\sum_{i=1}^{n} X_i$ is a Poisson random variable with mean $\sum_{i=1}^{n} \mu_i$ and variance $\sum_{i=1}^{n} \mu_i$.

17.2 Consider the following data taken on subgroups of size 5. The data contain 20 averages and ranges on the diameter (in millimeters) of an important component part of an engine. Display $\bar{X}$- and R-charts. Does the process appear to be in control?

Sample	$\bar{X}$	R
1	2.3972	0.0052
2	2.4191	0.0117
3	2.4215	0.0062
4	2.3917	0.0089
5	2.4151	0.0095
6	2.4027	0.0101
7	2.3921	0.0091
8	2.4171	0.0059
9	2.3951	0.0068
10	2.4215	0.0048
11	2.3887	0.0082
12	2.4107	0.0032
13	2.4009	0.0077
14	2.3992	0.0107
15	2.3889	0.0025
16	2.4107	0.0138
17	2.4109	0.0037
18	2.3944	0.0052
19	2.3951	0.0038
20	2.4015	0.0017

17.3 Suppose for Review Exercise 17.2 that the buyer has set specifications for the part. The specifications require that the diameter fall in the range covered by 2.40000 ± 0.0100 mm. What proportion of units produced by this process will not conform to specifications?

17.4 For the situation of Review Exercise 17.2, give numerical estimates of the mean and standard devia-

Review Exercises

tion of the diameter for the part being manufactured in the process.

17.5 Consider the data of Table 17.1. Suppose that additional samples of size 5 are taken and tensile strength recorded. The sampling produces the following results (in pounds per square inch).

Sample	$\bar{X}$	R
1	1511	22
2	1508	14
3	1522	11
4	1488	18
5	1519	6
6	1524	11
7	1519	8
8	1504	7
9	1500	8
10	1519	14

(a) Plot the data, using the $\bar{X}$- and R-charts for the preliminary data of Table 17.1.
(b) Does the process appear to be in control? If not, explain why.

17.6 Consider an in-control process with mean $\mu = 25$ and $\sigma = 1.0$. Suppose that subgroups of size 5 are used with control limits $\mu \pm 3\sigma/\sqrt{n}$, and centerline at μ. Suppose that a shift occurs in the mean, and the new mean is $\mu = 26.5$.

(a) What is the average number of samples required (following the shift) to detect the out-of-control situation?
(b) What is the standard deviation of the number of runs required?

17.7 Consider the situation of Example 17.2. The following data are taken on additional samples of size 5. Plot the $\bar{X}$- and S-values on the $\bar{X}$- and S-charts that were produced with the data in the preliminary sample. Does the process appear to be in control? Explain why or why not.

Sample	$\bar{X}$	S_i
1	62.280	0.062
2	62.319	0.049
3	62.297	0.077
4	62.318	0.042
5	62.315	0.038
6	62.389	0.052
7	62.401	0.059
8	62.315	0.042
9	62.298	0.036
10	62.337	0.068

17.8 Samples of size 50 are taken every hour from a process producing a certain type of item that is considered either defective or not defective. Twenty samples are taken.

(a) Construct a control chart for control of proportion defective.
(b) Does the process appear to be in control? Explain.

Sample	Number of Defective Items	Sample	Number of Defective Items
1	4	11	2
2	3	12	4
3	5	13	1
4	3	14	2
5	2	15	3
6	2	16	1
7	2	17	1
8	1	18	2
9	4	19	3
10	3	20	1

17.9 For the situation of Review Exercise 17.8, suppose that additional data are collected as follows:

Sample	Number of Defective Items
1	3
2	4
3	2
4	2
5	3
6	1
7	3
8	5
9	7
10	7

Does the process appear to be in control? Explain.

17.10 A quality control effort is being undertaken for a process where large steel plates are manufactured and surface defects are of concern. The goal is to set up a quality control chart for the number of defects per plate. The data are given below. Set up the appropriate control chart, using this sample information. Does the process appear to be in control?

Sample	Number of Defects	Sample	Number of Defects
1	4	11	1
2	2	12	2
3	1	13	2
4	3	14	3
5	0	15	1
6	4	16	4
7	5	17	3
8	3	18	2
9	2	19	1
10	2	20	3

Chapter 18

Bayesian Statistics

18.1 Bayesian Concepts

The classical methods of estimation that we have studied in this text are based solely on information provided by the random sample. These methods essentially interpret probabilities as relative frequencies. For example, in arriving at a 95% confidence interval for μ, we interpret the statement

$$P(-1.96 < Z < 1.96) = 0.95$$

to mean that 95% of the time in repeated experiments Z will fall between -1.96 and 1.96. Since

$$Z = \frac{\bar{X} - \mu}{\sigma/\sqrt{n}}$$

for a normal sample with known variance, the probability statement here means that 95% of the random intervals $(\bar{X} - 1.96\sigma/\sqrt{n}, \bar{X} + 1.96\sigma/\sqrt{n})$ contain the true mean μ. Another approach to statistical methods of estimation is called **Bayesian methodology**. The main idea of the method comes from Bayes' rule, described in Section 2.7. The key difference between the Bayesian approach and the classical or frequentist approach is that in Bayesian concepts, the parameters are viewed as random variables.

Subjective Probability

Subjective probability is the foundation of Bayesian concepts. In Chapter 2, we discussed two possible approaches to probability, namely the relative frequency and the indifference approaches. The first one determines a probability as a consequence of repeated experiments. For instance, to decide the free-throw percentage of a basketball player, we can record the number of shots made and the total number of attempts this player has made. The probability of hitting a free-throw for this player can be calculated as the ratio of these two numbers. On the other hand, if we have no knowledge of any bias in a die, the probability that a 3 will appear in the next throw will be 1/6. Such an approach to probability interpretation is based on the indifference rule.

However, in many situations, the preceding probability interpretations cannot be applied. For instance, consider the questions "What is the probability that it will rain tomorrow?" "How likely is it that this stock will go up by the end of the month?" and "What is the likelihood that two companies will be merged together?" They can hardly be interpreted by the aforementioned approaches, and the answers to these questions may be different for different people. Yet these questions are constantly asked in daily life, and the approach used to explain these probabilities is called *subjective probability*, which reflects one's subjective opinion.

Conditional Perspective

Recall that in Chapters 9 through 17, all statistical inferences were based on the fact that the parameters are unknown but fixed quantities, apart from those in Section 9.14, in which the parameters were treated as variables and the maximum likelihood estimates (MLEs) were calculated conditioning on the observed sample data. In Bayesian statistics, not only are the parameters treated as variables as in MLE calculation, but also they are treated as random.

Because the observed data are the only experimental results for the practitioner, statistical inference is based on the actual observed data from a given experiment. Such a view is called a *conditional perspective*. Furthermore, in Bayesian concepts, since the parameters are treated as random, a probability distribution can be specified, generally by using the *subjective probability* for the parameter. Such a distribution is called a *prior distribution* and it usually reflects the experimenter's prior belief about the parameter. In the Bayesian perspective, once an experiment is conducted and data are observed, all knowledge about the parameter is contained in the actual observed data and in the prior information.

Bayesian Applications

Although Bayes' rule is credited to Thomas Bayes, Bayesian applications were first introduced by French scientist Pierre Simon Laplace, who published a paper on using Bayesian inference on the unknown binomial proportions (for binomial distribution, see Section 5.2).

Since the introduction of the Markov chain Monte Carlo (MCMC) computational tools for Bayesian analysis in the early 1990s, Bayesian statistics has become more and more popular in statistical modeling and data analysis. Meanwhile, methodology developments using Bayesian concepts have progressed dramatically, and they are applied in fields such as bioinformatics, biology, business, engineering, environmental and ecology science, life science and health, medicine, and many others.

18.2 Bayesian Inferences

Consider the problem of finding a point estimate of the parameter θ for the population with distribution $f(x|\theta)$, given θ. Denote by $\pi(\theta)$ the prior distribution of θ. Suppose that a random sample of size n, denoted by $\mathbf{x} = (x_1, x_2, \ldots, x_n)$, is observed.

18.2 Bayesian Inferences

Definition 18.1: The distribution of θ, given $\mathbf{x}$, which is called the posterior distribution, is given by

$$\pi(\theta|\mathbf{x}) = \frac{f(\mathbf{x}|\theta)\pi(\theta)}{g(\mathbf{x})},$$

where $g(\mathbf{x})$ is the marginal distribution of $\mathbf{x}$.

The marginal distribution of $\mathbf{x}$ in the above definition can be calculated using the following formula:

$$g(\mathbf{x}) = \begin{cases} \sum_\theta f(\mathbf{x}|\theta)\pi(\theta), & \theta \text{ is discrete,} \\ \int_{-\infty}^{\infty} f(\mathbf{x}|\theta)\pi(\theta)\, d\theta, & \theta \text{ is continuous.} \end{cases}$$

Example 18.1: Assume that the prior distribution for the proportion of defectives produced by a machine is

p	0.1	0.2
$\pi(p)$	0.6	0.4

Denote by x the number of defectives among a random sample of size 2. Find the posterior probability distribution of p, given that x is observed.

Solution: The random variable X follows a binomial distribution

$$f(x|p) = b(x; 2, p) = \binom{2}{x} p^x q^{2-x}, \quad x = 0, 1, 2.$$

The marginal distribution of x can be calculated as

$$g(x) = f(x|0.1)\pi(0.1) + f(x|0.2)\pi(0.2)$$
$$= \binom{2}{x}[(0.1)^x(0.9)^{2-x}(0.6) + (0.2)^x(0.8)^{2-x}(0.4)].$$

Hence, for $x = 0, 1, 2$, we obtain the marginal probabilities as

x	0	1	2
$g(x)$	0.742	0.236	0.022

The posterior probability of $p = 0.1$, given x, is

$$\pi(0.1|x) = \frac{f(x|0.1)\pi(0.1)}{g(x)} = \frac{(0.1)^x(0.9)^{2-x}(0.6)}{(0.1)^x(0.9)^{2-x}(0.6) + (0.2)^x(0.8)^{2-x}(0.4)},$$

and $\pi(0.2|x) = 1 - \pi(0.1|x)$.

Suppose that $x = 0$ is observed.

$$\pi(0.1|0) = \frac{f(0|0.1)\pi(0.1)}{g(0)} = \frac{(0.1)^0(0.9)^{2-0}(0.6)}{0.742} = 0.6550,$$

and $\pi(0.2|0) = 0.3450$. If $x = 1$ is observed, $\pi(0.1|1) = 0.4576$, and $\pi(0.2|1) = 0.5424$. Finally, $\pi(0.1|2) = 0.2727$, and $\pi(0.2|2) = 0.7273$.

The prior distribution for Example 18.1 is discrete, although the natural range of p is from 0 to 1. Consider the following example, where we have a prior distribution covering the whole space for p.

Example 18.2: Suppose that the prior distribution of p is uniform (i.e., $\pi(p) = 1$, for $0 < p < 1$). Use the same random variable X as in Example 18.1 to find the posterior distribution of p.

Solution: As in Example 18.1, we have

$$f(x|p) = b(x; 2, p) = \binom{2}{x} p^x q^{2-x}, \quad x = 0, 1, 2.$$

The marginal distribution of x can be calculated as

$$g(x) = \int_0^1 f(x|p)\pi(p)\, dp = \binom{2}{x} \int_0^1 p^x (1-p)^{2-x}\, dp.$$

The integral above can be evaluated at each x directly as $g(0) = 1/3$, $g(1) = 1/3$, and $g(2) = 1/3$. Therefore, the posterior distribution of p, given x, is

$$\pi(p|x) = \frac{\binom{2}{x} p^x (1-p)^{2-x}}{1/3} = 3\binom{2}{x} p^x (1-p)^{2-x}, \quad 0 < p < 1.$$

The posterior distribution above is actually a beta distribution (see Section 6.8) with parameters $\alpha = x+1$ and $\beta = 3-x$. So, if $x = 0$ is observed, the posterior distribution of p is a beta distribution with parameters $(1,3)$. The posterior mean is $\mu = \frac{1}{1+3} = \frac{1}{4}$ and the posterior variance is $\sigma^2 = \frac{(1)(3)}{(1+3)^2(1+3+1)} = \frac{3}{80}$.

Using the posterior distribution, we can estimate the parameter(s) in a population in a straightforward fashion. In computing posterior distributions, it is very helpful if one is familiar with the distributions in Chapters 5 and 6. Note that in Definition 18.1, the *variable* in the posterior distribution is θ, while $\mathbf{x}$ is given. Thus, we can treat $g(\mathbf{x})$ as a constant as we calculate the posterior distribution of θ. Then the posterior distribution can be expressed as

$$\pi(\theta|\mathbf{x}) \propto f(\mathbf{x}|\theta)\pi(\theta),$$

where the symbol "$\propto$" stands for *is proportional to*. In the calculation of the posterior distribution above, we can leave the factors that do not depend on θ out of the normalization constant, i.e., the marginal density $g(\mathbf{x})$.

Example 18.3: Suppose that random variables $X_1, \ldots, X_n$ are independent and from a Poisson distribution with mean λ. Assume that the prior distribution of λ is exponential with mean 1. Find the posterior distribution of λ when $\bar{x} = 3$ with $n = 10$.

Solution: The density function of $\mathbf{X} = (X_1, \ldots, X_n)$ is

$$f(\mathbf{x}|\lambda) = \prod_{i=1}^n e^{-\lambda} \frac{\lambda^{x_i}}{x_i!} = e^{-n\lambda} \frac{\lambda^{\sum_{i=1}^n x_i}}{\prod_{i=1}^n x_i!},$$

and the prior distribution is

$$\pi(\theta) = e^{-\lambda}, \text{ for } \lambda > 0.$$

18.2 Bayesian Inferences

Hence, using Definition 18.1 we obtain the posterior distribution of λ as

$$\pi(\lambda|\mathbf{x}) \propto f(\mathbf{x}|\lambda)\pi(\lambda) = e^{-n\lambda}\frac{\lambda^{\sum_{i=1}^{n} x_i}}{\prod_{i=1}^{n} x_i!}e^{-\lambda} \propto e^{-(n+1)\lambda}\lambda^{\sum_{i=1}^{n} x_i}.$$

Referring to the gamma distribution in Section 6.6, we conclude that the posterior distribution of λ follows a gamma distribution with parameters $1 + \sum_{i=1}^{n} x_i$ and $\frac{1}{n+1}$.
Hence, we have the posterior mean and variance of λ as $\frac{\sum_{i=1}^{n} x_i + 1}{n+1}$ and $\frac{\sum_{i=1}^{n} x_i + 1}{(n+1)^2}$.
So, when $\bar{x} = 3$ with $n = 10$, we have $\sum_{i=1}^{10} x_i = 30$. Hence, the posterior distribution of λ is a gamma distribution with parameters 31 and $1/11$.

From Example 18.3 we observe that sometimes it is quite convenient to use the "proportional to" technique in calculating the posterior distribution, especially when the result can be formed to a commonly used distribution as described in Chapters 5 and 6.

Point Estimation Using the Posterior Distribution

Once the posterior distribution is derived, we can easily use the summary of the posterior distribution to make inferences on the population parameters. For instance, the posterior mean, median, and mode can all be used to estimate the parameter.

Example 18.4: Suppose that $x = 1$ is observed for Example 18.2. Find the posterior mean and the posterior mode.

***Solution*:** When $x = 1$, the posterior distribution of p can be expressed as

$$\pi(p|1) = 6p(1-p), \quad \text{for} \quad 0 < p < 1.$$

To calculate the mean of this distribution, we need to find

$$\int_0^1 6p^2(1-p)\, dp = 6\left(\frac{1}{3} - \frac{1}{4}\right) = \frac{1}{2}.$$

To find the posterior mode, we need to obtain the value of p such that the posterior distribution is maximized. Taking derivative of $\pi(p)$ with respect to p, we obtain $6 - 12p$. Solving for p in $6 - 12p = 0$, we obtain $p = 1/2$. The second derivative is -12, which implies that the posterior mode is achieved at $p = 1/2$.

Bayesian methods of estimation concerning the mean μ of a normal population are based on the following example.

Example 18.5: If $\bar{x}$ is the mean of a random sample of size n from a normal population with known variance σ^2, and the prior distribution of the population mean is a normal distribution with known mean μ_0 and known variance σ_0^2, then show that the posterior distribution of the population mean is also a normal distribution with mean μ^* and standard deviation σ^*, where

$$\mu^* = \frac{\sigma_0^2}{\sigma_0^2 + \sigma^2/n}\bar{x} + \frac{\sigma^2/n}{\sigma_0^2 + \sigma^2/n}\mu_0 \quad \text{and} \quad \sigma^* = \sqrt{\frac{\sigma_0^2 \sigma^2}{n\sigma_0^2 + \sigma^2}}.$$

Solution: The density function of our sample is

$$f(x_1, x_2, \ldots, x_n \mid \mu) = \frac{1}{(2\pi)^{n/2}\sigma^n} \exp\left[-\frac{1}{2}\sum_{i=1}^{n}\left(\frac{x_i - \mu}{\sigma}\right)^2\right],$$

for $-\infty < x_i < \infty$ and $i = 1, 2, \ldots, n$, and the prior is

$$\pi(\mu) = \frac{1}{\sqrt{2\pi}\sigma_0} \exp\left[-\frac{1}{2}\left(\frac{\mu - \mu_0}{\sigma_0}\right)^2\right], \quad -\infty < \mu < \infty.$$

Then the posterior distribution of μ is

$$\pi(\mu \mid \mathbf{x}) \propto \exp\left\{-\frac{1}{2}\left[\sum_{i=1}^{n}\left(\frac{x_i - \mu}{\sigma}\right)^2 + \left(\frac{\mu - \mu_0}{\sigma_0}\right)^2\right]\right\}$$

$$\propto \exp\left\{-\frac{1}{2}\left[\frac{n(\bar{x} - \mu)^2}{\sigma^2} + \frac{(\mu - \mu_0)^2}{\sigma_0^2}\right]\right\},$$

due to

$$\sum_{i=1}^{n}(x_i - \mu)^2 = \sum_{i=1}^{n}(x_i - \bar{x})^2 + n(\bar{x} - \mu)^2$$

from Section 8.5. Completing the squares for μ yields the posterior distribution

$$\pi(\mu \mid \mathbf{x}) \propto \exp\left[-\frac{1}{2}\left(\frac{\mu - \mu^*}{\sigma^*}\right)^2\right],$$

where

$$\mu^* = \frac{n\bar{x}\sigma_0^2 + \mu_0\sigma^2}{n\sigma_0^2 + \sigma^2}, \quad \sigma^* = \sqrt{\frac{\sigma_0^2\sigma^2}{n\sigma_0^2 + \sigma^2}}.$$

This is a normal distribution with mean μ^* and standard deviation σ^*.

The Central Limit Theorem allows us to use Example 18.5 also when we select sufficiently large random samples ($n \geq 30$ for many engineering experimental cases) from nonnormal populations (the distribution is not very far from symmetric), and when the prior distribution of the mean is approximately normal.

Several comments need to be made about Example 18.5. The posterior mean μ^* can also be written as

$$\mu^* = \frac{\sigma_0^2}{\sigma_0^2 + \sigma^2/n}\bar{x} + \frac{\sigma^2/n}{\sigma_0^2 + \sigma^2/n}\mu_0,$$

which is a weighted average of the sample mean $\bar{x}$ and the prior mean μ_0. Since both coefficients are between 0 and 1 and they sum to 1, the posterior mean μ^* is always between $\bar{x}$ and μ_0. This means that the posterior estimation of μ is influenced by both $\bar{x}$ and μ_0. Furthermore, the weight of $\bar{x}$ depends on the prior variance as well as the variance of the sample mean. For a large sample problem ($n \to \infty$), the posterior mean $\mu^* \to \bar{x}$. This means that the prior mean does not play any role in estimating the population mean μ using the posterior distribution. This is very reasonable since it indicates that when the amount of data is substantial, information from the data will dominate the information on μ provided by the prior. On the other hand, when the prior variance is large ($\sigma_0^2 \to \infty$), the posterior mean μ^* also goes to $\bar{x}$. Note that for a normal distribution, the larger the variance, the flatter the density function. The flatness of the normal distribution in this case means that there is almost no subjective prior information available on the parameter μ before the data are collected. Thus, it is reasonable that the posterior estimation μ^* only depends on the data value $\bar{x}$.

Now consider the posterior standard deviation σ^*. This value can also be written as

$$\sigma^* = \sqrt{\frac{\sigma_0^2 \sigma^2/n}{\sigma_0^2 + \sigma^2/n}}.$$

It is obvious that the value σ^* is smaller than both σ_0 and $\sigma/\sqrt{n}$, the prior standard deviation and the standard deviation of $\bar{x}$, respectively. This suggests that the posterior estimation is more accurate than both the prior and the sample data. Hence, incorporating both the data and prior information results in better posterior information than using any of the data or prior alone. This is a common phenomenon in Bayesian inference. Furthermore, to compute μ^* and σ^* by the formulas in Example 18.5, we have assumed that σ^2 is known. Since this is generally not the case, we shall replace σ^2 by the sample variance s^2 whenever $n \geq 30$.

Bayesian Interval Estimation

Similar to the classical confidence interval, in Bayesian analysis we can calculate a $100(1-\alpha)\%$ Bayesian interval using the posterior distribution.

Definition 18.2: The interval $a < \theta < b$ will be called a $100(1-\alpha)\%$ **Bayesian interval** for θ if

$$\int_{-\infty}^{a} \pi(\theta|x) \, d\theta = \int_{b}^{\infty} \pi(\theta|x) \, d\theta = \frac{\alpha}{2}.$$

Recall that under the frequentist approach, the probability of a confidence interval, say 95%, is interpreted as a coverage probability, which means that if an experiment is repeated again and again (with considerable unobserved data), the probability that the intervals calculated according to the rule will cover the true parameter is 95%. However, in Bayesian interval interpretation, say for a 95% interval, we can state that the probability of the unknown parameter falling into the calculated interval (which only depends on the observed data) is 95%.

Example 18.6: Supposing that $X \sim b(x; n, p)$, with known $n = 2$, and the prior distribution of p is uniform $\pi(p) = 1$, for $0 < p < 1$, find a 95% Bayesian interval for p.

Solution: As in Example 18.2, when $x = 0$, the posterior distribution is a beta distribution with parameters 1 and 3, i.e., $\pi(p|0) = 3(1-p)^2$, for $0 < p < 1$. Thus, we need to solve for a and b using Definition 18.2, which yields the following:

$$0.025 = \int_0^a 3(1-p)^2 \, dp = 1 - (1-a)^3$$

and

$$0.025 = \int_b^1 3(1-p)^2 \, dp = (1-b)^3.$$

The solutions to the above equations result in $a = 0.0084$ and $b = 0.7076$. Therefore, the probability that p falls into $(0.0084, 0.7076)$ is 95%.

For the normal population and normal prior case described in Example 18.5, the posterior mean μ^* is the Bayes estimate of the population mean μ, and a $100(1-\alpha)\%$ **Bayesian interval** for μ can be constructed by computing the interval

$$\mu^* - z_{\alpha/2}\sigma^* < \mu < \mu^* + z_{\alpha/2}\sigma^*,$$

which is centered at the posterior mean and contains $100(1-\alpha)\%$ of the posterior probability.

Example 18.7: An electrical firm manufactures light bulbs that have a length of life that is approximately normally distributed with a standard deviation of 100 hours. Prior experience leads us to believe that μ is a value of a normal random variable with a mean $\mu_0 = 800$ hours and a standard deviation $\sigma_0 = 10$ hours. If a random sample of 25 bulbs has an average life of 780 hours, find a 95% Bayesian interval for μ.

Solution: According to Example 18.5, the posterior distribution of the mean is also a normal distribution with mean

$$\mu^* = \frac{(25)(780)(10)^2 + (800)(100)^2}{(25)(10)^2 + (100)^2} = 796$$

and standard deviation

$$\sigma^* = \sqrt{\frac{(10)^2(100)^2}{(25)(10)^2 + (100)^2}} = \sqrt{80}.$$

The 95% Bayesian interval for μ is then given by

$$796 - 1.96\sqrt{80} < \mu < 796 + 1.96\sqrt{80},$$

or

$$778.5 < \mu < 813.5.$$

Hence, we are 95% sure that μ will be between 778.5 and 813.5.

On the other hand, ignoring the prior information about μ, we could proceed as in Section 9.4 and construct the classical 95% confidence interval

$$780 - (1.96)\left(\frac{100}{\sqrt{25}}\right) < \mu < 780 + (1.96)\left(\frac{100}{\sqrt{25}}\right),$$

or $740.8 < \mu < 819.2$, which is seen to be wider than the corresponding Bayesian interval.

18.3 Bayes Estimates Using Decision Theory Framework

Using Bayesian methodology, the posterior distribution of a parameter can be obtained. Bayes estimates can also be derived using the posterior distribution and a loss function when a loss is incurred. A loss function is a function that describes the cost of a decision associated with an event of interest. Here we only list a few commonly used loss functions and their associated Bayes estimates.

Squared-Error Loss

Definition 18.3: The **squared-error loss function** is

$$L(\theta, a) = (\theta - a)^2,$$

where θ is the parameter (or state of nature) and a an action (or estimate).

A Bayes estimate minimizes the posterior expected loss, given on the observed sample data.

Theorem 18.1: The mean of the posterior distribution $\pi(\theta|x)$, denoted by θ^*, is the **Bayes estimate of θ** under the squared-error loss function.

Example 18.8: Find the Bayes estimates of p, for all the values of x, for Example 18.1 when the squared-error loss function is used.

Solution: When $x = 0$, $p^* = (0.1)(0.6550) + (0.2)(0.3450) = 0.1345$.
When $x = 1$, $p^* = (0.1)(0.4576) + (0.2)(0.5424) = 0.1542$.
When $x = 2$, $p^* = (0.1)(0.2727) + (0.2)(0.7273) = 0.1727$.

Note that the classical estimate of p is $\hat{p} = x/n = 0, 1/2$, and 1, respectively, for the x values at 0, 1, and 2. These classical estimates are very different from the corresponding Bayes estimates.

Example 18.9: Repeat Example 18.8 in the situation of Example 18.2.

Solution: Since the posterior distribution of p is a $B(x+1, 3-x)$ distribution (see Section 6.8 on page 201), the Bayes estimate of p is

$$p^* = E^{\pi(p|x)}(p) = 3\binom{2}{x} \int_0^1 p^{x+1}(1-p)^{2-x}\, dp,$$

which yields $p^* = 1/4$ for $x = 0$, $p^* = 1/2$ for $x = 1$, and $p^* = 3/4$ for $x = 2$, respectively. Notice that when $x = 1$ is observed, the Bayes estimate and the classical estimate $\hat{p}$ are equivalent.

For the normal situation as described in Example 18.5, the Bayes estimate of μ under the squared-error loss will be the posterior mean μ^*.

Example 18.10: Suppose that the sampling distribution of a random variable, X, is Poisson with parameter λ. Assume that the prior distribution of λ follows a gamma distribution with parameters (α, β). Find the Bayes estimate of λ under the squared-error loss function.

Solution: Using Example 18.3, we conclude that the posterior distribution of λ follows a gamma distribution with parameters $(x+\alpha, (1+1/\beta)^{-1})$. Using Theorem 6.4, we obtain the posterior mean

$$\hat{\lambda} = \frac{x+\alpha}{1+1/\beta}.$$

Since the posterior mean is the Bayes estimate under the squared-error loss, $\hat{\lambda}$ is our Bayes estimate.

Absolute-Error Loss

The squared-error loss described above is similar to the least-squares concept we discussed in connection with regression in Chapters 11 and 12. In this section, we introduce another loss function as follows.

Definition 18.4: The **absolute-error loss function** is defined as

$$L(\theta, a) = |\theta - a|,$$

where θ is the parameter and a an action.

Theorem 18.2: The median of the posterior distribution $\pi(\theta|x)$, denoted by θ^*, is the **Bayes estimate** of θ under the absolute-error loss function.

Example 18.11: Under the absolute-error loss, find the Bayes estimator for Example 18.9 when $x = 1$ is observed.

Solution: Again, the posterior distribution of p is a $B(x+1, 3-x)$. When $x = 1$, it is a beta distribution with density $\pi(p \mid x = 1) = 6x(1-x)$ for $0 < x < 1$ and 0 otherwise. The median of this distribution is the value of p^* such that

$$\frac{1}{2} = \int_0^{p^*} 6p(1-p)\, dp = 3p^{*2} - 2p^{*3},$$

which yields the answer $p^* = \frac{1}{2}$. Hence, the Bayes estimate in this case is 0.5.

Exercises

18.1 Estimate the proportion of defectives being produced by the machine in Example 18.1 if the random sample of size 2 yields 2 defectives.

18.2 Let us assume that the prior distribution for the proportion p of drinks from a vending machine that overflow is

p	0.05	0.10	0.15
$\pi(p)$	0.3	0.5	0.2

If 2 of the next 9 drinks from this machine overflow, find

(a) the posterior distribution for the proportion p;

(b) the Bayes estimate of p.

Exercises

18.3 Repeat Exercise 18.2 when 1 of the next 4 drinks overflows and the uniform prior distribution is

$$\pi(p) = 10, \quad 0.05 < p < 0.15.$$

18.4 Service calls come to a maintenance center according to a Poisson process with λ calls per minute. A data set of 20 one-minute periods yields an average of 1.8 calls. If the prior for λ follows an exponential distribution with mean 2, determine the posterior distribution of λ.

18.5 A previous study indicates that the percentage of chain smokers, p, who have lung cancer follows a beta distribution (see Section 6.8) with mean 70% and standard deviation 10%. Suppose a new data set collected shows that 81 out of 120 chain smokers have lung cancer.
(a) Determine the posterior distribution of the percentage of chain smokers who have lung cancer by combining the new data and the prior information.
(b) What is the posterior probability that p is larger than 50%?

18.6 The developer of a new condominium complex claims that 3 out of 5 buyers will prefer a two-bedroom unit, while his banker claims that it would be more correct to say that 7 out of 10 buyers will prefer a two-bedroom unit. In previous predictions of this type, the banker has been twice as reliable as the developer. If 12 of the next 15 condominiums sold in this complex are two-bedroom units, find
(a) the posterior probabilities associated with the claims of the developer and banker;
(b) a point estimate of the proportion of buyers who prefer a two-bedroom unit.

18.7 The burn time for the first stage of a rocket is a normal random variable with a standard deviation of 0.8 minute. Assume a normal prior distribution for μ with a mean of 8 minutes and a standard deviation of 0.2 minute. If 10 of these rockets are fired and the first stage has an average burn time of 9 minutes, find a 95% Bayesian interval for μ.

18.8 The daily profit from a juice vending machine placed in an office building is a value of a normal random variable with unknown mean μ and variance σ^2. Of course, the mean will vary somewhat from building to building, and the distributor feels that these average daily profits can best be described by a normal distribution with mean $\mu_0 = \$30.00$ and standard deviation $\sigma_0 = \$1.75$. If one of these juice machines, placed in a certain building, showed an average daily profit of $\bar{x} = \$24.90$ during the first 30 days with a standard deviation of $s = \$2.10$, find

(a) a Bayes estimate of the true average daily profit for this building;
(b) a 95% Bayesian interval of μ for this building;
(c) the probability that the average daily profit from the machine in this building is between $24.00 and $26.00.

18.9 The mathematics department of a large university is designing a placement test to be given to incoming freshman classes. Members of the department feel that the average grade for this test will vary from one freshman class to another. This variation of the average class grade is expressed subjectively by a normal distribution with mean $\mu_0 = 72$ and variance $\sigma_0^2 = 5.76$.
(a) What prior probability does the department assign to the actual average grade being somewhere between 71.8 and 73.4 for next year's freshman class?
(b) If the test is tried on a random sample of 100 students from the next incoming freshman class, resulting in an average grade of 70 with a variance of 64, construct a 95% Bayesian interval for μ.
(c) What posterior probability should the department assign to the event of part (a)?

18.10 Suppose that in Example 18.7 the electrical firm does not have enough prior information regarding the population mean length of life to be able to assume a normal distribution for μ. The firm believes, however, that μ is surely between 770 and 830 hours, and it is thought that a more realistic Bayesian approach would be to assume the prior distribution

$$\pi(\mu) = \frac{1}{60}, \quad 770 < \mu < 830.$$

If a random sample of 25 bulbs gives an average life of 780 hours, follow the steps of the proof for Example 18.5 to find the posterior distribution

$$\pi(\mu \mid x_1, x_2, \ldots, x_{25}).$$

18.11 Suppose that the time to failure T of a certain hinge is an exponential random variable with probability density

$$f(t) = \theta e^{-\theta t}, \quad t > 0.$$

From prior experience we are led to believe that θ is a value of an exponential random variable with probability density

$$\pi(\theta) = 2e^{-2\theta}, \quad \theta > 0.$$

If we have a sample of n observations on T, show that the posterior distribution of Θ is a gamma distribution with parameters

$$\alpha = n+1 \quad \text{and} \quad \beta = \left(\sum_{i=1}^{n} t_i + 2\right)^{-1}.$$

18.12 Suppose that a sample consisting of 5, 6, 6, 7, 5, 6, 4, 9, 3, and 6 comes from a Poisson population with mean λ. Assume that the parameter λ follows a gamma distribution with parameters $(3,2)$. Under the squared-error loss function, find the Bayes estimate of λ.

18.13 A random variable X follows a negative binomial distribution with parameters $k = 5$ and p [i.e., $b^*(x; 5, p)$]. Furthermore, we know that p follows a uniform distribution on the interval $(0, 1)$. Find the Bayes estimate of p under the squared-error loss function.

18.14 A random variable X follows an exponential distribution with mean $1/\beta$. Assume the prior distribution of β is another exponential distribution with mean 2.5. Determine the Bayes estimate of β under the absolute-error loss function.

18.15 A random sample $X_1, \ldots, X_n$ comes from a uniform distribution (see Section 6.1) population $U(0, \theta)$ with unknown θ. The data are given below:

0.13, 1.06, 1.65, 1.73, 0.95, 0.56, 2.14, 0.33, 1.22, 0.20, 1.55, 1.18, 0.71, 0.01, 0.42, 1.03, 0.43, 1.02, 0.83, 0.88

Suppose the prior distribution of θ has the density

$$\pi(\theta) = \begin{cases} \frac{1}{\theta^2}, & \theta > 1, \\ 0, & \theta \leq 1. \end{cases}$$

Determine the Bayes estimator under the absolute-error loss function.

Bibliography

[1] Bartlett, M. S., and Kendall, D. G. (1946). "The Statistical Analysis of Variance Heterogeneity and Logarithmic Transformation," *Journal of the Royal Statistical Society*, Ser. B. **8**, 128–138.

[2] Bowker, A. H., and Lieberman, G. J. (1972). *Engineering Statistics*, 2nd ed. Upper Saddle River, N.J.: Prentice Hall.

[3] Box, G. E. P. (1988). "Signal to Noise Ratios, Performance Criteria and Transformations (with discussion)," *Technometrics*, **30**, 1–17.

[4] Box, G. E. P., and Fung, C. A. (1986). "Studies in Quality Improvement: Minimizing Transmitted Variation by Parameter Design," Report 8. University of Wisconsin-Madison, Center for Quality and Productivity Improvement.

[5] Box, G. E. P., Hunter, W. G., and Hunter, J. S. (1978). *Statistics for Experimenters*. New York: John Wiley & Sons.

[6] Brownlee, K. A. (1984). *Statistical Theory and Methodology: In Science and Engineering*, 2nd ed. New York: John Wiley & Sons.

[7] Carroll, R. J., and Ruppert, D. (1988). *Transformation and Weighting in Regression*. New York: Chapman and Hall.

[8] Chatterjee, S., Hadi, A. S., and Price, B. (1999). *Regression Analysis by Example*, 3rd ed. New York: John Wiley & Sons.

[9] Cook, R. D., and Weisberg, S. (1982). *Residuals and Influence in Regression*. New York: Chapman and Hall.

[10] Daniel, C. and Wood, F. S. (1999). *Fitting Equations to Data: Computer Analysis of Multifactor Data*, 2nd ed. New York: John Wiley & Sons.

[11] Daniel, W. W. (1989). *Applied Nonparametric Statistics*, 2nd ed. Belmont, Calif.: Wadsworth Publishing Company.

[12] Devore, J. L. (2003). *Probability and Statistics for Engineering and the Sciences*, 6th ed. Belmont, Calif: Duxbury Press.

[13] Dixon, W. J. (1983). *Introduction to Statistical Analysis*, 4th ed. New York: McGraw-Hill.

[14] Draper, N. R., and Smith, H. (1998). *Applied Regression Analysis*, 3rd ed. New York: John Wiley & Sons.

[15] Duncan, A. (1986). *Quality Control and Industrial Statistics*, 5th ed. Homewood, Ill.: Irwin.

[16] Dyer, D. D., and Keating, J. P. (1980). "On the Determination of Critical Values for Bartlett's Test," *Journal of the American Statistical Association*, **75**, 313–319.

[17] Ewan, W. D., and Kemp, K. W. (1960). "Sampling Inspection of Continuous Processes with No Autocorrelation between Successive Results," *Biometrika*, **47**, 363–380.

[18] Geary, R. C. (1947). "Testing for Normality," *Biometrika*, **34**, 209–242.

[19] Gunst, R. F., and Mason, R. L. (1980). *Regression Analysis and Its Application: A Data-Oriented Approach*. New York: Marcel Dekker.

[20] Guttman, I., Wilks, S. S., and Hunter, J. S. (1971). *Introductory Engineering Statistics*. New York: John Wiley & Sons.

[21] Harville, D. A. (1977). "Maximum Likelihood Approaches to Variance Component Estimation and to Related Problems," *Journal of the American Statistical Association*, **72**, 320–338.

[22] Hicks, C. R., and Turner, K. V. (1999). *Fundamental Concepts in the Design of Experiments*, 5th ed. Oxford: Oxford University Press.

[23] Hoaglin, D. C., Mosteller, F., and Tukey, J. W. (1991). *Fundamentals of Exploratory Analysis of Variance*. New York: John Wiley & Sons.

[24] Hocking, R. R. (1976). "The Analysis and Selection of Variables in Linear Regression," *Biometrics*, **32**, 1–49.

[25] Hodges, J. L., and Lehmann, E. L. (2005). *Basic Concepts of Probability and Statistics*, 2nd ed. Philadelphia: Society for Industrial and Applied Mathematics.

[26] Hoerl, A. E., and Wennard, R. W. (1970). "Ridge Regression: Applications to Nonorthogonal Problems," *Technometrics*, **12**, 55–67.

[27] Hogg, R. V., and Ledolter, J. (1992). *Applied Statistics for Engineers and Physical Scientists*, 2nd ed. Upper Saddle River, N.J.: Prentice Hall.

[28] Hogg, R. V., McKean, J. W., and Craig, A. (2005). *Introduction to Mathematical Statistics*, 6th ed. Upper Saddle River, N.J.: Prentice Hall.

[29] Hollander, M., and Wolfe, D. (1999). *Nonparametric Statistical Methods*. New York: John Wiley & Sons.

[30] Johnson, N. L., and Leone, F. C. (1977). *Statistics and Experimental Design: In Engineering and the Physical Sciences*, 2nd ed. Vols. I and II, New York: John Wiley & Sons.

[31] Kackar, R. (1985). "Off-Line Quality Control, Parameter Design, and the Taguchi Methods," *Journal of Quality Technology*, **17**, 176–188.

[32] Koopmans, L. H. (1987). *An Introduction to Contemporary Statistics*, 2nd ed. Boston: Duxbury Press.

[33] Kutner, M. H., Nachtsheim, C. J., Neter, J., and Li, W. (2004). *Applied Linear Regression Models*, 5th ed. New York: McGraw-Hill/Irwin.

[34] Larsen, R. J., and Morris, M. L. (2000). *An Introduction to Mathematical Statistics and Its Applications*, 3rd ed. Upper Saddle River, N.J.: Prentice Hall.

[35] Lehmann, E. L., and D'Abrera, H. J. M. (1998). *Nonparametrics: Statistical Methods Based on Ranks*, rev. ed. Upper Saddle River, N.J.: Prentice Hall.

[36] Lentner, M., and Bishop, T. (1986). *Design and Analysis of Experiments*, 2nd ed. Blacksburg, Va.: Valley Book Co.

[37] Mallows, C. L. (1973). "Some Comments on C_p," *Technometrics*, **15**, 661–675.

[38] McClave, J. T., Dietrich, F. H., and Sincich, T. (1997). *Statistics*, 7th ed. Upper Saddle River, N.J.: Prentice Hall.

[39] Montgomery, D. C. (2008a). *Design and Analysis of Experiments*, 7th ed. New York: John Wiley & Sons.

[40] Montgomery, D. C. (2008b). *Introduction to Statistical Quality Control*, 6th ed. New York: John Wiley & Sons.

[41] Mosteller, F., and Tukey, J. (1977). *Data Analysis and Regression*. Reading, Mass.: Addison-Wesley Publishing Co.

[42] Myers, R. H. (1990). *Classical and Modern Regression with Applications*, 2nd ed. Boston: Duxbury Press.

[43] Myers, R. H., Khuri, A. I., and Vining, G. G. (1992). "Response Surface Alternatives to the Taguchi Robust Parameter Design Approach," *The American Statistician*, **46**, 131–139.

[44] Myers, R. H., Montgomery, D. C., and Anderson-Cook, C. M. (2009). *Response Surface Methodology: Process and Product Optimization Using Designed Experiments*, 3rd ed. New York: John Wiley & Sons.

[45] Myers, R. H., Montgomery, D. C., Vining, G. G., and Robinson, T. J. (2008). *Generalized Linear Models with Applications in Engineering and the Sciences*, 2nd ed., New York: John Wiley & Sons.

[46] Noether, G. E. (1976). *Introduction to Statistics: A Nonparametric Approach*, 2nd ed. Boston: Houghton Mifflin Company.

[47] Olkin, I., Gleser, L. J., and Derman, C. (1994). *Probability Models and Applications*, 2nd ed. New York: Prentice Hall.

[48] Ott, R. L., and Longnecker, M. T. (2000). *An Introduction to Statistical Methods and Data Analysis*, 5th ed. Boston: Duxbury Press.

[49] Pacansky, J., England, C. D., and Wattman, R. (1986). "Infrared Spectroscopic Studies of Poly (perfluoropropyleneoxide) on Gold Substrate: A Classical Dispersion Analysis for the Refractive Index." *Applied Spectroscopy*, **40**, 8–16.

[50] Plackett, R. L., and Burman, J. P. (1946). "The Design of Multifactor Experiments," *Biometrika*, **33**, 305–325.

[51] Ross, S. M. (2002). *Introduction to Probability Models*, 9th ed. New York: Academic Press, Inc.

[52] Satterthwaite, F. E. (1946). "An Approximate Distribution of Estimates of Variance Components," *Biometrics*, **2**, 110–114.

[53] Schilling, E. G., and Nelson, P. R. (1976). "The Effect of Nonnormality on the Control Limits of $\bar{X}$ Charts," *Journal of Quality Technology*, **8**, 347–373.

[54] Schmidt, S. R., and Launsby, R. G. (1991). *Understanding Industrial Designed Experiments*. Colorado Springs, Col. Air Academy Press.

[55] Shoemaker, A. C., Tsui, K.-L., and Wu, C. F. J. (1991). "Economical Experimentation Methods for Robust Parameter Design," *Technometrics*, **33**, 415–428.

[56] Snedecor, G. W., and Cochran, W. G. (1989). *Statistical Methods*, 8th ed. Allies, Iowa: The Iowa State University Press.

[57] Steel, R. G. D., Torrie, J. H., and Dickey, D. A. (1996). *Principles and Procedures of Statistics: A Biometrical Approach*, 3rd ed. New York: McGraw-Hill.

[58] Taguchi, G. (1991). *Introduction to Quality Engineering*. White Plains, N.Y.: Unipub/Kraus International.

[59] Taguchi, G., and Wu, Y. (1985). *Introduction to Off-Line Quality Control*. Nagoya, Japan: Central Japan Quality Control Association.

[60] Thompson, W. O., and Cady, F. B. (1973). *Proceedings of the University of Kentucky Conference on Regression with a Large Number of Predictor Variables*. Lexington, Ken.: University of Kentucky Press.

[61] Tukey, J. W. (1977). *Exploratory Data Analysis*. Reading, Mass.: Addison-Wesley Publishing Co.

[62] Vining, G. G., and Myers, R. H. (1990). "Combining Taguchi and Response Surface Philosophies: A Dual Response Approach," *Journal of Quality Technology*, **22**, 38–45.

[63] Welch, W. J., Yu, T. K., Kang, S. M., and Sacks, J. (1990). "Computer Experiments for Quality Control by Parameter Design," *Journal of Quality Technology*, **22**, 15–22.

[64] Winer, B. J. (1991). *Statistical Principles in Experimental Design*, 3rd ed. New York: McGraw-Hill.

Appendix A
Statistical Tables and Proofs

Table A.1 Binomial Probability Sums $\sum_{x=0}^{r} b(x; n, p)$

						p					
n	r	0.10	0.20	0.25	0.30	0.40	0.50	0.60	0.70	0.80	0.90
1	0	0.9000	0.8000	0.7500	0.7000	0.6000	0.5000	0.4000	0.3000	0.2000	0.1000
	1	1.0000	1.0000	1.0000	1.0000	1.0000	1.0000	1.0000	1.0000	1.0000	1.0000
2	0	0.8100	0.6400	0.5625	0.4900	0.3600	0.2500	0.1600	0.0900	0.0400	0.0100
	1	0.9900	0.9600	0.9375	0.9100	0.8400	0.7500	0.6400	0.5100	0.3600	0.1900
	2	1.0000	1.0000	1.0000	1.0000	1.0000	1.0000	1.0000	1.0000	1.0000	1.0000
3	0	0.7290	0.5120	0.4219	0.3430	0.2160	0.1250	0.0640	0.0270	0.0080	0.0010
	1	0.9720	0.8960	0.8438	0.7840	0.6480	0.5000	0.3520	0.2160	0.1040	0.0280
	2	0.9990	0.9920	0.9844	0.9730	0.9360	0.8750	0.7840	0.6570	0.4880	0.2710
	3	1.0000	1.0000	1.0000	1.0000	1.0000	1.0000	1.0000	1.0000	1.0000	1.0000
4	0	0.6561	0.4096	0.3164	0.2401	0.1296	0.0625	0.0256	0.0081	0.0016	0.0001
	1	0.9477	0.8192	0.7383	0.6517	0.4752	0.3125	0.1792	0.0837	0.0272	0.0037
	2	0.9963	0.9728	0.9492	0.9163	0.8208	0.6875	0.5248	0.3483	0.1808	0.0523
	3	0.9999	0.9984	0.9961	0.9919	0.9744	0.9375	0.8704	0.7599	0.5904	0.3439
	4	1.0000	1.0000	1.0000	1.0000	1.0000	1.0000	1.0000	1.0000	1.0000	1.0000
5	0	0.5905	0.3277	0.2373	0.1681	0.0778	0.0313	0.0102	0.0024	0.0003	0.0000
	1	0.9185	0.7373	0.6328	0.5282	0.3370	0.1875	0.0870	0.0308	0.0067	0.0005
	2	0.9914	0.9421	0.8965	0.8369	0.6826	0.5000	0.3174	0.1631	0.0579	0.0086
	3	0.9995	0.9933	0.9844	0.9692	0.9130	0.8125	0.6630	0.4718	0.2627	0.0815
	4	1.0000	0.9997	0.9990	0.9976	0.9898	0.9688	0.9222	0.8319	0.6723	0.4095
	5	1.0000	1.0000	1.0000	1.0000	1.0000	1.0000	1.0000	1.0000	1.0000	1.0000
6	0	0.5314	0.2621	0.1780	0.1176	0.0467	0.0156	0.0041	0.0007	0.0001	0.0000
	1	0.8857	0.6554	0.5339	0.4202	0.2333	0.1094	0.0410	0.0109	0.0016	0.0001
	2	0.9842	0.9011	0.8306	0.7443	0.5443	0.3438	0.1792	0.0705	0.0170	0.0013
	3	0.9987	0.9830	0.9624	0.9295	0.8208	0.6563	0.4557	0.2557	0.0989	0.0159
	4	0.9999	0.9984	0.9954	0.9891	0.9590	0.8906	0.7667	0.5798	0.3446	0.1143
	5	1.0000	0.9999	0.9998	0.9993	0.9959	0.9844	0.9533	0.8824	0.7379	0.4686
	6	1.0000	1.0000	1.0000	1.0000	1.0000	1.0000	1.0000	1.0000	1.0000	1.0000
7	0	0.4783	0.2097	0.1335	0.0824	0.0280	0.0078	0.0016	0.0002	0.0000	
	1	0.8503	0.5767	0.4449	0.3294	0.1586	0.0625	0.0188	0.0038	0.0004	0.0000
	2	0.9743	0.8520	0.7564	0.6471	0.4199	0.2266	0.0963	0.0288	0.0047	0.0002
	3	0.9973	0.9667	0.9294	0.8740	0.7102	0.5000	0.2898	0.1260	0.0333	0.0027
	4	0.9998	0.9953	0.9871	0.9712	0.9037	0.7734	0.5801	0.3529	0.1480	0.0257
	5	1.0000	0.9996	0.9987	0.9962	0.9812	0.9375	0.8414	0.6706	0.4233	0.1497
	6		1.0000	0.9999	0.9998	0.9984	0.9922	0.9720	0.9176	0.7903	0.5217
	7			1.0000	1.0000	1.0000	1.0000	1.0000	1.0000	1.0000	1.0000

Table A.1 Binomial Probability Table

Table A.1 (continued) Binomial Probability Sums $\sum_{x=0}^{r} b(x; n, p)$

n	r	\multicolumn{10}{c}{p}
		0.10	0.20	0.25	0.30	0.40	0.50	0.60	0.70	0.80	0.90
8	0	0.4305	0.1678	0.1001	0.0576	0.0168	0.0039	0.0007	0.0001	0.0000	
	1	0.8131	0.5033	0.3671	0.2553	0.1064	0.0352	0.0085	0.0013	0.0001	
	2	0.9619	0.7969	0.6785	0.5518	0.3154	0.1445	0.0498	0.0113	0.0012	0.0000
	3	0.9950	0.9437	0.8862	0.8059	0.5941	0.3633	0.1737	0.0580	0.0104	0.0004
	4	0.9996	0.9896	0.9727	0.9420	0.8263	0.6367	0.4059	0.1941	0.0563	0.0050
	5	1.0000	0.9988	0.9958	0.9887	0.9502	0.8555	0.6846	0.4482	0.2031	0.0381
	6		0.9999	0.9996	0.9987	0.9915	0.9648	0.8936	0.7447	0.4967	0.1869
	7		1.0000	1.0000	0.9999	0.9993	0.9961	0.9832	0.9424	0.8322	0.5695
	8				1.0000	1.0000	1.0000	1.0000	1.0000	1.0000	1.0000
9	0	0.3874	0.1342	0.0751	0.0404	0.0101	0.0020	0.0003	0.0000		
	1	0.7748	0.4362	0.3003	0.1960	0.0705	0.0195	0.0038	0.0004	0.0000	
	2	0.9470	0.7382	0.6007	0.4628	0.2318	0.0898	0.0250	0.0043	0.0003	0.0000
	3	0.9917	0.9144	0.8343	0.7297	0.4826	0.2539	0.0994	0.0253	0.0031	0.0001
	4	0.9991	0.9804	0.9511	0.9012	0.7334	0.5000	0.2666	0.0988	0.0196	0.0009
	5	0.9999	0.9969	0.9900	0.9747	0.9006	0.7461	0.5174	0.2703	0.0856	0.0083
	6	1.0000	0.9997	0.9987	0.9957	0.9750	0.9102	0.7682	0.5372	0.2618	0.0530
	7		1.0000	0.9999	0.9996	0.9962	0.9805	0.9295	0.8040	0.5638	0.2252
	8			1.0000	1.0000	0.9997	0.9980	0.9899	0.9596	0.8658	0.6126
	9					1.0000	1.0000	1.0000	1.0000	1.0000	1.0000
10	0	0.3487	0.1074	0.0563	0.0282	0.0060	0.0010	0.0001	0.0000		
	1	0.7361	0.3758	0.2440	0.1493	0.0464	0.0107	0.0017	0.0001	0.0000	
	2	0.9298	0.6778	0.5256	0.3828	0.1673	0.0547	0.0123	0.0016	0.0001	
	3	0.9872	0.8791	0.7759	0.6496	0.3823	0.1719	0.0548	0.0106	0.0009	0.0000
	4	0.9984	0.9672	0.9219	0.8497	0.6331	0.3770	0.1662	0.0473	0.0064	0.0001
	5	0.9999	0.9936	0.9803	0.9527	0.8338	0.6230	0.3669	0.1503	0.0328	0.0016
	6	1.0000	0.9991	0.9965	0.9894	0.9452	0.8281	0.6177	0.3504	0.1209	0.0128
	7		0.9999	0.9996	0.9984	0.9877	0.9453	0.8327	0.6172	0.3222	0.0702
	8		1.0000	1.0000	0.9999	0.9983	0.9893	0.9536	0.8507	0.6242	0.2639
	9				1.0000	0.9999	0.9990	0.9940	0.9718	0.8926	0.6513
	10					1.0000	1.0000	1.0000	1.0000	1.0000	1.0000
11	0	0.3138	0.0859	0.0422	0.0198	0.0036	0.0005	0.0000			
	1	0.6974	0.3221	0.1971	0.1130	0.0302	0.0059	0.0007	0.0000		
	2	0.9104	0.6174	0.4552	0.3127	0.1189	0.0327	0.0059	0.0006	0.0000	
	3	0.9815	0.8389	0.7133	0.5696	0.2963	0.1133	0.0293	0.0043	0.0002	
	4	0.9972	0.9496	0.8854	0.7897	0.5328	0.2744	0.0994	0.0216	0.0020	0.0000
	5	0.9997	0.9883	0.9657	0.9218	0.7535	0.5000	0.2465	0.0782	0.0117	0.0003
	6	1.0000	0.9980	0.9924	0.9784	0.9006	0.7256	0.4672	0.2103	0.0504	0.0028
	7		0.9998	0.9988	0.9957	0.9707	0.8867	0.7037	0.4304	0.1611	0.0185
	8		1.0000	0.9999	0.9994	0.9941	0.9673	0.8811	0.6873	0.3826	0.0896
	9			1.0000	1.0000	0.9993	0.9941	0.9698	0.8870	0.6779	0.3026
	10					1.0000	0.9995	0.9964	0.9802	0.9141	0.6862
	11						1.0000	1.0000	1.0000	1.0000	1.0000

Table A.1 (continued) Binomial Probability Sums $\sum_{x=0}^{r} b(x; n, p)$

		p									
n	r	0.10	0.20	0.25	0.30	0.40	0.50	0.60	0.70	0.80	0.90
12	0	0.2824	0.0687	0.0317	0.0138	0.0022	0.0002	0.0000			
	1	0.6590	0.2749	0.1584	0.0850	0.0196	0.0032	0.0003	0.0000		
	2	0.8891	0.5583	0.3907	0.2528	0.0834	0.0193	0.0028	0.0002	0.0000	
	3	0.9744	0.7946	0.6488	0.4925	0.2253	0.0730	0.0153	0.0017	0.0001	
	4	0.9957	0.9274	0.8424	0.7237	0.4382	0.1938	0.0573	0.0095	0.0006	0.0000
	5	0.9995	0.9806	0.9456	0.8822	0.6652	0.3872	0.1582	0.0386	0.0039	0.0001
	6	0.9999	0.9961	0.9857	0.9614	0.8418	0.6128	0.3348	0.1178	0.0194	0.0005
	7	1.0000	0.9994	0.9972	0.9905	0.9427	0.8062	0.5618	0.2763	0.0726	0.0043
	8		0.9999	0.9996	0.9983	0.9847	0.9270	0.7747	0.5075	0.2054	0.0256
	9		1.0000	1.0000	0.9998	0.9972	0.9807	0.9166	0.7472	0.4417	0.1109
	10				1.0000	0.9997	0.9968	0.9804	0.9150	0.7251	0.3410
	11					1.0000	0.9998	0.9978	0.9862	0.9313	0.7176
	12						1.0000	1.0000	1.0000	1.0000	1.0000
13	0	0.2542	0.0550	0.0238	0.0097	0.0013	0.0001	0.0000			
	1	0.6213	0.2336	0.1267	0.0637	0.0126	0.0017	0.0001	0.0000		
	2	0.8661	0.5017	0.3326	0.2025	0.0579	0.0112	0.0013	0.0001		
	3	0.9658	0.7473	0.5843	0.4206	0.1686	0.0461	0.0078	0.0007	0.0000	
	4	0.9935	0.9009	0.7940	0.6543	0.3530	0.1334	0.0321	0.0040	0.0002	
	5	0.9991	0.9700	0.9198	0.8346	0.5744	0.2905	0.0977	0.0182	0.0012	0.0000
	6	0.9999	0.9930	0.9757	0.9376	0.7712	0.5000	0.2288	0.0624	0.0070	0.0001
	7	1.0000	0.9988	0.9944	0.9818	0.9023	0.7095	0.4256	0.1654	0.0300	0.0009
	8		0.9998	0.9990	0.9960	0.9679	0.8666	0.6470	0.3457	0.0991	0.0065
	9		1.0000	0.9999	0.9993	0.9922	0.9539	0.8314	0.5794	0.2527	0.0342
	10			1.0000	0.9999	0.9987	0.9888	0.9421	0.7975	0.4983	0.1339
	11				1.0000	0.9999	0.9983	0.9874	0.9363	0.7664	0.3787
	12					1.0000	0.9999	0.9987	0.9903	0.9450	0.7458
	13						1.0000	1.0000	1.0000	1.0000	1.0000
14	0	0.2288	0.0440	0.0178	0.0068	0.0008	0.0001	0.0000			
	1	0.5846	0.1979	0.1010	0.0475	0.0081	0.0009	0.0001			
	2	0.8416	0.4481	0.2811	0.1608	0.0398	0.0065	0.0006	0.0000		
	3	0.9559	0.6982	0.5213	0.3552	0.1243	0.0287	0.0039	0.0002		
	4	0.9908	0.8702	0.7415	0.5842	0.2793	0.0898	0.0175	0.0017	0.0000	
	5	0.9985	0.9561	0.8883	0.7805	0.4859	0.2120	0.0583	0.0083	0.0004	
	6	0.9998	0.9884	0.9617	0.9067	0.6925	0.3953	0.1501	0.0315	0.0024	0.0000
	7	1.0000	0.9976	0.9897	0.9685	0.8499	0.6047	0.3075	0.0933	0.0116	0.0002
	8		0.9996	0.9978	0.9917	0.9417	0.7880	0.5141	0.2195	0.0439	0.0015
	9		1.0000	0.9997	0.9983	0.9825	0.9102	0.7207	0.4158	0.1298	0.0092
	10			1.0000	0.9998	0.9961	0.9713	0.8757	0.6448	0.3018	0.0441
	11				1.0000	0.9994	0.9935	0.9602	0.8392	0.5519	0.1584
	12					0.9999	0.9991	0.9919	0.9525	0.8021	0.4154
	13					1.0000	0.9999	0.9992	0.9932	0.9560	0.7712
	14						1.0000	1.0000	1.0000	1.0000	1.0000

Table A.1 Binomial Probability Table

Table A.1 (continued) Binomial Probability Sums $\sum_{x=0}^{r} b(x; n, p)$

						p					
n	r	0.10	0.20	0.25	0.30	0.40	0.50	0.60	0.70	0.80	0.90
15	0	0.2059	0.0352	0.0134	0.0047	0.0005	0.0000				
	1	0.5490	0.1671	0.0802	0.0353	0.0052	0.0005	0.0000			
	2	0.8159	0.3980	0.2361	0.1268	0.0271	0.0037	0.0003	0.0000		
	3	0.9444	0.6482	0.4613	0.2969	0.0905	0.0176	0.0019	0.0001		
	4	0.9873	0.8358	0.6865	0.5155	0.2173	0.0592	0.0093	0.0007	0.0000	
	5	0.9978	0.9389	0.8516	0.7216	0.4032	0.1509	0.0338	0.0037	0.0001	
	6	0.9997	0.9819	0.9434	0.8689	0.6098	0.3036	0.0950	0.0152	0.0008	
	7	1.0000	0.9958	0.9827	0.9500	0.7869	0.5000	0.2131	0.0500	0.0042	0.0000
	8		0.9992	0.9958	0.9848	0.9050	0.6964	0.3902	0.1311	0.0181	0.0003
	9		0.9999	0.9992	0.9963	0.9662	0.8491	0.5968	0.2784	0.0611	0.0022
	10		1.0000	0.9999	0.9993	0.9907	0.9408	0.7827	0.4845	0.1642	0.0127
	11			1.0000	0.9999	0.9981	0.9824	0.9095	0.7031	0.3518	0.0556
	12				1.0000	0.9997	0.9963	0.9729	0.8732	0.6020	0.1841
	13					1.0000	0.9995	0.9948	0.9647	0.8329	0.4510
	14						1.0000	0.9995	0.9953	0.9648	0.7941
	15							1.0000	1.0000	1.0000	1.0000
16	0	0.1853	0.0281	0.0100	0.0033	0.0003	0.0000				
	1	0.5147	0.1407	0.0635	0.0261	0.0033	0.0003	0.0000			
	2	0.7892	0.3518	0.1971	0.0994	0.0183	0.0021	0.0001			
	3	0.9316	0.5981	0.4050	0.2459	0.0651	0.0106	0.0009	0.0000		
	4	0.9830	0.7982	0.6302	0.4499	0.1666	0.0384	0.0049	0.0003		
	5	0.9967	0.9183	0.8103	0.6598	0.3288	0.1051	0.0191	0.0016	0.0000	
	6	0.9995	0.9733	0.9204	0.8247	0.5272	0.2272	0.0583	0.0071	0.0002	
	7	0.9999	0.9930	0.9729	0.9256	0.7161	0.4018	0.1423	0.0257	0.0015	0.0000
	8	1.0000	0.9985	0.9925	0.9743	0.8577	0.5982	0.2839	0.0744	0.0070	0.0001
	9		0.9998	0.9984	0.9929	0.9417	0.7728	0.4728	0.1753	0.0267	0.0005
	10		1.0000	0.9997	0.9984	0.9809	0.8949	0.6712	0.3402	0.0817	0.0033
	11			1.0000	0.9997	0.9951	0.9616	0.8334	0.5501	0.2018	0.0170
	12				1.0000	0.9991	0.9894	0.9349	0.7541	0.4019	0.0684
	13					0.9999	0.9979	0.9817	0.9006	0.6482	0.2108
	14					1.0000	0.9997	0.9967	0.9739	0.8593	0.4853
	15						1.0000	0.9997	0.9967	0.9719	0.8147
	16							1.0000	1.0000	1.0000	1.0000

Table A.1 (continued) Binomial Probability Sums $\sum_{x=0}^{r} b(x; n, p)$

						p					
n	r	0.10	0.20	0.25	0.30	0.40	0.50	0.60	0.70	0.80	0.90
17	0	0.1668	0.0225	0.0075	0.0023	0.0002	0.0000				
	1	0.4818	0.1182	0.0501	0.0193	0.0021	0.0001	0.0000			
	2	0.7618	0.3096	0.1637	0.0774	0.0123	0.0012	0.0001			
	3	0.9174	0.5489	0.3530	0.2019	0.0464	0.0064	0.0005	0.0000		
	4	0.9779	0.7582	0.5739	0.3887	0.1260	0.0245	0.0025	0.0001		
	5	0.9953	0.8943	0.7653	0.5968	0.2639	0.0717	0.0106	0.0007	0.0000	
	6	0.9992	0.9623	0.8929	0.7752	0.4478	0.1662	0.0348	0.0032	0.0001	
	7	0.9999	0.9891	0.9598	0.8954	0.6405	0.3145	0.0919	0.0127	0.0005	
	8	1.0000	0.9974	0.9876	0.9597	0.8011	0.5000	0.1989	0.0403	0.0026	0.0000
	9		0.9995	0.9969	0.9873	0.9081	0.6855	0.3595	0.1046	0.0109	0.0001
	10		0.9999	0.9994	0.9968	0.9652	0.8338	0.5522	0.2248	0.0377	0.0008
	11		1.0000	0.9999	0.9993	0.9894	0.9283	0.7361	0.4032	0.1057	0.0047
	12			1.0000	0.9999	0.9975	0.9755	0.8740	0.6113	0.2418	0.0221
	13				1.0000	0.9995	0.9936	0.9536	0.7981	0.4511	0.0826
	14					0.9999	0.9988	0.9877	0.9226	0.6904	0.2382
	15					1.0000	0.9999	0.9979	0.9807	0.8818	0.5182
	16						1.0000	0.9998	0.9977	0.9775	0.8332
	17							1.0000	1.0000	1.0000	1.0000
18	0	0.1501	0.0180	0.0056	0.0016	0.0001	0.0000				
	1	0.4503	0.0991	0.0395	0.0142	0.0013	0.0001				
	2	0.7338	0.2713	0.1353	0.0600	0.0082	0.0007	0.0000			
	3	0.9018	0.5010	0.3057	0.1646	0.0328	0.0038	0.0002			
	4	0.9718	0.7164	0.5187	0.3327	0.0942	0.0154	0.0013	0.0000		
	5	0.9936	0.8671	0.7175	0.5344	0.2088	0.0481	0.0058	0.0003		
	6	0.9988	0.9487	0.8610	0.7217	0.3743	0.1189	0.0203	0.0014	0.0000	
	7	0.9998	0.9837	0.9431	0.8593	0.5634	0.2403	0.0576	0.0061	0.0002	
	8	1.0000	0.9957	0.9807	0.9404	0.7368	0.4073	0.1347	0.0210	0.0009	
	9		0.9991	0.9946	0.9790	0.8653	0.5927	0.2632	0.0596	0.0043	0.0000
	10		0.9998	0.9988	0.9939	0.9424	0.7597	0.4366	0.1407	0.0163	0.0002
	11		1.0000	0.9998	0.9986	0.9797	0.8811	0.6257	0.2783	0.0513	0.0012
	12			1.0000	0.9997	0.9942	0.9519	0.7912	0.4656	0.1329	0.0064
	13				1.0000	0.9987	0.9846	0.9058	0.6673	0.2836	0.0282
	14					0.9998	0.9962	0.9672	0.8354	0.4990	0.0982
	15					1.0000	0.9993	0.9918	0.9400	0.7287	0.2662
	16						0.9999	0.9987	0.9858	0.9009	0.5497
	17						1.0000	0.9999	0.9984	0.9820	0.8499
	18							1.0000	1.0000	1.0000	1.0000

Table A.1 Binomial Probability Table

Table A.1 (continued) Binomial Probability Sums $\sum_{x=0}^{r} b(x;n,p)$

						p					
n	r	0.10	0.20	0.25	0.30	0.40	0.50	0.60	0.70	0.80	0.90
19	0	0.1351	0.0144	0.0042	0.0011	0.0001					
	1	0.4203	0.0829	0.0310	0.0104	0.0008	0.0000				
	2	0.7054	0.2369	0.1113	0.0462	0.0055	0.0004	0.0000			
	3	0.8850	0.4551	0.2631	0.1332	0.0230	0.0022	0.0001			
	4	0.9648	0.6733	0.4654	0.2822	0.0696	0.0096	0.0006	0.0000		
	5	0.9914	0.8369	0.6678	0.4739	0.1629	0.0318	0.0031	0.0001		
	6	0.9983	0.9324	0.8251	0.6655	0.3081	0.0835	0.0116	0.0006		
	7	0.9997	0.9767	0.9225	0.8180	0.4878	0.1796	0.0352	0.0028	0.0000	
	8	1.0000	0.9933	0.9713	0.9161	0.6675	0.3238	0.0885	0.0105	0.0003	
	9		0.9984	0.9911	0.9674	0.8139	0.5000	0.1861	0.0326	0.0016	
	10		0.9997	0.9977	0.9895	0.9115	0.6762	0.3325	0.0839	0.0067	0.0000
	11		1.0000	0.9995	0.9972	0.9648	0.8204	0.5122	0.1820	0.0233	0.0003
	12			0.9999	0.9994	0.9884	0.9165	0.6919	0.3345	0.0676	0.0017
	13			1.0000	0.9999	0.9969	0.9682	0.8371	0.5261	0.1631	0.0086
	14				1.0000	0.9994	0.9904	0.9304	0.7178	0.3267	0.0352
	15					0.9999	0.9978	0.9770	0.8668	0.5449	0.1150
	16					1.0000	0.9996	0.9945	0.9538	0.7631	0.2946
	17						1.0000	0.9992	0.9896	0.9171	0.5797
	18							0.9999	0.9989	0.9856	0.8649
	19							1.0000	1.0000	1.0000	1.0000
20	0	0.1216	0.0115	0.0032	0.0008	0.0000					
	1	0.3917	0.0692	0.0243	0.0076	0.0005	0.0000				
	2	0.6769	0.2061	0.0913	0.0355	0.0036	0.0002				
	3	0.8670	0.4114	0.2252	0.1071	0.0160	0.0013	0.0000			
	4	0.9568	0.6296	0.4148	0.2375	0.0510	0.0059	0.0003			
	5	0.9887	0.8042	0.6172	0.4164	0.1256	0.0207	0.0016	0.0000		
	6	0.9976	0.9133	0.7858	0.6080	0.2500	0.0577	0.0065	0.0003		
	7	0.9996	0.9679	0.8982	0.7723	0.4159	0.1316	0.0210	0.0013	0.0000	
	8	0.9999	0.9900	0.9591	0.8867	0.5956	0.2517	0.0565	0.0051	0.0001	
	9	1.0000	0.9974	0.9861	0.9520	0.7553	0.4119	0.1275	0.0171	0.0006	
	10		0.9994	0.9961	0.9829	0.8725	0.5881	0.2447	0.0480	0.0026	0.0000
	11		0.9999	0.9991	0.9949	0.9435	0.7483	0.4044	0.1133	0.0100	0.0001
	12		1.0000	0.9998	0.9987	0.9790	0.8684	0.5841	0.2277	0.0321	0.0004
	13			1.0000	0.9997	0.9935	0.9423	0.7500	0.3920	0.0867	0.0024
	14				1.0000	0.9984	0.9793	0.8744	0.5836	0.1958	0.0113
	15					0.9997	0.9941	0.9490	0.7625	0.3704	0.0432
	16					1.0000	0.9987	0.9840	0.8929	0.5886	0.1330
	17						0.9998	0.9964	0.9645	0.7939	0.3231
	18						1.0000	0.9995	0.9924	0.9308	0.6083
	19							1.0000	0.9992	0.9885	0.8784
	20								1.0000	1.0000	1.0000

Table A.2 Poisson Probability Sums $\sum_{x=0}^{r} p(x;\mu)$

					μ				
r	0.1	0.2	0.3	0.4	0.5	0.6	0.7	0.8	0.9
0	0.9048	0.8187	0.7408	0.6703	0.6065	0.5488	0.4966	0.4493	0.4066
1	0.9953	0.9825	0.9631	0.9384	0.9098	0.8781	0.8442	0.8088	0.7725
2	0.9998	0.9989	0.9964	0.9921	0.9856	0.9769	0.9659	0.9526	0.9371
3	1.0000	0.9999	0.9997	0.9992	0.9982	0.9966	0.9942	0.9909	0.9865
4		1.0000	1.0000	0.9999	0.9998	0.9996	0.9992	0.9986	0.9977
5				1.0000	1.0000	1.0000	0.9999	0.9998	0.9997
6							1.0000	1.0000	1.0000

					μ				
r	1.0	1.5	2.0	2.5	3.0	3.5	4.0	4.5	5.0
0	0.3679	0.2231	0.1353	0.0821	0.0498	0.0302	0.0183	0.0111	0.0067
1	0.7358	0.5578	0.4060	0.2873	0.1991	0.1359	0.0916	0.0611	0.0404
2	0.9197	0.8088	0.6767	0.5438	0.4232	0.3208	0.2381	0.1736	0.1247
3	0.9810	0.9344	0.8571	0.7576	0.6472	0.5366	0.4335	0.3423	0.2650
4	0.9963	0.9814	0.9473	0.8912	0.8153	0.7254	0.6288	0.5321	0.4405
5	0.9994	0.9955	0.9834	0.9580	0.9161	0.8576	0.7851	0.7029	0.6160
6	0.9999	0.9991	0.9955	0.9858	0.9665	0.9347	0.8893	0.8311	0.7622
7	1.0000	0.9998	0.9989	0.9958	0.9881	0.9733	0.9489	0.9134	0.8666
8		1.0000	0.9998	0.9989	0.9962	0.9901	0.9786	0.9597	0.9319
9			1.0000	0.9997	0.9989	0.9967	0.9919	0.9829	0.9682
10				0.9999	0.9997	0.9990	0.9972	0.9933	0.9863
11				1.0000	0.9999	0.9997	0.9991	0.9976	0.9945
12					1.0000	0.9999	0.9997	0.9992	0.9980
13						1.0000	0.9999	0.9997	0.9993
14							1.0000	0.9999	0.9998
15								1.0000	0.9999
16									1.0000

Table A.2 Poisson Probability Table

Table A.2 (continued) Poisson Probability Sums $\sum_{x=0}^{r} p(x;\mu)$

r	\multicolumn{9}{c}{μ}								
	5.5	6.0	6.5	7.0	7.5	8.0	8.5	9.0	9.5
0	0.0041	0.0025	0.0015	0.0009	0.0006	0.0003	0.0002	0.0001	0.0001
1	0.0266	0.0174	0.0113	0.0073	0.0047	0.0030	0.0019	0.0012	0.0008
2	0.0884	0.0620	0.0430	0.0296	0.0203	0.0138	0.0093	0.0062	0.0042
3	0.2017	0.1512	0.1118	0.0818	0.0591	0.0424	0.0301	0.0212	0.0149
4	0.3575	0.2851	0.2237	0.1730	0.1321	0.0996	0.0744	0.0550	0.0403
5	0.5289	0.4457	0.3690	0.3007	0.2414	0.1912	0.1496	0.1157	0.0885
6	0.6860	0.6063	0.5265	0.4497	0.3782	0.3134	0.2562	0.2068	0.1649
7	0.8095	0.7440	0.6728	0.5987	0.5246	0.4530	0.3856	0.3239	0.2687
8	0.8944	0.8472	0.7916	0.7291	0.6620	0.5925	0.5231	0.4557	0.3918
9	0.9462	0.9161	0.8774	0.8305	0.7764	0.7166	0.6530	0.5874	0.5218
10	0.9747	0.9574	0.9332	0.9015	0.8622	0.8159	0.7634	0.7060	0.6453
11	0.9890	0.9799	0.9661	0.9467	0.9208	0.8881	0.8487	0.8030	0.7520
12	0.9955	0.9912	0.9840	0.9730	0.9573	0.9362	0.9091	0.8758	0.8364
13	0.9983	0.9964	0.9929	0.9872	0.9784	0.9658	0.9486	0.9261	0.8981
14	0.9994	0.9986	0.9970	0.9943	0.9897	0.9827	0.9726	0.9585	0.9400
15	0.9998	0.9995	0.9988	0.9976	0.9954	0.9918	0.9862	0.9780	0.9665
16	0.9999	0.9998	0.9996	0.9990	0.9980	0.9963	0.9934	0.9889	0.9823
17	1.0000	0.9999	0.9998	0.9996	0.9992	0.9984	0.9970	0.9947	0.9911
18		1.0000	0.9999	0.9999	0.9997	0.9993	0.9987	0.9976	0.9957
19			1.0000	1.0000	0.9999	0.9997	0.9995	0.9989	0.9980
20						0.9999	0.9998	0.9996	0.9991
21						1.0000	0.9999	0.9998	0.9996
22							1.0000	0.9999	0.9999
23								1.0000	0.9999
24									1.0000

Table A.2 (continued) Poisson Probability Sums $\sum_{x=0}^{r} p(x;\mu)$

r	10.0	11.0	12.0	13.0	14.0	15.0	16.0	17.0	18.0
0	0.0000	0.0000	0.0000						
1	0.0005	0.0002	0.0001	0.0000	0.0000				
2	0.0028	0.0012	0.0005	0.0002	0.0001	0.0000	0.0000		
3	0.0103	0.0049	0.0023	0.0011	0.0005	0.0002	0.0001	0.0000	0.0000
4	0.0293	0.0151	0.0076	0.0037	0.0018	0.0009	0.0004	0.0002	0.0001
5	0.0671	0.0375	0.0203	0.0107	0.0055	0.0028	0.0014	0.0007	0.0003
6	0.1301	0.0786	0.0458	0.0259	0.0142	0.0076	0.0040	0.0021	0.0010
7	0.2202	0.1432	0.0895	0.0540	0.0316	0.0180	0.0100	0.0054	0.0029
8	0.3328	0.2320	0.1550	0.0998	0.0621	0.0374	0.0220	0.0126	0.0071
9	0.4579	0.3405	0.2424	0.1658	0.1094	0.0699	0.0433	0.0261	0.0154
10	0.5830	0.4599	0.3472	0.2517	0.1757	0.1185	0.0774	0.0491	0.0304
11	0.6968	0.5793	0.4616	0.3532	0.2600	0.1848	0.1270	0.0847	0.0549
12	0.7916	0.6887	0.5760	0.4631	0.3585	0.2676	0.1931	0.1350	0.0917
13	0.8645	0.7813	0.6815	0.5730	0.4644	0.3632	0.2745	0.2009	0.1426
14	0.9165	0.8540	0.7720	0.6751	0.5704	0.4657	0.3675	0.2808	0.2081
15	0.9513	0.9074	0.8444	0.7636	0.6694	0.5681	0.4667	0.3715	0.2867
16	0.9730	0.9441	0.8987	0.8355	0.7559	0.6641	0.5660	0.4677	0.3751
17	0.9857	0.9678	0.9370	0.8905	0.8272	0.7489	0.6593	0.5640	0.4686
18	0.9928	0.9823	0.9626	0.9302	0.8826	0.8195	0.7423	0.6550	0.5622
19	0.9965	0.9907	0.9787	0.9573	0.9235	0.8752	0.8122	0.7363	0.6509
20	0.9984	0.9953	0.9884	0.9750	0.9521	0.9170	0.8682	0.8055	0.7307
21	0.9993	0.9977	0.9939	0.9859	0.9712	0.9469	0.9108	0.8615	0.7991
22	0.9997	0.9990	0.9970	0.9924	0.9833	0.9673	0.9418	0.9047	0.8551
23	0.9999	0.9995	0.9985	0.9960	0.9907	0.9805	0.9633	0.9367	0.8989
24	1.0000	0.9998	0.9993	0.9980	0.9950	0.9888	0.9777	0.9594	0.9317
25		0.9999	0.9997	0.9990	0.9974	0.9938	0.9869	0.9748	0.9554
26		1.0000	0.9999	0.9995	0.9987	0.9967	0.9925	0.9848	0.9718
27			0.9999	0.9998	0.9994	0.9983	0.9959	0.9912	0.9827
28			1.0000	0.9999	0.9997	0.9991	0.9978	0.9950	0.9897
29				1.0000	0.9999	0.9996	0.9989	0.9973	0.9941
30					0.9999	0.9998	0.9994	0.9986	0.9967
31					1.0000	0.9999	0.9997	0.9993	0.9982
32						1.0000	0.9999	0.9996	0.9990
33							0.9999	0.9998	0.9995
34							1.0000	0.9999	0.9998
35								1.0000	0.9999
36									0.9999
37									1.0000

Table A.3   Normal Probability Table

Table A.3 Areas under the Normal Curve

z	.00	.01	.02	.03	.04	.05	.06	.07	.08	.09
−3.4	0.0003	0.0003	0.0003	0.0003	0.0003	0.0003	0.0003	0.0003	0.0003	0.0002
−3.3	0.0005	0.0005	0.0005	0.0004	0.0004	0.0004	0.0004	0.0004	0.0004	0.0003
−3.2	0.0007	0.0007	0.0006	0.0006	0.0006	0.0006	0.0006	0.0005	0.0005	0.0005
−3.1	0.0010	0.0009	0.0009	0.0009	0.0008	0.0008	0.0008	0.0008	0.0007	0.0007
−3.0	0.0013	0.0013	0.0013	0.0012	0.0012	0.0011	0.0011	0.0011	0.0010	0.0010
−2.9	0.0019	0.0018	0.0018	0.0017	0.0016	0.0016	0.0015	0.0015	0.0014	0.0014
−2.8	0.0026	0.0025	0.0024	0.0023	0.0023	0.0022	0.0021	0.0021	0.0020	0.0019
−2.7	0.0035	0.0034	0.0033	0.0032	0.0031	0.0030	0.0029	0.0028	0.0027	0.0026
−2.6	0.0047	0.0045	0.0044	0.0043	0.0041	0.0040	0.0039	0.0038	0.0037	0.0036
−2.5	0.0062	0.0060	0.0059	0.0057	0.0055	0.0054	0.0052	0.0051	0.0049	0.0048
−2.4	0.0082	0.0080	0.0078	0.0075	0.0073	0.0071	0.0069	0.0068	0.0066	0.0064
−2.3	0.0107	0.0104	0.0102	0.0099	0.0096	0.0094	0.0091	0.0089	0.0087	0.0084
−2.2	0.0139	0.0136	0.0132	0.0129	0.0125	0.0122	0.0119	0.0116	0.0113	0.0110
−2.1	0.0179	0.0174	0.0170	0.0166	0.0162	0.0158	0.0154	0.0150	0.0146	0.0143
−2.0	0.0228	0.0222	0.0217	0.0212	0.0207	0.0202	0.0197	0.0192	0.0188	0.0183
−1.9	0.0287	0.0281	0.0274	0.0268	0.0262	0.0256	0.0250	0.0244	0.0239	0.0233
−1.8	0.0359	0.0351	0.0344	0.0336	0.0329	0.0322	0.0314	0.0307	0.0301	0.0294
−1.7	0.0446	0.0436	0.0427	0.0418	0.0409	0.0401	0.0392	0.0384	0.0375	0.0367
−1.6	0.0548	0.0537	0.0526	0.0516	0.0505	0.0495	0.0485	0.0475	0.0465	0.0455
−1.5	0.0668	0.0655	0.0643	0.0630	0.0618	0.0606	0.0594	0.0582	0.0571	0.0559
−1.4	0.0808	0.0793	0.0778	0.0764	0.0749	0.0735	0.0721	0.0708	0.0694	0.0681
−1.3	0.0968	0.0951	0.0934	0.0918	0.0901	0.0885	0.0869	0.0853	0.0838	0.0823
−1.2	0.1151	0.1131	0.1112	0.1093	0.1075	0.1056	0.1038	0.1020	0.1003	0.0985
−1.1	0.1357	0.1335	0.1314	0.1292	0.1271	0.1251	0.1230	0.1210	0.1190	0.1170
−1.0	0.1587	0.1562	0.1539	0.1515	0.1492	0.1469	0.1446	0.1423	0.1401	0.1379
−0.9	0.1841	0.1814	0.1788	0.1762	0.1736	0.1711	0.1685	0.1660	0.1635	0.1611
−0.8	0.2119	0.2090	0.2061	0.2033	0.2005	0.1977	0.1949	0.1922	0.1894	0.1867
−0.7	0.2420	0.2389	0.2358	0.2327	0.2296	0.2266	0.2236	0.2206	0.2177	0.2148
−0.6	0.2743	0.2709	0.2676	0.2643	0.2611	0.2578	0.2546	0.2514	0.2483	0.2451
−0.5	0.3085	0.3050	0.3015	0.2981	0.2946	0.2912	0.2877	0.2843	0.2810	0.2776
−0.4	0.3446	0.3409	0.3372	0.3336	0.3300	0.3264	0.3228	0.3192	0.3156	0.3121
−0.3	0.3821	0.3783	0.3745	0.3707	0.3669	0.3632	0.3594	0.3557	0.3520	0.3483
−0.2	0.4207	0.4168	0.4129	0.4090	0.4052	0.4013	0.3974	0.3936	0.3897	0.3859
−0.1	0.4602	0.4562	0.4522	0.4483	0.4443	0.4404	0.4364	0.4325	0.4286	0.4247
−0.0	0.5000	0.4960	0.4920	0.4880	0.4840	0.4801	0.4761	0.4721	0.4681	0.4641

Table A.3 (continued) Areas under the Normal Curve

z	.00	.01	.02	.03	.04	.05	.06	.07	.08	.09
0.0	0.5000	0.5040	0.5080	0.5120	0.5160	0.5199	0.5239	0.5279	0.5319	0.5359
0.1	0.5398	0.5438	0.5478	0.5517	0.5557	0.5596	0.5636	0.5675	0.5714	0.5753
0.2	0.5793	0.5832	0.5871	0.5910	0.5948	0.5987	0.6026	0.6064	0.6103	0.6141
0.3	0.6179	0.6217	0.6255	0.6293	0.6331	0.6368	0.6406	0.6443	0.6480	0.6517
0.4	0.6554	0.6591	0.6628	0.6664	0.6700	0.6736	0.6772	0.6808	0.6844	0.6879
0.5	0.6915	0.6950	0.6985	0.7019	0.7054	0.7088	0.7123	0.7157	0.7190	0.7224
0.6	0.7257	0.7291	0.7324	0.7357	0.7389	0.7422	0.7454	0.7486	0.7517	0.7549
0.7	0.7580	0.7611	0.7642	0.7673	0.7704	0.7734	0.7764	0.7794	0.7823	0.7852
0.8	0.7881	0.7910	0.7939	0.7967	0.7995	0.8023	0.8051	0.8078	0.8106	0.8133
0.9	0.8159	0.8186	0.8212	0.8238	0.8264	0.8289	0.8315	0.8340	0.8365	0.8389
1.0	0.8413	0.8438	0.8461	0.8485	0.8508	0.8531	0.8554	0.8577	0.8599	0.8621
1.1	0.8643	0.8665	0.8686	0.8708	0.8729	0.8749	0.8770	0.8790	0.8810	0.8830
1.2	0.8849	0.8869	0.8888	0.8907	0.8925	0.8944	0.8962	0.8980	0.8997	0.9015
1.3	0.9032	0.9049	0.9066	0.9082	0.9099	0.9115	0.9131	0.9147	0.9162	0.9177
1.4	0.9192	0.9207	0.9222	0.9236	0.9251	0.9265	0.9279	0.9292	0.9306	0.9319
1.5	0.9332	0.9345	0.9357	0.9370	0.9382	0.9394	0.9406	0.9418	0.9429	0.9441
1.6	0.9452	0.9463	0.9474	0.9484	0.9495	0.9505	0.9515	0.9525	0.9535	0.9545
1.7	0.9554	0.9564	0.9573	0.9582	0.9591	0.9599	0.9608	0.9616	0.9625	0.9633
1.8	0.9641	0.9649	0.9656	0.9664	0.9671	0.9678	0.9686	0.9693	0.9699	0.9706
1.9	0.9713	0.9719	0.9726	0.9732	0.9738	0.9744	0.9750	0.9756	0.9761	0.9767
2.0	0.9772	0.9778	0.9783	0.9788	0.9793	0.9798	0.9803	0.9808	0.9812	0.9817
2.1	0.9821	0.9826	0.9830	0.9834	0.9838	0.9842	0.9846	0.9850	0.9854	0.9857
2.2	0.9861	0.9864	0.9868	0.9871	0.9875	0.9878	0.9881	0.9884	0.9887	0.9890
2.3	0.9893	0.9896	0.9898	0.9901	0.9904	0.9906	0.9909	0.9911	0.9913	0.9916
2.4	0.9918	0.9920	0.9922	0.9925	0.9927	0.9929	0.9931	0.9932	0.9934	0.9936
2.5	0.9938	0.9940	0.9941	0.9943	0.9945	0.9946	0.9948	0.9949	0.9951	0.9952
2.6	0.9953	0.9955	0.9956	0.9957	0.9959	0.9960	0.9961	0.9962	0.9963	0.9964
2.7	0.9965	0.9966	0.9967	0.9968	0.9969	0.9970	0.9971	0.9972	0.9973	0.9974
2.8	0.9974	0.9975	0.9976	0.9977	0.9977	0.9978	0.9979	0.9979	0.9980	0.9981
2.9	0.9981	0.9982	0.9982	0.9983	0.9984	0.9984	0.9985	0.9985	0.9986	0.9986
3.0	0.9987	0.9987	0.9987	0.9988	0.9988	0.9989	0.9989	0.9989	0.9990	0.9990
3.1	0.9990	0.9991	0.9991	0.9991	0.9992	0.9992	0.9992	0.9992	0.9993	0.9993
3.2	0.9993	0.9993	0.9994	0.9994	0.9994	0.9994	0.9994	0.9995	0.9995	0.9995
3.3	0.9995	0.9995	0.9995	0.9996	0.9996	0.9996	0.9996	0.9996	0.9996	0.9997
3.4	0.9997	0.9997	0.9997	0.9997	0.9997	0.9997	0.9997	0.9997	0.9997	0.9998

Table A.4   Student t-Distribution Probability Table

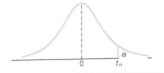

Table A.4 Critical Values of the t-Distribution

v	α						
	0.40	0.30	0.20	0.15	0.10	0.05	0.025
1	0.325	0.727	1.376	1.963	3.078	6.314	12.706
2	0.289	0.617	1.061	1.386	1.886	2.920	4.303
3	0.277	0.584	0.978	1.250	1.638	2.353	3.182
4	0.271	0.569	0.941	1.190	1.533	2.132	2.776
5	0.267	0.559	0.920	1.156	1.476	2.015	2.571
6	0.265	0.553	0.906	1.134	1.440	1.943	2.447
7	0.263	0.549	0.896	1.119	1.415	1.895	2.365
8	0.262	0.546	0.889	1.108	1.397	1.860	2.306
9	0.261	0.543	0.883	1.100	1.383	1.833	2.262
10	0.260	0.542	0.879	1.093	1.372	1.812	2.228
11	0.260	0.540	0.876	1.088	1.363	1.796	2.201
12	0.259	0.539	0.873	1.083	1.356	1.782	2.179
13	0.259	0.538	0.870	1.079	1.350	1.771	2.160
14	0.258	0.537	0.868	1.076	1.345	1.761	2.145
15	0.258	0.536	0.866	1.074	1.341	1.753	2.131
16	0.258	0.535	0.865	1.071	1.337	1.746	2.120
17	0.257	0.534	0.863	1.069	1.333	1.740	2.110
18	0.257	0.534	0.862	1.067	1.330	1.734	2.101
19	0.257	0.533	0.861	1.066	1.328	1.729	2.093
20	0.257	0.533	0.860	1.064	1.325	1.725	2.086
21	0.257	0.532	0.859	1.063	1.323	1.721	2.080
22	0.256	0.532	0.858	1.061	1.321	1.717	2.074
23	0.256	0.532	0.858	1.060	1.319	1.714	2.069
24	0.256	0.531	0.857	1.059	1.318	1.711	2.064
25	0.256	0.531	0.856	1.058	1.316	1.708	2.060
26	0.256	0.531	0.856	1.058	1.315	1.706	2.056
27	0.256	0.531	0.855	1.057	1.314	1.703	2.052
28	0.256	0.530	0.855	1.056	1.313	1.701	2.048
29	0.256	0.530	0.854	1.055	1.311	1.699	2.045
30	0.256	0.530	0.854	1.055	1.310	1.697	2.042
40	0.255	0.529	0.851	1.050	1.303	1.684	2.021
60	0.254	0.527	0.848	1.045	1.296	1.671	2.000
120	0.254	0.526	0.845	1.041	1.289	1.658	1.980
∞	0.253	0.524	0.842	1.036	1.282	1.645	1.960

Table A.4 (continued) Critical Values of the t-Distribution

v	\multicolumn{7}{c	}{α}
	0.02	0.015	0.01	0.0075	0.005	0.0025	0.0005
1	15.894	21.205	31.821	42.433	63.656	127.321	636.578
2	4.849	5.643	6.965	8.073	9.925	14.089	31.600
3	3.482	3.896	4.541	5.047	5.841	7.453	12.924
4	2.999	3.298	3.747	4.088	4.604	5.598	8.610
5	2.757	3.003	3.365	3.634	4.032	4.773	6.869
6	2.612	2.829	3.143	3.372	3.707	4.317	5.959
7	2.517	2.715	2.998	3.203	3.499	4.029	5.408
8	2.449	2.634	2.896	3.085	3.355	3.833	5.041
9	2.398	2.574	2.821	2.998	3.250	3.690	4.781
10	2.359	2.527	2.764	2.932	3.169	3.581	4.587
11	2.328	2.491	2.718	2.879	3.106	3.497	4.437
12	2.303	2.461	2.681	2.836	3.055	3.428	4.318
13	2.282	2.436	2.650	2.801	3.012	3.372	4.221
14	2.264	2.415	2.624	2.771	2.977	3.326	4.140
15	2.249	2.397	2.602	2.746	2.947	3.286	4.073
16	2.235	2.382	2.583	2.724	2.921	3.252	4.015
17	2.224	2.368	2.567	2.706	2.898	3.222	3.965
18	2.214	2.356	2.552	2.689	2.878	3.197	3.922
19	2.205	2.346	2.539	2.674	2.861	3.174	3.883
20	2.197	2.336	2.528	2.661	2.845	3.153	3.850
21	2.189	2.328	2.518	2.649	2.831	3.135	3.819
22	2.183	2.320	2.508	2.639	2.819	3.119	3.792
23	2.177	2.313	2.500	2.629	2.807	3.104	3.768
24	2.172	2.307	2.492	2.620	2.797	3.091	3.745
25	2.167	2.301	2.485	2.612	2.787	3.078	3.725
26	2.162	2.296	2.479	2.605	2.779	3.067	3.707
27	2.158	2.291	2.473	2.598	2.771	3.057	3.689
28	2.154	2.286	2.467	2.592	2.763	3.047	3.674
29	2.150	2.282	2.462	2.586	2.756	3.038	3.660
30	2.147	2.278	2.457	2.581	2.750	3.030	3.646
40	2.123	2.250	2.423	2.542	2.704	2.971	3.551
60	2.099	2.223	2.390	2.504	2.660	2.915	3.460
120	2.076	2.196	2.358	2.468	2.617	2.860	3.373
∞	2.054	2.170	2.326	2.432	2.576	2.807	3.290

Table A.5 Chi-Squared Distribution Probability Table

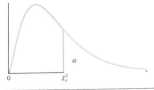

Table A.5 Critical Values of the Chi-Squared Distribution

v	α									
	0.995	0.99	0.98	0.975	0.95	0.90	0.80	0.75	0.70	0.50
1	0.0^4393	0.0^3157	0.0^3628	0.0^3982	0.00393	0.0158	0.0642	0.102	0.148	0.455
2	0.0100	0.0201	0.0404	0.0506	0.103	0.211	0.446	0.575	0.713	1.386
3	0.0717	0.115	0.185	0.216	0.352	0.584	1.005	1.213	1.424	2.366
4	0.207	0.297	0.429	0.484	0.711	1.064	1.649	1.923	2.195	3.357
5	0.412	0.554	0.752	0.831	1.145	1.610	2.343	2.675	3.000	4.351
6	0.676	0.872	1.134	1.237	1.635	2.204	3.070	3.455	3.828	5.348
7	0.989	1.239	1.564	1.690	2.167	2.833	3.822	4.255	4.671	6.346
8	1.344	1.647	2.032	2.180	2.733	3.490	4.594	5.071	5.527	7.344
9	1.735	2.088	2.532	2.700	3.325	4.168	5.380	5.899	6.393	8.343
10	2.156	2.558	3.059	3.247	3.940	4.865	6.179	6.737	7.267	9.342
11	2.603	3.053	3.609	3.816	4.575	5.578	6.989	7.584	8.148	10.341
12	3.074	3.571	4.178	4.404	5.226	6.304	7.807	8.438	9.034	11.340
13	3.565	4.107	4.765	5.009	5.892	7.041	8.634	9.299	9.926	12.340
14	4.075	4.660	5.368	5.629	6.571	7.790	9.467	10.165	10.821	13.339
15	4.601	5.229	5.985	6.262	7.261	8.547	10.307	11.037	11.721	14.339
16	5.142	5.812	6.614	6.908	7.962	9.312	11.152	11.912	12.624	15.338
17	5.697	6.408	7.255	7.564	8.672	10.085	12.002	12.792	13.531	16.338
18	6.265	7.015	7.906	8.231	9.390	10.865	12.857	13.675	14.440	17.338
19	6.844	7.633	8.567	8.907	10.117	11.651	13.716	14.562	15.352	18.338
20	7.434	8.260	9.237	9.591	10.851	12.443	14.578	15.452	16.266	19.337
21	8.034	8.897	9.915	10.283	11.591	13.240	15.445	16.344	17.182	20.337
22	8.643	9.542	10.600	10.982	12.338	14.041	16.314	17.240	18.101	21.337
23	9.260	10.196	11.293	11.689	13.091	14.848	17.187	18.137	19.021	22.337
24	9.886	10.856	11.992	12.401	13.848	15.659	18.062	19.037	19.943	23.337
25	10.520	11.524	12.697	13.120	14.611	16.473	18.940	19.939	20.867	24.337
26	11.160	12.198	13.409	13.844	15.379	17.292	19.820	20.843	21.792	25.336
27	11.808	12.878	14.125	14.573	16.151	18.114	20.703	21.749	22.719	26.336
28	12.461	13.565	14.847	15.308	16.928	18.939	21.588	22.657	23.647	27.336
29	13.121	14.256	15.574	16.047	17.708	19.768	22.475	23.567	24.577	28.336
30	13.787	14.953	16.306	16.791	18.493	20.599	23.364	24.478	25.508	29.336
40	20.707	22.164	23.838	24.433	26.509	29.051	32.345	33.66	34.872	39.335
50	27.991	29.707	31.664	32.357	34.764	37.689	41.449	42.942	44.313	49.335
60	35.534	37.485	39.699	40.482	43.188	46.459	50.641	52.294	53.809	59.335

Table A.5 (continued) Critical Values of the Chi-Squared Distribution

v	\multicolumn{10}{c}{α}
	0.30	0.25	0.20	0.10	0.05	0.025	0.02	0.01	0.005	0.001
1	1.074	1.323	1.642	2.706	3.841	5.024	5.412	6.635	7.879	10.827
2	2.408	2.773	3.219	4.605	5.991	7.378	7.824	9.210	10.597	13.815
3	3.665	4.108	4.642	6.251	7.815	9.348	9.837	11.345	12.838	16.266
4	4.878	5.385	5.989	7.779	9.488	11.143	11.668	13.277	14.860	18.466
5	6.064	6.626	7.289	9.236	11.070	12.832	13.388	15.086	16.750	20.515
6	7.231	7.841	8.558	10.645	12.592	14.449	15.033	16.812	18.548	22.457
7	8.383	9.037	9.803	12.017	14.067	16.013	16.622	18.475	20.278	24.321
8	9.524	10.219	11.030	13.362	15.507	17.535	18.168	20.090	21.955	26.124
9	10.656	11.389	12.242	14.684	16.919	19.023	19.679	21.666	23.589	27.877
10	11.781	12.549	13.442	15.987	18.307	20.483	21.161	23.209	25.188	29.588
11	12.899	13.701	14.631	17.275	19.675	21.920	22.618	24.725	26.757	31.264
12	14.011	14.845	15.812	18.549	21.026	23.337	24.054	26.217	28.300	32.909
13	15.119	15.984	16.985	19.812	22.362	24.736	25.471	27.688	29.819	34.527
14	16.222	17.117	18.151	21.064	23.685	26.119	26.873	29.141	31.319	36.124
15	17.322	18.245	19.311	22.307	24.996	27.488	28.259	30.578	32.801	37.698
16	18.418	19.369	20.465	23.542	26.296	28.845	29.633	32.000	34.267	39.252
17	19.511	20.489	21.615	24.769	27.587	30.191	30.995	33.409	35.718	40.791
18	20.601	21.605	22.760	25.989	28.869	31.526	32.346	34.805	37.156	42.312
19	21.689	22.718	23.900	27.204	30.144	32.852	33.687	36.191	38.582	43.819
20	22.775	23.828	25.038	28.412	31.410	34.170	35.020	37.566	39.997	45.314
21	23.858	24.935	26.171	29.615	32.671	35.479	36.343	38.932	41.401	46.796
22	24.939	26.039	27.301	30.813	33.924	36.781	37.659	40.289	42.796	48.268
23	26.018	27.141	28.429	32.007	35.172	38.076	38.968	41.638	44.181	49.728
24	27.096	28.241	29.553	33.196	36.415	39.364	40.270	42.980	45.558	51.179
25	28.172	29.339	30.675	34.382	37.652	40.646	41.566	44.314	46.928	52.619
26	29.246	30.435	31.795	35.563	38.885	41.923	42.856	45.642	48.290	54.051
27	30.319	31.528	32.912	36.741	40.113	43.195	44.140	46.963	49.645	55.475
28	31.391	32.620	34.027	37.916	41.337	44.461	45.419	48.278	50.994	56.892
29	32.461	33.711	35.139	39.087	42.557	45.722	46.693	49.588	52.335	58.301
30	33.530	34.800	36.250	40.256	43.773	46.979	47.962	50.892	53.672	59.702
40	44.165	45.616	47.269	51.805	55.758	59.342	60.436	63.691	66.766	73.403
50	54.723	56.334	58.164	63.167	67.505	71.420	72.613	76.154	79.490	86.660
60	65.226	66.981	68.972	74.397	79.082	83.298	84.58	88.379	91.952	99.608

Table A.6 F-Distribution Probability Table

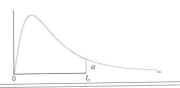

Table A.6 Critical Values of the F-Distribution

$$f_{0.05}(v_1, v_2)$$

v_2	v_1=1	2	3	4	5	6	7	8	9
1	161.45	199.50	215.71	224.58	230.16	233.99	236.77	238.88	240.54
2	18.51	19.00	19.16	19.25	19.30	19.33	19.35	19.37	19.38
3	10.13	9.55	9.28	9.12	9.01	8.94	8.89	8.85	8.81
4	7.71	6.94	6.59	6.39	6.26	6.16	6.09	6.04	6.00
5	6.61	5.79	5.41	5.19	5.05	4.95	4.88	4.82	4.77
6	5.99	5.14	4.76	4.53	4.39	4.28	4.21	4.15	4.10
7	5.59	4.74	4.35	4.12	3.97	3.87	3.79	3.73	3.68
8	5.32	4.46	4.07	3.84	3.69	3.58	3.50	3.44	3.39
9	5.12	4.26	3.86	3.63	3.48	3.37	3.29	3.23	3.18
10	4.96	4.10	3.71	3.48	3.33	3.22	3.14	3.07	3.02
11	4.84	3.98	3.59	3.36	3.20	3.09	3.01	2.95	2.90
12	4.75	3.89	3.49	3.26	3.11	3.00	2.91	2.85	2.80
13	4.67	3.81	3.41	3.18	3.03	2.92	2.83	2.77	2.71
14	4.60	3.74	3.34	3.11	2.96	2.85	2.76	2.70	2.65
15	4.54	3.68	3.29	3.06	2.90	2.79	2.71	2.64	2.59
16	4.49	3.63	3.24	3.01	2.85	2.74	2.66	2.59	2.54
17	4.45	3.59	3.20	2.96	2.81	2.70	2.61	2.55	2.49
18	4.41	3.55	3.16	2.93	2.77	2.66	2.58	2.51	2.46
19	4.38	3.52	3.13	2.90	2.74	2.63	2.54	2.48	2.42
20	4.35	3.49	3.10	2.87	2.71	2.60	2.51	2.45	2.39
21	4.32	3.47	3.07	2.84	2.68	2.57	2.49	2.42	2.37
22	4.30	3.44	3.05	2.82	2.66	2.55	2.46	2.40	2.34
23	4.28	3.42	3.03	2.80	2.64	2.53	2.44	2.37	2.32
24	4.26	3.40	3.01	2.78	2.62	2.51	2.42	2.36	2.30
25	4.24	3.39	2.99	2.76	2.60	2.49	2.40	2.34	2.28
26	4.23	3.37	2.98	2.74	2.59	2.47	2.39	2.32	2.27
27	4.21	3.35	2.96	2.73	2.57	2.46	2.37	2.31	2.25
28	4.20	3.34	2.95	2.71	2.56	2.45	2.36	2.29	2.24
29	4.18	3.33	2.93	2.70	2.55	2.43	2.35	2.28	2.22
30	4.17	3.32	2.92	2.69	2.53	2.42	2.33	2.27	2.21
40	4.08	3.23	2.84	2.61	2.45	2.34	2.25	2.18	2.12
60	4.00	3.15	2.76	2.53	2.37	2.25	2.17	2.10	2.04
120	3.92	3.07	2.68	2.45	2.29	2.18	2.09	2.02	1.96
∞	3.84	3.00	2.60	2.37	2.21	2.10	2.01	1.94	1.88

Reproduced from Table 18 of *Biometrika Tables for Statisticians*, Vol. I, by permission of E.S. Pearson and the Biometrika Trustees.

Table A.6 (continued) Critical Values of the F-Distribution

$$f_{0.05}(v_1, v_2)$$

v_2	\multicolumn{10}{c}{v_1}									
	10	12	15	20	24	30	40	60	120	∞
1	241.88	243.91	245.95	248.01	249.05	250.10	251.14	252.20	253.25	254.31
2	19.40	19.41	19.43	19.45	19.45	19.46	19.47	19.48	19.49	19.50
3	8.79	8.74	8.70	8.66	8.64	8.62	8.59	8.57	8.55	8.53
4	5.96	5.91	5.86	5.80	5.77	5.75	5.72	5.69	5.66	5.63
5	4.74	4.68	4.62	4.56	4.53	4.50	4.46	4.43	4.40	4.36
6	4.06	4.00	3.94	3.87	3.84	3.81	3.77	3.74	3.70	3.67
7	3.64	3.57	3.51	3.44	3.41	3.38	3.34	3.30	3.27	3.23
8	3.35	3.28	3.22	3.15	3.12	3.08	3.04	3.01	2.97	2.93
9	3.14	3.07	3.01	2.94	2.90	2.86	2.83	2.79	2.75	2.71
10	2.98	2.91	2.85	2.77	2.74	2.70	2.66	2.62	2.58	2.54
11	2.85	2.79	2.72	2.65	2.61	2.57	2.53	2.49	2.45	2.40
12	2.75	2.69	2.62	2.54	2.51	2.47	2.43	2.38	2.34	2.30
13	2.67	2.60	2.53	2.46	2.42	2.38	2.34	2.30	2.25	2.21
14	2.60	2.53	2.46	2.39	2.35	2.31	2.27	2.22	2.18	2.13
15	2.54	2.48	2.40	2.33	2.29	2.25	2.20	2.16	2.11	2.07
16	2.49	2.42	2.35	2.28	2.24	2.19	2.15	2.11	2.06	2.01
17	2.45	2.38	2.31	2.23	2.19	2.15	2.10	2.06	2.01	1.96
18	2.41	2.34	2.27	2.19	2.15	2.11	2.06	2.02	1.97	1.92
19	2.38	2.31	2.23	2.16	2.11	2.07	2.03	1.98	1.93	1.88
20	2.35	2.28	2.20	2.12	2.08	2.04	1.99	1.95	1.90	1.84
21	2.32	2.25	2.18	2.10	2.05	2.01	1.96	1.92	1.87	1.81
22	2.30	2.23	2.15	2.07	2.03	1.98	1.94	1.89	1.84	1.78
23	2.27	2.20	2.13	2.05	2.01	1.96	1.91	1.86	1.81	1.76
24	2.25	2.18	2.11	2.03	1.98	1.94	1.89	1.84	1.79	1.73
25	2.24	2.16	2.09	2.01	1.96	1.92	1.87	1.82	1.77	1.71
26	2.22	2.15	2.07	1.99	1.95	1.90	1.85	1.80	1.75	1.69
27	2.20	2.13	2.06	1.97	1.93	1.88	1.84	1.79	1.73	1.67
28	2.19	2.12	2.04	1.96	1.91	1.87	1.82	1.77	1.71	1.65
29	2.18	2.10	2.03	1.94	1.90	1.85	1.81	1.75	1.70	1.64
30	2.16	2.09	2.01	1.93	1.89	1.84	1.79	1.74	1.68	1.62
40	2.08	2.00	1.92	1.84	1.79	1.74	1.69	1.64	1.58	1.51
60	1.99	1.92	1.84	1.75	1.70	1.65	1.59	1.53	1.47	1.39
120	1.91	1.83	1.75	1.66	1.61	1.55	1.50	1.43	1.35	1.25
∞	1.83	1.75	1.67	1.57	1.52	1.46	1.39	1.32	1.22	1.00

Table A.6 F-Distribution Probability Table

Table A.6 (continued) Critical Values of the F-Distribution

$f_{0.01}(v_1, v_2)$

v_2	v_1								
	1	2	3	4	5	6	7	8	9
1	4052.18	4999.50	5403.35	5624.58	5763.65	5858.99	5928.36	5981.07	6022.47
2	98.50	99.00	99.17	99.25	99.30	99.33	99.36	99.37	99.39
3	34.12	30.82	29.46	28.71	28.24	27.91	27.67	27.49	27.35
4	21.20	18.00	16.69	15.98	15.52	15.21	14.98	14.80	14.66
5	16.26	13.27	12.06	11.39	10.97	10.67	10.46	10.29	10.16
6	13.75	10.92	9.78	9.15	8.75	8.47	8.26	8.10	7.98
7	12.25	9.55	8.45	7.85	7.46	7.19	6.99	6.84	6.72
8	11.26	8.65	7.59	7.01	6.63	6.37	6.18	6.03	5.91
9	10.56	8.02	6.99	6.42	6.06	5.80	5.61	5.47	5.35
10	10.04	7.56	6.55	5.99	5.64	5.39	5.20	5.06	4.94
11	9.65	7.21	6.22	5.67	5.32	5.07	4.89	4.74	4.63
12	9.33	6.93	5.95	5.41	5.06	4.82	4.64	4.50	4.39
13	9.07	6.70	5.74	5.21	4.86	4.62	4.44	4.30	4.19
14	8.86	6.51	5.56	5.04	4.69	4.46	4.28	4.14	4.03
15	8.68	6.36	5.42	4.89	4.56	4.32	4.14	4.00	3.89
16	8.53	6.23	5.29	4.77	4.44	4.20	4.03	3.89	3.78
17	8.40	6.11	5.18	4.67	4.34	4.10	3.93	3.79	3.68
18	8.29	6.01	5.09	4.58	4.25	4.01	3.84	3.71	3.60
19	8.18	5.93	5.01	4.50	4.17	3.94	3.77	3.63	3.52
20	8.10	5.85	4.94	4.43	4.10	3.87	3.70	3.56	3.46
21	8.02	5.78	4.87	4.37	4.04	3.81	3.64	3.51	3.40
22	7.95	5.72	4.82	4.31	3.99	3.76	3.59	3.45	3.35
23	7.88	5.66	4.76	4.26	3.94	3.71	3.54	3.41	3.30
24	7.82	5.61	4.72	4.22	3.90	3.67	3.50	3.36	3.26
25	7.77	5.57	4.68	4.18	3.85	3.63	3.46	3.32	3.22
26	7.72	5.53	4.64	4.14	3.82	3.59	3.42	3.29	3.18
27	7.68	5.49	4.60	4.11	3.78	3.56	3.39	3.26	3.15
28	7.64	5.45	4.57	4.07	3.75	3.53	3.36	3.23	3.12
29	7.60	5.42	4.54	4.04	3.73	3.50	3.33	3.20	3.09
30	7.56	5.39	4.51	4.02	3.70	3.47	3.30	3.17	3.07
40	7.31	5.18	4.31	3.83	3.51	3.29	3.12	2.99	2.89
60	7.08	4.98	4.13	3.65	3.34	3.12	2.95	2.82	2.72
120	6.85	4.79	3.95	3.48	3.17	2.96	2.79	2.66	2.56
∞	6.63	4.61	3.78	3.32	3.02	2.80	2.64	2.51	2.41

Table A.6 (continued) Critical Values of the F-Distribution

$$f_{0.01}(v_1, v_2)$$

v_2	\multicolumn{10}{c}{v_1}									
	10	12	15	20	24	30	40	60	120	∞
1	6055.85	6106.32	6157.28	6208.73	6234.63	6260.65	6286.78	6313.03	6339.39	6365.86
2	99.40	99.42	99.43	99.45	99.46	99.47	99.47	99.48	99.49	99.50
3	27.23	27.05	26.87	26.69	26.60	26.50	26.41	26.32	26.22	26.13
4	14.55	14.37	14.20	14.02	13.93	13.84	13.75	13.65	13.56	13.46
5	10.05	9.89	9.72	9.55	9.47	9.38	9.29	9.20	9.11	9.02
6	7.87	7.72	7.56	7.40	7.31	7.23	7.14	7.06	6.97	6.88
7	6.62	6.47	6.31	6.16	6.07	5.99	5.91	5.82	5.74	5.65
8	5.81	5.67	5.52	5.36	5.28	5.20	5.12	5.03	4.95	4.86
9	5.26	5.11	4.96	4.81	4.73	4.65	4.57	4.48	4.40	4.31
10	4.85	4.71	4.56	4.41	4.33	4.25	4.17	4.08	4.00	3.91
11	4.54	4.40	4.25	4.10	4.02	3.94	3.86	3.78	3.69	3.60
12	4.30	4.16	4.01	3.86	3.78	3.70	3.62	3.54	3.45	3.36
13	4.10	3.96	3.82	3.66	3.59	3.51	3.43	3.34	3.25	3.17
14	3.94	3.80	3.66	3.51	3.43	3.35	3.27	3.18	3.09	3.00
15	3.80	3.67	3.52	3.37	3.29	3.21	3.13	3.05	2.96	2.87
16	3.69	3.55	3.41	3.26	3.18	3.10	3.02	2.93	2.84	2.75
17	3.59	3.46	3.31	3.16	3.08	3.00	2.92	2.83	2.75	2.65
18	3.51	3.37	3.23	3.08	3.00	2.92	2.84	2.75	2.66	2.57
19	3.43	3.30	3.15	3.00	2.92	2.84	2.76	2.67	2.58	2.49
20	3.37	3.23	3.09	2.94	2.86	2.78	2.69	2.61	2.52	2.42
21	3.31	3.17	3.03	2.88	2.80	2.72	2.64	2.55	2.46	2.36
22	3.26	3.12	2.98	2.83	2.75	2.67	2.58	2.50	2.40	2.31
23	3.21	3.07	2.93	2.78	2.70	2.62	2.54	2.45	2.35	2.26
24	3.17	3.03	2.89	2.74	2.66	2.58	2.49	2.40	2.31	2.21
25	3.13	2.99	2.85	2.70	2.62	2.54	2.45	2.36	2.27	2.17
26	3.09	2.96	2.81	2.66	2.58	2.50	2.42	2.33	2.23	2.13
27	3.06	2.93	2.78	2.63	2.55	2.47	2.38	2.29	2.20	2.10
28	3.03	2.90	2.75	2.60	2.52	2.44	2.35	2.26	2.17	2.06
29	3.00	2.87	2.73	2.57	2.49	2.41	2.33	2.23	2.14	2.03
30	2.98	2.84	2.70	2.55	2.47	2.39	2.30	2.21	2.11	2.01
40	2.80	2.66	2.52	2.37	2.29	2.20	2.11	2.02	1.92	1.80
60	2.63	2.50	2.35	2.20	2.12	2.03	1.94	1.84	1.73	1.60
120	2.47	2.34	2.19	2.03	1.95	1.86	1.76	1.66	1.53	1.38
∞	2.32	2.18	2.04	1.88	1.79	1.70	1.59	1.47	1.32	1.00

Table A.7 Tolerance Factors for Normal Distributions

	Two-Sided Intervals							One-Sided Intervals						
	$\gamma = 0.05$			$\gamma = 0.01$				$\gamma = 0.05$			$\gamma = 0.01$			
	$1-\alpha$			$1-\alpha$				$1-\alpha$			$1-\alpha$			
n	0.90	0.95	0.99	0.90	0.95	0.99		0.90	0.95	0.99	0.90	0.95	0.99	
2	32.019	37.674	48.430	160.193	188.491	242.300		20.581	26.260	37.094	103.029	131.426	185.617	
3	8.380	9.916	12.861	18.930	22.401	29.055		6.156	7.656	10.553	13.995	17.170	23.896	
4	5.369	6.370	8.299	9.398	11.150	14.527		4.162	5.144	7.042	7.380	9.083	12.387	
5	4.275	5.079	6.634	6.612	7.855	10.260		3.407	4.203	5.741	5.362	6.578	8.939	
6	3.712	4.414	5.775	5.337	6.345	8.301		3.006	3.708	5.062	4.411	5.406	7.335	
7	3.369	4.007	5.248	4.613	5.488	7.187		2.756	3.400	4.642	3.859	4.728	6.412	
8	3.136	3.732	4.891	4.147	4.936	6.468		2.582	3.187	4.354	3.497	4.285	5.812	
9	2.967	3.532	4.631	3.822	4.550	5.966		2.454	3.031	4.143	3.241	3.972	5.389	
10	2.839	3.379	4.433	3.582	4.265	5.594		2.355	2.911	3.981	3.048	3.738	5.074	
11	2.737	3.259	4.277	3.397	4.045	5.308		2.275	2.815	3.852	2.898	3.556	4.829	
12	2.655	3.162	4.150	3.250	3.870	5.079		2.210	2.736	3.747	2.777	3.410	4.633	
13	2.587	3.081	4.044	3.130	3.727	4.893		2.155	2.671	3.659	2.677	3.290	4.472	
14	2.529	3.012	3.955	3.029	3.608	4.737		2.109	2.615	3.585	2.593	1.189	4.337	
15	2.480	2.954	3.878	2.945	3.507	4.605		2.068	2.566	3.520	2.522	3.102	4.222	
16	2.437	2.903	3.812	2.872	3.421	4.492		2.033	2.524	3.464	2.460	3.028	4.123	
17	2.400	2.858	3.754	2.808	3.345	4.393		2.002	2.486	3.414	2.405	2.963	4.037	
18	2.366	2.819	3.702	2.753	3.279	4.307		1.974	2.453	3.370	2.357	2.905	3.960	
19	2.337	2.784	3.656	2.703	3.221	4.230		1.949	2.423	3.331	2.314	2.854	3.892	
20	2.310	2.752	3.615	2.659	3.168	4.161		1.926	2.396	3.295	2.276	2.808	1.832	
25	2.208	2.631	3.457	2.494	2.972	3.904		1.838	2.292	3.158	2.129	2.633	3.601	
30	2.140	2.549	3.350	2.385	2.841	3.733		1.777	2.220	3.064	2.030	2.516	3.447	
35	2.090	2.490	3.272	2.306	2.748	3.611		1.732	2.167	2.995	1.957	2.430	3.334	
40	2.052	2.445	3.213	2.247	2.677	3.518		1.697	2.126	2.941	1.902	2.364	3.249	
45	2.021	2.408	3.165	2.200	2.621	3.444		1.669	2.092	2.898	1.857	2.312	3.180	
50	1.996	2.379	3.126	2.162	2.576	3.385		1.646	2.065	2.863	1.821	2.269	3.125	
60	1.958	2.333	3.066	2.103	2.506	3.293		1.609	2.022	2.807	1.764	2.202	3.038	
70	1.929	2.299	3.021	2.060	2.454	3.225		1.581	1.990	2.765	1.722	2.153	2.974	
80	1.907	2.272	2.986	2.026	2.414	3.173		1.559	1.965	2.733	1.688	2.114	2.924	
90	1.889	2.251	2.958	1.999	2.382	3.130		1.542	1.944	2.706	1.661	2.082	2.883	
100	1.874	2.233	2.934	1.977	2.355	3.096		1.527	1.927	2.684	1.639	2.056	2.850	
150	1.825	2.175	2.859	1.905	2.270	2.983		1.478	1.870	2.611	1.566	1.971	2.741	
200	1.798	2.143	2.816	1.865	2.222	2.921		1.450	1.837	2.570	1.524	1.923	2.679	
250	1.780	2.121	2.788	1.839	2.191	2.880		1.431	1.815	2.542	1.496	1.891	2.638	
300	1.767	2.106	2.767	1.820	2.169	2.850		1.417	1.800	2.522	1.476	1.868	2.608	
∞	1.645	1.960	2.576	1.645	1.960	2.576		1.282	1.645	2.326	1.282	1.645	2.326	

Adapted from C. Eisenhart, M. W. Hastay, and W. A. Wallis, *Techniques of Statistical Analysis*, Chapter 2, McGraw-Hill Book Company, New York, 1947. Used with permission of McGraw-Hill Book Company.

Table A.8 Sample Size for the t-Test of the Mean

	Level of t-Test																			
Single-Sided Test	$\alpha = 0.005$					$\alpha = 0.01$					$\alpha = 0.025$					$\alpha = 0.05$				
Double-Sided Test	$\alpha = 0.01$					$\alpha = 0.02$					$\alpha = 0.05$					$\alpha = 0.1$				
$\beta = 0.1$	.01	.05	.1	.2	.5	.01	.05	.1	.2	.5	.01	.05	.1	.2	.5	.01	.05	.1	.2	.5
0.05																				
0.10																				
0.15																				122
0.20									139					99						70
0.25			110					90				128	64				139	101	45	
0.30		134	78				115	63			119	90	45			122	97	71	32	
0.35	125	99	58			109	85	47			109	88	67	34		90	72	52	24	
0.40	115	97	77	45		101	85	66	37	117	84	68	51	26	101	70	55	40	19	
0.45	92	77	62	37	110	81	68	53	30	93	67	54	41	21	80	55	44	33	15	
0.50	100	75	63	51	30	90	66	55	43	25	76	54	44	34	18	65	45	36	27	13
0.55	83	63	53	42	26	75	55	46	36	21	63	45	37	28	15	54	38	30	22	11
0.60	71	53	45	36	22	63	47	39	31	18	53	38	32	24	13	46	32	26	19	9
0.65	61	46	39	31	20	55	41	34	27	16	46	33	27	21	12	39	28	22	17	8
0.70	53	40	34	28	17	47	35	30	24	14	40	29	24	19	10	34	24	19	15	8
0.75	47	36	30	25	16	42	31	27	21	13	35	26	21	16	9	30	21	17	13	7
0.80	41	32	27	22	14	37	28	24	19	12	31	22	19	15	9	27	19	15	12	6
0.85	37	29	24	20	13	33	25	21	17	11	28	21	17	13	8	24	17	14	11	6
0.90	34	26	22	18	12	29	23	19	16	10	25	19	16	12	7	21	15	13	10	5
0.95	31	24	20	17	11	27	21	18	14	9	23	17	14	11	7	19	14	11	9	5
1.00	28	22	19	16	10	25	19	16	13	9	21	16	13	10	6	18	13	11	8	5
1.1	24	19	16	14	9	21	16	14	12	8	18	13	11	9	6	15	11	9	7	
1.2	21	16	14	12	8	18	14	12	10	7	15	12	10	8	5	13	10	8	6	
1.3	18	15	13	11	8	16	13	11	9	6	13	10	9	7		11	8	7	6	
1.4	16	13	12	10	7	14	11	10	9	6	12	9	8	7		10	8	7	5	
1.5	15	12	11	9	7	13	10	9	8	6	11	8	7	6		9	7	6		
1.6	13	11	10	8	6	12	10	9	7	5	10	8	7	6		8	6	6		
1.7	12	10	9	8	6	11	9	8	7		9	7	6	5		8	6	5		
1.8	12	10	9	8	6	10	8	7	7		8	7	6			7	6			
1.9	11	9	8	7	6	10	8	7	6		8	6	6			7	5			
2.0	10	8	8	7	5	9	7	7	6		7	6	5			6				
2.1	10	8	7	7		8	7	6	6		7	6				6				
2.2	9	8	7	6		8	7	6	5		7	6				6				
2.3	9	7	7	6		8	6	6			6	5				5				
2.4	8	7	7	6		7	6	6			6									
2.5	8	7	6	6		7	6	6			6									
3.0	7	6	6	5		6	5	5			5									
3.5	6	5	5			5														
4.0	6																			

Value of $\Delta = |\delta|/\sigma$

Reproduced with permission from O. L. Davies, ed., *Design and Analysis of Industrial Experiments*, Oliver & Boyd, Edinburgh, 1956.

Table A.9 Table of Sample Sizes for the Test of the Difference between Two Means

Table A.9 Sample Size for the t-Test of the Difference between Two Means

			Level of t-Test															
Single-Sided Test	$\alpha = 0.005$				$\alpha = 0.01$				$\alpha = 0.025$				$\alpha = 0.05$					
Double-Sided Test	$\alpha = 0.01$				$\alpha = 0.02$				$\alpha = 0.05$				$\alpha = 0.1$					
$\beta = 0.1$	.01	.05	.1	.2	.5	.01	.05	.1	.2	.5	.01	.05	.1	.2	.5	.01	.05	.1	.2	.5

$\Delta = \|\delta\|/\sigma$	.01	.05	.1	.2	.5	.01	.05	.1	.2	.5	.01	.05	.1	.2	.5	.01	.05	.1	.2	.5
0.05																				
0.10																				
0.15																				
0.20																				137
0.25														124						88
0.30										123				87						61
0.35			110							90				64				102		45
0.40			85							70			100	50			108	78		35
0.45		118	68				101	55			105	79	39	108	86	62	28
0.50		96	55			106	82	45		106	86	64	32		88	70	51	23
0.55	101	79	46		106	88	68	38		87	71	53	27	112	73	58	42	19
0.60	101	85	67	39		90	74	58	32	104	74	60	45	23	89	61	49	36	16
0.65	87	73	57	34	104	77	64	49	27	88	63	51	39	20	76	52	42	30	14
0.70	100	75	63	50	29	90	66	55	43	24	76	55	44	34	17	66	45	36	26	12
0.75	88	66	55	44	26	79	58	48	38	21	67	48	39	29	15	57	40	32	23	11
0.80	77	58	49	39	23	70	51	43	33	19	59	42	34	26	14	50	35	28	21	10
0.85	69	51	43	35	21	62	46	38	30	17	52	37	31	23	12	45	31	25	18	9
0.90	62	46	39	31	19	55	41	34	27	15	47	34	27	21	11	40	28	22	16	8
0.95	55	42	35	28	17	50	37	31	24	14	42	30	25	19	10	36	25	20	15	7
1.00	50	38	32	26	15	45	33	28	22	13	38	27	23	17	9	33	23	18	14	7
1.1	42	32	27	22	13	38	28	23	19	11	32	23	19	14	8	27	19	15	12	6
1.2	36	27	23	18	11	32	24	20	16	9	27	20	16	12	7	23	16	13	10	5
1.3	31	23	20	16	10	28	21	17	14	8	23	17	14	11	6	20	14	11	9	5
1.4	27	20	17	14	9	24	18	15	12	8	20	15	12	10	6	17	12	10	8	4
1.5	24	18	15	13	8	21	16	14	11	7	18	13	11	9	5	15	11	9	7	4
1.6	21	16	14	11	7	19	14	12	10	6	16	12	10	8	5	14	10	8	6	4
1.7	19	15	13	10	7	17	13	11	9	6	14	11	9	7	4	12	9	7	6	3
1.8	17	13	71	10	6	15	12	10	8	5	13	10	8	6	4	11	8	7	5	
1.9	16	12	11	9	6	14	11	9	8	5	12	9	7	6	4	10	7	6	5	
2.0	14	11	10	8	6	13	10	9	7	5	11	8	7	6	4	9	7	6	4	
2.1	13	10	9	8	5	12	9	8	7	5	10	8	6	5	3	8	6	5	4	
2.2	12	10	8	7	5	11	9	7	6	4	9	7	6	5		8	6	5	4	
2.3	11	9	8	7	5	10	8	7	6	4	9	7	6	5		7	5	5	4	
2.4	11	9	8	6	5	10	8	7	6	4	8	6	5	4		7	5	4	4	
2.5	10	8	7	6	4	9	7	6	5	4	8	6	5	4		6	5	4	3	
3.0	8	6	6	5	4	7	6	5	4	3	6	5	4	4		5	4	3		
3.5	6	5	5	4	3	6	5	4	4		5	4	4	3	3	4	3			
4.0	6	5	4	4		5	4	3	4		4	3				4				

Reproduced with permission from O. L. Davies, ed., *Design and Analysis of Industrial Experiments*, Oliver & Boyd, Edinburgh, 1956.

Table A.10 Critical Values for Bartlett's Test

$b_k(0.01; n)$

	Number of Populations, k								
n	2	3	4	5	6	7	8	9	10
3	0.1411	0.1672							
4	0.2843	0.3165	0.3475	0.3729	0.3937	0.4110			
5	0.3984	0.4304	0.4607	0.4850	0.5046	0.5207	0.5343	0.5458	0.5558
6	0.4850	0.5149	0.5430	0.5653	0.5832	0.5978	0.6100	0.6204	0.6293
7	0.5512	0.5787	0.6045	0.6248	0.6410	0.6542	0.6652	0.6744	0.6824
8	0.6031	0.6282	0.6518	0.6704	0.6851	0.6970	0.7069	0.7153	0.7225
9	0.6445	0.6676	0.6892	0.7062	0.7197	0.7305	0.7395	0.7471	0.7536
10	0.6783	0.6996	0.7195	0.7352	0.7475	0.7575	0.7657	0.7726	0.7786
11	0.7063	0.7260	0.7445	0.7590	0.7703	0.7795	0.7871	0.7935	0.7990
12	0.7299	0.7483	0.7654	0.7789	0.7894	0.7980	0.8050	0.8109	0.8160
13	0.7501	0.7672	0.7832	0.7958	0.8056	0.8135	0.8201	0.8256	0.8303
14	0.7674	0.7835	0.7985	0.8103	0.8195	0.8269	0.8330	0.8382	0.8426
15	0.7825	0.7977	0.8118	0.8229	0.8315	0.8385	0.8443	0.8491	0.8532
16	0.7958	0.8101	0.8235	0.8339	0.8421	0.8486	0.8541	0.8586	0.8625
17	0.8076	0.8211	0.8338	0.8436	0.8514	0.8576	0.8627	0.8670	0.8707
18	0.8181	0.8309	0.8429	0.8523	0.8596	0.8655	0.8704	0.8745	0.8780
19	0.8275	0.8397	0.8512	0.8601	0.8670	0.8727	0.8773	0.8811	0.8845
20	0.8360	0.8476	0.8586	0.8671	0.8737	0.8791	0.8835	0.8871	0.8903
21	0.8437	0.8548	0.8653	0.8734	0.8797	0.8848	0.8890	0.8926	0.8956
22	0.8507	0.8614	0.8714	0.8791	0.8852	0.8901	0.8941	0.8975	0.9004
23	0.8571	0.8673	0.8769	0.8844	0.8902	0.8949	0.8988	0.9020	0.9047
24	0.8630	0.8728	0.8820	0.8892	0.8948	0.8993	0.9030	0.9061	0.9087
25	0.8684	0.8779	0.8867	0.8936	0.8990	0.9034	0.9069	0.9099	0.9124
26	0.8734	0.8825	0.8911	0.8977	0.9029	0.9071	0.9105	0.9134	0.9158
27	0.8781	0.8869	0.8951	0.9015	0.9065	0.9105	0.9138	0.9166	0.9190
28	0.8824	0.8909	0.8988	0.9050	0.9099	0.9138	0.9169	0.9196	0.9219
29	0.8864	0.8946	0.9023	0.9083	0.9130	0.9167	0.9198	0.9224	0.9246
30	0.8902	0.8981	0.9056	0.9114	0.9159	0.9195	0.9225	0.9250	0.9271
40	0.9175	0.9235	0.9291	0.9335	0.9370	0.9397	0.9420	0.9439	0.9455
50	0.9339	0.9387	0.9433	0.9468	0.9496	0.9518	0.9536	0.9551	0.9564
60	0.9449	0.9489	0.9527	0.9557	0.9580	0.9599	0.9614	0.9626	0.9637
80	0.9586	0.9617	0.9646	0.9668	0.9685	0.9699	0.9711	0.9720	0.9728
100	0.9669	0.9693	0.9716	0.9734	0.9748	0.9759	0.9769	0.9776	0.9783

Reproduced from D. D. Dyer and J. P. Keating, "On the Determination of Critical Values for Bartlett's Test," *J. Am. Stat. Assoc.*, **75**, 1980, by permission of the Board of Directors.

Table A.10 Table for Bartlett's Test

Table A.10 (continued) Critical Values for Bartlett's Test

$b_k(0.05; n)$

n	\multicolumn{9}{c	}{Number of Populations, k}
	2	3	4	5	6	7	8	9	10
3	0.3123	0.3058	0.3173	0.3299					
4	0.4780	0.4699	0.4803	0.4921	0.5028	0.5122	0.5204	0.5277	0.5341
5	0.5845	0.5762	0.5850	0.5952	0.6045	0.6126	0.6197	0.6260	0.6315
6	0.6563	0.6483	0.6559	0.6646	0.6727	0.6798	0.6860	0.6914	0.6961
7	0.7075	0.7000	0.7065	0.7142	0.7213	0.7275	0.7329	0.7376	0.7418
8	0.7456	0.7387	0.7444	0.7512	0.7574	0.7629	0.7677	0.7719	0.7757
9	0.7751	0.7686	0.7737	0.7798	0.7854	0.7903	0.7946	0.7984	0.8017
10	0.7984	0.7924	0.7970	0.8025	0.8076	0.8121	0.8160	0.8194	0.8224
11	0.8175	0.8118	0.8160	0.8210	0.8257	0.8298	0.8333	0.8365	0.8392
12	0.8332	0.8280	0.8317	0.8364	0.8407	0.8444	0.8477	0.8506	0.8531
13	0.8465	0.8415	0.8450	0.8493	0.8533	0.8568	0.8598	0.8625	0.8648
14	0.8578	0.8532	0.8564	0.8604	0.8641	0.8673	0.8701	0.8726	0.8748
15	0.8676	0.8632	0.8662	0.8699	0.8734	0.8764	0.8790	0.8814	0.8834
16	0.8761	0.8719	0.8747	0.8782	0.8815	0.8843	0.8868	0.8890	0.8909
17	0.8836	0.8796	0.8823	0.8856	0.8886	0.8913	0.8936	0.8957	0.8975
18	0.8902	0.8865	0.8890	0.8921	0.8949	0.8975	0.8997	0.9016	0.9033
19	0.8961	0.8926	0.8949	0.8979	0.9006	0.9030	0.9051	0.9069	0.9086
20	0.9015	0.8980	0.9003	0.9031	0.9057	0.9080	0.9100	0.9117	0.9132
21	0.9063	0.9030	0.9051	0.9078	0.9103	0.9124	0.9143	0.9160	0.9175
22	0.9106	0.9075	0.9095	0.9120	0.9144	0.9165	0.9183	0.9199	0.9213
23	0.9146	0.9116	0.9135	0.9159	0.9182	0.9202	0.9219	0.9235	0.9248
24	0.9182	0.9153	0.9172	0.9195	0.9217	0.9236	0.9253	0.9267	0.9280
25	0.9216	0.9187	0.9205	0.9228	0.9249	0.9267	0.9283	0.9297	0.9309
26	0.9246	0.9219	0.9236	0.9258	0.9278	0.9296	0.9311	0.9325	0.9336
27	0.9275	0.9249	0.9265	0.9286	0.9305	0.9322	0.9337	0.9350	0.9361
28	0.9301	0.9276	0.9292	0.9312	0.9330	0.9347	0.9361	0.9374	0.9385
29	0.9326	0.9301	0.9316	0.9336	0.9354	0.9370	0.9383	0.9396	0.9406
30	0.9348	0.9325	0.9340	0.9358	0.9376	0.9391	0.9404	0.9416	0.9426
40	0.9513	0.9495	0.9506	0.9520	0.9533	0.9545	0.9555	0.9564	0.9572
50	0.9612	0.9597	0.9606	0.9617	0.9628	0.9637	0.9645	0.9652	0.9658
60	0.9677	0.9665	0.9672	0.9681	0.9690	0.9698	0.9705	0.9710	0.9716
80	0.9758	0.9749	0.9754	0.9761	0.9768	0.9774	0.9779	0.9783	0.9787
100	0.9807	0.9799	0.9804	0.9809	0.9815	0.9819	0.9823	0.9827	0.9830

Table A.11 Critical Values for Cochran's Test

$\alpha = 0.01$

k \ n	2	3	4	5	6	7	8	9	10	11	17	37	145	∞
2	0.9999	0.9950	0.9794	0.9586	0.9373	0.9172	0.8988	0.8823	0.8674	0.8539	0.7949	0.7067	0.6062	0.5000
3	0.9933	0.9423	0.8831	0.8335	0.7933	0.7606	0.7335	0.7107	0.6912	0.6743	0.6059	0.5153	0.4230	0.3333
4	0.9676	0.8643	0.7814	0.7212	0.6761	0.6410	0.6129	0.5897	0.5702	0.5536	0.4884	0.4057	0.3251	0.2500
5	0.9279	0.7885	0.6957	0.6329	0.5875	0.5531	0.5259	0.5037	0.4854	0.4697	0.4094	0.3351	0.2644	0.2000
6	0.8828	0.7218	0.6258	0.5635	0.5195	0.4866	0.4608	0.4401	0.4229	0.4084	0.3529	0.2858	0.2229	0.1667
7	0.8376	0.6644	0.5685	0.5080	0.4659	0.4347	0.4105	0.3911	0.3751	0.3616	0.3105	0.2494	0.1929	0.1429
8	0.7945	0.6152	0.5209	0.4627	0.4226	0.3932	0.3704	0.3522	0.3373	0.3248	0.2779	0.2214	0.1700	0.1250
9	0.7544	0.5727	0.4810	0.4251	0.3870	0.3592	0.3378	0.3207	0.3067	0.2950	0.2514	0.1992	0.1521	0.1111
10	0.7175	0.5358	0.4469	0.3934	0.3572	0.3308	0.3106	0.2945	0.2813	0.2704	0.2297	0.1811	0.1376	0.1000
12	0.6528	0.4751	0.3919	0.3428	0.3099	0.2861	0.2680	0.2535	0.2419	0.2320	0.1961	0.1535	0.1157	0.0833
15	0.5747	0.4069	0.3317	0.2882	0.2593	0.2386	0.2228	0.2104	0.2002	0.1918	0.1612	0.1251	0.0934	0.0667
20	0.4799	0.3297	0.2654	0.2288	0.2048	0.1877	0.1748	0.1646	0.1567	0.1501	0.1248	0.0960	0.0709	0.0500
24	0.4247	0.2871	0.2295	0.1970	0.1759	0.1608	0.1495	0.1406	0.1338	0.1283	0.1060	0.0810	0.0595	0.0417
30	0.3632	0.2412	0.1913	0.1635	0.1454	0.1327	0.1232	0.1157	0.1100	0.1054	0.0867	0.0658	0.0480	0.0333
40	0.2940	0.1915	0.1508	0.1281	0.1135	0.1033	0.0957	0.0898	0.0853	0.0816	0.0668	0.0503	0.0363	0.0250
60	0.2151	0.1371	0.1069	0.0902	0.0796	0.0722	0.0668	0.0625	0.0594	0.0567	0.0461	0.0344	0.0245	0.0167
120	0.1225	0.0759	0.0585	0.0489	0.0429	0.0387	0.0357	0.0334	0.0316	0.0302	0.0242	0.0178	0.0125	0.0083
∞	0	0	0	0	0	0	0	0	0	0	0	0	0	0

Reproduced from C. Eisenhart, M. W. Hastay, and W. A. Wallis, *Techniques of Statistical Analysis*, Chapter 15, McGraw-Hill Book Company, New York, 1947. Used with permission of McGraw-Hill Book Company.

Table A.11 (continued) Critical Values for Cochran's Test

$\alpha = 0.05$

k	2	3	4	5	6	7	8	9	10	11	17	37	145	∞
2	0.9985	0.9750	0.9392	0.9057	0.8772	0.8534	0.8332	0.8159	0.8010	0.7880	0.7341	0.6602	0.5813	0.5000
3	0.9669	0.8709	0.7977	0.7457	0.7071	0.6771	0.6530	0.6333	0.6167	0.6025	0.5466	0.4748	0.4031	0.3333
4	0.9065	0.7679	0.6841	0.6287	0.5895	0.5598	0.5365	0.5175	0.5017	0.4884	0.4366	0.3720	0.3093	0.2500
5	0.8412	0.6838	0.5981	0.5441	0.5065	0.4783	0.4564	0.4387	0.4241	0.4118	0.3645	0.3066	0.2513	0.2000
6	0.7808	0.6161	0.5321	0.4803	0.4447	0.4184	0.3980	0.3817	0.3682	0.3568	0.3135	0.2612	0.2119	0.1667
7	0.7271	0.5612	0.4800	0.4307	0.3974	0.3726	0.3535	0.3384	0.3259	0.3154	0.2756	0.2278	0.1833	0.1429
8	0.6798	0.5157	0.4377	0.3910	0.3595	0.3362	0.3185	0.3043	0.2926	0.2829	0.2462	0.2022	0.1616	0.1250
9	0.6385	0.4775	0.4027	0.3584	0.3286	0.3067	0.2901	0.2768	0.2659	0.2568	0.2226	0.1820	0.1446	0.1111
10	6.6020	0.4450	0.3733	0.3311	0.3029	0.2823	0.2666	0.2541	0.2439	0.2353	0.2032	0.1655	0.1308	0.1000
12	0.5410	0.3924	0.3264	0.2880	0.2624	0.2439	0.2299	0.2187	0.2098	0.2020	0.1737	0.1403	0.1100	0.0833
15	0.4709	0.3346	0.2758	0.2419	0.2195	0.2034	0.1911	0.1815	0.1736	0.1671	0.1429	0.1144	0.0889	0.0667
20	0.3894	0.2705	0.2205	0.1921	0.1735	0.1602	0.1501	0.1422	0.1357	0.1303	0.1108	0.0879	0.0675	0.0500
24	0.3434	0.2354	0.1907	0.1656	0.1493	0.1374	0.1286	0.1216	0.1160	0.1113	0.0942	0.0743	0.0567	0.0417
30	0.2929	0.1980	0.1593	0.1377	0.1237	0.1137	0.1061	0.1002	0.0958	0.0921	0.0771	0.0604	0.0457	0.0333
40	0.2370	0.1576	0.1259	0.1082	0.0968	0.0887	0.0827	0.0780	0.0745	0.0713	0.0595	0.0462	0.0347	0.0250
60	0.1737	0.1131	0.0895	0.0765	0.0682	0.0623	0.0583	0.0552	0.0520	0.0497	0.0411	0.0316	0.0234	0.0167
120	0.0998	0.0632	0.0495	0.0419	0.0371	0.0337	0.0312	0.0292	0.0279	0.0266	0.0218	0.0165	0.0120	0.0083
∞	0	0	0	0	0	0	0	0	0	0	0	0	0	0

Table A.12 Upper Percentage Points of the Studentized Range Distribution: Values of $q(0.05; k, v)$

Degrees of Freedom, v	Number of Treatments k								
	2	3	4	5	6	7	8	9	10
1	18.0	27.0	32.8	37.2	40.5	43.1	15.1	47.1	49.1
2	6.09	5.33	9.80	10.89	11.73	12.43	13.03	13.54	13.99
3	4.50	5.91	6.83	7.51	8.04	8.47	8.85	9.18	9.46
4	3.93	5.04	5.76	6.29	6.71	7.06	7.35	7.60	7.83
5	3.64	4.60	5.22	5.67	6.03	6.33	6.58	6.80	6.99
6	3.46	4.34	4.90	5.31	5.63	5.89	6.12	6.32	6.49
7	3.34	4.16	4.68	5.06	5.35	5.59	5.80	5.99	6.15
8	3.26	4.04	4.53	4.89	5.17	5.40	5.60	5.77	5.92
9	3.20	3.95	4.42	4.76	5.02	5.24	5.43	5.60	5.74
10	3.15	3.88	4.33	4.66	4.91	5.12	5.30	5.46	5.60
11	3.11	3.82	4.26	4.58	4.82	5.03	5.20	5.35	5.49
12	3.08	3.77	4.20	4.51	4.75	4.95	5.12	5.27	5.40
13	3.06	3.73	4.15	4.46	4.69	4.88	5.05	5.19	5.32
14	3.03	3.70	4.11	4.41	4.65	4.83	4.99	5.13	5.25
15	3.01	3.67	4.08	4.37	4.59	4.78	4.94	5.08	5.20
16	3.00	3.65	4.05	4.34	4.56	4.74	4.90	5.03	5.05
17	2.98	3.62	4.02	4.31	4.52	4.70	4.86	4.99	5.11
18	2.97	3.61	4.00	4.28	4.49	4.67	4.83	4.96	5.07
19	2.96	3.59	3.98	4.26	4.47	4.64	4.79	4.92	5.04
20	2.95	3.58	3.96	4.24	4.45	4.62	4.77	4.90	5.01
24	2.92	3.53	3.90	4.17	4.37	4.54	4.68	4.81	4.92
30	2.89	3.48	3.84	4.11	4.30	4.46	4.60	4.72	4.83
40	2.86	3.44	3.79	4.04	4.23	4.39	4.52	4.63	4.74
60	2.83	3.40	3.74	3.98	4.16	4.31	4.44	4.55	4.65
120	2.80	3.36	3.69	3.92	4.10	4.24	4.36	4.47	4.56
∞	2.77	3.32	3.63	3.86	4.03	4.17	4.29	4.39	4.47

Table A.13 Table for Duncan's Test

Table A.13 Least Significant Studentized Ranges $r_p(0.05; p, v)$

$\alpha = 0.05$

v	\multicolumn{9}{c}{p}								
	2	3	4	5	6	7	8	9	10
1	17.97	17.97	17.97	17.97	17.97	17.97	17.97	17.97	17.97
2	6.085	6.085	6.085	6.085	6.085	6.085	6.085	6.085	6.085
3	4.501	4.516	4.516	4.516	4.516	4.516	4.516	4.516	4.516
4	3.927	4.013	4.033	4.033	4.033	4.033	4.033	4.033	4.033
5	3.635	3.749	3.797	3.814	3.814	3.814	3.814	3.814	3.814
6	3.461	3.587	3.649	3.68	3.694	3.697	3.697	3.697	3.697
7	3.344	3.477	3.548	3.588	3.611	3.622	3.626	3.626	3.626
8	3.261	3.399	3.475	3.521	3.549	3.566	3.575	3.579	3.579
9	3.199	3.339	3.420	3.470	3.502	3.523	3.536	3.544	3.547
10	3.151	3.293	3.376	3.430	3.465	3.489	3.505	3.516	3.522
11	3.113	3.256	3.342	3.397	3.435	3.462	3.48	3.493	3.501
12	3.082	3.225	3.313	3.370	3.410	3.439	3.459	3.474	3.484
13	3.055	3.200	3.289	3.348	3.389	3.419	3.442	3.458	3.470
14	3.033	3.178	3.268	3.329	3.372	3.403	3.426	3.444	3.457
15	3.014	3.160	3.25	3.312	3.356	3.389	3.413	3.432	3.446
16	2.998	3.144	3.235	3.298	3.343	3.376	3.402	3.422	3.437
17	2.984	3.130	3.222	3.285	3.331	3.366	3.392	3.412	3.429
18	2.971	3.118	3.210	3.274	3.321	3.356	3.383	3.405	3.421
19	2.960	3.107	3.199	3.264	3.311	3.347	3.375	3.397	3.415
20	2.950	3.097	3.190	3.255	3.303	3.339	3.368	3.391	3.409
24	2.919	3.066	3.160	3.226	3.276	3.315	3.345	3.370	3.390
30	2.888	3.035	3.131	3.199	3.250	3.290	3.322	3.349	3.371
40	2.858	3.006	3.102	3.171	3.224	3.266	3.300	3.328	3.352
60	2.829	2.976	3.073	3.143	3.198	3.241	3.277	3.307	3.333
120	2.800	2.947	3.045	3.116	3.172	3.217	3.254	3.287	3.314
∞	2.772	2.918	3.017	3.089	3.146	3.193	3.232	3.265	3.294

Abridged from H. L. Harter, "Critical Values for Duncan's New Multiple Range Test," *Biometrics*, **16**, No. 4, 1960, by permission of the author and the editor.

Table A.13 (continued) Least Significant Studentized Ranges $r_p(0.01; p, v)$

$\alpha = 0.01$

v	\multicolumn{9}{c}{p}								
	2	3	4	5	6	7	8	9	10
1	90.03	90.03	90.03	90.03	90.03	90.03	90.03	90.03	90.03
2	14.04	14.04	14.04	14.04	14.04	14.04	14.04	14.04	14.04
3	8.261	8.321	8.321	8.321	8.321	8.321	8.321	8.321	8.321
4	6.512	6.677	6.740	6.756	6.756	6.756	6.756	6.756	6.756
5	5.702	5.893	5.989	6.040	6.065	6.074	6.074	6.074	6.074
6	5.243	5.439	5.549	5.614	5.655	5.680	5.694	5.701	5.703
7	4.949	5.145	5.260	5.334	5.383	5.416	5.439	5.454	5.464
8	4.746	4.939	5.057	5.135	5.189	5.227	5.256	5.276	5.291
9	4.596	4.787	4.906	4.986	5.043	5.086	5.118	5.142	5.160
10	4.482	4.671	4.790	4.871	4.931	4.975	5.010	5.037	5.058
11	4.392	4.579	4.697	4.780	4.841	4.887	4.924	4.952	4.975
12	4.320	4.504	4.622	4.706	4.767	4.815	4.852	4.883	4.907
13	4.260	4.442	4.560	4.644	4.706	4.755	4.793	4.824	4.850
14	4.210	4.391	4.508	4.591	4.654	4.704	4.743	4.775	4.802
15	4.168	4.347	4.463	4.547	4.610	4.660	4.700	4.733	4.760
16	4.131	4.309	4.425	4.509	4.572	4.622	4.663	4.696	4.724
17	4.099	4.275	4.391	4.475	4.539	4.589	4.630	4.664	4.693
18	4.071	4.246	4.362	4.445	4.509	4.560	4.601	4.635	4.664
19	4.046	4.220	4.335	4.419	4.483	4.534	4.575	4.610	4.639
20	4.024	4.197	4.312	4.395	4.459	4.510	4.552	4.587	4.617
24	3.956	4.126	4.239	4.322	4.386	4.437	4.480	4.516	4.546
30	3.889	4.056	4.168	4.250	4.314	4.366	4.409	4.445	4.477
40	3.825	3.988	4.098	4.180	4.244	4.296	4.339	4.376	4.408
60	3.762	3.922	4.031	4.111	4.174	4.226	4.270	4.307	4.340
120	3.702	3.858	3.965	4.044	4.107	4.158	4.202	4.239	4.272
∞	3.643	3.796	3.900	3.978	4.040	4.091	4.135	4.172	4.205

Table A.14 Table for Dunnett's Two-Sided Test

Table A.14 Values of $d_{\alpha/2}(k, v)$ for Two-Sided Comparisons between k Treatments and a Control

	$\alpha = 0.05$								
	$k =$ Number of Treatment Means (excluding control)								
v	1	2	3	4	5	6	7	8	9
5	2.57	3.03	3.29	3.48	3.62	3.73	3.82	3.90	3.97
6	2.45	2.86	3.10	3.26	3.39	3.49	3.57	3.64	3.71
7	2.36	2.75	2.97	3.12	3.24	3.33	3.41	3.47	3.53
8	2.31	2.67	2.88	3.02	3.13	3.22	3.29	3.35	3.41
9	2.26	2.61	2.81	2.95	3.05	3.14	3.20	3.26	3.32
10	2.23	2.57	2.76	2.89	2.99	3.07	3.14	3.19	3.24
11	2.20	2.53	2.72	2.84	2.94	3.02	3.08	3.14	3.19
12	2.18	2.50	2.68	2.81	2.90	2.98	3.04	3.09	3.14
13	2.16	2.48	2.65	2.78	2.87	2.94	3.00	3.06	3.10
14	2.14	2.46	2.63	2.75	2.84	2.91	2.97	3.02	3.07
15	2.13	2.44	2.61	2.73	2.82	2.89	2.95	3.00	3.04
16	2.12	2.42	2.59	2.71	2.80	2.87	2.92	2.97	3.02
17	2.11	2.41	2.58	2.69	2.78	2.85	2.90	2.95	3.00
18	2.10	2.40	2.56	2.68	2.76	2.83	2.89	2.94	2.98
19	2.09	2.39	2.55	2.66	2.75	2.81	2.87	2.92	2.96
20	2.09	2.38	2.54	2.65	2.73	2.80	2.86	2.90	2.95
24	2.06	2.35	2.51	2.61	2.70	2.76	2.81	2.86	2.90
30	2.04	2.32	2.47	2.58	2.66	2.72	2.77	2.82	2.86
40	2.02	2.29	2.44	2.54	2.62	2.68	2.73	2.77	2.81
60	2.00	2.27	2.41	2.51	2.58	2.64	2.69	2.73	2.77
120	1.98	2.24	2.38	2.47	2.55	2.60	2.65	2.69	2.73
∞	1.96	2.21	2.35	2.44	2.51	2.57	2.61	2.65	2.69

Reproduced from Charles W. Dunnett, "New Tables for Multiple Comparison with a Control," *Biometrics*, **20**, No. 3, 1964, by permission of the author and the editor.

Table A.14 (continued) Values of $d_{\alpha/2}(k, v)$ for Two-Sided Comparisons between k Treatments and a Control

	$\alpha = 0.01$								
	k = Number of Treatment Means (excluding control)								
v	1	2	3	4	5	6	7	8	9
5	4.03	4.63	4.98	5.22	5.41	5.56	5.69	5.80	5.89
6	3.71	4.21	4.51	4.71	4.87	5.00	5.10	5.20	5.28
7	3.50	3.95	4.21	4.39	4.53	4.64	4.74	4.82	4.89
8	3.36	3.77	4.00	4.17	4.29	4.40	4.48	4.56	4.62
9	3.25	3.63	3.85	4.01	4.12	4.22	4.30	4.37	4.43
10	3.17	3.53	3.74	3.88	3.99	4.08	4.16	4.22	4.28
11	3.11	3.45	3.65	3.79	3.89	3.98	4.05	4.11	4.16
12	3.05	3.39	3.58	3.71	3.81	3.89	3.96	4.02	4.07
13	3.01	3.33	3.52	3.65	3.74	3.82	3.89	3.94	3.99
14	2.98	3.29	3.47	3.59	3.69	3.76	3.83	3.88	3.93
15	2.95	3.25	3.43	3.55	3.64	3.71	3.78	3.83	3.88
16	2.92	3.22	3.39	3.51	3.60	3.67	3.73	3.78	3.83
17	2.90	3.19	3.36	3.47	3.56	3.63	3.69	3.74	3.79
18	2.88	3.17	3.33	3.44	3.53	3.60	3.66	3.71	3.75
19	2.86	3.15	3.31	3.42	3.50	3.57	3.63	3.68	3.72
20	2.85	3.13	3.29	3.40	3.48	3.55	3.60	3.65	3.69
24	2.80	3.07	3.22	3.32	3.40	3.47	3.52	3.57	3.61
30	2.75	3.01	3.15	3.25	3.33	3.39	3.44	3.49	3.52
40	2.70	2.95	3.09	3.19	3.26	3.32	3.37	3.41	3.44
60	2.66	2.90	3.03	3.12	3.19	3.25	3.29	3.33	3.37
120	2.62	2.85	2.97	3.06	3.12	3.18	3.22	3.26	3.29
∞	2.58	2.79	2.92	3.00	3.06	3.11	3.15	3.19	3.22

Table A.15 Table for Dunnett's One-Sided Test

Table A.15 Values of $d_\alpha(k, v)$ for One-Sided Comparisons between k Treatments and a Control

	$\alpha = 0.05$								
	k = Number of Treatment Means (excluding control)								
v	1	2	3	4	5	6	7	8	9
5	2.02	2.44	2.68	2.85	2.98	3.08	3.16	3.24	3.30
6	1.94	2.34	2.56	2.71	2.83	2.92	3.00	3.07	3.12
7	1.89	2.27	2.48	2.62	2.73	2.82	2.89	2.95	3.01
8	1.86	2.22	2.42	2.55	2.66	2.74	2.81	2.87	2.92
9	1.83	2.18	2.37	2.50	2.60	2.68	2.75	2.81	2.86
10	1.81	2.15	2.34	2.47	2.56	2.64	2.70	2.76	2.81
11	1.80	2.13	2.31	2.44	2.53	2.60	2.67	2.72	2.77
12	1.78	2.11	2.29	2.41	2.50	2.58	2.64	2.69	2.74
13	1.77	2.09	2.27	2.39	2.48	2.55	2.61	2.66	2.71
14	1.76	2.08	2.25	2.37	2.46	2.53	2.59	2.64	2.69
15	1.75	2.07	2.24	2.36	2.44	2.51	2.57	2.62	2.67
16	1.75	2.06	2.23	2.34	2.43	2.50	2.56	2.61	2.65
17	1.74	2.05	2.22	2.33	2.42	2.49	2.54	2.59	2.64
18	1.73	2.04	2.21	2.32	2.41	2.48	2.53	2.58	2.62
19	1.73	2.03	2.20	2.31	2.40	2.47	2.52	2.57	2.61
20	1.72	2.03	2.19	2.30	2.39	2.46	2.51	2.56	2.60
24	1.71	2.01	2.17	2.28	2.36	2.43	2.48	2.53	2.57
30	1.70	1.99	2.15	2.25	2.33	2.40	2.45	2.50	2.54
40	1.68	1.97	2.13	2.23	2.31	2.37	2.42	2.47	2.51
60	1.67	1.95	2.10	2.21	2.28	2.35	2.39	2.44	2.48
120	1.66	1.93	2.08	2.18	2.26	2.32	2.37	2.41	2.45
∞	1.64	1.92	2.06	2.16	2.23	2.29	2.34	2.38	2.42

Reproduced from Charles W. Dunnett, "A Multiple Comparison Procedure for Comparing Several Treatments with a Control," *J. Am. Stat. Assoc.*, **50**, 1955, 1096–1121, by permission of the author and the editor.

Table A.15 (continued) Values of $d_\alpha(k,v)$ for One-Sided Comparisons between k Treatments and a Control

	$\alpha = 0.01$								
	k = Number of Treatment Means (excluding control)								
v	1	2	3	4	5	6	7	8	9
5	3.37	3.90	4.21	4.43	4.60	4.73	4.85	4.94	5.03
6	3.14	3.61	3.88	4.07	4.21	4.33	4.43	4.51	4.59
7	3.00	3.42	3.66	3.83	3.96	4.07	4.15	4.23	4.30
8	2.90	3.29	3.51	3.67	3.79	3.88	3.96	4.03	4.09
9	2.82	3.19	3.40	3.55	3.66	3.75	3.82	3.89	3.94
10	2.76	3.11	3.31	3.45	3.56	3.64	3.71	3.78	3.83
11	2.72	3.06	3.25	3.38	3.48	3.56	3.63	3.69	3.74
12	2.68	3.01	3.19	3.32	3.42	3.50	3.56	3.62	3.67
13	2.65	2.97	3.15	3.27	3.37	3.44	3.51	3.56	3.61
14	2.62	2.94	3.11	3.23	3.32	3.40	3.46	3.51	3.56
15	2.60	2.91	3.08	3.20	3.29	3.36	3.42	3.47	3.52
16	2.58	2.88	3.05	3.17	3.26	3.33	3.39	3.44	3.48
17	2.57	2.86	3.03	3.14	3.23	3.30	3.36	3.41	3.45
18	2.55	2.84	3.01	3.12	3.21	3.27	3.33	3.38	3.42
19	2.54	2.83	2.99	3.10	3.18	3.25	3.31	3.36	3.40
20	2.53	2.81	2.97	3.08	3.17	3.23	3.29	3.34	3.38
24	2.49	2.77	2.92	3.03	3.11	3.17	3.22	3.27	3.31
30	2.46	2.72	2.87	2.97	3.05	3.11	3.16	3.21	3.24
40	2.42	2.68	2.82	2.92	2.99	3.05	3.10	3.14	3.18
60	2.39	2.64	2.78	2.87	2.94	3.00	3.04	3.08	3.12
120	2.36	2.60	2.73	2.82	2.89	2.94	2.99	3.03	3.06
∞	2.33	2.56	2.68	2.77	2.84	2.89	2.93	2.97	3.00

Table A.16 Table for the Signed-Rank Test

Table A.16 Critical Values for the Signed-Rank Test

n	One-Sided $\alpha = 0.01$ Two-Sided $\alpha = 0.02$	One-Sided $\alpha = 0.025$ Two-Sided $\alpha = 0.05$	One-Sided $\alpha = 0.05$ Two-Sided $\alpha = 0.1$
5			1
6		1	2
7	0	2	4
8	2	4	6
9	3	6	8
10	5	8	11
11	7	11	14
12	10	14	17
13	13	17	21
14	16	21	26
15	20	25	30
16	24	30	36
17	28	35	41
18	33	40	47
19	38	46	54
20	43	52	60
21	49	59	68
22	56	66	75
23	62	73	83
24	69	81	92
25	77	90	101
26	85	98	110
27	93	107	120
28	102	117	130
29	111	127	141
30	120	137	152

Reproduced from F. Wilcoxon and R. A. Wilcox, *Some Rapid Approximate Statistical Procedures, American Cyanamid Company*, Pearl River, N.Y., 1964, by permission of the American Cyanamid Company.

Table A.17 Critical Values for the Wilcoxon Rank-Sum Test

One-Tailed Test at $\alpha = 0.001$ or Two-Tailed Test at $\alpha = 0.002$

n_1	6	7	8	9	10	11	12	13	14	15	16	17	18	19	20
1															
2															
3												0	0	0	0
4					0	0	0	1	1	1	2	2	3	3	3
5		0	0	1	1	2	2	3	3	4	5	5	6	7	7
6	0	1	2	2	3	4	4	5	6	7	8	9	10	11	12
7		2	3	3	5	6	7	8	9	10	11	13	14	15	16
8			5	5	6	8	9	11	12	14	15	17	18	20	21
9				7	8	10	12	14	15	17	19	21	23	25	26
10					10	12	14	17	19	21	23	25	27	29	32
11						15	17	20	22	24	27	29	32	34	37
12							20	23	25	28	31	34	37	40	42
13								26	29	32	35	38	42	45	48
14									32	36	39	43	46	50	54
15										40	43	47	51	55	59
16											48	52	56	60	65
17												57	61	66	70
18													66	71	76
19														77	82
20															88

One-Tailed Test at $\alpha = 0.01$ or Two-Tailed Test at $\alpha = 0.02$

n_1	5	6	7	8	9	10	11	12	13	14	15	16	17	18	19	20
1																
2									0	0	0	0	0	0	1	1
3			0	0	1	1	1	2	2	2	3	3	4	4	4	5
4	0	1	1	2	3	3	4	5	5	6	7	7	8	9	9	10
5	1	2	3	4	5	6	7	8	9	10	11	12	13	14	15	16
6		3	4	6	7	8	9	11	12	13	15	16	18	19	20	22
7			6	8	9	11	12	14	16	17	19	21	23	24	26	28
8				10	11	13	15	17	20	22	24	26	28	30	32	34
9					14	16	18	21	23	26	28	31	33	36	38	40
10						19	22	24	27	30	33	36	38	41	44	47
11							25	28	31	34	37	41	44	47	50	53
12								31	35	38	42	46	49	53	56	60
13									39	43	47	51	55	59	63	67
14										47	51	56	60	65	69	73
15											56	61	66	70	75	80
16												66	71	76	82	87
17													77	82	88	93
18														88	94	100
19															101	107
20																114

Based in part on Tables 1, 3, 5, and 7 of D. Auble, "Extended Tables for the Mann-Whitney Statistic," *Bulletin of the Institute of Educational Research at Indiana University*, **1**, No. 2, 1953, by permission of the director.

Table A.17   Table for the Rank-Sum Test

Table A.17 (continued) Critical Values for the Wilcoxon Rank-Sum Test

One-Tailed Test at $\alpha = 0.025$ or Two-Tailed Test at $\alpha = 0.05$

n_1	4	5	6	7	8	9	10	11	12	13	14	15	16	17	18	19	20
1																	
2					0	0	0	0	1	1	1	1	1	2	2	2	2
3		0	1	1	2	2	3	3	4	4	5	5	6	6	7	7	8
4	0	1	2	3	4	4	5	6	7	8	9	10	11	11	12	13	13
5		2	3	5	6	7	8	9	11	12	13	14	15	17	18	19	20
6			5	6	8	10	11	13	14	16	17	19	21	22	24	25	27
7				8	10	12	14	16	18	20	22	24	26	28	30	32	34
8					13	15	17	19	22	24	26	29	31	34	36	38	41
9						17	20	23	26	28	31	34	37	39	42	45	48
10							23	26	29	33	36	39	42	45	48	52	55
11								30	33	37	40	44	47	51	55	58	62
12									37	41	45	49	53	57	61	65	69
13										45	50	54	59	63	67	72	76
14											55	59	64	67	74	78	83
15												64	70	75	80	85	90
16													75	81	86	92	98
17														87	93	99	105
18															99	106	112
19																113	119
20																	127

One-Tailed Test at $\alpha = 0.05$ or Two-Tailed Test at $\alpha = 0.1$

n_1	3	4	5	6	7	8	9	10	11	12	13	14	15	16	17	18	19	20
1																	0	0
2			0	0	0	1	1	1	1	2	2	3	3	3	3	4	4	4
3	0	0	1	2	2	3	4	4	5	5	6	7	7	8	9	9	10	11
4		1	2	3	4	5	6	7	8	9	10	11	12	14	15	16	17	18
5			4	5	6	8	9	11	12	13	15	16	18	19	20	22	23	25
6				7	8	10	12	14	16	17	19	21	23	25	26	28	30	32
7					11	13	15	17	19	21	24	26	28	30	33	35	37	39
8						15	18	20	23	26	28	31	33	36	39	41	44	47
9							21	24	27	30	33	36	39	42	45	48	51	54
10								27	31	34	37	41	44	48	51	55	58	62
11									34	38	42	46	50	54	57	61	65	69
12										42	47	51	55	60	64	68	72	77
13											51	56	61	65	70	75	80	84
14												61	66	71	77	82	87	92
15													72	77	83	88	94	100
16														83	89	95	101	107
17															96	102	109	115
18																109	116	123
19																	123	130
20																		138

Table A.18 $P(V \leq v^*$ when H_0 is true) in the Runs Test

(n_1, n_2)	v^* 2	3	4	5	6	7	8	9	10
(2, 3)	0.200	0.500	0.900	1.000					
(2, 4)	0.133	0.400	0.800	1.000					
(2, 5)	0.095	0.333	0.714	1.000					
(2, 6)	0.071	0.286	0.643	1.000					
(2, 7)	0.056	0.250	0.583	1.000					
(2, 8)	0.044	0.222	0.533	1.000					
(2, 9)	0.036	0.200	0.491	1.000					
(2, 10)	0.030	0.182	0.455	1.000					
(3, 3)	0.100	0.300	0.700	0.900	1.000				
(3, 4)	0.057	0.200	0.543	0.800	0.971	1.000			
(3, 5)	0.036	0.143	0.429	0.714	0.929	1.000			
(3, 6)	0.024	0.107	0.345	0.643	0.881	1.000			
(3, 7)	0.017	0.083	0.283	0.583	0.833	1.000			
(3, 8)	0.012	0.067	0.236	0.533	0.788	1.000			
(3, 9)	0.009	0.055	0.200	0.491	0.745	1.000			
(3, 10)	0.007	0.045	0.171	0.455	0.706	1.000			
(4, 4)	0.029	0.114	0.371	0.629	0.886	0.971	1.000		
(4, 5)	0.016	0.071	0.262	0.500	0.786	0.929	0.992	1.000	
(4, 6)	0.010	0.048	0.190	0.405	0.690	0.881	0.976	1.000	
(4, 7)	0.006	0.033	0.142	0.333	0.606	0.833	0.954	1.000	
(4, 8)	0.004	0.024	0.109	0.279	0.533	0.788	0.929	1.000	
(4, 9)	0.003	0.018	0.085	0.236	0.471	0.745	0.902	1.000	
(4, 10)	0.002	0.014	0.068	0.203	0.419	0.706	0.874	1.000	
(5, 5)	0.008	0.040	0.167	0.357	0.643	0.833	0.960	0.992	1.000
(5, 6)	0.004	0.024	0.110	0.262	0.522	0.738	0.911	0.976	0.998
(5, 7)	0.003	0.015	0.076	0.197	0.424	0.652	0.854	0.955	0.992
(5, 8)	0.002	0.010	0.054	0.152	0.347	0.576	0.793	0.929	0.984
(5, 9)	0.001	0.007	0.039	0.119	0.287	0.510	0.734	0.902	0.972
(5, 10)	0.001	0.005	0.029	0.095	0.239	0.455	0.678	0.874	0.958
(6, 6)	0.002	0.013	0.067	0.175	0.392	0.608	0.825	0.933	0.987
(6, 7)	0.001	0.008	0.043	0.121	0.296	0.500	0.733	0.879	0.966
(6, 8)	0.001	0.005	0.028	0.086	0.226	0.413	0.646	0.821	0.937
(6, 9)	0.000	0.003	0.019	0.063	0.175	0.343	0.566	0.762	0.902
(6, 10)	0.000	0.002	0.013	0.047	0.137	0.288	0.497	0.706	0.864
(7, 7)	0.001	0.004	0.025	0.078	0.209	0.383	0.617	0.791	0.922
(7, 8)	0.000	0.002	0.015	0.051	0.149	0.296	0.514	0.704	0.867
(7, 9)	0.000	0.001	0.010	0.035	0.108	0.231	0.427	0.622	0.806
(7, 10)	0.000	0.001	0.006	0.024	0.080	0.182	0.355	0.549	0.743
(8, 8)	0.000	0.001	0.009	0.032	0.100	0.214	0.405	0.595	0.786
(8, 9)	0.000	0.001	0.005	0.020	0.069	0.157	0.319	0.500	0.702
(8, 10)	0.000	0.000	0.003	0.013	0.048	0.117	0.251	0.419	0.621
(9, 9)	0.000	0.000	0.003	0.012	0.044	0.109	0.238	0.399	0.601
(9, 10)	0.000	0.000	0.002	0.008	0.029	0.077	0.179	0.319	0.510
(10, 10)	0.000	0.000	0.001	0.004	0.019	0.051	0.128	0.242	0.414

Reproduced from C. Eisenhart and R. Swed, "Tables for Testing Randomness of Grouping in a Sequence of Alternatives," *Ann. Math. Stat.*, **14**, 1943, by permission of the editor.

Table A.18 Table for the Runs Test

Table A.18 (continued) $P(V \leq v^*$ when H_0 is true) in the Runs Test

(n_1, n_2)	11	12	13	14	15	16	17	18	19	20
(2,3)										
(2,4)										
(2,5)										
(2,6)										
(2,7)										
(2,8)										
(2,9)										
(2,10)										
(3,3)										
(3,4)										
(3,5)										
(3,6)										
(3,7)										
(3,8)										
(3,9)										
(3,10)										
(4,4)										
(4,5)										
(4,6)										
(4,7)										
(4,8)										
(4,9)										
(4,10)										
(5,5)										
(5,6)	1.000									
(5,7)	1.000									
(5,8)	1.000									
(5,9)	1.000									
(5,10)	1.000									
(6,6)	0.998	1.000								
(6,7)	0.992	0.999	1.000							
(6,8)	0.984	0.998	1.000							
(6,9)	0.972	0.994	1.000							
(6,10)	0.958	0.990	1.000							
(7,7)	0.975	0.996	0.999	1.000						
(7,8)	0.949	0.988	0.998	1.000	1.000					
(7,9)	0.916	0.975	0.994	0.999	1.000					
(7,10)	0.879	0.957	0.990	0.998	1.000					
(8,8)	0.900	0.968	0.991	0.999	1.000	1.000				
(8,9)	0.843	0.939	0.980	0.996	0.999	1.000	1.000			
(8,10)	0.782	0.903	0.964	0.990	0.998	1.000	1.000			
(9,9)	0.762	0.891	0.956	0.988	0.997	1.000	1.000	1.000		
(9,10)	0.681	0.834	0.923	0.974	0.992	0.999	1.000	1.000	1.000	
(10,10)	0.586	0.758	0.872	0.949	0.981	0.996	0.999	1.000	1.000	1.000

Table A.19 Sample Size for Two-Sided Nonparametric Tolerance Limits

$1-\alpha$	$1-\gamma$					
	0.50	0.70	0.90	0.95	0.99	0.995
0.995	336	488	777	947	1325	1483
0.99	168	244	388	473	662	740
0.95	34	49	77	93	130	146
0.90	17	24	38	46	64	72
0.85	11	16	25	30	42	47
0.80	9	12	18	22	31	34
0.75	7	10	15	18	24	27
0.70	6	8	12	14	20	22
0.60	4	6	9	10	14	16
0.50	3	5	7	8	11	12

Table A–25d of Wilfrid J. Dixon and Frank J. Massey, Jr., *Introduction to Statistical Analysis*, 3rd ed. McGraw-Hill, 1969. Reprinted with permission by The McGraw-Hill Companies, Inc.

Table A.20 Sample Size for One-Sided Nonparametric Tolerance Limits

$1-\alpha$	$1-\gamma$				
	0.50	0.70	0.95	0.99	0.995
0.995	139	241	598	919	1379
0.99	69	120	299	459	688
0.95	14	24	59	90	135
0.90	7	12	29	44	66
0.85	5	8	19	29	43
0.80	4	6	14	21	31
0.75	3	5	11	7	25
0.70	2	4	9	13	20
0.60	2	3	6	10	14
0.50	1	2	5	7	10

Table A–25e of Wilfrid J. Dixon and Frank J. Massey, Jr., *Introduction to Statistical Analysis*, 3rd ed. McGraw-Hill, 1969. Reprinted with permission by The McGraw-Hill Companies, Inc.

Table A.21 Table for Spearman's Rank Correlation Coefficients

Table A.21 Critical Values for Spearman's Rank Correlation Coefficients

n	$\alpha = 0.05$	$\alpha = 0.025$	$\alpha = 0.01$	$\alpha = 0.005$
5	0.900			
6	0.829	0.886	0.943	
7	0.714	0.786	0.893	
8	0.643	0.738	0.833	0.881
9	0.600	0.683	0.783	0.833
10	0.564	0.648	0.745	0.794
11	0.523	0.623	0.736	0.818
12	0.497	0.591	0.703	0.780
13	0.475	0.566	0.673	0.745
14	0.457	0.545	0.646	0.716
15	0.441	0.525	0.623	0.689
16	0.425	0.507	0.601	0.666
17	0.412	0.490	0.582	0.645
18	0.399	0.476	0.564	0.625
19	0.388	0.462	0.549	0.608
20	0.377	0.450	0.534	0.591
21	0.368	0.438	0.521	0.576
22	0.359	0.428	0.508	0.562
23	0.351	0.418	0.496	0.549
24	0.343	0.409	0.485	0.537
25	0.336	0.400	0.475	0.526
26	0.329	0.392	0.465	0.515
27	0.323	0.385	0.456	0.505
28	0.317	0.377	0.448	0.496
29	0.311	0.370	0.440	0.487
30	0.305	0.364	0.432	0.478

Reproduced from E. G. Olds, "Distribution of Sums of Squares of Rank Differences for Small Samples," *Ann. Math. Stat.*, **9**, 1938, by permission of the editor.

Table A.22 Factors for Constructing Control Charts

Obs. in Sample	Chart for Averages		Chart for Standard Deviations							Chart for Ranges					
	Factors for Control Limits		Factors for Centerline		Factors for Control Limits					Factors for Centerline		Factors for Control Limits			
n	A_2	A_3	c_4	$1/c_4$	B_3	B_4	B_5	B_6		d_2	$1/d_2$	d_3	D_3	D_4	
2	1.880	2.659	0.7979	1.2533	0	3.267	0	2.606		1.128	0.8865	0.853	0	3.267	
3	1.023	1.954	0.8862	1.1284	0	2.568	0	2.276		1.693	0.5907	0.888	0	2.574	
4	0.729	1.628	0.9213	1.0854	0	2.266	0	2.088		2.059	0.4857	0.880	0	2.282	
5	0.577	1.427	0.9400	1.0638	0	2.089	0	1.964		2.326	0.4299	0.864	0	2.114	
6	0.483	1.287	0.9515	1.0510	0.030	1.970	0.029	1.874		2.534	0.3946	0.848	0	2.004	
7	0.419	1.182	0.9594	1.0423	0.118	1.882	0.113	1.806		2.704	0.3698	0.833	0.076	1.924	
8	0.373	1.099	0.9650	1.0363	0.185	1.815	0.179	1.751		2.847	0.3512	0.820	0.136	1.864	
9	0.337	1.032	0.9693	1.0317	0.239	1.761	0.232	1.707		2.970	0.3367	0.808	0.184	1.816	
10	0.308	0.975	0.9727	1.0281	0.284	1.716	0.276	1.669		3.078	0.3249	0.797	0.223	1.777	
11	0.285	0.927	0.9754	1.0252	0.321	1.679	0.313	1.637		3.173	0.3152	0.787	0.256	1.744	
12	0.266	0.886	0.9776	1.0229	0.354	1.646	0.346	1.610		3.258	0.3069	0.778	0.283	1.717	
13	0.249	0.850	0.9794	1.0210	0.382	1.618	0.374	1.585		3.336	0.2998	0.770	0.307	1.693	
14	0.235	0.817	0.9810	1.0194	0.406	1.594	0.399	1.563		3.407	0.2935	0.763	0.328	1.672	
15	0.223	0.789	0.9823	1.0180	0.428	1.572	0.421	1.544		3.472	0.2880	0.756	0.347	1.653	
16	0.212	0.763	0.9835	1.0168	0.448	1.552	0.440	1.526		3.532	0.2831	0.750	0.363	1.637	
17	0.203	0.739	0.9845	1.0157	0.466	1.534	0.458	1.511		3.588	0.2787	0.744	0.378	1.622	
18	0.194	0.718	0.9854	1.0148	0.482	1.518	0.475	1.496		3.640	0.2747	0.739	0.391	1.608	
19	0.187	0.698	0.9862	1.0140	0.497	1.503	0.490	1.483		3.689	0.2711	0.734	0.403	1.597	
20	0.180	0.680	0.9869	1.0133	0.510	1.490	0.504	1.470		3.735	0.2677	0.729	0.415	1.585	
21	0.173	0.663	0.9876	1.0126	0.523	1.477	0.516	1.459		3.778	0.2647	0.724	0.425	1.575	
22	0.167	0.647	0.9882	1.0119	0.534	1.466	0.528	1.448		3.819	0.2618	0.720	0.434	1.566	
23	0.162	0.633	0.9887	1.0114	0.545	1.455	0.539	1.438		3.858	0.2592	0.716	0.443	1.557	
24	0.157	0.619	0.9892	1.0109	0.555	1.445	0.549	1.429		3.895	0.2567	0.712	0.451	1.548	
25	0.153	0.606	0.9896	1.0105	0.565	1.435	0.559	1.420		3.931	0.2544	0.708	0.459	4.541	

Section A.24 Proof of Mean of the Hypergeometric Distribution

Table A.23 The Incomplete Gamma Function: $F(x; \alpha) = \int_0^x \frac{1}{\Gamma(\alpha)} y^{\alpha-1} e^{-y} \, dy$

x	\multicolumn{10}{c}{α}
	1	2	3	4	5	6	7	8	9	10
1	0.6320	0.2640	0.0800	0.0190	0.0040	0.0010	0.0000	0.0000	0.0000	0.0000
2	0.8650	0.5940	0.3230	0.1430	0.0530	0.0170	0.0050	0.0010	0.0000	0.0000
3	0.9500	0.8010	0.5770	0.3530	0.1850	0.0840	0.0340	0.0120	0.0040	0.0010
4	0.9820	0.9080	0.7620	0.5670	0.3710	0.2150	0.1110	0.0510	0.0210	0.0080
5	0.9930	0.9600	0.8750	0.7350	0.5600	0.3840	0.2380	0.1330	0.0680	0.0320
6	0.9980	0.9830	0.9380	0.8490	0.7150	0.5540	0.3940	0.2560	0.1530	0.0840
7	0.9990	0.9930	0.9700	0.9180	0.8270	0.6990	0.5500	0.4010	0.2710	0.1700
8	1.0000	0.9970	0.9860	0.9580	0.9000	0.8090	0.6870	0.5470	0.4070	0.2830
9		0.9990	0.9940	0.9790	0.9450	0.8840	0.7930	0.6760	0.5440	0.4130
10		1.0000	0.9970	0.9900	0.9710	0.9330	0.8700	0.7800	0.6670	0.5420
11			0.9990	0.9950	0.9850	0.9620	0.9210	0.8570	0.7680	0.6590
12			1.0000	0.9980	0.9920	0.9800	0.9540	0.9110	0.8450	0.7580
13				0.9990	0.9960	0.9890	0.9740	0.9460	0.9000	0.8340
14				1.0000	0.9980	0.9940	0.9860	0.9680	0.9380	0.8910
15					0.9990	0.9970	0.9920	0.9820	0.9630	0.9300

A.24 Proof of Mean of the Hypergeometric Distribution

To find the mean of the hypergeometric distribution, we write

$$E(X) = \sum_{x=0}^{n} x \frac{\binom{k}{x}\binom{N-k}{n-x}}{\binom{N}{n}} = k \sum_{x=1}^{n} \frac{(k-1)!}{(x-1)!(k-x)!} \cdot \frac{\binom{N-k}{n-x}}{\binom{N}{n}}$$

$$= k \sum_{x=1}^{n} \frac{\binom{k-1}{x-1}\binom{N-k}{n-x}}{\binom{N}{n}}.$$

Since

$$\binom{N-k}{n-1-y} = \binom{(N-1)-(k-1)}{n-1-y} \quad \text{and} \quad \binom{N}{n} = \frac{N!}{n!(N-n)!} = \frac{N}{n}\binom{N-1}{n-1},$$

letting $y = x - 1$, we obtain

$$E(X) = k \sum_{y=0}^{n-1} \frac{\binom{k-1}{y}\binom{N-k}{n-1-y}}{\binom{N}{n}}$$

$$= \frac{nk}{N} \sum_{y=0}^{n-1} \frac{\binom{k-1}{y}\binom{(N-1)-(k-1)}{n-1-y}}{\binom{N-1}{n-1}} = \frac{nk}{N},$$

since the summation represents the total of all probabilities in a hypergeometric experiment when $N-1$ items are selected at random from $N-1$, of which $k-1$ are labeled success.

A.25 Proof of Mean and Variance of the Poisson Distribution

Let $\mu = \lambda t$.

$$E(X) = \sum_{x=0}^{\infty} x \cdot \frac{e^{-\mu}\mu^x}{x!} = \sum_{x=1}^{\infty} x \cdot \frac{e^{-\mu}\mu^x}{x!} = \mu \sum_{x=1}^{\infty} \frac{e^{-\mu}\mu^{x-1}}{(x-1)!}.$$

Since the summation in the last term above is the total probability of a Poisson random variable with mean μ, which can be easily seen by letting $y = x - 1$, it equals 1. Therefore, $E(X) = \mu$. To calculate the variance of X, note that

$$E[X(X-1)] = \sum_{x=0}^{\infty} x(x-1) \frac{e^{-\mu}\mu^x}{x!} = \mu^2 \sum_{x=2}^{\infty} \frac{e^{-\mu}\mu^{x-2}}{(x_2)!}.$$

Again, letting $y = x - 2$, the summation in the last term above is the total probability of a Poisson random variable with mean μ. Hence, we obtain

$$\sigma^2 = E(X^2) - [E(X)]^2 = E[X(X-1)] + E(X) - [E(X)]^2 = \mu^2 + \mu - \mu^2 = \mu = \lambda t.$$

A.26 Proof of Mean and Variance of the Gamma Distribution

To find the mean and variance of the gamma distribution, we first calculate

$$E(X^k) = \frac{1}{\beta^\alpha \Gamma(\alpha)} \int_0^\infty x^{\alpha+k-1} e^{-x/\beta}\, dx = \frac{\beta^{k+\alpha}\Gamma(\alpha+k)}{\beta^\alpha \Gamma(\alpha)} \int_0^\infty \frac{x^{\alpha+k-1} e^{-x/\beta}}{\beta^{k+\alpha}\Gamma(\alpha+k)}\, dx,$$

for $k = 0, 1, 2, \ldots$. Since the integrand in the last term above is a gamma density function with parameters $\alpha + k$ and β, it equals 1. Therefore,

$$E(X^k) = \beta^k \frac{\Gamma(k+\alpha)}{\Gamma(\alpha)}.$$

Using the recursion formula of the gamma function from page 194, we obtain

$$\mu = \beta \frac{\Gamma(\alpha+1)}{\Gamma(\alpha)} = \alpha\beta \quad \text{and} \quad \sigma^2 = E(X^2) - \mu^2 = \beta^2 \frac{\Gamma(\alpha+2)}{\Gamma(\alpha)} - \mu^2 = \beta^2 \alpha(\alpha+1) - (\alpha\beta)^2 = \alpha\beta^2.$$

Appendix B
Answers to Odd-Numbered Non-Review Exercises

Chapter 1

1.1 (a) Sample size = 15
  (b) Sample mean = 3.787
  (c) Sample median = 3.6
  (e) $\bar{x}_{tr(20)} = 3.678$
  (f) They are about the same.

1.3 (b) Yes, the aging process has reduced the tensile strength.
  (c) $\bar{x}_{\text{Aging}} = 209.90$, $\bar{x}_{\text{No aging}} = 222.10$
  (d) $\tilde{x}_{\text{Aging}} = 210.00$, $\tilde{x}_{\text{No aging}} = 221.50$. The means and medians are similar for each group.

1.5 (b) Control: $\bar{x} = 5.60$, $\tilde{x} = 5.00$, $\bar{x}_{tr(10)} = 5.13$.
  Treatment: $\bar{x} = 7.60$, $\tilde{x} = 4.50$, $\bar{x}_{tr(10)} = 5.63$.
  (c) The extreme value of 37 in the treatment group plays a strong leverage role for the mean calculation.

1.7 Sample variance = 0.943
  Sample standard deviation = 0.971

1.9 (a) No aging: sample variance = 23.66, sample standard deviation = 4.86.
  Aging: sample variance = 42.10, sample standard deviation = 6.49.
  (b) Based on the numbers in (a), the variation in "Aging" is smaller than the variation in "No aging," although the difference is not so apparent in the plot.

1.11 Control: sample variance = 69.38, sample standard deviation = 8.33.
  Treatment: sample variance = 128.04, sample standard deviation = 11.32.

1.13 (a) Mean = 124.3, median = 120
  (b) 175 is an extreme observation.

1.15 Yes, P-value = 0.03125, probability of obtaining $HHHHH$ with a fair coin.

1.17 (a) The sample means for nonsmokers and smokers are 30.32 and 43.70, respectively.
  (b) The sample standard deviations for nonsmokers and smokers are 7.13 and 16.93, respectively.
  (d) Smokers appear to take a longer time to fall asleep. For smokers the time to fall asleep is more variable.

1.19 (a)

Stem	Leaf	Frequency
0	22233457	8
1	023558	6
2	035	3
3	03	2
4	057	3
5	0569	4
6	0005	4

(b)

Class Interval	Class Midpoint	Freq.	Rel. Freq.
0.0–0.9	0.45	8	0.267
1.0–1.9	1.45	6	0.200
2.0–2.9	2.45	3	0.100
3.0–3.9	3.45	2	0.067
4.0–4.9	4.45	3	0.100
5.0–5.9	5.45	4	0.133
6.0–6.9	6.45	4	0.133

(c) Sample mean = 2.7967
Sample range = 6.3
Sample standard deviation = 2.2273

1.21 (a) $\bar{x} = 74.02$ and $\tilde{x} = 78$
(b) $s = 39.26$

1.23 (b) $\bar{x}_{1980} = 395.10$, $\bar{x}_{1990} = 160.15$
(c) The mean emissions dropped between 1980 and 1990; the variability also decreased because there were no longer extremely large emissions.

1.25 (a) Sample mean = 33.31
(b) Sample median = 26.35
(d) $\bar{x}_{\text{tr}(10)} = 30.97$

Chapter 2

2.1 (a) $S = \{8, 16, 24, 32, 40, 48\}$
(b) $S = \{-5, 1\}$
(c) $S = \{T, HT, HHT, HHH\}$
(d) $S = \{$Africa, Antarctica, Asia, Australia, Europe, North America, South America$\}$
(e) $S = \phi$

2.3 $A = C$

2.5 Using the tree diagram, we obtain
$S = \{1HH, 1HT, 1TH, 1TT, 2H, 2T, 3HH, 3HT, 3TH, 3TT, 4H, 4T, 5HH, 5HT, 5TH, 5TT, 6H, 6T\}$

2.7 $S_1 = \{MMMM, MMMF, MMFM, MFMM, FMMM, MMFF, MFMF, MFFM, FMFM, FFMM, FMMF, MFFF, FMFF, FFMF, FFFM, FFFF\}$;
$S_2 = \{0, 1, 2, 3, 4\}$

2.9 (a) $A = \{1HH, 1HT, 1TH, 1TT, 2H, 2T\}$
(b) $B = \{1TT, 3TT, 5TT\}$
(c) $A' = \{3HH, 3HT, 3TH, 3TT, 4H, 4T, 5HH, 5HT, 5TH, 5TT, 6H, 6T\}$
(d) $A' \cap B = \{3TT, 5TT\}$
(e) $A \cup B = \{1HH, 1HT, 1TH, 1TT, 2H, 2T, 3TT, 5TT\}$

2.11 (a) $S = \{M_1M_2, M_1F_1, M_1F_2, M_2M_1, M_2F_1, M_2F_2, F_1M_1, F_1M_2, F_1F_2, F_2M_1, F_2M_2, F_2F_1\}$
(b) $A = \{M_1M_2, M_1F_1, M_1F_2, M_2M_1, M_2F_1, M_2F_2\}$
(c) $B = \{M_1F_1, M_1F_2, M_2F_1, M_2F_2, F_1M_1, F_1M_2, F_2M_1, F_2M_2\}$
(d) $C = \{F_1F_2, F_2F_1\}$
(e) $A \cap B = \{M_1F_1, M_1F_2, M_2F_1, M_2F_2\}$
(f) $A \cup C = \{M_1M_2, M_1F_1, M_1F_2, M_2M_1, M_2F_1, M_2F_2, F_1F_2, F_2F_1\}$

2.15 (a) {nitrogen, potassium, uranium, oxygen}
(b) {copper, sodium, zinc, oxygen}
(c) {copper, sodium, nitrogen, potassium, uranium, zinc}
(d) {copper, uranium, zinc}
(e) ϕ
(f) {oxygen}

2.19 (a) The family will experience mechanical problems but will receive no ticket for a traffic violation and will not arrive at a campsite that has no vacancies.
(b) The family will receive a traffic ticket and arrive at a campsite that has no vacancies but will not experience mechanical problems.
(c) The family will experience mechanical problems and will arrive at a campsite that has no vacancies.
(d) The family will receive a traffic ticket but will not arrive at a campsite that has no vacancies.
(e) The family will not experience mechanical problems.

2.21 18

2.23 156

2.25 20

2.27 48

2.29 210

2.31 72

2.33 (a) 1024; (b) 243

2.35 362,880

2.37 2880

2.39 (a) 40,320; (b) 336

2.41 360

2.43 24

2.45 3360

2.47 56

2.49 (a) Sum of the probabilities exceeds 1.
(b) Sum of the probabilities is less than 1.
(c) A negative probability
(d) Probability of both a heart and a black card is zero.

2.51 $S = \{\$10, \$25, \$100\}$; $P(10) = \frac{11}{20}$, $P(25) = \frac{3}{10}$, $P(100) = \frac{15}{100}$; $\frac{17}{20}$

2.53 (a) 0.3; (b) 0.2

2.55 10/117

2.57 (a) 5/26; (b) 9/26; (c) 19/26

2.59 (a) 94/54,145; (b) 143/39,984

2.61 (a) 22/25; (b) 3/25; (c) 17/50

2.63 (a) 0.32; (b) 0.68; (c) office or den

2.65 (a) 0.8; (b) 0.45; (c) 0.55

2.67 (a) 0.31; (b) 0.93; (c) 0.31

2.69 (a) 0.009; (b) 0.999; (c) 0.01

2.71 (a) 0.048; (b) $50,000; (c) $12,500

2.73 (a) The probability that a convict who pushed dope also committed armed robbery.
(b) The probability that a convict who committed armed robbery did not push dope.
(c) The probability that a convict who did not push dope also did not commit armed robbery.

2.75 (a) 14/39; (b) 95/112

2.77 (a) 5/34; (b) 3/8

2.79 (a) 0.018; (b) 0.614; (c) 0.166; (d) 0.479

2.81 (a) 0.35; (b) 0.875; (c) 0.55

2.83 (a) 9/28; (b) 3/4; (c) 0.91

2.85 0.27

2.87 5/8

2.89 (a) 0.0016; (b) 0.9984

2.91 (a) 91/323; (b) 91/323

2.93 (a) 0.75112; (b) 0.2045

2.95 0.0960

2.97 0.40625

2.99 0.1124

2.101 0.857

Chapter 3

3.1 Discrete; continuous; continuous; discrete; discrete; continuous

3.3
Sample Space	w
HHH	3
HHT	1
HTH	1
THH	1
HTT	−1
THT	−1
TTH	−1
TTT	−3

3.5 (a) 1/30; (b) 1/10

3.7 (a) 0.68; (b) 0.375

3.9 (b) 19/80

3.11
x	0	1	2
$f(x)$	$\frac{2}{7}$	$\frac{4}{7}$	$\frac{1}{7}$

3.13 $F(x) = \begin{cases} 0, & \text{for } x < 0, \\ 0.41, & \text{for } 0 \leq x < 1, \\ 0.78, & \text{for } 1 \leq x < 2, \\ 0.94, & \text{for } 2 \leq x < 3, \\ 0.99, & \text{for } 3 \leq x < 4, \\ 1, & \text{for } x \geq 4 \end{cases}$

3.15 $F(x) = \begin{cases} 0, & \text{for } x < 0, \\ \frac{2}{7}, & \text{for } 0 \leq x < 1, \\ \frac{6}{7}, & \text{for } 1 \leq x < 2, \\ 1, & \text{for } x \geq 2 \end{cases}$
(a) 4/7; (b) 5/7

3.17 (b) 1/4; (c) 0.3

3.19 $F(x) = \begin{cases} 0, & x < 1 \\ \frac{x-1}{2}, & 1 \leq x < 3; \ 1/4 \\ 1, & x \geq 3 \end{cases}$

3.21 (a) 3/2; (b) $F(x) = \begin{cases} 0, & x < 0 \\ x^{3/2}, & 0 \leq x < 1 \\ 1, & x \geq 1 \end{cases}$; 0.3004

3.23
$$F(w) = \begin{cases} 0, & \text{for } w < -3, \\ \frac{1}{27}, & \text{for } -3 \leq w < -1, \\ \frac{7}{27}, & \text{for } -1 \leq w < 1, \\ \frac{19}{27}, & \text{for } 1 \leq w < 3, \\ 1, & \text{for } w \geq 3 \end{cases}$$
(a) 20/27; (b) 2/3

3.25

t	20	25	30
$P(T=t)$	$\frac{1}{5}$	$\frac{3}{5}$	$\frac{1}{5}$

3.27 (a)
$$F(x) = \begin{cases} 0, & x < 0, \\ 1 - \exp(-x/2000), & x \geq 0 \end{cases}$$
(b) 0.6065; (c) 0.6321

3.29 (b)
$$F(x) = \begin{cases} 0, & x < 1, \\ 1 - x^{-3}, & x \geq 1 \end{cases}$$
(c) 0.0156

3.31 (a) 0.2231; (b) 0.2212

3.33 (a) $k = 280$; (b) 0.3633; (c) 0.0563

3.35 (a) 0.1528; (b) 0.0446

3.37 (a) 1/36; (b) 1/15

3.39 (a)

$f(x,y)$		x		
	0	1	2	3
y 0	0	$\frac{3}{70}$	$\frac{9}{70}$	$\frac{3}{70}$
1	$\frac{2}{70}$	$\frac{18}{70}$	$\frac{18}{70}$	$\frac{2}{70}$
2	$\frac{3}{70}$	$\frac{9}{70}$	$\frac{3}{70}$	0

(b) 1/2

3.41 (a) 1/16; (b) $g(x) = 12x(1-x)^2$, for $0 \leq x \leq 1$; (c) 1/4

3.43 (a) 3/64; (b) 1/2

3.45 0.6534

3.47 (a) Dependent; (b) 1/3

3.49 (a)

x	1	2	3
$g(x)$	0.10	0.35	0.55

(b)

y	1	2	3
$h(y)$	0.20	0.50	0.30

(c) 0.2857

3.51 (a)

$f(x,y)$		x			
		0	1	2	3
y	0	$\frac{1}{55}$	$\frac{6}{55}$	$\frac{6}{55}$	$\frac{1}{55}$
	1	$\frac{6}{55}$	$\frac{16}{55}$	$\frac{6}{55}$	0
	2	$\frac{6}{55}$	$\frac{6}{55}$	0	0
	3	$\frac{1}{55}$	0	0	0

(b) 42/55

3.53 5/8

3.55 Independent

3.57 (a) 3; (b) 21/512

3.59 Dependent

Chapter 4

4.1 0.88

4.3 25¢

4.5 $1.23

4.7 $500

4.9 $6900

4.11 $(\ln 4)/\pi$

4.13 100 hours

4.15 0

4.17 209

4.19 $1855

4.21 $833.33

4.23 (a) 35.2; (b) $\mu_X = 3.20$, $\mu_Y = 3.00$

4.25 2

4.27 2000 hours

4.29 (b) 3/2

4.31 (a) 1/6; (b) $(5/6)^5$

4.33 $5,250,000

4.35 0.74

4.37 1/18; in terms of actual profit, the variance is $\frac{1}{18}(5000)^2$

4.39 1/6

4.41 118.9

4.43 $\mu_Y = 10$; $\sigma_Y^2 = 144$

4.45 0.01

Answers to Chapter 5

4.47 -0.0062

4.49 $\sigma_X^2 = 0.8456$, $\sigma_X = 0.9196$

4.51 $-1/\sqrt{5}$

4.53 $\mu_{g(X)} = 10.33$, $\sigma_{g(X)} = 6.66$

4.55 $0.80

4.57 209

4.59 $\mu = 7/2$, $\sigma^2 = 15/4$

4.61 $3/14$

4.63 52

4.65 (a) 7; (b) 0; (c) 12.25

4.67 $46/63$

4.69 (a) $E(X) = E(Y) = 1/3$ and $\text{Var}(X) = \text{Var}(Y) = 4/9$; (b) $E(Z) = 2/3$ and $\text{Var}(Z) = 8/9$

4.71 (a) 4; (b) 32; 16

4.73 By direct calculation, $E(e^Y) = 1884.32$. Using the second-order approximation, $E(e^Y) \approx 1883.38$, which is very close to the true value.

4.75 0.03125

4.77 (a) At most $4/9$; (b) at least $5/9$; (c) at least $21/25$; (d) 10

Chapter 5

5.1 $\mu = \frac{1}{k}\sum_{i=1}^{k} x_i$, $\sigma^2 = \frac{1}{k}\sum_{i=1}^{k}(x_i - \mu)^2$

5.3 $f(x) = \frac{1}{10}$, for $x = 1, 2, \ldots, 10$, and $f(x) = 0$ elsewhere; $3/10$

5.5 (a) 0.0480; (b) 0.2375; (c) $P(X = 5 \mid p = 0.3) = 0.1789$, $P = 0.3$ is reasonable.

5.7 (a) 0.0474; (b) 0.0171

5.9 (a) 0.7073; (b) 0.4613; (c) 0.1484

5.11 0.1240

5.13 0.8369

5.15 (a) 0.0778; (b) 0.3370; (c) 0.0870

5.17 $\mu = 3.5$, $\sigma^2 = 1.05$

5.19 $f(x_1, x_2, x_3) = \binom{n}{x_1, x_2, x_3} 0.35^{x_1} 0.05^{x_2} 0.60^{x_3}$

5.21 0.0095

5.23 0.0077

5.25 0.8670

5.27 (a) 0.2852; (b) 0.9887; (c) 0.6083

5.29 $5/14$

5.31 $h(x; 6, 3, 4) = \dfrac{\binom{4}{x}\binom{2}{3-x}}{\binom{6}{3}}$, for $x = 1, 2, 3$; $P(2 \leq X \leq 3) = 4/5$

5.33 (a) 0.3246; (b) 0.4496

5.35 0.9517

5.37 (a) 0.6815; (b) 0.1153

5.39 0.9453

5.41 0.6077

5.43 (a) $4/33$; (b) $8/165$

5.45 0.2315

5.47 (a) 0.3991; (b) 0.1316

5.49 0.0515

5.51 $63/64$

5.53 (a) 0.3840; (b) 0.0067

5.55 (a) 0.0630; (b) 0.9730

5.57 (a) 0.1429; (b) 0.1353

5.59 (a) 0.1638; (b) 0.032

5.61 0.2657

5.63 $\mu = 6$, $\sigma^2 = 6$

5.65 (a) 0.2650; (b) 0.9596

5.67 (a) 0.8243; (b) 14

5.69 4

5.71 5.53×10^{-4}; $\mu = 7.5$

5.73 (a) 0.0137; (b) 0.0830

5.75 0.4686

Chapter 6

6.3 (a) 0.6; (b) 0.7; (c) 0.5

6.5 (a) 0.0823; (b) 0.0250; (c) 0.2424; (d) 0.9236; (e) 0.8133; (f) 0.6435

6.7 (a) 0.54; (b) -1.72; (c) 1.28

6.9 (a) 0.1151; (b) 16.1; (c) 20.275; (d) 0.5403

6.11 (a) 0.0548; (b) 0.4514; (c) 23 cups; (d) 189.95 milliliters

6.13 (a) 0.8980; (b) 0.0287; (c) 0.6080

6.15 (a) 0.0571; (b) 99.11%; (c) 0.3974; (d) 27.952 minutes; (e) 0.0092

6.17 6.24 years

6.19 (a) 51%; (b) $18.37

6.21 (a) 0.0401; (b) 0.0244

6.23 26 students

6.25 (a) 0.3085; (b) 0.0197

6.27 (a) 0.9514; (b) 0.0668

6.29 (a) 0.1171; (b) 0.2049

6.31 0.1357

6.33 (a) 0.0778; (b) 0.0571; (c) 0.6811

6.35 (a) 0.8749; (b) 0.0059

6.37 (a) 0.0228; (b) 0.3974

6.41 $2.8e^{-1.8} - 3.4e^{-2.4} = 0.1545$

6.43 (a) $\mu = 6$; $\sigma^2 = 18$; (b) between 0 and 14.485 million liters

6.45 $\sum_{x=4}^{6} \binom{6}{x}(1-e^{-3/4})^x (e^{-3/4})^{6-x} = 0.3968$

6.47 (a) $\sqrt{\pi/2} = 1.2533$ years; (b) e^{-2}

6.49 (a) Mean $= 0.25$, median $= 0.206$; (b) variance $= 0.0375$; (c) 0.2963

6.51 $e^{-4} = 0.0183$

6.53 (a) $\mu = \alpha\beta = 50$; (b) $\sigma^2 = \alpha\beta^2 = 500$; $\sigma = \sqrt{500}$; (c) 0.815

6.55 (a) 0.1889; (b) 0.0357

6.57 Mean $= e^6$, variance $= e^{12}(e^4 - 1)$

6.59 (a) e^{-5}; (b) $\beta = 0.2$

Chapter 7

7.1 $g(y) = 1/3$, for $y = 1, 3, 5$

7.3 $g(y_1, y_2) = \binom{2}{\frac{y_1+y_2}{2}, \frac{y_1-y_2}{2}, 2-y_1}$

$\times \left(\frac{1}{4}\right)^{(y_1+y_2)/2} \left(\frac{1}{3}\right)^{(y_1-y_2)/2} \left(\frac{5}{12}\right)^{2-y_1}$;

for $y_1 = 0, 1, 2$; $y_2 = -2, -1, 0, 1, 2$; $y_2 \leq y_1$; $y_1 + y_2 = 0, 2, 4$

7.7 Gamma distribution with $\alpha = 3/2$ and $\beta = m/2b$

7.9 (a) $g(y) = 32/y^3$, for $y > 4$; (b) $1/4$

7.11 $h(z) = 2(1-z)$, for $0 < z < 1$

7.13 $h(w) = 6 + 6w - 12w^{1/2}$, for $0 < w < 1$

7.15 $g(y) = \begin{cases} \frac{2}{9\sqrt{y}}, & 0 < y < 1, \\ \frac{\sqrt{y}+1}{9\sqrt{y}}, & 1 < y < 4 \end{cases}$

7.19 Both equal μ.

7.23 (a) Gamma(2, 1); (b) Uniform(0, 1)

Chapter 8

8.1 (a) Responses of all people in Richmond who have a telephone;
(b) Outcomes for a large or infinite number of tosses of a coin;
(c) Length of life of such tennis shoes when worn on the professional tour;
(d) All possible time intervals for this lawyer to drive from her home to her office.

8.3 (a) $\bar{x} = 3.2$ seconds; (b) $\tilde{x} = 3.1$ seconds

8.5 (a) $\bar{x} = 2.4$; (b) $\tilde{x} = 2$; (c) $m = 3$

8.7 (a) 53.75; (b) 75 and 100

8.9 (a) Range is 10; (b) $s = 3.307$

8.11 (a) 2.971; (b) 2.971

8.13 $s = 0.585$

8.15 (a) 45.9; (b) 5.1

8.17 0.3159

8.19 (a) Variance is reduced from 0.49 to 0.16.
(b) Variance is increased from 0.04 to 0.64.

8.21 Yes

8.23 (a) $\mu = 5.3$; $\sigma^2 = 0.81$;
(b) $\mu_{\bar{X}} = 5.3$; $\sigma^2_{\bar{X}} = 0.0225$;
(c) 0.9082

8.25 (a) 0.6898; (b) 7.35

8.29 0.5596

8.31 (a) The chance that the difference in mean drying time is larger than 1.0 is 0.0013; (b) 13

8.33 (a) 1/2; (b) 0.3085

8.35 $P(\bar{X} \leq 775 \mid \mu = 760) = 0.9332$

8.37 (a) 27.488; (b) 18.475; (c) 36.415

8.39 (a) 0.297; (b) 32.852; (c) 46.928

8.41 (a) 0.05; (b) 0.94

8.45 (a) 0.975; (b) 0.10; (c) 0.875; (d) 0.99

8.47 (a) 2.500; (b) 1.319; (c) 1.714

8.49 No; $\mu > 20$

8.51 (a) 2.71; (b) 3.51; (c) 2.92;
(d) 0.47; (e) 0.34

8.53 The F-ratio is 1.44. The variances are not significantly different.

Chapter 9

9.1 56

9.3 $0.3097 < \mu < 0.3103$

9.5 (a) $22,496 < \mu < 24,504$; (b) error ≤ 1004

9.7 35

9.9 $10.15 < \mu < 12.45$

9.11 $0.978 < \mu < 1.033$

9.13 $47.722 < \mu < 49.278$

9.15 $(13,075, 33,925)$

9.17 $(6.05, 16.55)$

9.19 323.946 to 326.154

9.21 Upper prediction limit: 9.42;
upper tolerance limit: 11.72

9.25 Yes, the value of 6.9 is outside of the prediction interval.

9.27 (a) $(0.9876, 1.0174)$;
(b) $(0.9411, 1.0639)$;
(c) $(0.9334, 1.0716)$

9.35 $2.9 < \mu_1 - \mu_2 < 7.1$

9.37 $2.80 < \mu_1 - \mu_2 < 3.40$

9.39 $1.5 < \mu_1 - \mu_2 < 12.5$

9.41 $0.70 < \mu_1 - \mu_2 < 3.30$

9.43 $-6536 < \mu_1 - \mu_2 < 2936$

9.45 $(-0.74, 6.30)$

9.47 $(-6.92, 36.70)$

9.49 $0.54652 < \mu_B - \mu_A < 1.69348$

9.51 Method 1: $0.194 < p < 0.262$; method 2: $0.1957 < p < 0.2639$

9.53 (a) $0.498 < p < 0.642$; (b) error ≤ 0.072

9.55 (a) $0.739 < p < 0.961$; (b) no

9.57 (a) $0.644 < p < 0.690$; (b) error ≤ 0.023

9.59 2576

9.61 160

9.63 9604

9.65 $-0.0136 < p_F - p_M < 0.0636$

9.67 $0.0011 < p_1 - p_2 < 0.0869$

9.69 $(-0.0849, 0.0013)$; not significantly different

9.71 $0.293 < \sigma^2 < 6.736$; valid claim

9.73 $3.472 < \sigma^2 < 12.804$

9.75 $9.27 < \sigma < 34.16$

9.77 $0.549 < \sigma_1/\sigma_2 < 2.690$

9.79 $0.016 < \sigma_1^2/\sigma_2^2 < 0.454$; no

9.81 $\frac{1}{n}\sum_{i=1}^{n} x_i$

9.83 $\hat{\beta} = \bar{x}/5$

9.85 $\hat{\theta} = \max\{x_1, \ldots, x_n\}$

9.87 $x \ln p + (1-x)\ln(1-p)$. Set the derivative with respect to $p = 0$; $\hat{p} = x = 1.0$

Chapter 10

10.1 (a) Conclude that less than 30% of the public is allergic to some cheese products when, in fact, 30% or more is allergic.

(b) Conclude that at least 30% of the public is allergic to some cheese products when, in fact, less than 30% is allergic.

10.3 (a) The firm is not guilty;

(b) the firm is guilty.

10.5 (a) 0.0559;

(b) $\beta = 0.0017$; $\beta = 0.00968$; $\beta = 0.5557$

10.7 (a) 0.1286;

(b) $\beta = 0.0901$; $\beta = 0.0708$.

(c) The probability of a type I error is somewhat large.

10.9 (a) $\alpha = 0.0850$; (b) $\beta = 0.3410$

10.11 (a) $\alpha = 0.1357$; (b) $\beta = 0.2578$

10.13 $\alpha = 0.0094$; $\beta = 0.0122$

10.15 (a) $\alpha = 0.0718$; (b) $\beta = 0.1151$

10.17 (a) $\alpha = 0.0384$; (b) $\beta = 0.5$; $\beta = 0.2776$

10.19 $z = -2.76$; yes, $\mu < 40$ months; P-value $= 0.0029$

10.21 $z = -1.64$; P-value $= 0.10$

10.23 $t = 0.77$; fail to reject H_0.

10.25 $z = 8.97$; yes, $\mu > 20{,}000$ kilometers; P-value < 0.001

10.27 $t = 12.72$; P-value < 0.0005; reject H_0.

10.29 $t = -1.98$; P-value $= 0.0312$; reject H_0

10.31 $z = -2.60$; conclude $\mu_A - \mu_B \leq 12$ kilograms.

10.33 $t = 1.50$; there is not sufficient evidence to conclude that the increase in substrate concentration would cause an increase in the mean velocity of more than 0.5 micromole per 30 minutes.

10.35 $t = 0.70$; there is not sufficient evidence to support the conclusion that the serum is effective.

10.37 $t = 2.55$; reject H_0: $\mu_1 - \mu_2 > 4$ kilometers.

10.39 $t' = 0.22$; fail to reject H_0.

10.41 $t' = 2.76$; reject H_0.

10.43 $t = -2.53$; reject H_0; the claim is valid.

10.45 $t = 2.48$; P-value < 0.02; reject H_0.

10.47 $n = 6$

10.49 $78.28 \approx 79$

10.51 5

10.53 (a) H_0: $M_{\text{hot}} - M_{\text{cold}} = 0$, H_1: $M_{\text{hot}} - M_{\text{cold}} \neq 0$;

(b) paired t, $t = 0.99$; P-value > 0.30; fail to reject H_0.

10.55 P-value $= 0.4044$ (with a one-tailed test); the claim is not refuted.

10.57 $z = 1.44$; fail to reject H_0.

10.59 $z = -5.06$ with P-value ≈ 0; conclude that fewer than one-fifth of the homes are heated by oil.

10.61 $z = 0.93$ with P-value $= P(Z > 0.93) = 0.1762$; there is not sufficient evidence to conclude that the new medicine is effective.

10.63 $z = 2.36$ with P-value $= 0.0182$; yes, the difference is significant.

10.65 $z = 1.10$ with P-value $= 0.1357$; we do not have sufficient evidence to conclude that breast cancer is more prevalent in the urban community.

10.67 $\chi^2 = 18.13$ with P-value $= 0.0676$ (from computer output); do not reject H_0: $\sigma^2 = 0.03$.

10.69 $\chi^2 = 63.75$ with P-value $= 0.8998$ (from computer output); do not reject H_0.

10.71 $\chi^2 = 42.37$ with P-value $= 0.0117$ (from computer output); machine is out of control.

10.73 $f = 1.33$ with P-value $= 0.3095$ (from computer output); fail to reject H_0: $\sigma_1 = \sigma_2$.

Answers to Chapter 11

10.75 $f = 0.086$ with P-value $= 0.0328$ (from computer output); reject H_0: $\sigma_1 = \sigma_2$ at level greater than 0.0328.

10.77 $f = 19.67$ with P-value $= 0.0008$ (from computer output); reject H_0: $\sigma_1 = \sigma_2$.

10.79 $\chi^2 = 10.14$; reject H_0, the ratio is not 5:2:2:1.

10.81 $\chi^2 = 4.47$; there is not sufficient evidence to claim that the die is unbalanced.

10.83 $\chi^2 = 3.125$; do not reject H_0: geometric distribution.

10.85 $\chi^2 = 5.19$; do not reject H_0: normal distribution.

10.87 $\chi^2 = 5.47$; do not reject H_0.

10.89 $\chi^2 = 124.59$; yes, occurrence of these types of crime is dependent on the city district.

10.91 $\chi^2 = 5.92$ with P-value $= 0.4332$; do not reject H_0.

10.93 $\chi^2 = 31.17$ with P-value < 0.0001; attitudes are not homogeneous.

10.95 $\chi^2 = 1.84$; do not reject H_0.

Chapter 11

11.1 (a) $b_0 = 64.529$, $b_1 = 0.561$;
(b) $\hat{y} = 81.4$

11.3 (a) $\hat{y} = 5.8254 + 0.5676x$;
(c) $\hat{y} = 34.205$ at $50°C$

11.5 (a) $\hat{y} = 6.4136 + 1.8091x$;
(b) $\hat{y} = 9.580$ at temperature 1.75

11.7 (b) $\hat{y} = 31.709 + 0.353x$

11.9 (b) $\hat{y} = 343.706 + 3.221x$;
(c) $\hat{y} = \$456$ at advertising costs $= \$35$

11.11 (b) $\hat{y} = -1847.633 + 3.653x$

11.13 (a) $\hat{y} = 153.175 - 6.324x$;
(b) $\hat{y} = 123$ at $x = 4.8$ units

11.15 (a) $s^2 = 176.4$;
(b) $t = 2.04$; fail to reject H_0: $\beta_1 = 0$.

11.17 (a) $s^2 = 0.40$;
(b) $4.324 < \beta_0 < 8.503$;
(c) $0.446 < \beta_1 < 3.172$

11.19 (a) $s^2 = 6.626$;
(b) $2.684 < \beta_0 < 8.968$;
(c) $0.498 < \beta_1 < 0.637$

11.21 $t = -2.24$; reject H_0 and conclude $\beta < 6$

11.23 (a) $24.438 < \mu_{Y|24.5} < 27.106$;
(b) $21.88 < y_0 < 29.66$

11.25 $7.81 < \mu_{Y|1.6} < 10.81$

11.27 (a) 17.1812 mpg;
(b) no, the 95% confidence interval on mean mpg is $(27.95, 29.60)$;
(c) miles per gallon will likely exceed 18.

11.29 (b) $\hat{y} = 3.4156x$

11.31 The f-value for testing the lack of fit is 1.58, and the conclusion is that H_0 is not rejected. Hence, the lack-of-fit test is insignificant.

11.33 (a) $\hat{y} = 2.003x$;
(b) $t = 1.40$, fail to reject H_0.

11.35 $f = 1.71$ and P-value $= 0.2517$; the regression is linear.

11.37 (a) $b_0 = 10.812$, $b_1 = -0.3437$;
(b) $f = 0.43$; the regression is linear.

11.39 (a) $\hat{P} = -11.3251 - 0.0449T$;
(b) yes;
(c) $R^2 = 0.9355$;
(d) yes

11.41 (b) $\hat{N} = -175.9025 + 0.0902Y$; $R^2 = 0.3322$

11.43 $r = 0.240$

11.45 (a) $r = -0.979$;
(b) P-value $= 0.0530$; do not reject H_0 at 0.025 level;
(c) 95.8%

11.47 (a) $r = 0.784$;
(b) reject H_0 and conclude that $\rho > 0$;
(c) 61.5%.

Chapter 12

12.1 $\hat{y} = 0.5800 + 2.7122x_1 + 2.0497x_2$

12.3 (a) $\hat{y} = 27.547 + 0.922x_1 + 0.284x_2$;
(b) $\hat{y} = 84$ at $x_1 = 64$ and $x_2 = 4$

12.5 (a) $\hat{y} = -102.7132 + 0.6054x_1 + 8.9236x_2 + 1.4374x_3 + 0.0136x_4$;
(b) $\hat{y} = 287.6$

12.7 $\hat{y} = 141.6118 - 0.2819x + 0.0003x^2$

12.9 (a) $\hat{y} = 56.4633 + 0.1525x - 0.00008x^2$;
(b) $\hat{y} = 86.7\%$ when temperature is at $225°C$

12.11 $\hat{y} = -6.5122 + 1.9994x_1 - 3.6751x_2 + 2.5245x_3 + 5.1581x_4 + 14.4012x_5$

12.13 (a) $\hat{y} = 350.9943 - 1.2720x_1 - 0.1539x_2$;
(b) $\hat{y} = 140.9$

12.15 $\hat{y} = 3.3205 + 0.4210x_1 - 0.2958x_2 + 0.0164x_3 + 0.1247x_4$

12.17 0.1651

12.19 242.72

12.21 (a) $\hat{\sigma}^2_{B_2} = 28.0955$; (b) $\hat{\sigma}_{B_1 B_2} = -0.0096$

12.23 $t = 5.91$ with P-value $= 0.0002$. Reject H_0 and claim that $\beta_1 \neq 0$.

12.25 $0.4516 < \mu_{Y|x_1=900, x_2=1} < 1.2083$ and $-0.1640 < y_0 < 1.8239$

12.27 $263.7879 < \mu_{Y|x_1=75, x_2=24, x_3=90, x_4=98} < 311.3357$ and $243.7175 < y_0 < 331.4062$

12.29 (a) $t = -1.09$ with P-value $= 0.3562$;
(b) $t = -1.72$ with P-value $= 0.1841$;
(c) yes; not sufficient evidence to show that x_1 and x_2 are significant

12.31 $R^2 = 0.9997$

12.33 $f = 5.106$ with P-value $= 0.0303$; the regression is not significant at level 0.01.

12.35 $f = 34.90$ with P-value $= 0.0002$; reject H_0 and conclude $\beta_1 > 0$.

12.37 $f = 10.18$ with P-value < 0.01; x_1 and x_2 are significant in the presence of x_3 and x_4.

12.39 The two-variable model is better.

12.41 First model: $R^2_{\text{adj}} = 92.7\%$, C.V. $= 9.0385$.
Second model: $R^2_{\text{adj}} = 98.1\%$, C.V. $= 4.6287$.
The partial F-test shows P-value $= 0.0002$; model 2 is better.

12.43 Using x_2 alone is not much different from using x_1 and x_2 together since the R^2_{adj} are 0.7696 versus 0.7591, respectively.

12.45 (a) $\widehat{\text{mpg}} = 5.9593 - 0.00003773$ odometer $+ 0.3374$ octane $- 12.6266z_1 - 12.9846z_2$;
(b) sedan;
(c) they are not significantly different.

12.47 (b) $\hat{y} = 4.690$ seconds;
(c) $4.450 < \mu_{Y|\{180, 260\}} < 4.930$

12.49 $\hat{y} = 2.1833 + 0.9576x_2 + 3.3253x_3$

12.51 (a) $\hat{y} = -587.211 + 428.433x$;
(b) $\hat{y} = 1180 - 191.691x + 35.20945x^2$;
(c) quadratic model

12.53 $\hat{\sigma}^2_{B_1} = 20{,}588$; $\hat{\sigma}^2_{B_{11}} = 62.6502$;
$\hat{\sigma}_{B_1, B_{11}} = -1103.5$

12.55 (a) Intercept model is the best.

12.57 (a) $\hat{y} = 3.1368 + 0.6444x_1 - 0.0104x_2 + 0.5046x_3 - 0.1197x_4 - 2.4618x_5 + 1.5044x_6$;
(b) $\hat{y} = 4.6563 + 0.5133x_3 - 0.1242x_4$;
(c) C_p criterion: variables x_1 and x_2 with $s^2 = 0.7317$ and $R^2 = 0.6476$; s^2 criterion: variables x_1, x_3 and x_4 with $s^2 = 0.7251$ and $R^2 = 0.6726$;
(d) $\hat{y} = 4.6563 + 0.5133x_3 - 0.1242x_4$; This one does not lose much in s^2 and R^2;
(e) two observations have large R-Student values and should be checked.

12.59 (a) $\hat{y} = 125.8655 + 7.7586x_1 + 0.0943x_2 - 0.0092x_1 x_2$;
(b) the model with x_2 alone is the best.

12.61 (a) $\hat{p} = (1 + e^{2.9949 - 0.0308x})^{-1}$;
(b) 1.8515

Chapter 13

13.1 $f = 0.31$; not sufficient evidence to support the hypothesis that there are differences among the 6 machines.

13.3 $f = 14.52$; yes, the difference is significant.

13.5 $f = 8.38$; the average specific activities differ significantly.

13.7 $f = 2.25$; not sufficient evidence to support the hypothesis that the different concentrations of $MgNH_4PO_4$ significantly affect the attained height of chrysanthemums.

13.9 $b = 0.79 > b_4(0.01, 4, 4, 4, 9) = 0.4939$. Do not reject H_0. There is not sufficent evidence to claim that variances are different.

13.11 $b = 0.7822 < b_4(0.05, 9, 8, 15) = 0.8055$. The variances are significantly different.

13.13 (a) P-value < 0.0001, significant,
(b) for contrast 1 vs. 2, P-value $< 0.0001Z$, significantly different; for contrast 3 vs. 4, P-value $= 0.0648$, not significantly different

13.15 Results of Tukey's tests are given below.

$\bar{y}_4.$	$\bar{y}_3.$	$\bar{y}_1.$	$\bar{y}_5.$	$\bar{y}_2.$
2.98	4.30	5.44	6.96	7.90

13.17 (a) P-value $= 0.0121$; yes, there is a significant difference.

(b)

	Substrate Removal			
Depletion	Modified Hess	Kicknet	Surber	Kicknet

13.19 $f = 70.27$ with P-value < 0.0001; reject H_0.

$\bar{x}_0$	$\bar{x}_{25}$	$\bar{x}_{100}$	$\bar{x}_{75}$	$\bar{x}_{50}$
55.167	60.167	64.167	70.500	72.833

Temperature is important; both 75° and 50°(C) yielded batteries with significantly longer activated life.

13.21 The mean absorption is significantly lower for aggregate 4 than for the other aggregates.

13.23 Comparing the control to 1 and 2: significant; comparing the control to 3 and 4: insignificant

13.25 f(fertilizer) $= 6.11$; there is significant difference among the fertilizers.

13.27 $f = 5.99$; percent of foreign additives is not the same for all three brands of jam; brand A

13.29 P-value < 0.0001; significant

13.31 P-value $= 0.0023$; significant

13.33 P-value $= 0.1250$; not significant

13.35 P-value < 0.0001;
$f = 122.37$; the amount of dye has an effect on the color of the fabric.

13.37 (a) $y_{ij} = \mu + A_i + \epsilon_{ij}$, $A_i \sim n(x; 0, \sigma_\alpha)$, $\epsilon_{ij} \sim n(x; 0, \sigma)$;
(b) $\hat{\sigma}_\alpha^2 = 0$ (the estimated variance component is -0.00027); $\hat{\sigma}^2 = 0.0206$.

13.39 (a) $f = 14.9$; operators differ significantly;
(b) $\hat{\sigma}_\alpha^2 = 28.91$; $s^2 = 8.32$.

13.41 (a) $y_{ij} = \mu + A_i + \epsilon_{ij}$, $A_i \sim n(x; 0, \sigma_\alpha)$;
(b) yes; $f = 5.63$ with P-value $= 0.0121$;
(c) there is a significant loom variance component.

Chapter 14

14.1 (a) $f = 8.13$; significant;
(b) $f = 5.18$; significant;
(c) $f = 1.63$; insignificant

14.3 (a) $f = 14.81$; significant;
(b) $f = 9.04$; significant;
(c) $f = 0.61$; insignificant

14.5 (a) $f = 34.40$; significant;
(b) $f = 26.95$; significant;
(c) $f = 20.30$; significant

14.7 Test for effect of temperature: $f_1 = 10.85$ with P-value $= 0.0002$;
Test for effect of amount of catalyst: $f_2 = 46.63$ with P-value < 0.0001;
Test for effect of interaction: $f = 2.06$ with P-value $= 0.074$.

14.9 (a)

Source of Variation	df	Sum of Squares	Mean Squares	f	P
Cutting speed	1	12.000	12.000	1.32	0.2836
Tool geometry	1	675.000	675.000	74.31	< 0.0001
Interaction	1	192.000	192.000	21.14	0.0018
Error	8	72.667	9.083		
Total	11	951.667			

(b) The interaction effect masks the effect of cutting speed;

(c) $f_{\text{tool geometry}=1} = 16.51$ and P-value $= 0.0036$;
$f_{\text{tool geometry}=2} = 5.94$ and P-value $= 0.0407$.

14.11 (a)

Source of Variation	df	Sum of Squares	Mean Squares	f	P
Method	1	0.000104	0.000104	6.57	0.0226
Laboratory	6	0.008058	0.001343	84.70	< 0.0001
Interaction	6	0.000198	0.000033	2.08	0.1215
Error	14	0.000222	0.000016		
Total	27	0.008582			

(b) The interaction is not significant;

(c) Both main effects are significant;

(e) $f_{\text{laboratory}=1} = 0.01576$ and P-value = 0.9019; no significant difference between the methods in laboratory 1; $f_{\text{tool geometry}=2} = 9.081$ and P-value = 0.0093.

14.13 (b)

Source of Variation	df	Sum of Squares	Mean Squares	f	P
Time	1	0.060208	0.060208	157.07	< 0.0001
Treatment	1	0.060208	0.060208	157.07	< 0.0001
Interaction	1	0.000008	0.000008	.02	0.8864
Error	8	0.003067	0.000383		
Total	11	0.123492			

(c) Both time and treatment influence the magnesium uptake significantly, although there is no significant interaction between them.

(d) $Y = \mu + \beta_T \text{Time} + \beta_Z Z + \beta_{TZ} \text{Time } Z + \epsilon$, where $Z = 1$ when treatment = 1 and $Z = 0$ when treatment = 2;

(e) $f = 0.02$ with P-value = 0.8864; the interaction in the model is insignificant.

14.15 (a) Interaction is significant at a level of 0.05, with P-value of 0.0166.

(b) Both main effects are significant.

14.17 (a) AB: $f = 3.83$; significant;
AC: $f = 3.79$; significant;
BC: $f = 1.31$; not significant;
ABC: $f = 1.63$; not significant;

(b) A: $f = 0.54$; not significant;
B: $f = 6.85$; significant;
C: $f = 2.15$; not significant;

(c) The presence of AC interaction masks the main effect C.

14.19 (a) Stress: $f = 45.96$ with P-value < 0.0001;
coating: $f = 0.05$ with P-value = 0.8299;
humidity: $f = 2.13$ with P-value = 0.1257;
coating × humidity: $f = 3.41$ with P-value = 0.0385;
coating × stress: $f = 0.08$ with P-value = 0.9277;
humidity × stress: $f = 3.15$ with P-value = 0.0192;
coating × humidity × stress: $f = 1.93$ with P-value = 0.1138.

(b) The best combination appears to be uncoated, medium humidity, and a stress level of 20.

14.21

Effect	f	P
Temperature	14.22	< 0.0001
Surface	6.70	0.0020
HRC	1.67	0.1954
T × S	5.50	0.0006
T × HRC	2.69	0.0369
S × HRC	5.41	0.0007
T × S × HRC	3.02	0.0051

14.23 (a) Yes; brand × type; brand × temperature;

(b) yes;

(c) brand Y, powdered detergent, hot temperature.

14.25 (a)

Effect	f	P
Time	543.53	< 0.0001
Temp	209.79	< 0.0001
Solvent	4.97	0.0457
Time × Temp	2.66	0.1103
Time × Solvent	2.04	0.1723
Temp × Solvent	0.03	0.8558
Time × Temp × Solvent	6.22	0.0140

Although three two-way interactions are shown to be insignificant, they may be masked by the significant three-way interaction.

14.27 (a) $f = 1.49$; no significant interaction;

(b) $f(\text{operators}) = 12.45$; significant;
$f(\text{filters}) = 8.39$; significant;

(c) $\hat{\sigma}_\alpha^2 = 0.1777$ (filters);
$\hat{\sigma}_\beta^2 = 0.3516$ (operators);
$s^2 = 0.185$

14.29 (a) $\hat{\sigma}_\beta^2, \hat{\sigma}_\gamma^2, \hat{\sigma}_{\alpha\gamma}^2$ are significant;

(b) $\hat{\sigma}_\gamma^2$ and $\hat{\sigma}_{\alpha\gamma}^2$ are significant.

14.31 (a) Mixed model;

(b) Material: $f = 47.42$ with P-value $<$ 0.0001;
brand: $f = 1.73$ with P-value $= 0.2875$;
material $\times$ brand: $f = 16.06$ with P-value $= 0.0004$;

(c) no

Chapter 15

15.1 B and C are significant at level 0.05.

15.3 Factors A, B, and C have negative effects on the phosphorus compound, and factor D has a positive effect. However, the interpretation of the effect of individual factors should involve the use of interaction plots.

15.5 Significant effects:
A: $f = 9.98$; BC: $f = 19.03$.
Insignificant effects:
B: $f = 0.20$; C: $f = 6.54$; D: $f = 0.02$; AB: $f = 1.83$;
AC: $f = 0.20$; AD: $f = 0.57$; BD: $f = 1.83$;
CD: $f = 0.02$. Since the BC interaction is significant, both B and C would be investigated further.

15.9 (a) $b_A = 5.5$, $b_B = -3.25$, and $b_{AB} = 2.5$;

(b) the values of the coefficients are one-half those of the effects;

(c) $t_A = 5.99$ with P-value $= 0.0039$;
$t_B = -3.54$ with P-value $= 0.0241$;
$t_{AB} = 2.72$ with P-value $= 0.0529$;
$t^2 = F$.

15.11 (a) $A = -0.8750$, $B = 5.8750$, $C = 9.6250$, $AB = -3.3750$, $AC = -9.6250$, $BC = 0.1250$, and $ABC = -1.1250$;
B, C, AB, and AC appear important based on their magnitude.

(b)
Effects	P-Value
A	0.7528
B	0.0600
C	0.0071
AB	0.2440
AC	0.0071
BC	0.9640
ABC	0.6861

(c) Yes;

(d) At a high level of A, C essentially has no effect. At a low level of A, C has a positive effect.

15.13 (a)

Machine			
1	2	3	4
(1)	c	a	ac
ab	d	b	ad
cd	e	acd	ae
ce	abc	ace	bc
de	abd	ade	bd
abcd	abe	bcd	be
abce	cde	bce	acde
abde	abcde	bde	bcde

(b) ABD, CDE, $ABCDE$ (one possible design)

15.15 (a) x_2, x_3, x_1x_2, and x_1x_3;

(b) Curvature: P-value $= 0.0038$;

(c) One additional design point different from the original ones

15.17 $(0, -1)$, $(0, 1)$, $(-1, 0)$, $(1, 0)$ might be used.

15.19 (a) With BCD as the defining contrast, the principal block contains (1), a, bc, abc, bd, abd, cd, acd;

(b)
Block 1	Block 2
(1)	a
bc	abc
abd	bd
acd	cd

confounded by ABC;

(c) Defining contrast BCD produces the following aliases: $A \equiv ABCD$, $B \equiv CD$, $C \equiv BD$, $D \equiv BC$, $AB \equiv ACD$, $AC \equiv ABD$, and $AD \equiv ABC$. Since AD and ABC are confounded with blocks, there are only 2 degrees of freedom for error from the interactions not confounded.

Source of Variation	Degrees of Freedom
A	1
B	1
C	1
D	1
Blocks	1
Error	2
Total	7

15.21 (a) With the defining contrasts $ABCE$ and $ABDF$, the principal block contains (1), ab, acd, bcd, ce, $abce$, ade, bde, acf, bcf, df, $abdf$, aef, bef, $cdef$, $abcdef$;

(b) $A \equiv BCE \equiv BDF \equiv ACDEF$,
$AD \equiv BCDE \equiv BF \equiv ACEF$,
$B \equiv ACE \equiv ADF \equiv BCDEF$,
$AE \equiv BC \equiv BDEF \equiv ACDF$,
$C \equiv ABE \equiv ABCDF \equiv DEF$,
$AF \equiv BCEF \equiv BD \equiv ACDE$,
$D \equiv ABCDE \equiv ABF \equiv CEF$,
$CE \equiv AB \equiv ABCDEF \equiv DF$,
$E \equiv ABC \equiv ABDEF \equiv CDF$,
$DE \equiv ABCD \equiv ABEF \equiv CF$,
$F \equiv ABCEF \equiv ABD \equiv CDE$,
$BCD \equiv ADE \equiv ACF \equiv BEF$,
$AB \equiv CE \equiv DF \equiv ABCDEF$,
$BCF \equiv AEF \equiv ACD \equiv BDE$,
$AC \equiv BE \equiv BCDF \equiv ADEF$;

Source of Variation	Degrees of Freedom
A	1
B	1
C	1
D	1
E	1
F	1
AB	1
AC	1
AD	1
BC	1
BD	1
CD	1
CF	1
Error	2
Total	15

15.23

Source	df	SS	MS	f	P
A	1	6.1250	6.1250	5.81	0.0949
B	1	0.6050	0.6050	0.57	0.5036
C	1	4.8050	4.8050	4.56	0.1223
D	1	0.2450	0.2450	0.23	0.6626
Error	3	3.1600	1.0533		
Total	7	14.9400			

15.25

Source	df	SS	MS	f	P
A	1	388,129.00	388,129.00	3585.49	0.0001
B	1	277,202.25	277,202.25	2560.76	0.0001
C	1	4692.25	4692.25	43.35	0.0006
D	1	9702.25	9702.25	89.63	0.0001
E	1	1806.25	1806.25	16.69	0.0065
AD	1	1406.25	1406.25	12.99	0.0113
AE	1	462.25	462.25	4.27	0.0843
BD	1	1156.00	1156.00	10.68	0.0171
BE	1	961.00	961.00	8.88	0.0247
Error	6	649.50	108.25		
Total	15	686,167.00			

All main effects are significant at the 0.05 level; AD, BD, and BE are also significant at the 0.05 level.

15.27 The principal block contains af, be, cd, abd, ace, bcf, def, $abcdef$.

15.29 $A \equiv BD \equiv CE \equiv CDF \equiv BEF \equiv ABCF \equiv ADEF \equiv ABCDE$;
$B \equiv AD \equiv CF \equiv CDE \equiv AEF \equiv ABCE \equiv BDEF \equiv ABCDF$;
$C \equiv AE \equiv BF \equiv BDE \equiv ADF \equiv CDEF \equiv ABCD \equiv ABCEF$;
$D \equiv AB \equiv EF \equiv BCE \equiv ACF \equiv BCDF \equiv ACDE \equiv ABDEF$;
$E \equiv AC \equiv DF \equiv ABF \equiv BCD \equiv ABDE \equiv BCEF \equiv ACDEF$;
$F \equiv BC \equiv DE \equiv ACD \equiv ABE \equiv ACEF \equiv ABDF \equiv BCDEF$.

15.31 $x_1 = 1$ and $x_2 = 1$

15.33 (a) Yes;
(b) (i) $E(\hat{y}) = 79.00 + 5.281A$;
(ii) $\text{Var}(\hat{y}) = 6.22^2 \sigma_Z^2 + 5.70^2 A^2 \sigma_Z^2 + 2(6.22)(5.70) A \sigma_Z^2$;
(c) velocity at low level;
(d) velocity at low level;
(e) yes

15.35 $\hat{y} = 12.7519 + 4.7194x_1 + 0.8656x_2 - 1.4156x_3$; units are centered and scaled; test for lack of fit, $F = 81.58$, with P-value < 0.0001.

15.37 AFG, BEG, CDG, DEF, $CEFG$, $BDFG$, $BCDE$, $ADEG$, $ACDF$, $ABEF$, and $ABCDEFG$

Chapter 16

16.1 $x = 7$ with P-value $= 0.1719$; fail to reject H_0.

16.3 $x = 3$ with P-value $= 0.0244$; reject H_0.

16.5 $x = 4$ with P-value $= 0.3770$; fail to reject H_0.

16.7 $x = 4$ with P-value $= 0.1335$; fail to reject H_0.

16.9 $w = 43$; fail to reject H_0.

16.11 $w_+ = 17.5$; fail to reject H_0.

16.13 $w_+ = 15$ with $n = 13$; reject H_0 in favor of $\tilde{\mu}_1 - \tilde{\mu}_2 < 8$.

16.15 $u_1 = 4$; claim is not valid.

16.17 $u_2 = 5$; A operates longer.

16.19 $u = 15$; fail to reject H_0.

16.21 $h = 10.58$; operating times are different.

16.23 $v = 7$ with P-value $= 0.910$; random sample.

16.25 $v = 6$ with P-value $= 0.044$; fail to reject H_0.

16.27 $v = 4$; random sample.

16.29 0.70

16.31 0.995

16.33 (a) $r_s = 0.39$; (b) fail to reject H_0.

16.35 (a) $r_s = 0.72$; (b) reject H_0, so $\rho > 0$.

16.37 (a) $r_s = 0.71$; (b) reject H_0, so $\rho > 0$.

Chapter 18

18.1 $p^* = 0.173$

18.3 (a) $\pi(p \mid x = 1) = 40p(1-p)^3/0.2844$; $0.05 < p < 0.15$;
(b) $p^* = 0.106$

18.5 (a) $beta(95, 45)$; (b) 1

18.7 $8.077 < \mu < 8.692$

18.9 (a) 0.2509; (b) $68.71 < \mu < 71.69$; (c) 0.0174

18.13 $p^* = \frac{6}{x+2}$

18.15 2.21

Index

Acceptable quality level, 705
Acceptance sampling, 153
Additive rule, 56
Adjusted R^2, 464
Analysis of variance (ANOVA), 254, 507
    one-factor, 509
    table, 415
    three-factor, 579
    two-factor, 565
Approximation
    binomial to hypergeometric, 155
    normal to binomial, 187, 188
    Poisson to binomial, 163
Average, 111

Backward elimination, 479
Bartlett's test, 516
Bayes estimates, 717
    under absolute-error loss, 718
    under square-error loss, 717
Bayes' rule, 72, 75
Bayesian
    inference, 710
    interval, 715
    methodology, 265, 709
    perspective, 710
    posterior interval, 317
Bernoulli
    process, 144
    random variable, 83
    trial, 144
Beta distribution, 201
Bias, 227
Binomial distribution, 104, 145, 153, 155
    mean of, 147
    variance of, 147
Blocks, 509
Box plot, 3, 24, 25

Categorical variable, 472
Central composite design, 640
Central limit theorem, 233, 234, 238
Chebyshev's theorem, 135–137, 148, 155, 180, 186
Chi-squared distribution, 200
Cochran's test, 518
Coefficient of determination, 407, 433, 462
    adjusted, 464
Coefficient of variation, 471
Combination, 50
Complement of an event, 39
Completely randomized design, 8, 509
Conditional distribution, 99
    joint, 103
Conditional perspective, 710
Conditional probability, 62–66, 68, 75, 76
Confidence
    coefficient, 269
    degree of, 269
    limits, 269, 271
Confidence interval, 269, 270, 281, 317
    for difference of two means, 285–288, 290
    for difference of two proportions, 300, 301
    interpretation of, 289
    of large sample, 276
    for paired observations, 293
    for ratio of standard deviations, 306
    for ratio of variances, 306
    for single mean, 269–272, 275
        one-sided, 273
    for single proportion, 297
    for single variance, 304
    for standard deviation, 304
Contingency table, 373
    marginal frequency, 374
Continuity correction, 190
Continuous distribution
    beta, 201

chi-squared, 200
exponential, 195
gamma, 195
lognormal, 201
normal, 172
uniform, 171
Weibull, 203, 204
Control chart
 for attributes, 697
 Cusum chart, 705
 p-chart, 697
 R-chart, 688
 S-chart, 695
 U-chart, 704
 for variable, 684
 $\bar{X}$-chart, 686
Correlation coefficient, 125, 431
 Pearson product-moment, 432
 population, 432
 sample, 432
Covariance, 119, 123
C_p statistic, 491
Cross validation, 487
Cumulative distribution function, 85, 90

Degrees of freedom, 15, 16, 200, 244, 246
 Satterthwaite approximation of, 289
Descriptive statistics, 3, 9
Design of experiment
 blocking, 532
 central composite, 640
 completely randomized, 532
 contrast, 599
 control factors, 644
 defining relation, 627
 fractional factorial, 598, 612, 626, 627
 noise factor, 644
 orthogonal, 617
 randomized block, 533
 resolution, 637
Deviation, 120
Discrete distribution
 binomial, 143, 144
 geometric, 158, 160
 hypergeometric, 152, 153
 multinomial, 143, 149
 negative binomial, 158, 159
 Poisson, 161, 162

Distribution, 23
 beta, 201
 binomial, 104, 143–145, 175, 188
 bivariate normal, 431
 chi-squared, 200
 continuous uniform, 171
 empirical, 254
 Erlang, 207
 exponential, 104, 194, 195
 gamma, 194, 195
 Gaussian, 19, 172
 geometric, 143, 158, 160
 hypergeometric, 152–154, 175
 lognormal, 201
 multinomial, 143, 149
 multivariate hypergeometric, 156
 negative binomial, 143, 158–160
 normal, 19, 172, 173, 188
 Poisson, 143, 161, 162
 posterior, 711
 prior, 710
 skewed, 23
 standard normal, 177
 symmetric, 23
 t-, 246, 247
 variance ratio, 253
 Weibull, 203
Distribution-free method, 655
Distributional parameter, 104
Dot plot, 3, 8, 32
Dummy variable, 472
Duncan's multiple-range test, 527
Dunnett's test, 528

Erlang distribution, 207
Error
 in estimating the mean, 272
 experimental, 509
 sum of squares, 402
 type I, 322
 type II, 323
Estimate, 12
 of single mean, 269
Estimation, 12, 142, 266
 difference of two sample means, 285
 maximum likelihood, 307, 308, 312
 paired observations, 291
 proportion, 296

of the ratio of variances, 305
of single variance, 303
two proportions, 300
Estimator, 266
efficient, 267
maximum likelihood, 308–310
method of moments, 314, 315
point, 266, 268
unbiased, 266, 267
Event, 38
Expectation
mathematical, 111, 112, 115
Expected mean squares
ANOVA model, 548
Expected value, 112–115
Experiment-wise error rate, 525
Experimental error, 509
Experimental unit, 9, 286, 292, 562
Exponential distribution, 104, 194, 195
mean of, 196
memoryless property of, 197
relationship to Poisson process, 196
variance of, 196

F-distribution, 251–254
Factor, 28, 507
Factorial, 47
Factorial experiment, 561
in blocks, 583
factor, 507
interaction, 562
level, 507
main effects, 562
masking effects, 563
mixed model, 591
pooling mean squares, 583
random effects, 589
three-factor ANOVA, 579
treatment, 507
two-factor ANOVA, 565
Failure rate, 204, 205
Fixed effects experiment, 547
Forward selection, 479

Gamma distribution, 194, 195
mean of, 196
relationship to Poisson process, 196
variance of, 196

Gamma function, 194
incomplete, 199
Gaussian distribution, 19, 172
Geometric distribution, 158, 160
mean of, 160
variance of, 160
Goodness-of-fit test, 210, 255, 317, 370, 371

Histogram, 22
probability, 86
Historical data, 30
Hypergeometric distribution, 152–154
mean of, 154
variance of, 154
Hypothesis, 320
alternative, 320
null, 320
statistical, 319
testing, 320, 321

Independence, 62, 65, 67, 68
statistical, 101–103
Indicator variable, 472
Inferential statistics, 1
Interaction, 28, 562
Interquartile range, 24, 25
Intersection of events, 39
Interval estimate, 268
Bayesian, 715

Jacobian, 213
matrix, 214

Kruskall-Wallis test, 668

Lack of fit, 418
Least squares method, 394, 396
Level of significance, 323
Likelihood function, 308
Linear model, 133
Linear predictor, 498
Linear regression
ANOVA, 414
categorical variable, 472
coefficient of determination, 407
correlation, 430
data transformation, 424
dependent variable, 389
empirical model, 391

error sum of squares, 415
fitted line, 392
fitted value, 416
independent variable, 389
lack of fit, 418
least squares, 394
mean response, 394, 409
model selection, 476, 487
multiple, 390, 443
normal equation, 396
through the origin, 413
overfitting, 408
prediction, 408
prediction interval, 410, 411
pure experimental error, 419
random error, 391
regression coefficient, 392
regression sum of squares, 461
regressor, 389
residual, 395
simple, 389, 390
statistical model, 391
test of linearity, 416
total sum of squares, 414
Logistic regression, 497
effective dose, 500
odds ratio, 500
Lognormal distribution, 201
mean of, 202
variance of, 202
Loss function
absolute-error, 718
squared-error, 717

Marginal distribution, 97, 101, 102
joint, 103
Markov chain Monte Carlo, 710
Masking effect, 563
Maximum likelihood estimation, 307, 308, 710
residual, 550
restricted, 550
Mean, 19, 111, 112, 114, 115
population, 12, 16
trimmed, 12
Mean squared error, 284
Mean squares, 415
Mode, 713
normal distribution, 174

Model selection, 476
backward elimination, 480
C_p statistic, 491
forward selection, 479
PRESS, 487, 488
sequential methods, 476
stepwise regression, 480
Moment, 218
Moment-generating function, 218
Multicollinearity, 476
Multinomial distribution, 149
Multiple comparison test, 523
Duncan's, 527
Dunnett's, 528
experiment-wise error rate, 525
Tukey's, 526
Multiple linear regression, 443
adjusted R^2, 464
ANOVA, 455
error sum of squares, 460
HAT matrix, 483
inference, 455
multicollinearity, 476
normal equations, 444
orthogonal variables, 467
outlier, 484
polynomial, 446
R-student residuals, 483
regression sum of squares, 460
studentized residuals, 483
variable screening, 456
variance-covariance matrix, 453
Multiplication rule, 44
Multiplicative rule, 65
Multivariate hypergeometric distribution, 156
Mutually exclusive
events, 40

Negative binomial distribution, 158, 159
Negative binomial experiment, 158
Negative exponential distribution, 196
Nonlinear regression, 496
binary response, 497
count data, 497
logistic, 497
Nonparametric methods, 655
Kruskall-Wallis test, 668
runs test, 671

sign test, 656
signed-rank test, 660
tolerance limits, 674
Wilcoxon rank-sum test, 665
Normal distribution, 172, 173
mean of, 175
normal curve, 172–175
standard, 177
standard deviation of, 175
variance of, 175
Normal equations for linear regression, 444
Normal probability plot, 254
Normal quantile-quantile plot, 256, 257

Observational study, 3, 29
OC curve, 335
One-sided confidence bound, 273
One-way ANOVA, 509
contrast, 520
contrast sum of squares, 521
grand mean, 510
single-degree-of-freedom contrast, 520
treatment, 509
treatment effect, 510
Orthogonal contrasts, 522
Orthogonal variables, 467
Outlier, 24, 279, 484

p-chart, 697
P-value, 4, 109, 331–333
Paired observations, 291
Parameter, 12, 142
Partial F-test, 466
Permutation, 47
circular, 49
Plot
box, 24
normal quantile-quantile, 256, 257
probability, 254
quantile, 254, 255
stem-and-leaf, 21
Point estimate, 266, 268
standard error, 276
Points of inflection, normal distribution, 174
Poisson distribution, 143, 161, 162
mean of, 162
variance of, 162
Poisson experiment, 161

Poisson process, 161, 196
relationship to gamma distribution, 196
Polynomial regression, 443, 446
Pooled estimate of variance, 287
Pooled sample variance, 287
Population, 2, 4, 225, 226
mean of, 226
parameter, 16, 104
size of, 226
variance of, 226
Posterior distribution, 711
Power of a test, 329
Prediction interval, 277, 278, 281
for future observation, 278, 279
one-sided, 279
Prediction sum of squares, 487, 488
Prior distribution, 710
Probability, 35, 52, 53
additive rule, 56
coverage, 715
of an event, 52
indifference, 55, 709
mass function, 84
relative frequency, 55, 709
subjective, 55
subjective approach, 709
Probability density function, 88, 89
joint, 96
Probability distribution, 84
conditional, 99
continuous, 87
discrete, 84
joint, 94, 95, 102
marginal, 97
mean of, 111
variance of, 119
Probability function, 84
Probability mass function, 84
joint, 95
Product rule, 65

Quality control, 681
chart, 681, 682
in control, 682
out of control, 682
limits, 683
Quantile, 255
Quantile plot, 254, 255

R-chart, 688
R^2, 407, 462
    Adjusted, 464
Random effects experiment
    variance components, 549
Random effects model, 547, 548
Random sample, 227
    simple, 7
Random sampling, 225
Random variable, 81
    Bernoulli, 83, 147
    binomial, 144, 147
    chi-squared, 244
    continuous, 84
    continuous uniform, 171
    discrete, 83, 84
    discrete uniform, 150
    hypergeometric, 143, 153
    mean of, 111, 114
    multinomial, 149
    negative binomial, 158
    nonlinear function of, 133
    normal, 173
    Poisson, 161, 162
    transformation, 211
    variance of, 119, 122
Randomized complete block design, 533
Rank correlation coefficient, 675
    Spearman, 674
Rectangular distribution, 171
Regression, 20
Rejectable quality level, 705
Relative frequency, 22, 31, 111
Residual, 395, 427
Response surface, 642, 648
    robust parameter design, 644
Response surface methodology, 447, 639, 640
    control factor, 644
    control factors, 644
    noise factor, 644
    second order model, 640
Retrospective study, 30
Rule method, 37
Rule of elimination, 73–75
Runs test, 671

S-chart, 695
Sample, 1, 2, 225, 226
    biased, 7
    mean, 3, 11, 12, 19, 30–32, 225, 228
    median, 3, 11, 12, 30, 31, 228
    mode, 228
    random, 227
    range, 15, 30, 31, 229
    standard deviation, 3, 15, 16, 30, 31, 229, 230
    variance, 15, 16, 30, 225, 229
Sample mean, 111
Sample size, 7
    in estimating a mean, 272
    in estimating a proportion, 298
    in hypothesis testing, 351
Sample space, 35
    continuous, 83
    discrete, 83
    partition, 57
Sampling distribution, 232
    of mean, 233
Satterthwaite approximation of degrees of freedom, 289
Scatter plot, 3
Sign test, 656
Signed-rank test, 660
Significance level, 332
Single proportion test, 360
Standard deviation, 120, 122, 135
    sample, 15, 16
Standard error of mean, 277
Standard normal distribution, 177
Statistic, 228
Statistical independence, 101–103
Statistical inference, 3, 225, 265
Stem-and-leaf plot, 3, 21, 22, 31
Stepwise regression, 479
Subjective probability, 709, 710
Sum of squares
    error, 402, 415
    identity, 510, 536, 567
    lack-of-fit, 419
    regression, 415
    total, 407
    treatment, 511, 522, 536

t-distribution, 246–250
Test statistic, 322
Tests for equality of variances, 516
    Bartlett's, 516

INDEX

Cochran's, 518
Tests of hypotheses, 19, 266, 319
    choice of sample size, 349, 352
    critical region, 322
    critical value, 322
    goodness-of-fit, 210, 255, 370, 371
    important properties, 329
    one-tailed, 330
    P-value, 331, 333
    paired observations, 345
    partial F, 466
    single proportion, 360
    single sample, 336
    single sample, variance known, 336
    single sample, variance unknown, 340
    single variance, 366
    size of test, 323
    test for homogeneity, 376
    test for independence, 373
    test for several proportions, 377
    test statistics, 326
    on two means, 342
    two means with unknown and unequal variances, 345
    two means with unknown but equal variances, 343
    two-tailed, 330
    two variances, 366
Tolerance interval, 280, 281
Tolerance limits, 280
    of nonparametric method, 674
    one-sided, 281
Total probability, 72, 73
Treatment
    negative effect, 563
    positive effect, 563
Tree diagram, 36
Trimmed mean, 12
Tukey's test, 526
2^k factorial experiment, 597
    aliases, 628
    center runs, 620
    defining relation, 627
    design generator, 627
    diagnostic plotting, 604
    factor screening, 598
    fractional factorial, 626
    orthogonal design, 617
    Plackett-Burman designs, 638
    regression setting, 612
    resolution, 637

U-chart, 704
Unbiased estimator, 267
Uniform distribution, 171
Union of events, 40

Variability, 8, 9, 14–16, 119, 135, 228, 251, 253
    between/within samples, 253, 254
Variable transformation
    continuous, 213, 214
    discrete, 212
Variance, 119, 120, 122
    population, 16
    sample, 16
Variance ratio distribution, 253
Venn diagram, 40

Weibull distribution, 203
    cumulative distribution function for, 204
    failure rate of, 204, 205
    mean of, 203
    variance of, 203
Wilcoxon rank-sum test, 665

$\bar{X}$-chart, 686
    operating characteristic function, 691